STUDENT'S
SOLUTIONS MANUAL

Calculus for the
Life Sciences

Greenwell ■ Ritchey ■ Lial

Addison
Wesley

Boston San Francisco New York
London Toronto Sydney Tokyo Singapore Madrid
Mexico City Munich Paris Cape Town Hong Kong Montreal

Reproduced by Addison-Wesley from camera-ready copy.

Printed in the United States of America.

ISBN   0-201-77016-4

4 5 6 7 8 9 10 QWC 05 04 03

PREFACE

This book provides solutions for many of the exercises in *Calculus for the Life Sciences,* First Edition, by Raymond N. Greenwell, Nathan P. Ritchey, and Margaret L. Lial. Solutions are included for all of the exercises in Chapter R and for odd–numbered exercises in Chapters 1-13. Solutions are not provided for exercises with open–response answers. Sample tests are provided at the end of each chapter to help you determine if you have mastered the concepts in a given chapter.

This book should be used as an aid as you work to master your coursework. Try to solve the exercises that your instructor assigns before you refer to the solutions in this book. Then, if you have difficulty, read these solutions to guide you in solving the exercises. The solutions have been written so that they are consistent with the methods used in the textbook.

You may find that some of the solutions are presented in greater detail than others. Thus, if you cannot find an explanation for a difficulty that you encountered in one exercise, you may find the explanation in the solution for a similar exercise elsewhere in the exercise set.

In addition to solutions, you will find a list of suggestions on how to be successful in mathematics. A careful reading will be helpful for many students.

The following people have made valuable contributions to the production of this *Student's Solutions Manual:* LaurelTech Integrated Publishing Services, editors; Judy Martinez and Sheri Minkner, typists.

We also want to thank Tommy Thompson of Cedar Valley Community College for the essay "To the Student: Success in Mathematics."

CONTENTS

CHAPTER 3   THE DERIVATIVE

CHAPTER 4   CALCULATING THE DERIVATIVE

CHAPTER 5   GRAPHS AND THE DERIVATIVE

CHAPTER 6 APPLICATIONS OF THE DERIVATIVE

CHAPTER 7 INTEGRATION

CHAPTER 8 FURTHER TECHNIQUES AND APPLICATIONS OF INTEGRATION

CHAPTER 9 MULTIVARIABLE CALCULUS

CHAPTER 10 MATRICES

CHAPTER 11 DIFFERENTIAL EQUATIONS

CHAPTER 12 PROBABILITY

CHAPTER 13 PROBABILITY AND CALCULUS

TO THE STUDENT: SUCCESS IN MATHEMATICS

The main reason students have difficulty with mathematics is that they don't know how to study it. Studying mathematics *is* different from studying subjects like English or history. The key to success is regular practice.

This should not be surprising. After all, can you learn to play the piano or to ski well without a lot of regular practice? The same thing is true for learning mathematics. Working problems nearly every day is the key to becoming successful. Here is a list of things you can do to help you succeed in studying mathematics.

1. *Attend class regularly.* Pay attention in class to what your instructor says and does, and make careful notes. In particular, note the problems the instructor works on the board and copy the complete solutions. Keep these notes separate from your homework to avoid confusion when you read them over later.

2. Don't hesitate to ask questions in class. It is not a sign of weakness, but of strength. There are always other students with the same question who are too shy to ask.

3. *Read your text carefully.* Many students read only enough to get by, usually only the examples. Reading the complete section will help you to be successful with the homework problems. Most exercises are keyed to specific examples or objectives that will explain the procedures for working them.

4. Before you start on your homework assignment, rework the problems the instructor worked in class. This will reinforce what you have learned. Many students say, "I understand it perfectly when you do it, but I get stuck when I try to work the problem myself."

5. Do your homework assignment only *after* reading the text and reviewing your notes from class. Check your work with the answers in the back of the book. If you get a problem wrong and are unable to see why, mark that problem and ask your instructor about it. Then practice working additional problems of the same type to reinforce what you have learned.

6. Work as neatly as you can. Write your symbols clearly, and make sure the problems are clearly separated from each other. Working neatly will help you to think clearly and also make it easier to review the homework before a test.

7. After you have completed a homework assignment, look over the text again. Try to decide what the main ideas are in the lesson. Often they are clearly highlighted or boxed in the text.

8. Chapters 1–13 are each followed by a chapter test with questions from each section of the chapter. Use these as practice tests. Work through the problems under test conditions, without referring to the text or the answers until you are finished. You may want to time yourself to see how long it takes you. When you have finished, check your answers against the answers that follow each test and study those problems that you missed.

9. Keep any quizzes and tests that are returned to you and use them when you study for future tests and the final exam. These quizzes and tests indicate what your instructor considers most important. Be sure to correct any problems on these tests that you missed, so you will have the corrected work to study.

10. Don't worry if you do not understand a new topic right away. As you read more about it and work through the problems, you will gain understanding. Each time you look back at a topic you will understand it a little better. No one understands each topic completely right from the start.

ALGEBRA REFERENCE

R.1 Polynomials

1. $(2x^2 - 6x + 11) + (-3x^2 + 7x - 2)$
$= 2x^2 - 6x + 11 - 3x^2 + 7x - 2$
$= (2-3)x^2 + (7-6)x + (11-2)$
$= -x^2 + x + 9$

2. $(-4y^2 - 3y + 8) - (2y^2 - 6y - 2)$
$= (-4y^2 - 3y + 8) + (-2y^2 + 6y + 2)$
$= -4y^2 - 3y + 8 - 2y^2 + 6y + 2$
$= (-4y^2 - 2y^2) + (-3y + 6y)$
$\quad + (8 + 2)$
$= -6y^2 + 3y + 10$

3. $-3(4q^2 - 3q + 2) + 2(-q^2 + q - 4)$
$= -12q^2 + 9q - 6 - 2q^2 + 2q - 8$
$= -14q^2 + 11q - 14$

4. $2(3r^2 + 4r + 2) - 3(-r^2 + 4r - 5)$
$= (6r^2 + 8r + 4) + (3r^2 - 12r + 15)$
$= (6r^2 + 3r^2) + (8r - 12r)$
$\quad + (4 + 15)$
$= 9r^2 - 4r + 19$

5. $(0.613x^2 - 4.215x + 0.892)$
$\quad - 0.47(2x^2 - 3x + 5)$
$= 0.613x^2 - 4.215x + 0.892$
$\quad - 0.94x^2 + 1.41x - 2.35$
$= -0.327x^2 - 2.805x - 1.458$

6. $0.83(5r^2 - 2r + 7) - (7.12r^2 + 6.423r - 2)$
$= (4.15r^2 - 1.66r + 5.81)$
$\quad + (-7.12r^2 - 6.423r + 2)$
$= (4.15r^2 - 7.12r^2)$
$\quad + (-1.66r - 6.423r) + (5.81 + 2)$
$= -2.97r^2 - 8.083r + 7.81$

7. $-9m(2m^2 + 3m - 1)$
$= -9m(2m^2) - 9m(3m) - 9m(-1)$
$= -18m^3 - 27m^2 + 9m$

8. $(6k - 1)(2k - 3)$
$= (6k)(2k) + (6k)(-3) + (-1)(2k)$
$\quad + (-1)(-3)$
$= 12k^2 - 18k - 2k + 3$
$= 12k^2 - 20k + 3$

9. $(5r - 3s)(5r + 4s)$

Use the FOIL method to find this product.

$(5r - 3s)(5r + 4s)$
$= (5r)(5r) + (5r)(4s) + (-3s)(5r) + (-3s)(4s)$
$= 25r^2 + 20rs - 15rs - 12s^2$
$= 25r^2 + 5rs - 12s^2$

10. $(9k + q)(2k - q)$
$= (9k)(2k) + (9k)(-q) + (q)(2k)$
$\quad + (q)(-q)$
$= 18k^2 - 9kq + 2kq - q^2$
$= 18k^2 - 7kq - q^2$

11. $\left(\dfrac{2}{5}y + \dfrac{1}{8}z\right)\left(\dfrac{3}{5}y + \dfrac{1}{2}z\right)$

$= \left(\dfrac{2}{5}y\right)\left(\dfrac{3}{5}y\right) + \left(\dfrac{2}{5}y\right)\left(\dfrac{1}{2}z\right) + \left(\dfrac{1}{8}z\right)\left(\dfrac{3}{5}y\right)$
$\quad + \left(\dfrac{1}{8}z\right)\left(\dfrac{1}{2}z\right)$
$= \dfrac{6}{25}y^2 + \dfrac{1}{5}yz + \dfrac{3}{40}yz + \dfrac{1}{16}z^2$
$= \dfrac{6}{25}y^2 + \left(\dfrac{8}{40} + \dfrac{3}{40}\right)yz + \dfrac{1}{16}z^2$
$= \dfrac{6}{25}y^2 + \dfrac{11}{40}yz + \dfrac{1}{16}z^2$

12. $\left(\dfrac{3}{4}r - \dfrac{2}{3}s\right)\left(\dfrac{5}{4}r + \dfrac{1}{3}s\right)$

$= \left(\dfrac{3}{4}r\right)\left(\dfrac{5}{4}r\right) + \left(\dfrac{3}{4}r\right)\left(\dfrac{1}{3}s\right) + \left(-\dfrac{2}{3}s\right)\left(\dfrac{5}{4}r\right)$
$\quad + \left(-\dfrac{2}{3}s\right)\left(\dfrac{1}{3}s\right)$
$= \dfrac{15}{16}r^2 + \dfrac{1}{4}rs - \dfrac{5}{6}rs - \dfrac{2}{9}s^2$
$= \dfrac{15}{16}r^2 - \dfrac{7}{12}rs - \dfrac{2}{9}s^2$

13. $(12x - 1)(12x + 1)$
$= (12x)^2 - 1^2$
$= 144x^2 - 1$

14. $(6m + 5)(6m - 5)$
$= (6m)(6m) + (6m)(-5) + (5)(6m)$
$\quad + (5)(-5)$
$= 36m^2 - 30m + 30m - 25$
$= 36m^2 - 25$

15. $(3p - 1)(9p^2 + 3p + 1)$
$$= (3p - 1)(9p^2) + (3p - 1)(3p)$$
$$+ (3p - 1)(1)$$
$$= 3p(9p^2) - 1(9p^2) + 3p(3p)$$
$$- 1(3p) + 3p(1) - 1(1)$$
$$= 27p^3 - 9p^2 + 9p^2 - 3p + 3p - 1$$
$$= 27p^3 - 1$$

16. $(2p - 1)(3p^2 - 4p + 5)$
$$= (2p)(3p^2) + (2p)(-4p) + (2p)(5)$$
$$+ (-1)(3p^2) + (-1)(-4p) + (-1)(5)$$
$$= 6p^3 - 8p^2 + 10p - 3p^2 + 4p - 5$$
$$= 6p^3 - 11p^2 + 14p - 5$$

17. $(2m + 1)(4m^2 - 2m + 1)$
$$= 2m(4m^2 - 2m + 1) + 1(4m^2 - 2m + 1)$$
$$= 8m^3 - 4m^2 + 2m + 4m^2 - 2m + 1$$
$$= 8m^3 + 1$$

18. $(k + 2)(12k^3 - 3k^2 + k + 1)$
$$= k(12k^3) + k(-3k^2) + k(k) + k(1)$$
$$+ 2(12k^3) + 2(-3k^2) + 2(k) + 2(1)$$
$$= 12k^4 - 3k^3 + k^2 + k + 24k^3 - 6k^2$$
$$+ 2k + 2$$
$$= 12k^4 + 21k^3 - 5k^2 + 3k + 2$$

19. $(m - n + k)(m + 2n - 3k)$
$$= m(m + 2n - 3k) - n(m + 2n - 3k)$$
$$+ k(m + 2n - 3k)$$
$$= m^2 + 2mn - 3km - mn - 2n^2 + 3kn$$
$$+ km + 2kn - 3k^2$$
$$= m^2 + mn - 2n^2 - 2km + 5kn - 3k^2$$

20. $(r - 3s + t)(2r - s + t)$
$$= r(2r) + r(-s) + r(t) - 3s(2r)$$
$$- 3s(-s) - 3s(t) + t(2r) + t(-s)$$
$$+ t(t)$$
$$= 2r^2 - rs + rt - 6rs + 3s^2 - 3st$$
$$+ 2rt - st + t^2$$
$$= 2r^2 - 7rs + 3s^2 + 3rt - 4st + t^2$$

21. $(x + 1)(x + 2)(x + 3)$
$$= [x(x + 2) + 1(x + 2)](x + 3)$$
$$= [x^2 + 2x + x + 2](x + 3)$$
$$= [x^2 + 3x + 2](x + 3)$$
$$= x^2(x + 3) + 3x(x + 3) + 2(x + 3)$$
$$= x^3 + 3x^2 + 3x^2 + 9x + 2x + 6$$
$$= x^3 + 6x^2 + 11x + 6$$

22. $(x - 1)(x + 2)(x - 3)$
$$= [x(x + 2) + (-1)(x + 2)](x - 3)$$
$$= (x^2 + 2x - x - 2)(x - 3)$$
$$= (x^2 + x - 2)(x - 3)$$
$$= x^2(x - 3) + x(x - 3) + (-2)(x - 3)$$
$$= x^3 - 3x^2 + x^2 - 3x - 2x + 6$$
$$= x^3 - 2x^2 - 5x + 6$$

23. $(3a + b)^2$
$$= (3a + b)(3a + b)$$
$$= 3a(3a + b) + b(3a + b)$$
$$= 9a^2 + 3ab + 3ab + b^2$$
$$= 9a^2 + 6ab + b^2$$

24. $(x - 2y)^3$
$$= [(x - 2y)(x - 2y)](x - 2y)$$
$$= (x^2 - 2xy - 2xy + 4y^2)(x - 2y)$$
$$= (x^2 - 4xy + 4y^2)(x - 2y)$$
$$= (x^2 - 4xy + 4y^2)x + (x^2 - 4xy + 4y^2)(-2y)$$
$$= x^3 - 4x^2y + 4xy^2 - 2x^2y + 8xy^2 - 8y^3$$
$$= x^3 - 6x^2y + 12xy^2 - 8y^3$$

R.2   Factoring

1. $8a^3 - 16a^2 + 24a$
$$= 8a \cdot a^2 - 8a \cdot 2a + 8a \cdot 3$$
$$= 8a(a^2 - 2a + 3)$$

2. $3y^3 + 24y^2 + 9y$
$$= 3y \cdot y^2 + 3y \cdot 8y + 3y \cdot 3$$
$$= 3y(y^2 + 8y + 3)$$

3. $25p^4 - 20p^3q + 100p^2q^2$
$$= 5p^2 \cdot 5p^2 - 5p^2 \cdot 4pq + 5p^2 \cdot 20q^2$$
$$= 5p^2(5p^2 - 4pq + 20q^2)$$

4. $60m^4 - 120m^3n + 50m^2n^2$
$$= 10m^2 \cdot 6m^2 - 10m^2 \cdot 12mn$$
$$+ 10m^2 \cdot 5n^2$$
$$= 10m^2(6m^2 - 12mn + 5n^2)$$

5. $m^2 + 9m + 14 = (m + 2)(m + 7)$

since $2 \cdot 7 = 14$ and $2 + 7 = 9$.

6. $x^2 + 4x - 5 = (x + 5)(x - 1)$

since $5 \cdot (-1) = -5$ and $-1 + 5 = 4$.

7. $z^2 + 9z + 20 = (z + 4)(z + 5)$

since $4 \cdot 5 = 20$ and $4 + 5 = 9$.

8. $b^2 - 8b + 7 = (b - 7)(b - 1)$

since $(-7) \cdot (-1) = 7$ and $-7 + (-1) = -8$.

9. $a^2 - 6ab + 5b^2 = (a - b)(a - 5b)$

since $(-b)(-5b) = 5b^2$ and
$-b + (-5b) = -6b$.

10. $s^2 + 2st - 35t^2 = (s - 5t)(s + 7t)$

since $(-5t) \cdot (7t) = -35t^2$ and $(7t) + (-5t) = 2t$.

11. $y^2 - 4yz - 21z^2 = (y + 3z)(y - 7z)$

since $(3z)(-7z) = -21z^2$ and
$3z + (-7z) = -4z$.

12. $6a^2 - 48a - 120$
$= 6(a^2 - 8a - 20)$
$= 6(a - 10)(a + 2)$

13. $3m^3 + 12m^2 + 9m$
$= 3m(m^2 + 4m + 3)$
$= 3m(m + 1)(m + 3)$

14. $2x^2 - 5x - 3$

The possible factors of $2x^2$ are $2x$ and x and the possible factors of -3 are -3 and 1, or 3 and -1. Try various combinations until one works.

$$2x^2 - 5x - 3 = (2x + 1)(x - 3)$$

15. $3a^2 + 10a + 7$

The possible factors of $3a^2$ are $3a$ and a and the possible factors of 7 are 7 and 1. Try various combinations until one works.

$$3a^2 + 10a + 7 = (a + 1)(3a + 7)$$

16. $2a^2 - 17a + 30 = (2a - 5)(a - 6)$

17. $15y^2 + y - 2 = (5y + 2)(3y - 1)$

18. $21m^2 + 13mn + 2n^2$
$= (7m + 2n)(3m + n)$

19. $24a^4 + 10a^3b - 4a^2b^2$
$= 2a^2(12a^2 + 5ab - 2b^2)$
$= 2a^2(4a - b)(3a + 2b)$

20. $32z^5 - 20z^4a - 12z^3a^2$
$= 4z^3(8z^2 - 5za - 3a^2)$
$= 4z^3(8z + 3a)(z - a)$

21. $x^2 - 64 = x^2 - 8^2$
$= (x + 8)(x - 8)$

22. $9m^2 - 25 = (3m)^2 - (5)^2$
$= (3m + 5)(3m - 5)$

23. $121a^2 - 100$
$= (11a)^2 - 10^2$
$= (11a + 10)(11a - 10)$

24. $9x^2 + 64$ is the *sum* of two perfect squares. It cannot be factored. It is prime.

25. $z^2 + 14zy + 49y^2$
$= z^2 + 2 \cdot 7zy + 7^2y^2$
$= (z + 7y)^2$

26. $m^2 - 6mn + 9n^2$
$= m^2 - 2(3mn) + (3n)^2$
$= (m - 3n)^2$

27. $9p^2 - 24p + 16$
$= (3p)^2 - 2 \cdot 3p \cdot 4 + 4^2$
$= (3p - 4)^2$

28. $a^3 - 216$
$= a^3 - 6^3$
$= (a - 6)[(a)^2 + (a)(6) + (6)^2]$
$= (a - 6)(a^2 + 6a + 36)$

29. $8r^3 - 27s^3$
$= (2r)^3 - (3s)^3$
$= (2r - 3s)(4r^2 + 6rs + 9s^2)$

30. $64m^3 + 125$
$= (4m)^3 + 5^3$
$= (4m + 5)[(4m)^2 - (4m)(5) + (5)^2]$
$= (4m + 5)(16m^2 - 20m + 25)$

31. $x^4 - y^4 = (x^2)^2 - (y^2)^2$
$= (x^2 + y^2)(x^2 - y^2)$
$= (x^2 + y^2)(x + y)(x - y)$

32. $16a^4 - 81b^4$
$= (4a^2)^2 - (9b^2)^2$
$= (4a^2 + 9b^2)(4a^2 - 9b^2)$
$= (4a^2 + 9b^2)[(2a)^2 - (3b)^2]$
$= (4a^2 + 9b^2)(2a + 3b)(2a - 3b)$

R.3   Rational Expressions

1. $\dfrac{7z^2}{14z} = \dfrac{7z \cdot z}{2(7)z} = \dfrac{z}{2}$

2. $\dfrac{25p^3}{10p^2} = \dfrac{5 \cdot 5 \cdot p \cdot p \cdot p}{2 \cdot 5 \cdot p \cdot p} = \dfrac{5p}{2}$

3. $\dfrac{8k + 16}{9k + 18} = \dfrac{8(k + 2)}{9(k + 2)} = \dfrac{8}{9}$

4. $\dfrac{3(t+5)}{(t+5)(t-3)} = \dfrac{3}{t-3}$

5. $\dfrac{8x^2+16x}{4x^2} = \dfrac{8x(x+2)}{4x^2}$

$\qquad\qquad = \dfrac{2(x+2)}{x}$

6. $\dfrac{36y^2+72y}{9y} = \dfrac{36y(y+2)}{9y}$

$\qquad\qquad\quad = \dfrac{9\cdot 4\cdot y(y+2)}{9\cdot y}$

$\qquad\qquad\quad = 4(y+2)$

7. $\dfrac{m^2-4m+4}{m^2+m-6}$

$\quad = \dfrac{(m-2)(m-2)}{(m-2)(m+3)}$

$\quad = \dfrac{m-2}{m+3}$

8. $\dfrac{r^2-r-6}{r^2+r-12} = \dfrac{(r-3)(r+2)}{(r+4)(r-3)}$

$\qquad\qquad\quad = \dfrac{r+2}{r+4}$

9. $\dfrac{x^2+3x-4}{x^2-1}$

$\quad = \dfrac{(x-1)(x+4)}{(x-1)(x+1)}$

$\quad = \dfrac{x+4}{x+1}$

10. $\dfrac{z^2-5z+6}{z^2-4} = \dfrac{(z-3)(z-2)}{(z+2)(z-2)}$

$\qquad\qquad\quad = \dfrac{z-3}{z+2}$

11. $\dfrac{8m^2+6m-9}{16m^2-9}$

$\quad = \dfrac{(4m-3)(2m+3)}{(4m-3)(4m+3)}$

$\quad = \dfrac{2m+3}{4m+3}$

12. $\dfrac{6y^2+11y+4}{3y^2+7y+4}$

$\quad = \dfrac{(3y+4)(2y+1)}{(3y+4)(y+1)}$

$\quad = \dfrac{2y+1}{y+1}$

13. $\dfrac{9k^2}{25}\cdot\dfrac{5}{3k} = \dfrac{3\cdot 3\cdot 5k^2}{5\cdot 5\cdot 3k} = \dfrac{3k^2}{5k} = \dfrac{3k}{5}$

14. $\dfrac{15p^3}{9p^2}\div\dfrac{6p}{10p^2}$

$\quad = \dfrac{15p^3}{9p^2}\cdot\dfrac{10p^2}{6p}$

$\quad = \dfrac{150p^5}{54p^3}$

$\quad = \dfrac{25\cdot 6p^5}{9\cdot 6p^3}$

$\quad = \dfrac{25p^2}{9}$

15. $\dfrac{a+b}{2p}\cdot\dfrac{12}{5(a+b)}$

$\quad = \dfrac{(a+b)12}{2p(5)(a+b)} = \dfrac{12}{10p}$

$\quad = \dfrac{6}{5p}$

16. $\dfrac{a-3}{16}\div\dfrac{a-3}{32} = \dfrac{a-3}{16}\cdot\dfrac{32}{a-3}$

$\qquad\qquad\qquad = \dfrac{a-3}{16}\cdot\dfrac{16\cdot 2}{a-3}$

$\qquad\qquad\qquad = \dfrac{2}{1} = 2$

17. $\dfrac{2k+8}{6}\div\dfrac{3k+12}{2}$

$\quad = \dfrac{2(k+4)}{6}\cdot\dfrac{2}{3(k+4)}$

$\quad = \dfrac{2(k+4)(2)}{2\cdot 3\cdot 3(k+4)}$

$\quad = \dfrac{2}{9}$

18. $\dfrac{9y-18}{6y+12}\cdot\dfrac{3y+6}{15y-30}$

$\quad = \dfrac{9(y-2)}{6(y+2)}\cdot\dfrac{3(y+2)}{15(y-2)}$

$\quad = \dfrac{27}{90} = \dfrac{3\cdot 3}{10\cdot 3} = \dfrac{3}{10}$

19. $\dfrac{4a+12}{2a-10}\div\dfrac{a^2-9}{a^2-a-20}$

$\quad = \dfrac{4(a+3)}{2(a-5)}\cdot\dfrac{(a-5)(a+4)}{(a-3)(a+3)}$

$\quad = \dfrac{2(a+4)}{a-3}$

20. $\dfrac{6r-18}{9r^2+6r-24}\cdot\dfrac{12r-16}{4r-12}$

$\qquad=\dfrac{6(r-3)}{3(3r^2+2r-8)}\cdot\dfrac{4(3r-4)}{4(r-3)}$

$\qquad=\dfrac{6(r-3)}{3(3r-4)(r+2)}\cdot\dfrac{4(3r-4)}{4(r-3)}$

$\qquad=\dfrac{6}{3(r+2)}$

$\qquad=\dfrac{2}{r+2}$

21. $\dfrac{k^2-k-6}{k^2+k-12}\cdot\dfrac{k^2+3k-4}{k^2+2k-3}$

$\qquad=\dfrac{(k-3)(k+2)(k-1)(k+4)}{(k+4)(k-3)(k+3)(k-1)}$

$\qquad=\dfrac{k+2}{k+3}$

22. $\dfrac{m^2+3m+2}{m^2+5m+4}\div\dfrac{m^2+5m+6}{m^2+10m+24}$

$\qquad=\dfrac{m^2+3m+2}{m^2+5m+4}\cdot\dfrac{m^2+10m+24}{m^2+5m+6}$

$\qquad=\dfrac{(m+1)(m+2)}{(m+4)(m+1)}\cdot\dfrac{(m+6)(m+4)}{(m+3)(m+2)}$

$\qquad=\dfrac{m+6}{m+3}$

23. $\dfrac{2m^2-5m-12}{m^2-10m+24}\div\dfrac{4m^2-9}{m^2-9m+18}$

$\qquad=\dfrac{2m^2-5m-12}{m^2-10m+24}\cdot\dfrac{m^2-9m+18}{4m^2-9}$

$\qquad=\dfrac{(2m+3)(m-4)(m-6)(m-3)}{(m-6)(m-4)(2m-3)(2m+3)}$

$\qquad=\dfrac{m-3}{2m-3}$

24. $\dfrac{6n^2-5n-6}{6n^2+5n-6}\cdot\dfrac{12n^2-17n+6}{12n^2-n-6}$

$\qquad=\dfrac{(2n-3)(3n+2)}{(2n+3)(3n-2)}\cdot\dfrac{(3n-2)(4n-3)}{(3n+2)(4n-3)}$

$\qquad=\dfrac{2n-3}{2n+3}$

25. $\dfrac{a+1}{2}-\dfrac{a-1}{2}$

$\qquad=\dfrac{(a+1)-(a-1)}{2}$

$\qquad=\dfrac{a+1-a+1}{2}$

$\qquad=\dfrac{2}{2}=1$

26. $\dfrac{3}{p}+\dfrac{1}{2}$

Multiply the first term by $\frac{2}{2}$ and the second by $\frac{p}{p}$.

$\qquad\dfrac{2\cdot3}{2\cdot p}+\dfrac{p\cdot1}{p\cdot2}=\dfrac{6}{2p}+\dfrac{p}{2p}$

$\qquad\qquad=\dfrac{6+p}{2p}$

27. $\dfrac{2}{y}-\dfrac{1}{4}=\left(\dfrac{4}{4}\right)\dfrac{2}{y}-\left(\dfrac{y}{y}\right)\dfrac{1}{4}$

$\qquad=\dfrac{8-y}{4y}$

28. $\dfrac{1}{6m}+\dfrac{2}{5m}+\dfrac{4}{m}$

$\qquad=\dfrac{5\cdot1}{5\cdot6m}+\dfrac{6\cdot2}{6\cdot5m}+\dfrac{30\cdot4}{30\cdot m}$

$\qquad=\dfrac{5}{30m}+\dfrac{12}{30m}+\dfrac{120}{30m}$

$\qquad=\dfrac{5+12+120}{30m}$

$\qquad=\dfrac{137}{30m}$

29. $\dfrac{1}{m-1}+\dfrac{2}{m}$

$\qquad=\dfrac{m}{m}\left(\dfrac{1}{m-1}\right)+\dfrac{m-1}{m-1}\left(\dfrac{2}{m}\right)$

$\qquad=\dfrac{m+2m-2}{m(m-1)}$

$\qquad=\dfrac{3m-2}{m(m-1)}$

30. $\dfrac{6}{r}-\dfrac{5}{r-2}$

$\qquad=\dfrac{6(r-2)}{r(r-2)}-\dfrac{5r}{r(r-2)}$

$\qquad=\dfrac{6(r-2)-5r}{r(r-2)}$

$\qquad=\dfrac{6r-12-5r}{r(r-2)}$

$\qquad=\dfrac{r-12}{r(r-2)}$

31. $\dfrac{8}{3(a-1)}+\dfrac{2}{a-1}$

$\qquad=\dfrac{8}{3(a-1)}+\dfrac{3}{3}\left(\dfrac{2}{a-1}\right)$

$\qquad=\dfrac{8+6}{3(a-1)}$

$\qquad=\dfrac{14}{3(a-1)}$

32. $\dfrac{2}{5(k-2)} + \dfrac{3}{4(k-2)}$

$= \dfrac{4 \cdot 2}{4 \cdot 5(k-2)} + \dfrac{5 \cdot 3}{5 \cdot 4(k-2)}$

$= \dfrac{8}{20(k-2)} + \dfrac{15}{20(k-2)}$

$= \dfrac{8+15}{20(k-2)}$

$= \dfrac{23}{20(k-2)}$

33. $\dfrac{2}{x^2 - 2x - 3} + \dfrac{5}{x^2 - x - 6}$

$= \dfrac{2}{(x-3)(x+1)} + \dfrac{5}{(x-3)(x+2)}$

$= \left(\dfrac{x+2}{x+2}\right)\dfrac{2}{(x-3)(x+1)}$

$\quad + \left(\dfrac{x+1}{x+1}\right)\dfrac{5}{(x-3)(x+2)}$

$= \dfrac{2x+4+5x+5}{(x+2)(x-3)(x+1)}$

$= \dfrac{7x+9}{(x+2)(x-3)(x+1)}$

34. $\dfrac{2y}{y^2 + 7y + 12} - \dfrac{y}{y^2 + 5y + 6}$

$= \dfrac{2y}{(y+4)(y+3)} - \dfrac{y}{(y+3)(y+2)}$

$= \dfrac{2y(y+2)}{(y+4)(y+3)(y+2)}$

$\quad - \dfrac{y(y+4)}{(y+3)(y+2)(y+4)}$

$= \dfrac{2y(y+2) - y(y+4)}{(y+4)(y+3)(y+2)}$

$= \dfrac{2y^2 + 4y - y^2 - 4y}{(y+4)(y+3)(y+2)}$

$= \dfrac{y^2}{(y+4)(y+3)(y+2)}$

35. $\dfrac{3k}{2k^2 + 3k - 2} - \dfrac{2k}{2k^2 - 7k + 3}$

$= \dfrac{3k}{(2k-1)(k+2)} - \dfrac{2k}{(2k-1)(k-3)}$

$\left(\dfrac{k-3}{k-3}\right)\dfrac{3k}{(2k-1)(k+2)}$

$\quad - \left(\dfrac{k+2}{k+2}\right)\dfrac{2k}{(2k-1)(k-3)}$

$= \dfrac{(3k^2 - 9k) - (2k^2 + 4k)}{(2k-1)(k+2)(k-3)}$

$= \dfrac{k^2 - 13k}{(2k-1)(k+2)(k-3)}$

$= \dfrac{k(k-13)}{(2k-1)(k+2)(k-3)}$

36. $\dfrac{4m}{3m^2 + 7m - 6} - \dfrac{m}{3m^2 - 14m + 8}$

$= \dfrac{4m}{(3m-2)(m+3)} - \dfrac{m}{(3m-2)(m-4)}$

$= \dfrac{4m(m-4)}{(3m-2)(m+3)(m-4)}$

$\quad - \dfrac{m(m+3)}{(3m-2)(m-4)(m+3)}$

$= \dfrac{4m(m-4) - m(m+3)}{(3m-2)(m-4)(m+3)}$

$= \dfrac{4m^2 - 16m - m^2 - 3m}{(3m-2)(m+3)(m-4)}$

$= \dfrac{3m^2 - 19m}{(3m-2)(m+3)(m-4)}$

$= \dfrac{m(3m-19)}{(3m-2)(m+3)(m-4)}$

37. $\dfrac{2}{a+2} + \dfrac{1}{a} + \dfrac{a-1}{a^2 + 2a}$

$= \dfrac{2}{a+2} + \dfrac{1}{a} + \dfrac{a-1}{a(a+2)}$

$= \left(\dfrac{a}{a}\right)\dfrac{2}{a+2} + \left(\dfrac{a+2}{a+2}\right)\dfrac{1}{a} + \dfrac{a-1}{a(a+2)}$

$= \dfrac{2a + a + 2 + a - 1}{a(a+2)}$

$= \dfrac{4a+1}{a(a+2)}$

38. $\dfrac{5x+2}{x^2 - 1} + \dfrac{3}{x^2 + x} - \dfrac{1}{x^2 - x}$

$= \dfrac{5x+2}{(x+1)(x-1)} + \dfrac{3}{x(x+1)} - \dfrac{1}{x(x-1)}$

$= \left(\dfrac{x}{x}\right)\left(\dfrac{5x+2}{(x+1)(x-1)}\right) + \left(\dfrac{x-1}{x-1}\right)\left(\dfrac{3}{x(x+1)}\right)$

$\quad - \left(\dfrac{x+1}{x+1}\right)\left(\dfrac{1}{x(x-1)}\right)$

$= \dfrac{x(5x+2) + (x-1)(3) - (x+1)(1)}{x(x+1)(x-1)}$

$= \dfrac{5x^2 + 2x + 3x - 3 - x - 1}{x(x+1)(x-1)}$

$= \dfrac{5x^2 + 4x - 4}{x(x+1)(x-1)}$

R.4   Equations

1. $0.2m - 0.5 = 0.1m + 0.7$
$$10(0.2m - 0.5) = 10(0.1m + 0.7)$$
$$2m - 5 = m + 7$$
$$m - 5 = 7$$
$$m = 12$$

The solution is 12.

2. $\dfrac{5}{6}k - 2k + \dfrac{1}{3} = \dfrac{2}{3}$

Multiply both sides of the equation by 6.

$$6\left(\frac{5}{6}k\right) - 6(2k) + 6\left(\frac{1}{3}\right) = 6\left(\frac{2}{3}\right)$$
$$5k - 12k + 2 = 4$$
$$-7k + 2 = 4$$
$$-7k = 2$$
$$k = -\frac{2}{7}$$

The solution is $-\frac{2}{7}$.

3. $2x + 8 = x - 4$
$$x + 8 = -4$$
$$x = -12$$

The solution is -12.

4. $5x + 2 = 8 - 3x$
$$8x + 2 = 8$$
$$8x = 6$$
$$x = \frac{3}{4}$$

The solution is $\frac{3}{4}$.

5. $3r + 2 - 5(r + 1) = 6r + 4$
$$3r + 2 - 5r - 5 = 6r + 4$$
$$-3 - 2r = 6r + 4$$
$$-3 = 8r + 4$$
$$-7 = 8r$$
$$-\frac{7}{8} = r$$

The solution is $-\frac{7}{8}$.

6. $5(a + 3) + 4a - 5 = -(2a - 4)$
$$5a + 15 + 4a - 5 = -2a + 4$$
$$9a + 10 = -2a + 4$$
$$11a + 10 = 4$$
$$11a = -6$$
$$a = -\frac{6}{11}$$

The solution is $-\frac{6}{11}$.

7. $2[m - (4 + 2m) + 3] = 2m + 2$
$$2[m - 4 - 2m + 3] = 2m + 2$$
$$2[-m - 1] = 2m + 2$$
$$-2m - 2 = 2m + 2$$
$$-2m = 2m + 4$$
$$-4m = 4$$
$$m = -1$$

The solution is -1.

8. $4[2p - (3 - p) + 5] = -7p - 2$
$$4[2p - 3 + p + 5] = -7p - 2$$
$$4[3p + 2] = -7p - 2$$
$$12p + 8 = -7p - 2$$
$$19p + 8 = -2$$
$$19p = -10$$
$$p = -\frac{10}{19}$$

The solution is $-\frac{10}{19}$.

9. $x^2 + 5x + 6 = 0$
$$(x + 3)(x + 2) = 0$$
$$x + 3 = 0 \quad \text{or} \quad x + 2 = 0$$
$$x = -3 \quad \text{or} \quad x = -2$$

The solutions are -3 and -2.

10. $x^2 = 3 + 2x$
$$x^2 - 2x - 3 = 0$$
$$(x - 3)(x + 1) = 0$$
$$x - 3 = 0 \quad \text{or} \quad x + 1 = 0$$
$$x = 3 \quad \text{or} \quad x = -1$$

The solutions are 3 and -1.

11. $m^2 + 16 = 8m$
$$m^2 - 8m + 16 = 0$$
$$(m)^2 - 2(4m) + (4)^2 = 0$$
$$(m - 4)^2 = 0$$
$$m - 4 = 0$$
$$m = 4$$

The solution is 4.

12. $2k^2 - k = 10$
$$2k^2 - k - 10 = 0$$
$$(2k - 5)(k + 2) = 0$$
$$2k - 5 = 0 \quad \text{or} \quad k + 2 = 0$$
$$k = \frac{5}{2} \quad \text{or} \quad k = -2$$

The solutions are $\frac{5}{2}$ and -2.

13.
$$6x^2 - 5x = 4$$
$$6x^2 - 5x - 4 = 0$$
$$(3x - 4)(2x + 1) = 0$$
$$3x - 4 = 0 \quad \text{or} \quad 2x + 1 = 0$$
$$3x = 4 \qquad\qquad 2x = -1$$
$$x = \frac{4}{3} \quad \text{or} \qquad x = -\frac{1}{2}$$

The solutions are $\frac{4}{3}$ and $-\frac{1}{2}$.

14.
$$m(m - 7) = -10$$
$$m^2 - 7m + 10 = 0$$
$$(m - 5)(m - 2) = 0$$
$$m - 5 = 0 \quad \text{or} \quad m - 2 = 0$$
$$m = 5 \quad \text{or} \qquad m = 2$$

The solutions are 5 and 2.

15.
$$9x^2 - 16 = 0$$
$$(3x)^2 - (4)^2 = 0$$
$$(3x + 4)(3x - 4) = 0$$
$$3x + 4 = 0 \qquad \text{or} \quad 3x - 4 = 0$$
$$3x = -4 \qquad\qquad 3x = 4$$
$$x = -\frac{4}{3} \quad \text{or} \qquad x = \frac{4}{3}$$

The solutions are $-\frac{4}{3}$ and $\frac{4}{3}$.

16.
$$z(2z + 7) = 4$$
$$2z^2 + 7z - 4 = 0$$
$$(2z - 1)(z + 4) = 0$$
$$2z - 1 = 0 \quad \text{or} \quad z + 4 = 0$$
$$z = \frac{1}{2} \quad \text{or} \qquad z = -4$$

The solutions are $\frac{1}{2}$ and -4.

17.
$$12y^2 - 48y = 0$$
$$12y(y) - 12y(4) = 0$$
$$12y(y - 4) = 0$$
$$12y = 0 \quad \text{or} \quad y - 4 = 0$$
$$y = 0 \quad \text{or} \qquad y = 4$$

The solutions are 0 and 4.

18. $3x^2 - 5x + 1 = 0$

Use the quadratic formula.

$$x = \frac{-(-5) \pm \sqrt{(-5)^2 - 4(3)(1)}}{2(3)}$$

$$= \frac{5 \pm \sqrt{25 - 12}}{6}$$

$$x = \frac{5 + \sqrt{13}}{6} \quad \text{or} \quad x = \frac{5 - \sqrt{13}}{6}$$

$$\approx 1.434 \qquad\qquad \approx 0.232$$

The solutions are $\frac{5+\sqrt{13}}{6} \approx 1.434$ and $\frac{5-\sqrt{13}}{6} \approx 0.232$.

19.
$$2m^2 = m + 4$$
$$2m^2 - m - 4 = 0$$

Use the quadratic formula.

$$x = \frac{-(-1) \pm \sqrt{(-1)^2 - 4(2)(-4)}}{2(2)}$$

$$x = \frac{1 \pm \sqrt{1 + 32}}{4}$$

$$x = \frac{1 \pm \sqrt{33}}{4}$$

The solutions are $\frac{1+\sqrt{33}}{4} \approx 1.686$ and $\frac{1-\sqrt{33}}{4} \approx -1.186$.

20. $p^2 + p - 1 = 0$

$$p = \frac{-1 \pm \sqrt{1^2 - 4(1)(-1)}}{2(1)}$$

$$= \frac{-1 \pm \sqrt{5}}{2}$$

The solutions are $\frac{-1+\sqrt{5}}{2} \approx 0.618$ and $\frac{-1-\sqrt{5}}{2} \approx -1.618$.

21.
$$k^2 - 10k = -20$$
$$k^2 - 10k + 20 = 0$$

$$k = \frac{-(-10) \pm \sqrt{(-10)^2 - 4(1)(20)}}{2(1)}$$

$$k = \frac{10 \pm \sqrt{100 - 80}}{2}$$

$$k = \frac{10 \pm \sqrt{20}}{2}$$

$$k = \frac{10 \pm \sqrt{4}\sqrt{5}}{2}$$

$$k = \frac{10 \pm 2\sqrt{5}}{2}$$

$$k = \frac{2(5 \pm \sqrt{5})}{2}$$

$$k = 5 \pm \sqrt{5}$$

The solutions are $5 + \sqrt{5} \approx 7.236$ and $5 - \sqrt{5} \approx 2.764$.

22. $2x^2 + 12x + 5 = 0$

$$x = \frac{-12 \pm \sqrt{(12)^2 - 4(2)(5)}}{2(2)}$$

$$= \frac{-12 \pm \sqrt{104}}{4} = \frac{-12 \pm \sqrt{4 \cdot 26}}{4}$$

$$= \frac{-12 \pm \sqrt{4}\sqrt{26}}{4} = \frac{-12 \pm 2\sqrt{26}}{4}$$

$$= \frac{2(-6 \pm \sqrt{26})}{2 \cdot 2} = \frac{-6 \pm \sqrt{26}}{2}$$

The solutions are $\frac{-6+\sqrt{26}}{2} \approx -0.450$ and

$\frac{-6-\sqrt{26}}{2} \approx -5.550$.

23. $2r^2 - 7r + 5 = 0$

$(2r - 5)(r - 1) = 0$

$2r - 5 = 0 \quad$ or $\quad r - 1 = 0$

$2r = 5$

$r = \frac{5}{2} \quad$ or $\qquad r = 1$

The solutions are $\frac{5}{2}$ and 1.

24. $2x^2 - 7x + 30 = 0$

$$x = \frac{-(-7) \pm \sqrt{(-7)^2 - 4(2)(30)}}{2(2)}$$

$$x = \frac{7 \pm \sqrt{49 - 240}}{4}$$

$$x = \frac{7 \pm \sqrt{-191}}{4}$$

Since there is a negative number under the radical sign, $\sqrt{-191}$ is not a real number. Thus, there are no real-number solutions.

25. $3k^2 + k = 6$

$3k^2 + k - 6 = 0$

$$k = \frac{-1 \pm \sqrt{1 - 4(3)(-6)}}{2(3)}$$

$$= \frac{-1 \pm \sqrt{73}}{6}$$

The solutions are $\frac{-1+\sqrt{73}}{6} \approx 1.257$ and $\frac{-1-\sqrt{73}}{6} \approx -1.591$.

26. $5m^2 + 5m = 0$

$5m(m + 1) = 0$

$5m = 0 \quad$ or $\quad m + 1 = 0$

$m = 0 \quad$ or $\qquad m = -1$

The solutions are 0 and -1.

27. $$\frac{3x - 2}{7} = \frac{x + 2}{5}$$

$$35\left(\frac{3x - 2}{7}\right) = 35\left(\frac{x + 2}{5}\right)$$

$$5(3x - 2) = 7(x + 2)$$

$$15x - 10 = 7x + 14$$

$$8x = 24$$

$$x = 3$$

The solution is $x = 3$.

28. $\frac{x}{3} - 7 = 6 - \frac{3x}{4}$

Multiply both sides by 12, the least common denominator of 3 and 4.

$$12\left(\frac{x}{3} - 7\right) = 12\left(6 - \frac{3x}{4}\right)$$

$$12\left(\frac{x}{3}\right) - (12)(7) = (12)(6) - (12)\left(\frac{3x}{4}\right)$$

$$4x - 84 = 72 - 9x$$

$$13x - 84 = 72$$

$$13x = 156$$

$$x = 12$$

The solution is 12.

29. $\frac{4}{x - 3} - \frac{8}{2x + 5} + \frac{3}{x - 3} = 0$

$$\frac{4}{x - 3} + \frac{3}{x - 3} - \frac{8}{2x + 5} = 0$$

$$\frac{7}{x - 3} - \frac{8}{2x + 5} = 0$$

Multiply both sides by $(x - 3)(2x + 5)$. Note that $x \neq 3$ and $x \neq -\frac{5}{2}$.

$$(x-3)(2x+5)\left(\frac{7}{x - 3} - \frac{8}{2x + 5}\right) = (x-3)(2x+5)(0)$$

$$7(2x + 5) - 8(x - 3) = 0$$

$$14x + 35 - 8x + 24 = 0$$

$$6x + 59 = 0$$

$$6x = -59$$

$$x = -\frac{59}{6}$$

Note: It is especially important to check solutions of equations that involve rational expressions. Here, a check shows that $-\frac{59}{6}$ is a solution.

30. $\dfrac{5}{2p+3} - \dfrac{3}{p-2} = \dfrac{4}{2p+3}$

Multiply both sides by $(2p+3)(p-2)$.

Note that $p \neq -\dfrac{3}{2}$ and $p \neq 2$.

$(2p+3)(p-2)\left(\dfrac{5}{2p+3} - \dfrac{3}{p-2}\right)$

$\quad = (2p+3)(p-2)\left(\dfrac{4}{2p+3}\right)$

$(2p+3)(p-2)\left(\dfrac{5}{2p+3}\right) - (2p+3)(p-2)\left(\dfrac{3}{p-2}\right)$

$\quad = (2p+3)(p-2)\left(\dfrac{4}{2p+3}\right)$

$(p-2)(5) - (2p+3)(3) = (p-2)(4)$

$5p - 10 - 6p - 9 = 4p - 8$

$-p - 19 = 4p - 8$

$-5p - 19 = -8$

$-5p = 11$

$p = -\dfrac{11}{5}$

The solutions is $-\dfrac{11}{5}$.

31. $\dfrac{2}{m} + \dfrac{m}{m+3} = \dfrac{3m}{m^2 + 3m}$

$\dfrac{2}{m} + \dfrac{m}{m+3} = \dfrac{3m}{m(m+3)}$

Multiply both sides by $m(m+3)$.
Note that $m \neq 0$, and $m \neq -3$.

$m(m+3)\left(\dfrac{2}{m} + \dfrac{m}{m+3}\right) = m(m+3)\left(\dfrac{3m}{m(m+3)}\right)$

$2(m+3) + m(m) = 3m$

$2m + 6 + m^2 = 3m$

$m^2 - m + 6 = 0$

$m = \dfrac{-(-1) \pm \sqrt{(-1)^2 - 4(1)(6)}}{2(1)}$

$\quad = \dfrac{1 \pm \sqrt{1 - 24}}{2}$

$\quad = \dfrac{1 \pm \sqrt{-23}}{2}$

There are no real number solutions.

32. $\dfrac{2y}{y-1} = \dfrac{5}{y} + \dfrac{10 - 8y}{y^2 - y}$

$\dfrac{2y}{y-1} = \dfrac{5}{y} + \dfrac{10 - 8y}{y(y-1)}$

Multiply both sides by $y(y-1)$.
Note that $y \neq 0$ and $y \neq 1$.

$y(y-1)\left(\dfrac{2y}{y-1}\right) = y(y-1)\left[\dfrac{5}{y} + \dfrac{10 - 8y}{y(y-1)}\right]$

$y(y-1)\left(\dfrac{2y}{y-1}\right) = y(y-1)\left(\dfrac{5}{y}\right)$

$\quad + y(y-1)\left[\dfrac{10 - 8y}{y(y-1)}\right]$

$y(2y) = (y-1)(5) + (10 - 8y)$

$2y^2 = 5y - 5 + 10 - 8y$

$2y^2 = 5 - 3y$

$2y^2 + 3y - 5 = 0$

$(2y + 5)(y - 1) = 0$

$2y + 5 = 0 \quad \text{or} \quad y - 1 = 0$

$y = -\dfrac{5}{2} \quad \text{or} \qquad y = 1$

Since $y \neq 1$, 1 is not a solution.
The solution is $-\dfrac{5}{2}$.

33. $\dfrac{1}{x-2} - \dfrac{3x}{x-1} = \dfrac{2x+1}{x^2 - 3x + 2}$

$\dfrac{1}{x-2} - \dfrac{3x}{x-1} = \dfrac{2x+1}{(x-2)(x-1)}$

Multiply both sides by $(x-2)(x-1)$.
Note that $x \neq 2$ and $x \neq 1$.

$(x-2)(x-1)\left(\dfrac{1}{x-2} - \dfrac{3x}{x-1}\right) = (x-2)(x-1)$

$\quad \cdot \left[\dfrac{2x+1}{(x-2)(x-1)}\right]$

$(x-2)(x-1)\left(\dfrac{1}{x-2}\right)$

$\quad - (x-2)(x-1)\cdot\left(\dfrac{3x}{x-1}\right) = \dfrac{(x-2)(x-1)(2x+1)}{(x-2)(x-1)}$

$(x-1) - (x-2)(3x) = 2x+1$

$x - 1 - 3x^2 + 6x = 2x + 1$

$-3x^2 + 7x - 1 = 2x + 1$

$-3x^2 + 5x - 2 = 0$

$3x^2 - 5x + 2 = 0$

$(3x - 2)(x - 1) = 0$

$3x - 2 = 0 \quad \text{or} \quad x - 1 = 0$

$x = \dfrac{2}{3} \quad \text{or} \qquad x = 1$

1 is not a solution since $x \neq 1$.
The solution is $\dfrac{2}{3}$.

34.
$$\frac{5}{a} + \frac{-7}{a+1} = \frac{a^2 - 2a + 4}{a^2 + a}$$

$$a(a+1)\left(\frac{5}{a} + \frac{-7}{a+1}\right) = a(a+1)\left(\frac{a^2 - 2a + 4}{a^2 + a}\right)$$

Note that $a \neq 0$ and $a \neq -1$.

$$5(a+1) + (-7)(a) = a^2 - 2a + 4$$
$$5a + 5 - 7a = a^2 - 2a + 4$$
$$5 - 2a = a^2 - 2a + 4$$
$$5 = a^2 + 4$$
$$0 = a^2 - 1$$
$$0 = (a+1)(a-1)$$
$$a + 1 = 0 \quad \text{or} \quad a - 1 = 0$$
$$a = -1 \quad \text{or} \quad a = 1$$

Since -1 would make two denominators zero, 1 is the only solution.

35.
$$\frac{2b^2 + 5b - 8}{b^2 + 2b} + \frac{5}{b+2} = -\frac{3}{b}$$

$$\frac{2b^2 + 5b - 8}{b(b+2)} + \frac{5}{b+2} = \frac{-3}{b}$$

Multiply both sides by $b(b+2)$.
Note that $b \neq 0$ and $b \neq -2$.

$$b(b+2)\left(\frac{2b^2 + 5b - 8}{b^2 + 2b}\right)$$

$$+ b(b+2)\left(\frac{5}{b+2}\right) = b(b+2)\left(-\frac{3}{b}\right)$$
$$2b^2 + 5b - 8 + 5b = (b+2)(-3)$$
$$2b^2 + 10b - 8 = -3b - 6$$
$$2b^2 + 13b - 2 = 0$$

$$b = \frac{-13 \pm \sqrt{(13)^2 - 4(2)(-2)}}{2(2)} = \frac{-13 \pm \sqrt{169 + 16}}{4}$$

$$b = \frac{-13 \pm \sqrt{185}}{4}$$

The solutions are $\frac{-13+\sqrt{185}}{4} \approx 0.150$ and $\frac{-13-\sqrt{185}}{4} \approx -6.650$.

36.
$$\frac{2}{x^2 - 2x - 3} + \frac{5}{x^2 - x - 6} = \frac{1}{x^2 + 3x + 2}$$
$$\frac{2}{(x-3)(x+1)} + \frac{5}{(x-3)(x+2)} = \frac{1}{(x+2)(x+1)}$$

Multiply both sides by $(x-3)(x+1)(x+2)$.
Note that $x \neq 3$, $x \neq -1$, and $x \neq -2$.

$$(x-3)(x+1)(x+2)\left(\frac{2}{(x-3)(x+1)}\right)$$

$$+ (x-3)(x+1)(x+2)\left(\frac{5}{(x-3)(x+2)}\right)$$

$$= (x-3)(x+1)(x+2)\left(\frac{1}{(x+2)(x+1)}\right)$$

$$2(x+2) + 5(x+1) = x - 3$$
$$2x + 4 + 5x + 5 = x - 3$$
$$7x + 9 = x - 3$$
$$6x + 9 = -3$$
$$6x = -12$$
$$x = -2$$

However, $x \neq -2$. Therefore there is no solution.

37.
$$\frac{2}{y^2 + 7y + 12} - \frac{1}{y^2 + 5y + 6} = \frac{5}{y^2 + 6y + 8}$$
$$\frac{2}{(y+4)(y+3)} - \frac{1}{(y+3)(y+2)} = \frac{5}{(y+4)(y+2)}$$

Multiply both sides by $(y+4)(y+3)(y+2)$.
Note that $y \neq -4$, $y \neq -3$, and $y \neq -2$.

$$(y+4)(y+3)(y+2)\left(\frac{2}{(y+4)(y+3)}\right)$$

$$- (y+4)(y+3)(y+2)\left(\frac{1}{(y+3)(y+2)}\right)$$

$$= (y+4)(y+3)(y+2)\left(\frac{5}{(y+4)(y+2)}\right)$$

$$2(y+2) - (y+4) = 5(y+3)$$
$$2y + 4 - y - 4 = 5y + 15$$
$$y = 5y + 15$$
$$-4y = 15$$
$$y = -\frac{15}{4}$$

The solution is $-\frac{15}{4}$.

R.5   Inequalities

1. $x < 0$

Because the inequality symbol means "less than," the endpoint at 0 is not included. This inequality is written in interval notation is $(-\infty, 0)$. To graph this interval on a number line, place an open circle at 0 and draw a heavy arrow pointing to the left.

2. $x \geq -3$

Because the inequality sign means "greater than or equal to," the endpoint at -3 is included. This inequality is written in interval notation as $[-3, \infty)$. To graph this interval on a number line, place a closed circle at -3 and draw a heavy arrow pointing to the right.

3. $-1 \leq x < 2$

The endpoint at -1 is included, but the endpoint at 2 is not. This inequality is written in interval notation as $[-1, 2)$. To graph this interval, place a closed circle at -1 and an open circle at 2; then draw a heavy line segment between them.

4. $-5 < x \leq -4$

The endpoint at -4 is included, but the endpoint at -5 is not. This inequality is written in interval notation as $(-5, -4]$. To graph this interval, place an open circle at -5 and a closed circle at -4; then draw a heavy line segment between them.

5. $-9 > x$

This inequality may be rewritten as $x < -9$, and is written in interval notation as $(-\infty, -9)$. Note that the endpoint at -9 is not included. To graph this interval, place an open circle at -9 and draw a heavy arrow pointing to the left.

6. $6 \leq x$

This inequality may be written as $x \geq 6$, and is written in interval notation as $[6, \infty)$. Note that the endpoint at 6 is included. To graph this interval, place a closed circle at 6 and draw a heavy arrow pointing to the right.

7. $(-4, 3)$

This represents all the numbers between -4 and 3, not including the endpoints. This interval can be written as the inequality $-4 < x < 3$.

8. $[2, 7)$

This represents all the numbers between 2 and 7, including 2 but not including 7. This interval can be written as the inequality $2 \leq x < 7$.

9. $(-\infty, -1]$

This represents all the numbers to the left of -1 on the number line and includes the endpoint. This interval can be written as the inequality $x \leq -1$.

10. $(3, \infty)$

This represents all the numbers to the right of 3, and does not include the endpoint. This interval can be written as the inequality $x > 3$.

11. Notice that the endpoint -2 is included, but 6 is not. The interval show in the graph can be written as the inequality $-2 \leq x < 6$.

12. Notice that neither endpoint is included. The interval shown in the graph can be written as $0 < x < 8$.

13. Notice that both endpoints are included. The interval shown in the graph can be written as $x \leq -4$ or $x \geq 4$.

14. Notice that the endpoint 0 is not included, but 3 is included. The interval shown in the graph can be written as $x < 0$ or $x \geq 3$.

15. $-3p - 2 \geq 1$

$-3p \geq 3$

$\left(-\dfrac{1}{3}\right)(-3p) \leq \left(-\dfrac{1}{3}\right)(3)$

$p \leq -1$

The solution in interval notation is $(-\infty, -1]$.

16. $6k - 4 < 3k - 1$

$6k < 3k + 3$

$3k < 3$

$k < 1$

The solution in interval notation is $(-\infty, 1)$.

17. $m - (4 + 2m) + 3 < 2m + 2$

$m - 4 - 2m + 3 < 2m + 2$

$-m - 1 < 2m + 2$

$-3m - 1 < 2$

$-3m < 3$

$-\dfrac{1}{3}(-3m) > -\dfrac{1}{3}(3)$

$m > -1$

The solution is $(-1, \infty)$.

18. $-2(3y - 8) \geq 5(4y - 2)$

$-6y + 16 \geq 20y - 10$

$-6y + 16 + (-16) \geq 20y - 10 + (-16)$

$-6y \geq 20y - 26$

$-6y + (-20y) \geq 20y + (-20y) - 26$

$-26y \geq -26$

$-\dfrac{1}{26}(-26)y \leq -\dfrac{1}{26}(-26)$

$y \leq 1$

The solution is $(-\infty, 1]$.

19. $3p - 1 < 6p + 2(p - 1)$

$3p - 1 < 6p + 2p - 2$

$3p - 1 < 8p - 2$

$-5p - 1 < -2$

$-5p < -1$

$-\dfrac{1}{5}(-5p) > -\dfrac{1}{5}(-1)$

$p > \dfrac{1}{5}$

The solution is $\left(\dfrac{1}{5}, \infty\right)$.

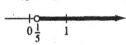

20. $x + 5(x + 1) > 4(2 - x) + x$

$x + 5x + 5 > 8 - 4x + x$

$6x + 5 > 8 - 3x$

$6x > 3 - 3x$

$9x > 3$

$x > \dfrac{1}{3}$

The solution is $\left(\dfrac{1}{3}, \infty\right)$.

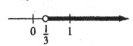

21. $-7 < y - 2 < 4$

$-7 + 2 < y - 2 + 2 < 4 + 2$

$-5 < y < 6$

The solution is $(-5, 6)$.

22. $8 \leq 3r + 1 \leq 13$

$8 + (-1) \leq 3r + 1 + (-1) \leq 13 + (-1)$

$7 \leq 3r \leq 12$

$\dfrac{1}{3}(7) \leq \dfrac{1}{3}(3r) \leq \dfrac{1}{3}(12)$

$\dfrac{7}{3} \leq r \leq 4$

The solution is $\left[\dfrac{7}{3}, 4\right]$.

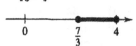

23. $-4 \leq \dfrac{2k - 1}{3} \leq 2$

$3(-4) \leq 3\left(\dfrac{2k - 1}{3}\right) \leq 3(2)$

$-12 \leq 2k - 1 \leq 6$

$-11 \leq 2k \leq 7$

$-\dfrac{11}{2} \leq k \leq \dfrac{7}{2}$

The solution is $\left[-\dfrac{11}{2}, \dfrac{7}{2}\right]$.

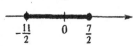

24. $-1 \leq \dfrac{5y+2}{3} \leq 4$

$3(-1) \leq 3\left(\dfrac{5y+2}{3}\right) \leq 3(4)$

$-3 \leq 5y + 2 \leq 12$
$-5 \leq 5y \leq 10$
$-1 \leq y \leq 2$

The solution is $[-1, 2]$.

25. $\dfrac{3}{5}(2p+3) \geq \dfrac{1}{10}(5p+1)$

$10\left(\dfrac{3}{5}\right)(2p+3) \geq 10\left(\dfrac{1}{10}\right)(5p+1)$

$6(2p+3) \geq 5p + 1$
$12p + 18 \geq 5p + 1$
$7p \geq -17$
$p \geq -\dfrac{17}{7}$

The solution is $\left[-\frac{17}{7}, \infty\right)$.

26. $\dfrac{8}{3}(z-4) \leq \dfrac{2}{9}(3z+2)$

$(9)\dfrac{8}{3}(z-4) \leq (9)\dfrac{2}{9}(3z+2)$

$24(z-4) \leq 2(3z+2)$
$24z - 96 \leq 6z + 4$
$24z \leq 6z + 100$
$18z \leq 100$
$z \leq \dfrac{100}{18}$
$z \leq \dfrac{50}{9}$

The solution is $\left(-\infty, \frac{50}{9}\right]$.

27. $(m+2)(m-4) < 0$

Solve $(m+2)(m-4) = 0$.

$$m = -2 \quad \text{or} \quad m = 4$$

Intervals: $(-\infty, -2), \ (-2, 4), (4, \infty)$

For $(-\infty, -2)$, choose -3 to test for m.

$$(-3+2)(-3-4) = -1(-7) = 7 \not< 0$$

For $(-2, 4)$, choose 0.

$$(0+2)(0-4) = 2(-4) = -8 < 0$$

For $(4, \infty)$, choose 5.

$$(5+2)(5-4) = 7(1) = 7 \not< 0$$

The solution is $(-2, 4)$.

28. $(t+6)(t-1) \geq 0$

Solve $(t+6)(t-1) = 0$.

$$(t+6)(t-1) = 0$$
$$t = -6 \quad \text{or} \quad t = 1$$

Intervals: $(-\infty, -6), \ (-6, 1), \ (1, \infty)$

For $(-\infty, -6)$, choose -7 to test for t.

$$(-7+6)(-7-1) = (-1)(-8) = 8 \geq 0$$

For $(-6, 1)$, choose 0.

$$(0+6)(0-1) = (6)(-1) = -6 \not\geq 0$$

For $(1, \infty)$, choose 2.

$$(2+6)(2-1) = (8)(1) = 8 \geq 0$$

Because the symbol $\geq$ is used, the endpoints -6 and 1 are included in the solution, $(-\infty, -6] \cup [1, \infty)$.

29. $y^2 - 3y + 2 < 0$
$(y - 2)(y - 1) < 0$

Solve $(y - 2)(y - 1) = 0$.

$$y = 2 \quad \text{or} \quad y = 1$$

Intervals: $(-\infty, 1), \ (1, 2), (2, \infty)$

For $(-\infty, 1)$, choose $y = 0$.

$$0^2 - 3(0) + 2 = 2 \not< 0$$

For $(1, 2)$, choose $y = \frac{3}{2}$.

$$\left(\frac{3}{2}\right)^2 - 3\left(\frac{3}{2}\right) + 2 = \frac{9}{4} - \frac{9}{2} + 2$$

$$= \frac{9 - 18 + 8}{4}$$

$$= -\frac{1}{4} < 0$$

For $(2, \infty)$, choose 3.

$$3^2 - 3(3) + 2 = 2 \not< 0$$

The solution is $(1, 2)$.

30. $2k^2 + 7k - 4 > 0$

Solve $2k^2 + 7k - 4 = 0$.

$$2k^2 + 7k - 4 = 0$$
$$(2k - 1)(k + 4) = 0$$

$$k = \frac{1}{2} \quad \text{or} \quad k = -4$$

Intervals: $(-\infty, -4), \left(-4, \frac{1}{2}\right), \left(\frac{1}{2}, \infty\right)$

For $(-\infty, -4)$, choose -5.

$$2(-5)^2 + 7(-5) - 4 = 11 > 0$$

For $\left(-4, \frac{1}{2}\right)$, choose 0.

$$2(0)^2 + 7(0) - 4 = -4 \not> 0$$

For $\left(\frac{1}{2}, \infty\right)$, choose 1.

$$2(1)^2 + 7(1) - 4 = 5 > 0$$

The solution is $(-\infty, -4) \cup \left(\frac{1}{2}, \infty\right)$.

31. $q^2 - 7q + 6 \le 0$

Solve $q^2 - 7q + 6 = 0$.

$$(q - 1)(q - 6) = 0$$

$$q = 1 \quad \text{or} \quad q = 6$$

These solutions are also solutions of the given inequality because the symbol $\le$ indicates that the endpoints are included.

Intervals $(-\infty, 1), \ (1, 6), \ (6, \infty)$

For $(-\infty, 1)$, choose 0.

$$0^2 - 7(0) + 6 = 6 \not\le 0$$

For $(1, 6)$, choose 2.

$$2^2 - 7(2) + 6 = -4 \le 0$$

For $(6, \infty)$, choose 7.

$$7^2 - 7(7) + 6 = 6 \not\le 0$$

The solution is $[1, 6]$.

32. $2k^2 - 7k - 15 \le 0$

Solve $2k^2 - 7k - 15 = 0$.

$$2k^2 - 7k - 15 = 0$$
$$(2k + 3)(k - 5) = 0$$

$$k = -\frac{3}{2} \quad \text{or} \quad k = 5$$

Intervals: $\left(-\infty, -\frac{3}{2}\right), \left(-\frac{3}{2}, 5\right), (5, \infty)$

For $\left(-\infty, -\frac{3}{2}\right)$, choose -2.

$$2(-2)^2 - 7(-2) - 15 = 7 \not\le 0$$

For $\left(-\frac{3}{2}, 5\right)$, choose 0.

$$2(0)^2 - 7(0) - 15 = -15 \le 0$$

For $(5, \infty)$, choose 6.

$$2(6)^2 - 7(6) - 15 \not\le 0$$

The solution is $\left[-\frac{3}{2}, 5\right]$.

33. $6m^2 + m > 1$

Solve $6m^2 + m = 1$.

$$6m^2 + m - 1 = 0$$
$$(2m + 1)(3m - 1) = 0$$

$$m = -\frac{1}{2} \quad \text{or} \quad m = \frac{1}{3}$$

Intervals: $\left(-\infty, -\frac{1}{2}\right), \left(-\frac{1}{2}, \frac{1}{3}\right), \left(\frac{1}{3}, \infty\right)$

For $\left(-\infty, -\frac{1}{2}\right)$, choose -1.

$$6(-1)^2 + (-1) = 5 > 1$$

For $\left(-\frac{1}{2}, \frac{1}{3}\right)$, choose 0.

$$6(0)^2 + 0 = 0 \not> 1$$

For $\left(\frac{1}{3}, \infty\right)$ choose 1.

$$6(1)^2 + 1 = 7 > 1$$

The solution is $\left(-\infty, -\frac{1}{2}\right) \cup \left(\frac{1}{3}, \infty\right)$.

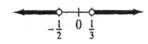

34. $10r^2 + r \leq 2$

Solve $10r^2 + r = 2$.

$$10r^2 + r = 2$$
$$10r^2 + r - 2 = 0$$
$$(5r - 2)(2r + 1) = 0$$

$$r = \frac{2}{5} \quad \text{or} \quad r = -\frac{1}{2}$$

Intervals: $\left(-\infty, -\frac{1}{2}\right), \left(-\frac{1}{2}, \frac{2}{5}\right), \left(\frac{2}{5}, \infty\right)$

For $\left(-\infty, -\frac{1}{2}\right)$, choose -1.

$$10(-1)^2 + (-1) = 9 \not\leq 2$$

For $\left(-\frac{1}{2}, \frac{2}{5}\right)$, choose 0.

$$10(0)^2 + 0 = 0 \leq 2$$

For $\left(\frac{2}{5}, \infty\right)$, choose 1.

$$10(1)^2 + 1 = 11 \not\leq 2$$

The solution is $\left[-\frac{1}{2}, \frac{2}{5}\right]$.

35. $2y^2 + 5y \leq 3$

Solve $2y^2 + 5y = 3$.

$$2y^2 + 5y - 3 = 0$$
$$(y + 3)(2y - 1) = 0$$

$$y = -3 \quad \text{or} \quad y = \frac{1}{2}$$

Intervals: $\left(-\infty, -3\right), \left(-3, \frac{1}{2}\right), \left(\frac{1}{2}, \infty\right)$

For $\left(-\infty, -3\right)$, choose -4.

$$2(-4)^2 + 5(-4) = 12 \not\leq 3$$

For $\left(-3, \frac{1}{2}\right)$, choose 0.

$$2(0)^2 + 5(0) = 0 \leq 3$$

For $\left(\frac{1}{2}, \infty\right)$, choose 1.

$$2(1)^2 + 5(1) = 7 \not\leq 3$$

The solution is $\left[-3, \frac{1}{2}\right]$.

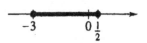

36. $3a^2 + a > 10$

Solve $3a^2 + a = 10$.

$$3a^2 + a = 10$$
$$3a^2 + a - 10 = 0$$
$$(3a - 5)(a + 2) = 0$$

$$a = \frac{5}{3} \quad \text{or} \quad a = -2$$

Intervals: $\left(-\infty, -2\right), \left(-2, \frac{5}{3}\right), \left(\frac{5}{3}, \infty\right)$

For $\left(-\infty, -2\right)$, choose -3.

$$3(-3)^2 + (-3) = 24 > 10$$

For $\left(-2, \frac{5}{3}\right)$, choose 0.

$$3(0)^2 + 0 = 0 \not> 10$$

For $\left(\frac{5}{3}, \infty\right)$, choose 2.

$$3(2)^2 + 2 = 14 > 10$$

The solution is $\left(-\infty, -2\right) \cup \left(\frac{5}{3}, \infty\right)$.

37. $x^2 \leq 25$

Solve $x^2 = 25$.

$$x = -5 \quad \text{or} \quad x = 5$$

Intervals: $(-\infty, -5), (-5, 5), (5, \infty)$

For $(-\infty, -5)$, choose -6.

$$(-6)^2 = 36 \not\leq 25$$

For $(-5, 5)$, choose 0.

$$0^2 = 0 \leq 25$$

For $(5, \infty)$, choose 6.

$$6^2 = 36 \not\leq 25$$

The solution is $[-5, 5]$.

38. $p^2 - 16p > 0$

Solve $p^2 - 16p = 0$.

$$p^2 - 16p = 0$$
$$p(p - 16) = 0$$
$$p = 0 \quad \text{or} \quad p = 16$$

Intervals: $(-\infty, 0), (0, 16), (16, \infty)$

For $(-\infty, 0)$, choose -1.

$$(-1)^2 - 16(-1) = 17 > 0$$

For $(0, 16)$, choose 1.

$$(1)^2 - 16(1) = -15 \not> 0$$

For $(16, \infty)$, choose 17.

$$(17)^2 - 16(17) = 17 > 0$$

The solution is $(-\infty, 0) \cup (16, \infty)$.

39. $\dfrac{m - 3}{m + 5} \leq 0$

Solve $\dfrac{m - 3}{m + 5} = 0$.

$$(m + 5)\frac{m - 3}{m + 5} = (m + 5)(0)$$
$$m - 3 = 0$$
$$m = 3$$

Set the denominator equal to 0 and solve.

$$m + 5 = 0$$
$$m = -5$$

Intervals: $(-\infty, -5), (-5, 3), (3, \infty)$

For $(-\infty, -5)$, choose -6.

$$\frac{-6 - 3}{-6 + 5} = 9 \not\leq 0$$

For $(-5, 3)$, choose 0.

$$\frac{0 - 3}{0 + 5} = -\frac{3}{5} \leq 0$$

For $(3, \infty)$, choose 4.

$$\frac{4 - 3}{4 + 5} = \frac{1}{9} \not\leq 0$$

Although the $\leq$ symbol is used, including -5 in the solution would cause the denominator to be zero.

The solution is $(-5, 3]$.

40. $\dfrac{r + 1}{r - 1} > 0$

Solve the equation $\dfrac{r + 1}{r - 1} = 0$.

$$\frac{r + 1}{r - 1} = 0$$
$$(r - 1)\frac{r + 1}{r - 1} = (r - 1)(0)$$
$$r + 1 = 0$$
$$r = -1$$

Find the value for which the denominator equals zero.

$$r - 1 = 0$$
$$r = 1$$

Intervals: $(-\infty, -1), (-1, 1), (1, \infty)$

For $(-\infty, -1)$, choose -2.

$$\frac{-2 + 1}{-2 - 1} = \frac{-1}{-3} = \frac{1}{3} > 0$$

For $(-1, 1)$, choose 0.

$$\frac{0 + 1}{0 - 1} = \frac{1}{-1} = -1 \not> 0$$

For $(1, \infty)$, choose 2.

$$\frac{2 + 1}{2 - 1} = \frac{3}{1} = 3 > 0$$

The solution is $(-\infty, -1) \cup (1, \infty)$.

41. $\dfrac{k-1}{k+2} > 1$

Solve $\dfrac{k-1}{k+2} = 1$.

$$k - 1 = k + 2$$
$$-1 \neq 2$$

The equation has no solution.
Solve $k + 2 = 0$.

$$k = -2$$

Intervals: $(-\infty, -2)$, $(-2, \infty)$

For $(-\infty, -2)$, choose -3.

$$\frac{-3-1}{-3+2} = 4 > 1$$

For $(-2, \infty)$, choose 0.

$$\frac{0-1}{0+2} = -\frac{1}{2} \not> 1$$

The solution is $(-\infty, -2)$.

42. $\dfrac{a-5}{a+2} < -1$

Solve the equation $\dfrac{a-5}{a+2} = -1$.

$$\frac{a-5}{a+2} = -1$$
$$a - 5 = -1(a+2)$$
$$a - 5 = -a - 2$$
$$2a = 3$$

$$a = \frac{3}{2}$$

Set the denominator equal to zero and solve for a.

$$a + 2 = 0$$
$$a = -2$$

Intervals: $(-\infty, -2), \left(-2, \frac{3}{2}\right), \left(\frac{3}{2}, \infty\right)$

For $(-\infty, -2)$, choose -3.

$$\frac{-3-5}{-3+2} = \frac{-8}{-1} = 8 \not< -1$$

For $\left(-2, \frac{3}{2}\right)$, choose 0.

$$\frac{0-5}{0+2} = \frac{-5}{2} = -\frac{5}{2} < -1$$

For $\left(\frac{3}{2}, \infty\right)$, choose 2.

$$\frac{2-5}{2+2} = \frac{-3}{4} = -\frac{3}{4} \not< -1$$

The solution is $\left(-2, \frac{3}{2}\right)$.

43. $\dfrac{2y+3}{y-5} \leq 1$

Solve $\dfrac{2y+3}{y-5} = 1$.

$$2y + 3 = y - 5$$
$$y = -8$$

Solve $y - 5 = 0$.

$$y = 5$$

Intervals: $(-\infty, -8)$, $(-8, 5)$, $(5, \infty)$

For $(-\infty, , -8)$, choose $y = -10$.

$$\frac{2(-10)+3}{-10-5} = \frac{17}{15} \not\leq 1$$

For $(-8, 5)$, choose $y = 0$.

$$\frac{2(0)+3}{0-5} = -\frac{3}{5} \leq 1$$

For $(5, \infty)$, choose $y = 6$.

$$\frac{2(6)+3}{6-5} = \frac{15}{1} \not\leq 1$$

The solution is $[-8, 5)$.

44. $\dfrac{a+2}{3+2a} \leq 5$

For the equation $\dfrac{a+2}{3+2a} = 5$.

$$\frac{a+2}{3+2a} = 5$$
$$a + 2 = 5(3 + 2a)$$
$$a + 2 = 15 + 10a$$
$$-9a = 13$$

$$a = -\frac{13}{9}$$

Set the denominator equal to zero and solve for a.

$$3 + 2a = 0$$
$$2a = -3$$

$$a = -\frac{3}{2}$$

Intervals: $\left(-\infty, -\frac{3}{2}\right), \left(-\frac{3}{2}, -\frac{13}{9}\right), \left(-\frac{13}{9}, \infty\right)$

For $\left(-\infty, -\frac{3}{2}\right)$, choose -2.

$$\frac{-2+2}{3+2(-2)} = \frac{0}{-1} = 0 \leq 5$$

For $\left(-\frac{3}{2}, -\frac{13}{9}\right)$, choose -1.46.

$$\frac{-1.46 + 2}{3 + 2(-1.46)} = \frac{0.54}{0.08} = 6.75 \not\leq 5$$

For $\left(-\frac{13}{9}, \infty\right)$, choose 0.

$$\frac{0 + 2}{3 + 2(0)} = \frac{2}{3} \leq 5$$

The value $-\frac{3}{2}$ cannot be included in the solution since it would make the denominator zero. The solution is $\left(-\infty, -\frac{3}{2}\right) \cup \left[-\frac{13}{9}, \infty\right)$.

45. $\dfrac{7}{k+2} \geq \dfrac{1}{k+2}$

Solve $\dfrac{7}{k+2} = \dfrac{1}{k+2}$.

$$\frac{7}{k+2} - \frac{1}{k+2} = 0$$

$$\frac{6}{k+2} = 0$$

The equation has no solution.

Solve $k + 2 = 0$.

$$k = -2$$

Intervals: $(-\infty, -2), (-2, \infty)$

For $(-\infty, -2)$, choose $k = -3$.

$$\frac{6}{-3+2} = -6 \not\geq 0$$

For $(-2, \infty)$, choose $k = 0$.

$$\frac{6}{0+2} = 3 \geq 0$$

The solution is $(-2, \infty)$.

46. $\dfrac{5}{p+1} > \dfrac{12}{p+1}$

Solve the equation $\dfrac{5}{p+1} = \dfrac{12}{p+1}$.

$$\frac{5}{p+1} = \frac{12}{p+1}$$

$$5 = 12$$

The equation has no solution.
Set the denominator equal to zero and solve for p.

$$p + 1 = 0$$

$$p = -1$$

Intervals: $(-\infty, -1), (-1, \infty)$

For $(-\infty, -1)$, choose -2.

$$\frac{5}{-2+1} = -5 \text{ and } \frac{12}{-2+1} = -12, \text{ so}$$

$$\frac{5}{-2+1} > \frac{12}{-2+1}.$$

For $(-1, \infty)$, choose 0.

$$\frac{5}{0+1} = 5 \text{ and } \frac{12}{0+1} = 12, \text{ so}$$

$$\frac{5}{0+1} \not> \frac{12}{0+1}.$$

The solution is $(-\infty, -1)$.

47. $\dfrac{3x}{x^2 - 1} < 2$

Solve

$$\frac{3x}{x^2 - 1} = 2.$$
$$3x = 2x^2 - 2$$
$$-2x^2 + 3x + 2 = 0$$
$$(2x + 1)(-x + 2) = 0$$

$$x = -\frac{1}{2} \quad \text{or} \quad x = 2$$

Set $x^2 - 1 = 0$.

$$x = 1 \quad \text{or} \quad x = -1$$

Intervals: $(-\infty, -1), \left(-1, -\frac{1}{2}\right), \left(-\frac{1}{2}, 1\right),$ $(1, 2), (2, \infty)$

For $(-\infty, -1)$, choose $x = -2$.

$$\frac{3(-2)}{(-2)^2 - 1} = -\frac{6}{3} = -2 < 2$$

For $\left(-1, -\frac{1}{2}\right)$, choose $x = -\frac{3}{4}$.

$$\frac{3\left(-\frac{3}{4}\right)}{\left(-\frac{3}{4}\right)^2 - 1} = \frac{-\frac{9}{4}}{\frac{9}{16} - 1} = \frac{36}{7} \not< 2$$

For $\left(-\frac{1}{2}, 1\right)$, choose $x = 0$.

$$\frac{3(0)}{0^2 - 1} = 0 < 2$$

For $(1, 2)$, choose $x = \frac{3}{2}$.

$$\frac{3\left(\frac{3}{2}\right)}{\left(\frac{3}{2}\right)^2 - 1} = \frac{\frac{9}{2}}{\frac{5}{4}} = \frac{18}{5} \not< 2$$

For $(2, \infty)$, choose $x = 3$.

$$\frac{3(3)}{3^2 - 1} = \frac{9}{8} < 2$$

The solution is $(-\infty, -1) \cup \left(-\frac{1}{2}, 1\right) \cup (2, \infty)$.

48. $\dfrac{8}{p^2 + 2p} > 1$

Solve the equation $\dfrac{8}{p^2 + 2p} = 1$.

$$\frac{8}{p^2 + 2p} = 1$$
$$8 = p^2 + 2p$$
$$0 = p^2 + 2p - 8$$
$$0 = (p + 4)(p - 2)$$
$$p + 4 = 0 \quad \text{or} \quad p - 2 = 0$$
$$p = -4 \quad \text{or} \quad p = 2$$

Set the denominator equal to zero and solve for p.

$$p^2 + 2p = 0$$
$$p(p + 2) = 0$$
$$p = 0 \quad \text{or} \quad p + 2 = 0$$
$$p = -2$$

Intervals: $(-\infty, -4), \ (-4, -2), \ (-2, 0),$
$(0, 2), \ (2, \infty)$

For $(-\infty, -4)$, choose -5.

$$\frac{8}{(-5)^2 + 2(-5)} = \frac{8}{15} \not> 1$$

For $(-4, -2)$, choose -3.

$$\frac{8}{(-3)^2 + 2(-3)} = \frac{8}{9 - 6} = \frac{8}{3} > 1$$

For $(-2, 0)$, choose -1.

$$\frac{8}{(-1)^2 + 2(-1)} = \frac{8}{-1} = -8 \not> 1$$

For $(0, 2)$, choose 1.

$$\frac{8}{(1)^2 + 2(1)} = \frac{8}{3} > 1$$

For $(2, \infty)$, choose 3.

$$\frac{8}{(3)^2 + (2)(3)} = \frac{8}{15} \not> 1$$

The solution is $(-4, -2) \cup (0, 2)$.

49. $\dfrac{z^2 + z}{z^2 - 1} \geq 3$

Solve

$$\frac{z^2 + z}{z^2 - 1} = 3.$$
$$z^2 + z = 3z^2 - 3$$
$$-2z^2 + z + 3 = 0$$
$$-1(2z^2 - z - 3) = 0$$
$$-1(z + 1)(2z - 3) = 0$$
$$z = -1 \quad \text{or} \quad z = \frac{3}{2}$$

Set $z^2 - 1 = 0$.

$$z^2 = 1$$
$$z = -1 \quad \text{or} \quad z = 1$$

Intervals: $(-\infty, -1), \ (-1, 1), \ \left(1, \frac{3}{2}\right), \left(\frac{3}{2}, \infty\right)$

For $(-\infty, -1)$, choose $x = -2$.

$$\frac{(-2)^2 + 3}{(-2)^2 - 1} = \frac{7}{3} \not\geq 3$$

For $(-1, 1)$, choose $x = 0$.

$$\frac{0^2 + 3}{0^2 - 1} = -3 \not\geq 3$$

For $\left(1, \frac{3}{2}\right)$, choose $x = \frac{3}{2}$.

$$\frac{\left(\frac{3}{2}\right)^2 + 3}{\left(\frac{3}{2}\right)^2 - 1} = \frac{21}{5} \geq 3$$

For $\left(\frac{3}{2}, \infty\right)$, choose $x = 2$.

$$\frac{2^2 + 3}{2^2 - 1} = \frac{7}{3} \not\geq 3$$

The solution is $\left(1, \frac{3}{2}\right]$.

50. $\dfrac{a^2 + 2a}{a^2 - 4} \leq 2$

Solve the equation $\dfrac{a^2 + 2a}{a^2 - 4} = 2$.

$$\frac{a^2 + 2a}{a^2 - 4} = 2$$
$$a^2 + 2a = 2(a^2 - 4)$$
$$a^2 + 2a = 2a^2 - 8$$
$$0 = a^2 - 2a - 8$$
$$0 = (a - 4)(a + 2)$$
$$a - 4 = 0 \quad \text{or} \quad a + 2 = 0$$
$$a = 4 \quad \text{or} \quad a = -2$$

But -2 is not a possible solution.
Set the denominator equal to zero and solve for a.

$$a^2 - 4 = 0$$
$$(a + 2)(a - 2) = 0$$
$$a + 2 = 0 \quad \text{or} \quad a - 2 = 0$$
$$a = -2 \quad \text{or} \quad a = 2$$

Intervals: $(-\infty, -2),\ (-2, 2),$
$(2, 4),\ (4, \infty)$

For $(-\infty, -2)$, choose -3.

$$\frac{(-3)^2 + 2(-3)}{(-3)^2 - 4} = \frac{9 - 6}{9 - 4} = \frac{3}{5} \le 2$$

For $(-2, 2)$, choose 0.

$$\frac{(0)^2 + 2(0)}{0 - 4} = \frac{0}{-4} = 0 \le 2$$

For $(2, 4)$, choose 3.

$$\frac{(3)^2 + 2(3)}{(3)^2 - 4} = \frac{9 + 6}{9 - 5} = \frac{15}{4} \not\le 2$$

For $(4, \infty)$, choose 5.

$$\frac{(5)^2 + 2(5)}{(5)^2 - 4} = \frac{25 + 10}{25 - 4} = \frac{35}{21} \le 2$$

The value 4 will satisfy the original inequality, but the values -2 and 2 will not since they make the denominator zero. The solution is $(-\infty, -2) \cup (-2, 2) \cup [4, \infty)$.

R.6   Exponents

1. $8^{-2} = \dfrac{1}{8^2} = \dfrac{1}{64}$

2. $3^{-4} = \dfrac{1}{3^4} = \dfrac{1}{81}$

3. $5^0 = 1$, by definition.

4. $(-12)^0 = 1$, by definition.

5. $-(-3)^{-2} = -\dfrac{1}{(-3)^2} = -\dfrac{1}{9}$

6. $-(-3^{-2}) = -\left(-\dfrac{1}{3^2}\right) = -\left(-\dfrac{1}{9}\right) = \dfrac{1}{9}$

7. $\left(\dfrac{2}{7}\right)^{-2} = \dfrac{1}{\left(\frac{2}{7}\right)^2} = \dfrac{1}{\frac{4}{49}} = \dfrac{49}{4}$

8. $\left(\dfrac{4}{3}\right)^{-3} = \dfrac{1}{\left(\frac{4}{3}\right)^3} = \dfrac{1}{\frac{64}{27}} = \dfrac{27}{64}$

9. $\dfrac{3^{-4}}{3^2} = 3^{(-4)-2} = 3^{-4-2} = 3^{-6} = \dfrac{1}{3^6}$

10. $\dfrac{8^9 \cdot 8^{-7}}{8^{-3}} = 8^{9+(-7)-(-3)} = 8^{9-7+3} = 8^5$

11. $\dfrac{10^8 \cdot 10^{-10}}{10^4 \cdot 10^2}$

$$= \frac{10^{8+(-10)}}{10^{4+2}} = \frac{10^{-2}}{10^6}$$

$$= 10^{-2-6} = 10^{-8}$$

$$= \frac{1}{10^8}$$

12. $\left(\dfrac{5^{-6} \cdot 5^3}{5^{-2}}\right)^{-1} = (5^{-6+3-(-2)})^{-1}$

$$= (5^{-6+3+2})^{-1} = (5^{-1})^{-1}$$

$$= 5^{(-1)(-1)} = 5^1 = 5$$

13. $\dfrac{x^4 \cdot x^3}{x^5} = \dfrac{x^{4+3}}{x^5} = \dfrac{x^7}{x^5} = x^{7-5} = x^2$

14. $\dfrac{y^9 \cdot y^7}{y^{13}} = y^{9+7-13} = y^3$

15. $\dfrac{(4k^{-1})^2}{2k^{-5}} = \dfrac{4^2 k^{-2}}{2k^{-5}} = \dfrac{16k^{-2-(-5)}}{2}$

$$= 8k^{-2+5} = 8k^3$$

$$= 2^3 k^3$$

16. $\dfrac{(3z^2)^{-1}}{z^5} = \dfrac{3^{-1}(z^2)^{-1}}{z^5} = \dfrac{3^{-1}z^{2(-1)}}{z^5}$

$$= \frac{3^{-1}z^{-2}}{z^5} = 3^{-1}z^{-2-5}$$

$$= 3^{-1}z^{-7} = \frac{1}{3} \cdot \frac{1}{z^7} = \frac{1}{3z^7}$$

17. $\dfrac{2^{-1}x^3 y^{-3}}{xy^{-2}} = 2^{-1}x^{3-1}y^{-3-(-2)}$

$$= 2^{-1}x^2 y^{-3+2} = 2^{-1}x^2 y^{-1}$$

$$= \frac{1}{2}x^2 \cdot \frac{1}{y} = \frac{x^2}{2y}$$

18. $\dfrac{5^{-2}m^2 y^{-2}}{5^2 m^{-1}y^{-2}} = \dfrac{5^{-2}}{5^2} \cdot \dfrac{m^2}{m^{-1}} \cdot \dfrac{y^{-2}}{y^{-2}}$

$$= 5^{-2-2}m^{2-(-1)}y^{-2-(-2)}$$

$$= 5^{-2-2}m^{2+1}y^{-2+2}$$

$$= 5^{-4}m^3 y^0 = \frac{1}{5^4} \cdot m^3 \cdot 1$$

$$= \frac{m^3}{5^4}$$

19. $\left(\dfrac{a^{-1}}{b^2}\right)^{-3} = \dfrac{(a^{-1})^{-3}}{(b^2)^{-3}} = \dfrac{a^{(-1)(-3)}}{b^{2(-3)}}$

$$= \frac{a^3}{b^{-6}} = a^3 b^6$$

20. $\left(\dfrac{2c^2}{d^3}\right)^{-2} = \dfrac{2^{-2}(c^2)^{-2}}{(d^3)^{-2}}$

$\qquad = \dfrac{2^{-2}c^{(2)(-2)}}{d^{(3)(-2)}} = \dfrac{2^{-2}c^{-4}}{d^{-6}}$

$\qquad = \dfrac{d^6}{2^2 c^4}$

21. $\left(\dfrac{x^6 y^{-3}}{x^{-2} y^5}\right)^{1/2} = (x^{6-(-2)} y^{-3-5})^{1/2}$

$\qquad = (x^8 y^{-8})^{1/2}$

$\qquad = (x^8)^{1/2}(y^{-8})^{1/2}$

$\qquad = x^4 y^{-4}$

$\qquad = \dfrac{x^4}{y^4}$

22. $\left(\dfrac{a^{-7} b^{-1}}{b^{-4} a^2}\right)^{1/3} = \left(a^{-7-2} b^{-1-(-4)}\right)^{1/3}$

$\qquad = (a^{-9} b^3)^{1/3}$

$\qquad = (a^{-9})^{1/3} (b^3)^{1/3}$

$\qquad = a^{-3} b^1$

$\qquad = \dfrac{b}{a^3}$

23. $a^{-1} + b^{-1} = \dfrac{1}{a} + \dfrac{1}{b}$

$\qquad = \left(\dfrac{b}{b}\right)\left(\dfrac{1}{a}\right) + \left(\dfrac{a}{a}\right)\left(\dfrac{1}{b}\right)$

$\qquad = \dfrac{b}{ab} + \dfrac{a}{ab}$

$\qquad = \dfrac{b+a}{ab}$

$\qquad = \dfrac{a+b}{ab}$

24. $b^{-2} - a = \dfrac{1}{b^2} - a$

$\qquad = \dfrac{1}{b^2} - a\left(\dfrac{b^2}{b^2}\right)$

$\qquad = \dfrac{1}{b^2} - \dfrac{ab^2}{b^2}$

$\qquad = \dfrac{1-ab^2}{b^2}$

25. $\dfrac{2n^{-1} - 2m^{-1}}{m+n^2} = \dfrac{\frac{2}{n} - \frac{2}{m}}{m+n^2}$

$\qquad = \dfrac{\frac{2}{n}\cdot\frac{m}{m} - \frac{2}{m}\cdot\frac{n}{n}}{(m+n^2)}$

$\qquad = \dfrac{2m-2n}{mn(m+n^2)} \quad \text{or} \quad \dfrac{2(m-n)}{mn(m+n^2)}$

26. $\left(\dfrac{m}{3}\right)^{-1} + \left(\dfrac{n}{2}\right)^{-2} = \left(\dfrac{3}{m}\right)^1 + \left(\dfrac{2}{n}\right)^2$

$\qquad = \dfrac{3}{m} + \dfrac{4}{n^2}$

$\qquad = \left(\dfrac{3}{m}\right)\left(\dfrac{n^2}{n^2}\right) + \left(\dfrac{4}{n^2}\right)\left(\dfrac{m}{m}\right)$

$\qquad = \dfrac{3n^2}{mn^2} + \dfrac{4m}{mn^2}$

$\qquad = \dfrac{3n^2 + 4m}{mn^2}$

27. $(x^{-1} - y^{-1})^{-1} = \dfrac{1}{\frac{1}{x} - \frac{1}{y}}$

$\qquad = \dfrac{1}{\frac{1}{x}\cdot\frac{y}{y} - \frac{1}{y}\cdot\frac{x}{x}}$

$\qquad = \dfrac{1}{\frac{y}{xy} - \frac{x}{xy}}$

$\qquad = \dfrac{1}{\frac{y-x}{xy}}$

$\qquad = \dfrac{xy}{y-x}$

28. $(x^{-2} + y^{-2})^{-2} = \left(\dfrac{1}{x^2} + \dfrac{1}{y^2}\right)^{-2}$

$\qquad = \left[\left(\dfrac{1}{x^2}\right)\left(\dfrac{y^2}{y^2}\right) + \left(\dfrac{x^2}{x^2}\right)\left(\dfrac{1}{y^2}\right)\right]^{-2}$

$\qquad = \left(\dfrac{y^2}{x^2 y^2} + \dfrac{x^2}{x^2 y^2}\right)^{-2}$

$\qquad = \left(\dfrac{y^2 + x^2}{x^2 y^2}\right)^{-2} = \left(\dfrac{x^2 y^2}{y^2 + x^2}\right)^2$

$\qquad = \dfrac{(x^2)^2(y^2)^2}{(x^2 + y^2)^2} = \dfrac{x^4 y^4}{(x^2 + y^2)^2}$

29. $81^{1/2} = (9^2)^{1/2} = 9^{2(1/2)} = 9^1 = 9$

30. $27^{1/3} = \sqrt[3]{27} = 3$

31. $32^{2/5} = (32^{1/5})^2 = 2^2 = 4$

32. $-125^{2/3} = -(125^{1/3})^2 = -5^2 = -25$

33. $\left(\dfrac{4}{9}\right)^{1/2} = \dfrac{4^{1/2}}{9^{1/2}} = \dfrac{2}{3}$

34. $\left(\dfrac{64}{27}\right)^{1/3} = \dfrac{64^{1/3}}{27^{1/3}} = \dfrac{4}{3}$

35. $16^{-5/4} = (16^{1/4})^{-5} = 2^{-5} = \dfrac{1}{2^5} \quad \text{or} \quad \dfrac{1}{32}$

36. $625^{-1/4} = \dfrac{1}{625^{1/4}} = \dfrac{1}{5}$

37. $\left(\dfrac{27}{64}\right)^{-1/3} = \dfrac{27^{-1/3}}{64^{-1/3}} = \dfrac{64^{1/3}}{27^{1/3}}$

$\qquad = \dfrac{4}{3}$

38. $\left(\dfrac{121}{100}\right)^{-3/2} = \dfrac{1}{\left(\frac{121}{100}\right)^{3/2}} = \dfrac{1}{\left[\left(\frac{121}{100}\right)^{1/2}\right]^3}$

$\qquad = \dfrac{1}{\left(\frac{11}{10}\right)^3} = \dfrac{1}{\frac{1,331}{1,000}} = \dfrac{1,000}{1,331}$

39. $2^{1/2} \cdot 2^{3/2} = 2^{1/2+3/2} = 2^{4/2}$

$\qquad = 2^2 = 4$

40. $27^{2/3} \cdot 27^{-1/3} = 27^{(2/3)+(-1/3)}$

$\qquad = 27^{2/3-1/3}$

$\qquad = 27^{1/3} = 3$

41. $\dfrac{4^{2/3} \cdot 4^{5/3}}{4^{1/3}} = \dfrac{4^{2/3+5/3}}{4^{1/3}}$

$\qquad = 4^{7/3-1/3} = 4^{6/3}$

$\qquad = 4^2 = 16$

42. $\dfrac{3^{-5/2} \cdot 3^{3/2}}{3^{7/2} \cdot 3^{-9/2}}$

$\qquad = 3^{(-5/2)+(3/2)-(7/2)-(-9/2)}$

$\qquad = 3^{-5/2+3/2-7/2+9/2}$

$\qquad = 3^0 = 1$

43. $\dfrac{7^{-1/3} \cdot 7r^{-3}}{7^{2/3} \cdot (r^{-2})^2}$

$\qquad = \dfrac{7^{-1/3+1}r^{-3}}{7^{2/3} \cdot r^{-4}}$

$\qquad = 7^{-1/3+3/3-2/3}r^{-3-(-4)}$

$\qquad = 7^0 r^{-3+4} = 1 \cdot r^1 = r$

44. $\dfrac{12^{3/4} \cdot 12^{5/4} \cdot y^{-2}}{12^{-1} \cdot (y^{-3})^{-2}}$

$\qquad = \dfrac{12^{3/4+5/4} \cdot y^{-2}}{12^{-1} \cdot y^{(-3)(-2)}} = \dfrac{12^{8/4} \cdot y^{-2}}{12^{-1} \cdot y^6}$

$\qquad = \dfrac{12^2 \cdot y^{-2}}{12^{-1}y^6}$

$\qquad = 12^{2-(-1)} \cdot y^{-2-6} = 12^3 y^{-8}$

$\qquad = \dfrac{12^3}{y^8}$

45. $\dfrac{6k^{-4} \cdot (3k^{-1})^{-2}}{2^3 \cdot k^{1/2}}$

$\qquad = \dfrac{2 \cdot 3k^{-4}(3^{-2})(k^2)}{2^3 k^{1/2}}$

$\qquad = 2^{1-3}3^{1+(-2)}k^{-4+2-1/2}$

$\qquad = 2^{-2}3^{-1}k^{-5/2}$

$\qquad = \dfrac{1}{2^2} \cdot \dfrac{1}{3} \cdot \dfrac{1}{k^{5/2}} = \dfrac{1}{2^2 3 k^{5/2}}$

or $\dfrac{1}{12k^{5/2}}$

46. $\dfrac{8p^{-3}(4p^2)^{-2}}{p^{-5}} = \dfrac{8p^{-3} \cdot 4^{-2}p^{(2)(-2)}}{p^{-5}}$

$\qquad = \dfrac{8p^{-3}4^{-2}p^{-4}}{p^{-5}}$

$\qquad = 8 \cdot 4^{-2}p^{(-3)+(-4)-(-5)}$

$\qquad = 8 \cdot 4^{-2}p^{-3-4+5}$

$\qquad = 8 \cdot 4^{-2}p^{-2}$

$\qquad = 8 \cdot \dfrac{1}{4^2} \cdot \dfrac{1}{p^2}$

$\qquad = 8 \cdot \dfrac{1}{16} \cdot \dfrac{1}{p^2}$

$\qquad = \dfrac{8}{16p^2} = \dfrac{1}{2p^2}$

47. $\dfrac{a^{4/3}}{a^{2/3}} \cdot \dfrac{b^{1/2}}{b^{-3/2}} = a^{4/3-2/3}b^{1/2-(-3/2)}$

$\qquad = a^{2/3}b^2$

48. $\dfrac{x^{1/3} \cdot y^{2/3} \cdot z^{1/4}}{x^{5/3} \cdot y^{-1/3} \cdot z^{3/4}}$

$\qquad = x^{1/3-(5/3)}y^{(2/3)-(-1/3)}z^{1/4-(3/4)}$

$\qquad = x^{1/3-5/3}y^{2/3+1/3}z^{1/4-3/4}$

$\qquad = x^{-4/3}y^{3/3}z^{-2/4}$

$\qquad = \dfrac{y}{x^{4/3}z^{2/4}}$

$\qquad = \dfrac{y}{x^{4/3}z^{1/2}}$

49. $\dfrac{k^{-3/5} \cdot h^{-1/3} \cdot t^{2/5}}{k^{-1/5} \cdot h^{-2/3} \cdot t^{1/5}}$

$\qquad = k^{-3/5-(-1/5)}h^{-1/3-(-2/3)}t^{2/5-1/5}$

$\qquad = k^{-3/5+1/5}h^{-1/3+2/3}t^{2/5-1/5}$

$\qquad = k^{-2/5}h^{1/3}t^{1/5}$

$\qquad = \dfrac{h^{1/3}t^{1/5}}{k^{2/5}}$

50. $\dfrac{m^{7/3} \cdot n^{-2/5} \cdot p^{3/8}}{m^{-2/3} \cdot n^{3/5} \cdot p^{-5/8}}$

$= m^{7/3-(-2/3)}n^{-2/5-(3/5)}p^{3/8-(-5/8)}$

$= m^{7/3+2/3}n^{-2/5-3/5}p^{3/8+5/8}$

$= m^{9/3}n^{-5/5}p^{8/8}$

$= m^3 n^{-1} p^1$

$= \dfrac{m^3 p}{n}$

51. $12x^2(x^2+2)^2 - 4x(4x^3+1)(x^2+2)$

$= (4x)(3x)(x^2+2)(x^2+2)$
$\quad - 4x(4x^3+1)(x^2+2)$

$= 4x(x^2+2)[3x(x^2+2) - (4x^3+1)]$

$= 4x(x^2+2)(3x^3+6x-4x^3-1)$

$= 4x(x^2+2)(-x^3+6x-1)$

52. $6x(x^3+7)^2 - 6x^2(3x^2+5)(x^3+7)$

$= 6x(x^3+7)(x^3+7) - 6x(x)(3x^2+5)(x^3+7)$

$= 6x(x^3+7)[(x^3+7) - x(3x^2+5)]$

$= 6x(x^3+7)(x^3+7-3x^3-5x)$

$= 6x(x^3+7)(-2x^3-5x+7)$

53. $(x^2+2)(x^2-1)^{-1/2}(x) + (x^2-1)^{1/2}(2x)$

$= (x^2+2)(x^2-1)^{-1/2}(x)$
$\quad + (x^2-1)^1(x^2-1)^{-1/2}(2x)$

$= x(x^2-1)^{-1/2}[(x^2+2) + (x^2-1)(2)]$

$= x(x^2-1)^{-1/2}(x^2+2+2x^2-2)$

$= x(x^2-1)^{-1/2}(3x^2)$

$= 3x^3(x^2-1)^{-1/2}$

54. $9(6x+2)^{1/2} + 3(9x-1)(6x+2)^{-1/2}$

$= 3 \cdot 3(6x+2)^{-1/2}(6x+2)^1$
$\quad + 3(9x-1)(6x+2)^{-1/2}$

$= 3(6x+2)^{-1/2}[3(6x+2) + (9x-1)]$

$= 3(6x+2)^{-1/2}(18x+6+9x-1)$

$= 3(6x+2)^{-1/2}(27x+5)$

55. $x(2x+5)^2(x^2-4)^{-1/2} + 2(x^2-4)^{1/2}(2x+5)$

$= (2x+5)^2(x^2-4)^{-1/2}(x)$
$\quad + (x^2-4)^1(x^2-4)^{-1/2}(2)(2x+5)$

$= (2x+5)(x^2-4)^{-1/2}$
$\quad \cdot [(2x+5)(x) + (x^2-4)(2)]$

$= (2x+5)(x^2-4)^{-1/2}$
$\quad \cdot (2x^2+5x+2x^2-8)$

$= (2x+5)(x^2-4)^{-1/2}(4x^2+5x-8)$

56. $(4x^2+1)^2(2x-1)^{-1/2} + 16x(4x^2+1)(2x-1)^{1/2}$

$= (4x^2+1)(4x^2+1)(2x-1)^{-1/2}$
$\quad + 16x(4x^2+1)(2x-1)^{-1/2}(2x-1)$

$= (4x^2+1)(2x-1)^{-1/2}$
$\quad \cdot [(4x^2+1) + 16x(2x-1)]$

$= (4x^2+1)(2x-1)^{-1/2}(4x^2+1+32x^2-16x)$

$= (4x^2+1)(2x-1)^{-1/2}(36x^2-16x+1)$

R.7   Radicals

1. $\sqrt[3]{125} = 5$ because $5^3 = 125$.

2. $\sqrt[4]{1{,}296} = \sqrt[4]{6^4} = 6$

3. $\sqrt[5]{-3{,}125} = -5$ because $(-5)^5 = -3{,}125$.

4. $\sqrt{50} = \sqrt{25 \cdot 2} = \sqrt{25}\sqrt{2} = 5\sqrt{2}$

5. $\sqrt{2{,}000} = \sqrt{4 \cdot 100 \cdot 5} = 2 \cdot 10\sqrt{5}$
$\quad = 20\sqrt{5}$

6. $\sqrt{32y^5} = \sqrt{(16y^4)(2y)} = \sqrt{16y^4}\sqrt{2y}$
$\quad = 4y^2\sqrt{2y}$

7. $7\sqrt{2} - 8\sqrt{18} + 4\sqrt{72}$
$\quad = 7\sqrt{2} - 8\sqrt{9 \cdot 2} + 4\sqrt{36 \cdot 2}$
$\quad = 7\sqrt{2} - 8(3)\sqrt{2} + 4(6)\sqrt{2}$
$\quad = 7\sqrt{2} - 24\sqrt{2} + 24\sqrt{2}$
$\quad = 7\sqrt{2}$

8. $4\sqrt{3} - 5\sqrt{12} + 3\sqrt{75}$
$\quad = 4\sqrt{3} - 5(\sqrt{4}\sqrt{3}) + 3(\sqrt{25}\sqrt{3})$
$\quad = 4\sqrt{3} - 5(2\sqrt{3}) + 3(5\sqrt{3})$
$\quad = 4\sqrt{3} - 10\sqrt{3} + 15\sqrt{3}$
$\quad = (4 - 10 + 15)\sqrt{3} = 9\sqrt{3}$

9. $2\sqrt{5} - 3\sqrt{20} + 2\sqrt{45}$
$\quad = 2\sqrt{5} - 3\sqrt{4 \cdot 5} + 2\sqrt{9 \cdot 5}$
$\quad = 2\sqrt{5} - 3(2)\sqrt{5} + 2(3)\sqrt{5}$
$\quad = 2\sqrt{5} - 6\sqrt{5} + 6\sqrt{5}$
$\quad = 2\sqrt{5}$

10. $3\sqrt{28} - 4\sqrt{63} + \sqrt{112}$
$\quad = 3(\sqrt{4}\sqrt{7}) - 4(\sqrt{9}\sqrt{7}) + (\sqrt{16}\sqrt{7})$
$\quad = 3(2\sqrt{7}) - 4(3\sqrt{7}) + (4\sqrt{7})$
$\quad = 6\sqrt{7} - 12\sqrt{7} + 4\sqrt{7}$
$\quad = (6 - 12 + 4)\sqrt{7}$
$\quad = -2\sqrt{7}$

11. $\sqrt[3]{2} - \sqrt[3]{16} + 2\sqrt[3]{54}$

$= \sqrt[3]{2} - (\sqrt[3]{8 \cdot 2}) + 2(\sqrt[3]{27 \cdot 2})$

$= \sqrt[3]{2} - \sqrt[3]{8}\sqrt[3]{2} + 2(\sqrt[3]{27}\sqrt[3]{2})$

$= \sqrt[3]{2} - 2\sqrt[3]{2} + 2(3\sqrt[3]{2})$

$= \sqrt[3]{2} - 2\sqrt[3]{2} + 6\sqrt[3]{2}$

$= 5\sqrt[3]{2}$

12. $2\sqrt[3]{3} + 4\sqrt[3]{24} - \sqrt[3]{81}$

$= 2\sqrt[3]{3} + 4\sqrt[3]{8 \cdot 3} - \sqrt[3]{27 \cdot 3}$

$= 2\sqrt[3]{3} + 4(2)\sqrt[3]{3} - 3\sqrt[3]{3}$

$= 2\sqrt[3]{3} + 8\sqrt[3]{3} - 3\sqrt[3]{3}$

$= 7\sqrt[3]{3}$

13. $\sqrt[3]{32} - 5\sqrt[3]{4} + 2\sqrt[3]{108}$

$= \sqrt[3]{8 \cdot 4} - 5\sqrt[3]{4} + 2\sqrt[3]{27 \cdot 4}$

$= \sqrt[3]{8}\sqrt[3]{4} - 5\sqrt[3]{4} + 2\sqrt[3]{27}\sqrt[3]{4}$

$= 2\sqrt[3]{4} - 5\sqrt[3]{4} + 2(3\sqrt[3]{4})$

$= 2\sqrt[3]{4} - 5\sqrt[3]{4} + 6\sqrt[3]{4}$

$= 3\sqrt[3]{4}$

14. $\sqrt{2x^3y^2z^4} = \sqrt{x^2y^2z^4 \cdot 2x}$

$= xyz^2\sqrt{2x}$

15. $\sqrt{98r^3s^4t^{10}}$

$= \sqrt{(49 \cdot 2)(r^2 \cdot r)(s^4)(t^{10})}$

$= \sqrt{(49r^2s^4t^{10})(2r)}$

$= \sqrt{49r^2s^4t^{10}}\sqrt{2r}$

$= 7rs^2t^5\sqrt{2r}$

16. $\sqrt[3]{16z^5x^8y^4} = \sqrt[3]{8z^3x^6y^3 \cdot 2z^2x^2y}$

$= 2zx^2y\sqrt[3]{2z^2x^2y}$

17. $\sqrt[4]{x^8y^7z^{11}} = \sqrt[4]{(x^8)(y^4 \cdot y^3)(z^8z^3)}$

$= \sqrt[4]{(x^8y^4z^8)(y^3z^3)}$

$= \sqrt[4]{x^8y^4z^8}\sqrt[4]{y^3z^3}$

$= x^2yz^2\sqrt[4]{y^3z^3}$

18. $\sqrt{a^3b^5} - 2\sqrt{a^7b^3} + \sqrt{a^3b^9}$

$= \sqrt{a^2b^4ab} - 2\sqrt{a^6b^2ab} + \sqrt{a^2b^8ab}$

$= ab^2\sqrt{ab} - 2a^3b\sqrt{ab} + ab^4\sqrt{ab}$

$= (ab^2 - 2a^3b + ab^4)\sqrt{ab}$

$= ab\sqrt{ab}(b - 2a^2 + b^3)$

19. $\sqrt{p^7q^3} - \sqrt{p^5q^9} + \sqrt{p^9q}$

$= \sqrt{(p^6p)(q^2q)} - \sqrt{(p^4p)(q^8q)}$
$\quad + \sqrt{(p^8p)q}$

$= \sqrt{(p^6q^2)(pq)} - \sqrt{(p^4q^8)(pq)}$
$\quad + \sqrt{(p^8)(pq)}$

$= \sqrt{p^6q^2}\sqrt{pq} - \sqrt{p^4q^8}\sqrt{pq} + \sqrt{p^8}\sqrt{pq}$

$= p^3q\sqrt{pq} - p^2q^4\sqrt{pq} + p^4\sqrt{pq}$

$= p^2pq\sqrt{pq} - p^2q^4\sqrt{pq} + p^2p^2\sqrt{pq}$

$= p^2\sqrt{pq}(pq - q^4 + p^2)$

20. $\dfrac{5}{\sqrt{7}} = \dfrac{5}{\sqrt{7}} \cdot \dfrac{\sqrt{7}}{\sqrt{7}} = \dfrac{5\sqrt{7}}{7}$

21. $\dfrac{-2}{\sqrt{3}} = \dfrac{-2}{\sqrt{3}} \cdot \dfrac{\sqrt{3}}{\sqrt{3}} = \dfrac{-2\sqrt{3}}{\sqrt{9}} = -\dfrac{2\sqrt{3}}{3}$

22. $\dfrac{-3}{\sqrt{12}} = \dfrac{-3}{\sqrt{4 \cdot 3}} = \dfrac{-3}{2\sqrt{3}} \cdot \dfrac{\sqrt{3}}{\sqrt{3}}$

$= \dfrac{-3\sqrt{3}}{6}$

$= -\dfrac{\sqrt{3}}{2}$

23. $\dfrac{4}{\sqrt{8}} = \dfrac{4}{\sqrt{8}} \cdot \dfrac{\sqrt{2}}{\sqrt{2}} = \dfrac{4\sqrt{2}}{\sqrt{16}} = \dfrac{4\sqrt{2}}{4} = \sqrt{2}$

24. $\dfrac{3}{1 - \sqrt{5}} = \dfrac{3}{1 - \sqrt{5}} \cdot \dfrac{1 + \sqrt{5}}{1 + \sqrt{5}}$

$= \dfrac{3(1 + \sqrt{5})}{1 - 5}$

$= \dfrac{-3(1 + \sqrt{5})}{4}$

25. $\dfrac{5}{2 - \sqrt{6}} = \dfrac{5}{2 - \sqrt{6}} \cdot \dfrac{2 + \sqrt{6}}{2 + \sqrt{6}}$

$= \dfrac{5(2 + \sqrt{6})}{4 + 2\sqrt{6} - 2\sqrt{6} - \sqrt{36}}$

$= \dfrac{5(2 + \sqrt{6})}{4 - \sqrt{36}} = \dfrac{5(2 + \sqrt{6})}{4 - 6}$

$= \dfrac{5(2 + \sqrt{6})}{-2}$

$= -\dfrac{5(2 + \sqrt{6})}{2}$

26. $\dfrac{-2}{\sqrt{3} - \sqrt{2}}$

$= \dfrac{-2}{\sqrt{3} - \sqrt{2}} \cdot \dfrac{\sqrt{3} + \sqrt{2}}{\sqrt{3} + \sqrt{2}}$

$= \dfrac{-2(\sqrt{3} + \sqrt{2})}{3 - 2} = \dfrac{-2(\sqrt{3} + \sqrt{2})}{1}$

$= -2(\sqrt{3} + \sqrt{2})$

27. $\dfrac{1}{\sqrt{10} + \sqrt{3}} = \dfrac{1}{\sqrt{10} + \sqrt{3}} \cdot \dfrac{\sqrt{10} - \sqrt{3}}{\sqrt{10} - \sqrt{3}}$

$= \dfrac{\sqrt{10} - \sqrt{3}}{\sqrt{100} - \sqrt{30} + \sqrt{30} - \sqrt{9}}$

$= \dfrac{\sqrt{10} - \sqrt{3}}{\sqrt{100} - \sqrt{9}} = \dfrac{\sqrt{10} - \sqrt{3}}{10 - 3}$

$= \dfrac{\sqrt{10} - \sqrt{3}}{7}$

28. $\dfrac{1}{\sqrt{r}-\sqrt{3}} = \dfrac{1}{\sqrt{r}-\sqrt{3}} \cdot \dfrac{\sqrt{r}+\sqrt{3}}{\sqrt{r}+\sqrt{3}}$

$\qquad = \dfrac{\sqrt{r}+\sqrt{3}}{r-3}$

29. $\dfrac{5}{\sqrt{m}-\sqrt{5}} = \dfrac{5}{\sqrt{m}-\sqrt{5}} \cdot \dfrac{\sqrt{m}+\sqrt{5}}{\sqrt{m}+\sqrt{5}}$

$\qquad = \dfrac{5(\sqrt{m}+\sqrt{5})}{\sqrt{m^2}+\sqrt{5m}-\sqrt{5m}-\sqrt{25}}$

$\qquad = \dfrac{5(\sqrt{m}+\sqrt{5})}{\sqrt{m^2}-\sqrt{25}} = \dfrac{5(\sqrt{m}+\sqrt{5})}{m-5}$

30. $\dfrac{y-5}{\sqrt{y}-\sqrt{5}} = \dfrac{y-5}{\sqrt{y}-\sqrt{5}} \cdot \dfrac{\sqrt{y}+\sqrt{5}}{\sqrt{y}+\sqrt{5}}$

$\qquad = \dfrac{(y-5)(\sqrt{y}+\sqrt{5})}{y-5}$

$\qquad = \sqrt{y}+\sqrt{5}$

31. $\dfrac{z-11}{\sqrt{z}-\sqrt{11}} = \dfrac{z-11}{\sqrt{z}-\sqrt{11}} \cdot \dfrac{\sqrt{z}+\sqrt{11}}{\sqrt{z}+\sqrt{11}}$

$\qquad = \dfrac{(z-11)(\sqrt{z}+\sqrt{11})}{\sqrt{z^2}+\sqrt{11z}-\sqrt{11z}-\sqrt{121}}$

$\qquad = \dfrac{(z-11)(\sqrt{z}+\sqrt{11})}{\sqrt{z^2}-\sqrt{121}}$

$\qquad = \dfrac{(z-11)(\sqrt{z}+\sqrt{11})}{(z-11)}$

$\qquad = \sqrt{z}+\sqrt{11}$

32. $\dfrac{\sqrt{x}+\sqrt{x+1}}{\sqrt{x}-\sqrt{x+1}} = \dfrac{\sqrt{x}+\sqrt{x+1}}{\sqrt{x}-\sqrt{x+1}} \cdot \dfrac{\sqrt{x}+\sqrt{x+1}}{\sqrt{x}+\sqrt{x+1}}$

$\qquad = \dfrac{x+2\sqrt{x(x+1)}+(x+1)}{x-(x+1)}$

$\qquad = \dfrac{2x+2\sqrt{x(x+1)}+1}{-1}$

$\qquad = -2x-2\sqrt{x(x+1)}-1$

33. $\dfrac{\sqrt{p}+\sqrt{p^2-1}}{\sqrt{p}-\sqrt{p^2-1}}$

$\qquad = \dfrac{\sqrt{p}+\sqrt{p^2-1}}{\sqrt{p}-\sqrt{p^2-1}} \cdot \dfrac{\sqrt{p}+\sqrt{p^2-1}}{\sqrt{p}+\sqrt{p^2-1}}$

$\qquad = \dfrac{(\sqrt{p})^2+2\sqrt{p}\sqrt{p^2-1}+(\sqrt{p^2-1})^2}{\sqrt{p^2}+\sqrt{p}\sqrt{p^2-1}-\sqrt{p}\sqrt{p^2-1}-(\sqrt{p^2-1})^2}$

$\qquad = \dfrac{p+2\sqrt{p}\sqrt{p^2-1}+(p^2-1)}{p-(p^2-1)}$

$\qquad = \dfrac{p^2+p+2\sqrt{p(p^2-1)}-1}{-p^2+p+1}$

34. $\dfrac{1+\sqrt{2}}{2} = \dfrac{(1+\sqrt{2})(1-\sqrt{2})}{2(1-\sqrt{2})}$

$\qquad = \dfrac{1-2}{2(1-\sqrt{2})}$

$\qquad = -\dfrac{1}{2(1-\sqrt{2})}$

35. $\dfrac{1-\sqrt{3}}{3} = \dfrac{1-\sqrt{3}}{3} \cdot \dfrac{1+\sqrt{3}}{1+\sqrt{3}}$

$\qquad = \dfrac{1^2-(\sqrt{3})^2}{3(1+\sqrt{3})}$

$\qquad = \dfrac{1-3}{3(1+\sqrt{3})}$

$\qquad = -\dfrac{2}{3(1+\sqrt{3})}$

36. $\dfrac{\sqrt{x}+\sqrt{x+1}}{\sqrt{x}-\sqrt{x+1}}$

$\qquad = \dfrac{\sqrt{x}+\sqrt{x+1}}{\sqrt{x}-\sqrt{x+1}} \cdot \dfrac{\sqrt{x}-\sqrt{x+1}}{\sqrt{x}-\sqrt{x+1}}$

$\qquad = \dfrac{x-(x+1)}{x-2\sqrt{x}\cdot\sqrt{x+1}+(x+1)}$

$\qquad = \dfrac{-1}{2x-2\sqrt{x(x+1)}+1}$

37. $\dfrac{\sqrt{p}+\sqrt{p^2-1}}{\sqrt{p}-\sqrt{p^2-1}}$

$\qquad = \dfrac{\sqrt{p}+\sqrt{p^2-1}}{\sqrt{p}-\sqrt{p^2-1}} \cdot \dfrac{\sqrt{p}-\sqrt{p^2-1}}{\sqrt{p}-\sqrt{p^2-1}}$

$\qquad = \dfrac{(\sqrt{p})^2-(\sqrt{p^2-1})^2}{(\sqrt{p})^2-2\sqrt{p}\sqrt{p^2-1}+(\sqrt{p^2-1})^2}$

$\qquad = \dfrac{p-(p^2-1)}{p-2\sqrt{p(p^2-1)}+p^2-1}$

$\qquad = \dfrac{p-p^2+1}{p-2\sqrt{p(p^2-1)}+p^2-1}$

$\qquad = \dfrac{-p^2+p+1}{p^2+p-2\sqrt{p(p^2-1)}-1}$

38. $\sqrt{16-8x+x^2}$

$\qquad = \sqrt{(4-x)^2}$

$\qquad = |4-x|$

Since $\sqrt{}$ denotes the nonnegative root, we must have $4-x \geq 0$.

39. $\sqrt{4y^2 + 4y + 1}$
$$= \sqrt{(2y + 1)^2}$$
$$= |2y + 1|$$

Since $\sqrt{}$ denotes the nonnegative root, we must have $2y + 1 \geq 0$.

40. $\sqrt{4 - 25z^2} = \sqrt{(2 + 5z)(2 - 5z)}$

This factorization does not produce a perfect square, so the expression $\sqrt{4 - 25z^2}$ cannot be simplified.

41. $\sqrt{9k^2 + h^2}$

The expression $9k^2 + h^2$ is the sum of two squares and cannot be factored. Therefore, $\sqrt{9k^2 + h^2}$ cannot be simplified.

Chapter 1

FUNCTIONS

1.1 Lines and Linear Functions

1. Find the slope of the line through $(4, 5)$ and $(-1, 2)$.

$$m = \frac{5 - 2}{4 - (-1)}$$
$$= \frac{3}{5}$$

3. Find the slope of the line through $(8, 4)$ and $(8, -7)$.

$$m = \frac{4 - (-7)}{8 - 8}$$
$$= \frac{11}{0}$$

The slope is undefined; the line is vertical.

5. $y = 2x$

Using the slope-intercept form, $y = mx + b$, we see that the slope is 2.

7. $5x - 9y = 11$

Rewrite the equation in slope-intercept form.

$$9y = 5x - 11$$
$$y = \frac{5}{9}x - \frac{11}{9}$$

The slope is $\frac{5}{9}$.

9. $x = -6$

This is a vertical line; the slope is undefined.

11. $y = 8$

This is a horizontal line, which has a slope of 0.

13. Find the slope of a line parallel to $2y - 4x = 7$.
Rewrite the equation in slope-intercept form.

$$2y = 4x + 7$$
$$y = 2x + \frac{7}{2}$$

This slope is 2, so a parallel line will also have slope 2.

15. The line goes through $(1, 3)$, with slope $m = -2$. Use the point-slope form.

$$y - 3 = -2(x - 1)$$
$$y = -2x + 2 + 3$$
$$y = -2x + 5$$

17. The line goes through $(6, 1)$, with slope $m = 0$. Use the point-slope form.

$$y - 1 = 0(x - 6)$$
$$y - 1 = 0$$
$$y = 1$$

19. The line goes through $(4, 2)$ and $(1, 3)$. Find the slope, then use the point-slope form with either of the two given points.

$$m = \frac{3 - 2}{1 - 4}$$
$$= -\frac{1}{3}$$

Use the point $(1, 3)$.

$$y - 3 = -\frac{1}{3}(x - 1)$$
$$y = -\frac{1}{3}x + \frac{1}{3} + 3$$
$$y = -\frac{1}{3}x + \frac{10}{3}$$

21. The line goes through $\left(\frac{1}{2}, \frac{5}{3}\right)$ and $\left(3, \frac{1}{6}\right)$.

$$m = \frac{\frac{1}{6} - \frac{5}{3}}{3 - \frac{1}{2}} = \frac{\frac{1}{6} - \frac{10}{6}}{\frac{6}{2} - \frac{1}{2}}$$
$$= \frac{-\frac{9}{6}}{\frac{5}{2}} = -\frac{18}{30} = -\frac{3}{5}$$

$$y - \frac{5}{3} = -\frac{3}{5}\left(x - \frac{1}{2}\right)$$
$$y - \frac{5}{3} = -\frac{3}{5}x + \frac{3}{10}$$
$$y = -\frac{3}{5}x + \frac{3}{10} + \frac{5}{3}$$
$$y = -\frac{3}{5}x + \frac{9}{30} + \frac{50}{30}$$
$$y = -\frac{3}{5}x + \frac{59}{30}$$

23. The line goes through $(-8, 4)$ and $(-8, 6)$.

$$m = \frac{4-6}{-8-(-8)} = \frac{-2}{0}$$

which is undefined.

This is a vertical line; the value of x is always -8.
The equation of this line is $x = -8$.

25. The line has x-intercept 3 and y-intercept -2.
Two points on the line are $(3, 0)$ and $(0, -2)$. Find the slope, then use the slope-intercept form.

$$m = \frac{0-(-2)}{3-0}$$
$$= \frac{2}{3}$$
$$b = -2$$
$$y = \frac{2}{3}x - 2$$

27. The vertical line through $(-6, 5)$ goes through the point $(-6, 0)$, so the equation is $x = -6$.

29. Write an equation of the line through $(-1, 4)$, parallel to $x + 3y = 5$.
Rewrite the equation of the given line in slope-intercept form.

$$x + 3y = 5$$
$$3y = -x + 5$$
$$y = -\frac{1}{3}x + \frac{5}{3}$$

The slope is $-\frac{1}{3}$.

Use $m = -\frac{1}{3}$ and the point $(-1, 4)$ in the point-slope form.

$$y - 4 = -\frac{1}{3}[x - (-1)]$$
$$y = -\frac{1}{3}(x + 1) + 4$$
$$y = -\frac{1}{3}x - \frac{1}{3} + 4$$
$$y = -\frac{1}{3}x + \frac{11}{3}$$

31. Write an equation of the line through $(3, -4)$, perpendicular to $x + y = 4$.
Rewrite the equation of the given line as

$$y = -x + 4.$$

The slope of this line is -1. To find the slope of a perpendicular line, solve

$$-1m = -1.$$
$$m = 1$$

Use $m = 1$ and $(3, -4)$ in the point-slope form.

$$y - (-4) = 1(x - 3)$$
$$y = x - 3 - 4$$
$$y = x - 7$$

33. Write an equation of the line with y-intercept 2, perpendicular to $3x + 2y = 6$.
Find the slope of the given line.

$$3x + 2y = 6$$
$$2y = -3x + 6$$
$$y = -\frac{3}{2}x + 3$$

The slope is $-\frac{3}{2}$, so the slope of the perpendicular line will be $\frac{2}{3}$. If the y-intercept is 2, then using the slope-intercept form, we have

$$y = mx + b$$
$$y = \frac{2}{3}x + 2$$

35. Do the points $(4, 3), (2, 0)$, and $(-18, -12)$ lie on the same line?
Find the slope between $(4, 3)$ and $(2, 0)$.

$$m = \frac{0-3}{2-4} = \frac{-3}{-2} = \frac{3}{2}$$

Find the slope between $(4, 3)$ and $(-18, -12)$.

$$m = \frac{-12-3}{-18-4} = \frac{-15}{-22} = \frac{15}{22}$$

Since these slopes are not the same, the points do not lie on the same line.

37. A parallelogram has 4 sides, with opposite sides parallel. The slope of the line through $(1, 3)$ and $(2, 1)$ is

$$m = \frac{3-1}{1-2} = \frac{2}{-1} = -2.$$

The slope of the line through $\left(-\frac{5}{2}, 2\right)$ and $\left(-\frac{7}{2}, 4\right)$ is

$$m = \frac{2-4}{-\frac{5}{2}-\left(-\frac{7}{2}\right)} = \frac{-2}{1} = -2.$$

Since these slopes are equal, these two sides are parallel.

The slope of the line through $\left(-\frac{7}{2}, 4\right)$ and $(1, 3)$ is

$$m = \frac{4 - 3}{-\frac{7}{2} - 1} = \frac{1}{-\frac{9}{2}} = -\frac{2}{9}.$$

The slope of the line through $\left(-\frac{5}{2}, 2\right)$ and $(2, 1)$ is

$$m = \frac{2 - 1}{-\frac{5}{2} - 2} = \frac{1}{-\frac{9}{2}} = -\frac{2}{9}.$$

Since these slopes are equal, these two sides are parallel.

Since both pairs of opposite sides are parallel, the quadrilateral is a parallelogram.

39. The line goes through $(0, 2)$ and $(-2, 0)$.

$$m = \frac{2 - 0}{0 - (-2)} = \frac{2}{2} = 1$$

The correct choice is (a).

41. The line appears to go through $(0, 0)$ and $(-1, 4)$.

$$m = \frac{4 - 0}{-1 - 0} = \frac{4}{-1} = -4$$

43. (a) See the figure in the textbook.

Segment MN is drawn perpendicular to segment PQ. Recall that MQ is the length of segment MQ.

$$m_1 = \frac{\Delta y}{\Delta x} = \frac{MQ}{PQ}$$

From the diagram, we know that $PQ = 1$. Thus, $m_1 = \frac{MQ}{1}$, so MQ has length m_1.

(b) $\quad m_2 = \dfrac{\Delta y}{\Delta x} = \dfrac{-QN}{PQ} = \dfrac{-QN}{1}$

$\qquad QN = -m_2$

(c) Triangles MPQ, PNQ, and MNP are right triangles by construction. In triangles MPQ and MNP,

$$\text{angle } M = \text{angle } M,$$

and in the right triangles PNQ and MNP,

$$\text{angle } N = \text{angle } N.$$

Since all right angles are equal, and since triangles with two equal angles are similar, triangle MPQ is similar to triangle MNP and triangle PNQ is similar to triangle MNP.

Therefore, triangles MPQ and PNQ are similar to each other.

(d) Since corresponding sides in similar triangles are proportional,

$$MQ = k \cdot PQ \quad \text{and} \quad PQ = k \cdot QN.$$

$$\frac{MQ}{PQ} = \frac{k \cdot PQ}{k \cdot QN}$$

$$\frac{MQ}{PQ} = \frac{PQ}{QN}$$

From the diagram, we know that $PQ = 1$.

$$MQ = \frac{1}{QN}$$

From (a) and (b), $m_1 = MQ$ and $-m_2 = QN$.

Substituting, we get

$$m_1 = \frac{1}{-m_2}.$$

Multiplying both sides by m_2, we have

$$m_1 m_2 = -1.$$

45. $y = 2x + 3$

Three ordered pairs that satisfy this equation are $(-2, -1)$, $(0, 3)$, and $(1, 5)$. Plot these points and draw the line through them.

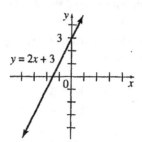

47. $y = -6x + 12$

Three ordered pairs that satisfy this equation are $(0, 12)$, $(1, 6)$, and $(2, 0)$. Plot these points and draw the line through them.

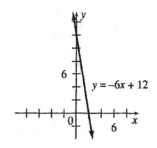

49. $3x - y = -9$

Find the intercepts.

If $y = 0$, then

$$3x - 0 = -9$$
$$3x = -9$$
$$x = -3,$$

so the x-intercept is -3.

If $x = 0$, then

$$3(0) - y = -9$$
$$-y = -9$$
$$y = 9,$$

so the y-intercept is 9.
Plot the ordered pairs $(-3, 0)$ and $(0, 9)$ and draw the line through these points. (A third point may be used as a check.)

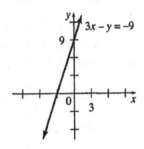

51. $4y + 5x = 10$

Find the intercepts.

If $y = 0$, then

$$4(0) + 5x = 10$$
$$5x = 10$$
$$x = 2,$$

so the x-intercept is 2.

If $x = 0$, then

$$4y + 5(0) = 10$$
$$4y = 10$$
$$y = \frac{10}{4} = \frac{5}{2},$$

so the y-intercept is $\frac{5}{2}$.

Plot the ordered pairs $(2, 0)$ and $\left(0, \frac{5}{2}\right)$ and draw the line through these points. (A third point may be used as a check.)

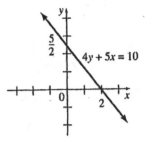

53. $x = 4$

For any value of y, the x-value is 4. Because all ordered pairs that satisfy this equation have the same first number, this equation does not represent a function. The graph is the vertical line with x-intercept 4.

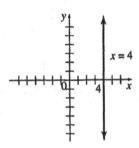

55. $y - 4 = 0$

This equation may be rewritten as $y = 4$, or, equivalently, $y = 0x + 4$. The y-value is 4 for any value of x. The graph is the horizontal line with y-intercept 4.

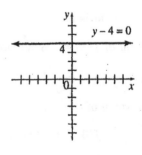

57. $y = -5x$

Three ordered pairs that satisfy this equation are $(0,0)$, $(-1,5)$, and $(1,-5)$. Use these points to draw the graph.

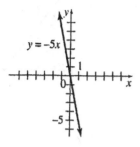

59. $x - 3y = 0$

If $y = 0$, then $x = 0$, so the x-intercept is 0. If $x = 0$, then $y = 0$, so the y-intercept is 0. Both intercepts give the same ordered pair, $(0,0)$.

To get a second point, choose some other value of x (or y). For example, if $x = 3$, then

$$x - 3y = 0$$
$$3 - 3y = 0$$
$$-3y = -3$$
$$y = 1,$$

giving the ordered pair $(3,1)$. Graph the line through $(0,0)$ and $(3,1)$.

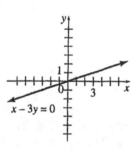

61. This statement is false.

The graph of $f(x) = -3$ is a horizontal line.

63. This statement is true.

For any value of a,

$$f(0) = a \cdot 0 = 0,$$

so the point $(0,0)$, which is the origin, lies on the line.

65. (a) Let $x =$ age.

$$u = 0.85(220 - x) = 187 - 0.85x$$
$$l = 0.7(220 - x) = 154 - 0.7x$$

(b) $u = 187 - 085(20) = 170$
$l = 154 - 0.7(20) = 140$

The target heart rate zone is 140 to 170 beats per minute.

(c) $u = 187 - 0.85(40) = 153$
$l = 154 - 0.7(40) = 126$

The target heart rate zone is 126 to 153 beats per minute.

(d) $154 - 0.7x = 187 - 0.85(x + 36)$
$154 - 0.7x = 187 - 0.85x - 30.6$
$154 - 0.7x = 156.4 - 0.85x$
$0.15x = 2.4$
$x = 16$

The younger woman is 16; the older is $16 + 36 = 52$.
$l = 0.7(220 - 16) \approx 143$ beats per minute.

67. Let x be the number of years after 1900. Then the "life expectancy from birth" line contains the points $(0,46)$ and $(75,75)$.

$$m = \frac{75 - 46}{75 - 0} = \frac{29}{75}$$

Use the point-slope form.

$$y - 46 = \frac{29}{75}(x - 0)$$
$$y = \frac{29}{75}x + 46$$

The "life expectancy from age 65" line contains the points $(0,76)$ and $(75,80)$.

$$m = \frac{80 - 76}{75 - 0} = \frac{4}{75}$$
$$y - 76 = \frac{4}{75}(x - 0)$$
$$y = \frac{4}{75}x + 76$$

Set the two equations equal to determine where the lines intersect. At this point, life expectancy should increase no further.

$$\frac{29}{75}x + 46 = \frac{4}{75}x + 76$$
$$29x + 3,450 = 4x + 5,700$$
$$25x = 2,250$$
$$x = 90$$

Determine the y-value when $x = 90$.

$$y = \frac{4}{75}(90) + 76$$

$$y = 4.8 + 76$$
$$= 80.8$$

Thus, the maximum life expectancy for humans is about 81 years.

69. (a) Since the years of healthy life is going up linearly 28 million years every 10 years, the slope of the line is $m = \frac{28}{10} = 2.8$. Since 35 million years of healthy life was lost in 1990, it follows that $y = 35$ when $x = 0$. So, the y-intercept is $b = 35$. Therefore, $y = mx + b = 2.8x + 35$ is the number of years (in millions) of healthy life lost globally to tobacco x years after 1990.

(b) Since the number of years lost to diarrhea is going down linearly 22 million years every 10 years, the slope of the line is $m = \frac{-22}{10} = -2.2$. Since 100 million years of healthy life was lost in 1990, it follows that $y = 100$, when $x = 0$. So, the y-intercept is $b = 100$. Therefore, $y = -2.2x + 100$ is the number of years lost (in millions) to diarrhea x years after 1990.

(c) Find out when the number of years of healthy life lost to tobacco will equal that lost to diarrhea. We solve the equation

$$2.8x + 35 = -2.2x + 100.$$
$$5x = 65$$
$$x = 13$$

So, the number of years of healthy life lost to tobacco will first exceed that lost to diarrhea 13 years after 1990, or in 2003.

71. (a) The function is of the form $y = f(x) = mx + b$ and contains the points $(5, 8.2)$ and $(17, 33.4)$. The slope is

$$m = \frac{8.2 - 33.4}{5 - 17} = \frac{-25.2}{-12} = 2.1.$$

Using the point-slope form we have

$$y - 8.2 = 2.1(x - 5)$$
$$y = 2.1x - 10.5 + 8.2$$
$$= 2.1x - 2.3.$$

So $y = 2.1x - 2.3$ is the number of male alates in an ant colony that is x years old.

(b) $f(1) = 2.1(1) - 2.3 = -0.2 \approx 0$, so the function predicts 0 alates in a one year-old colony.

(c) Set $f(x) = 40$, and solve for x.

$$2.1x - 2.3 = 40$$
$$x = \frac{40 + 2.3}{2.1} \approx 20.1$$

Therefore, this model predicts that a colony that is about 20 years old would have 40 male alates.

73. The cost to use the first thermometer on x patients is $y = 2x + 10$ dollars. The cost to use the second thermometer or x patients is $y = 0.75x + 10$ dollars. If these two costs are equal, then

$$2x + 10 = 0.75x + 120$$
$$1.25x = 110$$
$$x = 88$$

So, the costs of the these two thermometers are equal; after each is used on 88 patients.

75. (a) Since the average temperature rises $0.3°C$ every ten years, the slope is

$$m = \frac{0.3}{10} = 0.03.$$

$b = 15$, since a point is $(0, 15)$.
Let T be the average global temperature in degrees Celsius.

$$T = 0.03t + 15$$

(b) Let $T = 19$; find t.

$$19 = 0.03t + 15$$
$$4 = 0.03t$$
$$133.3 = t$$
$$133 \approx t$$
$$1970 + 133 = 2103$$

The temperature will rise to $19°C$ in about the year 2103.

77. (a) The line goes through $(0, 86,821)$ and $(22, 199,483)$.

$$m = \frac{199,483 - 86,821}{22 - 0} = \frac{112,662}{22}$$
$$m = 5,121$$
$$b = 86,821$$
$$y = 5,121x + 86,821$$

(b) The year 2010 corresponds to $x = 36$.

$$y = 5{,}121(36) + 86{,}821$$
$$y = 271{,}177$$

We predict that the number of immigrants to California in 2010 will be about 271,177.

79. (a) Use the points $(2, 9)$ and $(28, 17)$.

$$m = \frac{17 - 9}{28 - 2} = \frac{8}{26} \approx 0.31$$

Let p be the percentage of college students age 35 and older.

$$p - 9 = 0.31(t - 2)$$
$$p - 9 = 0.31t - .62$$
$$p = 0.31t + 8.38$$
$$p = 0.31t + 8.4$$

(b) Let $t = 40$; find p.

$$p = 0.31(40) + 8.4$$
$$p = 20.8$$

In 2010, the prediction is about 21%.

(c) Let $p = 31$; find t.

$$31 = 0.31t + 8.4$$
$$22.6 = 0.31t$$
$$t \approx 73$$

$$1970 + 73 = 2043$$

The predicted year is 2043.

81. If the temperatures are numerically equal, then $F = C$.

$$F = \frac{9}{5}C + 32$$

$$C = \frac{9}{5}C + 32$$

$$-\frac{4}{5}C = 32$$

$$C = -40$$

The Celsius and Fahrenheit temperatures are numerically equal at $-40°$.

1.2 The Least Squares Line

3.

x	y	xy	x^2	y^2
6.8	0.8	5.44	46.24	0.64
7.0	1.2	8.4	49.0	1.44
7.1	0.9	6.39	50.41	0.81
7.2	0.9	6.48	51.84	0.81
7.4	1.5	11.1	54.76	2.25
35.5	5.3	37.81	252.25	5.95

$$r = \frac{5(37.81) - (35.5)(5.3)}{\sqrt{5(252.25) - (35.5)^2} \cdot \sqrt{5(5.95) - (5.3)^2}}$$

$$\approx 0.6985$$
$$r^2 = (0.6985)^2 \approx 0.5$$

The answer is choice (c).

5. (a)

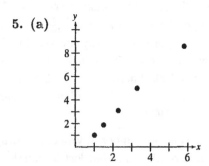

Yes, the data appear to be linear.

(b)

x	y	xy	x^2	y^2
5.8	8.6	49.88	33.64	73.96
1.5	1.9	2.85	2.25	3.61
2.3	3.1	7.13	5.29	9.61
1.0	1.0	1.0	1.0	1.0
3.3	5.0	16.5	10.89	25.0
13.9	19.6	77.36	53.07	113.18

Solve the system

$$5b + 13.9m = 19.6$$
$$13.9b + 53.07m = 77.36$$

for b and m.

$$5b + 13.9m = 19.6$$
$$5b = 19.6 - 13.9m$$

$$b = \frac{19.6 - 13.9m}{5}$$

$$13.9 \left(\frac{19.6 - 13.9m}{5} \right) + 53.07m = 77.36$$

$$13.9(19.6 - 13.9m) + 265.35m = 386.8$$
$$272.44 - 193.21m + 265.35m = 386.8$$
$$72.14m = 114.36$$

$$m = \frac{114.36}{72.14} \approx 1.585$$

$$b = \frac{19.6 - 13.9 \left(\frac{114.36}{72.14} \right)}{5}$$

$$\approx -0.487$$

Therefore, the least squares line is

$$Y = 1.585x - 0.487.$$

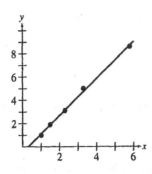

(c) No; for small values of w, for example, $x = 0.3$ or less, the least squares line gives negative y values, or negative egg lengths.

(d) $r = \dfrac{n(\sum xy) - (\sum x)(\sum y)}{\sqrt{n(\sum x^2) - (\sum x)^2} \cdot \sqrt{n(\sum y^2) - (\sum y)^2}}$

$$= \frac{5(77.36) - (13.9)(19.6)}{\sqrt{5(53.07) - 13.9^2} \sqrt{5(113.18) - 19.6^2}}$$

$$\approx 0.999$$

7. (a) Using the linear regression program of a graphing calculator we find that $r \approx 0.994$. Since r is close to 1, the data fits a straight line very well.

(b) Again with the aid of a graphing calculator we find the equation of the least squares line to be

$$Y = 1.3525x - 2.51.$$

(c)

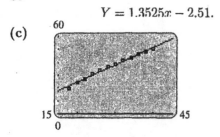

Yes, the line accurately fits the data.

(d) Since the slope of the least squares line is $m = 1.3525 = \frac{1.3525}{1}$, then the fetal stature is increasing by approximately 1.3525 cm each week.

(e) Using the formula for the least squares line, since $Y = 1.3525(45) - 2.51 \approx 58.35$, we would expect a 45-week-old fetus to have a stature of approximately 58.35 cm in length.

9. (a)

x	y	xy	x^2	y^2
88.6	20.0	1,772.00	7,849.96	400.00
71.6	16.0	1,145.60	5,126.56	256.00
93.3	19.8	1,847.34	8,704.89	392.04
84.3	18.4	1,551.12	7,106.49	338.56
80.6	17.1	1,378.26	6,496.36	292.41
75.2	15.5	1,165.60	5,655.04	240.25
69.7	14.7	1,024.59	4,858.09	216.09
82.0	17.1	1,402.20	6,724.00	292.41
69.4	15.4	1,068.76	4,816.36	237.16
83.3	16.2	1,349.46	6,938.89	262.44
79.6	15.0	1,194.00	6,336.16	225.00
82.6	17.2	1,420.72	6,822.76	295.84
80.6	16.0	1,289.60	6,496.36	256.00
83.5	17.0	1,419.50	6,972.25	289.00
76.3	14.4	1,098.72	5,821.69	207.36
1,200.6	249.8	20,127.47	96,725.86	4,200.56

$$15b + 1,200.6m = 249.8$$
$$1,200.6b + 96,725.86m = 20,127.47$$

$$15b = 249.8 - 1,200.6m$$
$$b = \frac{249.8 - 1,200.6m}{15}$$

$$1,200.6 \left(\frac{249.8 - 1,200.6m}{15} \right) + 96,725.86m = 20,127.47$$
$$1,200.6(249.8 - 1,200.6m) + 1,450,887.9m = 301,912.05$$
$$299,909.88 - 1,441,440.36m = 301,912.05$$
$$+ 1,450,887.9m$$
$$9,447.54m = 2,002.17$$
$$m \approx 0.211925$$
$$m \approx 0.212$$

$$b = \frac{249.8 - 1,200.6(0.211925)}{15}$$

$$\approx -0.309$$

$$Y = 0.212x - 0.309$$

(b) Let $x = 73$; find Y.

$$Y = 0.212(73) - 0.309$$

$$\approx 15.2$$

If the temperature were 73°F, you would expect to hear 15.2 chirps per second.

(c) Let $Y = 18$; find x.

$$18 = 0.212x - 0.315$$

$$18.315 = 0.212x$$

$$86.4 \approx x$$

When the crickets are chirping 18 times per second, the temperature is 86.4°F.

(d)

$$r = \frac{15(20,127.47) - (1,200.6)(249.8)}{\sqrt{15(96,725.86) - (1,200.6)^2} \cdot \sqrt{15(4,200.56) - (249.8)^2}}$$

$$\approx 0.835$$

11. (a) Convert the time from hours and minutes into hours.

$$\text{average speed} = \frac{\sum y}{\sum x}$$

$$= \frac{658}{195.70}$$

$$= 3.36$$

Hirst's average speed is 3.36 miles per hour.

(b)

Yes, the data appear to lie approximately on a straight line.

(c)

x	y
0.0000	0.0
2.9500	12.2
5.0500	19.0
7.3667	28.0
9.4333	33.4
10.1667	36.6
11.2667	43.8
16.0167	59.4
21.6667	69.6
24.0000	78.4
26.8000	85.0
28.9833	91.5
32.0000	101.1

Using a graphing calculator,

$$Y = 3.06x + 4.54$$

(d) Also, using a graphing calculator,

$$r \approx 0.9964$$

Yes, this value indicates a good fit of the least squares line to the data.

(e) A good value for Hirst's average speed would be

$$m = 3.06 \text{ miles per hour.}$$

13. $r = \dfrac{38(46,209,266) - (175,878)(9,989)}{\sqrt{38(872,066,218) - (175,878)^2} \cdot \sqrt{38(2,629,701) - (9,989)^2}} \approx -0.049$

No, there does not appear to be a trend.

15. (a)   **(b)**

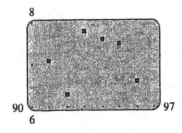

There is no linear pattern.

Yes, it has somewhat of a linear pattern,
but a line would not be a good fit.

(c)

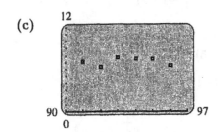

A line would be a fair fit for this set of points.

(d) Choosing a larger range of y makes the points appear to be closer to a line.

(e)

x	y	xy	x^2	y^2
91	7.1	646.1	8,281	50.41
92	6.3	579.6	8,464	39.69
93	7.8	725.4	8,649	60.84
94	7.6	714.4	8,836	57.76
95	7.5	712.5	9,025	56.25
96	6.6	633.6	9,216	43.56
561	42.9	4,011.6	52,471	308.51

$$r = \dfrac{6(4,011.6) - (561)(42.9)}{\sqrt{6(52,471) - 561^2} \cdot \sqrt{6(308.51) - 42.9^2}} \approx 0.081$$

17. (a)

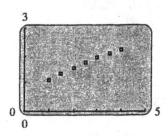

(b)

L	T	LT	L^2	T^2
1.0	1.11	1.11	1	1.2321
1.5	1.36	2.04	2.25	1.8496
2.0	1.57	3.14	4	2.4649
2.5	1.76	4.4	6.25	3.0976
3.0	1.92	5.76	9	3.6864
3.5	2.08	7.28	12.25	4.3264
4.0	2.22	8.88	16	4.9284
17.5	12.02	32.61	50.75	21.5854

$$7b + 17.5m = 12.02$$
$$17.5b + 50.75m = 32.61$$

$$7b + 17.5m = 12.02$$
$$7b = 12.02 - 17.5m$$
$$b = \frac{12.02 - 17.5m}{7}$$

$$17.5\left(\frac{12.02 - 17.5m}{7}\right) + 50.75m = 32.61$$

$$17.5\,(12.02 - 17.5m) + 355.25m = 228.27$$
$$210.35 - 306.25m + 355.25m = 228.27$$
$$49m = 17.92$$
$$m \approx 0.3657$$

$$b = \frac{12.02 - 17.5(0.3657)}{7} \approx 0.803$$
$$Y = 0.366x + 0.803$$

The line seems to fit the data.

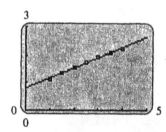

(c) $r = \dfrac{7(32.61) - (17.5)(12.02)}{\sqrt{7(50.75) - 17.5^2} \cdot \sqrt{7(21.5854) - 12.02^2}}$

≈ 0.995, which indicates a good fit and confirms the conclusion in part (b).

1.3 Properties of Functions

1. The x-value of 82 corresponds to two y-values, 93 and 14. In a function, each value of x must correspond to exactly one value of y.
The rule is not a function.

3. Each x-value corresponds to exactly one y-value.
The rule is a function.

5. $y = x^3$

Each x-value corresponds to exactly one y-value. The rule is a function.

7. $x = |y|$

Each value of x (except 0) corresponds to two y-values.
The rule is not a function.

9. $y = 2x + 3$

x	-2	-1	0	1	2	3
y	-1	1	3	5	7	9

Pairs: $(-2, -1)$, $(-1, 1)$, $(0, 3)$, $(1, 5)$, $(2, 7)$, $(3, 9)$

Range: $\{-1, 1, 3, 5, 7, 9\}$

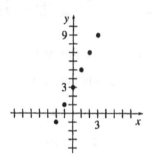

11. $2y - x = 5$
$$2y = 5 + x$$
$$y = \frac{1}{2}x + \frac{5}{2}$$

x	-2	-1	0	1	2	3
y	$\frac{3}{2}$	2	$\frac{5}{2}$	3	$\frac{7}{2}$	4

Pairs: $(-2, \frac{3}{2})$, $(-1, 2)$, $(0, \frac{5}{2})$, $(1, 3)$, $(2, \frac{7}{2})$, $(3, 4)$

Range: $\{\frac{3}{2}, 2, \frac{5}{2}, 3, \frac{7}{2}, 4\}$

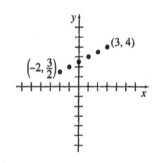

13. $y = x(x+1)$

x	-2	-1	0	1	2	3
y	2	0	0	2	6	12

Pairs: $(-2,2)$, $(-1,0)$, $(0,0)$,
$(1,2)$, $(2,6)$, $(3,12)$

Range: $\{0, 2, 6, 12\}$

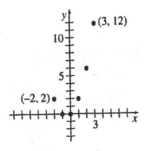

15. $y = x^2$

x	-2	-1	0	1	2	3
y	4	1	0	1	4	9

Pairs: $(-2,4)$, $(-1,1)$, $(0,0)$,
$(1,1)$, $(2,4)$, $(3,9)$

Range: $\{0, 1, 4, 9\}$

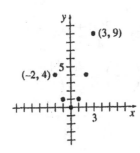

17. $y = \dfrac{1}{x+3}$

x	-2	-1	0	1	2	3
y	1	$\frac{1}{2}$	$\frac{1}{3}$	$\frac{1}{4}$	$\frac{1}{5}$	$\frac{1}{6}$

Pairs: $(-2,1)$, $(-1,\frac{1}{2})$, $(0,\frac{1}{3})$, $(1,\frac{1}{4})$, $(2,\frac{1}{5})$, $(3,\frac{1}{6})$

Range: $\{1, \frac{1}{2}, \frac{1}{3}, \frac{1}{4}, \frac{1}{5}, \frac{1}{6}\}$

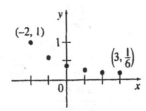

19. $y = \dfrac{3x - 3}{x + 5}$

x	-2	-1	0	1	2	3
y	-3	$-\frac{3}{2}$	$-\frac{3}{5}$	0	$\frac{3}{7}$	$\frac{3}{4}$

Pairs: $(-2,-3)$, $(-1,-\frac{3}{2})$, $(0,-\frac{3}{5})$, $(1,0)$, $(2,\frac{3}{7})$, $(3,\frac{3}{4})$

Range: $\{-3, -\frac{3}{2}, -\frac{3}{5}, 0, \frac{3}{7}, \frac{3}{4}\}$

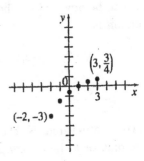

21. $f(x) = 2x$

x can take on any value, so the domain is the set of real numbers, $(-\infty, \infty)$.

23. $f(x) = x^4$

x can take on any value, so the domain is the set of real numbers, $(-\infty, \infty)$.

25. $f(x) = \sqrt{16 - x^2}$

For $f(x)$ to be a real number, $16 - x^2 \geq 0$.
Solve $16 - x^2 = 0$.

$$(4 - x)(4 + x) = 0$$
$$x = 4 \quad \text{or} \quad x = -4$$

The numbers form the intervals $(-\infty, -4)$, $(-4, 4)$, and $(4, \infty)$.
Only values in the interval $(-4, 4)$ satisfy the inequality. The domain is $[-4, 4]$.

27. $f(x) = (x - 3)^{1/2} = \sqrt{x - 3}$

For $f(x)$ to be a real number,

$$x - 3 \geq 0$$
$$x \geq 3.$$

The domain is $[3, \infty)$.

29. $f(x) = \dfrac{2}{x^2 - 4} = \dfrac{2}{(x-2)(x+2)}$

Since division by zero is not defined,
$(x - 2) \cdot (x + 2) \neq 0$.
When $(x - 2)(x + 2) = 0$,

$$x - 2 = 0 \quad \text{or} \quad x + 2 = 0$$
$$x = 2 \quad \text{or} \quad x = -2.$$

Thus, x can be any real number except ± 2.
The domain is

$$(-\infty, -2) \cup (-2, 2) \cup (2, \infty).$$

31. $f(x) = -\sqrt{\dfrac{2}{x^2 + 9}}$

x can take on any value. No choice for x will cause
the denominator to be zero. Also, no choice for x
will produce a negative number under the radical.
The domain is $(-\infty, \infty)$.

33. $f(x) = \sqrt{x^2 - 4x - 5} = \sqrt{(x-5)(x+1)}$

See the method used in Exercise 25.

$$(x - 5)(x + 1) \geq 0$$

when $x \geq 5$ and when $x \leq -1$.
The domain is $(-\infty, -1] \cup [5, \infty)$.

35. $f(x) = \dfrac{1}{\sqrt{x^2 - 6x + 8}} = \dfrac{1}{\sqrt{(x-4)(x-2)}}$

$(x - 4)(x - 2) > 0$, since the radicand cannot be
negative and the denominator of the function cannot be zero.
Solve $(x - 4)(x - 2) = 0$.

$$x - 4 = 0 \quad \text{or} \quad x - 2 = 0$$
$$x = 4 \quad \text{or} \quad x = 2$$

Use the values 2 and 4 to divide the number line
into 3 intervals, $(-\infty, 2)$, $(2, 4)$ and $(4, \infty)$.
Only the values in the intervals $(-\infty, 2)$ and $(4, \infty)$
satisfy the inequality.
The domain is $(-\infty, 2) \cup (4, \infty)$.

37. By reading the graph, the domain is all numbers
greater than or equal to -5 and less than 4. The
range is all numbers greater than or equal to -2
and less than or equal to 6.
Domain: $[-5, 4)$; range: $[-2, 6]$

39. By reading the graph, x can take on any value,
but y is less than or equal to 12.
Domain: $(-\infty, \infty)$; range: $(-\infty, 12]$

41. $f(x) = -x^2 + 5x + 1$

(a) $f(4) = -(4)^2 + 5(4) + 1$
$= -16 + 20 + 1$
$= 5$

(b) $f\left(-\dfrac{1}{2}\right) = -\left(-\dfrac{1}{2}\right)^2 + 5\left(-\dfrac{1}{2}\right) + 1$
$= -\dfrac{1}{4} - \dfrac{5}{2} + 1$
$= -\dfrac{7}{4}$

(c) $f(a) = -(a)^2 + 5(a) + 1$
$= -a^2 + 5a + 1$

(d) $f\left(\dfrac{2}{m}\right) = -\left(\dfrac{2}{m}\right)^2 + 5\left(\dfrac{2}{m}\right) + 1$
$= -\dfrac{4}{m^2} + \dfrac{10}{m} + 1$
$\text{or} \quad \dfrac{-4 + 10m + m^2}{m^2}$

(e)
$$f(x) = 1$$
$$-x^2 + 5x + 1 = 1$$
$$-x^2 + 5x = 0$$
$$x(-x + 5) = 0$$
$$x = 0 \quad \text{or} \quad x = 5$$

43. $f(x) = \dfrac{2x + 1}{x - 2}$

(a) $f(4) = \dfrac{2(4) + 1}{4 - 2} = \dfrac{9}{2}$

(b) $f\left(-\dfrac{1}{2}\right) = \dfrac{2(-\frac{1}{2}) + 1}{-\frac{1}{2} - 2}$
$= \dfrac{-1 + 1}{\frac{5}{2}}$
$= \dfrac{0}{\frac{5}{2}} = 0$

(c) $f(a) = \dfrac{2(a) + 1}{(a) - 2} = \dfrac{2a + 1}{a - 2}$

(d) $f\left(\dfrac{2}{m}\right) = \dfrac{2\left(\frac{2}{m}\right) + 1}{\frac{2}{m} - 2}$
$= \dfrac{\frac{4}{m} + \frac{m}{m}}{\frac{2}{m} - \frac{2m}{m}}$
$= \dfrac{\frac{4+m}{m}}{\frac{2-2m}{m}}$
$= \dfrac{4 + m}{m} \cdot \dfrac{m}{2 - 2m}$
$= \dfrac{4 + m}{2 - 2m}$

(e) $f(x) = 1$

$$\frac{2x+1}{x-2} = 1$$

$$2x + 1 = x - 2$$

$$x = -3$$

45. The domain is all real numbers between the end-points of the curve, or $[-2, 4]$.
The range is all real numbers between the minimum and maximum values of the function or $[0, 4]$.

(a) $f(-2) = 0$

(b) $f(0) = 4$

(c) $f\left(\frac{1}{2}\right) = 3$

(d) From the graph, $f(x) = 1$ when $x = -1.5$, 1.5, or 2.5.

47. The domain is all real numbers between the end-points of the curve, or $[-2, 4]$.
The range is all real numbers between the minimum and maximum values of the function or $[-3, 2]$.

(a) $f(-2) = -3$

(b) $f(0) = -2$

(c) $f\left(\frac{1}{2}\right) = -1$

(d) From the graph, $f(x) = 1$ when $x = 2.5$.

49. $f(x) = 6x^2 - 2$

$$\begin{aligned}
f(m - 3) &= 6(m - 3)^2 - 2 \\
&= 6(m^2 - 6m + 9) - 2 \\
&= 6m^2 - 36m + 54 - 2 \\
&= 6m^2 - 36m + 52
\end{aligned}$$

51. $g(r + h)$
$$\begin{aligned}
&= (r + h)^2 - 2(r + h) + 5 \\
&= r^2 + 2hr + h^2 - 2r - 2h + 5
\end{aligned}$$

53. $g\left(\frac{3}{q}\right) = \left(\frac{3}{q}\right)^2 - 2\left(\frac{3}{q}\right) + 5$

$$= \frac{9}{q^2} - \frac{6}{q} + 5$$

$$\text{or} \quad \frac{9 - 6q + 5q^2}{q^2}$$

55. A vertical line drawn anywhere through the graph will intersect the graph in only one place. The graph represents a function.

57. A vertical line drawn through the graph may intersect the graph in two places. The graph does not represent a function.

59. A vertical line drawn anywhere through the graph will intersect the graph in only one place. The graph represents a function.

61. $f(x) = x^2 - 4$

(a) $f(x + h) = (x + h)^2 - 4$
$$= x^2 + 2hx + h^2 - 4$$

(b) $f(x + h) - f(x)$
$$\begin{aligned}
&= [(x + h)^2 - 4] - (x^2 - 4) \\
&= x^2 + 2hx + h^2 - 4 - x^2 + 4 \\
&= 2hx + h^2
\end{aligned}$$

(c) $\dfrac{f(x + h) - f(x)}{h}$

$$= \frac{[(x + h)^2 - 4] - (x^2 - 4)}{h}$$

$$= \frac{x^2 + 2hx + h^2 - 4 - x^2 + 4}{h}$$

$$= \frac{2hx + h^2}{h}$$

$$= 2x + h$$

63. $f(x) = 2x^2 - 4x - 5$

(a) $f(x + h)$
$$\begin{aligned}
&= 2(x + h)^2 - 4(x + h) - 5 \\
&= 2(x^2 + 2hx + h^2) - 4x - 4h - 5 \\
&= 2x^2 + 4hx + 2h^2 - 4x - 4h - 5
\end{aligned}$$

(b) $f(x + h) - f(x)$
$$\begin{aligned}
&= 2x^2 + 4hx + 2h^2 - 4x - 4h - 5 \\
&\quad - (2x^2 - 4x - 5) \\
&= 2x^2 + 4hx + 2h^2 - 4x - 4h - 5 \\
&\quad - 2x^2 + 4x + 5 \\
&= 4hx + 2h^2 - 4h
\end{aligned}$$

(c) $\dfrac{f(x + h) - f(x)}{h}$

$$= \frac{4hx + 2h^2 - 4h}{h}$$

$$= \frac{h(4x + 2h - 4)}{h}$$

$$= 4x + 2h - 4$$

65. $f(x) = \dfrac{1}{x}$

(a) $f(x + h) = \dfrac{1}{x + h}$

(b) $f(x + h) - f(x)$

$= \dfrac{1}{x + h} - \dfrac{1}{x}$

$= \left(\dfrac{x}{x}\right)\dfrac{1}{x + h} - \dfrac{1}{x}\left(\dfrac{x + h}{x + h}\right)$

$= \dfrac{x - (x + h)}{x(x + h)}$

$= \dfrac{-h}{x(x + h)}$

(c) $\dfrac{f(x + h) - f(x)}{h}$

$= \dfrac{\frac{1}{x+h} - \frac{1}{x}}{h}$

$= \dfrac{\frac{1}{x+h}\left(\frac{x}{x}\right) - \frac{1}{x}\left(\frac{x+h}{x+h}\right)}{h}$

$= \dfrac{\frac{x-(x+h)}{(x+h)x}}{h}$

$= \dfrac{1}{h}\left[\dfrac{x - x - h}{(x + h)x}\right]$

$= \dfrac{1}{h}\left[\dfrac{-h}{(x + h)x}\right]$

$= \dfrac{-1}{x(x + h)}$

67. If $f(x) = 2x^2 + 5x + 1$ and $g(x) = 3x - 1$, then

$(f \circ g)(x) = f(g(x))$
$= 2g(x)^2 + 5g(x) + 1$
$= 2(3x - 1)^2 + 5(3x - 1) + 1$
$= 2(9x^2 - 6x + 1) + 15x - 5 + 1$
$= 18x^2 - 12x + 2 + 15x - 4$
$= 18x^2 + 3x - 2, \text{ and}$

$(g \circ f)(x) = g(f(x))$
$= 3f(x) - 1$
$= 3(2x^2 + 5x + 1) - 1$
$= 6x^2 + 15x + 3 - 1$
$= 6x^2 + 15x + 2.$

69. (a) The curve in the graph crosses the point with x-coordinate 17:37 and y-coordinate of approximately 140. So, at time 17 hours, 37 minutes the whole reaches a depth of about 140 m.

(b) The curve in the graph crosses the point with x-coordinate 17:39 and y-coordinate of approximately 240. So, at time 17 hours, 39 minutes the whole reaches a depth of about 240 m.

71. (a)-(i) By the given function f, a muskrat weighing 800 g expends

$$f(800) = 0.01(800)^{0.88}$$
$$\approx 3.6, \text{ or approximately}$$

3.6 kcal/km when swimming at the surface of the water.

(ii) A sea otter weighing 20,000 g expends

$$f(20,000) = 0.01(20,000)^{0.88}$$
$$\approx 61, \text{ or approximately}$$

61 kcal/km when swimming at the surface of the water.

(b) If z is the number of kilogram of an amimal's weight, then $x = g(z) = 1,000z$ is the number of grams since 1 kilogram equals 1,000 grams.

(c) Since the function $y = f(x)$ has input x grams and output y kcal/km, and the function $x = g(z)$ has imput z kilograms and output x grams, the only function of $f \circ g$ or $g \circ f$ that makes sence is $f \circ g$ defined by

$$y = (f \circ g)(z) = (f(g(z)))$$
$$= 0.01(g(z))^{0.88}$$
$$= 0.01(1,000z)^{0.88}$$
$$= 0.01(1,000^{0.88})z^{0.88}$$
$$\approx 4.4z^{0.88}$$

This function $(f \circ g)(z) = 4.4z^{0.88}$ gives the energy expenditure in kcal/km for an animal swimming at the surface of the water where z is the weight in kg.

73. (a) The independent variable is the years.

(b) The dependent variable is the number of Internet users.

(c) $f(1997) = 98$ million users

(d) The domain is $1995 \le x \le 1999$.
The range is $26,000,000 \le y \le 205,000,000$.

75. **(a)** Let $w =$ the width of the field;
$l =$ the length.

The perimeter of the field is 6,000 ft, so

$$2l + 2w = 6,000$$
$$l + w = 3,000$$
$$l = 3,000 - w.$$

Thus, the area of the field is given by

$$A = lw$$
$$A = (3,000 - w)w.$$

(b) Since $l = 3,000 - w$ and w cannot be negative, $0 \le w \le 3,000$.
The domain of A is $0 \le w \le 3,000$.

(c)

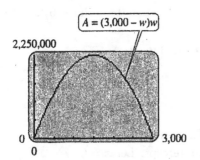

1.4   Quadratic Functions; Translation and Reflection

3. The graph of $y = x^2 - 3$ is the graph of $y = x^2$ translated 3 units downward.
This is graph D.

5. The graph of $y = (x+3)^2 + 2$ is the graph of $y = x^2$ translated 3 units to the left and 2 units upward.
This is graph B.

7. The graph of $y = -(x + 3)^2 + 2$ is the graph of $y = x^2$ reflected in the x-axis, translated three units to the left, and two units upward.
This is graph E.

9. $y = x^2 - 10x + 21$
$= (x - 7)(x - 3)$

Set $y = 0$ to find the x-intercepts.

$$0 = (x - 7)(x - 3)$$
$$x = 7 \text{ or } x = 3$$

The x-intercepts are 7 and 3.

Set $x = 0$ to find the y-intercept.

$$y = 0^2 - 10(0) + 21$$
$$y = 21$$

The y-intercept is 21.
The x-coordinate of the vertex is

$$x = \frac{-b}{2a} = \frac{10}{2} = 5.$$

Substitute to find the y-coordinate.

$$y = 5^2 - 10(5) + 21$$
$$= 25 - 50 + 21$$
$$= -4$$

The vertex is $(5, -4)$.

The axis is $x = 5$, the vertical line through the vertex.

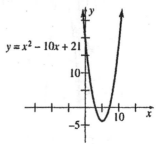

11. $y = -3x^2 + 12x - 11$

x-intercepts: Let $y = 0$.

The equation

$$0 = -3x^2 + 12x - 11$$

cannot be solved by factoring. Use the quadratic formula.

$$0 = -3x^2 + 12x - 11$$
$$x = \frac{-12 \pm \sqrt{12^2 - 4(-3)(-11)}}{2(-3)}$$
$$= \frac{-12 \pm \sqrt{144 - 132}}{-6}$$
$$= \frac{-12 \pm \sqrt{12}}{-6} = \frac{-12 \pm 2\sqrt{3}}{-6}$$
$$= 2 \pm \frac{\sqrt{3}}{3}$$

The x-intercepts are $2 + \frac{\sqrt{3}}{3} \approx 2.58$ and $2 - \frac{\sqrt{3}}{3} \approx 1.42$.

y-intercept: Let $x = 0$.

$$y = -3(0)^2 + 12(0) - 11$$
$$y = -11$$

The *y*-intercept is -11.

Vertex: $x = \dfrac{-b}{2a} = \dfrac{-12}{2(-3)} = \dfrac{-12}{-6} = 2$

$$y = -3(2)^2 + 12(2) - 11$$
$$= -12 + 24 - 11$$
$$= 1$$

The vertex is $(2, 1)$.

The axis is $x = 2$, the vertical line through the vertex.

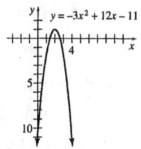

13. $f(x) = -x^2 + 6x - 6$

x-intercepts: Let $f(x) = 0$.

$0 = -x^2 + 6x - 6$

$$x = \frac{-6 \pm \sqrt{6^2 - 4(-1)(-6)}}{2(-1)}$$

$$= \frac{-6 \pm \sqrt{36 - 24}}{-2} = \frac{-6 \pm \sqrt{12}}{-2}$$

$$= \frac{-6 \pm 2\sqrt{3}}{-2} = 3 \pm \sqrt{3}$$

The *x*-intercepts are $3 + \sqrt{3} \approx 4.73$ and $3 - \sqrt{3} \approx 1.27$.

y-intercept: Let $x = 0$.

$$y = -0^2 + 6(0) - 6$$

The *y*-intercept is -6.

Vertex: $x = \dfrac{-b}{2a} = \dfrac{-6}{2(-1)} = \dfrac{-6}{-2} = 3$

$$y = -3^2 + 6(3) - 6$$
$$= -9 + 18 - 6 = 3$$

The vertex is $(3, 3)$.

The axis is $x = 3$.

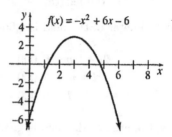

15. $f(x) = -3x^2 + 24x - 36$
$$= -3(x^2 - 8x + 12)$$
$$= -3(x - 6)(x - 2)$$

x-intercepts: Let $f(x) = 0$.

$$0 = -3(x - 6)(x - 2)$$

Divide by -3.

$$0 = (x - 6)(x - 2)$$
$$x = 6, \ x = 2$$

The *x*-intercepts are 6 and 2.

y-intercept: Let $x = 0$.

$$y = -3(0)^2 + 24(0) - 36$$

The *y*-intercept is -36.

Vertex: $x = \dfrac{-b}{2a} = \dfrac{-24}{2(-3)} = \dfrac{-24}{-6} = 4$

$$y = -3(4)^2 + 24(4) - 36$$
$$= -48 + 96 - 36 = 12$$

The vertex is $(4, 12)$.
The axis is $x = 4$.

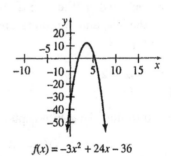

$f(x) = -3x^2 + 24x - 36$

17. $f(x) = \dfrac{5}{2}x^2 + 10x + 8$

x-intercepts: Let $f(x) = 0$.

$$0 = \frac{5}{2}x^2 + 10x - 8$$

$$x = \frac{-10 \pm \sqrt{10^2 - 4\left(\frac{5}{2}\right)(8)}}{2\left(\frac{5}{2}\right)}$$

$$= \frac{-10 \pm \sqrt{100 - 80}}{5}$$

$$= \frac{-10 \pm \sqrt{20}}{5}$$

$$= -2 \pm \frac{2\sqrt{5}}{5}$$

The x-intercepts are $-2 + \frac{2\sqrt{5}}{5} \approx -1.11$ and $-2 + \frac{2\sqrt{5}}{5} \approx -2.89$.

y-intercept: Let $x = 0$.

$$y = \frac{5}{2}(0)^2 + 10(0) + 8$$

The y-intercept is 8.

Vertex: $x = \dfrac{-b}{2a} = \dfrac{-10}{2\left(\frac{5}{2}\right)} = \dfrac{-10}{5} = -2$

$$y = \frac{5}{2}(-2)^2 + 10(-2) + 8$$

$$= \frac{5}{2}(4) - 20 + 8$$

$$= 10 - 20 + 8$$

$$= -2$$

The vertex is $(-2, -2)$.
The axis is $x = -2$.

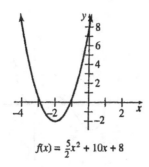

$$f(x) = \frac{5}{2}x^2 + 10x + 8$$

19. $y = -\dfrac{1}{2}x^2 - x - \dfrac{7}{2}$

x-intercepts: Let $y = 0$.

$$0 = -\frac{1}{2}x^2 - x - \frac{7}{2}$$

Multiply by -2 to clear fractions.

$$0 = x^2 + 2x + 7$$

Use the quadratic formula.

$$x = \frac{-2 \pm \sqrt{4 - 4(1)(7)}}{2}$$

$$= \frac{-2 \pm \sqrt{-24}}{2}$$

Since the radicand is negative, there are no x-intercepts.

y-intercept: Let $x = 0$.

$$y = -\frac{1}{2}(0)^2 - 0 - \frac{7}{2} = -\frac{7}{2}$$

The y-intercept is $-\frac{7}{2}$.

Vertex: $\dfrac{-b}{2a} = \dfrac{-(-1)}{2\left(-\frac{1}{2}\right)} = \dfrac{1}{-1} = -1$

$$y = -\frac{1}{2}(-1)^2 - (-1) - \frac{7}{2}$$

$$= -\frac{1}{2} + 1 - \frac{7}{2}$$

$$= -3$$

The vertex is $(-1, -3)$.
The axis is $x = -1$.

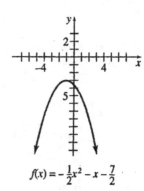

$$f(x) = -\frac{1}{2}x^2 - x - \frac{7}{2}$$

21. The graph of $y = \sqrt{x - 2} - 4$ is the graph of $y = \sqrt{x}$ translated 2 units to the right and 4 units downward.
This is graph A.

23. The graph of $y = \sqrt{-x-2} - 4$ is the same as the graph of $y = \sqrt{-(x+2)} - 4$. This is the graph of $y = \sqrt{x}$ reflected in the y-axis, translated 2 units to the left, and translated 4 units downward. This is graph B.

25. The graph of $y = -f(x)$ is the graph of $y = f(x)$ reflected in the x-axis.

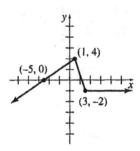

27. The graph of $y = f(-x)$ is the graph of $y = f(x)$ reflected in the y-axis.

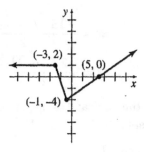

29. $f(x) = \sqrt{x-1} + 3$

Translate the graph of $f(x) = \sqrt{x}$ 1 unit right and 3 units upward.

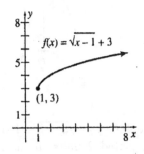

31. $f(x) = -\sqrt{-4-x} - 2$
$= -\sqrt{-(x+4)} - 2$

Translate the graph of $f(x) = \sqrt{x}$ 4 units left and 2 units downward.
Reflect vertically and horizontally.

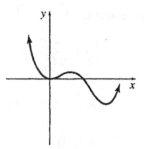

33. If $0 < a < 1$, the graph of $f(ax)$ will be wider than the graph of $f(x)$.
Multiplying x by a fraction makes the y-values less than the original y-values.

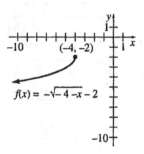

35. If $-1 < a < 0$, the graph of $f(ax)$ will be reflected horizontally, since a is negative. It will be flatter because multiplying x by a fraction decreases the corresponding y-values.

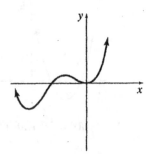

37. If $0 < a < 1$, the graph of $af(x)$ will be flatter than the graph of $f(x)$. Each y-value is only a fraction of the height of the original y-values.

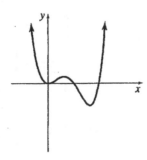

39. If $-1 < a < 0$, the graph will be reflected vertically, since a will be negative. Also, because a is a fraction, the graph will be flatter because each y-value will only be a fraction of its original height.

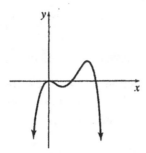

41. **(a)** Since the graph of $y = f(x)$ is reflected vertically to obtain the graph of $y = -f(x)$, the x-intercept is unchanged. The x-intercept of the graph of $y = f(x)$ is r.

   (b) Since the graph of $y = f(x)$ is reflected horizontally to obtain the graph of $y = f(-x)$, the x-intercept of the graph of $y = f(-x)$ is $-r$.

   (c) Since the graph of $y = f(x)$ is reflected both horizontally and vertically to obtain the graph of $y = -f(-x)$, the x-intercept of the graph of $y = -f(-x)$ is $-r$.

43. **(a)** The vertex of the quadratic function $y = 0.057x - 0.001x^2$ is at

$$x = -\frac{b}{2a} = -\frac{0.057}{2(-0.001)} = 28.5.$$

   Since the coefficient of the leading term, -0.001, is negative, then the graph of the function opens downward, so a maximum is reached at 28.5 weeks of gestation.

(b) The maximum splenic artery resistance reached at the vertex is

$$y = 0.057(28.5) - 0.001(28.5)^2$$
$$\approx 0.81.$$

(c) The splenic artery resistance equals 0, when $y = 0$.

$$0.057x - 0.001x^2 = 0 \quad \text{Substitute in the expression in } x \text{ for } y.$$
$$x(0.057 - 0.001x) = 0 \quad \text{Factor.}$$
$$x = 0 \text{ or } 0.057 - 0.001x = 0 \quad \text{Set each factor equal to 0.}$$
$$x = \frac{0.057}{0.001} = 57$$

So, the splenic artery resistance equals 0 at 0 weeks or 57 weeks of gestation.
No, this is not reasonable because at $x = 0$ or 57 weeks, the fetus does not exist.

45. **(a)** $E_0: y_0 = px_0^2 + qx_0 + r$

   $E_1: y_1 = px_1^2 + qx_1 + r$

   $E_1 - E_0:$

   $$y_1 - y_0 = px_1^2 + qx_1 + r - (px_0^2 + qx_0 + r)$$
   $$= p(x_1^2 - x_0^2) + q(x_1 - x_0) + r - r$$

   Therefore,
   $$y_1 - y_0 - q(x_1 - x_0) = p(x_1^2 - x_0^2), \text{ so}$$
   $$p = \frac{y_1 - y_0 - q(x_1 - x_0)}{x_1^2 - x_0^2}$$

(b) $E_0: y_0 = px_0^2 + qx_0 + r$

   $$= \frac{y_1 - y_0 - q(x_1 - x_0)}{x_1^2 - x_0^2} x_0^2 + qx_0 + r,$$

   so

   $$r = y_0 - x_0^2 \left(\frac{y_1 - y_0}{x_1^2 - x_0^2} \right) + qx_0^2 \left(\frac{x_1 - x_0}{x_1^2 - x_0^2} \right) - qx_0$$

(c) $y = px^2 + qx + r$

$$= \left(\frac{y_1 - y_0 - q(x_1 - x_0)}{x_1^2 - x_0^2} \right) x^2 + qx$$

$$+ \left(y_0 - x_0^2 \left(\frac{y_1 - y_0}{x_1^2 - x_0^2} \right) + qx_0^2 \left(\frac{x_1 - x_0}{x_1^2 - x_0^2} \right) - qx_0 \right)$$

$$= \left(\frac{y_1 - y_0}{x_1^2 - x_0^2} \right) x^2 - q \left(\frac{x_1 - x_0}{x_1^2 - x_0^2} \right) x^2 + qx + y_0$$

$$- \left(\frac{y_1 - y_0}{x_1^2 - x_0^2} \right) x_0^2 + q \left(\frac{x_1 - x_0}{x_1^2 - x_0^2} \right) x_0^2 - qx_0$$

$$= \left(\frac{y_1 - y_0}{x_1^2 - x_0^2} \right) x^2$$

$$+ q \left[x - x_0 - \left(\frac{x_1 - x_0}{x_1^2 - x_0^2} \right) x^2 + \left(\frac{x_1 - x_0}{x_1^2 - x_0^2} \right) x_0^2 \right]$$

$$+ y_0 - \left(\frac{y_1 - y_0}{x_1^2 - x_0^2} \right) x_0^2$$

(d) Define $K = \dfrac{y_1 - y_0}{x_1^2 - x_0^2}$,

$$L = Kx_0^2, \text{ and}$$

$$M = \frac{x_1 - x_0}{x_1^2 - x_0^2} = \frac{1}{x_1 + x_0}.$$

Substitute these into the result from part (c).

$$y = Kx^2 + q\left(x - x_0 - Mx^2 + Mx_0^2 \right) + y_0 - KLx_0^2$$

(e) Continue to substitute

$$A = L - y_0,$$
$$P = Mx_0^2, \text{ and}$$
$$R = x_0 - P,$$

into the result from part (d), and solve for q.

$$y = Kx^2 + q(x - x_0 - Mx^2 + P) - (L - y_0)$$
$$= Kx^2 + q(x - Mx^2 - R) - A, \text{ so}$$

$$q = \frac{y - Kx^2 + A}{x - R - Mx^2}$$

(f) If $(x_0, y_0) = (1, 2)$ and $m(x_1, y_1) = (3, 5)$, then

$$K = \frac{5 - 2}{3^2 - 1^2} = \frac{3}{8},$$

$$L = \frac{3}{8}(1)^2 = \frac{3}{8},$$

$$M = \frac{1}{3 + 1} = \frac{1}{4},$$

$$A = \frac{3}{8} - 2 = -\frac{13}{8},$$

$$P = \frac{1}{4} \cdot 1^2 = \frac{1}{4}, \text{ and}$$

$$R = 1 - \frac{1}{4} = \frac{3}{4}.$$

Therefore,

$$q = \frac{y - Kx^2 + A}{x - R - Mx^2}$$

$$= \frac{y - \frac{3}{8}x^2 - \frac{13}{8}}{x - \frac{3}{4} - \frac{1}{4}x^2}$$

47. $f(x) = -0.2369x^2 + 1.425x + 6.905$

This function defines a parabola opening downward, so the maximum percent is at the vertex. The x-coordinate of the vertex is

$$x = \frac{-b}{2a} = \frac{-1.425}{2(-0.2369)} \approx 3.0076$$

Since $x = 0$ represents the year 1992, the percent of freshmen reached its maximum in $1992 + 3$, or 1995.

The domain of $f(x)$ is $0 \le x \le 6$.

49. $h = 32t - 16t^2$
$$= -16t^2 + 32t$$

(a) Find the vertex.

$$x = \frac{-b}{2a} = \frac{-32}{-32} = 1$$

$$y = -16(1)^2 + 32(1)$$
$$= 16$$

The vertex is $(1, 16)$, so the maximum height is 16 ft.

(b) When the object hits the ground, $h = 0$, so

$$32t - 16t^2 = 0$$
$$16t(2 - t) = 0$$
$$t = 0 \quad \text{or} \quad t = 2.$$

When $t = 0$, the object is thrown upward. When $t = 2$, the object hits the ground; that is, after 2 sec.

51. (a) $f(x) = -19.321x^2 + 3,608.7x - 168,310$

$$\frac{-b}{2a} = \frac{-3,608.7}{2(-19.321)} \approx 93.388$$

The value $x = 93.388$ corresponds to 1993.

(b) $f(93.388)$
$$= -19.321(93.388)^2 + 3,608.7(93.388)$$
$$- 168,310$$
$$\approx 194.68$$

The maximum amount spent is approximately $195 million.

(c) Graph

$f(x) = -19.321x^2 + 3,608.7x - 168,310.$

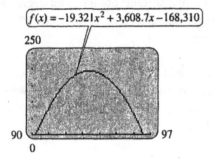

53. Let $x =$ the width.
Then $320 - 2x =$ the length.

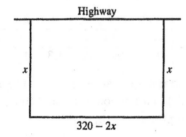

Area $= x(320 - 2x) = -2x^2 + 320x$

Find the vertex:

$$x = \frac{-b}{2a} = \frac{-320}{-4} = 80$$

$$y = -2(80)^2 + 320(80) = 12,800$$

The graph of the area function is a parabola with vertex (80, 12,800).
The maximum area of 12,800 sq ft occurs when the width is 80 ft and the length is

$$320 - 2x = 320 - 2(80) = 160 \text{ ft.}$$

55. Draw a sketch of the arch with the vertex at the origin.

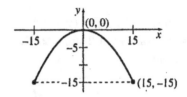

Since the arch is a parabola that opens downward, the equation of the parabola is the form $y = a(x - h)^2 + k$, where the vertex $(h, k) = (0, 0)$ and $a < 0$. That is, the equation is of the form $y = ax^2$.

Since the arch is 30 meters wide at the base and 15 meters high, the points $(15, -15)$ and $(-15, -15)$ are on the parabola. Use $(15, -15)$ as one point on the parabola.

$$-15 = a(15)^2$$

$$a = \frac{-15}{15^2} = -\frac{1}{15}$$

So, the equation is

$$y = -\frac{1}{15}x^2.$$

Ten feet from the ground (the base) is at $y = -5$. Substitute -5 for y and solve for x.

$$-5 = -\frac{1}{15}x^2$$
$$x^2 = -5(-15) = 75$$
$$x = \pm\sqrt{75} = \pm 5\sqrt{3}$$

The width of the arch ten feet from the ground is then

$$5\sqrt{3} - (-5\sqrt{3}) = 10\sqrt{3} \text{ meters}$$
$$\approx 17.32 \text{ meters.}$$

1.5   Polynomial and Rational Functions

3. The graph of $f(x) = (x+2)^3 - 5$ is the graph of $y = x^3$ translated 2 units to the left and 5 units downward.

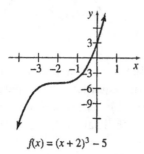

$$f(x) = (x+2)^3 - 5$$

5. The graph of $f(x) = -(x-3)^4 + 1$ is the graph of $y = x^4$ reflected vertically, translated 3 units to the right, and translated 1 unit upward.

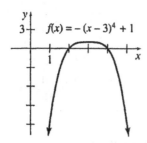

7. The graph of $y = x^3 - 7x - 9$ has the right end up, the left end down, at most two turning points, and a y-intercept of -9.
This is graph D.

9. The graph of $y = -x^3 - 4x^2 + x + 6$ has the right end down, the left end up, at most two turning points, and a y-intercept of 6.
This is graph E.

11. The graph of $y = x^4 - 5x^2 + 7$ has both ends up, at most three turning points, and a y-intercept of 7.
This is graph I.

13. The graph of $y = -x^4 + 2x^3 + 10x + 15$ has both ends down, at most three turning points, and a y-intercept of 15.
This is graph G.

15. The graph of $y = -x^5 + 4x^4 + x^3 - 16x^2 + 12x + 5$ has the right end down, the left end up, at most four turning points, and a y-intercept of 5.
This is graph A.

17. The graph of $y = \frac{2x^2+3}{x^2+1}$ has no vertical asymptote, the line $y = 2$ as a horizontal asymptote, and a y-intercept of 3.
This is graph D.

19. The graph $y = \frac{-2x^2-3}{x^2+1}$ has no vertical asymptote, the line $y = -2$ as a horizontal asymptote, and a y-intercept of -3.
This is graph E.

21. The right end is up and the left end is up. There are three turning points. The degree is an even integer equal to 4 or more. The x^n term has a $+$ sign.

23. The right end is up and the left end is down. There are four turning points. The degree is an odd integer equal to 5 or more. The x^n term has a $+$ sign.

25. The right end is down and the left end is up. There are six turning points. The degree is an odd integer equal to 7 or more. The x^n term has a $-$ sign.

27. $y = \dfrac{-4}{x-3}$

The function is undefined for $x = 3$, so the line $x = 3$ is a vertical asymptote.

x	-97	-7	-2	0	2	4	13	103
$x-3$	-100	-10	-5	-3	-1	1	10	100
y	0.04	0.4	0.8	1.3	4	-4	-0.4	-0.04

The graph approaches $y = 0$, so the line $y = 0$ (the x-axis) is a horizontal asymptote.
Asymptotes: $y = 0$, $x = 3$

x-intercept: none, because the x-axis is an asymptote

y-intercept: $\frac{4}{3}$, the value when $x = 0$

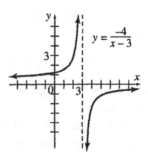

$$y = \frac{-4}{x-3}$$

29. $y = \dfrac{2}{3 + 2x}$

$3 + 2x = 0$ when $2x = -3$ or $x = -\frac{3}{2}$, so the line $x = -\frac{3}{2}$ is a vertical asymptote.

x	-51.5	-6.5	-2	-1	3.5	48.5
$3 + 2x$	-100	-10	-1	1	10	100
y	-0.02	-0.2	-2	2	0.2	0.02

The graph approaches $y = 0$, so the line $y = 0$ (the x-axis) is a horizontal asymptote.
Asymptotes: $y = 0$, $x = -\frac{3}{2}$

x-intercept: none, since the x-axis is an asymptote

y-intercept: $\frac{2}{3}$, the value when $x = 0$

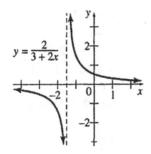

31. $y = \dfrac{3x}{x - 1}$

$x - 1 = 0$ when $x = 1$, so the line $x = 1$ is a vertical asymptote.

x	-99	-9	-1	0
$3x$	-297	-27	-3	0
$x - 1$	-100	-10	-2	-1
y	2.97	2.7	1.5	0

x	0.5	1.5	12	11	101
$3x$	1.5	4.5	6	33	303
$x - 1$	-0.5	0.5	1	10	100
y	-3	9	6	3.3	3.03

As x gets larger,

$$\frac{3x}{x - 1} \approx \frac{3x}{x} = 3.$$

Thus, $y = 3$ is a horizontal asymptote.
Asymptotes: $y = 3$, $x = 1$

x-intercept: 0, the value when $y = 0$

y-intercept: 0, the value when $x = 0$

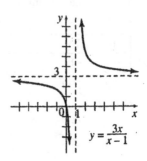

33. $y = \dfrac{x + 1}{x - 4}$

$x - 4 = 0$ when $x = 4$, so $x = 4$ is a vertical asymptote.

x	-96	-6	-1	0	3
$x + 1$	-95	-5	0	1	4
$x - 4$	-100	-10	-5	-4	-1
y	0.95	0.5	0	-0.25	-4

x	3.5	4.5	5	14	104
$x + 1$	4.5	5.5	6	15	105
$x - 4$	-0.5	0.5	1	10	100
y	-9	11	6	1.5	1.05

As x gets larger,

$$\frac{x + 1}{x - 4} \approx \frac{x}{x} = 1.$$

Thus, $y = 1$ is a horizontal asymptote.
Asymptotes: $y = 1$, $x = 4$

x-intercept: -1, the value when $y = 0$

y-intercept: $-\frac{1}{4}$, the value when $x = 0$

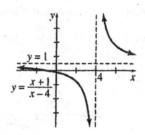

35. $y = \dfrac{1 - 2x}{5x + 20}$

$5x + 20 = 0$ when $5x = -20$ or $x = -4$, so the line $x = -4$ is a vertical asymptote.

x	-7	-6	-5	-3	-2	-1	0
$1 - 2x$	15	13	11	7	5	3	1
$5x + 20$	-15	-10	-5	5	10	15	20
y	-1	-1.3	-2.2	1.4	0.5	0.2	0.05

As x gets larger,

$$\frac{1 - 2x}{5x + 20} \approx \frac{-2x}{5x} = -\frac{2}{5}.$$

Thus, the line $y = -\frac{2}{5}$ is a horizontal asymptote.

Asymptotes: $x = -4$, $y = -\frac{2}{5}$

x-intercept: $\frac{1}{2}$, the value when $y = 0$

y-intercept: $\frac{1}{20}$, the value when $x = 0$

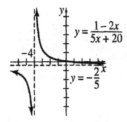

37. $y = \dfrac{-x - 4}{3x + 6}$

$3x + 6 = 0$ when $3x = -6$ or $x = -2$, so the line $x = -2$ is a vertical asymptote.

x	-5	-4	-3	-1	0	1
$-x - 4$	1	0	-1	-3	-4	-5
$3x + 6$	-9	-6	-3	3	6	9
y	-0.11	0	0.33	-1	-0.67	-0.56

As x gets larger,

$$\frac{-x - 4}{3x + 6} \approx \frac{-x}{3x} = -\frac{1}{3}.$$

The line $y = -\frac{1}{3}$ is a horizontal asymptote.
Asymptotes: $y = -\frac{1}{3}$, $x = -2$

x-intercept: -4, the value when $y = 0$

y-intercept: $-\frac{2}{3}$, the value when $x = 0$

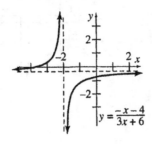

39. For a vertical asymptote at $x = 1$, put $x - 1$ in the denominator. For a horizontal asymptote at $y = 2$, the degree of the numerator must equal the degree of the denominator and the quotient of their leading terms must equal 2. So, $2x$ in the numerator would cause y to approach 2 as x gets larger.

So, one possible answer is $y = \dfrac{2x}{x - 1}$.

41. $f(x) = (x - 1)(x - 2)(x + 3)$,
$g(x) = x^3 + 2x^2 - x - 2$,
$h(x) = 3x^3 + 6x^2 - 3x - 6$

(a) $f(1) = (0)(-1)(4) = 0$

(b) $f(x)$ is zero when $x = 2$ and when $x = -3$.

(c) $g(-1) = (-1)^3 + 2(-1)^2 - (-1) - 2$
$ = -1 + 2 + 1 - 2 = 0$
$g(1) = (1)^3 + 2(1)^2 - (1) - 2$
$ = 1 + 2 - 1 - 2$
$ = 0$
$g(-2) = (-2)^3 + 2(-2)^2 - (-2) - 2$
$ = -8 + 8 + 2 - 2$
$ = 0$

(d) $g(x) = [x - (-1)](x - 1)[x - (-2)]$
$g(x) = (x + 1)(x - 1)(x + 2)$

(e) $h(x) = 3g(x)$
$ = 3(x + 1)(x - 1)(x + 2)$

(f) If f is a polynomial and $f(a) = 0$ for some number a, then one factor of the polynomial is $x - a$.

43. $f(x) = \dfrac{1}{x^5 - 2x^3 - 3x^2 + 6}$

(a) Two vertical asymptotes appear, one at $x = -1.4$ and one at $x = 1.4$.

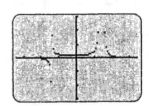

(b) Three vertical asymptotes appear, one at $x = -1.414$, one at $x = 1.414$, and one at $x = 1.442$.

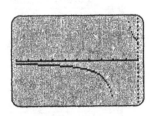

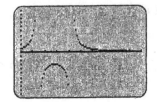

45. (c) $y = \dfrac{b(1-x) + cx}{1 - a(1-x)}$

$$= \dfrac{0.025(1-x) + 0.92x}{1 - 0.76(1-x)}$$

$$= \dfrac{0.025 - 0.025x + 0.92x}{1 - 0.76 + 0.76x}$$

$$= \dfrac{0.025 + 0.895x}{0.24 + 0.76x}$$

(d)

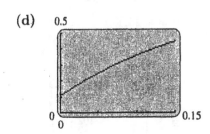

(e) Substitute 0.20 in for y and solve for x.

$$0.20 = \dfrac{0.025 + 0.895x}{0.24 + 0.76x}$$

$$0.20(0.24 + 0.76x) = 0.025 + 0.895x$$

$$0.048 + 0.152x = 0.025 + 0.895x$$

$$0.743x = 0.023$$

$$x = \dfrac{0.023}{0.743}$$

$$\approx 0.031$$

One method for checking this answer with a graphing calculator is to find the intersection point of the function graphed in part (d) with the line $y = 0.20$.

47. (a)

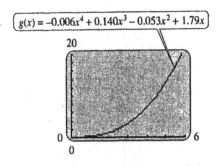

$g(x) = -0.006x^4 + 0.140x^3 - 0.053x^2 + 1.79x$

(b) Because the leading coefficient is negative and the degree of the polynomial is even, the graph will have right end down, so it cannot keep increasing forever.

49. $A(x) = -0.015x^3 + 1.058x$

(a)

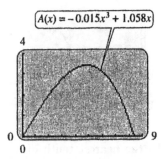

$A(x) = -0.015x^3 + 1.058x$

(b) Reading the graph, we find that concentration is maximum between 4 hours and 5 hours, but closer to 5 hours.

(c) Concentration exceeds 0.08% from less than 1 hr to about 8.4 hours.

51. $f(x) = \dfrac{\lambda x}{1 + (ax)^b}$

(a) A reasonable domain for the function is $[0, \infty)$. Populations are not measured using negative numbers and they may get extremely large.

(b) If $\lambda = a = b = 1$, the function becomes

$$f(x) = \dfrac{x}{1 + x}.$$

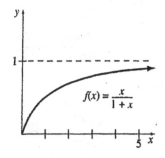

(c) If $\lambda = a = 1$ and $b = 2$, the function becomes

$$f(x) = \frac{x}{1 + x^2}.$$

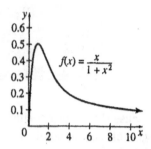

(d) As seen from the graphs, when b increases, the population of the next generation, $f(x)$, gets smaller when the current generation, x, is larger.

53. (a)

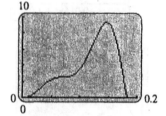

(b) To the nearest tenth of a second, $y = 0$ at $t \approx 0.0$ and $t \approx 0.2$. This function is valid only for values $y > 0$ and $t \geq 0$. For $t > 0.20$, $y < 0$, so this function is valid only when $0 \leq t \leq 0.20$.

(c) A maximum program of a graphing calculator can be used to approximate the highest maximum point on the graph in part (a). This point is approximately $(0.1465, 9.30)$, so at 0.1465 sec a maximum power of 9.30 horsepower is reached.

55. $y = \dfrac{6.5x}{102 - x}$

y = percent of pollutant
x = cost in thousands of dollars

(a) $x = 0$

$$y = \frac{6.5(0)}{102 - 0} = \frac{0}{102} = \$0$$

$x = 50$

$$y = \frac{6.5(50)}{102 - 50} = \frac{325}{52} = 6.25$$

$6.25(1{,}000) = \$6{,}250$

$x = 80$

$$y = \frac{6.5(80)}{102 - 80} = \frac{520}{22} = 23.636$$

$(23.636)(1{,}000) = \$23{,}636$

$x = 90$

$$y = \frac{6.5(90)}{102 - 90} = \frac{585}{12} = 48.75$$

$(48.75)(1{,}000) = \$48{,}750$

$x = 95$

$$y = \frac{6.5(95)}{102 - 95} = \frac{617.5}{7} = 88.214$$

$(88.214)(1{,}000) = 88{,}214$

$x = 99$

$$y = \frac{6.5(99)}{102 - 99} = \frac{643.5}{3} = 214.5$$

$(214.500)(1{,}000) = 214{,}500$

$x = 100$

$$y = \frac{6.5(100)}{102 - 100} = \frac{650}{2} = 325$$

$(325)(1{,}000) = \$325{,}000$

(b)

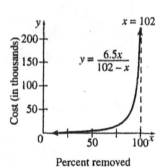

57. (a)

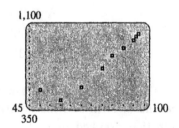

(b) $f(x) = 0.29195x^2 - 30.355x + 1,238.5$

(c)

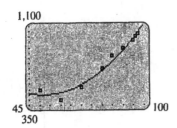

(d) $f(x) = -0.01382x^3 + 3.3697x^2 - 252.65x + 6,425.5$

(e)

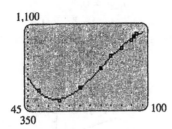

Chapter 1 Review Exercises

3. Through $(-2, 5)$ and $(4, 7)$

$$m = \frac{5 - 7}{-2 - 4}$$
$$= \frac{-2}{-6}$$
$$= \frac{1}{3}$$

5. Through the origin and $(11, -2)$

$$m = \frac{-2 - 0}{11 - 0} = -\frac{2}{11}$$

7. $2x + 3y = 15$
$$3y = -2x + 15$$
$$y = -\frac{2}{3}x + 5$$
$$m = -\frac{2}{3}$$

9. $y + 4 = 9$
$$y = 5$$
$$y = 0x + 5$$
$$m = 0$$

11. $y = -3x$
$$y = -3x + 0$$
$$m = -3$$

13. Through $(5, -1)$; slope $\frac{2}{3}$

Use point-slope form.

$$y - (-1) = \frac{2}{3}(x - 5)$$

$$y + 1 = \frac{2}{3}(x - 5)$$

$$3(y + 1) = 2(x - 5)$$
$$3y + 3 = 2x - 10$$
$$3y = 2x - 13$$

$$y = \frac{2}{3}x - \frac{13}{3}$$

15. Through $(5, -2)$ and $(1, 3)$

$$m = \frac{-2 - 3}{5 - 1} = \frac{-5}{4} = -\frac{5}{4}$$

Use point-slope form.

$$y - (-2) = \frac{-5}{4}(x - 5)$$

$$y + 2 = -\frac{5}{4}x + \frac{25}{4}$$

$$4y + 8 = -5x + 25$$
$$4y = -5x + 17$$

$$y = -\frac{5}{4}x + \frac{17}{4}$$

17. Through $(-1, 4)$; undefined slope

Undefined slope means the line is vertical. The equation of the vertical line through $(-1, 4)$ is $x = -1$.

19. Through $(2, -1)$, parallel to $3x - y = 1$

Solve $3x - y = 1$ for y.

$$-y = 1 - 3x$$
$$y = -1 + 3x$$
$$m = 3$$

The desired line has the same slope. Use the point-slope form.

$$y - (-1) = 3(x - 2)$$
$$y + 1 = 3x - 6$$
$$y = 3x - 7$$

21. Through $(2, -10)$, perpendicular to a line with un-defined slope

A line with undefined slope is a vertical line. A line perpendicular to a vertical line is a horizontal line with equation of the form $y = k$. The desired line passed through $(2, -10)$, so $k = -10$. Thus, an equation of the desired line is $y = -10$.

23. Through $(-7, 4)$, perpendicular to $y = 8$

The given line, $y = 8$, is a horizontal line. A line perpendicular to a horizontal line is a vertical line with equation of the form $x = h$.

The desired line passes through $(-7, 4)$, so $h = -7$. Thus, an equation of the desired line is $x = -7$.

25. $y = 6 - 2x$

Find the intercepts.
Let $x = 0$.
$$y = 6 - 2(0) = 6$$
The y-intercept is 6.
Let $y = 0$.
$$0 = 6 - 2x$$
$$2x = 6$$
$$x = 3$$
The x-intercept is 3.
Draw the line through $(0, 6)$ and $(3, 0)$.

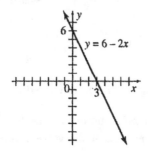

27. $2x + 7y = 14$

Find the intercepts.
When $x = 0$, $y = 2$, so the y-intercept is 2.
When $y = 0$, $x = 7$, so the x-intercept is 7.
Draw the line through $(0, 2)$ and $(7, 0)$.

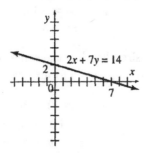

29. $y = 1$

This is the horizontal line passing through $(0, 1)$.

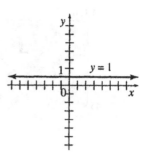

31. $x + 3y = 0$

When $x = 0$, $y = 0$.
When $x = 3$, $y = -1$.
Draw the line through $(0, 0)$ and $(3, -1)$.

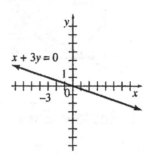

35. $y = (2x + 1)(x - 1)$
$\quad = 2x^2 - x - 1$

x	-3	-2	-1	0	1	2	3
y	20	9	2	-1	0	5	14

Pairs: $(-3, 20), (-2, 9), (-1, 2), (0, -1), (1, 0),$ $(2, 5), (3, \ 14)$
Range: $\{-1, 0, 2, 5, 9, 14, 20\}$

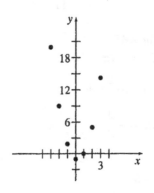

37. $f(x) = -x^2 + 2x - 4$

   (a) $f(6) = -(6)^2 + 2(6) - 4$
$$= -36 + 12 - 4$$
$$= -28$$

   (b) $f(-2) = -(-2)^2 + 2(-2) - 4$
$$= -4 - 4 - 4$$
$$= -12$$

   (c) $f(-4) = -(-4)^2 + 2(-4) - 4$
$$= -16 - 8 - 4$$
$$= -28$$

   (d) $f(r + 1)$
$$= -(r + 1)^2 + 2(r + 1) - 4$$
$$= -(r^2 + 2r + 1) + 2r + 2 - 4$$
$$= -r^2 - 3$$

39. $f(x) = 5x^2 - 3$ and $g(x) = -x^2 + 4x + 1$

   (a) $f(-2) = 5(-2)^2 - 3 = 17$

   (b) $g(3) = -(3)^2 + 4(3) + 1 = 4$

   (c) $f(-k) = 5(-k)^2 - 3 = 5k^2 - 3$

   (d) $g(3m) = -(3m)^2 + 4(3m) + 1$
$$= -9m^2 + 12m + 1$$

   (e) $f(x + h) = 5(x + h)^2 - 3$
$$= 5(x^2 + 2xh + h^2) - 3$$
$$= 5x^2 + 10xh + 5h^2 - 3$$

   (f) $g(x + h) = -(x + h)^2 + 4(x + h) + 1$
$$= -(x^2 + 2xh + h^2) + 4x + 4h + 1$$
$$= -x^2 - 2xh - h^2 + 4x + 4h + 1$$

   (g) $\dfrac{f(x + h) - f(x)}{h}$
$$= \dfrac{5(x + h)^2 - 3 - (5x^2 - 3)}{h}$$
$$= \dfrac{5(x^2 + 2hx + h^2) - 3 - 5x^2 + 3}{h}$$
$$= \dfrac{5x^2 + 10hx + 5h^2 - 5x^2}{h}$$
$$= \dfrac{10hx + 5h^2}{h}$$
$$= 10x + 5h$$

   (h) $\dfrac{g(x + h) - g(x)}{h}$
$$= \dfrac{-(x + h)^2 + 4(x + h) + 1 - (-x^2 + 4x + 1)}{h}$$
$$= \dfrac{-(x^2 + 2xh + h^2) + 4x + 4h + 1 + x^2 - 4x - 1}{h}$$
$$= \dfrac{-x^2 - 2xh - h^2 + 4h + x^2}{h}$$
$$= \dfrac{-2xh - h^2 + 4h}{h}$$
$$= -2x - h + 4$$

41. $y = \dfrac{3x - 4}{x}$

$x \neq 0$

Domain: $(-\infty, 0) \cup (0, \infty)$

43. $y = -\dfrac{1}{4}x^2 + x + 2$

The graph is a parabola.
Let $y = 0$.
$$0 = -\dfrac{1}{4}x^2 + x + 2$$

Multiply by 4.
$$0 = -x^2 + 4x + 8$$
$$x = \dfrac{-4 \pm \sqrt{4^2 - 4(-1)(8)}}{2(-1)}$$
$$= \dfrac{-4 \pm \sqrt{48}}{-2}$$
$$= 2 \pm 2\sqrt{3}$$

The x-intercepts are $2 + 2\sqrt{3} \approx 5.46$
and $2 - 2\sqrt{3} \approx -1.46$.
Let $x = 0$.
$$y = -\dfrac{1}{4}(0)^2 + 0 + 2$$

$y = 2$ is the y-intercept.

Vertex: $x = \dfrac{-b}{2a} = \dfrac{-1}{2\left(-\frac{1}{4}\right)} = 2$

$$y = -\dfrac{1}{4}(2)^2 + 2 + 2$$
$$= -1 + 4$$
$$= 3$$

The vertex is $(2,3)$.

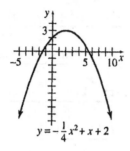

$$y = -\frac{1}{4}x^2 + x + 2$$

45. $y = -3x^2 - 12x - 1$

x-intercepts: -3.91 and -0.09
y-intercept: -1
Vertex: $(-2, 11)$

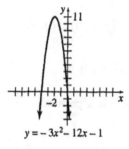

$$y = -3x^2 - 12x - 1$$

47. $f(x) = 1 - x^4$
$\qquad = -x^4 + 1$

Reflect the graph of $f(x) = x^4$ vertically and translate 1 unit upward.

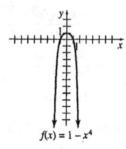

$$f(x) = 1 - x^4$$

49. $y = -(x+2)^4 - 2$

Translate the graph of $y = x^4$ 2 units to the left and 2 units downward. Reflect vertically.

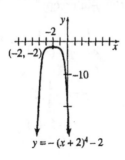

$$y = -(x+2)^4 - 2$$

51. $f(x) = \dfrac{2}{3x - 1}$

Vertical asymptote: $3x - 1 = 0$ or $x = \frac{1}{3}$
Horizontal asymptote: $y = 0$, since $\frac{2}{3x-1}$ approaches zero as x gets larger.

x	-2	-1	0	1	2
y	$-\frac{2}{7}$	$-\frac{1}{2}$	-2	1	$\frac{2}{5}$

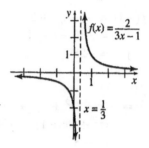

53. $f(x) = \dfrac{6x}{x + 2}$

Vertical asymptote: $x = -2$
Horizontal asymptote: $y = 6$

x	-5	-4	-3	-1	0	1	2
y	10	12	18	-6	0	2	3

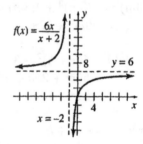

55. If $f(x) = 3x + 4$ and $g(x) = x^2 - 6x - 7$, then

$$\begin{aligned}
(f \circ g)(x) &= f(g(x)) \\
&= 3g(x) + 4 \\
&= 3(x^2 - 6x - 7) + 4 \\
&= 3x^2 - 18x - 21 + 4 \\
&= 3x^2 - 18x - 17, \text{ and}
\end{aligned}$$

$$\begin{aligned}
(g \circ f)(x) &= g(f(x)) \\
&= f(x)^2 - 6f(x) - 7 \\
&= (3x + 4)^2 - 6(3x + 4) - 7 \\
&= 9x^2 + 24x + 16 - 18x - 24 - 7 \\
&= 9x^2 + 6x - 15
\end{aligned}$$

57. (a) $y = \dfrac{x - 1.1}{0.124} = \dfrac{1}{0.124}x - \dfrac{1.1}{0.124}$

$\approx 8.06x - \dfrac{1.1}{0.124}$,

so the slope is about 8.06. This means each additional local species per 0.5 square meter leads to an additional 8.06 regional species per 0.5 hectare.

(b) If $x = 6$, then $y = \dfrac{6 - 1.1}{0.124} \approx 39.5$. So, if the average local diversity is 6 species per 0.5 square meter, then the regional diviersity must be 39.5 species per 0.5 hectare.

(c) If $y = 70$, then

$$70 = \dfrac{x - 1.1}{0.124}$$

$$70(0.124) = x - 1.1$$

$$x = 8.68 + 1.1$$

$$= 9.78$$

Therefore, a regional diversity of 70 species per 0.5 hectare means there is a local diversity of 9.78 species per 0.5 meter.

59. (a)(i) Substitute the value for RCV of one formula for RCV of the other formula and solve for S.

$$1,486S^2 - 4,106S + 4,514 = 1,486S - 825$$
$$1,486S^2 - 5,592S + 5,339 = 0$$

Find the discriminant of the quadratic formula $b^2 - 4ac$.

$$b^2 - 4ac = (-5,592)^2 - 4(1,486)(5,339)$$
$$= -464,552$$

Since the discriminant is negative, this quadratic equation has no real solution. Therefore, there is no value for S that will give the same value for RCV in the two formiulas.

(ii) This result can also be found with a graphing calculator by graphing the two functions defined by these two formulas and noting that the graphs of these functions never intersect.

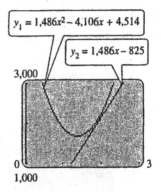

$y_1 = 1,486x^2 - 4,106x + 4,514$

$y_2 = 1,486x - 825$

(b) Substitute PV of one formula in for PV in the other formula and solve for S.

$$1,278S^{1.289} = 1,395S$$
$$1,278S^{1.289} - 1,395S = 0$$
$$9S(142S^{0.289} - 155) = 0, \text{so}$$
$$9S = 0 \quad \text{or} \quad 142S^{0.289} - 155 = 0$$

$$S = 0 \quad \text{or} \quad S = \left(\dfrac{155}{142}\right)^{\frac{1}{0.289}}$$

$$\approx 1.35$$

Therefore, these two formulas give the same answer when $S = 0$ square meters or $S \approx 1.35$ square meters.

$PV|_{S=0} = 1,395(0) = 0 = 1,278(0)^{1.289}$, and
$PV|_{S=c} = 1,395c \approx 1,889 \approx 1,278c^{1.289}$, where

$$c = \left(\dfrac{155}{142}\right)^{\frac{1}{0.289}} \approx 1.35$$

Therefore, the predicted plasma volumes are $PV = 0$ milliliter when $S = 0$ and $PV \approx 1,889$ milliliters when $S \approx 1.35$.

61. (a)

x	y	xy	x^2	y^2
1,523	44	67,012	2,319,529	1,936
2,670	74	197,580	7,128,900	5,476
2,833	76	215,308	8,025,889	5,776
3,465	79	273,735	12,006,225	6,241
2,395	61	146,095	5,736,025	3,721
3,181	72	229,032	10,118,761	5,184
2,834	71	201,214	8,031,556	5,041
1,883	67	126,161	3,545,689	4,489
2,960	78	230,880	8,761,600	6,084
3,671	76	278,996	13,476,241	5,776
27,415	698	1,966,013	79,150,415	49,724

$$r = \frac{10(1{,}966{,}013) - (27{,}415)(698)}{\sqrt{10(79{,}150{,}415) - (27{,}415)^2} \cdot \sqrt{10(49{,}724) - (698)^2}}$$

$$r \approx 0.8286$$

Yes, the data seem to fit a straight line, but not as close as in some other examples.

(b)

80

1,000   4,000
40

Yes, the data seem to fit a straight line.

(c)

$$10b + 27{,}415m = 698$$
$$27{,}415b + 79{,}150{,}415m = 1{,}966{,}013$$
$$10b = 698 - 27{,}415m$$
$$b = 69.8 - 2{,}741.5m$$
$$27{,}415(69.8 - 2{,}741.5m) + 79{,}150{,}415m = 1{,}966{,}013$$
$$1{,}913{,}567 - 75{,}158{,}222.5m + 79{,}150{,}415m = 1{,}966{,}013$$
$$3{,}992{,}192.5m = 52{,}446$$
$$m \approx 0.013137$$
$$m \approx 0.0131$$

$$b = 69.8 - 2{,}741.5(0.013137)$$
$$b \approx 33.8$$

$$Y = 0.0131x + 33.8$$

(d) Let $x = 3{,}149$.

$$Y = 0.0131(3{,}149) + 33.8$$
$$Y \approx 75$$

The predicted life expectancy in the United Kingdom, with a daily calorie supply of 3,149, is about 75 years.
This is slightly lower than the actual figure of 77 years.

63. $F(x) = -\dfrac{2}{3}x^2 + \dfrac{14}{3}x + 96$

The maximum fever occurs at the vertex of the parabola.

$$x = \frac{-b}{2a} = \frac{-\frac{14}{3}}{-\frac{4}{3}} = \frac{7}{2}$$

$$y = -\frac{2}{3}\left(\frac{7}{2}\right)^2 + \frac{14}{3}\left(\frac{7}{2}\right) + 96$$

$$= -\frac{2}{3}\left(\frac{49}{4}\right) + \frac{49}{3} + 96$$

$$= -\frac{49}{6} + \frac{49}{3} + 96$$

$$= -\frac{49}{6} + \frac{98}{6} + \frac{576}{6} = \frac{625}{6} \approx 104.2$$

The maximum fever occurs on the third day. It is about 104.2°F.

65. (a) $100\% - 87.5\% = 12.5\%$

$$12.5\% = 0.125 = \frac{125}{1{,}000} = \frac{1}{8}$$

The fraction let in is 1 over the SPF rating.

(b)

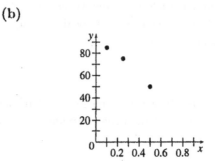

(c) $UVB = 1 - \dfrac{1}{SPF}$

(d) $1 - \dfrac{1}{8} = 87.5\%$

$$1 - \frac{1}{4} = 75.0\%$$

The increase is 12.5%.

(e) $1 - \dfrac{1}{30} = 96.\overline{6}\%$

$$1 - \frac{1}{15} = 93.\overline{3}\%$$

The increase is $3.\overline{3}\%$ or about 3.3%.

(f) The increase in percent protection decreases to zero.

67. $p(t) = \dfrac{1.79 \cdot 10^{11}}{(2{,}026.87 - t)^{0.99}}$

(a) $p(1999) \approx 6.64$ billion

This is about 644 million more than the estimate of 5,996 million.

(b) $p(2020) \approx 26.56$ billion
$p(2025) \approx 96.32$ billion

69. $y = \dfrac{7x}{100 - x}$

(a) $y = \dfrac{7(80)}{100 - 80} = \dfrac{560}{20} = 28$

The cost is $28,000.

(b) $y = \dfrac{7(50)}{100 - 50} = \dfrac{350}{50} = 7$

The cost is $7,000.

(c) $\dfrac{7(90)}{100 - 90} = \dfrac{630}{10} = 63$

The cost is $63,000.

(d) Plot the points $(80, 28)$, $(50, 7)$, and $(90, 63)$.

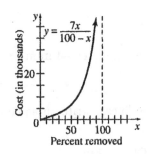

(e) No, because all of the pollutant would be removed when $x = 100$, at which point the denominator of the function would be zero.

71. **(a)** The general trend has been to decrease from 9.125 to 7.454 parts per million.

(b) Replace x with $x - 1982$.

$g(x) = -0.012053(x - 1982)^2$
$\qquad\quad - 0.046607(x - 1982) + 9.125$

73. Using a graphing calculator. Follow the steps outlined in Examples 3-5 in Section 1.3 of the textbook.

(a) $r = 0.494$

The data do not seem to fit a straight line very closely.

(b)

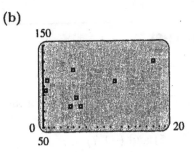

The data points do not lie on a straight line. This agrees with part a.

(c) $Y = 1.70x + 87.8$

(d) The slope is 1.70 thousand (or 1,700). On average, the governor's salary increases $1,700 for each additional million in population.

75. **(a)** $x = 0.9$ means the speed is 10% slower on the return trip.
$x = 1.1$ means the speed is 10% faster on the return trip.

(b) $v_{aver} = \dfrac{2d}{\frac{d}{v} + \frac{d}{xv}} = \dfrac{2d}{\frac{d}{v} + \frac{d}{xv}} \cdot \dfrac{xv}{xv}$

$\qquad = \dfrac{2dxv}{dx + d} = \dfrac{2dxv}{d(x + 1)}$

$\qquad = \dfrac{2xv}{x + 1} = \left(\dfrac{2x}{x + 1}\right)v$

(c) The formula for v_{aver} is a rational function with a horizontal asymptote at $v_{aver} = 2v$. This means that as the return velocity becomes greater and greater, the average velocity approaches twice the velocity on the first part of the trip, and can never exceed twice that velocity.

Chapter 1 Test

[1.1]

Find the slope of each line that has a slope.

1. Through $(2, -5)$ and $(-1, 7)$

2. Through $(9, 5)$ and $(9, 2)$

3. $3x - 7y = 9$

4. Perpendicular to $2x + 5y = 7$

Find an equation in the form $y = mx + b$ (where possible) for each line.

5. Through $(-1, 6)$ and $(5, -3)$

6. x–intercept 6, y–intercept -5

7. Through $(0, 3)$, parallel to $2x - 4y = 1$

8. Through $(1, -4)$, perpendicular to $3x + y = 1$

9. Through $(2, 5)$, perpendicular to $x = 5$

Graph each of the following.

10. $3x + 5y = 15$

11. $2x + y = 0$

12. $x - 3 = 0$

13. British CPI The Consumer Price Index (CPI) in the United Kingdom increased from 78.5 in 1980 to 200.1 in 2000.*

(a) If the change in the United Kingdom's CPI is considered to be linear, write an equation expressing the CPI, y, in terms of the number of years after 1980, x.

(b) Use the result from part (a) to predict the United Kingdom's CPI in 2010.

[1.2]

14. An electronics firm was planning to expand its product line and wanted to get an idea of the salary picture for technicians it would hire in this field. The following data was collected.

$$\sum x = 176 \qquad \sum x^2 = 3,162$$
$$\sum y = 356 \qquad \sum y^2 = 10,870$$
$$\sum xy = 5,629 \qquad n = 12$$

(a) Find an equation for the least squares line.

(b) Find the coefficient of correlation.

*U.S. Department of Labor, Bureau of Labor Statistics, Foreign Labor Statistics (www.bls.gov).

15. An economist was interested in the production costs for companies supplying chemicals for use in fertilizers. The data below represents the relationship between the number of tons produced during a given year (x) and the production cost per ton (y) for seven companies.

Number of Tons (in thousands)	Cost per Ton (in dollars)
3.0	40
4.0	50
2.4	50
5.0	35
2.6	55
4.0	35
5.5	30

 (a) Find the equation for the least squares line.

 (b) Find the coefficient of correlation.

16. Gold Medal Times The following table shows the time (in seconds) of the gold medal winner for the men's 100 meter freestyle swimming in the Olympics for selected years.

Year	Time
1912	63.4
1920	61.4
1928	58.6
1936	57.6
1952	57.4
1960	55.2
1968	52.2
1976	49.99
1984	49.80
1992	49.02

 (a) Find the equation for the least squares line for the time, Y, in terms of the number of years after 1912, x.

 (b) Use the result from part (a) to estimate the time of the gold medal winner in the 2000 Olympics.

[1.3]

17. List the ordered pairs obtained from each equation, given $\{-3, -2, -1, 0, 1, 2, 3\}$ as the domain. Give the range.

 (a) $2x + 3y = 6$ (b) $y = \dfrac{1}{x^2 - 2}$

18. Let $f(x) = 3x - 4$ and $g(x) = -x^2 + 5x$. Find each of the following.

 (a) $f(-3)$ (b) $g(-2)$ (c) $f(2m)$ (d) $g(k-1)$ (e) $f(x+h)$

19. Given the functions $f(x) = (x-2)^2$ and $g(x) = 7 - 3x$, find each of the following.

 (a) $h(x) = (f \circ g)(x)$ (b) $k(x) = (g \circ f)(x)$

20. The volume V of a sphere of radius r is given by

$$V = f(r) = \frac{4}{3}\pi r^3.$$

A balloon is being inflated in such a way that its radius (in inches) is given by

$$r = g(t) = 2t$$

after t seconds. Only one of the two functions $f \circ g$ and $g \circ f$ makes physical sense. Calculate that function and describe what it does.

[1.4]

21. Graph the parabola

$$y = 2x^2 - 4x - 2$$

and give its vertex, axis, x–intercept, and y–intercept.

22. The manufacturer of a certain product has determined that his profit in dollars for making x units of this product is given by the equation

$$p = -2x^2 + 120x + 3,000.$$

(a) Find the number of units that will maximize the profit.

(b) What is the maximum profit for making this product?

23. Traffic Volume Due to a major road construction project, traffic patterns will change considerably. City planners anticipate the volume, v, in thousands of cars per week, around what is, for the moment, a quiet neighborhood, will be given by

$$v = 4.77 + 1.49w - 0.02w^2, \ 0 \le w \le 52$$

w weeks after the project is begun.

(a) When can the residents expect maximum traffic volume?

(b) What will be the maximum volume?

Use translations and reflections to graph the following functions.

24. $y = 4 - x^2$ **25.** $f(x) = \frac{1}{2}x^3 + 1$

[1.5]

Graph each function.

26. $f(x) = x^3 - 2x^2 - x + 2$ **27.** $f(x) = \frac{3}{2x - 1}$ **28.** $f(x) = \frac{x - 1}{3x + 6}$

29. Find the horizontal and vertical asymptotes and x– and y–intercepts, if any, for the following function.

$$f(x) = \frac{3x - 5}{5x + 10}$$

30. Suppose a cost-benefit model is given by the equation

$$y = \frac{20x}{110 - x},$$

where y is the cost in thousands of dollars of removing x percent of a certain pollutant. Find the cost in thousands of dollars of removing each of the following percents of pollution.

(a) 70%   (b) 90%   (c) 100%

Chapter 1 Test Answers

1. -4

2. Undefined

3. $\frac{3}{7}$

4. $\frac{5}{2}$

5. $y = -\frac{3}{2}x + \frac{9}{2}$

6. $y = \frac{5}{6}x - 5$

7. $y = \frac{1}{2}x + 3$

8. $y = \frac{1}{3}x - \frac{13}{3}$

9. $y = 5$

10.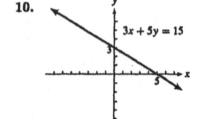

11.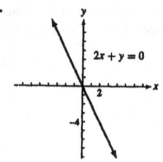

12.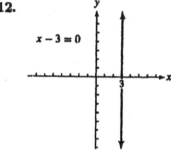

13. (a) $y = 6.08x + 78.50$ (b) About 260.9

14. (a) $Y = 0.70x + 19.37$ (b) $r = 0.96$

15. (a) $Y = -6.37x + 66.24$ (b) $r = -0.79$

16. (a) $Y = -0.1781x + 62.7289$
    (b) About 47.05 seconds

17. (a) $(-3,4)$, $\left(-2,\frac{10}{3}\right)$, $\left(-1,\frac{8}{3}\right)$, $(0,2)$, $\left(1,\frac{4}{3}\right)$, $\left(2,\frac{2}{3}\right)$, $(3,0)$; range: $\left\{0,\frac{2}{3},\frac{4}{3},2,\frac{8}{3},\frac{10}{3},4\right\}$

    (b) $\left(-3,\frac{1}{7}\right)$, $\left(-2,\frac{1}{2}\right)$, $(-1,-1)$, $\left(0,-\frac{1}{2}\right)$, $(1,-1)$, $\left(2,\frac{1}{2}\right)$, $\left(3,\frac{1}{7}\right)$; range: $\left\{-1,-\frac{1}{2},\frac{1}{7},\frac{1}{2}\right\}$

18. (a) -13 (b) -14 (c) $6m - 4$ (d) $-k^2 + 7k - 6$ (e) $3x + 3h - 4$

19. (a) $9x^2 - 30x + 25$ (b) $-3x^2 + 12x - 5$

20. $(f \circ g)(t) = \frac{32}{3}\pi t^3$ gives the volume in cubic inches where t is the time in seconds.

21. Vertex: $(1, -4)$; axis: $x = 1$;
x-intercepts: $1 - \sqrt{2} \approx -0.414$,
$1 + \sqrt{2} \approx 2.414$; y-intercept: -2

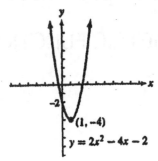

22. (a) 30   (b) $4,800

23. (a) During week 37   (b) About 32,500 cars

24.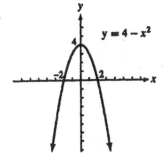

25.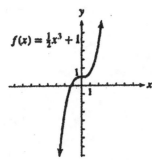

26.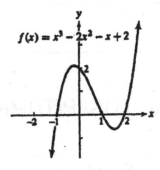

27.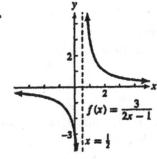

28.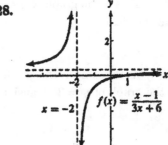

29. Horizontal: $y = \frac{3}{5}$; vertical: $x = -2$;
x-intercept: $\frac{5}{3}$; y-intercept: $-\frac{1}{2}$

30. (a) $35,000   (b) $90,000   (c) $200,000

EXPONENTIAL, LOGARITHMIC, AND TRIGONOMETRIC FUNCTIONS

2.1 Exponential Functions

1.

number of folds	1 2 3 4 5 ... 10 ... 50
layers of paper	2 4 8 16 32 ... 1,024 ... 2^{50}

$2^{50} = 1.125899907 \times 10^{15}$

3. The graph of $y = 3^x$ is the graph of an exponential function $y = a^x$ with $a > 1$.
This is graph E.

5. The graph of $y = \left(\frac{1}{3}\right)^{1-x}$ is the graph of $y = (3^{-1})^{1-x}$ or $y = 3^{x-1}$. This is the graph of $y = 3^x$ translated 1 unit to the right.
This is graph C.

7. The graph of $y = 3(3)^x$ is the same as the graph of $y = 3^{x+1}$. This is the graph of $y = 3^x$ translated 1 unit to the left.
This is graph F.

9. The graph of $y = 2 - 3^{-x}$ is the same as the graph of $y = -3^{-x} + 2$. This is the graph of $y = 3^x$ reflected in the x-axis, reflected in the y-axis, and translated up 2 units.
This is graph A.

11. The graph of $y = 3^{x-1}$ is the graph of $y = 3^x$ translated 1 unit to the right.
This is graph C.

13. $2^x = \dfrac{1}{8}$

$2^x = \dfrac{1}{2^3}$

$2^x = 2^{-3}$

$x = -3$

15. $e^x = \dfrac{1}{e^2}$

$e^x = e^{-2}$

$x = -2$

17. $25^x = 125^{x-2}$

$(5^2)^x = (5^3)^{x-2}$

$5^{2x} = 5^{3x-6}$

$2x = 3x - 6$

$6 = x$

19. $16^{x+2} = 64^{2x-1}$

$(2^4)^{x+2} = (2^6)^{2x-1}$

$2^{4x+8} = 2^{12x-6}$

$4x + 8 = 12x - 6$

$14 = 8x$

$\dfrac{7}{4} = x$

21. $e^{-x} = (e^2)^{x+3}$

$e^{-x} = e^{2x+6}$

$-x = 2x + 6$

$-3x = 6$

$x = -2$

23. $5^{-|x|} = \dfrac{1}{25}$

$5^{-|x|} = 5^{-2}$

$-|x| = -2$

$|x| = 2$

$x = 2 \quad \text{or} \quad x = -2$

25. $5^{x^2+x} = 1$

$5^{x^2+x} = 5^0$

$x^2 + x = 0$

$x(x + 1) = 0$

$x = 0 \quad \text{or} \quad x + 1 = 0$

$x = 0 \quad \text{or} \qquad x = -1$

31.

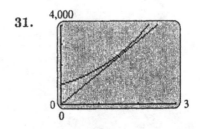

33. (a) $t = 10$

$$N = 100 \exp \{9.8901 \exp [-\exp (2.5420 - 0.2167(10))]\}$$
$$\approx 1,006$$

$t = 20$

$$N = 100 \exp \{9.8901 \exp [-\exp (2.5420 - 0.2167(20))]\}$$
$$\approx 432,500$$

$t = 30$

$$N = 100 \exp \{9.8901 \exp [-\exp (2.5420 - 0.2167(30))]\}$$
$$= 1,637,000$$

$t = 40$

$$N = 100 \exp \{9.8901 \exp [-\exp (2.5420 - 0.2167(40))]\}$$
$$= 1,931,000$$

$t = 50$

$$N = 100 \exp \{9.8901 \exp [-\exp (2.5420 - 0.2167(50))]\}$$
$$= 1,969,000$$

(b)

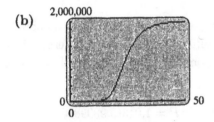

(c) According to parts a and b, the number of bacteria levels off at around 2,000,000.

35. $A(t) = 2,600e^{0.018t}$

(a) 1970: $t = 20$

$$A(20) = 2,600e^{0.018(20)}$$
$$= 2,600e^{0.36}$$
$$\approx 3,727$$

The function gives a population of 3,727 million in 1970.
This is very close to the actual population of about 3,700 million.

(b) 1990: $t = 40$

$$A(40) = 2,600e^{0.018(40)}$$
$$= 2,600e^{0.72}$$
$$\approx 5,341$$

The function gives a population of 5,341 million in 1990.

(c) 2010: $t = 60$

$$= 2,600e^{0.018(60)}$$
$$= 2,600e^{1.08}$$
$$= 7,656$$

From the function, we estimate that the world population in 2010 will be 7,656 million.

37. (a) $\dfrac{C(1975)}{C(1950)} \approx \dfrac{0.13}{0.05} = 2.6$

$\dfrac{C(2000)}{C(1975)} \approx \dfrac{0.35}{0.13} \approx 2.7$

An exponential function seems to find the data.

(b)

$$C(x) = C_0 a^{x-1950}$$
$$C(1950) = 0.05$$
$$0.05 = C_0 a^{1950-1950}$$
$$C_0 = 0.05$$
$$C(x) = 0.05 a^{x-1950}$$
$$C(2000) = 0.35$$
$$0.35 = 0.05 a^{2000-1950}$$
$$a^{50} = \frac{0.35}{0.05}$$
$$a = \sqrt[50]{\frac{0.35}{0.05}} \approx 1.040$$
$$C(x) = 0.05(1.040)^{x-1950}$$

(c)

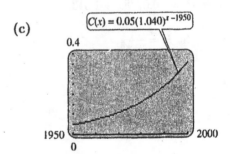

$$C(1975) = 0.05(1.040)^{25} \approx 0.133$$

From the graph, $C(1975) \approx 0.133$.
The values appear to be identical.

(d) The annual percent increase is $1.040 - 1$, which is 0.040 or 4.0%.

(e) Keep the graph from part (c), and add the graph of the horizontal line $y = 0.5$ on the same screen. Find the intersection point of the line and the curve at $(2008.7084, 0.5)$. This indicates that if the increase were to continue, the concentration would reach 0.50 ppb during the year 2008.

39. (a) When $x = 0$, $P = 1{,}013$.
When $x = 10{,}000$, $P = 265$.
First we fit $P = ae^{kx}$.

$$1{,}013 = ae^0$$
$$a = 1{,}013$$
$$P = 1{,}013e^{kx}$$
$$265 = 1{,}013e^{k(10{,}000)}$$

$$\frac{265}{1{,}013} = e^{10{,}000k}$$
$$10{,}000k = \ln\left(\frac{265}{1{,}013}\right)$$

$$k = \frac{\ln\left(\frac{265}{1{,}013}\right)}{10{,}000} \approx -1.34 \times 10^{-4}$$

Therefore $P = 1{,}013e^{(\cdot -1.34 \times 10^{-4})x}$.
Next we fit $P = mx + b$.
We use the points $(0, 1{,}013)$ and $(10{,}000, 265)$.

$$m = \frac{265 - 1{,}013}{10{,}000 - 0} = -0.0748$$

$$b = 1{,}013$$

Therefore $P = -0.0748x + 1{,}013$.

Finally, we fit $P = \frac{1}{ax+b}$.

$$1{,}013 = \frac{1}{a(0) + b}$$

$$b = \frac{1}{1{,}013} \approx 9.87 \times 10^{-4}$$

$$P = \frac{1}{ax + \frac{1}{1{,}013}}$$

$$265 = \frac{1}{10{,}000a + \frac{1}{1{,}013}}$$

$$\frac{1}{265} = 10{,}000a + \frac{1}{1{,}013}$$

$$10{,}000a = \frac{1}{265} - \frac{1}{1{,}013}$$

$$a = \frac{\frac{1}{265} - \frac{1}{1{,}013}}{10{,}000} \approx 2.79 \times 10^{-7}$$

Therefore,

$$P = \frac{1}{(2.79 \times 10^{-7})x + (9.87 \times 10^{-4})}.$$

(b)

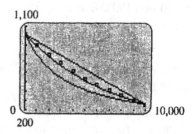

$P = 1{,}013e^{(-1.34 \times 10^{-4})x}$ is the best fit.

(c) $P(1{,}500) = 1{,}013e^{-1.34 \times 10^{-4}(1{,}500)} \approx 829$
$P(11{,}000) = 1{,}013e^{-1.34 \times 10^{-4}(11{,}000)} \approx 232$

We predict that the pressure at 1,500 meters will be 829 millibars, and at 11,000 meters will be 232 millibars.

(d) Using exponential regression, we obtain $P = 1{,}038(0.99998661)^x$ which differs slightly from the function found in part (b) which can be rewritten as

$$P = 1{,}013(0.99998660)^x.$$

41. (a) $f(x) = f_0 a^x$
$f_0 = 26$ million

Use the point $(4, 205)$ to find a

$$205 = 26a^4$$
$$a^4 = \frac{205}{26}$$
$$a = \sqrt[4]{\frac{205}{26}}$$
$$\approx 1.6757$$
$$f(x) = 26(1.6757)^x$$

(b)

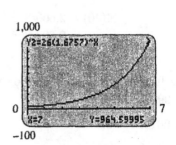

The number of users in 2002 is about 965 million.

(c) 1995 − 1996:

$$\frac{f(1) - f(0)}{f(0)} \approx \frac{43.568 - 26}{26}$$

$$\approx 0.68 \approx 68\%$$

1996 − 1997:

$$\frac{f(2) - f(1)}{f(1)} \approx \frac{73.0 - 43.568}{43.568}$$

$$\approx 0.68 \approx 68\%$$

1997 − 1998:

$$\frac{f(3) - f(2)}{f(2)} \approx \frac{122.34 - 73.0}{73.0}$$

$$\approx 0.68 \approx 68\%$$

1998 − 1999:

$$\frac{f(4) - f(3)}{f(3)} \approx \frac{205 - 122.34}{122.34}$$

$$\approx 0.68 \approx 68\%$$

The average yearly percent increase in users was about 68%.

(d) There were 122 million users in 1998.

43. $A = P\left(1 + \frac{r}{m}\right)^{tm}$, $P = 26{,}000$, $r = 0.12$, $t = 3$

(a) annually, $m = 1$

$$A = 26{,}000\left(1 + \frac{0.12}{1}\right)^{3(1)}$$

$$= 26{,}000(1.12)^3$$

$$= 36{,}528.13$$

Interest = $36,528.13 − $26,000
$$= \$10{,}528.13$$

(b) semiannually, $m = 2$

$$A = 26{,}000\left(1 + \frac{0.12}{2}\right)^{3(2)}$$

$$= 26{,}000(1.06)^6$$

$$= 36{,}881.50$$

Interest = $36,881.50 − $26,000 = $10,881.50

(c) quarterly, $m = 4$

$$A = 26{,}000\left(1 + \frac{0.12}{4}\right)^{3(4)}$$

$$= 26{,}000(1.03)^{12}$$

$$= 37{,}069.78$$

Interest = $37,069.78 − $26,000 = $11,069.78

(d) monthly, $m = 12$

$$A = 26{,}000\left(1 + \frac{0.12}{12}\right)^{3(12)}$$

$$= 26{,}000(1.01)^{36}$$

$$= 37{,}199.99$$

Interest = $37,199.99 − $26,000 = $11,199.99

45. $A = P\left(1 + \frac{r}{m}\right)^{tm}$, $P = 5000$, $A = 8000$, $t = 4$

(a) $m = 1$

$$8000 = 5000\left(1 + \frac{r}{1}\right)^{4(1)}$$

$$\frac{8}{5} = (1 + r)^4$$

$$\left(\frac{8}{5}\right)^{1/4} - 1 = r$$

$$0.125 = r$$

The interest rate is 12.5%.

(b) $m = 4$

$$8000 = 5000\left(1 + \frac{r}{4}\right)^{4(4)}$$

$$\frac{8}{5} = \left(1 + \frac{r}{4}\right)^{16}$$

$$\left(\frac{8}{5}\right)^{1/16} - 1 = \frac{r}{4}$$

$$4\left[\left(\frac{8}{5}\right)^{1/16} - 1\right] = r$$

$$0.119 = r$$

The interest rate is 11.9%.

47. (a) $30{,}000 = 10{,}500\left(1 + \frac{r}{4}\right)^{(4)(12)}$

$$\frac{300}{105} = \left(1 + \frac{r}{4}\right)^{48}$$

$$1 + \frac{r}{4} = \left(\frac{300}{105}\right)^{1/48}$$

$$4 + r = 4\left(\frac{300}{105}\right)^{1/48}$$

$$r = 4\left(\frac{300}{105}\right)^{1/48} - 4$$

$$r \approx 0.0884$$

The required interest rate is 8.84%.

(b) $30,000 = 10,500e^{12r}$

$$\frac{300}{105} = e^{12r}$$

$$12r = \ln\left(\frac{300}{105}\right)$$

$$r = \frac{\ln\left(\frac{300}{105}\right)}{12}$$

$$r \approx 0.0875$$

The required interest rate is 8.75%.

2.2 Logarithmic Functions

1. $2^3 = 8$

Since $a^y = x$ means $y = \log_a x$, the equation in logarithmic form is

$$\log_2 8 = 3.$$

3. $3^4 = 81$

The equation in logarithmic form is

$$\log_3 81 = 4.$$

5. $3^{-2} = \frac{1}{9}$

The equation in logarithmic form is

$$\log_3 \frac{1}{9} = -2.$$

7. $\log_2 128 = 7$

Since $y = \log_a x$ means $a^y = x$, the equation in exponential form is

$$2^7 \quad = \quad 128.$$

9. $\log_{25} \frac{1}{25} = -1$

The equation in exponential form is

$$25^{-1} = \frac{1}{25}.$$

11. $\quad \log 10,000 = 4$
$$\log_{10} 10,000 = 4$$
$$10^4 = 10,000$$

When no base is written, $\log_{10}$ is understood.

13. Let $\log_5 25 = x.$

Then, $\quad 5^x = 25$
$$5^x = 5^2$$
$$x = 2.$$

Thus, $\log_5 25 = 2.$

15. $\log_4 64 = x$
$$4^x = 64$$
$$4^x = 4^3$$
$$x = 3$$

17. $\log_2 \frac{1}{4} = x$
$$2^x = \frac{1}{4}$$
$$2^x = 2^{-2}$$
$$x = -2$$

19. $\log_2 \sqrt[3]{\frac{1}{4}} = x$
$$2^x = \left(\frac{1}{4}\right)^{1/3}$$
$$2^x = \left(\frac{1}{2^2}\right)^{1/3}$$
$$2^x = 2^{-2/3}$$
$$x = -\frac{2}{3}$$

21. $\ln e = x$

Recall that ln means $\log_e$.

$$e^x = e$$
$$x = 1$$

23. $\ln e^{5/3} = x$
$$e^x = e^{5/3}$$
$$x = \frac{5}{3}$$

25. The logarithm to the base 3 of 4 is written $\log_3 4$. The subscript denotes the base.

27. $\log_9 7m = \log_9 7 + \log_9 m$

29. $\log_3 \frac{3p}{5k}$

$$= \log_3 3p - \log_3 5k$$
$$= (\log_3 3 + \log_3 p) - (\log_3 5 + \log_3 k)$$
$$= 1 + \log_3 p - \log_3 5 - \log_3 k$$

31. $\log_3 \dfrac{5\sqrt{2}}{\sqrt[4]{7}}$

$\qquad = \log_3 5\sqrt{2} - \log_3 \sqrt[4]{7}$
$\qquad = \log_3 5 \cdot 2^{1/2} - \log_3 7^{1/4}$
$\qquad = \log_3 5 + \log_3 2^{1/2} - \log_3 7^{1/4}$
$\qquad = \log_3 5 + \dfrac{1}{2} \log_3 2 - \dfrac{1}{4} \log_3 7$

33. $\log_b 8 = \log_b 2^3$
$\qquad = 3 \log_b 2$
$\qquad = 3a$

35. $\log_b 72b = \log_b 72 + \log_b b$
$\qquad = \log_b 72 + 1$
$\qquad = \log_b 2^3 \cdot 3^3 + 1$
$\qquad = \log_b 2^3 + \log_b 3^2 + 1$
$\qquad = 3 \log_b 2 + 2 \log_b 3 + 1$
$\qquad = 3a + 2c + 1$

37. $\log_5 20 = \dfrac{\ln 20}{\ln 5}$
$\qquad \approx \dfrac{3}{1.61}$
$\qquad \approx 1.86$

39. $\log_{1.2} 0.55 = \dfrac{\ln 0.55}{\ln 1.2}$
$\qquad \approx -3.28$

41. $\log_x 25 = -2$
$\qquad x^{-2} = 25$
$\qquad (x^{-2})^{-1/2} = 25^{-1/2}$
$\qquad x = \dfrac{1}{5}$

43. $\log_8 4 = z$
$\qquad 8^z = 4$
$\qquad (2^3)^z = 2^2$
$\qquad 2^{3z} = 2^2$
$\qquad 3z = 2$
$\qquad z = \dfrac{2}{3}$

45. $\log_r 7 = \dfrac{1}{2}$
$\qquad r^{1/2} = 7$
$\qquad (r^{1/2})^2 = 7^2$
$\qquad r = 49$

47. $\log_5 (9x - 4) = 1$
$\qquad 5^1 = 9x - 4$
$\qquad 9 = 9x$
$\qquad 1 = x$

49. $\log_9 m - \log_9 (m - 4) = -2$
$\qquad \log_9 \dfrac{m}{m-4} = -2$
$\qquad 9^{-2} = \dfrac{m}{m-4}$
$\qquad \dfrac{1}{81} = \dfrac{m}{m-4}$
$\qquad m - 4 = 81m$
$\qquad -4 = 80m$
$\qquad -0.05 = m$

This value is not possible since $\log_9 (-0.05)$ does not exist.

Thus, there is no solution to the original equation.

51. $\log_3 (x - 2) + \log_3 (x + 6) = 2$
$\qquad \log_3 [(x - 2)(x + 6)] = 2$
$\qquad (x - 2)(x + 6) = 3^2$
$\qquad x^2 + 4x - 12 = 9$
$\qquad x^2 + 4x - 21 = 0$
$\qquad (x + 7)(x - 3) = 0$
$\qquad x = -7 \quad \text{or} \quad x = 3$

$x = -7$ does not check in the original equation. The only solution is 3.

53. $\qquad 4^x = 12$
$\qquad x \log 4 = \log 12$
$\qquad x = \dfrac{\log 12}{\log 4}$
$\qquad \approx 1.79$

55. $\qquad e^{2y} = 12$
$\qquad \ln e^{2y} = \ln 12$
$\qquad 2y \ln e = \ln 12$
$\qquad 2y(1) = \ln 12$
$\qquad y = \dfrac{\ln 12}{2}$
$\qquad \approx 1.24$

57. $\qquad 10e^{3z-7} = 5$
$\qquad \ln 10e^{3z-7} = \ln 5$
$\qquad \ln 10 + \ln e^{3z-7} = \ln 5$
$\qquad \ln 10 + (3z - 7) \ln e = \ln 5$
$\qquad 3z - 7 = \ln 5 - \ln 10 + 7$
$\qquad 3z = \ln 5 - \ln 10 + 7$
$\qquad z = \dfrac{\ln 5 - \ln 10 + 7}{3}$
$\qquad \approx 2.10$

59.
$$1.5(1.05)^x = 2(1.01)^x$$
$$\ln[1.5(1.05)^x] = \ln[2(1.01)^x]$$
$$\ln 1.5 + x \ln 1.05 = \ln 2 + x \ln 1.01$$
$$x(\ln 1.05 - \ln 1.01) = \ln 2 - \ln 1.5$$
$$x = \frac{\ln 2 - \ln 1.5}{\ln 1.05 - \ln 1.01}$$
$$\approx 7.41$$

61. $f(x) = \ln(x^2 - 4)$

Since the domain of $f(x) = \ln x$ is $(0, \infty)$, the domain of $f(x) = \ln(x^2 - 4)$ is the set of all real numbers x for which
$$x^2 - 4 > 0.$$

To solve this quadratic inequality, first solve the corresponding quadratic equation.
$$x^2 - 4 = 0$$
$$(x + 2)(x - 2) = 0$$
$$x + 2 = 0 \quad \text{or} \quad x - 2 = 0$$
$$x = -2 \quad \text{or} \quad x = 2$$

These two solutions determine three intervals on the number line: $(-\infty, -2), (-2, 2)$, and $(2, \infty)$.

If $x = -3$, $(-3 + 2)(-3 - 2) > 0$.
If $x = 0$, $(0 + 2)(0 - 2) \not> 0$.
If $x = 3$, $(3 + 2)(3 - 2) > 0$.

The domain is $x < -2$ or $x > 2$, which is written in interval notation as $(-\infty, -2) \cup (2, \infty)$.

63. Let $m = \log_a \dfrac{x}{y}$, $n = \log_a x$, and $p = \log_a y$.

Then $a^m = \dfrac{x}{y}$, $a^n = x$, and $a^p = y$.

Substituting gives
$$a^m = \frac{x}{y} = \frac{a^n}{a^p} = a^{n-p}.$$

So $m = n - p$.
Therefore,
$$\log_a \frac{x}{y} = \log_a x - \log_a y.$$

65. (a) $t = \dfrac{\ln 2}{\ln(1 + 0.03)}$

≈ 23.4 years

(b) $t = \dfrac{\ln 2}{\ln(1 + 0.06)}$

≈ 11.9 years

(c) $t = \dfrac{\ln 2}{\ln(1 + 0.08)}$

$= 9.0$ years

67. Set the exponential growth functions $y = 4,500(1.04)^t$ and $y = 300(1.06)^t$ equal to each other and solve for t.

$$4,500(1.04)^t = 3,000(1.06)^t$$
$$\frac{3}{2}(1.04)^t = (1.06)^t$$
$$\ln\left(\frac{3}{2}(1.04)^t\right) = \ln(1.06)^t$$
$$\ln\frac{3}{2} + \ln(1.04)^t = t\ln(1.06)^t$$
$$\ln\frac{3}{2} + t\ln 1.04 = t\ln(1.06)^t$$
$$\ln\frac{3}{2} = t(\ln 1.06 - \ln 1.04)$$
$$t = \frac{\ln\frac{3}{2}}{\ln 1.06 - \ln 1.04}$$
$$t \approx 21.29$$

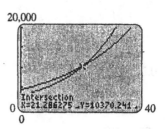

After about 21.3 years the black squirrels will outnumber the gray squirrels.

69. (a) $w = 4,000$ gm
$A = 4.688(4,000)^{0.8168 - 0.0154 \log 4,000}$
$\approx 2,590$ square centimeters

(b) $w = 8,000$ gm
$A \approx 4.688(8,000)^{0.8168 - 0.0154 \log 8,000}$
$\approx 4,211$ square centimeters

(c)

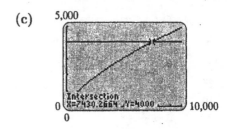

The graph shows than an infant with a surface area of 4,000 square centimeters has a weight of 7,430 gm.

71. (a) The total number of individuals in the community is $50 + 50$, or 100.

Let $P_1 = \dfrac{50}{100} = 0.5$, $P_2 = 0.5$.

$$
\begin{aligned}
H &= -1[P_1 \ln P_1 + P_2 \ln P_2] \\
&= -1[0.5 \ln 0.5 + 0.5 \ln 0.5] \\
&\approx 0.693
\end{aligned}
$$

(b) For 2 species, the maximum diversity is $\ln 2$.

(c) Yes, $\ln 2 \approx 0.693$.

73. (a) 3 species, $\frac{1}{3}$ each:

$$P_1 = P_2 = P_3 = \frac{1}{3}$$

$$
\begin{aligned}
H &= -(P_1 \ln P_1 + P_2 \ln P_2 + P_3 \ln P_3) \\
&= -3\left(\frac{1}{3} \ln \frac{1}{3}\right) \\
&= -\ln \frac{1}{3} \\
&\approx 1.099
\end{aligned}
$$

(b) 4 species, $\frac{1}{4}$ each:

$$P_1 = P_2 = P_3 = P_4 = \frac{1}{4}$$

$$
\begin{aligned}
H &= (P_1 \ln P_1 + P_2 \ln P_2 + P_3 \ln P_3 + P_4 \ln P_4) \\
&= -4\left(\frac{1}{4} \ln \frac{1}{4}\right) \\
&= -\ln \frac{1}{4} \\
&\approx 1.386
\end{aligned}
$$

(c) Notice that

$$-\ln \frac{1}{3} = \ln(3^{-1})^{-1} = \ln 3 \approx 1.099$$

and

$$-\ln \frac{1}{4} = \ln(4^{-1})^{-1} = \ln 4 \approx 1.386$$

by Property c of logarithms, so the populations are at a maximum index of diversity.

75. $C(t) = C_0 e^{-kt}$

When $t = 0$, $C(t) = 2$, and when $t = 3$, $C(t) = 1$.

$$
\begin{aligned}
2 &= C_0 e^{-k(0)} \\
C_0 &= 2 \\
1 &= 2e^{-3k} \\
\frac{1}{2} &= e^{-3k} \\
-3k &= \ln \frac{1}{2} = \ln 2^{-1} = -\ln 2 \\
k &= \frac{\ln 2}{3}
\end{aligned}
$$

$$
\begin{aligned}
T &= \frac{1}{k} \ln \frac{C_2}{C_1} \\
T &= \frac{1}{\frac{\ln 2}{3}} \ln \frac{5 C_1}{C_1} \\
T &= \frac{3 \ln 5}{\ln 2} \\
T &\approx 7.0
\end{aligned}
$$

The drug should be given about every 7 hours.

77. Let $x = 0$ represent 1994.

In x years, the population of New York is given by $18.2(1 + 0.001)^x$ or $18.2(1.001)^x$, and the population of Florida is given by $14.0(1 + 0.017)^x$ or $14.0(1.017)^x$.

First we find when the populations will be equal.

$$
\begin{aligned}
18.2 \,(1.001)^x &= 14.0 \,(1.017)^x \\
\frac{(1.001)^x}{(1.017)^x} &= \frac{14.0}{18.2} \\
\left(\frac{1.001}{1.017}\right)^x &= \frac{14.0}{18.2} \\
\ln \left(\frac{1.001}{1.017}\right)^x &= \ln \left(\frac{14.0}{18.2}\right) \\
x \ln \left(\frac{1.001}{1.017}\right) &= \ln \left(\frac{14.0}{18.2}\right) \\
x &= \frac{\ln \left(\frac{14.0}{18.2}\right)}{\ln \left(\frac{1.001}{1.017}\right)} \\
x &\approx 16.54
\end{aligned}
$$

$1994 + 16.54 = 2,010.54$

By graphing $y_1 = 18.2(1.001)^x$ and $y_2 = 14.0(1.017)^x$ on the same screen, we find that the two functions intersect at approximately $(16.54, 18.5)$, which verifies our answer.

The population of Florida will exceed that of New York in 2011.

79. $N(r) = -5,000 \ln r$

(a) $N(0.9) = -5,000 \ln (0.9) \approx 530$

(b) $N(0.5) = -5,000 \ln (0.5) \approx 3,500$

(c) $N(0.3) = -5,000 \ln (0.3) \approx 6,000$

(d) $N(0.7) = -5,000 \ln (0.7) \approx 1,800$

(e) $-5,000 \ln r = 1,000$

$$\ln r = \frac{1,000}{-5,000}$$

$$\ln r = -\frac{1}{5}$$

$$r = e^{-1/5}$$

$$r \approx 0.8$$

81. $R(I) = \log \dfrac{I}{I_0}$

(a) $R(1,000,000\, I_0)$

$= \log \dfrac{1,000,000\, I_0}{I_0}$

$= \log 1,000,000$

$= 6$

(b) $R(100,000,000\, I_0)$

$= \log \dfrac{100,000,000\, I_0}{I_0}$

$= \log 100,000,000$

$= 8$

(c) $R(I) = \log \dfrac{I}{I_0}$

$6.7 = \log \dfrac{I}{I_0}$

$10^{6.7} = \dfrac{I}{I_0}$

$I \approx 5,000,000 I_0$

(d) $R(I) = \log \dfrac{I}{I_0}$

$8.1 = \log \dfrac{I}{I_0}$

$10^{8.1} = \dfrac{I}{I_0}$

$I \approx 126,000,000 I_0$

(e) $\dfrac{1985 \text{ quake}}{1999 \text{ quake}} = \dfrac{126,000,000 I_0}{5,000,000 I_0} \approx 25$

The 1985 earthquake had an amplitude more than 25 times that of the 1999 earthquake.

(f) $R(E) = \dfrac{2}{3} \log \dfrac{E}{E_0}$

For the 1999 earthquake

$$6.7 = \frac{2}{3} \log \frac{E}{E_0}$$

$$10.05 = \log \frac{E}{E_0}$$

$$\frac{E}{E_0} = 10^{10.05}$$

$$E = 10^{10.05} E_0$$

For the 1985 earthquake,

$$8.1 = \frac{2}{3} \log \frac{E}{E_0}$$

$$12.15 = \log \frac{E}{E_0}$$

$$\frac{E}{E_0} = 10^{12.15}$$

$$E = 10^{12.15} E_0$$

The ratio of their energies is

$$\frac{10^{12.15} E_0}{10^{10.05} E_0} = 10^{2.1} \approx 126$$

The 1985 earthquake had an energy about 126 times that of the 1999 earthquake.

(g) Find the energy of a magnitude 6.7 earthquake. Using the formula from part f,

$$6.7 = \frac{2}{3} \log \frac{E}{E_0}$$

$$\log \frac{E}{E_0} = 10.05$$

$$\frac{E}{E_0} = 10^{10.05}$$

$$E = E_0 10^{10.05}$$

For an earthquake that releases 15 times this much energy, $E = E_0(15)10^{10.05}$.

$$R(E_0(15)10^{10.05}) = \frac{2}{3} \log \left(\frac{E_0(15)10^{10.05}}{E_0} \right)$$

$$= \frac{2}{3} \log(15 \cdot 10^{10.05})$$

$$\approx 7.5$$

So, it's true that a magnitude 7.5 earthquake releases 15 times more energy than one of magnitude 6.7.

2.3   Applications: Growth and Decay

1. y_0 represents the initial quantity; k represents the rate of growth or decay.

3. The half-life of a quantity is the time period for the quantity to decay to one-half of the initial amount.

5. Assume that $y = y_0 e^{kt}$ is the amount left of a radioactive substance decaying with a half-life of T. From Exercise 5, we know $k = \frac{-\ln 2}{T}$, so

$$y = y_0 e^{(-\ln 2/T)t} = y_0 e^{-t/T \ln 2} = y_0 e^{\ln(2^{-t/T})}$$

$$= y_0 2^{-t/T} = y_0 \left[\left(\frac{1}{2} \right)^{-1} \right]^{-t/T} = y_0 \left(\frac{1}{2} \right)^{t/T}$$

7. Solve for r. Let $t = 1997 - 1980 = 17$.

$$\frac{1}{84} = \frac{1}{250}(1 + r)^{17}$$

$$\frac{250}{84} = (1 + r)^{17}$$

$$\left(\frac{250}{84} \right)^{1/17} = 1 + r$$

$$r = \left(\frac{250}{84} \right)^{1/17} - 1$$

$$r \approx 0.066 \text{ or about } 6.6\%$$

No, these numbers do not represent a 4% increase annually. They represent a 6.6% increase.

9. $y = y_0 e^{kt}$
$y = 40{,}000$, $y_0 = 25{,}000$, $t = 10$

(a) $40{,}000 = 25{,}000 e^{k(10)}$
$1.6 = e^{10k}$
$\ln 1.6 = 10k$
$0.047 = k$

The equation is
$$y = 25{,}000 e^{0.047t}.$$

(b) $y = 60{,}000$
$60{,}000 = 25{,}000 e^{0.047t}$
$2.4 = e^{0.047t}$
$\ln 2.4 = 0.047t$
$18.6 = t$

There will be 60,000 bacteria in about 18.6 hours.

11. $f(t) = 500 e^{0.1t}$

(a) $f(t) = 3{,}000$
$3{,}000 = 500 e^{0.1t}$
$6 = e^{0.1t}$
$\ln 6 = 0.1t$
$17.9 \approx t$

It will take 17.9 days.

(b) If $t = 0$ corresponds to January 1, the date January 17 should be placed on the product. January 18 would be more than 17.9 days.

13. (a) From the graph, the risks of chromosomal abnormality per 1,000 at ages 20, 35, 42, and 49 are 2, 5, 29, and 125, respectively.

(Note: It is difficult to read the graph accurately. If you read different values from the graph, your answers to parts (b)-(e) may differ from those given here.)

(b) $y = Ce^{kt}$

When $t = 20$, $y = 2$, and when $t = 35$, $y = 5$.

$$2 = Ce^{20k}$$
$$5 = Ce^{35k}$$

$$\frac{5}{2} = \frac{Ce^{35k}}{Ce^{20k}}$$
$$2.5 = e^{15k}$$
$$15k = \ln 2.5$$
$$k = \frac{\ln 2.5}{15}$$
$$k \approx 0.061$$

(c) $y = Ce^{kt}$

When $t = 42$, $y = 29$, and when $t = 49$, $y = 125$.

$$29 = Ce^{42k}$$
$$125 = Ce^{49k}$$

$$\frac{125}{29} = \frac{Ce^{49k}}{Ce^{42k}}$$
$$\frac{125}{29} = e^{7k}$$
$$7k = \ln \left(\frac{125}{29} \right)$$
$$k = \frac{\ln \left(\frac{125}{29} \right)}{7}$$
$$k \approx 0.21$$

(d) Since the values of k are different, we cannot assume the graph is of the form $y = Ce^{kt}$.

(e) The results are summarized in the following table.

n	Value of k for [20, 35]	Value of k for [42, 49]
2	0.0011	0.0023
3	2.6×10^{-5}	3.4×10^{-5}
4	6.8×10^{-7}	5.5×10^{-7}

The value of n should be somewhere between 3 and 4.

15. Find k.

$$A(t) = A_0 e^{kt}$$

Let $A(t) = \left(\dfrac{1}{2}\right) A_0$ and $t = 1.25$.

$$\frac{1}{2} A_0 = A_0 e^{1.25k}$$

$$\frac{1}{2} = e^{1.25k}$$

$$\ln \frac{1}{2} = \ln e^{1.25k}$$

$$\ln \frac{1}{2} = 1.25k$$

$$\frac{\ln \frac{1}{2}}{1.25} = k$$

Let $A_0 = 1$ and let $t = 0.25$.

$$A(0.25) = 1 \cdot e^{\left(\frac{\ln \frac{1}{2}}{1.25} \cdot 0.25\right)}$$

$$\approx 0.87$$

Therefore, 87 percent of the potassium 40 remains from a creature that died 250 million years ago.

17.
$$A(t) = A_0 e^{kt}$$
$$0.60\, A_0 = A_0 e^{(-\ln 2/5,600)t}$$
$$0.60 = e^{(-\ln 2/5,600)t}$$
$$\ln 0.60 = -\frac{\ln 2}{5,600} t$$
$$\frac{5,600(\ln 0.60)}{-\ln 2} = t$$
$$4,127 \approx t$$

The sample was about 4,100 years old.

19.
$$\frac{1}{2} A_0 = A_0 e^{-0.00043t}$$
$$\frac{1}{2} = e^{-0.00043t}$$
$$\ln \frac{1}{2} = -0.00043t$$
$$\ln 1 - \ln 2 = -0.00043t$$
$$\frac{0 - \ln 2}{-0.00043} = t$$
$$1{,}612 \approx t$$

The half-life of radium 226 is about 1,600 years.

21. (a) $\quad A(t) = A_0 \left(\dfrac{1}{2}\right)^{t/1,620}$

$$A(100) = 2.0 \left(\frac{1}{2}\right)^{100/1,620}$$

$$A(100) \approx 1.9$$

After 100 years, about 1.9 grams will remain.

(b) $\quad 0.1 = 2.0 \left(\dfrac{1}{2}\right)^{t/1,620}$

$$\frac{0.1}{2} = \left(\frac{1}{2}\right)^{t/1,620}$$

$$\ln 0.05 = \frac{t}{1,620} \ln \frac{1}{2}$$

$$t = \frac{1,620 \ln 0.05}{\ln \left(\frac{1}{2}\right)}$$

$$t \approx 7,000$$

The half-life is about 7,000 years.

23. (a) $y = y_0 e^{kt}$

When $t = 0$, $y = 25.0$, so $y_0 = 25.0$.
When $t = 50$, $y = 19.5$.

$$19.5 = 25.0 e^{50k}$$

$$\frac{19.5}{25.0} = e^{50k}$$

$$50k = \ln \left(\frac{19.5}{25.0}\right)$$

$$k = \frac{\ln \left(\frac{19.5}{25.0}\right)}{50}$$

$$k \approx -0.00497$$
$$y = 25.0 e^{-0.00497t}$$

(b)
$$\frac{1}{2}y_0 = y_0 e^{-0.00497t}$$

$$\frac{1}{2} = e^{-0.00497t}$$

$$-0.00497t = \ln\left(\frac{1}{2}\right)$$

$$t = \frac{\ln\left(\frac{1}{2}\right)}{-0.00497}$$

$$t \approx 139$$

The half-life is about 139 days.

25. (a) Let $t =$ the number of degrees Celsius.

$y = y_0 \cdot e^{kt}$
$y_0 = 10$ when $t = 0°$.
To find k, let $y = 11$ when $t = 10°$.

$$11 = 10e^{10k}$$

$$e^{10k} = \frac{11}{10}$$

$$10k = \ln 1.1$$

$$k = \frac{\ln 1.1}{10}$$

$$\approx 0.0095$$

The equation is

$$y = 10e^{0.0095t}.$$

(b) Let $y = 15$; solve for t.

$$15 = 10e^{0.0095t}$$
$$\ln 1.5 = 0.0095t$$

$$t = \frac{\ln 1.5}{0.0095}$$

$$\approx 42.7$$

15 grams will dissolve at 42.7°C.

27. $f(t) = T_0 + Ce^{-kt}$

$$25 = 20 + 100e^{-0.1t}$$
$$5 = 100e^{-0.1t}$$
$$e^{-0.1t} = 0.05$$
$$-0.1t = \ln 0.05$$

$$t = \frac{\ln 0.05}{-0.1}$$

$$\approx 30$$

It will take about 30 min.

29. The figure is not correct.

$$(1 + .054)(1 + .06)(1 + .049) = 1.1720$$

This is a 17.2% increase.

2.4   Trigonometric Functions

1. $60° = 60\left(\dfrac{\pi}{180}\right) = \dfrac{\pi}{3}$

3. $150° = 150\left(\dfrac{\pi}{180}\right) = \dfrac{5\pi}{6}$

5. $210° = 210\left(\dfrac{\pi}{180}\right) = \dfrac{7\pi}{6}$

7. $390° = 390\left(\dfrac{\pi}{180}\right) = \dfrac{13\pi}{6}$

9. $\dfrac{7\pi}{4} = \dfrac{7\pi}{4}\left(\dfrac{180°}{\pi}\right) = 315°$

11. $\dfrac{11\pi}{6} = \dfrac{11\pi}{6}\left(\dfrac{180°}{\pi}\right) = 330°$

13. $\dfrac{8\pi}{5} = \dfrac{8\pi}{5}\left(\dfrac{180°}{\pi}\right) = 288°$

15. $\dfrac{4\pi}{15} = \dfrac{4\pi}{15}\left(\dfrac{180°}{\pi}\right) = 48°$

17. Let $\alpha =$ the angle with terminal side through $(-3, 4)$. Then $x = -3$, $y = 4$, and

$$r = \sqrt{x^2 + y^2} = \sqrt{(-3)^2 + (4)^2}$$
$$= \sqrt{25} = 5.$$

$$\sin\alpha = \frac{y}{r} = \frac{4}{5} \qquad \cot\alpha = \frac{x}{y} = -\frac{3}{4}$$

$$\cos\alpha = \frac{x}{r} = -\frac{3}{5} \qquad \sec\alpha = \frac{r}{x} = -\frac{5}{3}$$

$$\tan\alpha = \frac{y}{x} = -\frac{4}{3} \qquad \csc\alpha = \frac{r}{y} = \frac{5}{4}$$

19. Let $\alpha =$ the angle with terminal side through $(6, 8)$. Then $x = 6$, $y = 8$, and

$$r = \sqrt{x^2 + y^2} = \sqrt{36 + 64}$$
$$= \sqrt{100} = 10.$$

$$\sin\alpha = \frac{y}{r} = \frac{4}{5} \qquad \cot\alpha = \frac{x}{y} = \frac{3}{4}$$

$$\cos\alpha = \frac{x}{r} = \frac{3}{5} \qquad \sec\alpha = \frac{r}{x} = \frac{5}{3}$$

$$\tan\alpha = \frac{y}{x} = \frac{4}{3} \qquad \csc\alpha = \frac{r}{y} = \frac{5}{4}$$

21. In quadrant I, all six trigonometric functions are positive, so their sign is +.

23. In quadrant III, $x < 0$ and $y < 0$. Furthermore, $r > 0$.

$\sin \theta = \dfrac{y}{r} < 0$, so the sign is $-$.

$\cos \theta = \dfrac{x}{r} < 0$, so the sign is $-$.

$\tan \theta = \dfrac{y}{x} > 0$, so the sign is $+$.

$\cot \theta = \dfrac{x}{y} > 0$, so the sign is $+$.

$\sec \theta = \dfrac{r}{x} < 0$, so the sign is $-$.

$\csc \theta = \dfrac{r}{y} < 0$, so the sign is $-$.

25. When an angle θ of 30° is drawn in standard position, one choice of a point on its terminal side is $(x, y) = (\sqrt{3}, 1)$. Then

$$r = \sqrt{x^2 + y^2} = \sqrt{3 + 1} = 2.$$

$$\tan \theta = \frac{y}{x} = \frac{1}{\sqrt{3}} = \frac{\sqrt{3}}{3}$$

$$\cot \theta = \frac{x}{y} = \sqrt{3}$$

$$\csc \theta = \frac{r}{y} = 2$$

27. When an angle θ of 60° is drawn in standard position, one choice of a point on its terminal side is $(x, y) = (1, \sqrt{3})$. Then

$$r = \sqrt{x^2 + y^2} = \sqrt{1 + 3} = 2.$$

$$\sin \theta = \frac{y}{r} = \frac{\sqrt{3}}{2}$$

$$\cot \theta = \frac{x}{y} = \frac{1}{\sqrt{3}} = \frac{\sqrt{3}}{3}$$

$$\csc \theta = \frac{r}{y} = \frac{2}{\sqrt{3}} = \frac{2\sqrt{3}}{3}$$

29. When an angle θ of 135° is drawn in standard position, one choice of a point on its terminal side is $(x, y) = (-1, 1)$. Then

$$r = \sqrt{x^2 + y^2} = \sqrt{1 + 1} = \sqrt{2}.$$

$$\tan \theta = \frac{y}{x} = -1$$

$$\cot \theta = \frac{x}{y} = -1$$

31. When an angle θ of 210° is drawn in standard position, one choice of a point on its terminal side is $(x, y) = (-\sqrt{3}, -1)$. Then

$$r = \sqrt{x^2 + y^2} = \sqrt{3 + 1} = 2.$$

$$\cos \theta = \frac{x}{r} = -\frac{\sqrt{3}}{2}$$

$$\sec \theta = \frac{r}{x} = \frac{2}{-\sqrt{3}} = -\frac{2\sqrt{3}}{3}$$

33. When an angle of $\frac{\pi}{3}$ is drawn in standard position, one choice of a point on its terminal side is $(x, y) = (1, \sqrt{3})$. Then

$$r = \sqrt{x^2 + y^2} = \sqrt{1 + 3} = 2.$$

$$\sin \frac{\pi}{3} = \frac{y}{r} = \frac{\sqrt{3}}{2}$$

35. When an angle of $\frac{\pi}{4}$ is drawn in standard position, one choice of a point on its terminal side is $(x, y) = (1, 1)$.

$$\tan \frac{\pi}{4} = \frac{y}{x} = 1$$

37. When an angle of $\frac{\pi}{6}$ is drawn in standard position, one choice of a point on its terminal side is $(x, y) = (\sqrt{3}, 1)$ Then

$$r = \sqrt{x^2 + y^2} = \sqrt{3 + 1} = 2.$$

$$\sec \frac{\pi}{6} = \frac{r}{x} = \frac{2}{\sqrt{3}} = \frac{2\sqrt{3}}{3}$$

39. When an angle of 3π is drawn in standard position, one choice of a point on its terminal side is $(x, y) = (-1, 0)$. Then

$$r = \sqrt{x^2 + y^2} = \sqrt{1} = 1.$$

$$\cos 3\pi = \frac{x}{r} = -1$$

41. When an angle of $\frac{4\pi}{3}$ is drawn in standard position, one choice of a point on its terminal side is $(x, y) = (-1, -\sqrt{3})$. Then

$$r = \sqrt{x^2 + y^2} = \sqrt{1 + 3} = 2.$$

$$\sin \frac{4\pi}{3} = \frac{y}{r} = -\frac{\sqrt{3}}{2}$$

43. When an angle of $\frac{5\pi}{4}$ is drawn in standard position, one choice of a point on its terminal side is $(x, y) = (-1, -1)$. Then

$$r = \sqrt{x^2 + y^2} = \sqrt{1 + 1} = \sqrt{2}.$$

$$\csc \frac{5\pi}{4} = \frac{r}{y} = -\sqrt{2}$$

45. When an angle of $-\frac{\pi}{3}$ is drawn in standard position, one choice of a point on its terminal side is $(x, y) = (1, -\sqrt{3})$. Then

$$\tan\left(-\frac{\pi}{3}\right) = \frac{y}{x} = -\sqrt{3}$$

47. When an angle of $-\frac{7\pi}{6}$ is drawn in standard position, one choice of a point on its terminal side is $(x, y) = (-\sqrt{3}, 1)$. Then

$$r = \sqrt{x^2 + y^2} = \sqrt{3 + 1} = 2.$$

$$\sin\left(-\frac{7\pi}{6}\right) = \frac{y}{r} = \frac{1}{2}$$

49. $\sin 39° = 0.6293$

51. $\tan 82° = 7.1154$

53. $\sin 0.4014 = 0.3907$

55. $\cos 1.4137 = 0.1564$

57. $f(x) = \cos(3x)$ is of the form $f(x) = a\cos(bx)$ where $a = 1$ and $b = 3$. Thus, $a = 1$ and $T = \frac{2\pi}{b} = \frac{2\pi}{3}$.

59. $s(x) = 3\sin(880\pi t - 7)$ is of the form $s(x) = a\sin(bt + c)$ where $a = 3$, $b = 880\pi$, and $c = -7$. Thus, $a = 3$ and $T = \frac{2\pi}{b} = \frac{2\pi}{880\pi} = \frac{1}{440}$.

61. The graph of $y = 2\cos x$ is similar to the graph of $y = \cos x$ except that it has twice the amplitude. (That is, its height is twice as great.)

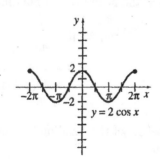

$y = 2\cos x$

63. The graph of $y = -\frac{1}{2}\cos x$ is similar to the graph of $y = \cos x$ except that it has half the amplitude and is reflected about the x-axis.

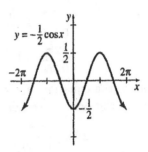

65. $y = 4\sin\left(\frac{1}{2}x + \pi\right) + 2$ has amplitude $a = 4$, period $T = \frac{2\pi}{b} = \frac{2\pi}{\frac{1}{2}} = 4\pi$, place shift $\frac{c}{b} = \frac{\pi}{\frac{1}{2}} = 2\pi$, and vertical shift $d = 2$. Thus, the graph of $y = 4\sin\left(\frac{1}{2}x + \pi\right) + 2$ is similar to the graph of $f(x) = \sin t$ except that it has 4 times the amplitude, twice the period, and is shifted up 2 units vertically. Also, $y = \sin\left(\frac{1}{2}x + \pi\right) + 2$ is shifted 2π units to the left relative to the graph of $g(x) = \sin\left(\frac{1}{2}x\right)$.

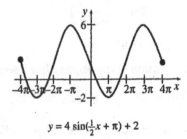

$y = 4\sin\left(\frac{1}{2}x + \pi\right) + 2$

67. The graph of $y = -3\tan x$ is similar to the graph of $y = \tan x$ except that it is reflected about the x-axis and each ordinate value is three times larger in absolute value. Note that the points $\left(-\frac{\pi}{4}, 3\right)$ and $\left(\frac{\pi}{4}, -3\right)$ lie on the graph.

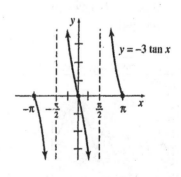

$y = -3\tan x$

69. (a) Since the amplitude is 2 and the period is 0.350,

$$0.350 = \frac{2\pi}{b}$$

$$\frac{1}{0.350} = \frac{b}{2\pi}$$

$$\frac{2\pi}{0.350} = b.$$

Therefore, the equation is $y = 2\sin(2\pi x/0.350)$, where x is the time in seconds.

(b)
$$2 = 2\sin\left(\frac{2\pi x}{0.350}\right)$$

$$1 = \sin\left(\frac{2\pi x}{0.350}\right)$$

$$\frac{\pi}{2} = \frac{2\pi x}{0.350}$$

$$\frac{0.350\pi}{4\pi} = x$$

$$0.0875 = x$$

The image reaches its maximum amplitude after 0.0875 seconds.

(c) $x = 2$

$$y = 2\sin\left(\frac{2\pi(2)}{0.350}\right)$$

$$\approx -1.95$$

The position of the object after 2 seconds is $-1.95°$.

71. (a) $\theta_2 = 23°, a = 10°, B_2 = 170°$

$$\tan\theta = (\tan 23°)(\cos 10° - \cot 170° \sin 10°)$$
$$\approx 0.83605$$
$$\theta \approx \tan^{-1}(0.83605)$$
$$\theta \approx 40°$$

(b) $\theta_2 = 20°, a = 10°, B_2 = 160°$

$$\tan\theta = (\tan 20°)(\cos 10° - \cot 160° \sin 10°)$$
$$\approx 0.53209$$
$$\theta \approx \tan^{-1}(0.53209)$$
$$\theta \approx 28°$$

73.

$P(t) = 7\,[1 - \cos(2\,\pi)](t + 10) + 100e^{0.2t}$

75. Solving $\dfrac{c_1}{c_2} = \dfrac{\sin\theta_1}{\sin\theta_2}$ for c_2 gives

$$c_2 = \frac{c_1 \sin\theta_2}{\sin\theta_1}.$$

$c_1 = 3 \cdot 10^8$, $\theta_1 = 46°$, and $\theta_2 = 31°$ so

$$c_2 = \frac{3 \cdot 10^8(\sin 31°)}{\sin 46°}$$

$$= 214,796,150$$

$$\approx 2.1 \times 10^8 \text{ m/sec}.$$

77. On the horizontal scale, one whole period clearly spans four squares, so $4 \cdot 30° = 120°$ is the period.

79. $T(x) = 37\sin\left[\dfrac{2\pi}{365}(x - 101)\right] + 25$

(a) $T(60) = 37\sin\left[\dfrac{2\pi}{365}(-41)\right] + 25$

$$\approx 1°C$$

(b) $T(91) = 37\sin\left[\dfrac{2\pi}{365}(-10)\right] + 25$

$$= 19°C$$

(c) $T(101) = 37\sin\left[\dfrac{2\pi}{365}(0)\right] + 25$

$$= 25°C$$

(d) $T(150) = 37\sin\left[\dfrac{2\pi}{365}(49)\right] + 25$

$$\approx 53°C$$

(e) The maximum and minimum values of the sine function are 1 and -1, respectively. Thus, the maximum value of T is

$$37(1) + 25 = 62°C$$

and the minimum value of T is

$$37(-1) + 25 = -12°C.$$

(f) The period is $\dfrac{2\pi}{\frac{2\pi}{365}} = 365$.

81. Let $h = $ the height of the building.

$$\tan 37.4° = \frac{h}{48}$$

$$h = 48\tan 37.4°$$

$$\approx 48(0.7646)$$

$$\approx 36.7$$

The height of the building is approximately 36.7 m.

Chapter 2 Review Exercises

3. $y = \dfrac{1}{e^x - 1}$

$e^x - 1 \neq 0$

$e^x \neq 1$

$x \neq 0$

Domain: $(-\infty, 0) \cup (0, \infty)$

5. $y = \tan x$

$= \dfrac{\sin x}{\cos x}$

$\cos x \neq 0$

$x \neq \pm\dfrac{\pi}{2}, \pm\dfrac{3\pi}{2}, \pm\dfrac{5\pi}{2}, \cdots$

Domain: $\left\{ x \mid x \neq \pm\dfrac{\pi}{2}, \pm\dfrac{3\pi}{2}, \pm\dfrac{5\pi}{2}, \cdots \right\}$

7. $y = 5^{-x} + 1$

x	-2	-1	0	1	2
y	26	6	2	$\frac{6}{5}$	$\frac{26}{5}$

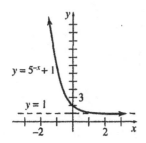

9. $y = \left(\dfrac{1}{2}\right)^{x-1}$

x	-2	-1	0	1	2
y	8	4	2	1	$\frac{1}{2}$

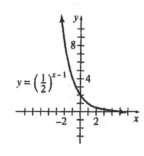

11. $y = 1 + \log_3 x$

$y - 1 = \log_3 x$

$3^{y-1} = x$

x	$\frac{1}{9}$	$\frac{1}{3}$	1	3	9
y	-1	0	1	2	3

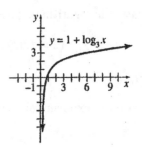

13. $y = 2 - \ln x^2$

x	-4	-3	-2	-1	1	2	3	4
y	-0.8	-0.2	0.6	2	2	0.6	-0.2	-0.8

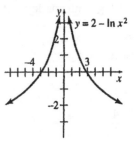

15. $\left(\dfrac{9}{16}\right)^x = \dfrac{3}{4}$

$\left(\dfrac{3}{4}\right)^{2x} = \left(\dfrac{3}{4}\right)^1$

$2x = 1$

$x = \dfrac{1}{2}$

17. $\left(\dfrac{1}{2}\right) = \left(\dfrac{b}{4}\right)^{1/4}$

$\left(\dfrac{1}{2}\right)^4 = \dfrac{b}{4}$

$4\left(\dfrac{1}{2}\right)^4 = b$

$4\left(\dfrac{1}{16}\right) = b$

$\dfrac{1}{4} = b$

19. $3^{1/2} = \sqrt{3}$

The equation in logarithmic form is

$$\log_3 \sqrt{3} = \frac{1}{2}.$$

21. $10^{1.07918} = 12$

The equation in logarithmic form is

$$\log_{10} 12 = 1.07918.$$

23. $\log_{10} 100 = 2$

The equation in exponential form is

$$10^2 = 100.$$

25. $\log 15.46 = 1.18921$

The equation in exponential form is

$$10^{1.18921} = 15.46.$$

Recall that $\log x$ means $\log_{10} x$.

27. $\log_{32} 16 = x$

$$32^x = 16$$
$$2^{5x} = 2^4$$
$$5x = 4$$
$$x = \frac{4}{5}$$

29. $\log_{100} 1{,}000 = x$

$$100^x = 1{,}000$$
$$(10^2)^x = 10^3$$
$$2x = 3$$
$$x = \frac{3}{2}$$

31. $\log_3 2y^3 - \log_3 8y^2$

$$= \log_3 \frac{2y^3}{8y^2}$$

$$= \log_3 \frac{y}{4}$$

33. $5 \log_4 r - 3 \log_4 r^2$

$$= \log_4 r^5 - \log_4 (r^2)^3$$

$$= \log_4 \frac{r^5}{r^6}$$

$$= \log_4 \frac{1}{r}$$

$$= \log_4 r^{-1}$$

$$= -\log_4 r$$

35. $\quad 3^{z-2} = 11$

$$\ln 3^{z-2} = \ln 11$$
$$(z-2) \ln 3 = \ln 11$$
$$z - 2 = \frac{\ln 11}{\ln 3}$$
$$z = \frac{\ln 11}{\ln 3} + 2$$
$$\approx 4.183$$

37. $\quad 15^{-k} = 9$

$$\ln 15^{-k} = \ln 9$$
$$-k \ln 15 = \ln 9$$
$$k = -\frac{\ln 9}{\ln 15}$$
$$\approx -0.811$$

39. $\quad e^{3x-1} = 12$

$$\ln (e^{3x-1}) = \ln 12$$
$$3x - 1 = \ln 12$$
$$3x = 1 + \ln 12$$
$$x = \frac{1 + \ln 12}{3}$$
$$\approx 1.162$$

41. $\left(1 + \dfrac{2p}{5}\right)^2 = 3$

$$1 + \frac{2p}{5} = \pm\sqrt{3}$$
$$5 + 2p = \pm 5\sqrt{3}$$
$$2p = -5 \pm 5\sqrt{3}$$
$$p = \frac{-5 \pm 5\sqrt{3}}{2}$$
$$p = \frac{-5 + 5\sqrt{3}}{2}$$
$$\approx 1.830$$
$$\text{or} \quad p = \frac{-5 - 5\sqrt{3}}{2}$$
$$\approx 6.830$$

43. $\log_3(2x + 5) = 2$

$$3^2 = 2x + 5$$
$$9 = 2x + 5$$
$$4 = 2x$$
$$x = 2$$

45. $\log_2(5m - 2) - \log_2(m + 3) = 2$

$$\left(\log_2 \frac{5m - 2}{m + 3}\right) = 2$$
$$\frac{5m - 2}{m + 3} = 2^2$$
$$5m - 2 = 4(m + 3)$$
$$5m - 2 = 4m + 12$$
$$m = 14$$

47. $f(x) = \log_a x;\ a > 0,\ a \neq 1$

(a) The domain is $(0, \infty)$.

(b) The range is $(-\infty, \infty)$.

(c) The x-intercept is 1.

(d) There are no discontinues..

(e) The y-axis, $x = 0$, is a vertical asymptote.

(f) f is increasing if $a > 1$.

(g) f is decreasing if $0 < a < 1$.

53. $90° = 90\left(\dfrac{\pi}{180}\right) = \dfrac{90\pi}{180} = \dfrac{\pi}{2}$

55. $210° = 210\left(\dfrac{\pi}{180}\right) = \dfrac{210\pi}{180} = \dfrac{7\pi}{6}$

57. $360° = 2\pi$

59. $7\pi = 7\pi\left(\dfrac{180}{\pi}\right)^° = 1{,}260°$

61. $\dfrac{9\pi}{20} = \dfrac{9\pi}{20}\left(\dfrac{180}{\pi}\right)^° = 81°$

63. $\dfrac{13\pi}{20} = \dfrac{13\pi}{20}\left(\dfrac{180}{\pi}\right)^° = 117°$

65. When an angle of 60° is drawn in standard position, one choice of a point on its terminal side is $(x, y) = (1, \sqrt{3})$. Then

$$r = \sqrt{x^2 + y^2} = \sqrt{1 + 3} = 2,$$

so

$$\sin 60° = \frac{y}{r} = \frac{\sqrt{3}}{2}.$$

67. When an angle of $-30°$ is drawn in standard position, one choice of a point on its terminal side is $(x, y) = (\sqrt{3}, -1)$. Then

$$r = \sqrt{x^2 + y^2} = \sqrt{3 + 1} = 2,$$

so

$$\cos(-30°) = \frac{x}{r} = \frac{\sqrt{3}}{2}.$$

69. When an angle of 120° is drawn in standard position, one choice of a point on its terminal side is $(x, y) = (-1, \sqrt{3})$. Then

$$r = \sqrt{x^2 + y^2} = \sqrt{1 + 3} = 2,$$

so

$$\csc 120° = \frac{r}{y} = \frac{2}{\sqrt{3}} = \frac{2\sqrt{3}}{3}.$$

71. When an angle of $\frac{\pi}{6}$ is drawn in standard position, one choice of a point on its terminal side is $(x, y) = (\sqrt{3}, 1)$. Then

$$r = \sqrt{x^2 + y^2} = \sqrt{3 + 1} = 2,$$

so

$$\sin \frac{\pi}{6} = \frac{y}{r} = \frac{1}{2}.$$

73. When an angle of $\frac{5\pi}{4}$ is drawn in standard position, one choice of a point on its terminal side is $(x, y) = (-1, -1)$. Then

$$r = \sqrt{x^2 + y^2} = \sqrt{1 + 1} = \sqrt{2},$$

so

$$\sec \frac{5\pi}{4} = \frac{r}{x} = \frac{\sqrt{2}}{-1} = -\sqrt{2}.$$

75. $\sin 47° \approx 0.7314$

77. $\tan 81° = 6.314$

79. $\sin 1.4661 = 0.9945$

81. $\cos .5934 = 0.8290$

83. The graph of $y = \cos x$ appears in Figure 14 in Section 13.1 in the textbook. To get $y = 3\cos x$, each value of y in $y = \cos x$ must be multiplied by 3. This gives a graph going through $(0, 3)$, $(\pi, -3)$ and $(2\pi, 3)$.

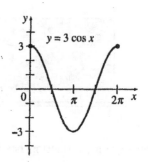

85. The graph of $y = \tan x$ appears in Figure 15 in Section 13.1. The difference between the graph of $y = \tan x$ and $y = -\tan x$ is that the y-values of points on the graph of $y = -\tan x$ are the opposites of the y-values of the corresponding points on the graph of $y = \tan x$.

A sample calculation:

When $x = \frac{\pi}{4}$,

$$y = -\tan \frac{\pi}{4} = -1.$$

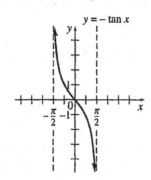

87. (a) Let $(x_1, y_1) = (66, 1.1)$ and $(x_2, y_2) = (97, 2.1)$. Find the slope, m.

$$m = \frac{2.1 - 1.1}{97 - 66} \approx -0.0323$$

Find the equation of the line using the point $(66, 1.1)$.

$$y = mx + b$$
$$1.1 = 0.0323(66) + b$$
$$-1.03 \approx b$$

The equation is $y = 0.0323x - 1.03$.

(b) Per capita grain production

$$= \frac{\text{world grain production}}{\text{population}}$$

$$= \frac{0.0323x - 1.03}{\left(\dfrac{9.803}{1 + 27.28e^{-0.03391x}} + 0.99\right)}$$

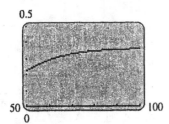

Per capita grain production has been increasing, but at a slower rate, and is leveling off at around 0.36 ton per person.

89. (a) The volume of a sphere is $\frac{4}{3}\pi r^3$.

The radius of a cancer cell is $\frac{2(10)^{-5}}{2}$ meters or 10^{-5} meters.

$$V = \frac{4}{3}\pi(10^{-5})^3$$
$$\approx 4.19 \times 10^{-15} \text{ cubic meters.}$$

(b) The formula for the total volulme of the cancer cells after t days is $V = (4.19 \times 10^{-15})2^t$ cubic meters.

(c) The radius of a cancer cell with a diameter of 1 centimeter is $\frac{10^{-2}\text{m}}{2}$ or 0.005 m.

$$V = \frac{4}{3}(\pi)(0.005)^3$$

Set the volume equal to the formula found in part (b) and solve for t.

$$\frac{4}{3}\pi(0.005)^3 = (4.19 \times 10^{-15})2^t$$
$$124{,}963{,}908.3 = 2^t$$
$$\ln 124{,}963{,}908.3 = t\ln 2$$
$$27 \approx t$$

In about 27 days, the cancer cell will grow to a tumor with a diameter of 1 cm.

91. $y = 17{,}000,\ y_0 = 15{,}000,\ t = 4$

(a)
$$y = y_0 e^{kt}$$
$$17{,}000 = 15{,}000e^{4k}$$
$$\frac{17}{15} = e^{4k}$$
$$\ln\left(\frac{17}{15}\right) = 4k$$
$$\frac{0.125}{4} = k$$
$$0.0313 = k$$

So, $y = 15{,}000e^{0.0313t}$.

(b)
$$45{,}000 = 15{,}000e^{0.0313t}$$
$$3 = e^{0.0313t}$$
$$\ln 3 = 0.0313t$$
$$\frac{\ln 3}{0.0313} = k$$
$$35 = k$$

It would take about 35 years.

93. Graph

$$y = c(t) = e^{-t} - e^{-2t}$$

on a graphing calculator and locate the maximum point. A calculator shows that the x-coordinate of the maximum point is about 0.69, and the y-coordinate is exactly 0.25. Thus, the maximum concentration of 0.25 occurs at about 0.69 minutes.

95. $y = \dfrac{7x}{100 - x}$

(a) $y = \dfrac{7(80)}{100 - 80} = \dfrac{560}{20} = 28$

The cost is $28,000.

(b) $y = \dfrac{7(50)}{100 - 50} = \dfrac{350}{50} = 7$

The cost is $7,000.

(c) $\dfrac{7(90)}{100 - 90} = \dfrac{630}{10} = 63$

The cost is $63,000.

(d) Plot the points $(80, 28)$, $(50, 7)$, and $(90, 63)$.

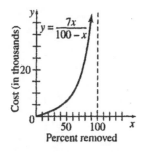

(e) No, because all of the pollutant would be removed when $x = 100$, at which point the denominator of the function would be zero.

97. $P(t) = 90 + 15 \sin 144\pi t$

The maximum possible value of $\sin \alpha$, is 1, while the minimum possible value is -1. Replacing α with $144\pi t$ gives

$$-1 \leq \sin 144\pi t \leq 1,$$
$$90 + 15(-1) \leq P(t) \leq 90 + 15(1)$$
$$75 \leq P(t) \leq 105.$$

Therefore, the minimum value of $P(t)$ is 75 and the maximum value of $P(t)$ is 105.

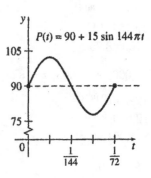

99. $P = \$2,781.36$, $r = 8\%$, $t = 6$, $m = 4$

$$A = P\left(1 + \frac{r}{m}\right)^{tm}$$

$$A = 2,781.36\left(1 + \frac{0.08}{4}\right)^{(6)(4)}$$

$$= 2,781.36(1.02)^{24}$$
$$= \$4,473.64$$

$$\text{Interest} = \$4,473.64 - \$2,781.36$$
$$= \$1,692.28$$

101. $2,100 deposited at 4% compounded quarterly.

$$A = P\left(1 + \frac{r}{m}\right)^{tm}$$

to double:

$$2(2,100) = 2,100\left(1 + \frac{0.04}{4}\right)^{t \cdot 4}$$

$$2 = 1.01^{4t}$$
$$\ln 2 = 4t \ln 1.01$$

$$t = \frac{\ln 2}{4 \ln 1.01}$$

$$\approx 17.5 \text{ years}$$

to triple;

$$3(2,100) = 2,100\left(1 + \frac{0.04}{4}\right)^{t \cdot 4}$$

$$3 = 1.04^{4t}$$
$$\ln 3 = 4t \ln 1.01$$

$$t = \frac{\ln 3}{4 \ln 1.01}$$

$$\approx 27.6 \text{ years}$$

103. $P = \$12,104$, $r = 8\%$, $t = 4$

$A = Pe^{rt}$

$A = 12,104e^{0.08(4)}$

$\quad = 12,104e^{0.32}$

$\quad = \$16,668.75$

105. $P = \$12,000$, $r = 0.05$, $t = 8$

$A = 12,000e^{0.05(8)}$

$\quad = 12,000e^{0.40}$

$\quad = \$17,901.90$

107. (a)

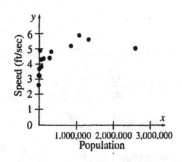

(b) Using a graphing calculator, $r = 0.63$.

(c)

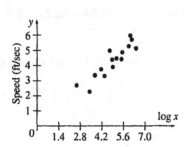

Yes, the data are now more linear than in part (a).

(d) Using a graphing calculator, $r = 0.91$. r is closer to 1.

(e) Using a graphing calculator,

$Y = 0.873 \log x - 0.0255$.

Chapter 2 Test

[2.1]

1. Solve the equation $8^{2y-1} = 4^{y+1}$.

2. Graph the function $f(x) = 4^{x-1}$.

3. \$315 is deposited in an account paying 6% compounded quarterly for 3 years. Find the following.

   (a) The amount in the account after 3 years.

   (b) The amount of interest earned by this deposit.

4. **California's Population** The population of the state of California from 1900 to 2000 can be approximated by the exponential function

$$A(t) = 1.854e^{0.033t}$$

   where t is the number of years since 1900.

   (a) The state's population was about 6,907,000 in 1940. How closely does this function approximate this value?

   (b) Use the function to estimate California's population in 1990. (The actual 1990 population was about 29,126,000.)

   (c) Predict the state's population in the year 2010 using this function.

[2.2]

5. Evaluate each logarithm without using a calculator.

   (a) $\log_8 16$ (b) $\ln e^{-3/4}$

6. Use properties of logarithms to simplify the following.

   (a) $\log_2 3k + \log_2 4k^2$ (b) $4\log_3 r - 3\log_3 m$

7. Solve each equation. Round to the nearest thousandth if necessary.

   (a) $2^{x+1} = 10$ (b) $\log_2(x+3) + \log_2(x-3) = 4$

8. **Infant Length** A formula used to approximate the length of a male infant based on his age is

$$L = 42.372 + 14.231 \ln t, \ 0 \le t \le 36^*$$

   where L is the length in centimeters and t is the child's age in months.

   (a) Find the length of a 12-month-old child using this formula.

   (b) Find the length of a 24-month-old child using this formula.

   (c) Using a graphing calculator, find the age of a child whose length is 82 cm.

---

*U.S. Department of Health and Human Services, Centers for Disease Control and Prevention, National Center for Health Statistics (http://www.cdc.gov/growthcharts).

[2.3]

9. Find the interest rate needed for \$5,000 to grow to \$10,000 in 8 years with continuous compounding.

10. How long will it take for \$1 to triple at an average rate of 7% compounded continuously?

11. Suppose sales of a certain item are given by $S(x) = 2,500 - 1,500e^{-2x}$, where x represents the number of years that the item has been on the market and $S(x)$ represents sales in thousands. Find the limit on sales.

12. Find the effective rate for the account described in Problem 3.

13. The population of Smalltown has grown exponentially from 14,000 in 1994 to 16,500 in 1997. At this rate, in what year will the population reach 17,400?

14. Potassium 42 decays exponentially. A sample which contained 1,000 grams 5 hours ago has decreased to 758 grams at present.

     (a) Write an exponential equation to express the amount, y, present after t hours.

     (b) What is the half-life of potassium 42?

15. **Half-Life** Among the transuranic elements – those that follow uranium in the periodic table of elements – perhaps the most useful is curium. Fourteen isotopes have been identified.

     (a) Curium 242 has a relatively short half-life but has a high power output and is useful for pacemakers and satellite power sources. It decays according to

$$A(t) = A_0 e^{-1.55405t}$$

     where t is the time in years. What is the half-life of curium 242?

     (b) The most stable isotope is curium 247 with decay equation

$$A(t) = A_0 e^{-0.000000044433t}$$

     where t is the time in years. What is the half-life of curium 247?

16. **Deforestation** An ecological organization has made a study of deforestation in a small country. It estimates that the disappearance of the nation's forests is progressing according to the formula

$$p(t) = \frac{27.8}{1 + 9e^{-0.25t}}$$

where p is the percentage deforested after t years.

     (a) Find the percent deforested at the beginning of the study ($t = 0$).

     (b) Find the percent deforested after 10, 20, and 30 years.

     (c) Does there appear to be a limit to this process? If so, what is it?

17. Find the present value of $15,000 at 6% compounded quarterly for 4 years.

18. Mr. Jones needs $20,000 for a down payment on a house in 5 years. How much must he deposit now at 5.8% compounded quarterly in order to have $20,000 in 5 years?

[2.4]

19. Convert 72° to radian measure. Express your answer as a multiple of π.

20. Convert $\frac{7\pi}{20}$ to degree measure.

21. Evaluate $\tan \frac{11\pi}{6}$ without using a calculator.

22. Use a calculator to find the value of $\sin 2.986$.

23. Graph $y = -4 \sin x$ over a one-period interval.

24. Graph $y = \frac{1}{2} \tan x$ over a one-period interval.

25. **Respiration Cycle** A normal adult at rest breathes in and out about a dozen times per minute for a respiration cycle of about 5 seconds. The volume v (in liters of air) in the lungs can be approximated by the function

$$v(t) = 0.48 - 0.32 \cos \frac{2\pi}{5} t, \ 0 \le t \le 5.$$

   (a) What is the period of this function? How does this compare to the respiration cycle?

   (b) Find and interpret $v(0)$, $v(2)$, and $v(4)$.

   (c) Consider the first respiration cycle. What is the maximum volume and when does it occur?

Chapter 2 Test Answers

1. $\frac{5}{4}$

2.

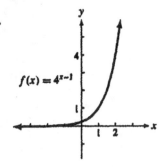

$f(x) = 4^{x-1}$

3. (a) $376.62   (b) $61.62

4. (a) 6.940 million; very closely
   (b) 36.138 million
   (c) 69.920 million

5. (a) $\frac{4}{3}$ (b) $-\frac{3}{4}$

6. (a) $\log_2 12k^3$ (b) $\log_3 \frac{r^4}{m^5}$

7. (a) 2.322   (b) 5

8. (a) 77.7 cm   (b) 87.6 cm   (c) 16 months

9. 8.66%

10. About 15.7 years

11. 2,500,000

12. 6.14%

13. 1998

14. (a) $y = 1,000e^{-0.0554t}$ (b) About 12.5 hours

15. (a) About 0.446 years or 163 days
    (b) About 15,600,000 years

16. (a) 2.8%   (b) 16%, 26.2%, 27.7%   (c) 27.8%

17. $11,820.47

18. $14,996.47

19. $\frac{2\pi}{5}$

20. 63°

21. $-\frac{\sqrt{3}}{3}$

22. 0.1550

23.

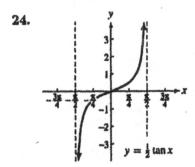

$y = -4 \sin x$

24.

$y = \frac{1}{2} \tan x$

25. (a) 10; two respiration cycles
    (b) 0.16, 0.74, 0.38; the volume of air in the
        lungs after 0, 2, and 4 seconds, respectively
    (c) 0.8 liters of air, after 2.5 seconds

Chapter 3

THE DERIVATIVE

3.1 Limits

1. Since $\lim\limits_{x \to 2^-} f(x)$ does not equal $\lim\limits_{x \to 2^+} f(x)$, $\lim\limits_{x \to 2} f(x)$ does not exist. The answer is C.

3. Since $\lim\limits_{x \to 4^-} f(x) = \lim\limits_{x \to 4^+} f(x) = 6$,

   $\lim\limits_{x \to 4} f(x) = 6$. The answer is b.

5. By reading the graph, as x gets closer to 3 from the left or the right, $f(x)$ gets closer to 3.

   $$\lim_{x \to 3} f(x) = 3$$

7. By reading the graph, as x approaches 0 from the left or the right, $f(x)$ approaches 0.

   $$\lim_{x \to 0} f(x) = 0$$

9. (a) (i) By reading the graph, as x gets closer to -2 from the left, $f(x)$ gets closer to -1.
   $$\lim_{x \to -2^-} f(x) = -1$$

   (ii) By reading the graph, as x gets closer to -2 from the right, $f(x)$ gets closer to $-\frac{1}{2}$.
   $$\lim_{x \to -2^+} f(x) = -\frac{1}{2}$$

   (iii) Since $\lim\limits_{x \to -2^-} f(x) = -1$ and $\lim\limits_{x \to -2^+} f(x)$
   $= -\frac{1}{2}$, $\lim\limits_{x \to -2} f(x)$ does not exist.

   (iv) $f(-2)$ does not exist since there is no point on the graph with an x-coordinate of -2.

(b) (i) By reading the graph, as x gets closer to -1 from the left, $f(x)$ gets closer to $-\frac{1}{2}$.
   $$\lim_{x \to -1^-} f(x) = -\frac{1}{2}$$

   (ii) By reading the graph, as x gets closer to -1 from the right, $f(x)$ gets closer to $-\frac{1}{2}$.
   $$\lim_{x \to -1^+} f(x) = -\frac{1}{2}$$

   (iii) Since $\lim\limits_{x \to -1^-} f(x) = -\frac{1}{2}$ and
   $$\lim_{x \to -1^+} f(x) = -\frac{1}{2}, \lim_{x \to -1} f(x) = -\frac{1}{2}.$$

   (iv) $f(-1) = -\frac{1}{2}$ since $\left(-1, -\frac{1}{2}\right)$ is a point of the graph.

11. By reading the graph, as x moves further to the right, $f(x)$ gets closer to 3. Therefore, $\lim\limits_{x \to \infty} f(x) = 3$.

13. $\lim\limits_{x \to 2} F(x)$ in Exercise 6 exists because $\lim\limits_{x \to 2^-} F(x) = 4$ and $\lim\limits_{x \to 2^+} F(x) = 4$.
   $\lim\limits_{x \to -2} f(x)$ in Exercise 9 does not exist since $\lim\limits_{x \to -2^-} f(x) = -1$, but $\lim\limits_{x \to -2^+} f(x) = -\frac{1}{2}$.

15. From the table, as x approaches 1 from the left or the right, $f(x)$ approaches 2.

   $$\lim_{x \to 1} f(x) = 2$$

17. $k(x) = \dfrac{x^3 - 2x - 4}{x - 2}$; find $\lim\limits_{x \to 2} k(x)$.

x	1.9	1.99	1.999
$k(x)$	9.41	9.9401	9.9941

x	2.001	2.01	2.1
$k(x)$	10.006	10.0601	10.61

   As x approaches 2 from the left or the right, $k(x)$ approaches 10.

   $$\lim_{x \to 2} k(x) = 10$$

19. $h(x) = \dfrac{\sqrt{x} - 2}{x - 1}$; find $\lim\limits_{x \to 1} h(x)$.

x	0.9	0.99	0.999
$h(x)$	10.51317	100.50126	1000.50013

x	1.001	1.01	1.1
$h(x)$	-999.50012	-99.50124	-9.51191

$\lim\limits_{x \to 1^-} h(x) = \infty$

$\lim\limits_{x \to 1^+} h(x) = -\infty$

Thus, $\lim\limits_{x \to 1} h(x)$ does not exist.

21. $\lim\limits_{x \to 4} \, [f(x) - g(x)] = \lim\limits_{x \to 4} f(x) - \lim\limits_{x \to 4} g(x)$
$$= 16 - 8 = 8$$

23. $\lim\limits_{x \to 4} \dfrac{f(x)}{g(x)} = \dfrac{\lim\limits_{x \to 4} f(x)}{\lim\limits_{x \to 4} g(x)} = \dfrac{16}{8} = 2$

25. $\lim\limits_{x \to 4} \sqrt{f(x)} = \lim\limits_{x \to 4} \, [f(x)]^{1/2}$
$$= [\lim\limits_{x \to 4} f(x)]^{1/2}$$
$$= 16^{1/2} = 4$$

27. $\lim\limits_{x \to 4} 2^{g(x)} = 2^{\lim\limits_{x \to 4} g(x)}$
$$= 2^8$$
$$= 256$$

29. $\lim\limits_{x \to 4} \dfrac{f(x) + g(x)}{2g(x)}$

$$= \dfrac{\lim\limits_{x \to 4} \, [f(x) + g(x)]}{\lim\limits_{x \to 4} 2g(x)}$$

$$= \dfrac{\lim\limits_{x \to 4} f(x) + \lim\limits_{x \to 4} g(x)}{2 \lim\limits_{x \to 4} g(x)}$$

$$= \dfrac{16 + 8}{2(8)} = \dfrac{24}{16} = \dfrac{3}{2}$$

31. $\lim\limits_{x \to 4} \left[\sin\left(\dfrac{\pi}{16} \cdot g(x) \right) \right] = \sin\left(\dfrac{\pi}{16} \cdot \lim\limits_{x \to 4} g(x) \right)$

$$= \sin\left(\dfrac{\pi}{16} \cdot 8 \right)$$

$$= \sin\left(\dfrac{\pi}{2} \right)$$

$$= 1$$

33. $\lim\limits_{x \to 3} \dfrac{x^2 - 9}{x - 3} = \lim\limits_{x \to 3} \dfrac{(x - 3)(x + 3)}{x - 3}$

$$= \lim\limits_{x \to 3} \, (x + 3)$$

$$= \lim\limits_{x \to 3} x + \lim\limits_{x \to 3} 3$$

$$= 3 + 3 = 6$$

35. $\lim\limits_{x \to 1} \dfrac{5x^2 - 7x + 2}{x^2 - 1} = \lim\limits_{x \to 1} \dfrac{(5x - 2)(x - 1)}{(x + 1)(x - 1)}$

$$= \lim\limits_{x \to 1} \dfrac{5x - 2}{x + 1}$$

$$= \dfrac{5 - 2}{2}$$

$$= \dfrac{3}{2}$$

37. $\lim\limits_{x \to -2} \dfrac{x^2 - x - 6}{x + 2} = \lim\limits_{x \to -2} \dfrac{(x - 3)(x + 2)}{x + 2}$

$$= \lim\limits_{x \to -2} \, (x - 3)$$

$$= \lim\limits_{x \to -2} x + \lim\limits_{x \to -2} \, (-3)$$

$$= -2 - 3 = -5$$

39. $\lim\limits_{x \to 0} \dfrac{\frac{1}{x+3} - \frac{1}{3}}{x}$

$$= \lim\limits_{x \to 0} \left(\dfrac{1}{x + 3} - \dfrac{1}{3} \right)\left(\dfrac{1}{x} \right)$$

$$= \lim\limits_{x \to 0} \left[\dfrac{3}{3(x + 3)} - \dfrac{x + 3}{3(x + 3)} \right]\left(\dfrac{1}{x} \right)$$

$$= \lim\limits_{x \to 0} \dfrac{3 - x - 3}{3(x + 3)(x)}$$

$$= \lim\limits_{x \to 0} \dfrac{-x}{3(x + 3)x}$$

$$= \lim\limits_{x \to 0} \dfrac{-1}{3(x + 3)}$$

$$= \dfrac{-1}{3(0 + 3)} = -\dfrac{1}{9}$$

41. $\lim\limits_{x \to 25} \dfrac{\sqrt{x} - 5}{x - 25}$

$$= \lim\limits_{x \to 25} \dfrac{\sqrt{x} - 5}{x - 25} \cdot \dfrac{\sqrt{x} + 5}{\sqrt{x} + 5}$$

$$= \lim\limits_{x \to 25} \dfrac{x - 25}{(x - 25)(\sqrt{x} + 5)}$$

$$= \lim\limits_{x \to 25} \dfrac{1}{\sqrt{x} + 5}$$

$$= \dfrac{1}{\sqrt{25} + 5}$$

$$= \dfrac{1}{10}$$

43. $\lim\limits_{h \to 0} \dfrac{(x+h)^2 - x^2}{h}$

$= \lim\limits_{h \to 0} \dfrac{x^2 + 2hx + h^2 - x^2}{h}$

$= \lim\limits_{h \to 0} \dfrac{2hx + h^2}{h}$

$= \lim\limits_{h \to 0} \dfrac{h(2x + h)}{h}$

$= \lim\limits_{h \to 0} (2x + h)$

$= 2x + 0 = 2x$

45. $\lim\limits_{x \to \infty} \dfrac{3x}{5x - 1} = \lim\limits_{x \to \infty} \dfrac{\frac{3x}{x}}{\frac{5x}{x} - \frac{1}{x}}$

$= \lim\limits_{x \to \infty} \dfrac{3}{5 - \frac{1}{x}}$

$= \dfrac{3}{5 - 0} = \dfrac{3}{5}$

47. $\lim\limits_{x \to \infty} \dfrac{x^2 + 2x}{2x^2 - 2x + 1}$

$= \lim\limits_{x \to \infty} \dfrac{\frac{x^2}{x^2} + \frac{2x}{x^2}}{\frac{2x^2}{x^2} - \frac{2x}{x^2} + \frac{1}{x^2}}$

$= \lim\limits_{x \to \infty} \dfrac{1 + \frac{2}{x}}{2 - \frac{2}{x} + \frac{1}{x^2}}$

$= \dfrac{1 + 0}{2 - 0 + 0} = \dfrac{1}{2}$

49. $\lim\limits_{x \to \infty} \dfrac{3x^3 + 2x - 1}{2x^4 - 3x^3 - 2}$

$= \lim\limits_{x \to \infty} \dfrac{\frac{3x^3}{x^4} + \frac{2x}{x^4} - \frac{1}{x^4}}{\frac{2x^4}{x^4} - \frac{3x^3}{x^4} - \frac{2}{x^4}}$

$= \lim\limits_{x \to \infty} \dfrac{\frac{3}{x} + \frac{2}{x^3} - \frac{1}{x^4}}{2 - \frac{3}{x} - \frac{2}{x^4}}$

$= \dfrac{0 + 0 - 0}{2 - 0 - 0} = 0$

51. $\lim\limits_{x \to \infty} \dfrac{2x^3 - x - 3}{6x^2 - x - 1}$

$= \lim\limits_{x \to \infty} \dfrac{\frac{2x^3}{x^3} - \frac{x}{x^3} - \frac{3}{x^3}}{\frac{6x^2}{x^3} - \frac{x}{x^3} - \frac{1}{x^3}}$

$= \lim\limits_{x \to \infty} \dfrac{2 - \frac{1}{x^2} - \frac{3}{x^3}}{\frac{6}{x} - \frac{1}{x^2} - \frac{1}{x^3}}$

$= \dfrac{2 - 0 - 0}{0 - 0 - 0} = \dfrac{2}{0}$

Therefore $\lim\limits_{x \to \infty} \dfrac{2x^3 - x - 3}{6x^2 - x - 1}$ does not exist.

53. $\lim\limits_{x \to 0} \dfrac{1 - \cos^2 x}{\sin^2 x} = \lim\limits_{x \to 0} \dfrac{\sin^2 x}{\sin^2 x}$

$= \lim\limits_{x \to 0} 1$

$= 1$

55. (a) $\lim\limits_{x \to -2} \dfrac{3x}{(x+2)^3}$ does not exist since

$\lim\limits_{x \to -2+} \dfrac{3x}{(x+2)^3} = -\infty$ and $\lim\limits_{x \to -2-} \dfrac{3x}{(x+2)^3} = \infty$.

(b) Since $(x + 2)^3 = 0$ when $x = -2$, $x = -2$ is the vertical asymptote of the graph of $F(x)$.

(c) The two answers are related. Since $x = -2$ is a vertical asymptote, we know that $\lim\limits_{x \to -2} F(x)$ does not exist.

59. (a) $\lim\limits_{x \to -\infty} e^x = 0$ since, as the graph goes further to the left, e^x gets closer to 0.

(b) The graph of e^x has a horizontal asymptote at $y = 0$ since $\lim\limits_{x \to -\infty} e^x = 0$.

61. (a) $\lim\limits_{x \to 0^+} \ln x = -\infty$ since, as the graph gets closer to $x = 0$, the value of $\ln x$ get smaller.

(b) The graph of $y = \ln x$ has a vertical asymptote at $x = 0$ since $\lim\limits_{x \to 0^+} \ln x = -\infty$.

65. $\lim\limits_{x \to 1} \dfrac{x^4 + 4x^3 - 9x^2 + 7x - 3}{x - 1}$

(a)

x	1.01	1.001	1.0001	0.99	0.999	0.9999
$f(x)$	5.0908	5.009	5.0009	4.9108	4.991	4.9991

As $x \to 1^-$ and as $x \to 1^+$, we see that $f(x) \to 5$.

(b) Graph

$$y = \dfrac{x^4 + 4x^3 - 9x^2 + 7x - 3}{x - 1}$$

on a graphing calculator. One suitable choice for the viewing window is $[-6, 6]$ by $[-10, 40]$ with Xscl $= 1$, Yscl $= 10$.

Because $x - 1 = 0$ when $x = 1$, we know that the function is undefined at this x-value. The graph does not show an asymptote at $x = 1$. This indicates that the rational expression that defines this function is not written in lowest terms, and that the graph should have an open circle to show a "hole" in the graph at $x = 1$. The graphing calculator doesn't show the hole, but if we try to find the value of the function at $x = 1$, we see that it is undefined. (Using the TABLE feature on a TI-83, we see that for $x = 1$, the y-value is listed as "ERROR.")

By viewing the function near $x = 1$ and using the ZOOM feature, we see that as x gets close to 1 from the left or the right, y gets close to 5, suggesting that

$$\lim_{x \to 1} \frac{x^4 + 4x^3 - 9x^2 + 7x - 3}{x - 1} = 5.$$

67. $\lim\limits_{x \to -1} \dfrac{x^{1/3} + 1}{x + 1}$

(a)

x	-1.01	-1.001	-1.0001	-0.99	-0.999	-0.9999
$f(x)$	0.33223	0.33322	.033332	0.33445	0.33344	0.33334

We see that as $x \to -1^-$ and as $x \to -1^+$,

$$f(x) \to 0.3333 \text{ or } \tfrac{1}{3}.$$

(b) Graph

$$y = \frac{x^{1/3} + 1}{x + 1}.$$

One suitable choice for the viewing window is $[-5, 5]$ by $[-2, 2]$.

Because $x + 1 = 0$ when $x = -1$, we know that the function is undefined at this x-value. The graph does not show an asymptote at $x = -1$. This indicates that the rational expression that defined this function is not written lowest terms, and that the graph should have an open circle to show a "hole" in the graph at $x = -1$. The graphing calculator doesn't show the hole, but if we try to find the value of the function at $x = -1$, we see that it is undefined. (Using the TABLE feature on a TI-83, we see that for $x = -1$, the y-value is listed as "ERROR.")

By viewing the function near $x = -1$ and using the ZOOM feature, we see that as x gets close to -1 from the left or right, y gets close to 0.3333, suggesting that

$$\lim_{x \to -1} \frac{x^{1/3} + 1}{x + 1} = 0.3333 \text{ or } \frac{1}{3}.$$

69. $\lim\limits_{x \to \infty} \dfrac{\sqrt{9x^2 + 5}}{2x}$

Graph the functions on a graphing calculator. A good choice for the viewing window is $[-10, 10]$ by $[-5, 5]$.

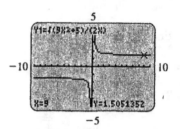

(a) The graph appears to have horizontal asymptotes at $y = \pm 1.5$. We see that as $x \to \infty$, $y \to 1.5$, so we determine that

$$\lim_{x \to \infty} \frac{\sqrt{9x^2 + 5}}{2x} = 1.5.$$

(b) As $x \to \infty$,

$$\sqrt{9x^2 + 5} \to \sqrt{9x^2} = 3|x|,$$

and

$$\frac{\sqrt{9x^2 + 5}}{2x} \to \frac{3|x|}{2x}.$$

Since $x > 0$, $|x| = x$, so

$$\frac{3|x|}{2x} = \frac{3x}{2x} = \frac{3}{2}.$$

Thus,

$$\lim_{x \to \infty} \frac{\sqrt{9x^2 + 5}}{2x} = \frac{3}{2} \text{ or } 1.5.$$

71. $\lim\limits_{x \to -\infty} \dfrac{\sqrt{36x^2 + 2x + 7}}{3x}$

Graph this function on a graphing calculator. A good choice for the viewing window is $[-10, 10]$ by $[-5, 5]$.

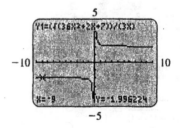

(a) The graph appears to have horizontal asymptotes at $y = \pm 2$. We see that as $x \to -\infty$, $y \to -2$, so we determine that

$$\lim_{x \to -\infty} \frac{\sqrt{36x^2 + 2x + 7}}{3x} = -2.$$

(b) As $x \to -\infty$,

$$\sqrt{36x^2 + 2x + 7} \to \sqrt{36x^2} = 6\,|x|$$

and

$$\frac{\sqrt{36x^2 + 2x + 7}}{3x} \to \frac{6\,|x|}{3x}.$$

Since $x < 0$, $|x| = -x$, so

$$\frac{6\,|x|}{3x} = \frac{6(-x)}{3x} = -2.$$

Thus,

$$\lim_{x \to -\infty} \frac{\sqrt{36x^2 + 2x + 7}}{3x} = -2.$$

73. $\displaystyle \lim_{x \to \infty} \frac{\left(1 + 5x^{1/3} + 2x^{5/3}\right)^3}{x^5}$

Graph this function on a graphing calculator. A good choice for the viewing window is $[-20, 20]$ by $[0, 20]$ with Xscl = 5, Yscl = 5.

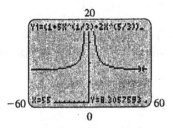

(a) The graph appears to have a horizontal asymptote at $y = 8$. We see that as $x \to \infty$, $y \to 8$, so we determine that

$$\lim_{x \to \infty} \frac{\left(1 + 5x^{1/3} + 2x^{5/3}\right)^3}{x^5} = 8.$$

(b) As $x \to \infty$, the highest power term dominates in the numerator, so

$$\left(1 + 5x^{1/3} + 2x^{5/3}\right)^3 \to \left(2x^{5/3}\right)^3 = 2^3 x^5$$
$$= 8x^5.$$

and

$$\frac{\left(1 + 5x^{1/3} + 2x^{5/3}\right)^3}{x^5} \to \frac{8x^5}{x^5} = 8.$$

Thus,

$$\lim_{x \to \infty} \frac{\left(1 + 5x^{1/3} + 2x^{5/3}\right)^3}{x^5} = 8.$$

77. (a) $\displaystyle \lim_{t \to 4} N(t) = 96$ patients because $N(t)$ is constant as t approaches 4 from the left or the right.

(b) $\displaystyle \lim_{t \to 6} T(x) = 92$ patients because as t approaches 6 from the right, $N(t)$ is constant at 92 patients.

(c) $\displaystyle \lim_{t \to 6^-} N(t) = 96$ patients because as t approaches 6 from the left, $N(t)$ is constant at 96 patients.

(d) $\displaystyle \lim_{t \to 6} N(t)$ does not exist because

$$\lim_{t \to 6^+} N(t) \neq \lim_{t \to 6^-} N(t).$$

(e) $N(6) = 92$ since $(6, 92)$ is a point on the graph.

79. $A(h) = \dfrac{0.17h}{h^2 + 2}$

$$\begin{aligned}
\lim_{x \to \infty} A(h) &= \lim_{x \to \infty} \frac{0.17h}{h^2 + 2} \\
&= \lim_{x \to \infty} \frac{\frac{0.17h}{h^2}}{\frac{h^2}{h^2} + \frac{2}{h^2}} \\
&= \lim_{x \to \infty} \frac{\frac{0.17}{h}}{1 + \frac{2}{h^2}} \\
&= \frac{0}{1 + 0} = 0
\end{aligned}$$

This means that the concentration of the drug in the bloodstream approaches 0 as the number of hours after injection increases.

81. (a)
$$\begin{aligned}
D(t) &= 155(1 - e^{-0.0133t}) \\
D(20) &= 155(1 - e^{-0.0133(20)}) \\
&= 155(1 - e^{-0.266}) \\
&\approx 36.2
\end{aligned}$$

The depth of the sediment layer deposited below the bottom of the lake in 1970 was 36.2 cm.

(b)
$$\begin{aligned}
\lim_{t \to \infty} D(t) &= \lim_{t \to \infty} 155(1 - e^{-0.0133t}) \\
&= 155 \lim_{t \to \infty} (1 - e^{-0.0133t}) \\
&= 155\left(\lim_{t \to \infty} 1 - \lim_{t \to \infty} e^{-0.0133t}\right) \\
&= 155(1) - 155 \lim_{t \to \infty} e^{-0.0133t} \\
&= 155 - 155 e^{\lim_{t \to \infty} -0.0133t}
\end{aligned}$$

$$\lim_{t \to \infty} -0.0133t = -\infty$$

As $t \to \infty$,

$$155 - 155 e^{\lim_{t \to \infty} -0.0133t} \to 155 - 155(0) = 155.$$

Thus,
$$\lim_{t \to \infty} D(t) = 155.$$

Going back in time (t is years before 1990), the depth of the sediment approaches 155 cm.

83.
$$
\begin{aligned}
\lim_{S \to \infty} I &= \lim_{S \to \infty} E[1 - e^{-a(S-h)/E}] \\
&= \lim_{S \to \infty} [E - Ee^{-a(S-h)/E}] \\
&= \lim_{S \to \infty} E - \lim_{S \to \infty} Ee^{-a(S-h)/E} \\
&= E - E \lim_{S \to \infty} e^{-a(S-h)/E} \\
&= E - Ee^{\lim_{S \to \infty} [-a(S-h)/E]} \\
&= E - E(0) \\
&= E
\end{aligned}
$$

3.2 Continuity

1. Discontinuous at $x = -1$

   (a) $\displaystyle \lim_{x \to -1^-} f(x) = \frac{1}{2}$

   (b) $\displaystyle \lim_{x \to -1^+} f(x) = \frac{1}{2}$

   (c) $\displaystyle \lim_{x \to -1} f(x) = \frac{1}{2}$ (since (a) and (b) have the same answers)

   (d) $f(-1)$ does not exist.

3. Discontinuous at $x = 1$

   (a) $\displaystyle \lim_{x \to 1^-} f(x) = -2$

   (b) $\displaystyle \lim_{x \to 1^+} f(x) = -2$

   (c) $\displaystyle \lim_{x \to 1} f(x) = -2$ (since (a) and (b) have the same answers)

   (d) $f(1) = 2$

5. Discontinuous at $x = -5$ and $x = 0$

   (a) $\displaystyle \lim_{x \to -5^-} f(x) = \infty$ $\displaystyle \lim_{x \to 0^-} f(x) = 0$

   (b) $\displaystyle \lim_{x \to -5^+} f(x) = -\infty$ $\displaystyle \lim_{x \to 0^+} f(x) = 0$

   (c) $\displaystyle \lim_{x \to -5} f(x)$ does not exist, since the answers to (a) and (b) are different.

   $\displaystyle \lim_{x \to 0} f(x) = 0$, since the answers to (a) and (b) are the same.

   (d) $f(-5)$ does not exist. $f(0)$ does not exist.

7. $f(x) = \dfrac{5 + x}{x(x - 2)}$

$f(x)$ is discontinuous at $x = 0$ and $x = 2$ since the denominator equals 0 at these two values.

$\displaystyle \lim_{x \to 0} f(x)$ does not exist since $\displaystyle \lim_{x \to 0^-} f(x) = \infty$ and $\displaystyle \lim_{x \to 0^+} f(x) = -\infty$.

$\displaystyle \lim_{x \to 2} f(x)$ does not exist since $\displaystyle \lim_{x \to 2^-} f(x) = -\infty$ and $\displaystyle \lim_{x \to 2^+} f(x) = \infty$.

9. $f(x) = \dfrac{x^2 - 4}{x - 2}$

$f(x)$ is discontinuous at $x = 2$ since the denominator equals zero at that value.

Since
$$\frac{x^2 - 4}{x - 2} = \frac{(x + 2)(x - 2)}{x - 2}$$
$$= x + 2,$$

$\displaystyle \lim_{x \to 2} f(x) = 2 + 2 = 4.$

11. $p(x) = x^2 - 4x + 11$

Since $p(x)$ is a polynomial function, it is continuous everywhere and thus discontinuous no where.

13. $p(x) = \dfrac{|x + 2|}{x + 2}$

$p(x)$ is discontinuous at $x = -2$ since the denominator is undefined at that value.

Since $\displaystyle \lim_{x \to -2^-} p(x) = -1$ and $\displaystyle \lim_{x \to -2^+} p(x) = 1$, $\displaystyle \lim_{x \to -2} p(x)$ does not exist.

15. $f(x) = \sin\left(\dfrac{x}{x + 2}\right)$

$f(x)$ is discontinuous at $x = -2$ since the denominator is undefined at that value.

$\displaystyle \lim_{x \to -2} \sin\left(\frac{x}{x + 2}\right)$ does not exist since

$\displaystyle \lim_{x \to -2^-} \sin\left(\frac{x}{x + 2}\right)$ does not exist and

$\displaystyle \lim_{x \to -2^+} \sin\left(\frac{x}{x + 2}\right)$ does not exist.

17. $k(x) = e^{\sqrt{x-1}}$

$k(x)$ is discontinuous where $x \leq 1$.

$\lim\limits_{x \to a} k(x)$ does not exist where $a \leq 1$ since $k(x)$ is not defined where $a < 1$ and $\lim\limits_{x \to a} k(x)$ does not exist where $a = 1$.

19. $r(x) = \ln\left(\dfrac{x}{x-1}\right)$

$r(x)$ is discontinuous where $0 \leq x \leq 1$ since the function is undefined at these values.

$\lim\limits_{x \to a} r(x)$ does not exist where $0 \leq a \leq 1$ since $r(x)$ is undefined at these values.

21. $f(x) = \begin{cases} 1 & \text{if} \quad x < 2 \\ x+3 & \text{if} \quad 2 \leq x \leq 4 \\ 7 & \text{if} \quad x > 4 \end{cases}$

(a)

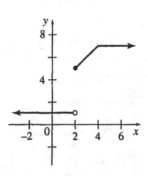

(b) $f(x)$ is discontinuous at $x = 2$.

(c) $\lim\limits_{x \to 2^-} f(x) = 1 \qquad\qquad \lim\limits_{x \to 2^+} f(x) = 5$

23. $g(x) = \begin{cases} 11 & \text{if} \quad x < -1 \\ x^2+2 & \text{if} \quad -1 \leq x \leq 3 \\ 11 & \text{if} \quad x > 3 \end{cases}$

(a)

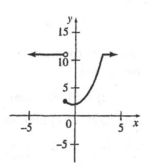

(b) $g(x)$ is discontinuous at $x = -1$.

(c) $\lim\limits_{x \to -1^-} g(x) = 11$

$\lim\limits_{x \to -1^+} g(x) = (-1)^2 + 2 = 3$

25. $h(x) = \begin{cases} 4x + 4 & \text{if} \quad x \leq 0 \\ x^2 - 4x + 4 & \text{if} \quad x > 0 \end{cases}$

(a)

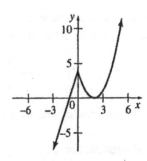

(b) There are no points of discontinuity.

29. $f(x) = \dfrac{x^2 + x + 2}{x^3 - 0.9x^2 + 4.14x - 5.4} = \dfrac{P(x)}{Q(x)}$

(a) Graph

$$Y_1 = \frac{P(x)}{Q(x)} = \frac{x^2 + x + 2}{x^3 - 0.9x^2 + 4.14x - 5.4}$$

on a graphing calculator. A good choice for the viewing window is $[-3, 3]$ by $[-10, 10]$.

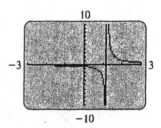

The graph has a vertical asymptote at $x = 1.2$, which indicates that f is discontinuous at $x = 1.2$.

(b) Graph

$$Y_2 = Q(x) = x^3 - 0.9x^2 + 4.14x - 5.4$$

using the same viewing window.

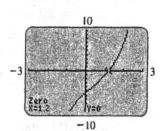

We see that this graph has one x-intercept, 1.2. This indicates that 1.2 is the only real solution of the equation $Q(x) = 0$.

This result verifies our answer from part (a) because a rational function of the form

$$f(x) = \frac{P(x)}{Q(x)}$$

will be discontinuous wherever $Q(x) = 0$.

31. (a) At diagnosis, $t = 40$

$$N(40) = 2^{40}$$
$$\approx 1.0995 \times 10^{12}$$

At diagnosis there are about 1.0995×10^{12} tumor cells.

(b) If 99.9% of the cells are killed instantaneously, 0.1% remain.

$$N(t) = 0.001 \cdot 2^{40} \approx 1.0995 \times 10^9$$
$$\log_2 1.0995 \times 10^9 \approx 30$$

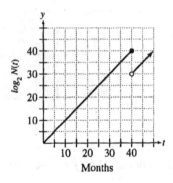

Months

(c) The function described in part (b) is discontinuous at $t = 40$ since $\lim_{t \to 40} N(t)$ does not exist.

$$\lim_{t \to 40^-} N(t) = \lim_{t \to 40^-} 2^t$$
$$= 2^{40} \text{ and}$$
$$\lim_{t \to 40^+} N(t) = \lim_{t \to 40^+} 0.99 \cdot 2^t$$
$$= 0.999 \cdot 2^{20}$$
$$\approx 1.0995 \times 10^{12}$$

Hence,

$$\lim_{t \to 40^-} N(t) \neq \lim_{t \to 40^+} N(t)$$

35. $W(t) = \begin{cases} 48 + 3.64t + 0.6363t^2 + 0.00963t^3, 1 \leq t \leq 28 \\ -1{,}004 + 65.8t, 28 < t \leq 56 \end{cases}$

(a) $W(25) = 48 + 3.64(25) + 0.6363(25)^2 + 0.00963(25)$
$$\approx 687.156$$

A male broiler at 25 days weighs about 687 grams.

(b) $W(t)$ is not a continuous function.
At $t = 28$

$$\lim_{t \to 28^-} W(t)$$
$$= \lim_{t \to 28^-} 48 + 3.64t + 0.6363t^2 + 0.00963t^3$$
$$= 48 + 3.64(28) + 0.6363(28)^2 + 0.00963(28)^3$$
$$\approx 860.18$$

and

$$\lim_{t \to 28^-} W(t) \neq \lim_{t \to 28^+} (-1004 + 65.8t)$$
$$= -1004 + 65.8(28)$$
$$= 838.4$$

so

$$\lim_{t \to 28^-} W(t) \neq \lim_{t \to 28^+} W(t)$$

Thus $W(t)$ is discontinuous.

(c)

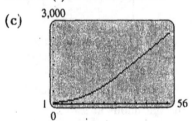

37. (a) $\lim_{x \to 6} P(x)$

As x approaches 6 from the left or the right, the value of $P(x)$ for the corresponding point on the graph approaches 500.

Thus, $\lim_{x \to 6} P(x) = \$500$.

(b) $\lim_{x \to -10} P(x) = \$1{,}500$

because as x approaches 10 from the left, $P(x)$ approaches $1,500.

(c) $\lim_{x \to 10^+} P(x) = \$1{,}000$ because as x approaches 10 from the right, $P(x)$ approaches $1,000.

(d) Since $\lim_{x \to 10^+} P(x) \neq \lim_{x \to 10^-} P(x)$,

$$\lim_{x \to 10} P(x) \text{ does not exist.}$$

(e) From the graph, the function is discontinuous at $x = 10$. This may be the result of a change of shifts.

(f) From the graph, the second shift will be as profitable as the first shift when 15 units are produced.

39. In dollars,

$$F(x) = \begin{cases} 1.20x & \text{if } 0 < x \le 100 \\ 1.00x & \text{if } x > 100. \end{cases}$$

(a) $F(80) = 1.20(80) = \$96$

(b) $F(150) = 1.00(150) = \$150$

(c) $F(100) = 1.20(100) = \$120$

(d) F is discontinuous at $x = 100$.

3.3   Rates of Change

1. $y = x^2 + 2x = f(x)$ between $x = 0$ and $x = 3$

Average rate of change

$$= \frac{f(3) - f(0)}{3 - 0}$$

$$= \frac{15 - 0}{3}$$

$$= 5$$

3. $y = 2x^3 - 4x^2 + 6x = f(x)$ between $x = -1$ and $x = 1$

Average rate of change

$$= \frac{f(1) - f(-1)}{1 - (-1)}$$

$$= \frac{4 - (-12)}{2} = \frac{16}{2}$$

$$= 8$$

5. $y = \sqrt{x} = f(x)$ between $x = 1$ and $x = 4$

Average rate of change

$$= \frac{f(4) - f(1)}{4 - 1}$$

$$= \frac{2 - 1}{3}$$

$$= \frac{1}{3}$$

7. $y = \dfrac{1}{x - 1} = f(x)$ between $x = -2$ and $x = 0$

Average rate of change

$$= \frac{f(0) - f(-2)}{0 - (-2)}$$

$$= \frac{-1 - \left(-\frac{1}{3}\right)}{2} = \frac{-\frac{2}{3}}{2}$$

$$= -\frac{1}{3}$$

9. $\lim\limits_{h \to 0} \dfrac{s(6 + h) - s(6)}{h}$

$$= \lim\limits_{h \to 0} \frac{(6+h)^2 + 5(6+h) + 2 - [6^2 + 5(6) + 2]}{h}$$

$$= \lim\limits_{h \to 0} \frac{h^2 + 17h + 68 - 68}{h} = \lim\limits_{h \to 0} \frac{h^2 + 17h}{h}$$

$$= \lim\limits_{h \to 0} \frac{h(h + 17)}{h} = \lim\limits_{h \to 0} (h + 17) = 17$$

The instantaneous velocity at $t = 6$ is 17.

11. $s(t) = t^3 + 2t + 9$

$$\lim\limits_{h \to 0} \frac{s(1 + h) - s(1)}{h}$$

$$= \lim\limits_{h \to 0} \frac{[(1+h)^3 + 2(1+h) + 9] - [(1)^3 + 2(1) + 9]}{h}$$

$$= \lim\limits_{h \to 0} \frac{[1 + 3h + 3h^2 + h^3 + 2 + 2h + 9] - [1 + 2 + 9]}{h}$$

$$= \lim\limits_{h \to 0} \frac{h^3 + 3h^2 + 5h + 12 - 12}{h}$$

$$= \lim\limits_{h \to 0} \frac{h^3 + 3h^2 + 5h}{h} = \lim\limits_{h \to 0} \frac{h(h^2 + 3h + 5)}{h}$$

$$= \lim\limits_{h \to 0} (h^2 + 3h + 5) = 5$$

The instantaneous velocity at $t = 1$ is 5.

13. $f(x) = x^2 + 2x$ at $x = 0$

$$\lim\limits_{h \to 0} \frac{f(0 + h) - f(0)}{h}$$

$$= \lim\limits_{h \to 0} \frac{(0 + h)^2 + 2(0 + h) - [0^2 + 2(0)]}{h}$$

$$= \lim\limits_{h \to 0} \frac{h^2 + 2h}{h} = \lim\limits_{h \to 0} \frac{h(h + 2)}{h}$$

$$= \lim\limits_{h \to 0} h + 2 = 2$$

The instantaneous rate of change at $x = 0$ is 2.

15. $g(t) = 1 - t^2$ at $t = -1$

$$\lim_{h \to 0} \frac{g(-1+h) - g(-1)}{h}$$

$$= \lim_{h \to 0} \frac{1 - (-1+h)^2 - [1 - (-1)^2]}{h}$$

$$= \lim_{h \to 0} \frac{1 - (1 - 2h + h^2) - 1 + 1}{h}$$

$$= \lim_{h \to 0} \frac{2h - h^2}{h} = \lim_{h \to 0} \frac{h(2 - h)}{h}$$

$$= \lim_{h \to 0} (2 - h) = 2$$

The instantaneous rate of change at $t = -1$ is 2.

17. $f(x) = x^x$ at $x = 2$

h	
0.01	$\dfrac{f(2 + 0.01) - f(2)}{0.01}$ $= \dfrac{2.01^{2.01} - 2^2}{0.01}$ $= 6.84$
0.001	$\dfrac{f(2 + 0.001) - f(2)}{0.001}$ $= \dfrac{2.001^{2.001} - 2^2}{0.001}$ $= 6.779$
0.0001	$\dfrac{f(2 + 0.0001) - f(2)}{0.0001}$ $= \dfrac{2.0001^{2.0001} - 2^2}{0.0001}$ $= 6.773$
0.00001	$\dfrac{f(2 + 0.00001) - f(2)}{0.00001}$ $= \dfrac{2.00001^{2.00001} - 2^2}{0.00001}$ $= 6.7727$
0.000001	$\dfrac{f(2 + 0.000001) - f(2)}{0.000001}$ $= \dfrac{2.000001^{2.000001} - 2^2}{0.000001}$ $= 6.7726$

The instantaneous rate of change at $x = 2$ is 6.7726.

19. $f(x) = x^{\ln x}$ at $x = 2$

h	
0.01	$\dfrac{f(2 + 0.01) - f(2)}{0.01}$ $= \dfrac{2.01^{\ln 2.01} - 2^{\ln 2}}{0.01}$ $= 1.1258$
0.001	$\dfrac{f(2 + 0.001) - f(2)}{0.001}$ $= \dfrac{2.001^{\ln 2.001} - 22^{\ln 2}}{0.001}$ $= 1.1212$
0.0001	$\dfrac{f(2 + 0.0001) - f(2)}{0.0001}$ $= \dfrac{2.0001^{\ln 2.0001} - 2^{\ln 2}}{0.0001}$ $= 1.1207$
0.00001	$\dfrac{f(2 + 0.00001) - f(2)}{0.00001}$ $= \dfrac{2.00001^{\ln 2.00001} - 2^{\ln 2}}{0.00001}$ $= 1.1207$

The instantaneous rate of change at $x = 2$ is 1.1207.

23. $L(t) = -0.01t^2 + 0.788t - 7.048$

(a) $\dfrac{L(28) - L(22)}{28 - 22} = \dfrac{7.176 - 5.448}{6}$

$$= 0.288$$

The average rate of growth during weeks 22 through 28 is 0.288 mm per week.

(b) $\displaystyle\lim_{h\to 0}\frac{L(t+h)-L(t)}{h}$

$\displaystyle=\lim_{h\to 0}\frac{L(22+h)-L(22)}{h}$

$\displaystyle=\lim_{h\to 0}\frac{[-0.01(22+h)^2+0.788(22+h)-7.048]-5.448}{h}$

$\displaystyle=\lim_{h\to 0}\frac{-0.01(h^2+44h+484)+17.336+0.788h-12.496}{h}$

$\displaystyle=\lim_{h\to 0}\frac{-0.01h^2+0.348h}{h}$

$\displaystyle=\lim_{h\to 0}(-0.01h+0.348)$

$=0.348$

The instantaneous rate of growth at exactly 22 weeks is 0.348 mm per week.

(c)

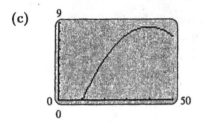

25. $p(t)=t^2+t$

(a) $p(1)=1^2+1=2$
$p(4)=4^2+4=20$

Average rate of change $=\dfrac{p(4)-p(1)}{4-1}$

$\qquad\qquad\qquad\quad=\dfrac{20-2}{3}$

$\qquad\qquad\qquad\quad=6$

The average rate of change is 6% per day.

(b) $\displaystyle\lim_{h\to 0}\frac{p(3+h)-p(3)}{h}$

$\displaystyle=\lim_{h\to 0}\frac{(3+h)^2+(3+h)-(3^2+3)}{h}$

$\displaystyle=\lim_{h\to 0}\frac{9+6h+h^2+3+h-12}{h}$

$\displaystyle=\lim_{h\to 0}\frac{h^2+7h}{h}$

$\displaystyle=\lim_{h\to 0}\frac{h(h+7)}{h}$

$\displaystyle=\lim_{h\to 0}(h+7)=7$

The instantaneous rate of change is 7% per day.

27. (a) $P(2)=5,\ P(1)=3$

Average rate of change $=\dfrac{P(2)-P(1)}{2-1}$

$\qquad\qquad\qquad\quad=\dfrac{5-3}{2-1}=\dfrac{2}{1}=2$

From 1 min to 2 min, the population of bacteria increases, on the average, 2 million per min.

(b) $P(3)=4.2,\ P(2)=5$

Average rate of change $=\dfrac{P(3)-P(2)}{3-2}$

$\qquad\qquad\qquad\quad=\dfrac{4.2-5}{3-2}=\dfrac{-0.8}{1}=-0.8$

From 2 min to 3 min, the population of bacteria decreases, on the average, -0.8 million or 800,000 per min.

(c) $P(3)=4.2,\ P(4)=2$

Average rate of change $=\dfrac{P(4)-P(3)}{4-3}$

$\qquad\qquad\qquad\quad=\dfrac{2-4.2}{4-3}=\dfrac{-2.2}{1}=-2.2$

From 3 min to 4 min, the population of bacteria decreases, on the average, -2.2 million per min.

(d) $P(4)=2,\ P(5)=1$

Average rate of change $=\dfrac{P(5)-P(4)}{5-4}$

$\qquad\qquad\qquad\quad=\dfrac{1-2}{5-4}=\dfrac{-1}{1}=-1$

From 4 min to 5 min, the population decreases, on the average, -1 million per min.

(e) The population increased up to 2 min after the bactericide was introduced, but decreased after 2 min.

(f) The rate of decrease of the population slows down at about 3 min.
The graph becomes less and less steep after that point.

29. Let $M(t)$ be marijuana production (in metric tons) for year t and $O(t)$ be opium production (in metric tons) for year t.

(a) $M(1994) = 13{,}500$
$M(1996) = 12{,}000$
$M(1998) = 9{,}500$

Average rate of change from 1994 – 1996

$$= \frac{M(1996) - M(1994)}{1996 - 1994}$$

$$= \frac{12{,}000 - 13{,}500}{2}$$

$$= -750$$

On average, the production decreased 750 metric tons per year.

Average rate of change from 1996 – 1998

$$= \frac{M(1998) - M(1996)}{1998 - 1996}$$

$$= \frac{9{,}500 - 12{,}000}{2}$$

$$= -1{,}250$$

On average, the production decreased 1,000 metric tons per year.

$$O(1994) = 3{,}600$$
$$O(1996) = 4{,}400$$
$$O(1998) = 3{,}600$$

Average rate of change from 1994 – 1996

$$= \frac{O(1996) - O(1994)}{1996 - 1994}$$

$$= \frac{4{,}400 - 3{,}600}{2}$$

$$= 400$$

On average, the production increased 400 metric tons per year.

Average rate of change from 1996 – 1998.

$$= \frac{O(1998) - O(1996)}{1998 - 1996}$$

$$= \frac{3{,}600 - 4{,}400}{2}$$

$$= -400$$

On average, the production decreased 400 metric tons per year.

(b) Marijuana had the greatest change in net production.

31. Let $B(t) =$ the amount in the Medicare Trust Fund for year t.

(a) $B(1994) = 152$
$B(1998) = 125$

Average change in fund

$$= \frac{125 - 152}{1998 - 1994}$$

$$= \frac{-27}{4}$$

$$= -6.75$$

On average, the amount in the fund decreased approximately $6.75 billion per year from 1994 – 1998.

(b) $B(1998) = 125$
$B(2010) = 300$

Average change in fund

$$= \frac{300 - 125}{2010 - 1998}$$

$$= \frac{175}{12}$$

$$\approx 14.58$$

On average, the amount in the fund increases approximately $14.58 billion per year from 1998 – 2010.

(c) $B(1990) = 125$
$B(1998) = 125$

Average change in fund

$$= \frac{125 - 125}{1998 - 1990}$$

$$= \frac{0}{8}$$

$$= 0$$

On average, the amount in the fund changes approximately $0 per year from 1990 to 1998.

33. (a) By looking at the graphs, we see that from 1991 to 1997 the curve representing the ratio of Internet messages to telephone calls is steeper than the curve representing the ratio of households with PCs to households without PCs; thus the curve representing the ratio of Internet messages to telephone calls has a greater rate of change over this time interval.

(b) The nonlinear curve, that is, the curve representing the ratio of Internet messages to telephone calls, has an increasing rate of change.

35. (a) $\dfrac{s(2)-s(0)}{2-0}=\dfrac{10-0}{2}=5$ ft/sec

(b) $\dfrac{s(4)-s(2)}{4-2}=\dfrac{14-10}{2}=2$ ft/sec

(c) $\dfrac{s(6)-s(4)}{6-4}=\dfrac{20-14}{2}=3$ ft/sec

(d) $\dfrac{s(8)-s(6)}{8-6}=\dfrac{30-20}{2}=5$ ft/sec

(e) (i) $\dfrac{f(x_0+h)-f(x_0-h)}{2h}$

$$=\dfrac{f(4+2)-f(4-2)}{(2)(2)}$$

$$=\dfrac{f(6)-f(2)}{4}$$

$$=\dfrac{20-10}{4}$$

$$=\dfrac{10}{4}=2.5 \text{ ft/sec}$$

(ii) $\dfrac{2+3}{2}=2.5$ ft/sec

(f) (i) $\dfrac{f(x_0+h)-f(x_0-h)}{2h}$

$$=\dfrac{f(6+2)-f(6-2)}{(2)(2)}$$

$$=\dfrac{f(8)-f(4)}{4}$$

$$=\dfrac{30-14}{4}$$

$$=\dfrac{16}{4}=4 \text{ ft/sec}$$

(ii) $\dfrac{3+5}{2}=4$ ft/sec

Definition of the Derivative

1. (a) $f(x)=5$ is a horizontal line and has slope 0; the derivative is 0.

(b) $f(x)=x$ has slope 1; the derivative is 1.

(c) $f(x)=-x$ has slope of -1; the derivative is -1.

(d) $x=3$ is vertical and has undefined slope; the derivative does not exist.

(e) $y=mx+b$ has slope m; the derivative is m.

3. $f(x)=\frac{x^2-1}{x+2}$ is not differentiable when $x+2=0$ or $x=-2$ because the function is undefined and a vertical asymptote occurs there.

5. Using the points $(5,3)$ and $(6,5)$, we have

$$m=\dfrac{5-3}{6-5}=\dfrac{2}{1}$$

$$=2.$$

7. Using the points $(-2,2)$ and $(2,3)$, we have

$$m=\dfrac{3-2}{2-(-2)}$$

$$=\dfrac{1}{4}.$$

9. Using the points $(-3,-3)$ and $(0,-3)$, we have

$$m=\dfrac{-3-(-3)}{0-3}=\dfrac{0}{-3}$$

$$=0.$$

11. $f(x)=-4x^2+11x$

Step 1 $f(x+h)$
$$=-4(x+h)^2+11(x+h)$$
$$=-4(x^2+2xh+h^2)+11x+11h$$
$$=-4x^2-8xh-4h^2+11x+11h$$

Step 2 $f(x+h)-f(x)$
$$=-4x^2-8xh-4h^2+11x+11h$$
$$\quad -(-4x^2+11x)$$
$$=-8xh-4h^2+11h$$
$$=h(-8x-4h+11)$$

Step 3 $\dfrac{f(x+h)-f(x)}{h}$
$$=\dfrac{h(-8x-4h+11)}{h}$$
$$=-8x-4h+11$$

Step 4 $f'(x)=\lim\limits_{h\to 0}\dfrac{f(x+h)-f(x)}{h}$
$$=\lim\limits_{h\to 0}(-8x-4h+11)$$
$$=-8x+11$$

$$f'(-2)=-8(-2)+11=27$$
$$f'(0)=-8(0)+11=11$$
$$f'(3)=-8(3)+11=-13$$

13. $f(x) = \dfrac{-2}{x}$

$$f(x + h) = \dfrac{-2}{x + h}$$

$$\begin{aligned} f(x + h) - f(x) &= \dfrac{-2}{x + h} - \dfrac{-2}{x} \\ &= \dfrac{-2x + 2(x + h)}{x(x + h)} \\ &= \dfrac{-2x + 2x + 2h}{x(x + h)} \\ &= \dfrac{2h}{x(x + h)} \end{aligned}$$

$$\begin{aligned} \dfrac{f(x + h) - f(x)}{h} &= \dfrac{2h}{hx(x + h)} \\ &= \dfrac{2}{x(x + h)} \\ &= \dfrac{2}{x^2 + xh} \end{aligned}$$

$$\begin{aligned} f'(x) &= \lim_{h \to 0} \dfrac{f(x + h) - f(x)}{h} \\ &= \lim_{h \to 0} \dfrac{2}{x^2 + xh} \\ &= \dfrac{2}{x^2} \end{aligned}$$

$$f'(-2) = \dfrac{2}{(-2)^2} = \dfrac{2}{4} = \dfrac{1}{2}$$

$f'(0) = \frac{2}{0^2}$ which is undefined so $f'(0)$ does not exist.

$$f'(3) = \dfrac{2}{3^2} = \dfrac{2}{9}$$

15. $f(x) = \sqrt{x}$

Steps 1-3 are combined.

$$\begin{aligned} \dfrac{f(x + h) - f(x)}{h} &= \dfrac{\sqrt{x + h} - \sqrt{x}}{h} \\ &= \dfrac{\sqrt{x + h} - \sqrt{x}}{h} \cdot \dfrac{\sqrt{x + h} + \sqrt{x}}{\sqrt{x + h} + \sqrt{x}} \\ &= \dfrac{x + h - x}{h(\sqrt{x + h} + \sqrt{x})} \\ &= \dfrac{1}{\sqrt{x + h} + \sqrt{x}} \end{aligned}$$

$$\begin{aligned} f'(x) &= \lim_{h \to 0} \dfrac{f(x + h) - f(x)}{h} \\ &= \lim_{h \to 0} \dfrac{1}{\sqrt{x + h} + \sqrt{x}} \\ &= \dfrac{1}{2\sqrt{x}} \end{aligned}$$

$f'(-2) = \dfrac{1}{2\sqrt{-2}}$ which is undefined so $f'(-2)$ does not exist.

$f'(0) = \dfrac{1}{2\sqrt{0}} = \dfrac{1}{0}$ which is undefined so $f'(0)$ does not exist.

$$f'(3) = \dfrac{1}{2\sqrt{3}}$$

17. $f(x) = x^2 + 2x;\ x = 3$

$$\begin{aligned} &\dfrac{f(x + h) - f(x)}{h} \\ &= \dfrac{[(x + h)^2 + 2(x + h)] - (x^2 + 2x)}{h} \\ &= \dfrac{(x^2 + 2hx + h^2 + 2x + 2h) - (x^2 + 2x)}{h} \\ &= \dfrac{2hx + h^2 + 2h}{h} = 2x + h + 2 \end{aligned}$$

$$f'(x) = \lim_{h \to 0}(2x + h + 2) = 2x + 2$$

$f'(3) = 2(3) + 2 = 8$ is the slope of the tangent line at $x = 3$.

Use $m = 8$ and $(3, 15)$ in the point-slope form.

$$\begin{aligned} y - 15 &= 8(x - 3) \\ y &= 8x - 9 \end{aligned}$$

19. $f(x) = \dfrac{5}{x};\ x = 2$

$$\begin{aligned} &\dfrac{f(x + h) - f(x)}{h} \\ &= \dfrac{\frac{5}{x+h} - \frac{5}{x}}{h} \\ &= \dfrac{\frac{5x - 5(x+h)}{(x+h)x}}{h} \\ &= \dfrac{5x - 5x - 5h}{h(x + h)(x)} \\ &= \dfrac{-5h}{h(x + h)x} \\ &= \dfrac{-5}{(x + h)x} \end{aligned}$$

$$f'(x) = \lim_{h \to 0} \dfrac{-5}{(x + h)(x)} = -\dfrac{5}{x^2}$$

$f'(2) = \frac{-5}{2^2} = -\frac{5}{4}$ is the slope of the tangent line at $x = 2$.

Now use $m = -\frac{5}{4}$ and $(2, \frac{5}{2})$ in the point-slope form.

$$y - \frac{5}{2} = -\frac{5}{4}(x - 2)$$

$$y - \frac{5}{2} = -\frac{5}{4}x + \frac{10}{4}$$

$$y = -\frac{5}{4}x + 5$$

$$5x + 4y = 20$$

21. $f(x) = 4\sqrt{x};\ x = 9$

$$\frac{f(x+h) - f(x)}{h}$$

$$= \frac{4\sqrt{x+h} - 4\sqrt{x}}{h} \cdot \frac{4\sqrt{x+h} + 4\sqrt{x}}{4\sqrt{x+h} + 4\sqrt{x}}$$

$$= \frac{16(x+h) - 16x}{h(4\sqrt{x+h} + 4\sqrt{x})}$$

$$f'(x) = \lim_{h \to 0} \frac{16(x+h) - 16x}{h(4\sqrt{x+h} + 4\sqrt{x})}$$

$$= \lim_{h \to 0} \frac{16h}{h(4\sqrt{x+h} + 4\sqrt{x})}$$

$$= \lim_{h \to 0} \frac{4}{(\sqrt{x+h} + \sqrt{x})} = \frac{4}{2\sqrt{x}}$$

$$= \frac{2}{\sqrt{x}}$$

$f'(9) = \frac{2}{\sqrt{9}} = \frac{2}{3}$ is the slope of the tangent line at $x = 9$.

Use $m = \frac{2}{3}$ and $(9, 12)$ in the point-slope form.

$$y - 12 = \frac{2}{3}(x - 9)$$

$$y = \frac{2}{3}x + 6$$

$$3y = 2x + 18$$

23. $f(x) = -4x^2 + 11x$

$$\frac{f(x+h) - f(x)}{h}$$

$$= \frac{-4(x+h)^2 + 11(x+h) - (-4x^2 + 11x)}{h}$$

$$= \frac{-8xh - 4h^2 + 11h}{h}$$

$$f'(x) = \lim_{h \to 0} (-8x - 4h + 11)$$

$$= -8x + 11$$

$$f'(2) = -8(2) + 11 = -5$$

$$f'(16) = -8(16) + 11 = -117$$

$$f'(-3) = -8(-3) + 11 = 35$$

25. $f(x) = 8x + 6$

$$\frac{f(x+h) - f(x)}{h} = \frac{[8(x+h) + 6] - (8x + 6)}{h}$$

$$= \frac{(8x + 8h + 6) - (8x + 6)}{h}$$

$$= \frac{8h}{h}$$

$$= 8$$

$$f'(x) = \lim_{h \to 0} 8$$

$$= 8$$

$$f'(2) = 8;\ f'(16) = 8;\ f'(-3) = 8$$

27. $f(x) = -\frac{2}{x}$

$$\frac{f(x+h) - f(x)}{h} = \frac{\frac{-2}{x+h} - \left(\frac{-2}{x}\right)}{h}$$

$$= \frac{\frac{-2x + 2(x+h)}{(x+h)x}}{h}$$

$$= \frac{2h}{h(x+h)x}$$

$$= \frac{2}{(x+h)x}$$

$$f'(x) = \lim_{h \to 0} \frac{2}{(x+h)x}$$

$$= \frac{2}{x^2}$$

$$f'(2) = \frac{2}{2^2}$$

$$= \frac{1}{2}$$

$$f'(16) = \frac{2}{16^2}$$

$$= \frac{2}{256}$$

$$= \frac{1}{128}.$$

$$f'(-3) = \frac{2}{(-3)^2}$$

$$= \frac{2}{9}$$

29. $f(x) = \sqrt{x}$

$$\frac{f(x+h) - f(x)}{h}$$

$$= \frac{\sqrt{x+h} - \sqrt{x}}{h} \cdot \frac{\sqrt{x+h} + \sqrt{x}}{\sqrt{x+h} + \sqrt{x}}$$

$$= \frac{(x+h) - x}{h(\sqrt{x+h} + \sqrt{x})}$$

$$= \frac{h}{h(\sqrt{x+h} + \sqrt{x})}$$

$$= \frac{1}{\sqrt{x+h} + \sqrt{x}}$$

$$f'(x) = \lim_{h \to 0} \frac{1}{\sqrt{x+h} + \sqrt{x}}$$

$$= \frac{1}{2\sqrt{x}}$$

$$f'(2) = \frac{1}{2\sqrt{2}}$$

$$f'(16) = \frac{1}{2\sqrt{16}}$$

$$= \frac{1}{8}$$

$f'(-3) = \dfrac{1}{2\sqrt{-3}}$ is not a real number, so

$f'(-3)$ does not exist.

31. At $x = 0$, the graph of $f(x)$ has a sharp point. Therefore, there is no derivative for $x = 0$.

33. For $x = -3$ and $x = 0$, the tangent to the graph of $f(x)$ is vertical. For $x = -1$, there is a gap in the graph $f(x)$. For $x = 2$, the function $f(x)$ does not exist. For $x = 3$ and $x = 5$, the graph $f(x)$ has sharp points. Therefore, no derivative exists for $x = -3$, $x = -1$, $x = 0$, $x = 2$, $x = 3$, and $x = 5$.

35. (a) The rate of change of $f(x)$ is positive when $f(x)$ is increasing, that is, on $(a, 0)$ and (b, c).

(b) The rate of change of $f(x)$ is negative when $f(x)$ is decreasing, that is, on $(0, b)$.

(c) The rate of change is zero when the tangent to the graph is horizontal, that is, at $x = 0$ and $x = b$.

37. The zeros of graph (b) correspond to the turning points of graph (a), the points where the derivative is zero. Graph (a) gives the distance, while graph (b) gives the velocity.

39. $f(x) = x^x$, $a = 3$

(a)

h	
0.01	$\dfrac{f(3 + 0.01) - f(3)}{0.01}$
	$= \dfrac{3.01^{3.01} - 3^3}{0.01}$
	$= 57.3072$
0.001	$\dfrac{f(3 + 0.001) - f(3)}{0.001}$
	$= \dfrac{3.001^{3.001} - 3^3}{0.001}$
	$= 56.7265$
0.00001	$\dfrac{f(3 + 0.00001) - f(3)}{0.00001}$
	$= \dfrac{3.00001^{3.00001} - 3^3}{0.00001}$
	$= 56.6632$
0.000001	$\dfrac{f(3 + 0.000001) - f(3)}{0.000001}$
	$= \dfrac{3.000001^{3.000001} - 3^3}{0.000001}$
	$= 56.6626$
0.0000001	$\dfrac{f(3 + 0.0000001) - f(3)}{0.0000001}$
	$= \dfrac{3.0000001^{3.0000001} - 3^3}{0.0000001}$
	$= 56.6625$

It appears that $f'(3) = 56.6625$.

(b) Graph the function on a graphing calculator and move the cursor to an x-value near $x = 3$. A good choice for the initial viewing window is $[0, 4]$ by $[0, 60]$ with Xscl $= 1$, Yscl $= 10$.

Now zoom in on the function several times. Each time you zoom in, the graph will look less like a curve and more like a straight line. Use the TRACE feature to select two points on the graph, and record their coordinates. Use these two points to compute the slope. The result will be close to the most accurate value found in part (a), which is 56.6625.

Note: In this exercise, the method used in part (a) gives more accurate results than the method used in part (b).

41. $f(x) = x^{1/x}$, $a = 3$

(a)

h	
0.01	$\dfrac{f(3+0.01) - f(3)}{0.01}$ $= \dfrac{3.01^{1/3.01} - 3^{1/3}}{0.01}$ $= -0.0160$
0.001	$\dfrac{f(3+0.001) - f(3)}{0.001}$ $= \dfrac{3.001^{1/3.001} - 3^{1/3}}{0.001}$ $= -0.0158$
0.0001	$\dfrac{f(3+0.0001) - f(3)}{0.0001}$ $= \dfrac{3.0001^{1/3.0001} - 3^{1/3}}{0.0001}$ $= -0.0158$

It appears that $f'(3) = -0.0158$.

(b) Graph the function on a graphing calculator and move the cursor to an x-value near $x = 3$. A good choice for the initial viewing window is $[0, 5]$ by $[0, 3]$.

Follow the procedure outlined in the solution for Exercise 39, part (b). Note that near $x = 3$, the graph is very close to a horizontal line, so we expect that it slope will be close to 0. The final result will be close to the value found in part (a) of this exercise, which is -0.0158.

43. $l(x) = -3.6 + 0.17x$

(b) $l(70) = -3.6 + 0.17(70)$
$= 8.3$

A snake of length 70 cm will prey on a catfish of length 8.3 cm.

(b) $l^1(x) = \lim\limits_{h \to 0} \dfrac{l(x+h) - l(x)}{h}$

$= \lim\limits_{h \to 0} \dfrac{-3.6 + 0.17x + 0.17h + 3.6 - 0.17x}{h}$

$= \lim\limits_{h \to 0} \dfrac{0.17h}{h}$

$= 0.17$

$l^1(x) = 0.17$

Longer snakes eat bigger catfish at the constant rate of 0.17 cm increase in length of prey to each 1 cm increase in size of snake.

(c) $\qquad l(x) = 0$
$-3.6 + 0.17x = 0$
$0.17x = 3.6$
$x \approx 21.176$

The smallest snake for which the function makes sense has a length of about 21 cm.

45. $I(t) = 27 + 72t - 1.5t^2$

(a)

$I'(t) = \lim\limits_{h \to 0} \dfrac{I(t+h) - I(t)}{h}$

$= \lim\limits_{h \to 0} \dfrac{27+72t+72h-1.5t^2-3th-1.5h^2-27-72t+1.5}{h}$

$= \lim\limits_{h \to 0} \dfrac{72h - 3th - 1.5h^2}{h}$

$= \lim\limits_{h \to 0} 72 - 3t - 1.5h$

$= 72 - 3t$

$I'(5) = 72 - 3(5)$
$= 72 - 15$
$= 57$

The rate of change of the intake of food 5 minutes into a meal is 57 grams per minute.

(b) $\qquad I'(24) \stackrel{?}{=} 0$
$72 - 3(24) \stackrel{?}{=} 0$
$0 = 0$

24 minutes after the meal starts the rate of food consumption is 0.

(c) After 24 minutes the rate of food consumption is negative according to the function where a rate of zero is more accurate. A logical range for this function is

$$0 \le t \le 24.$$

47. (a) From the graph, V_{mp} is just about at the turning point of the curve. Thus, the slope of the tangent line is approximately zero. The power expenditure is not changing.

(b) From the graph, the slope of the tangent line at V_{mr} is approximately 0.1. The power expended is increasing 0.1 unit per unit increase in speed.

(c) The slope of the tangent line at V_{opt} is a bit greater than that at V_{mr}, about 0.12. The power expended increases 0.12 units for each unit increase in speed.

(d) The power level first decreases to V_{mp}, then increases at greater rates.

(e) V_{mr} is the point which produces the smallest slope of a line.

49. The slope of the tangent line to the graph at the first point is found by finding two points on the tangent line.

$$(x_1, y_1) = (1{,}000, 13.5)$$
$$(x_2, y_2) = (0, 18.5)$$
$$m = \frac{18.5 - 13.5}{0 - 1{,}000} = \frac{5}{-1{,}000} = -0.005$$

At the second point, we have

$$(x_1, y_1) = (1{,}000, 13.5)$$
$$(x_2, y_2) = (2{,}000, 21.5).$$
$$m = \frac{21.5 - 13.5}{2{,}000 - 1{,}000}$$
$$= \frac{8}{1{,}000}$$
$$= 0.008$$

At the third point, we have

$$(x_1, y_1) = (5{,}000, 20)$$
$$(x_2, y_2) = (3{,}000, 22.5).$$
$$m = \frac{22.5 - 20}{3{,}000 - 5{,}000}$$
$$= \frac{2.5}{-2{,}000}$$
$$= -0.00125$$

At 500 ft, the temperature decreases 0.005° per foot. At about 1,500 ft, the temperature increases 0.008° per foot. At 5,000 ft, the temperature decreases 0.00125° per foot.

51. The slope of the graph at $x = 24$ can be estimated using the points $(24, 360)$ and $(33, 395)$.

$$\text{slope} = \frac{395 - 360}{33 - 24} = \frac{35}{9} = 3\tfrac{8}{9}$$

Thus, the derivative for a 24 ounce bat is about 4 ft per oz which means that the distance the ball travels is increasing 4 feet per ounce.

The slope of the graph at $x = 51$ can be estimated using the points $(15, 400)$ and $(51, 390)$.

$$\text{slope} = \frac{390 - 400}{51 - 15} = \frac{-10}{36} = -\frac{5}{18} \approx -0.3$$

Thus, the derivative for a 51 ounce bat is about -0.3 ft per oz which means that the distance the ball travels is decreasing 0.3 feet per ounce.

3.4 Graphical Differentiation

3. Since the x-intercepts of the graph of f' occur whenever the graph of f has a horizontal tangent line, Y_1 is the derivative of Y_2. Notice that Y_1 has 2 x-intercepts; each occurs at an x-value where the tangent line to Y_2 is horizontal.
Note also that Y_1 is positive whenever Y_2 is increasing, and that Y_1 is negative whenever Y_2 is decreasing.

5. Since the x-intercepts of the graph of f' occur whenever the graph of f has a horizontal tangent line, Y_2 is the derivative of Y_1. Notice that Y_2 has 1 x-intercept which occurs at the x-value where the tangent line to Y_1 is horizontal. Also notice that the range on which Y_1 is increasing, Y_2 is positive and the range on which it is decreasing, Y_2 is negative.

7. To graph f', observe the intervals where the slopes of tangent lines are positive and where they are negative to determine where the derivative is positive and where it is negative. Also, whenever f has a horizontal tangent, f' will be 0, so the graph of f' will have an x-intercept. The x-values of the three turning point on the graph of f become the three x-intercepts of the graph of f.

Estimate the magnitude of the slope at several points by drawing tangents to the graph of f.

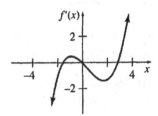

9. On the interval $(-\infty, -2)$, the graph of f is a horizontal line, so its slope is 0. Thus, on this interval, the graph of f' is $y = 0$ on $(-\infty, -2)$. On the interval $(-2, 0)$, the graph of f is a straight line, so its slope is constant. To find this slope, use the points $(-2, 2)$ and $(0, 0)$.

$$m = \frac{2-0}{-2-0} = \frac{2}{-2} = -1$$

On the interval $(0, 1)$, the slope is also constant. To find this slope, use the points $(0, 0)$ and $(1, 1)$.

$$m = \frac{1-0}{1-0} = 1$$

On the interval $(1, \infty)$, the graph is again a horizontal line, so $m = 0$. The graph of f' will be made up of portions of the y-axis and the lines $y = -1$ and $y = 1$.

Because the graph of f has "sharp points" or "corners" at $x = -2, x = 0$, and $x = 1$, we know that $f'(-2), f'(0)$, and $f'(1)$ do not exist. We show this on the graph of f' by using open circles at the endpoints of the portions of the graph.

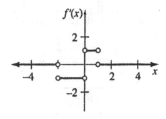

11. On the interval $(-\infty, -2)$, the graph of f is a straight line, so its slope is constant. To find this slope, use the points $(-4, 2)$ and $(-2, 0)$.

$$m = \frac{0-2}{-2-(-4)} = \frac{-2}{2} = -1$$

On the interval $(2, \infty)$, the slope of f is also constant. To find this slope, use the points $(2, 0)$ and $(3, 2)$.

$$m = \frac{2-0}{3-2} = \frac{2}{1} = 2$$

Thus, we have $f'(x) = -1$ on $(-\infty, -2)$ and $f'(x) = 2$ on $(2, \infty)$.

Because f is discontinuous at $x = -2$ and $x = 2$, we know that $f'(-2)$ and $f'(2)$ do not exist, which we indicate with open circles at $(-2, -1)$ and $(2, 2)$ on the graph of f'.

On the interval $(-2, 2)$, all tangent lines have positive slopes, so the graph of f' will be above the y-axis. Notice that the slope of f (and thus the y-value of f') decreases on $(-2, 0)$ and increases on $(0, 2)$, with a minimum value on this interval of 1 at $x = 0$.

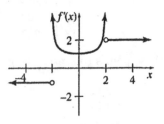

13. We observe that the slopes of tangent lines are positive on the interval $(-\infty, 0)$ and negative on the interval $(0, \infty)$, so the value of f' will be positive on $(-\infty, 0)$ and negative on $(0, \infty)$. Since f is undefined at $x = 0$, $f'(0)$ does not exist.

Notice that the graph of f becomes very flat when $|x| \to \infty$. The *value* of f approaches 0 and also the *slope* approaches 0. Thus, $y = 0$ (the x-axis) is a horizontal asymptote for both the graph of f and the graph of f'.

As $x \to 0^-$ and $x \to 0^+$, the graph of f gets very steep, so $|f'(x)| \to \infty$. Thus, $x = 0$ (the y-axis) is a vertical asymptote for both the graph of f and the graph of f'.

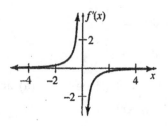

15. The growth rate of the function $y = f(x)$ is given by the derivative of this function $y' = f(x)$. We use the graph of f to sketch the graph of f'.

First, notice as x increase, y increases throughout the domain of f, but at a slower and slower rate. The slope of f is positive but always decreasing, and approaches 0 as t gets large. Thus, y' will always be positive and decreasing. It will approach but never reach 0.

To plot point on the graph of f', we need to estimate the slope of f at several points. From the graph of f, we obtain the values given in the following table.

t	y'
2	1,000
10	700
13	250

Use these points to sketch the graph.

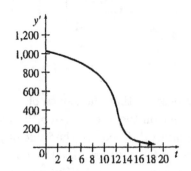

17.

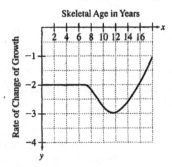

Chapter 3 Review Exercises

5. (a) $\lim_{x \to -3-} = 4$

(b) $\lim_{x \to -3+} = 4$

(c) $\lim_{x \to -3} = 4$ (since parts (a) and (b) have the same answer)

(d) $f(-3) = 4$, since $(-3, 4)$ is a point of the graph.

7. (a) $\lim_{x \to 4^-} f(x) = \infty$

(b) $\lim_{x \to 4^+} f(x) = -\infty$

(c) $\lim_{x \to 4} f(x)$ does not exist since the answers to parts (a) and (b) are different.

(d) $f(4)$ does not exist since the graph has no point with an x-value of 4.

9. $\lim_{x \to -\infty} g(x) = \infty$ since the y-value gets very large as the x-value gets very small.

11. $\lim_{x \to 6} \dfrac{2x + 5}{x - 3} = \dfrac{2(6) + 5}{6 + (-3)}$

$$= \frac{17}{3}$$

13. $\lim_{x \to 4} \dfrac{x^2 - 16}{x - 4} = \lim_{x \to 4} \dfrac{(x - 4)(x + 4)}{x - 4}$

$$= \lim_{x \to 4} (x + 4)$$

$$= 4 + 4$$

$$= 8$$

15. $\lim_{x \to -4} \dfrac{2x^2 + 3x - 20}{x + 4} = \lim_{x \to -4} \dfrac{(2x - 5)(x + 4)}{x + 4}$

$$= \lim_{x \to -4} (2x - 5)$$

$$= 2(-4) - 5$$

$$= -13$$

17. $\lim_{x \to 9} \dfrac{\sqrt{x} - 3}{x - 9} = \lim_{x \to 9} \dfrac{\sqrt{x} - 3}{x - 9} \cdot \dfrac{\sqrt{x} + 3}{\sqrt{x} + 3}$

$$= \lim_{x \to 9} \frac{x - 9}{(x - 9)(\sqrt{x} + 3)}$$

$$= \lim_{x \to 9} \frac{1}{\sqrt{x} + 3}$$

$$= \frac{1}{\sqrt{9} + 3}$$

$$= \frac{1}{6}$$

19. $\lim_{x \to \infty} \dfrac{x^2 + 5}{5x^2 - 1} = \lim_{x \to \infty} \dfrac{\frac{x^2}{x^2} + \frac{5}{x^2}}{\frac{5x^2}{x^2} - \frac{1}{x^2}}$

$$= \lim_{x \to \infty} \frac{1 + \frac{5}{x^2}}{5 - \frac{1}{x^2}}$$

$$= \frac{1 + 0}{5 - 0}$$

$$= \frac{1}{5}$$

21. $\lim\limits_{x\to-\infty}\left(\dfrac{3}{4}+\dfrac{2}{x}-\dfrac{5}{x^2}\right)$

$= \lim\limits_{x\to-\infty}\dfrac{3}{4}+\lim\limits_{x\to-\infty}\dfrac{2}{x}-\lim\limits_{x\to-\infty}\dfrac{5}{x^2}$

$= \dfrac{3}{4}+0-0=\dfrac{3}{4}$

23. As shown on the graph, $f(x)$ is discontinuous at x_2 and x_4.

25. $f(x)$ is discontinuous at $x=0$ and $x=\frac{1}{2}$ since that is where the denominator of $f(x)$ equals 0.
$f(0)$ and $f\left(\frac{1}{2}\right)$ do not exist.
$\lim\limits_{x\to0} f(x)$ does not exist since $\lim\limits_{x\to0^+} f(x)=\infty$,
but $\lim\limits_{x\to0^-} f(x)=-\infty$. $\lim\limits_{x\to\frac{1}{2}} f(x)$ does not exist
since $\lim\limits_{x\to\frac{1}{2}^-}=\infty$, but $\lim\limits_{x\to\frac{1}{2}^+} f(x)=-\infty$.

27. $f(x)$ is discontinuous at $x=-5$ since that is where the denominator of $f(x)$ equals 0.
$f(-5)$ does not exist.
$\lim\limits_{x\to-5} f(x)$ does not exist since $\lim\limits_{x\to-5^-} f(x)=\infty$,
but $\lim\limits_{x\to-5^+} f(x)=-\infty$.

29. $f(x)=x^2+3x-4$ is continuous everywhere since f is a polynomial function.

31. (a)

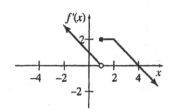

(b) The graph is discontinuous at $x=1$.

(c) $\lim\limits_{x\to1^-} f(x)=0$; $\lim\limits_{x\to1^+} f(x)=2$

33. $f(x)=\dfrac{x^4+2x^3+2x^2-10x+5}{x^2-1}$

(a) Find the values of $f(x)$ when x is close to 1.

x	y
1.1	2.6005
1.01	2.06
1.001	2.006
1.0001	2.0006
0.99	1.94
0.999	1.994
0.9999	1.9994

It appears that $\lim\limits_{x\to1} f(x)=2$.

(b) Graph

$$y=\dfrac{x^4+2x^3+2x^2-10x+5}{x^2-1}$$

on a graphing calculator. One suitable choice for the viewing window is $[-2,6]$ by $[-10,10]$.

Because $x^2-1=0$ when $x=-1$ or $x=1$, this function is discontinuous at these two x-values. The graph shows a vertical asymptote at $x=-1$ but not at $x=1$. The graph should have an open circle to show a "hole" in the graph at $x=1$. The graphing calculator doesn't show the hole, but trying to find the value of the function of $x=1$ will show that this value is undefined.

By viewing the function near $x=1$ and using the ZOOM feature, we see that as x gets close to 1 from the left or the right, y gets close to 2, suggesting that

$$\lim\limits_{x\to1}\dfrac{x^4+2x^3+2x^2-10x+5}{x^2-1}=2.$$

35. $y=6x^2+2=f(x)$; from $x=1$ to $x=4$

$f(4)=6(4)^2+2=98$
$f(1)=6(1)^2+2=8$

Average rate of change $=\dfrac{98-8}{4-1}=\dfrac{90}{3}=30$

$y'=12x$

Instantaneous rate of change at $x=1$:

$$f'(1)=12(1)=12$$

37. $y=\dfrac{-6}{3x-5}=f(x)$; from $x=4$ to $x=9$

$f(9)=\dfrac{-6}{3(9)-5}=\dfrac{-6}{22}=-\dfrac{3}{11}$

$f(4)=\dfrac{-6}{3(4)-5}=-\dfrac{6}{7}$

Average rate of change $=\dfrac{\frac{-3}{11}-\left(-\frac{6}{7}\right)}{9-4}$

$=\dfrac{\frac{-21+66}{77}}{5}$

$=\dfrac{45}{5(77)}=\dfrac{9}{77}$

$y'=\dfrac{(3x-5)(0)-(-6)(3)}{(3x-5)^2}=\dfrac{18}{(3x-5)^2}$

Instantaneous rate of change at $x=4$:

$$f'(4)=\dfrac{18}{(3\cdot4-5)^2}=\dfrac{18}{7^2}=\dfrac{18}{49}$$

39. $y = 4x + 3 = f(x)$

$$y' = \lim_{h \to 0} \frac{f(x+h) - f(x)}{h} = \lim_{h \to 0} \frac{[4(x+h)+3] - [4x+3]}{h}$$

$$= \lim_{h \to 0} \frac{4x + 4h + 3 - 4x - 3}{h} = \lim_{h \to 0} \frac{4h}{h} = \lim_{h \to 0} 4 = 4$$

41. $f(x) = (\ln x)^x, x_0 = 2$

(a)

h	
0.01	$\dfrac{f(2 + 0.01) - f(2)}{0.01}$ $= \dfrac{(\ln 2.01)^{2.01} - (\ln 2)^2}{0.01}$ $= 0.5191$
0.001	$\dfrac{f(2 + 0.001) - f(2)}{0.001}$ $= \dfrac{(\ln 2.001)^{2.001} - (\ln 2)^2}{0.001}$ $= 0.5173$
0.0001	$\dfrac{f(2 + 0.0001) - f(2)}{0.0001}$ $= \dfrac{(\ln 2.0001)^{2.0001} - (\ln 2)^2}{0.0001}$ $= 0.5171$
0.00001	$\dfrac{f(2 + 0.00001) - f(2)}{0.00001}$ $= \dfrac{(\ln 2.00001)^{2.00001} - (\ln 2)^2}{0.00001}$ $= 0.5171$

(b) Using a graphing calculator will confirm this result.

43. On the interval $(-\infty, 0)$, the graph of f is a straight line, so its slope is constant. To find this slope, use the points $(-2, 2)$ and $(0, 0)$.

$$m = \frac{0 - 2}{0 - (-2)} = \frac{-2}{2} = -1$$

Thus, the value of f' will be -1 on this interval. The graph of f has a sharp point at 0, so $f'(0)$ does not exist. To show this, we use an open circle on the graph of f' at $(0, -1)$.

We also observe that the slope of f is positive but decreasing from $x = 0$ to about $x = 1$, and then negative from there on. As $x \to \infty$, $f(x) \to 0$ and also $f'(x) = 0$.

Use this information to complete the graph of f'.

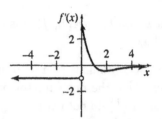

45. **(a)** The slope of the tangent line at $x = 100$ is 0.02. This means that the risk of heart attack is going up at a rate of 0.02 per 1,000 people for each increase in the blood cholesterol of 1 mg/dL.

(b) The slope of the tangent line at $x = 200$ is 0.15. This means that the risk of heart attack is going up at a rate of 0.15 per 1,000 people for each increase in the blood cholesterol of 1 mg/dL.

(c) $\dfrac{C(300) - C(100)}{300 - 100} = \dfrac{20 - 7}{200} = \dfrac{13}{200} = 0.065$

The average rate of change in the risk is 0.065 per 1,000 people per mg/dL of cholesterol.

47. $N(t) = 1,883.41 - 16.91t$

(a)

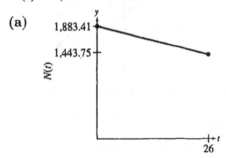

(b)

$$N'(t) = \lim_{h \to 0} \frac{N(t+h) - N(t)}{h}$$

$$= \lim_{h \to 0} \frac{1,883.41 - 16.91(t+h) - (1,883.41 - 16.91t)}{h}$$

$$= \lim_{h \to 0} \frac{-16.91h}{h} = -16.91$$

$N'(t) = -16.91$

Approximately 17 fewer wolves were harvested each year during the years 1964 to 1990.

49. (a) The rate can be estimated by estimating the slope of the tangent line to the curve at the given time. Answers will vary depending on the points used to deterimine the slope of the tangent line.

(i) The tangent line to the curve at 17:37 appears to be a depth of 40 meters at 17:36 and a depth of 140 meters at 17:37. The rate the whale is descending at 17:37 is about

$$\frac{140 - 40}{17:37 - 17:36} = \frac{100}{1} = 100 \text{ meters per minute.}$$

(ii) The tangent line to the curve at 17:39 appears to be a depth of 160 meters at 17:38 and a depth of 240 meters at 17:30/The rate the whale is descending at 17:39 is about.

$$\frac{240 - 160}{17:39 - 17:38} = \frac{80}{1} = 80 \text{ meters per minute.}$$

(b) The whale appears to have 5 distinct rates at which it is descending.

Interval	Rate (meters per minute)
17:35 − 17:35.5	$\frac{f(17:35.5) - f(17:35)}{17:35.5 - 17:35} = \frac{10 - 10}{0.5}$ $= 0$
17:36 − 17:37	$\frac{f(17:37) - f(17:36)}{17:37 - 17:36} = \frac{140 - 40}{1}$ $= 100$
17:37.5 − 17:38	$\frac{f(17:38) - f(17:37.5)}{17:38 - 17:37.5} = \frac{175 - 160}{0.5}$ $= 30$
17:38.5 − 17:39.5	$\frac{f(17:39.5) - f(17:38.5)}{17:39.5 - 17:38.5} = \frac{300 - 220}{1}$ $= 80$
17:40 − 17:41	$\frac{f(17:41) - f(17:40)}{17:41 - 17:40} = \frac{400 - 330}{1}$ $= 70$

Making smooth transitions between each interval, we get

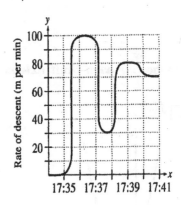

51.

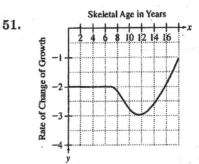

For a 10-year old girl, the remaining growth is about 14 cm and the rate of change is about −2.75 cm per year.

53. (a) The slope of the tangent line at $x = 100$ is 1. This means that the ball is rising 1 ft for each foot it travels horizontally.

(b) The slope of the tangent line at $x = 200$ is −2.7. This means that the ball is dropping 2.7 ft for each foot it travels horizontally.

Chapter 3 Test

[3.1]

Decide whether each limit exists. If a limit exists, find its value.

1. $\lim\limits_{x\to 1} f(x)$

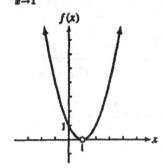

3. $\lim\limits_{x\to 2} f(x)$

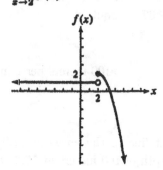

2. $\lim\limits_{x\to -1} f(x)$

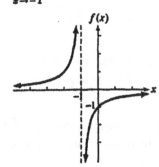

4. $\lim\limits_{x\to -2} f(x)$

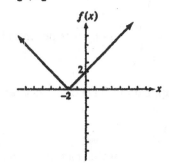

Use the properties of limits to help decide whether the following limits exist. If a limit exists, find its value.

5. $\lim\limits_{x\to 2}\left(\dfrac{1}{x}+1\right)(3x-2)$

7. $\lim\limits_{x\to 2}\dfrac{x^2-5x+6}{x-2}$

9. $\lim\limits_{x\to \pi/2}\dfrac{\sin^2 x-1}{\cos^2 x}$

6. $\lim\limits_{x\to 0}\dfrac{3x+5}{4x}$

8. $\lim\limits_{x\to 1}\dfrac{x-1}{\sqrt{x}-1}$

[3.2]

Find all points $x = a$ where the function is discontinuous. For each point of discontinuity, give $\lim\limits_{x\to a} f(x)$ if it exists.

10. $f(x) = \dfrac{x^2-16}{x+4}$

11. $g(x) = \dfrac{3+x}{(2x+3)(x-5)}$

12. $h(x) = \dfrac{3x^4+2x^2-7}{4}$

13. Is $f(x) = \frac{x-2}{x(3-x)(x+4)}$ continuous at the given values of x?

    (a) $x = 2$ **(b)** $x = 0$ **(c)** $x = 5$ **(d)** $x = 3$

14. Use the graph to answer the following questions.

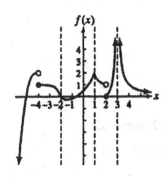

    (a) On which of the following intervals is the graph continuous?
        $(-5, -2)$, $(-3, 2)$, $(1, 4)$

    (b) Where is the function discontinuous?

    (c) Find $\lim\limits_{x \to -4^-} f(x)$.   **(d)** Find $\lim\limits_{x \to -4^+} f(x)$.

15. Consider the following function.

$$f(x) = \begin{cases} -3 & \text{if } x < -1 \\ 2x - 1 & \text{if } -1 \le x \le 2 \\ 6 & \text{if } x > 2 \end{cases}$$

    (a) Graph this function.   **(b)** Find any points of discontinuity.

    (c) Find the limit from the left and from the right at any point(s) of discontinuity.

16. Consider the following function.

$$f(x) = \begin{cases} \sin x & \text{if } x < \frac{\pi}{2} \\ 1 + \cos x & \text{if } \frac{\pi}{2} \le x < \pi \\ -\sin x & \text{if } \pi \le x < 2\pi \end{cases}$$

    (a) Find $\lim\limits_{x \to \pi/2} f(x)$.   **(b)** Is f continuous at $x = \frac{\pi}{2}$?

    (c) Find $\lim\limits_{x \to \pi} f(x)$.   **(d)** Is f continuous at $x = \pi$?

[3.3]

17. Find the average rate of change of $f(x) = x^3 - 5x$ between $x = 1$ and $x = 5$.

18. Use the graph to find the average rate of change of f on the given intervals.

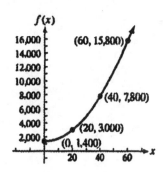

(a) From $x = 0$ to $x = 20$

(b) From $x = 20$ to $x = 60$

(c) From $x = 0$ to $x = 40$

(d) From $x = 0$ to $x = 60$

Find the instantaneous rate of change for each function at the given value.

19. $f(x) = 3x^2 - 7$ at $x = 1$

20. $g(t) = 4 - 2t^2$ at $t = -3$

21. Suppose the total profit in thousands of dollars from selling x units is given by

$$P(x) = 2x^2 - 6x + 9.$$

(a) Find the average rate of change of profit as x increases from 2 to 4.

(b) Find the marginal profit when 10 units are sold.

(c) Find the average rate of change of profit when sales are increased from 100 to 200 units.

22. Splenic Artery Resistance Researchers have found that the splenic artery resistance index in the fetus can be described by the function

$$y = 0.057x - 0.001x^2,$$

where x is the number of weeks of gestation.

(a) Find the average rate of change in the index as x increases from 9 to 10 weeks.

(b) Find the instantaneous rate of change in the index when the fetus is 10 weeks old.

(c) Find the instantaneous rate of change in the index when the fetus is 40 weeks old.

[3.4]

Use the definition of the derivative to find the derivative of each function.

23. $y = x^3 - 5x^2$

24. $y = \dfrac{3}{x}$

Find $f'(x)$ and use it to find $f'(3)$, $f'(0)$, $f'(-1)$.

25. $f(x) = \dfrac{-3}{x}$

26. $f(x) = 3x^2 + 4x$

27. $f(x) = -\dfrac{1}{2}\sqrt{x}$

28. Tooth Length The length (in mm) of the mesiodistal crown of the first molar for a human fetus can be approximated by

$$l = -0.01x^2 + 0.788x - 7.048.$$

where x is the number of weeks since conception.

(a) Use the definition of the derivative to find l'.

(b) Use algebra to find where l' is zero.

29. The function

$$f(x) = \frac{Kx}{A + x}$$

is used in biology to give the growth rate of a population in the presence of a quantity x of food. Let $K = 5$ and $A = 2$ and use the definition of the derivative to find $f'(x)$.

[3.5]

Sketch the graph of the derivative for each function shown.

30.

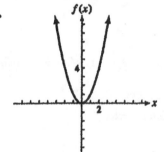

31.

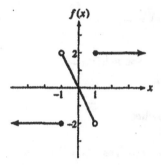

Chapter 3 Test Answers

1. 0

2. Does not exist

3. Does not exist

4. 0

5. 6

6. Does not exist

7. −1

8. 2

9. −1

10. $a = -4$, $\lim\limits_{x \to -4} f(x) = -8$

11. $a = -\frac{3}{2}$, limit does not exist; $a = 5$, limit does not exist.

12. Discontinuous nowhere

13. (a) Yes   (b) No   (c) Yes   (d) No

14. (a) $(-3, 2)$
    (b) At $x = -4$, $x = 2$, and $x = 3$
    (c) 2   (d) 1

15. (a)

$$f(x) = \begin{cases} -3 & \text{if } x < -1 \\ 2x - 1 & \text{if } -1 \le x \le 2 \\ 6 & \text{if } x > 2 \end{cases}$$

    (b) $x = 2$
    (c) From the left: 3;
        from the right: 6

16. (a) 1   (b) Yes   (c) 0   (d) Yes

17. 26

18. (a) 80   (b) 320   (c) 160   (d) 240

19. 6

30. 12

21. (a) \$6,000 per unit   (b) \$34,000 per unit
    (c) \$594,000 per unit

22. (a) 0.038   (b) 0.037   (c) −0.023

23. $y' = 3x^2 - 10x$

24. $y' = -\frac{3}{x^2}$

25. $f'(x) = \frac{3}{x^2}$, $\frac{1}{3}$, undefined, 3

26. $f'(x) = 6x + 4$, 22, 4, −2

27. $f'(x) = \frac{-1}{4\sqrt{x}}$, $\frac{-1}{4\sqrt{3}}$ or $\frac{-\sqrt{3}}{12}$, undefined, undefined

28. (a) $l' = -0.02x + 0.788$
    (b) $x = 39.4$ or 39.4 weeks

29. $f'(x) = \frac{10}{(2+x)^2}$

30.

31.

CALCULATING THE DERIVATIVE

4.1 Techniques For Finding Derivatives

1. $y = 10x^3 - 9x^2 + 6x + 5$

$\dfrac{dy}{dx} = 10(3x^{3-1}) - 9(2x^{2-1}) + 6x^{1-1} + 0$

$= 30x^2 - 18x + 6$

3. $y = x^4 - 5x^3 + \dfrac{y^2}{9} + 5$

$\dfrac{dy}{dx} = 4x^{4-1} - 5(3x^{3-1}) + \dfrac{1}{9}(2x^{2-1}) + 0$

$= 4x^3 - 15x^2 + \dfrac{2}{9}x$

5. $f(x) = 6x^{1.5} - 4x^{0.5}$

$f'(x) = 6(1.5x^{1.5-1}) - 4(0.5x^{0.5-1})$

$= 9x^{0.5} - 2x^{-0.5}$ or $9x^{0.5} - \dfrac{2}{x^{0.5}}$

7. $y = 8\sqrt{x} + 6x^{3/4}$

$= 8x^{1/2} + 6x^{3/4}$

$\dfrac{dy}{dx} = 8\left(\dfrac{1}{2}x^{1/2-1}\right) + 6\left(\dfrac{3}{4}x^{3/4-1}\right)$

$= 4x^{-1/2} + \dfrac{9}{2}x^{-1/4}$

or $\dfrac{4}{x^{1/2}} + \dfrac{9}{2x^{1/4}}$

9. $g(x) = 6x^{-5} - x^{-1}$

$g'(x) = 6(-5)x^{-5-1} - (-1)x^{-1-1}$

$= -30x^{-6} + x^{-2}$

or $\dfrac{-30}{x^6} + \dfrac{1}{x^2}$

11. $y = x^{-5} - x^{-2} + 5x^{-1}$

$\dfrac{dy}{dx} = -5x^{-5-1} - (-2x^{-2-1}) + 5(-1x^{-1-1})$

$= -5x^{-6} + 2x^{-3} - 5x^{-2}$

or $\dfrac{-5}{x^6} + \dfrac{2}{x^3} - \dfrac{5}{x^2}$

13. $f(t) = \dfrac{4}{t} + \dfrac{2}{t^3} + \sqrt{2}$

$= 4t^{-1} + 2t^{-3} + \sqrt{2}$

$f'(t) = 4(-1t^{-1-1}) + 2(-3t^{-3-1}) + 0$

$= -4t^{-2} - 6t^{-4}$ or $\dfrac{-4}{t^2} - \dfrac{6}{t^4}$

15. $y = \dfrac{3}{x^6} + \dfrac{1}{x^5} - \dfrac{7}{x^2}$

$= 3x^{-6} + x^{-5} - 7x^{-2}$

$\dfrac{dy}{dx} = 3(-6x^{-7}) + (-5x^{-6}) - 7(-2x^{-3})$

$= -18x^{-7} - 5x^{-6} + 14x^{-3}$

or $\dfrac{-18}{x^7} - \dfrac{5}{x^6} + \dfrac{14}{x^3}$

17. $h(x) = x^{-1/2} - 14x^{-3/2}$

$h'(x) = -\dfrac{1}{2}x^{-3/2} - 14\left(-\dfrac{3}{2}x^{-5/2}\right)$

$= \dfrac{-x^{-3/2}}{2} + 21x^{-5/2}$

or $\dfrac{-1}{2x^{3/2}} + \dfrac{21}{x^{5/2}}$

19. $y = \dfrac{-2}{\sqrt[3]{x}}$

$= \dfrac{-2}{x^{1/3}} = -2x^{-1/3}$

$\dfrac{dy}{dx} = -2\left(-\dfrac{1}{3}x^{-4/3}\right)$

$= \dfrac{2x^{-4/3}}{3}$ or $\dfrac{2}{3x^{4/3}}$

21. $g(x) = \dfrac{x^2 - 2x}{\sqrt{x}}$

$= \dfrac{x^2 - 2x}{x^{1/2}}$

$= \dfrac{x^2}{x^{1/2}} - \dfrac{2x}{x^{1/2}}$

$= x^{3/2} - 2x^{1/2}$

$g'(x) = \dfrac{3}{2}x^{3/2-1} - 2\left(\dfrac{1}{2}x^{1/2-1}\right)$

$= \dfrac{3}{2}x^{1/2} - x^{-1/2}$

or $\dfrac{3}{2}\sqrt{x} - \dfrac{1}{\sqrt{x}}$

23. $h(x) = (x^2 - 1)^3$
$$= x^6 - 3x^4 + 3x^2 - 1$$
$$h'(x) = 6x^{6-1} - 3(4x^{4-1}) + 3(2x^{2-1}) - 0$$
$$= 6x^5 - 12x^3 + 6x$$

27. $D_x \left[9x^{-1/2} + \dfrac{2}{x^{3/2}} \right]$

$$= D_x[9x^{-1/2} + 2x^{-3/2}]$$

$$= 9\left(-\frac{1}{2}x^{-3/2} \right) + 2\left(-\frac{3}{2}x^{-5/2} \right)$$

$$= -\frac{9}{2}x^{-3/2} - 3x^{-5/2}$$

or $\dfrac{-9}{2x^{3/2}} - \dfrac{3}{x^{5/2}}$

29. $f(x) = \dfrac{x^2}{6} - 4x$

$$= \frac{1}{6}x^2 - 4x$$

$$f'(x) = \frac{1}{6}(2x) - 4$$

$$= \frac{1}{3}x - 4$$

$$f'(-2) = \frac{1}{3}(-2) - 4$$

$$= -\frac{2}{3} - 4$$

$$= -\frac{14}{3}$$

31. $y = x^4 - 5x^3 + 2; \ x = 2$
$$y' = 4x^3 - 15x^2$$
$$y'(2) = 4(2)^3 - 15(2)^2$$
$$= -28$$

The slope of tangent line at $x = 2$ is -28.
Use $m = -28$ and $(x_1, y_1) = (2, -22)$ to obtain the equation.

$$y - (-22) = -28(x - 2)$$
$$y = -28x + 34$$
$$28x + y = 34$$

33. $y = -2x^{1/2} + x^{3/2}; x = 4$

$$y' = -2\left[\frac{1}{2}x^{-1/2} \right] + \frac{3}{2}x^{1/2}$$

$$= -x^{-1/2} + \frac{3}{2}x^{1/2}$$

$$= -\frac{1}{x^{1/2}} + \frac{3x^{1/2}}{2}$$

$$y'(4) = \frac{-1}{(4)^{1/2}} + \frac{3(4)^{1/2}}{2}$$

$$= \frac{-1}{2} + \frac{3(2)}{2}$$

$$= -\frac{1}{2} + 3 = \frac{5}{2}$$

The slope of the tangent line at $x = 4$ is $\frac{5}{2}$.

35. $f(x) = 9x^2 - 8x + 4$
$$f'(x) = 18x - 8$$

Let $f'(x) = 0$ to find the point where the slope of the tangent line is zero.

$$18x - 8 = 0$$
$$18x = 8$$
$$x = \frac{8}{18} = \frac{4}{9}$$

Find the y-coordinate.

$$f(x) = 9x^2 - 8x + 4$$

$$f\left(\frac{4}{9} \right) = 9\left(\frac{4}{9} \right)^2 - 8\left(\frac{4}{9} \right) + 4$$

$$= 9\left(\frac{16}{81} \right) - \frac{32}{9} + 4$$

$$= \frac{16}{9} - \frac{32}{9} + \frac{36}{9} = \frac{20}{9}$$

The slope of the tangent line is zero at one point, $\left(\frac{4}{9}, \frac{20}{9} \right)$.

37. $f(x) = x^3 + 15x^2 + 63x - 10$
$$f'(x) = 3x^2 + 30x + 63$$

If the tangent line is horizontal, then its slope is zero and $f'(x) = 0$.

$$3x^2 + 30x + 63 = 0$$
$$3(x^2 + 10x + 21) = 0$$
$$3(x + 3)(x + 7) = 0$$
$$x = -3 \ \text{ or } \ x = -7$$

Therefore, the tangent line is horizontal at $x = -3$ or $x = -7$.

39. $f(x) = x^3 - 5x^2 + 6x + 3$
$f'(x) = 3x^2 - 10x + 6$

If the tangent line is horizontal, then its slope is zero and $f'(x) = 0$.

$$3x^2 - 10x + 6 = 0$$

$$x = \frac{10 \pm \sqrt{100 - 72}}{6}$$

$$x = \frac{10 \pm \sqrt{28}}{6}$$

$$x = \frac{10 \pm 2\sqrt{7}}{6}$$

$$x = \frac{5 \pm \sqrt{7}}{3}$$

The tangent line is horizontal at $x = \frac{5 \pm \sqrt{7}}{3}$.

41. $f(x) = 2x^3 - 9x^2 - 12x + 5$
$f'(x) = 6x^2 - 18x - 12$

If the slope of the tangent line is 12, $f'(x) = 12$.

$$6x^2 - 18x - 12 = 12$$
$$6x^2 - 18x - 24 = 0$$
$$6(x^2 - 3x - 4) = 0$$
$$6(x - 4)(x + 1) = 0$$
$$x = 4 \quad \text{or} \quad x = -1$$

$f(4) = -59$ and $f(-1) = 6$
The slope of the tangent line is -12 at $(4, -59)$ and $(-1, 6)$.

43. $f(x) = 3g(x) - 2h(x) + 3$
$f'(x) = 3g'(x) - 2h'(x)$
$f'(5) = 3g'(5) - 2h'(5)$
$\quad\quad = 3(10) - 2(-4) = 38$

45. **(a)** From the graph, $f(1) = 2$, because the curve goes through $(1, 2)$.

(b) $f'(1)$ gives the slope of the tangent line to f at 1. The line goes through $(-1, 1)$ and $(1, 2)$.

$$m = \frac{2 - 1}{1 - (-1)} = \frac{1}{2},$$

so

$$f'(1) = \frac{1}{2}.$$

(c) The domain of f is $[-1, \infty)$ because the x-coordinates of the points of f start at $x = -1$ and continue infinitely through the positive real numbers.

(d) The range of f is $[0, \infty)$ because the y-coordinates of the points on f start at $y = 0$ and continue infinitely through the positive real numbers.

47. $\dfrac{f(x)}{k} = \dfrac{1}{k} \cdot f(x)$

Use the rule for the derivative of a constant times a function.

$$\frac{d}{dx}\left[\frac{f(x)}{k}\right] = \frac{d}{dx}\left[\frac{1}{k} \cdot f(x)\right]$$

$$= \frac{1}{k}f'(x) = \frac{f'(x)}{k}$$

49. $N(t) = 0.00437t^{3.2}$
$N'(t) = 0.013984t^{2.2}$

(a) $N'(5) \approx 0.4824$

(b) $N'(10) \approx 2.216$

51. $M(t) = 4t^{3/2} + 2t^{1/2}$

(a) $M(16) = 4 \cdot 16^{3/2} + 2 \cdot 16^{1/2}$
$\quad\quad = 256 + 8$
$\quad\quad = 264$

(b) $M(25) = 4 \cdot 25^{3/2} + 2 \cdot 25^{1/2}$
$\quad\quad = 500 + 10$
$\quad\quad = 510$

(c) $M'(t) = \dfrac{3}{2} \cdot 4t^{1/2} + \dfrac{1}{2} \cdot 2t^{-1/2}$

$\quad\quad\quad = 6t^{1/2} + t^{-1/2}$
$M'(16) = 6 \cdot 16^{1/2} + 16^{-1/2}$

$\quad\quad\quad = 6 \cdot 4 + \dfrac{1}{4}$

$\quad\quad\quad = 24 + \dfrac{1}{4}$

$\quad\quad\quad = \dfrac{97}{4}$ or 24.25

The rate of change is about $\frac{97}{4}$ or 24.25 matings per degree.

53. $V(t) = -2{,}159 + 1{,}313t - 60.82t^2$

(a) $V(3) = -2{,}159 + 1{,}313(3) - 60.82(3)^2$
$\quad\quad = 1{,}232.62$ cm^3

(b) $V'(t) = 1{,}313 - 121.64t$
$V'(3) = 1{,}313 - 121.64(3)$
$\quad\quad = 948.08$ cm^3/yr

55.
$$V = C(R_0 - R)R^2$$
$$= CR_0R^2 - CR^3$$

$$\frac{dV}{dR} = 2CR_0R - 3CR^2 = 0$$
$$CR(2R_0 - 3R) = 0$$

$$CR = 0 \quad \text{or} \quad 2R_0 - 3R = 0$$

$$R = 0 \qquad\qquad R = \frac{2}{3}R_0$$

Discard $R = 0$, since a closed windpipe produces no airflow. Velocity is maximized when $R = \frac{2}{3}R_0$.

57. $l(x) = -2.318 + 0.2356x - 0.002674x^2$

(a) The problem states that a fetus this formula concerns is at least 18 weeks old. So, the minimum x value should be 18. Considering the gestation time of a cow in general, a meaninful range for this funcion is $18 \le x \le 44$.

(b) $l(x) = 0.2356 - (2)0.002674x$
$$= 0.2356 - 0.005348x$$

(c) $l'(25) = 0.2356 - 0.005348(25)$
$$= 0.1019 \text{ cm/week}$$

59.
$$f(m) = 40m^{-0.6}$$
$$f'(m) = 40(-0.6)m^{-0.6-1} = -24m^{-1.6}$$
$$f'(0.01) = -24(0.01)^{-1.6}$$
$$\approx -38{,}037.43\ldots$$
$$\approx -38{,}000 \text{ species/g}$$

The number of species is decreasing by about 38,000 per gram the body mass increases when the body mass is 0.01g.

61. $v(x) = -35.98 + 12.09x - 0.4450x^2$

(a)
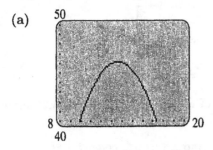

(b) Analyzing the graph, we see that the maximum value occurs at about $x = 13.5843\ldots \approx 13.58$.

(c) $v'(x) = 12.09 - 0.890x$
$$v'(13.58) = 12.09 - 0.890(13.58) \approx 0$$

63. $s(t) = 11t^2 + 4t + 2$

(a) $v(t) = s'(t) = 22t + 4$

(b) $v(0) = 22(0) + 4 = 4$
$$v(5) = 22(5) + 4 = 114$$
$$v(10) = 22(10) + 4 = 224$$

65. $s(t) = 4t^3 + 8t^2$

(a) $v(t) = s'(t) = 12t^2 + 16t$

(b) $v(0) = 12(0)^2 + 16(0) = 0$
$$v(5) = 12(5)^2 + 16(5)$$
$$= 300 + 80 = 380$$
$$v(10) = 12(10)^2 + 16(10)$$
$$= 1{,}200 + 160 = 1{,}360$$

67. $s(t) = -16t^2 + 144$

$$\text{velocity} = s'(t)$$
$$= -32t$$

(a) $s'(1) = -32 \cdot 1 = -32$
$$s'(2) = -32 \cdot 2$$
$$= -64$$

The rock's velocity is -32 ft/sec after 1 second and -64 ft/sec after 2 seconds.

(b) The rock will hit the ground when $s(t) = 0$.

$$-16t^2 + 144 = 0$$
$$t = \sqrt{\frac{144}{16}}$$
$$= 3$$

The rock will hit the ground after 3 seconds.

(c) The velocity at impact is the velocity at 3 seconds.

$$s'(3) = -32 \cdot 3$$
$$= -96$$

Its velocity on impact is -96 ft/sec.

69. (a) $d(x) = 1.66 - 0.90x + 0.47x^2$
$$d(.5) = 1.66 - 0.90(0.5) + 0.47(0.5)^2$$
$$= 1.3275 \text{ g/cm}^3$$

(b) $d'(x) = -0.90 + 0.47(2x)$
$$= -0.90 + 0.94x$$
$$d'(5) = -0.90 + 0.94(0.5)$$
$$= -0.43 \text{ g/cm}^3$$

71. $M(x) = 2.99x^3 - 382.12x^2 + 15,316.35x$
$\qquad - 172,148.37$
$M'(x) = 2.99(3x^{3-1}) - 382.12(2x^{2-1})$
$\qquad + 15,316.35x^{1-1} - 0$
$\qquad = 8.97x^2 - 764.24x + 15,316.35$

(a) $1930 - 1900 = 30$
$M'(30) = 462.15$

(b) $1935 - 1900 = 35$
$M'(35) = -443.8$

(c) $1940 - 1900 = 40$
$M'(40) = -901.25$

(d) $1955 - 1900 = 55$
$M'(55) = 417.4$

(e) $1980 - 1900 = 80$
$M'(80) = 11,585.15$

(f) $1999 - 1900 = 99$
$M'(99) = 27,571.56$

(g) The amount of money in circulation was increasing at the rate of \$462.15 million/year in 1930 but then began to decrease. The decrease of money continued through the early 1950's. Since then, the amount has again been increasing.

4.2 Derivatives of Products and Quotients

1. $y = (3x^2 + 2)(2x - 1)$

$\dfrac{dy}{dx} = (3x^2 + 2)(2) + (2x - 1)(6x)$
$\qquad = 6x^2 + 4 + 12x^2 - 6x$
$\qquad = 18x^2 - 6x + 4$

3. $y = (2x - 5)^2 = (2x - 5)(2x - 5)$

$\dfrac{dy}{dx} = (2x - 5)(2) + (2x - 5)(2)$
$\qquad = 4x - 10 + 4x - 10$
$\qquad = 8x - 20$

5. $k(t) = (t^2 - 1)^2$
$\qquad = (t^2 - 1)(t^2 - 1)$
$k'(t) = (t^2 - 1)(2t) + (t^2 - 1)(2t)$
$\qquad = 2t^3 - 2t + 2t^3 - 2t$
$\qquad = 4t^3 - 4t$

7. $y = (x + 1)(\sqrt{x} + 2)$
$\quad = (x + 1)(x^{1/2} + 2)$

$\dfrac{dy}{dx} = (x + 1)\left(\dfrac{1}{2}x^{-1/2}\right) + (x^{1/2} + 2)(1)$

$\quad = \dfrac{1}{2}x^{1/2} + \dfrac{1}{2}x^{-1/2} + x^{1/2} + 2$

$\quad = \dfrac{3}{2}x^{1/2} + \dfrac{1}{2}x^{-1/2} + 2$

or $\quad \dfrac{3x^{1/2}}{2} + \dfrac{1}{2x^{1/2}} + 2$

9. $p(y) = (y^{-1} + y^{-2})(2y^{-3} - 5y^{-4})$
$p'(y) = (y^{-1} + y^{-2})(-6y^{-4} + 20y^{-5})$
$\qquad + (-y^2 - 2y^{-3})(2y^{-3} - 5y^{-4})$
$\qquad = -6y^{-5} + 20y^{-6} - 6y^{-6} + 20y^{-7}$
$\qquad - 2y^{-5} + 5y^{-6} - 4y^{-6} + 10y^{-7}$
$\qquad = -8y^{-5} + 15y^{-6} + 30y^{-7}$

11. $f(x) = \dfrac{7x + 1}{3x + 8}$

$f'(x) = \dfrac{(3x + 8)(7) - (7x + 1)(3)}{(3x + 8)^2}$

$\qquad = \dfrac{21x + 56 - 21x - 3}{(3x + 8)^2}$

$\qquad = \dfrac{53}{(3x + 8)^2}$

13. $y = \dfrac{5 - 3t}{4 + t}$

$\dfrac{dy}{dx} = \dfrac{(4 + t)(-3) - (5 - 3t)(1)}{(4 + t)^2}$

$\qquad = \dfrac{-12 - 3t - 5 + 3t}{(4 + t)^2}$

$\qquad = \dfrac{-17}{(4 + t)^2}$

15. $y = \dfrac{x^2 + x}{x - 1}$

$\dfrac{dy}{dx} = \dfrac{(x - 1)(2x + 1) - (x^2 + x)(1)}{(x - 1)^2}$

$\qquad = \dfrac{2x^2 + x - 2x - 1 - x^2 - x}{(x - 1)^2}$

$\qquad = \dfrac{x^2 - 2x - 1}{(x - 1)^2}$

17. $f(t) = \dfrac{4t + 11}{t^2 - 3}$

$f'(t) = \dfrac{(t^2 - 3)(4) - (4t + 11)(2t)}{(t^2 - 3)^2}$

$= \dfrac{4t^2 - 12 - 8t^2 - 22t}{(t^2 - 3)^2}$

$= \dfrac{-4t^2 - 22t - 12}{(t^2 - 3)^2}$

19. $g(x) = \dfrac{x^2 - 4x + 2}{x + 3}$

$g'(x) = \dfrac{(x + 3)(2x - 4) - (x^2 - 4x + 2)(1)}{(x + 3)^2}$

$= \dfrac{2x^2 - 4x + 6x - 12 - x^2 + 4x - 2}{(x + 3)^2}$

$= \dfrac{x^2 + 6x - 14}{(x + 3)^2}$

21. $p(t) = \dfrac{\sqrt{t}}{t - 1}$

$= \dfrac{t^{1/2}}{t - 1}$

$p'(t) = \dfrac{(t - 1)\left(\frac{1}{2} t^{-1/2}\right) - t^{1/2}(1)}{(t - 1)^2}$

$= \dfrac{\frac{1}{2} t^{1/2} - \frac{1}{2} t^{-1/2} - t^{1/2}}{(t - 1)^2}$

$= \dfrac{-\frac{1}{2} t^{1/2} - \frac{1}{2} t^{-1/2}}{(t - 1)^2}$

$= \dfrac{-\frac{\sqrt{t}}{2} - \frac{1}{2\sqrt{t}}}{(t - 1)^2} \quad \text{or} \quad \dfrac{-t - 1}{2\sqrt{t}(t - 1)^2}$

23. $y = \dfrac{5x + 6}{\sqrt{x}} = \dfrac{5x + 6}{x^{1/2}}$

$\dfrac{dy}{dx} = \dfrac{(x^{1/2})(5) - (5x + 6)\left(\frac{1}{2} x^{-1/2}\right)}{(x^{1/2})^2}$

$= \dfrac{5x^{1/2} - \frac{5}{2} x^{1/2} - 3x^{-1/2}}{x}$

$= \dfrac{\frac{5}{2} x^{1/2} - 3x^{-1/2}}{x}$

$= \dfrac{\frac{5\sqrt{x}}{2} - \frac{3}{\sqrt{x}}}{x} \quad \text{or} \quad \dfrac{5x - 6}{2x\sqrt{x}}$

25. $g(y) = \dfrac{y^{1.4} + 1}{y^{2.5} + 2}$

$g'(y) = \dfrac{(y^{2.5} + 2)(1.4y^{0.4}) - (y^{1.4} + 1)(2.5y^{1.5})}{(y^{2.5} + 2)^2}$

$= \dfrac{1.4y^{2.9} + 2.8y^{0.4} - 2.5y^{2.9} - 2.5y^{1.5}}{(y^{2.5} + 2)^2}$

$= \dfrac{-1.1y^{2.9} - 2.5y^{1.5} + 2.8y^{0.4}}{(y^{2.5} + 2)^2}$

27. $h(x) = f(x)g(x)$
$h'(x) = f(x)g'(x) + g(x)f'(x)$
$h'(3) = f(3)g'(3) + g(3)f'(3)$
$\quad = 7(5) + 4(6) = 59$

29. In the first step, the two terms in the numerator are reversed. The correct work follows.

$D_x\left(\dfrac{2x + 5}{x^2 - 1}\right)$

$= \dfrac{(x^2 - 1)(2) - (2x + 5)(2x)}{(x^2 - 1)^2}$

$= \dfrac{2x^2 - 2 - 4x^2 - 10x}{(x^2 - 1)^2}$

$= \dfrac{-2x^2 - 10x - 2}{(x^2 - 1)^2}$

31. $f(x) = \dfrac{x}{x - 2}$, at $(3, 3)$

$m = f'(x) = \dfrac{(x - 2)(1) - x(1)}{(x - 2)^2}$

$= -\dfrac{2}{(x - 2)^2}$

At $(3, 3)$,

$m = -\dfrac{2}{(3 - 2)^2} = -2.$

Use the point-slope form.

$y - 3 = -2(x - 3)$
$y = -2x + 9$

33. $f(x)v(x) = u(x)$
$f'(x)v(x) + f(x)v'(x) = u'(x)$
$f'(x)v(x) = u'(x) - f(x)v'(x)$

$= u'(x) - \dfrac{u(x)}{v(x)} v'(x)$

$f'(x) = \dfrac{u'(x) - \frac{u(x)}{v(x)} v'(x)}{v(x)}$

$= \dfrac{u'(x)v(x) - u(x)v'(x)}{[v(x)]^2}$

35. Graph the numerical derivative of $f(x) = \frac{x-2}{x^2+4}$ for x ranging from -5 to 10. The derivative crosses the x-axis at approximately -0.828 and 4.828.

37. $f(x) = \dfrac{Kx}{A+x}$

(a) $f'(x) = \dfrac{(A+x)K - Kx(1)}{(A+x)^2}$

$f'(x) = \dfrac{AK}{(A+x)^2}$

(b) $f'(A) = \dfrac{AK}{(A+A)^2}$

$= \dfrac{AK}{4A^2}$

$= \dfrac{K}{4A}$

39. $R(w) = \dfrac{30(w-4)}{w-1.5}$

(a) $R(5) = \dfrac{30(5-4)}{5-1.5}$

≈ 8.57 min

(b) $R(7) = \dfrac{30(7-4)}{7-1.5}$

≈ 16.36 min

(c) $R'(w) = \dfrac{(w-1.5)(30) - 30(w-4)(1)}{(w-1.5)^2}$

$= \dfrac{30w - 45 - 30w + 120}{(w-1.5)^2}$

$= \dfrac{75}{(w-1.5)^2}$

$R'(5) = \dfrac{75}{(5-1.5)^2}$

$\approx 6.12 \dfrac{\text{min}^2}{\text{kcal}}$

$R'(7) = \dfrac{75}{(7-1.5)^2}$

$\approx 2.48 \dfrac{\text{min}^2}{\text{kcal}}$

41. $T(n) = \dfrac{an}{1+bn^2}$

$T'(n) = \dfrac{a(1+bn^2) - an(2bn)}{(1+bn^2)^2}$

$= \dfrac{a(1-bn^2)}{(1+bn^2)^2}$

The rate of change of the cell traction force per unit mass with respect to n, the number of cells, is given by $T'(n)$.

43. $L(C) = \dfrac{aC^n}{k+C^n}$

$L'(C) = \dfrac{a(k+C^n) - aC^n}{(k+C^n)^2}(nC^{n-1})$

$= \dfrac{aknC^{n-1}}{(k+C^n)^2}$

$L'(C)$ is the rate of change of the amount of neurotransmitter released with respect to C, the amount of intrcellular calcium.

45. $f(x) = \dfrac{x^2}{2(1-x)}$

$f'(x) = \dfrac{2(1-x)(2x) - x^2(-2)}{[2(1-x)]^2}$

$= \dfrac{4x - 4x^2 + 2x^2}{4(1-x)^2}$

$= \dfrac{4x - 2x^2}{4(1-x)^2}$

$= \dfrac{2x(2-x)}{4(1-x)^2}$

$= \dfrac{x(2-x)}{2(1-x)^2}$

(a) $f'(0.1) = \dfrac{0.1(2-0.1)}{2(1-0.1)^2} \approx 0.1173$

(b) $f'(0.6) = \dfrac{0.6(2-0.6)}{2(1-0.6)^2} = 2.625$

4.3   The Chain Rule

In Exercises 1-5, $f(x) = 4x^2 - 2x$ and $g(x) = 8x + 1$.

1. $g(2) = 8(2) + 1 = 17$
$f[g(2)] = f[17]$
$= 4(17)^2 - 2(17)$
$= 1{,}156 - 34 = 1{,}122$

3. $f(2) = 4(2)^2 - 2(2)$
$= 16 - 4 = 12$
$g[f(2)] = g[12]$
$= 8(12) + 1$
$= 96 + 1 = 97$

5. $f[g(k)] = 4(8k+1)^2 - 2(8k+1)$
$= 4(64k^2 + 16k + 1)$
$\quad - 16k - 2$
$= 256k^2 + 48k + 2$

7. $f(x) = \dfrac{x}{8} + 12$; $g(x) = 3x - 1$

$$f[g(x)] = \frac{3x - 1}{8} + 12$$

$$= \frac{3x - 1}{8} + \frac{96}{8}$$

$$= \frac{3x + 95}{8}$$

$$g[f(x)] = 3\left[\frac{x}{8} + 12\right] - 1$$

$$= \frac{3x}{8} + 36 - 1$$

$$= \frac{3x}{8} + 35$$

$$= \frac{3x}{8} + \frac{280}{8}$$

$$= \frac{3x + 280}{8}$$

9. $f(x) = \dfrac{1}{x}$; $g(x) = x^2$

$$f[g(x)] = \frac{1}{x^2}$$

$$g[f(x)] = \left(\frac{1}{x}\right)^2$$

$$= \frac{1}{x^2}$$

11. $f(x) = \sqrt{x + 2}$; $g(x) = 8x^2 - 6$

$$f[g(x)] = \sqrt{(8x^2 - 6) + 2}$$
$$= \sqrt{8x^2 - 4}$$
$$g[f(x)] = 8(\sqrt{x + 2})^2 - 6$$
$$= 8x + 16 - 6$$
$$= 8x + 10$$

13. $f(x) = \sqrt{x + 1}$; $g(x) = \dfrac{-1}{x}$

$$f[g(x)] = \sqrt{\frac{-1}{x} + 1}$$

$$= \sqrt{\frac{x - 1}{x}}$$

$$g[f(x)] = \frac{-1}{\sqrt{x + 1}}$$

17. $y = (5 - x)^{2/5}$

If $f(x) = x^{2/5}$ and
$\quad g(x) = 5 - x$,

then $y = f[g(x)] = (5 - x)^{2/5}$.

19. $y = -\sqrt{13 + 7x}$

If $f(x) = -\sqrt{x}$ and
$\quad g(x) = 13 + 7x$,

then $y = f[g(x)] = -\sqrt{13 + 7x}$.

21. $y = (x^2 + 5x)^{1/3} - 2(x^2 + 5x)^{2/3} + 7$

If $f(x) = x^{1/3} - 2x^{2/3} + 7$ and
$\quad g(x) = x^2 + 5x$,

then

$$y = f[g(x)] = (x^2 + 5x)^{1/3}$$
$$- 2(x^2 + 5x)^{2/3} + 7.$$

23. $y = (2x^3 + 9x)^5$

Let $f(x) = x^5$ and $g(x) = 2x^3 + 9x$. Then $(2x^3 + 9x)^5 = f[g(x)]$.

$$\frac{dy}{dx} = f'[g(x)] \cdot g'(x)$$

$$f'(x) = 5x^4$$
$$f'[g(x)] = 5[g(x)]^4$$
$$= 5(2x^3 + 9x)^4$$
$$g'(x) = 6x^2 + 9$$

$$\frac{dy}{dx} = 5(2x^3 + 9x)^4(6x^2 + 9)$$

25. $f(x) = -8(3x^4 + 2)^3$

Use the generalized power rule with
$u = 3x^4 + 2$, $n = 3$ and $u' = 12x^3$.

$$f'(x) = -8[3(3x^4 + 2)^{3-1} \cdot 12x^3]$$
$$= -8[36x^3(3x^4 + 2)^2]$$
$$= -288x^3(3x^4 + 2)^2$$

27. $s(t) = 12(2t^4 + 5)^{3/2}$

Use generalized power rule with $u = 2t^4 + 5$,
$n = \frac{3}{2}$, and $u' = 8t^3$.

$$s'(t) = 12\left[\frac{3}{2}(2t^4 + 5)^{1/2} \cdot 8t^3\right]$$

$$= 12[12t^3(2t^4 + 5)^{1/2}]$$
$$= 144t^3(2t^4 + 5)^{1/2}$$

29. $f(t) = 8\sqrt{4t^2 + 7}$
$= 8(4t^2 + 7)^{1/2}$

Use generalized power rule with $u = 4t^2 + 7$, $n = \frac{1}{2}$, and $u' = 8t$.

$$f'(t) = 8\left[\frac{1}{2}(4t^2 + 7)^{-1/2} \cdot 8t\right]$$

$$= 8[4t(4t^2 + 7)^{-1/2}]$$
$$= 32t(4t^2 + 7)^{-1/2}$$

$$= \frac{32t}{(4t^2 + 7)^{1/2}}$$

$$= \frac{32t}{\sqrt{4t^2 + 7}}$$

31. $r(t) = 4t(2t^5 + 3)^2$

Use the product rule and the power rule.

$$r'(t) = 4t[2(2t^5 + 3) \cdot 10t^4]$$
$$+ (2t^5 + 3)^2 \cdot 4$$
$$= 80t^5(2t^5 + 3) + 4(2t^5 + 3)^2$$
$$= 4(2t^5 + 3)[20t^5 + (2t^5 + 3)]$$
$$= 4(2t^5 + 3)(22t^5 + 3)$$

33. $y = (x^3 + 2)(x^2 - 1)^2$

Use the product rule and the power rule.

$$\frac{dy}{dx} = (x^3 + 2)[2(x^2 - 1) \cdot 2x]$$
$$+ (x^2 - 1)^2(3x^2)$$
$$= (x^3 + 2)[4x(x^2 - 1)]$$
$$+ 3x^2(x^2 - 1)^2$$
$$= (x^2 - 1)[4x(x^3 + 2) + 3x^2(x^2 - 1)]$$
$$= (x^2 - 1)(4x^4 + 8x + 3x^4 - 3x^2)$$
$$= (x^2 - 1)(7x^4 - 3x^2 + 8x)$$

35. $p(z) = z(6z + 1)^{4/3}$

Use the product rule and the power rule.

$$p'(z) = z \cdot \frac{4}{3}(6z + 1)^{1/3} \cdot 6$$

$$+ 1 \cdot (6z + 1)^{4/3}$$
$$= 8z(6z + 1)^{1/3} + (6z + 1)^{4/3}$$
$$= (6z + 1)^{1/3}[8z + (6z + 1)^1]$$
$$= (6z + 1)^{1/3}(14z + 1)$$

37. $y = \dfrac{1}{(3x^2 - 4)^5} = (3x^2 - 4)^{-5}$

$$\frac{dy}{dx} = -5(3x^2 - 4)^{-6} \cdot 6x$$
$$= -30x(3x^2 - 4)^{-6}$$

$$= \frac{-30x}{(3x^2 - 4)^6}$$

39. $p(t) = \dfrac{(2t + 3)^3}{4t^2 - 1}$

$$p'(t) = \frac{(4t^2 - 1)[3(2t + 3)^2 \cdot 2] - (2t + 3)^3(8t)}{(4t^2 - 1)^2}$$

$$= \frac{6(4t^2 - 1)(2t + 3)^2 - 8t(2t + 3)^3}{(4t^2 - 1)^2}$$

$$= \frac{(2t + 3)^2[6(4t^2 - 1) - 8t(2t + 3)]}{(4t^2 - 1)^2}$$

$$= \frac{(2t + 3)^2[24t^2 - 6 - 16t^2 - 24t]}{(4t^2 - 1)^2}$$

$$= \frac{(2t + 3)^2[8t^2 - 24t - 6]}{(4t^2 - 1)^2}$$

$$= \frac{2(2t + 3)^2(4t^2 - 12t - 3)}{(4t^2 - 1)^2}$$

41. $y = \dfrac{x^2 + 4x}{(3x^3 + 2)^4}$

$$\frac{dy}{dx} = \frac{(3x^3 + 2)^4(2x + 4) - (x^2 + 4x)[4(3x^3 + 2)^3 \cdot 9x^2]}{[(3x^3 + 2)^4]^2}$$

$$= \frac{(3x^3 + 2)^4(2x + 4) - 36x^2(x^2 + 4x)(3x^3 + 2)^3}{(3x^3 + 2)^8}$$

$$= \frac{2(3x^3 + 2)^3[(3x^3 + 2)(x + 2) - 18x^2(x^2 + 4x)]}{(3x^3 + 2)^8}$$

$$= \frac{2(3x^4 + 6x^3 + 2x + 4 - 18x^4 - 72x^3)}{(3x^3 + 2)^5}$$

$$= \frac{-30x^4 - 132x^3 + 4x + 8}{(3x^3 + 2)^5}$$

43. (a) $D_x(f[g(x)])$ at $x = 1$

$$= f'[g(1)] \cdot g'(1)$$

$$= f'(2) \cdot \left(\frac{2}{7}\right)$$

$$= -7\left(\frac{2}{7}\right)$$

$$= -2$$

(b) $D_x(f[g(x)])$ at $x = 2$

$$= f'[g(2)] \cdot g'(2)$$

$$= f'(3) \cdot \left(\frac{3}{7}\right)$$

$$= -8\left(\frac{3}{7}\right)$$

$$= -\frac{24}{7}$$

45. $f(x) = \sqrt{x^2 + 16}; x = 3$

$f(x) = (x^2 + 16)^{1/2}$

$f'(x) = \dfrac{1}{2}(x^2 + 16)^{-1/2}(2x)$

$f'(x) = \dfrac{x}{\sqrt{x^2 + 16}}$

$f'(3) = \dfrac{3}{\sqrt{3^2 + 16}} = \dfrac{3}{5}$

$f(3) = \sqrt{3^2 + 16} = 5$

We use $m = \frac{3}{5}$ and the point $P(3, 5)$ in the point-slope form.

$$y - 5 = \frac{3}{5}(x - 3)$$

$$y - 5 = \frac{3}{5}x - \frac{9}{5}$$

$$y = \frac{3}{5}x + \frac{16}{5}$$

47. $f(x) = x(x^2 - 4x + 5)^4; x = 2$

$f'(x) = x \cdot 4(x^2 - 4x + 5)^3 \cdot (2x - 4)$
$\qquad + 1 \cdot (x^2 - 4x + 5)^4$

$\quad = (x^2 - 4x + 5)^3$
$\qquad \cdot [4x(2x - 4) + (x^2 - 4x + 5)]$

$\quad = (x^2 - 4x + 5)^3(9x^2 - 20x + 5)$

$f'(2) = (1)^3(1) = 1$

$f(2) = 2(1)^4 = 2$

We use $m = 1$ and the point $P(2, 2)$.

$$y - 2 = 1(x - 2)$$
$$y - 2 = x - 2$$
$$y = x$$

49. $f(x) = \sqrt{x^3 - 6x^2 + 9x + 1}$

$f(x) = (x^3 - 6x^2 + 9x + 1)^{1/2}$

$f'(x) = \dfrac{1}{2}(x^3 - 6x^2 + 9x + 1)^{-1/2}$
$\qquad\qquad \cdot (3x^2 - 12x + 9)$

$f'(x) = \dfrac{3(x^2 - 4x + 3)}{2\sqrt{x^3 - 6x^2 + 9x + 1}}$

If the tangent line is horizontal, its slope is zero and $f'(x) = 0$.

$$\frac{3(x^2 - 4x + 3)}{2\sqrt{x^3 - 6x^2 + 9x + 1}} = 0$$

$$3(x^2 - 4x + 3) = 0$$

$$3(x - 1)(x - 3) = 0$$

$$x = 1 \quad \text{or} \quad x = 3$$

The tangent line is horizontal at $x = 1$ and $x = 3$.

53. $P(x) = 2x^2 + 1; \ x = f(a) = 3a + 2$

$P[f(a)] = 2(3a + 2)^2 + 1$
$\qquad\quad = 2(9a^2 + 12a + 4) + 1$
$\qquad\quad = 18a^2 + 24a + 9$

55. $L(w) = 2.472w^{2.571}, w(t) = 0.265 + 0.21t$

$L'(t) = L'(w) \cdot w'(t)$
$\qquad = (2.472)(2.571)w^{(2.571-1)} \cdot (0.21)$
$\qquad \approx 1.335w(t)^{1.571}$
$\qquad \approx 1.335(0.265 + 0.21t)^{1.571}$

$L'(25) \approx 1.335(0.265 + 0.21(25))^{1.571}$
$\qquad\quad \approx 19.5 \text{ mm/week}$

57. $N(t) = 2t(5t + 9)^{1/2} + 12$

$N'(t) = (2t)\left[\dfrac{1}{2}(5t + 9)^{-1/2}(5)\right]$
$\qquad\quad + 2(5t + 9)^{1/2} + 0$

$\quad = 5t(5t + 9)^{-1/2} + 2(5t + 9)^{1/2}$

$\quad = (5t + 9)^{-1/2}[5t + 2(5t + 9)]$

$\quad = (5t + 9)^{-1/2}(15t + 18)$

$\quad = \dfrac{15t + 18}{(5t + 9)^{1/2}}$

(a) $N'(0) = \dfrac{15(0) + 18}{[5(0) + 9]^{1/2}}$

$\qquad\qquad = \dfrac{18}{9^{1/2}} = 6$

(b) $N'\left(\dfrac{7}{5}\right) = \dfrac{15\left(\frac{7}{5}\right) + 18}{\left[5\left(\frac{7}{5}\right) + 9\right]^{1/2}}$

$\qquad\qquad = \dfrac{21 + 8}{(7 + 9)^{1/2}}$

$\qquad\qquad = \dfrac{39}{(16)^{1/2}}$

$\qquad\qquad = \dfrac{39}{4}$

$\qquad\qquad = 9.75$

(c) $N'(8) = \dfrac{15(8) + 18}{[5(8) + 9]^{1/2}}$

$\qquad\qquad = \dfrac{120 + 18}{(49)^{1/2}}$

$\qquad\qquad = \dfrac{138}{7}$

$\qquad\qquad \approx 19.71$

59. (a) $R(Q) = Q\left(C - \dfrac{Q}{3}\right)^{1/2}$

$R'(Q) = Q\left[\dfrac{1}{2}\left(C - \dfrac{Q}{3}\right)^{-1/2}\left(-\dfrac{1}{3}\right)\right]$

$\qquad + \left(C - \dfrac{Q}{3}\right)^{1/2}$ (1)

$\qquad = -\dfrac{1}{6}Q\left(C - \dfrac{Q}{3}\right)^{-1/2} + \left(C - \dfrac{Q}{3}\right)^{1/2}$

$\qquad = -\dfrac{Q}{6\left(C - \frac{Q}{3}\right)^{1/2}} + \left(C - \dfrac{Q}{3}\right)^{1/2}$

(b) $R'(Q) = -\dfrac{Q}{6\left(C - \frac{Q}{3}\right)^{1/2}} + \left(C - \dfrac{Q}{3}\right)^{1/2}$

If $Q = 87$ and $C = 59$, then

$R'(Q) = \left(59 - \dfrac{87}{3}\right)^{1/2} - \dfrac{87}{6\left(59 - \frac{87}{3}\right)^{1/2}}$

$\qquad = (30)^{1/2} - \dfrac{87}{6(30)^{1/2}}$

$\qquad = 5.48 - \dfrac{87}{32.88}$

$\qquad = 5.48 - 2.65$

$\qquad = 2.83.$

(c) Because $R'(Q)$ is positive, the patient's sensitivity to the drug is increasing.

61. $V(r) = \dfrac{4}{3}\pi r^3, S(r) = 4\pi r^2, r(t) = 6 - \dfrac{3}{17}t$

(a) $r(t) = 0$ when $6 - \dfrac{3}{17}t = 0;$

$t = \dfrac{17(6)}{3} = 34$ min.

(b) $\dfrac{dV}{dr} = 4\pi r^2, \dfrac{dS}{dr} = 8\pi r, \dfrac{dr}{dt} = -\dfrac{3}{17}$

$\dfrac{dV}{dt} = \dfrac{dV}{dr}\cdot\dfrac{dr}{dt} = -\dfrac{12}{17}\pi r^2$

$\qquad = -\dfrac{12}{17}\pi\left(6 - \dfrac{3}{17}t\right)^2$

$\dfrac{dS}{dt} = \dfrac{dS}{dr}\cdot\dfrac{dr}{dt} = -\dfrac{24}{17}\pi r$

$\qquad = -\dfrac{24}{17}\pi\left(6 - \dfrac{3}{17}t\right)$

When $t = 17$,

$\dfrac{dV}{dt} = -\dfrac{12}{17}\pi\left[6 - \dfrac{3}{17}(17)\right]^2$

$\qquad = -\dfrac{108}{17}\pi$ mm^3/mm

$\dfrac{dS}{dt} = -\dfrac{24}{17}\pi\left[6 - \dfrac{3}{17}(17)\right]$

$\qquad = -\dfrac{72}{17}\pi$ mm^2/mm

At $t = 17$ minutes, the volume is decreasing by $\dfrac{108}{17}\pi$ mm^3 per minute and the surface area is decreasing by $\dfrac{72}{17}\pi$ mm^2 per minute.

63. $V = \dfrac{6,000}{1 + 0.3t + 0.1t^2}$

The rate of change of the value is

$V'(t)$

$\qquad = \dfrac{(1 + 0.3t + 0.1t^2)(0) - 6,000(0.3 + 0.2t)}{(1 + 0.3t + 0.1t^2)^2}$

$\qquad = \dfrac{-6,000(0.3 + 0.2t)}{(1 + 0.3t + 0.1t^2)^2}.$

(a) 2 years after purchase, the rate of change in the value is

$V'(2) = \dfrac{-6,000[0.3 + 0.2(2)]}{[1 + 0.3(2) + 0.1(2)^2]^2}$

$\qquad = \dfrac{-6,000(0.3 + 0.4)}{(1 + 0.6 + 0.4)^2}$

$\qquad = \dfrac{-4,200}{4}$

$\qquad = -\$1,050.$

(b) 4 years after purchase, the rate of change in the value is

$V'(4) = \dfrac{-6,000[0.3 + 0.2(4)]}{[1 + 0.3(4) + 0.1(4)^2]^2}$

$\qquad = \dfrac{-6,600}{14.44}$

$\qquad = -\$457.06.$

4.4   Derivatives of Exponential Functions

1. $\qquad\qquad y = e^{4x}$

Let $\quad g(x) = 4x,$

with $\quad g'(x) = 4.$

$\dfrac{dy}{dx} = 4e^{4x}$

3. $y = -8e^{2x}$

$$\frac{dy}{dx} = -8(2e^{2x}) = -16e^{2x}$$

5. $y = -16e^{x+1}$

$g(x) = x + 1$

$g'(x) = 1$

$$\frac{dy}{dx} = -16[(1)e^{x+1}] = -16e^{x+1}$$

7. $y = e^{x^2}$

$g(x) = x^2$

$g'(x) = 2x$

$$\frac{dy}{dx} = 2xe^{x^2}$$

9. $y = 3e^{2x^2}$

$g(x) = 2x^2$

$g'(x) = 4x$

$$\frac{dy}{dx} = 3(4xe^{2x^2})$$
$$= 12xe^{2x^2}$$

11. $y = 4e^{2x^2-4}$

$g(x) = 2x^2 - 4$

$g'(x) = 4x$

$$\frac{dy}{dx} = 4[(4x)e^{2x^2-4}]$$
$$= 16xe^{2x^2-4}$$

13. $y = xe^x$

Use the product rule.

$$\frac{dy}{dx} = xe^x + e^x \cdot 1$$
$$= e^x(x + 1)$$

15. $y = (x - 3)^2 e^{2x}$

Use the product rule.

$$\frac{dy}{dx} = (x - 3)^2(2)e^{2x} + e^{2x}(2)(x - 3)$$
$$= 2(x - 3)^2 e^{2x} + 2(x - 3)e^{2x}$$
$$= 2(x - 3)e^{2x}[(x - 3) + 1]$$
$$= 2(x - 3)(x - 2)e^{2x}$$

17. $y = \dfrac{x^2}{e^x}$

Use the quotient rule.

$$\frac{dy}{dx} = \frac{e^x(2x) - x^2 e^x}{(e^x)^2}$$
$$= \frac{xe^x(2 - x)}{e^{2x}}$$
$$= \frac{x(2 - x)}{e^x}$$

19. $y = \dfrac{e^x + e^{-x}}{x}$

$$\frac{dy}{dx} = \frac{x(e^x - e^{-x}) - (e^x + e^{-x})}{x^2}$$

21. $p = \dfrac{10,000}{9 + 4e^{-0.2t}}$

$$\frac{dp}{dt} = \frac{(9 + 4e^{-0.2t}) \cdot 0 - 10,000[0 + 4(-0.2)e^{-0.2t}]}{(9 + 4e^{-0.2t})^2}$$
$$= \frac{8,000e^{-0.2t}}{(9 + 4e^{-0.2t})^2}$$

23. $f(z) = (2z + e^{-z^2})^2$

$f'(z) = 2(2z + e^{-z^2})^1(2 - 2ze^{-z^2})$

$\quad\; = 4(2z + e^{-z^2})(1 - ze^{-z^2})$

25. $y = 2^{-x} = e^{\ln 2^{-x}} = e^{-x \ln 2}$

$$\frac{dy}{dx} = -(\ln 2)e^{-x \ln 2} = -(\ln 2)2^{-x}$$

27. $y = -10^{3x^2-4} = -e^{\ln (10^{3x^2-4})}$

$\quad = -e^{(3x^2-4)\ln 10}$

$$\frac{dy}{dx} = -\ln 10(6x)e^{(3x^2-4)\ln 10}$$
$$= -6x(\ln 10)(10^{3x^2-4})$$
$$= -6x(10^{3x^2-4})\ln 10$$

29. $s = 5 \cdot 7^{\sqrt{t-2}} = 5 \cdot e^{\ln 7^{\sqrt{t-2}}} = 5e^{\sqrt{t-1}(\ln 7)}$

$$\frac{ds}{dt} = 5(\ln 7)\frac{1}{2}(t - 2)^{-1/2}e^{\sqrt{t-2}\ln 7}$$
$$= \frac{5(\ln 7)e^{\sqrt{t-2}\ln 7}}{2(t - 2)^{1/2}} = \frac{(5 \ln 7)(7^{\sqrt{t-2}})}{2\sqrt{t - 2}}$$

31. Graph

$$y = \frac{e^{x+0.0001} - e^x}{0.0001}$$

on a graphing calculator. A good choice for the viewing window is $[-1, 4]$ by $[-1, 16]$ with Xscl $=$ 1, Yscl $= 2$.

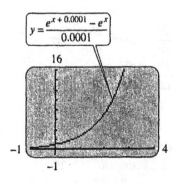

If we graph $y = e^x$ on the same screen, we see that the two graphs coincide. They are close enough to being identical that they are indistinguishable. By the definition of the derivative, if $f(x) = e^x$,

$$f'(x) = \lim_{h \to 0} \frac{f(x+h) - f(x)}{h}$$

$$= \lim_{h \to 0} \frac{e^{x+h} - e^x}{h},$$

and $h = 0.0001$ is very close to 0. Comparing the two graphs provides graphical evidence that

$$f'(x) = e^x.$$

33. $G(t) = \dfrac{m\, G_0}{G_0 + (m - G_0)e^{-kmt}};$

where $G_0 = 200; m = 10{,}000;$ and $k = 0.00001.$

(a) $G(t) = \dfrac{10{,}000(200)}{200 + (10{,}000 - 200)e^{-(0.00001)(10{,}000)t}}$

$G(t) = \dfrac{10{,}000}{1 + 49e^{-0.1t}}$

(b) $G(t) = 10{,}000(1 + 49e^{-0.1t})^{-1}$
  $G'(t) = -10{,}000(1 + 49e^{-0.1t})^{-2}(-4.9e^{-0.1t})$

$$= \frac{49{,}000e^{-0.1t}}{(1 + 49e^{-0.1t})^2}$$

$G(6) = \dfrac{10{,}000}{1 + 49e^{-0.6}} \approx 359$

$G'(6) = \dfrac{49{,}000e^{-0.6}}{(1 + 49e^{-0.6})^2} \approx 34.6$

(c) After 3 years, $t = 36.$

$G(36) = \dfrac{10{,}000}{1 + 49e^{-3.6}} \approx 4{,}276$

$G'(36) = \dfrac{49{,}000e^{-3.6}}{(1 + 49e^{-3.6})^2} \approx 245$

(d) After 7 years, $t = 84.$

$G(84) = \dfrac{10{,}000}{1 + 49e^{-8.4}} \approx 9{,}891$

$G'(84) = \dfrac{49{,}000e^{-8.4}}{(1 + 49e^{-8.4})^2} \approx 10.78$

(e) It increases for a while and then gradually decreases to 0.

35. $P(x) = 0.04e^{-4x}$

(a) $P(.5) = 0.04e^{-4(0.5)}$
  $= 0.04e^{-2}$
  ≈ 0.005

(b) $P(1) = 0.04e^{-4(1)}$
  $= 0.04e^{-4}$
  ≈ 0.0007

(c) $P(2) = 0.04e^{-4(2)}$
  $= 0.04e^{-8}$
  ≈ 0.000013

$P'(x) = 0.04(-4)e^{-4x}$
  $= -0.16e^{-4x}$

(d) $P'(0.5) = -0.16e^{-4(0.5)}$
  $= -0.16e^{-2}$
  ≈ -0.022

(e) $P'(1) = -0.16e^{-4(1)}$
  $= -0.16e^{-4}$
  ≈ -0.0029

(f) $P'(2) = -0.16e^{-4(2)}$
  $= -0.16e^{-8}$
  ≈ -0.000054

37. $P(t) = 0.00239e^{0.0957t}$

(a) $P(25) = 0.00239e^{0.0957(25)}$
  $\approx 0.026\%$
  $P(50) = 0.00239e^{0.0957(50)}$
  $\approx 0.286\%$
  $P(75) = 0.00239e^{0.0957(75)}$
  $\approx 3.130\%$

(b) $P'(t) = 0.00239e^{0.0957t}(0.0957)$
  $= 0.000228723e^{0.0957t}$
  $P'(25) = 0.000228723e^{0.0957(25)}$
  $\approx 0.0025\%/\text{year}$
  $P'(50) = 0.000228723e^{0.0957(50)}$
  $\approx 0.0274\%/\text{year}$
  $P'(75) = 0.000228723e^{0.0957(75)}$
  $\approx 0.300\%/\text{year}$

39. $p(x) = 0.001131e^{0.1268x}$

    (a) $p(25) = 0.001131e^{0.1268(25)} \approx 0.027$

    (b) When $p(x) = 1$,

$$0.001131e^{0.1268x} = 1$$

$$e^{0.1268x} = \frac{1}{0.001131}$$

$$0.1268x = \ln\frac{1}{0.001131}$$

$$x = \frac{1}{0.1268}\ln\frac{1}{0.001131}$$

$$\approx 54$$

This represents the year 2024.

    (c) $p'(x) = 0.001131e^{0.1268x}(0.1268)$

$$= 0.0001434108e^{0.1268x}$$

$$p'(32) = 0.0001434108e^{0.1268(32)}$$

$$\approx 0.008$$

The marginal increase in the proportion per year in 2002 is approximately 0.008.

41. $L(t) = 589\{1 - e^{-0.168(t+2.682))}\}$

    (a) Over time, the length of the average cutlass fish approaches 589 mm assymptotically.

    (b) $(0.95)589 = 589\{1 - e^{(-0.168(t+2.682))}\}$

$$0.95 = 1 - e^{-0.168(t+2.682)}$$

$$\ln(0.05) = -0.168(t + 2.682)$$

$$t = 15.1497\ldots \approx 15 \text{ years old}$$

    (c) $L'(t) = 589(-e^{-0.168(t+2.682)})(-0.168)$

$$= 98.952e^{-0.168(t+2.682)})$$

$$L'(4) = 98.952e^{-0.168(4+2.682)}$$

$$\approx 32.2 \text{ mm/year}$$

When a cutlassfish is 4 yers old, it is growing in length at a rate of 32.2 millimeters per year.

    (d)

43. $W(t) = 619(1 - 0.905e^{-0.002t})^{1.2386}$

    (a) $W(t)$ is a strictly increasing function.

$$\lim_{t\to\infty} W(t) = \lim_{t\to\infty} 619(1 - 0.905e^{-0.002t})^{1.2386}$$
$$= 619(1 - 0)^{1.2386} = 619 \text{ kg}$$

    (b) $0.9(619) = 619(1 - 0.905e^{-0.002t})^{1.2386}$

$$0.9 = (1 - 0.905e^{-0.002t})^{1.2386}$$

$$\frac{1 - 0.9^{1/1.2386}}{0.905} = e^{-0.002t}$$

$$t = 1{,}203.379 \approx 1{,}203 \text{ days}$$

    (c) $W'(t) \approx 1.3877(1 - 0.905e^{-0.002t})^{0.2386}(e^{-0.002}$

$$W'(1{,}000) \approx 1.3877(1 - 0.905e^{-0.002(1{,}000)})^{0.2386}$$
$$\cdot (e^{-0.002(1{,}000)})$$
$$= 0.1820\ldots \approx 0.18 \text{ kg/day.}$$

When the female cow is 1,000 days old, it is gaining 0.18 kilograms per day.

45. $I = E[1 - e^{-a(S-h)/E}]$

$$\frac{dI}{dS} = E\left(\frac{-a}{E}\right)(-e^{-a(S-h)/E})$$

$$= ae^{-a(S-h)/E}$$

47. $t(r) = 218 + 31(0.933)^n$

    (a) $t(51) = 218 + 31(0.933)^{51}$

$$\approx 218.9 \text{ sec}$$

    (b) $t'(n) = (31\ln 0.933)(0.933)^n$

$$t'(51) = (31\ln 0.933)(0.933)^{51}$$

$$\approx -0.63$$

The record is decreasing by 0.063 seconds per year at the end of 2001.

    (c) As $n \to \infty$, $(0.933)^n \to 0$ and $t(n) \to 218$. If the estimate is correct, then this is the least amount of time that it will ever take a human to run a mile. At the end of 2000, the world record stood at 3:43.13, or 223.13 seconds.

49. $P(t) = 26.7e^{0.023t}$

    (a) $P(5) = 26.7e^{0.023(5)} \approx 29.95$

so, the U.S. Latino-American population in 2000 was approximately 29,950,000.

    (b) $P'(t) = 26.7e^{0.023t}(0.023)$

$$= 0.6141e^{0.023t}$$

$$P'(5) = 0.6141e^{0.023(5)}$$

$$\approx 0.689$$

The Latino-American population was increasing at the rate of 0.689 million/year at the end of the year 2000.

51. $Q(t) = CV(1 - e^{-t/RC})$

(a) $I_c = \dfrac{dQ}{dt} = CV\left[0 - e^{-t/RC}\left(-\dfrac{1}{RC}\right)\right]$

$\qquad = CV\left(\dfrac{1}{RC}\right)e^{-t/RC}$

$\qquad = \dfrac{V}{R}e^{-t/RC}$

(b) When $C = 10^{-5}$ farads, $R = 10^7$ ohms, and $V = 10$ volts, after 200 seconds

$I_c = \frac{10}{10^7}e^{-200/(10^7 \cdot 10^{-5})} \approx 1.35 \times 10^{-7}$ amps

4.5   Derivatives of Logarithmic Functions

1. $y = \ln(8x)$

$\dfrac{dy}{dx} = \dfrac{d}{dx}(\ln 8x)$

$\qquad = \dfrac{d}{dx}(\ln 8 + \ln x)$

$\qquad = \dfrac{d}{dx}(\ln 8) + \dfrac{d}{dx}(\ln x)$

$\qquad = 0 + \dfrac{1}{x}$

$\qquad = \dfrac{1}{x}$

3. $y = \ln(3 - x)$
$\quad g(x) = 3 - x$
$\quad g'(x) = -1$

$\dfrac{dy}{dx} = \dfrac{g'(x)}{g(x)}$

$\qquad = \dfrac{-1}{3 - x}$ or $\dfrac{1}{x - 3}$

5. $y = \ln\left|2x^2 - 7x\right|$
$\quad g(x) = 2x^2 - 7x$
$\quad g'(x) = 4x - 7$

$\dfrac{dy}{dx} = \dfrac{4x - 7}{2x^2 - 7x}$

7. $y = \ln\sqrt{x + 5}$
$\quad g(x) = \sqrt{x + 5}$
$\qquad = (x + 5)^{1/2}$

$\quad g'(x) = \dfrac{1}{2}(x + 5)^{-1/2}$

$\dfrac{dy}{dx} = \dfrac{\frac{1}{2}(x + 5)^{-1/2}}{(x + 5)^{1/2}}$

$\qquad = \dfrac{1}{2(x + 5)}$

9. $y = \ln(x^4 + 5x^2)^{3/2}$

$\qquad = \dfrac{3}{2}\ln(x^4 + 5x^2)$

$\dfrac{dy}{dx} = \dfrac{3}{2}D_x[\ln(x^4 + 5x^2)]$

$\quad g(x) = x^4 + 5x^2$
$\quad g'(x) = 4x^3 + 10x$

$\dfrac{dy}{dx} = \dfrac{3}{2}\left(\dfrac{4x^3 + 10x}{x^4 + 5x^2}\right)$

$\qquad = \dfrac{3}{2}\left[\dfrac{2x(2x^2 + 5)}{x^2(x^2 + 5)}\right]$

$\qquad = \dfrac{3(2x^2 + 5)}{x(x^2 + 5)}$

11. $y = -3x\ln(x + 2)$

Use the product rule.

$\dfrac{dy}{dx} = -3x\left[\dfrac{d}{dx}\ln(x + 2)\right]$

$\qquad + \ln(x + 2)\left[\dfrac{d}{dx}(-3x)\right]$

$\qquad = -3x\left(\dfrac{1}{x + 2}\right) + [\ln(x + 2)](-3)$

$\qquad = -\dfrac{3x}{x + 2} - 3\ln(x + 2)$

13. $s = t^2\ln|t|$

$\dfrac{ds}{dt} = t^2 \cdot \dfrac{1}{t} + 2t\ln|t|$

$\qquad = t + 2t\ln|t|$

$\qquad = t(1 + 2\ln|t|)$

15. $y = \dfrac{2 \ln (x+3)}{x^2}$

Use the quotient rule.

$$\frac{dy}{dx} = \frac{x^2 \left(\frac{2}{x+3}\right) - 2 \ln (x+3) \cdot 2x}{(x^2)^2}$$

$$= \frac{\frac{2x^2}{x+3} - 4x \ln (x+3)}{x^4}$$

$$= \frac{2x^2 - 4x(x+3) \ln (x+3)}{x^4(x+3)}$$

$$= \frac{x[2x - 4(x+3) \ln (x+3)]}{x^4(x+3)}$$

$$= \frac{2x - 4(x+3) \ln (x+3)}{x^3(x+3)}$$

17. $y = \dfrac{\ln x}{4x+7}$

Use the quotient rule.

$$\frac{dy}{dx} = \frac{(4x+7)\left(\frac{1}{x}\right) - (\ln x)(4)}{(4x+7)^2}$$

$$= \frac{\frac{4x+7}{x} - 4 \ln x}{(4x+7)^2}$$

$$= \frac{4x+7 - 4x \ln x}{x(4x+7)^2}$$

19. $y = \dfrac{3x^2}{\ln x}$

$$\frac{dy}{dx} = \frac{(\ln x)(6x) - 3x^2 \left(\frac{1}{x}\right)}{(\ln x)^2}$$

$$= \frac{6x \ln x - 3x}{(\ln x)^2}$$

21. $y = (\ln |x+1|)^4$

$$\frac{dy}{dx} = 4(\ln |x+1|)^3 \left(\frac{1}{x+1}\right)$$

$$= \frac{4(\ln |x+1|)^3}{x+1}$$

23. $y = \ln |\ln x|$
       $g(x) = \ln x$

$$g'(x) = \frac{1}{x}$$

$$\frac{dy}{dx} = \frac{g'(x)}{g(x)}$$

$$= \frac{\frac{1}{x}}{\ln x}$$

$$= \frac{1}{x \ln x}$$

25. $y = e^{x^2} \ln x, \; x > 0$

$$\frac{dy}{dx} = e^{x^2} \left(\frac{1}{x}\right) + (\ln x)(2x)e^{x^2}$$

$$= \frac{e^{x^2}}{x} + 2xe^{x^2} \ln x$$

27. $y = \dfrac{e^x}{\ln x}, \; x > 0$

Use the quotient rule.

$$\frac{dy}{dx} = \frac{(\ln x)e^x - e^x \left(\frac{1}{x}\right)}{(\ln x)^2} \cdot \frac{x}{x}$$

$$= \frac{xe^x \ln x - e^x}{x (\ln x)^2}$$

29. $g(z) = (e^{2z} + \ln z)^3$

$$g'(z) = 3(e^{2z} + \ln z)^2 \left(e^{2z} \cdot 2 + \frac{1}{z}\right)$$

$$= 3(e^{2z} + \ln z)^2 \left(\frac{2ze^{2z} + 1}{z}\right)$$

31. $y = \log (2x - 3) = \dfrac{\ln (2x-3)}{\ln 10}$

$$\frac{dy}{dx} = \frac{1}{\ln 10} \cdot \frac{d}{dx}[\ln (2x-3)]$$

$$= \frac{1}{\ln 10} \cdot \frac{2}{2x-3}$$

$$= \frac{2}{(\ln 10)(2x-3)}$$

33. $y = \log |3x| = \dfrac{\ln |3x|}{\ln 10}$

$$\frac{dy}{dx} = \frac{1}{\ln 10} \cdot \frac{3}{3x}$$

$$= \frac{1}{x \ln 10}$$

35. $y = \log_7 \sqrt{2x-3} = \dfrac{\ln \sqrt{2x-3}}{\ln 7}$

$$= \frac{1}{\ln 7} \cdot \ln (2x-3)^{1/2}$$

$$= \frac{1}{\ln 7} \cdot \frac{1}{2} \ln (2x-3)$$

$$\frac{dy}{dx} = \frac{1}{2 \ln 7} \cdot \frac{2}{2x-3}$$

$$= \frac{1}{(\ln 7)(2x-3)}$$

37. $y = \log_2 (2x^2 - x)^{5/2}$

$$= \frac{\ln (2x^2 - x)^{5/2}}{\ln 2}$$

$$= \frac{1}{\ln 2} \cdot \frac{5}{2} \ln (2x^2 - x)$$

$$\frac{dy}{dx} = \frac{5}{2 \ln 2} \cdot \frac{4x - 1}{2x^2 - x}$$

$$= \frac{5(4x - 1)}{2(\ln 2)(2x^2 - x)}$$

39. $z = 10^y \log y$

$$z' = 10^y \cdot \frac{1}{\ln 10} \cdot \frac{1}{y} + (\ln 10)10^y \log y$$

$$= \frac{10^y}{y \ln 10} + (\ln 10) \, 10^y \log y$$

41. Note that a is a constant.

$$\frac{d}{dx} \ln |ax| = \frac{d}{dx}(\ln |a| + \ln |x|)$$

$$= \frac{d}{dx} \ln |a| + \frac{d}{dx} \ln x$$

$$= 0 + \frac{d}{dx} \ln |x|$$

$$= \frac{d}{dx} \ln |x|$$

Therefore,

$$\frac{d}{dx} \ln |ax| = \frac{d}{dx} \ln |x|.$$

43. Graph

$$y = \frac{\ln |x + 0.0001| - \ln |x|}{0.0001}$$

on a graphing calculator. A good choice for the viewing window is $[-3, 3]$.

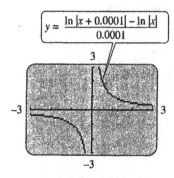

If we graph $y = \frac{1}{x}$ on the same screen, we see that the two graphs coincide.

By the definition of the derivative, if $f(x) = \ln |x|$,

$$f'(x) = \lim_{h \to 0} \frac{f(x + h) - f(x)}{h}$$

$$= \lim_{h \to 0} \frac{\ln |x + h| - \ln |x|}{h},$$

and $h = 0.0001$ is very close to 0.
Comparing the two graphs provides graphical evidence that

$$f'(x) = \frac{1}{x}.$$

45. $\ln \left(\dfrac{N(t)}{N_0} \right) = 9.8901 e^{-e^{2.54197 - 0.2167t}}$

(a) $\dfrac{N(t)}{1,000} = e^{9.8901 e^{-e^{2.54197 - 0.2167t}}}$

$N(t) = 1,000 e^{9.8901 e^{-e^{2.54197 - 0.2167t}}}$

(b) $N'(20) \approx 1,307,416$ bacteria/hour

Twenty hours into the experiment, the number of bacteria are increasing at a rate of 1,307,416 per hour.

(c) $S(t) = \ln \left(\dfrac{N(t)}{N_0} \right)$

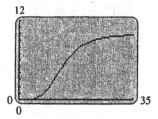

(d)

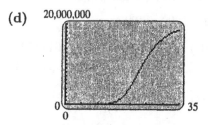

The two graphs have the same general shape, but $N(t)$ is scaled much larger.

(e) $\lim\limits_{t \to \infty} S(t) = \lim\limits_{t \to \infty} 9.8901 e^{-e^{2.54197 - 0.2167t}}$

$\qquad\qquad = 9.8901$

$$S(t) = \ln\left(\frac{N(t)}{N_0}\right)$$

$$N(t) = N_0 e^{S(t)}$$

$$\lim\limits_{t \to \infty} N(t) = N_0 e^{\lim\limits_{t \to \infty} S(t)} = 1{,}000 e^{9.8901}$$

$$\approx 19{,}734{,}033 \text{ bacteria}$$

47. $\log y = 1.54 - 0.008x - 0.658 \log x$

(a) $y(x) = 10^{(1.54 - 0.008x - 0.658 \log x)}$

$\qquad = 10^{1.54}(10^{-0.008x})(10^{-0.6581 \log x})$

$\qquad = 10^{1.54}(10^{0.008})^{-x}(10^{\log x})^{-0.6581}$

$\qquad \approx 34.7(1.0186)^{-x} x^{-0.6581}$

(b) **(i)** $\quad y(20) = 34.7(1.0186)^{-20} 20^{-0.6581}$

$\qquad\qquad = 3.3423\ldots$

$\qquad\qquad \approx 3.342$

$\qquad$ **(ii)** $\quad y(40) = 34.7(1.0186)^{-40} 40^{-0.6581}$

$\qquad\qquad = 1.4650\ldots$

$\qquad\qquad \approx 1.465$

(c) $\dfrac{dy}{dx} = -34.7(1.0186)^{-x} x^{-0.6581}\left(\ln(1.0186) + \dfrac{0.6581}{x}\right)$

$\qquad$ **(i)** $\quad \dfrac{dy}{dx}(20) = -0.17157\ldots \approx -0.172$

$\qquad$ **(ii)** $\quad \dfrac{dy}{dx}(40) = -0.051105\ldots \approx -0.0511$

49. $P(t) = (t + 100)\ln(t + 2)$

$$P'(t) = (t + 100)\left(\frac{1}{t + 2}\right) + \ln(t + 2)$$

$$P'(2) = (102)\left(\frac{1}{4}\right) + \ln(4)$$

$$\approx 26.9$$

$$P'(8) = (108)\left(\frac{1}{10}\right) + \ln(10)$$

$$\approx 13.1$$

51. $E(t) = -15{,}790.44 + 3{,}804.6 \ln(t)$

(a) $2002 - 1900 = 102$

$\qquad E(102) = -15{,}790.44 + 3{,}804.6 \ln 102$

$\qquad\qquad \approx 1{,}805.73 \text{ million metric tons}$

(b) $\quad E'(t) = 3{,}804.6\left(\dfrac{1}{t}\right)$

$\qquad E'(102) = 3{,}804.6\left(\dfrac{1}{102}\right)$

$\qquad\qquad = 37.3$

The greenhouse gas emissions are increasing at the rate of 37.3 million metric tons per year.

53. $f(x) = \dfrac{29{,}000(2.322 - \log x)}{x}$

$$f'(x) = \frac{x\left[29{,}000\left(-\dfrac{1}{(\ln 10)x}\right)\right] - 29{,}000(2.322 - \log x)(1)}{x^2}$$

$$= 29{,}000 \cdot \frac{-\dfrac{1}{\ln 10} - (2.322 - \log x)}{x^2}$$

$$= -29{,}000 \cdot \frac{1 + (\ln 10)(2.322 - \log x)}{x^2 \ln 10}$$

(a) $f(30) = \dfrac{29{,}000(2.322 - \log 30)}{30}$

$$\approx 817$$

$$f'(30) = -29{,}000 \frac{1 + (\ln 10)(2.322 - \log 30)}{(30)^2 \ln 10}$$

$$\approx -41.2$$

For a street 30 ft wide, the maximum traffic flow is 817 vehicles/hr, and the rate of change is -41.2 vehicles/hr per foot.

(b) $f(40) = \dfrac{29{,}000(2.322 - \log 40)}{40}$

$$\approx 522$$

$$f'(40) = -29{,}000 \cdot \frac{1 + (\ln 10)(2.322 - \log 40)}{(40)^2 \ln 10}$$

$$\approx -20.9$$

For a street 40 ft wide, the maximum traffic flow is 522 vehicles/hr, and the rate of change is -20.9 vehicles/hr per foot.

4.6 Derivatives of Trigonometric Functions

1. $y = 2\sin 6x$

$$\frac{dy}{dx} = 2(\cos 6x) \cdot D_x(6x)$$

$$= 2(\cos 6x) \cdot 6$$

$$= 12 \cos 6x$$

3. $y = 12 \tan(9x + 1)$

$$\frac{dy}{dx} = [12 \sec^2(9x + 1) \cdot D_x(9x + 1)$$

$$= [12 \sec(9x + 1)] \cdot 9$$

$$= 108 \sec^2(9x + 1)$$

5. $y = \cos^4 x$

$$\frac{dy}{dx} = [4(\cos x)^3]D_x(\cos x)$$
$$= (4\cos^3 x)(-\sin x)$$
$$= -4\sin x \cos^3 x$$

7. $y = \tan^5 x$

$$\frac{dy}{dx} = 5(\tan x)^4 \cdot D_x(\tan x)$$
$$= 5\tan^4 x \sec^2 x$$

9. $y = -5x \cdot \sin 4x$

$$\frac{dy}{dx} = -5x \cdot D_x(\sin 4x) + (\sin 4x) \cdot D_x(-5x)$$
$$= -5x(\cos 4x) \cdot D_x(4x) + (\sin 4x)(-5)$$
$$= -5x(\cos 4x) \cdot 4 - 5\sin 4x$$
$$= -5(4x\cos 4x + \sin 4x)$$

11. $y = \dfrac{\csc x}{x}$

$$\frac{dy}{dx} = \frac{x \cdot D_x(\csc x) - (\csc x) \cdot D_x x}{x^2}$$
$$= \frac{-x\csc x \cot x - \csc x}{x^2}$$

13. $y = \sin e^{5x}$

$$\frac{dy}{dx} = (\cos e^{5x}) \cdot D_x(e^{5x})$$
$$= (\cos e^{5x}) \cdot e^{5x} \cdot D_x(5x)$$
$$= (\cos e^{5x}) \cdot e^{5x} \cdot 5$$
$$= 5e^{5x}\cos e^{5x}$$

15. $y = e^{\sin x}$

$$\frac{dy}{dx} = (e^{\sin x}) \cdot D_x(\sin x) = (\cos x)e^{\sin x}$$

17. $y = \sin(\ln 4x^2)$

$$\frac{dy}{dx} = [\cos(\ln 4x^2)] \cdot D_x(\ln 4x^2)$$
$$= \cos(\ln 4x^2) \cdot \frac{D_x(4x^2)}{4x^2}$$
$$= \cos(\ln 4x^2)\frac{8x}{4x^2} = \cos(\ln 4x^2) \cdot \frac{2}{x}$$
$$= \frac{2}{x}\cos(\ln 4x^2)$$

19. $y = \ln\left|\sin x^2\right|$

$$\frac{dy}{dx} = \frac{D_x(\sin x^2)}{\sin x^2} = \frac{(\cos x^2) \cdot D_x(x^2)}{\sin x^2}$$
$$= \frac{(\cos x^2) \cdot 2x}{\sin x^2} = \frac{2x\cos x^2}{\sin x^2}$$
$$\text{or}\quad 2x\cot x^2$$

21. $y = \dfrac{2\sin x}{3 - 2\sin x}$

$$\frac{dy}{dx} = \frac{(3 - 2\sin x)D_x(2\sin x) - (2\sin x) \cdot D_x(3 - 2\sin x)}{(3 - 2\sin x)^2}$$
$$= \frac{(3 - 2\sin x) \cdot 2D_x(\sin x) - (2\sin x) \cdot [-2D_x(\sin x)]}{(3 - 2\sin x)^2}$$
$$= \frac{6\cos x - 4\sin x \cos x + 4\sin x \cos x}{(3 - 2\sin x)^2}$$
$$= \frac{6\cos x}{(3 - 2\sin x)^2}$$

23. $y = \sqrt{\dfrac{\sin x}{\sin 3x}} = \left(\dfrac{\sin x}{\sin 3x}\right)^{1/2}$

$$\frac{dy}{dx} = \frac{1}{2}\left(\frac{\sin x}{\sin 3x}\right)^{1/2} \cdot D_x\left(\frac{\sin x}{\sin 3x}\right)$$
$$= \frac{1}{2}\left(\frac{\sin 3x}{\sin x}\right)^{-1/2}$$
$$\cdot \left[\frac{(\sin 3x) \cdot D_x(\sin x) - (\sin x) \cdot D_x(\sin 3x)}{(\sin 3x)^2}\right]$$
$$= \frac{1}{2}\left(\frac{\sin 3x}{\sin x}\right)^{-1/2}$$
$$\cdot \left(\frac{(\sin 3x)(\cos x) - (\sin x)(\cos 3x) \cdot D_x(3x)}{\sin^2 3x}\right)$$
$$= \frac{1(\sin 3x)^{1/2}}{2(\sin x)^{1/2}} \cdot \frac{\sin 3x \cos x - 3\sin x \cos 3x}{\sin^2 3x}$$
$$= \frac{(\sin 3x)^{1/2}(\sin 3x \cos x - 3\sin x \cos 3x)}{2(\sin x)^{1/2}(\sin^2 3x)}$$
$$= \frac{\sqrt{\sin 3x}\,[\sin 3x \cos x - 3\sin x \cos 3x]}{2\sqrt{\sin x}\,(\sin^2 3x)}$$

25. $y = 2\csc x - 3\tan 4x - 7\cos\left(\dfrac{x}{8}\right) + e^{3x}$

$$\frac{dy}{dx} = 2(-\csc x \cot x) - 3\sec^2 4x \cdot D_x(4x)$$
$$-7\left[-\sin\left(\frac{x}{8}\right)\right] \cdot D_x\left(\frac{x}{8}\right) + e^{3x} \cdot D_x(3x)$$
$$= -2\csc x \cot x - 12\sec^2 4x + \frac{7}{8}\sin\left(\frac{x}{8}\right) + 3e^{3x}$$

27. $y = \sin x; x = 0$

Let $f(x) = \sin x$.
Then $f'(x) = \cos x$, so

$$f'(0) = \cos 0 = 1.$$

The slope of the tangent line to the graph of $y = \sin x$ at $x = 0$ is 1.

29. $y = \cos x; x = \dfrac{\pi}{2}$

Let $f(x) = \cos x$.
Then $f'(x) = -\sin x$, so

$$f'\left(\frac{\pi}{2}\right) = -\sin\frac{\pi}{2}$$
$$= -1.$$

The slope of the tangent line to the graph of $y = \cos x$ at $x = \frac{\pi}{2}$ is -1.

31. $y = \tan x; x = 0$

Let $f(x) = \tan x$.
Then $f'(x) = \sec^2 x$, so

$$f'(0) = \sec^2 0$$
$$= \frac{1}{\cos^2 0}$$
$$= 1.$$

The slope of the tangent line to the graph of $y = \tan x$ at $x = 0$ is 1.

33. Since $\cot x = \frac{\cos x}{\sin x}$, by using the quotient rule,

$$D_x(\cos x) = D_x\left(\frac{\cos x}{\sin x}\right)$$
$$= \frac{(\sin x)(-\sin x) - (\cos x)(\cos x)}{\sin^2 x}$$
$$= \frac{-\sin^2 x - \cos^2 x}{\sin^2 x}$$
$$= -\frac{\sin^2 x + \cos^2 x}{\sin^2 x}$$
$$= -\frac{1}{\sin^2 x}$$
$$= -\csc^2 x.$$

35. Since $\csc x = \frac{1}{\sin x} = (\sin x)^{-1}$,

$$D_x(\csc x) = D_x\left(\frac{1}{\sin x}\right) = D_x(\sin x)^{-1}$$
$$= -1(\sin x)^{-2}\cos x$$
$$= -\frac{\cos x}{\sin^2 x}$$
$$= -\frac{1}{\sin x}\cdot\frac{\cos x}{\sin x}$$
$$= -\csc x \cot x.$$

37. $y = \dfrac{\pi}{8}\cos 3\pi\left(t - \dfrac{1}{3}\right)$

(a) The graph should resemble the graph of $y = \cos x$ with the following difference: The maximum and minimum values of y are $\frac{\pi}{8}$ and $-\frac{\pi}{8}$. The period of the graph will be $\frac{2\pi}{3\pi} = \frac{2}{3}$ units. The graph will be shifted horizontal $\frac{1}{3}$ units to the right.

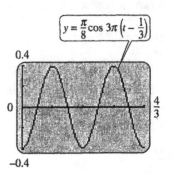

(b) velocity $= \dfrac{dy}{dt}$

$$= D_t\left[\frac{\pi}{8}\cos 3\pi\left(t - \frac{1}{3}\right)\right]$$
$$= \frac{\pi}{8}D_t\left[\cos 3\pi\left(t - \frac{1}{3}\right)\right]$$
$$= \frac{\pi}{8}\left[-\sin 3\pi\left(t - \frac{1}{3}\right)\right]D_t\left[3\pi\left(t - \frac{1}{3}\right)\right]$$
$$= \frac{\pi}{8}\left[-\sin 3\pi\left(t - \frac{1}{3}\right)\right]\cdot 3\pi$$
$$= -\frac{3\pi^2}{8}\sin 3\pi\left(t - \frac{1}{3}\right)$$

Acceleration $= \dfrac{d^2 y}{dt^2}$

$$= D_t\left[\frac{-3\pi^2\sin 3\pi\left(t - \frac{1}{3}\right)}{8}\right]$$
$$= \frac{-3\pi^2}{8}D_t\left[\sin 3\pi\left(t - \frac{1}{3}\right)\right]$$
$$= \frac{-3\pi^2}{8}\left[\cos 3\pi\left(t - \frac{1}{3}\right)\right]D_t\left[3\pi\left(t - \frac{1}{3}\right)\right]$$
$$= \frac{-3\pi^2}{8}\left[\cos 3\pi\left(t - \frac{1}{3}\right)\right]\cdot 3\pi$$
$$= -\frac{9\pi^3}{8}\cos 3\pi\left(t - \frac{1}{3}\right)$$

(c) $\dfrac{d^2y}{dt^2} + 9\pi^2 y = -\dfrac{9\pi^3}{8}\cos 3\pi\left(t - \dfrac{1}{3}\right) + 9\pi^2\left[\dfrac{\pi}{8}\cos 3\pi\left(t - \dfrac{1}{3}\right)\right]$

$\qquad\qquad\qquad\quad = -\dfrac{9\pi^3}{8}\cos 3\pi\left(t - \dfrac{1}{3}\right) + \dfrac{9\pi^3}{8}\cos 3\pi\left(t - \dfrac{1}{3}\right) = 0$

(d) $a(1) = -\dfrac{9\pi^3}{8}\cos 3\pi\left(t - \dfrac{1}{3}\right) = -\dfrac{9\pi^3}{8}\cos 2\pi = -\dfrac{9\pi^3}{8}\cdot 1 = -\dfrac{9\pi^3}{8} < 0$

$\quad\;\; y(1) = \dfrac{\pi}{8}\cos 3\pi\left(t - \dfrac{1}{3}\right) = \dfrac{\pi}{8}\cos 2\pi = \dfrac{\pi}{8}\cdot 1 = \dfrac{\pi}{8}$

Therefore, at $t = 1$ second, the force is clockwise and the arm makes an angle of $\frac{\pi}{8}$ radians forward from the vertical. The arm is moving clockwise.

$$a\left(\dfrac{4}{3}\right) = -\dfrac{9\pi^3}{8}\cos 3\pi\left(\dfrac{4}{3} - \dfrac{1}{3}\right) = -\dfrac{9\pi^3}{8}\cos(3\pi) = -\dfrac{9\pi^3}{8}(-1) = \dfrac{9\pi^3}{8} > 0$$

$$y\left(\dfrac{4}{3}\right) = \dfrac{\pi}{8}\cos 3\pi\left(\dfrac{4}{3} - \dfrac{1}{3}\right) = \dfrac{\pi}{8}\cos(3\pi) = \dfrac{\pi}{8}(-1) = -\dfrac{\pi}{8}$$

Therefore, at $t = \frac{4}{3}$ seconds, the force is counterclockwise and the arm makes an angle of $-\frac{\pi}{8}$ radians from the vertical. The arm is moving counterclockwise.

$$a\left(\dfrac{5}{3}\right) = -\dfrac{9\pi^3}{8}\cos 3\pi\left(\dfrac{5}{3} - \dfrac{1}{3}\right) = -\dfrac{9\pi^3}{8}\cos(4\pi) = -\dfrac{9\pi^3}{8}\cdot 1 = -\dfrac{9\pi^3}{8} < 0$$

$$y\left(\dfrac{5}{3}\right) = \dfrac{\pi}{8}\cos 3\pi\left(\dfrac{5}{3} - \dfrac{1}{3}\right) = \dfrac{\pi}{8}\cos(4\pi) = \dfrac{\pi}{8}\cdot 1 = \dfrac{\pi}{8}$$

Therefore, at $t = \frac{5}{3}$ second, the answer corresponds to $t = 1$ second. So the arm is moving clockwise and makes an angle of $\frac{\pi}{8}$ from the vertical.

39. (a) Using the calculator to graph $C(x) = 0.04x^2 + 0.6x + 330 + 7.5\sin(2\pi x)$ in a $[0, 25]$ by $[320, 380]$ viewing window, gives the following graph.

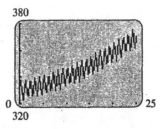

(b) $C(25) = 0.04(25)^2 + 0.6(25) + 330 + 7.5\sin(2\pi(25)) = 370$ parts per million
$\quad\; C(35.5) = 0.04(35.5)^2 + 0.6(35.5) + 330 + 7.5\sin(2\pi(35.5)) = 401.71$ parts per million
$\quad\; C(50.2) = 0.04(50.2)^2 + 0.6(50.2) + 330 + 7.5\sin(2\pi(50.2)) \approx 468.05$ parts per million

(c) Since $C'(x) = 0.08x + 0.6 + 15\pi\cos(2\pi x)$, $C'(50.2)$ is given by

$$C'(50.2) = 0.08(50.2) + 0.6 + 15\pi\cos(2\pi(50.2)) \approx 19.18 \text{ parts per million per year.}$$

The level of carbon dioxide will be increasing at the beginning of 2010 at 19.18 parts per million.

41. (a) $s(\theta) = 2.625\cos\theta + 2.625(15 + \cos^2\theta)^{1/2}$

$$\frac{ds}{dt} = \frac{ds}{d\theta} \cdot \frac{d\theta}{dt} = \left[-2.625\sin\theta + 1.3125(15+\cos^2\theta)^{-1/2} \cdot D_\theta(15+\cos^2\theta) \cdot \frac{d\theta}{dt}\right]$$

$$= \left[-2.625\sin\theta + \frac{1.3125}{\sqrt{15+\cos^2\theta}} \cdot (-2\sin\theta\cos\theta)\right] \cdot \frac{d\theta}{dt} = -2.625\sin\theta\left(1 + \frac{\cos\theta}{\sqrt{15+\cos^2\theta}}\right)\frac{d\theta}{dt}$$

(b) With $\theta = 4.944$ and $\dfrac{d\theta}{dt} = 505{,}168.1\ \dfrac{\text{radians}}{\text{hour}}$, we have

$$\frac{ds}{dt} = -2.625\sin(4.944)\left(1 + \frac{\cos(4.944)}{\sqrt{15+\cos^2(4.944)}}\right)(505{,}168.1) \approx 1{,}367{,}018.749\ \frac{\text{inches}}{\text{hour}}$$

$$= 1{,}367{,}018.749\ \frac{\text{inches}}{\text{hour}} \times \frac{1\text{ foot}}{12\text{ inches}} \times \frac{1\text{ mile}}{5{,}280\text{ feet}} \approx 21.6\text{ miles per hour}$$

43. $s(t) = \sin t + 2\cos t$

$v(t) = s'(t) = D_t(\sin t + 2\cos t) = \cos t - 2\sin t$

(a) $v(0) = 1 - 2(0) = 1$ **(b)** $v\left(\dfrac{\pi}{2}\right) = 0 - 2(1) = -2$ **(c)** $v(\pi) = -1 - 2(0) = -1$

$a(t) = v'(t) = s''(t) = D_t(\cos t - 2\sin t) = -\sin t - 2\cos t$

(d) $a(0) = 0 - 2(1) = -2$ **(e)** $a\left(\dfrac{\pi}{2}\right) = -1 - 2(0) = -1$ **(f)** $a(\pi) = 0 - 2(-1) = 2$

Chapter 4 Review Exercises

1. $y = 5x^2 - 7x - 9$
$y' = 5(2x) - 7$
$ = 10x - 7$

3. $y = 6x^{7/3}$

$y' = 6\left(\dfrac{7}{3}\right)x^{4/3} = \dfrac{42}{3}x^{4/3} = 14x^{4/3}$

5. $f(x) = x^{-3} + \sqrt{x}$

$ = x^{-3} + x^{1/2}$

$f'(x) = -3x^{-4} + \left(\dfrac{1}{2}\right)x^{-1/2}$

or $\dfrac{-3}{x^4} + \dfrac{1}{2x^{1/2}}$

7. $k(x) = \dfrac{3x}{x+5}$

$k'(x) = \dfrac{(x+5)(3) - (3x)(1)}{(x+5)^2}$

$ = \dfrac{3x + 15 - 3x}{(x+5)^2} = \dfrac{15}{(x+5)^2}$

9. $y = \dfrac{x^2 - x + 1}{x - 1}$

$y' = \dfrac{(x-1)(2x-1) - (x^2 - x + 1)(1)}{(x-1)^2} = \dfrac{2x^2 - 3x + 1 - x^2 + x - 1}{(x-1)^2} = \dfrac{x^2 - 2x}{(x-1)^2}$

11. $f(x) = (3x-2)^4$
$f'(x) = 4(3x-2)^3(3) = 12(3x-2)^3$

13. $y = \sqrt{2t-5}$
$y' = (2t-5)^{1/2} = \dfrac{1}{2}(2t-5)^{-1/2}(2) = (2t-5)^{-1/2}$ or $\dfrac{1}{(2t-5)^{1/2}}$

15. $y = 3x(2x+1)^3$

$\quad y' = 3x(3)(2x+1)^2(2) + (2x+1)^3(3)$

$\qquad = (18x)(2x+1)^2 + 3(2x+1)^3$

$\qquad = 3(2x+1)^2[6x + (2x+1)]$

$\qquad = 3(2x+1)^2(8x+1)$

17. $r(t) = \dfrac{5t^2 - 7t}{(3t+1)^3}$

$\quad r'(t) = \dfrac{(3t+1)^3(10t-7) - (5t^2-7t)(3)(3t+1)^2(3)}{[(3t+1)^3]^2}$

$\qquad = \dfrac{(3t+1)^3(10t-7) - 9(5t^2-7t)(3t+1)^2}{(3t+1)^6}$

$\qquad = \dfrac{(3t+1)(10t-7) - 9(5t^2-7t)}{(3t+1)^4}$

$\qquad = \dfrac{30t^2 - 11t - 7 - 45t^2 + 63t}{(3t+1)^4}$

$\qquad = \dfrac{-15t^2 + 52t - 7}{(3t+1)^4}$

19. $p(t) = t^2(t^2+1)^{5/2}$

$\quad p'(t) = t^2 \cdot \dfrac{5}{2}(t^2+1)^{3/2} \cdot 2t + 2t(t^2+1)^{5/2}$

$\qquad = 5t^3(t^2+1)^{3/2} + 2t(t^2+1)^{5/2}$

$\qquad = t(t^2+1)^{3/2}[5t^2 + 2(t^2+1)^1]$

$\qquad = t(t^2+1)^{3/2}(7t^2+2)$

21. $y = -6e^{2x}$

$\quad y' = -6(2e^{2x}) = -12e^{2x}$

23. $\quad y = e^{-2x^3}$

$\quad g(x) = -2x^3$

$\quad g'(x) = -6x^2$

$\qquad y' = -6x^2 e^{-2x^3}$

25. $\quad y = 5xe^{2x}$

Use the product rule.

$$y' = 5x(2e^{2x}) + e^{2x}(5)$$
$$= 10xe^{2x} + 5e^{2x}$$
$$= 5e^{2x}(2x+1)$$

27. $\quad y = \ln(2+x^2)$

$\quad g(x) = 2 + x^2$

$\quad g'(x) = 2x$

$$y' = \frac{2x}{2+x^2}$$

29. $y = \dfrac{\ln|3x|}{x-3}$

$\quad y' = \dfrac{(x-3)\left(\frac{1}{3x}\right)(3) - (\ln|3x|)(1)}{(x-3)^2}$

$\qquad = \dfrac{\frac{x-3}{x} - \ln|3x|}{(x-3)^2} \cdot \dfrac{x}{x}$

$\qquad = \dfrac{x - 3 - x\ln|3x|}{x(x-3)^2}$

31. $y = \dfrac{xe^x}{\ln(x^2-1)}$

$\quad y' = \dfrac{\ln(x^2-1)[xe^x + e^x] - xe^x\left(\frac{1}{x^2-1}\right)(2x)}{[\ln(x^2-1)]^2}$

$\qquad = \dfrac{e^x(x+1)\ln(x^2-1) - \frac{2x^2 e^x}{x^2-1}}{[\ln(x^2-1)]^2} \cdot \dfrac{x^2-1}{x^2-1}$

$\qquad = \dfrac{e^x(x+1)(x^2-1)\ln(x^2-1) - 2x^2 e^x}{(x^2-1)[\ln(x^2-1)]^2}$

33. $s = (t^2 + e^t)^2$

$\quad s' = 2(t^2+e^t)(2t + e^t)$

35. $y = 3 \cdot 10^{-x^2}$

$\quad y' = 3 \cdot (\ln 10)10^{-x^2}(-2x)$

$\qquad = -6x(\ln 10) \cdot 10^{-x^2}$

37. $g(z) = \log_2(z^3 + z + 1)$

$\quad g'(z) = \dfrac{1}{\ln 2} \cdot \dfrac{3z^2+1}{z^3+z+1}$

$\qquad = \dfrac{3z^2+1}{(\ln 2)(z^3+z+1)}$

41. $y = 6\tan 3x$

$\quad y' = 6\sec^2 3x \cdot D_x(3x)$

$\qquad = 18\sec^2 3x$

43. $y = \cot(9 - x^2)$

$\quad y' = [-\csc^2(9-x^2)] \cdot D_x(9-x^2)$

$\qquad = 2x\csc^2(9-x^2)$

45. $y = 2\sin^4(4x^2)$

$\quad y' = [8\sin^3(4x^2)] \cdot D_x[\sin(4x^2)]$

$\qquad = 8\sin^3(4x^2) \cdot \cos(4x^2) \cdot D_x(4x^2)$

$\qquad = 64x\sin^3(4x^2)\cos(4x^2)$

47. $y = \cos(1 + x^2)$

$\quad y' = [-\sin(1+x^2)] \cdot D_x(1+x^2)$

$\qquad = -2x\sin(1+x)^2$

49. $y = e^{-x} \sin x$

$y' = e^{-x} \cdot D_x (\sin x) + \sin x \cdot D_x (e^{-x})$

$\quad = e^{-x} \cos x - e^{-x} \sin x$

$\quad = e^{-x} (\cos x - \sin x)$

51. $y = \dfrac{\cos^2 x}{1 - \cos x}$

$y' = \dfrac{(1 - \cos x)(-2 \cos x \sin x) - (\cos^2 x)(\sin x)}{(1 - \cos x)^2}$

$\quad = \dfrac{-2 \cos x \sin x + \cos^2 x \sin x}{(1 - \cos x)^2}$

53. $y = \dfrac{\tan x}{1 + x}$

$y' = \dfrac{(1 + x)(\sec^2 x) - (\tan x)(1)}{(1 + x)^2}$

$\quad = \dfrac{\sec^2 x + x \sec^2 x - \tan x}{(1 + x)^2}$

55. $y = \ln |5 \sin x|$

$y' = \dfrac{1}{5 \sin x} \cdot D_x (5 \sin x)$

$\quad = \dfrac{\cos x}{\sin x}$

or $\cot x$

57. $y = 8 - x^2$; $x = 1$

$y = 8 - x^2$

$y' = -2x$

slope $= y'(1) = -2(1) = -2$

Use $(1, 7)$ and $m = -2$ in the point-slope form.

$$y - 7 = -2(x - 1)$$
$$y - 7 = -2x + 2$$
$$2x + y = 9$$
$$y = -2x + 9$$

59. $y = \dfrac{x}{x^2 - 1}$; $x = 2$

$\dfrac{dy}{dx} = \dfrac{(x^2 - 1) \cdot 1 - x(2x)}{(x^2 - 1)^2}$

$\quad = \dfrac{-x^2 - 1}{(x^2 - 1)^2}$

The value of $\frac{dy}{dx}$ when $x = 2$ is the slope.

$$m = \dfrac{-(2^2) - 1}{(2^2 - 1)^2} = \dfrac{-5}{9} = -\dfrac{5}{9}$$

When $x = 2$,

$$y = \dfrac{2}{4 - 1} = \dfrac{2}{3}.$$

Use $m = -\frac{5}{9}$ with $P\left(2, \frac{2}{3}\right)$.

$$y - \dfrac{2}{3} = -\dfrac{5}{9}(x - 2)$$
$$y - \dfrac{6}{9} = -\dfrac{5}{9}x + \dfrac{10}{9}$$
$$y = -\dfrac{5}{9}x + \dfrac{16}{9}$$

61. $y = -\sqrt{8x + 1}$; $x = 3$

$\quad\quad y = -(8x + 1)^{1/2}$

$\dfrac{dy}{dx} = -\dfrac{1}{2}(8x + 1)^{-1/2}(8)$

$\dfrac{dy}{dx} = -\dfrac{4}{(8x + 1)^{1/2}}$

The value of $\frac{dy}{dx}$ when $x = 3$ is the slope.

$$m = -\dfrac{4}{(24 + 1)^{1/2}} = -\dfrac{4}{5}$$

When $x = 3$,

$$y = -\sqrt{24 + 1} = -5.$$

Use $m = -\frac{4}{5}$ with $P(3, -5)$.

$$y + 5 = -\dfrac{4}{5}(x - 3)$$
$$y + \dfrac{25}{5} = -\dfrac{4}{5}x + \dfrac{12}{5}$$
$$y = -\dfrac{4}{5}x - \dfrac{13}{5}$$

63. $y = xe^x$; $x = 1$

$\dfrac{dy}{dx} = xe^x + 1 \cdot e^x$

$\quad = e^x(x + 1)$

The value of $\frac{dy}{dx}$ when $x = 1$ is the slope.

$$m = e^1(1 + 1) = 2e$$

When $x = 1$, $y = 1e^1 = e$. Use $m = 2e$ with $P(1, e)$.

$$y - e = 2e(x - 1)$$
$$y = 2ex - e$$

65. $y = x \ln x$; $x = e$

$$\frac{dy}{dx} = x \cdot \frac{1}{x} + 1 \cdot \ln x$$

$$= 1 + \ln x$$

The value of $\frac{dy}{dx}$ when $x = e$ is the slope

$$m = 1 + \ln e = 1 + 1 = 2.$$

When $x = e$, $y = e \ln e = e \cdot 1 = e$. Use $m = 2$ with $P(e, e)$.

$$y - e = 2(x - e)$$
$$y = 2x - e$$

67. $y = \tan(\pi x)$; $x = 1$

$$\frac{dy}{dx} = \pi \sec^2(\pi x)$$

The value of $\frac{dy}{dx}$ when $x = 1$ is the slope

$$m = \pi \sec^2(\pi) = \pi(-1)^2 = \pi.$$

When $x = 1$, $y = \tan(\pi) = 0$. Use $m = \pi$ with $P(1, 0)$.

$$y - 0 = \pi(x - 1)$$
$$y = \pi x - \pi$$

69. $c(x) = 0.19x$

(a) $c(20) = 0.19(20) = 3.8$ fish/hour

In two hours an angler will catch an estimted $2(3.8)$ or 7.6 fish.

(b) $c'(x) = 0.19$

The rate of change is constant. The estimated number of fish angler can catch in an hour increases by 0.19 walleye per hour as the density of walleye per acre increases.

71. $T(n) = 0.25 + (1.93 \times 10^{-2})n - (4.78 \times 10^{-5})n^2$

(a)

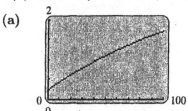

(b) Answers vary. Sample answer: 201 fish. At this point $T(n)$ has already begun to decrease with respect to n.

(c) $T'(n) = (1.93 \times 10^{-2}) - (9.56 \times 10^{-5})n$

$$T'(40) = (1.93 \times 10^{-2}) - (9.56 \times 10^{-5})40$$
$$= 0.015476 \approx 0.015 \text{ hour/fish}$$

When 40 fish are caught in the net, the processing time is increasing by 0.015 hours per additional fish caught.

73. $L(t) = 71.5(1 - e^{-0.1t})$ and
$W(L) = 0.01289 \cdot L^{2.9}$

(a) $L(5) = 71.5(1 - e^{-0.5})$
$$\approx 28.1$$

The approximate length of a 5-year-old monkey-face is 28.1 cm.

(b) $L'(t) = 71.5(0.1e^{-0.1t})$
$L'(5) = 71.5(0.1e^{-0.5})$

$$\approx 4.34$$

The length is growing by about 4.34 cm/year.

(c) $W[L(5)] \approx 0.01289(28.1)^{2.9}$
$$\approx 205$$

The approximate value of $W[L(5)]$ is 205 grams.

(d) $W'(L) = 0.01289(2.9)L^{1.9}$
$$= 0.037381L^{1.9}$$
$W'[L(5)] \approx .037381(28.1)^{1.9}$
$$\approx 21.2$$

The rate of change of the weight with respect to length is 21.2 grams/cm.

(e) $\frac{dW}{dt} = \frac{dW}{dL} \cdot \frac{dL}{dt}$

$$\approx (21.2)(4.34)$$

$$\approx 92.0$$

The weight is growing at about 92.0 grams/year.

75. $P(t) = 90 + 15 \sin 144\pi t$

The maximum possible value of $\sin \alpha$, is 1, while the minimum possible value is -1. Replacing α with $144\pi t$ gives

$$-1 \le \sin 144\pi t \le 1,$$
$$90 + 15(-1) \le P(t) \le 90 + 15(1)$$
$$75 \le P(t) \le 105.$$

Therefore, the minimum value of $P(t)$ is 75 and the maximum value of $P(t)$ is 105.

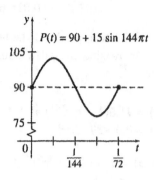

77. $\sin \theta = \dfrac{s}{L_2}$

$L_2 \sin \theta = s$

$L_2 = \dfrac{s}{\sin \theta}$

79. $\cot \theta = \dfrac{L_0 - L_1}{s}$

$s \cot \theta = L_0 - L_1$

$L_1 + s \cot \theta = L_0$

$L_1 = L_0 - s \cot \theta$

81. $R_2 = k \cdot \dfrac{L_2}{r_2{}^4}$ where R_2 is the resistance along DC.

83. $R = k \left(\dfrac{L_1}{r_1{}^4} + \dfrac{L_2}{r_2{}^4} \right)$

$= k \left(\dfrac{L_0 - s \cot \theta}{r_1{}^4} + \dfrac{s}{(\sin \theta) r_2{}^4} \right)$

$= \dfrac{k(L_0 - s \cot \theta)}{r_1{}^4} + \dfrac{ks}{r_2{}^4 \sin \theta}$

85. Using Exercise 84 gives

$\dfrac{ks \csc^2 \theta}{r_1{}^4} - \dfrac{ks \cos \theta}{r_2{}^4 \sin^2 \theta} = 0.$

87. Using Exercise 86 gives

$k \left(\dfrac{1}{r_1{}^4} - \dfrac{\cos \theta}{r_2{}^4} \right) = 0$

$\dfrac{1}{r_1{}^4} - \dfrac{\cos \theta}{r_2{}^4} = 0 \quad k \neq 0$

$\dfrac{1}{r_1{}^4} = \dfrac{\cos \theta}{r_2{}^4}$

$\cos \theta = \dfrac{r_2{}^4}{r_1{}^4}$

89. If $r_1 = 1.4$ and $r_2 = 0.8$, then

$$\cos \theta = \frac{(0.8)^4}{(1.4)^4} \approx 0.1066.$$

Thus,

$$\theta \approx 84°.$$

91. $C(x) = 0.05(1.040)^{x-1950}$
$C'(x) = 0.05(\ln\ 1.040)1.040^{x-1950}$
$C'(1998) = 0.05(\ln\ 1.040)1.040^{48}$
≈ 0.0129

The concentration of CFC-11 is increasing at a rate of 0.0129 ppb per year in 1998.

93. $N(t) = N_0 e^{-0.217t}$

(a)   1950 − 950 = 1000 years = 1 millennium;
$N_0 = 210$
$N(1) = 210e^{-0.217(1)} \approx 169$

This value is close to the number 167.

(b)   For 2050, let $N_0 = 167$ and $t = 0.1$.
$N(0.1) = 167e^{-0.217(.1)} \approx 163$

(c) If we use the function found in part (b), then $N'(t) = (-0.217)(167e^{-0.217t})$ and $N'(2) \approx -23.5$. This gives the rate at which the number of basic ancient Chinese words still in use is changing per millennium 2 millennia after 1950. That is, in 3950 the number of words still in use will be decreasing at a rate of 23.5 words per millennium.

95. $s(t) = A \cos(Bt + C)$
$s'(t) = -A \sin(Bt + C) \cdot D_t(Bt + C)$
$\quad = -A \sin(Bt + C) \cdot B$
$\quad = -AB \sin(Bt + C)$
$s''(t) = -AB \cos(Bt + C) \cdot D_t(Bt + C)$
$\quad = -AB \cos(Bt + C) \cdot B$
$\quad = -B^2 A \cos(Bt + C)$
$\quad = -B^2 s(t)$

Chapter 4 Test

[4.1]

Find the derivative of each function.

1. $y = 2x^4 - 3x^3 + 4x^2 - x + 1$ **2.** $f(x) = x^{3/4} - x^{-2/3} + x^{-1}$ **3.** $f(x) = -2x^{-2} - 3\sqrt{x}$

4. Find the slope of the tangent line to $y = -2x + \frac{1}{x} + \sqrt{x}$ at $x = 1$. Find the equation of the tangent line.

5. For the function

$$f(x) = x^3 + \frac{3}{2}x^2 - 60x + 18,$$

find all values of x where the tangent line is horizontal.

6. At the beginning of an experiment, a culture is determined to have 2×10^4 bacteria. Thereafter, the number of bacteria observed at time t (in hours) is given by the equation

$$B(t) = 10^4 \left(2 - 3\sqrt{t} + 2t + t^2\right).$$

How fast is the population growing at the end of 4 hours?

7. Bighorn Sheep The cumulative horn volume in years 2–9 for certain types of bighorn ram, found in the Rocky Mountains, can be described by the quadratic function

$$V(t) = -2,159 + 1,313t - 60.82t^2,$$

where $V(t)$ is the horn volume (in cm^3) and t is the year of growth.

(a) Find the horn volume for a four-year-old ram.

(b) Find the rate at which the horn volume of a four-year-old is changing.

8. Brain Weight The brain weight of a human fetus during the last trimester can be accurately estimated from the circumference of the head by

$$w(x) = \frac{x^3}{100} - \frac{1,500}{x},$$

where $w(x)$ is the weight of the brain (in g) and x is the circumference (in cm) of the head.*

(a) Estimate the brain weight of a fetus that has a head circumference of 28 cm.

(b) Find the rate of change of the brain weight for a fetus that has a head circumference of 28 cm.

---

*Dobbing, J. and J. Sands, "Head Circumference, Biparietal Diameter and Brain Growth in Fetal and Postnatal Life," Early Human Development, Vol. 2, No. 1, Apr 1978, pp. 81–87.

[4.2]

Find the derivative of each function.

9. $f(x) = (2x^2 - 5x)(3x^2 + 1)$

10. $y = \dfrac{x^2 + x}{x^3 - 1}$

11. Management has determined that the cost in thousands of dollars for producing x units is given by the equation

$$C(x) = \frac{3x^2}{x^2 + 1} + 200.$$

Find and interpret the marginal cost when $x = 3$.

[4.3]

12. Find $f[g(x)]$ and $g[f(x)]$ for the following pair of functions.

$$f(x) = \frac{3}{x^2}; \; g(x) = \frac{1}{x}$$

Find each of the following.

13. $\dfrac{dy}{dx}$ if $y = \dfrac{\sqrt{x-2}}{x+1}$

14. $D_x\left[\sqrt{(3x^2-1)^3}\right]$

15. $f'(2)$ if $f(t) = \dfrac{2t-1}{\sqrt{t+2}}$

Find the derivative of each function.

16. $y = -2x\sqrt{3x-1}$

17. $y = (5x^3 - 2x)^5$

18. Find the equation of the tangent line to the graph of the function $f(x) = \sqrt{x^2 - 9}$ at $x = 5$. (Write the equation in slope-intercept form.)

19. **African Wild Dog** The nomadic African wild dog is currently one of the most endangered carnivores in the world. After years of attempts to rescue the dog from extinction, scientists have collected a lot of data on the habits of this nomadic creature. The collected data shows the following mathematical relationship between the weight and the length of the dog.

$$L(w) = 2.472w^{2.571},$$

where w (in kg) is the mass and $L(w)$ (in mm) is the length of the mammal. Suppose that the weight of a particular wild dog can be estimated by

$$w(t) = 0.265 + 0.21t,$$

where $w(t)$ is the weight (in kg) and t is the age, in weeks of an African wild dog that is less than one year old. How fast is the length of a 30-week-old African wild dog changing?

[4.4]

Find the derivative of each function.

20. $y = 3x^2 e^{2x}$

21. $y = \left(e^{2x} - \ln |x|\right)^3$

22. The concentration of a certain drug in the bloodstream at time t in minutes is given by

$$c(t) = e^{-t} - e^{-3t}.$$

Find the rate of change in the concentration at each of the following times.

    (a) $t = 0$ **(b)** $t = 1$ **(c)** $t = 2$

23. Cutlassfish The cutlassfish is one of the most important resources of the commercial marine fishing industry in China. Researchers have developed a von Bertalanffy growth model that used the age of a certain species of cutlassfish to estimate preanal length such that

$$L(t) = 589 \left\{1 - e^{(-0.168(t+2.682))}\right\},$$

where $L(t)$ is the length of the fish (in mm) at time t (yr).

    (a) Determine the age of a fish that has grown to 90% of its maximum weight.

    (b) Find $L'(3)$.

[4.5]

Find the derivative of each function.

24. $y = \ln \left(2x^3 + 5\right)^{2/3}$

25. $y = \ln \left(x^2 + 8\right)^7$

26. Find the derivative of $y = \dfrac{\ln |3x - 1|}{x - 2}$.

27. Body Surface Area Boyd found that there is a mathematical relationship between a person's weight and the total body surface area (BSA). The relationship is

$$B(W) = 4.688 W^{(0.8168 - 0.0154 \log_{10} W)},$$

where W is the weight (in g) and $B(W)$ is the BSA in cm^2.

    (a) Find the BSA for a person who weighs 40 kg. Express your answer in m^2.

    (b) Find $B'(40,000)$.

[4.6]

Find the derivative of each of the following functions.

28. $y = -3 \cos 4x$

30. $y = 3x^2 \sin x$

32. $y = \ln |3 \cos x|$

29. $y = \cot^2 \dfrac{x}{2}$

31. $y = \dfrac{\sin^2 x}{\sin x + 1}$

33. $y = e^{\cos x}$

34. Find the slope of the tangent to the graph of $y = \tan x$ at $x = \frac{\pi}{3}$.

35. Find the maximum and minimum values of y for the simple harmonic motion given by $y = 3\sin(2x + 3)$.

36. Explain why the slope of $y = \tan x$ is never zero.

Chapter 4 Test Answers

1. $\frac{dy}{dx} = 8x^3 - 9x^2 + 8x - 1$

2. $f'(x) = \frac{3}{4x^{1/4}} + \frac{2}{3x^{5/3}} - \frac{1}{x^2}$

3. $f'(x) = \frac{4}{x^3} - \frac{3}{2\sqrt{x}}$

4. $-\frac{5}{2}$; $5x + 2y = 5$

5. $x = -5$, $x = 4$

6. 9.25×10^4 bacteria per hour

7. (a) 2,119.88 cm^3

   (b) 826.44 cm^3/yr

8. (a) 166 g   (b) 25.43 g/cm

9. $f'(x) = 24x^3 - 45x^2 + 4x - 5$

10. $\frac{dy}{dx} = \frac{-x^4 - 2x^3 - 2x - 1}{(x^3 - 1)^2}$

11. $C'(3) = 0.18$; after three units have been produced, the cost to produce one more unit will be approximately 0.18(1,000) or $180.

12. $f[g(x)] = 3x^2$; $g[f(x)] = \frac{x^2}{3}$

13. $\frac{5 - x}{2\sqrt{x - 2}(x + 1)^2}$

14. $9x\sqrt{3x^2 - 1}$

15. $\frac{13}{16}$

16. $\frac{dy}{dx} = \frac{2 - 9x}{\sqrt{3x - 1}}$

17. $\frac{dy}{dx} = 5\left(5x^3 - 2x\right)^4\left(15x^2 - 2\right)$

18. $y = \frac{5}{4}x - \frac{9}{4}$

19. About 25.7 mm/wk

20. $6x^2 e^{2x} + 6x e^{2x}$

21. $\frac{3\left(e^{2x} - \ln|x|\right)^2\left(2x e^{2x} - 1\right)}{x}$

22. (a) 2   (b) -0.219 (c) -0.128

23. (a) About 11 yr

    (b) About 38.1 mm/yr

24. $\frac{dy}{dx} = \frac{4x^2}{2x^3 + 5}$

25. $\frac{dy}{dx} = \frac{14x}{x^2 + 8}$

26. $\frac{3x - 6 - (3x - 1)\ln|3x - 1|}{(3x - 1)(x - 2)^2}$

27. (a) About 1.3 m^2

    (b) About 0.21 g/cm^2

28. $12\sin 4x$

29. $-\cot \frac{x}{2}\csc^2 \frac{x}{2}$

30. $6x\sin x + 3x^2\cos x$

31. $\frac{\sin^2 x \cos x + 2\sin x \cos x}{(\sin x + 1)^2}$

32. $-\tan x$

33. $-\sin x e^{\cos x}$

34. 4

35. 3, -3

36. The slope of $y = \tan x$ is given by $y' = \sec^2 x$. Since $\sec x$ is never zero, the slope of $y = \tan x$ is never zero.

Chapter 5

GRAPHS AND THE DERIVATIVE

5.1 Increasing and Decreasing Functions

1. By reading the graph, f is

   (a) increasing on $(1, \infty)$ and
   (b) decreasing on $(-\infty, 1)$.

3. By reading the graph, g is

   (a) increasing on $(-\infty, -2)$ and
   (b) decreasing on $(-2, \infty)$.

5. By reading the graph, h is

   (a) increasing on $(-\infty, -4)$ and $(-2, \infty)$ and
   (b) decreasing on $(-4, -2)$.

7. By reading the graph, f is

   (a) increasing on $(-7, -4)$ and $(-2, \infty)$ and
   (b) decreasing on $(-\infty, -7)$ and $(-4, -2)$.

9. $y = 2 + 3.6x - 1.2x^2$

   (a) $y' = 3.6 - 2.4x$

   y' is zero when

   $$3.6 - 2.4x = 0$$
   $$x = \frac{3}{2},$$

   and there are no values of x where y' does not exist, so $\frac{3}{2}$ is the only critical number.

   Test a point in each interval.

   (b) When $x = 1$, $y' = 1.2 > 0$, so the function is increasing on $\left(-\infty, \frac{3}{2}\right)$.

   (c) When $x = 2$, $y' = -1.2 < 0$, so the function is decreasing on $\left(\frac{3}{2}, \infty\right)$.

11. $f(x) = \frac{2}{3}x^3 - x^2 - 24x - 4$

    (a) $f'(x) = 2x^2 - 2x - 24$
    $$= 2(x^2 - x - 12)$$
    $$= 2(x + 3)(x - 4)$$

    $f'(x)$ is zero when $x = -3$ or $x = 4$, so the critical numbers are -3 and 4.

    Test a point in each interval.

    $$f'(-4) = 16 > 0$$
    $$f'(0) = -24 < 0$$
    $$f'(5) = 16 > 0$$

    (b) f is increasing on $(-\infty, -3)$ and $(4, \infty)$.

    (c) f is decreasing on $(-3, 4)$.

13. $f(x) = 4x^3 - 15x^2 - 72x + 5$

    (a) $f'(x) = 12x^2 - 30x - 72$
    $$= 6(2x^2 - 5x - 12)$$
    $$= 6(2x + 3)(x - 4)$$

    $f'(x)$ is zero when $x = -\frac{3}{2}$ or $x = 4$, so the critical numbers are $-\frac{3}{2}$ and 4.

    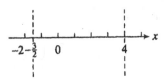

    $$f'(-2) = 36 > 0$$
    $$f'(0) = -72 < 0$$
    $$f'(5) = 78 > 0$$

    (b) f is increasing on $\left(-\infty, -\frac{3}{2}\right)$ and $(4, \infty)$.

    (c) f is decreasing on $\left(-\frac{3}{2}, 4\right)$.

15. $f(x) = x^4 + 4x^3 + 4x^2 + 1$

  (a) $f'(x) = 4x^3 + 12x^2 + 8x$

$$= 4x(x^2 + 3x + 2)$$
$$= 4x(x + 2)(x + 1)$$

$f'(x)$ is zero when $x = 0$, $x = -2$, or $x = -1$, so the critical numbers are 0, -2, and -1.

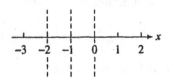

Test a point in each interval.

$$f'(-3) = -12(-1)(-2) = -24 < 0$$
$$f'(-1.5) = -6(0.5)(-0.5) = 1.5 > 0$$
$$f'(-0.5) = -2(1.5)(0.5) = -1.5 < 0$$
$$f'(1) = 4(3)(2) = 24 > 0$$

  (b) f is increasing on $(-2, -1)$ and $(0, \infty)$.

  (c) f is decreasing on $(-\infty, -2)$ and $(-1, 0)$.

17. $y = -3x + 6$

  (a) $y' = -3 < 0$

There are no critical numbers since y' is never 0 and always exists.

  (b) Since y' is always negative, the function is increasing on no interval.

  (c) y' is always negative, so the function is decreasing everywhere, or on the interval $(-\infty, \infty)$.

19. $f(x) = \dfrac{x + 2}{x + 1}$

  (a) $f'(x) = \dfrac{(x + 1)(1) - (x + 2)(1)}{(x + 1)^2}$

$$= \dfrac{-1}{(x + 1)^2}$$

The derivative is never 0, but it fails to exist at $x = -1$. Since -1 is not in the domain of f, however, -1 is not a critical number.

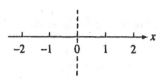

$$f'(-2) = -1 < 0$$
$$f'(0) = -1 < 0$$

  (b) f is increasing on no interval.

  (c) f is decreasing everywhere that it is defined, on $(-\infty, -1)$ and on $(-1, \infty)$.

21. $y = \sqrt{x^2 + 1}$

$$= (x^2 + 1)^{1/2}$$

  (a) $y' = \dfrac{1}{2}(x^2 + 1)^{-1/2}(2x)$

$$= x(x^2 + 1)^{-1/2}$$

$$= \dfrac{x}{\sqrt{x^2 + 1}}$$

$y' = 0$ when $x = 0$.
Since y does not fail to exist for any x, and since $y' = 0$ when $x = 0$, 0 is the only critical number.

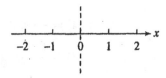

$$y'(1) = \dfrac{1}{\sqrt{2}} > 0$$

$$y'(-1) = \dfrac{-1}{\sqrt{2}} < 0$$

  (b) y is increasing on $(0, \infty)$.

  (c) y is decreasing on $(-\infty, 0)$.

23. $f(x) = x^{2/3}$

  (a) $f'(x) = \dfrac{2}{3}x^{-1/3} = \dfrac{2}{3x^{1/3}}$

$f'(x)$ is never zero, but fails to exist when $x = 0$, so 0 is the only critical number.

$$f'(-1) = -\dfrac{2}{3} < 0$$

$$f'(1) = \dfrac{2}{3} > 0$$

  (b) f is increasing on $(0, \infty)$.

  (c) f is decreasing on $(-\infty, 0)$.

25. $y = x - 4 \ln (3x - 9)$

(a) $y' = 1 - \dfrac{12}{3x - 9} = 1 - \dfrac{4}{x - 3}$

$= \dfrac{x - 7}{x - 3}$

y' is zero when $x = 7$. The derivative does not exist at $x = 3$, but note that the domain of f is $(3, \infty)$.

Thus, the only critical number is 7.

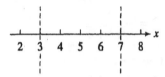

Choose values in the intervals $(3, 7)$ and $(7, \infty)$.

$$f'(4) = -3 < 0$$

$$f(8) = \frac{1}{5} > 0$$

(b) The function is increasing on $(7, \infty)$.

(c) The function is decreasing on $(3, 7)$.

27. $y = x^{2/3} - x^{5/3}$

(a) $y' = \dfrac{2}{3} x^{-1/3} - \dfrac{5}{3} x^{2/3} = \dfrac{2 - 5x}{3x^{1/3}}$

$y' = 0$ when $x = \frac{2}{5}$. The derivative does not exist at $x = 0$. So the critical numbers are 0 and $\frac{2}{5}$.

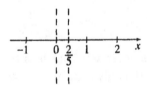

Test a point in each interval.

$$y'(-1) = \frac{7}{-3} < 0$$

$$y'\left(\frac{1}{5}\right) = \frac{1}{3\left(\frac{1}{5}\right)^{1/3}} = \frac{5^{1/3}}{3} > 0$$

$$y'(1) = \frac{-3}{3} = -1 < 0$$

(b) y is increasing on $\left(0, \frac{2}{5}\right)$.

(c) y is decreasing on $(-\infty, 0)$ and $\left(\frac{2}{5}, \infty\right)$.

29. $y = \sin x$

(a) $y' = \cos x$

y' is zero when $x = \pm \dfrac{\pi}{2}, \pm \dfrac{3\pi}{2}, \pm \dfrac{5\pi}{2}, \cdots$

Therefore, the critical numbers are $\frac{n\pi}{2}$, where n is an odd integer.

(b) The function is increasing on

$$\left(n\pi, \left(n + \frac{1}{2}\right)\pi\right) \cup \left(\left(n + \frac{3}{2}\right)\pi, (n + 2)\pi\right)$$

where n is an even integer.

(c) The function is decreasing on

$$\left(\left(n + \frac{1}{2}\right)\pi, \left(n + \frac{3}{2}\right)\pi\right)$$

where n is an even integer.

31. $y = 3 \sec x$

(a) $y' = 3 \sec x \tan x$

y' is zero when $x = 0, \pm \pi, \pm 2\pi, \ldots$
Therefore the critical numbers are $n\pi$, where n is an integer.

(b) The function is increasing on

$$\left((n + 1)\pi, \left(n + \frac{3}{2}\right)\pi\right) \cup \left(\left(n + \frac{3}{2}\right)\pi, (n + 2)\pi\right),$$

where n is an odd integer.

(c) The function is decreasing on

$$\left(n\pi, \left(n + \frac{1}{2}\right)\pi\right) \cup \left(\left(n + \frac{1}{2}\right)\pi, (n + 1)\pi\right)$$

where n is an odd integer.

33. $f(x) = ax^2 + bx + c, \ a > 0$
$f'(x) = 2ax + b$

Let $f'(x) = 0$ to find the critical number.

$$2ax + b = 0$$
$$2ax = -b$$
$$x = \frac{-b}{2a}$$

Choose a value in the interval $\left(-\infty, \frac{-b}{2a}\right)$. Since $a > 0$,

$$\frac{-b}{2a} - \frac{1}{2a} = \frac{-b - 1}{2a} < \frac{-b}{2a}.$$

$$f'\left(\frac{-b - 1}{2a}\right) = 2a\left(\frac{-b - 1}{2a}\right) + b$$

$$= -1 < 0$$

Choose a value in the interval $\left(\frac{-b}{2a}, \infty\right)$. Since $a > 0$,

$$\frac{-b}{2a} + \frac{1}{2a} = \frac{-b+1}{2a} > \frac{-b}{2a}.$$

$$f'\left(\frac{-b+1}{2a}\right) = 1 > 0.$$

$f(x)$ is increasing on $\left(\frac{-b}{2a}, \infty\right)$ and decreasing on $\left(-\infty, \frac{-b}{2a}\right)$.

This tells us that the curve opens upward and $x = \frac{-b}{2a}$ is the x-coordinate of the vertex.

$$f\left(\frac{-b}{2a}\right) = a\left(\frac{-b}{2a}\right)^2 + b\left(\frac{-b}{2a}\right) + c$$

$$= \frac{ab^2}{4a^2} - \frac{b^2}{2a} + c$$

$$= \frac{b^2}{4a} - \frac{2b^2}{4a} + \frac{4ac}{4a}$$

$$= \frac{4ac - b^2}{4a}$$

The vertex is $\left(\frac{-b}{2a}, \frac{4ac-b^2}{4a}\right)$ or $\left(-\frac{b}{2a}, \frac{4ac-b^2}{4a}\right)$.

35. $f(x) = e^x$
$f'(x) = e^x > 0$

$f(x) = e^x$ is increasing on $(-\infty, \infty)$.
$f(x) = e^x$ is decreasing nowhere.
Since $f'(x)$ is always positive, it is never equal zero. Therefore, the tangent line is horizontal nowhere.

37. (a) These curves are graphs of functions since they all pass the vertical line test.

(b) The graph for particulates increases from April to July; it decreases from July to November; it is constant from January to April and November to December.

(c) All graphs are constant from January to April and November to December. In these months, most trees do not have leaves.

39. $A(x) = -0.015x^3 + 1.058x$
$A'(x) = -0.045x^2 + 1.058$
$A'(x) = 0$ when

$$-0.045x^2 + 1.058 = 0$$
$$-0.045x^2 = -1.058$$
$$x^2 \approx 23.5$$
$$x \approx \pm 4.8.$$

The function only applies for the interval $[0, 8]$, so we disregard the solution -4.8.
Then, 4.8 divides $[0, 8]$ into two intervals.

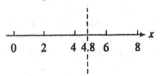

$A'(4) = 0.338 > 0$
$A'(5) = -0.067 < 0$

(a) A is increasing on the interval $(0, 4.8)$.

(b) A is decreasing on the interval $(4.8, 8)$.

41. $K(t) = \dfrac{5t}{t^2 + 1}$

$$K'(t) = \frac{5(t^2+1) - 2t(5t)}{(t^2+1)^2}$$

$$= \frac{5t^2 + 5 - 10t^2}{(t^2+1)^2}$$

$$= \frac{5 - 5t^2}{(t^2+1)^2}$$

$K'(t) = 0$ when

$$\frac{5 - 5t^2}{(t^2+1)^2} = 0$$
$$5 - 5t^2 = 0$$
$$5t^2 = 5$$
$$t = \pm 1.$$

Since t is the time after a drug is administered, the function applies only for $[0, \infty)$, so we discard $t = -1$. Then 1 divides the interval into two intervals.

$K'(.5) = 2.4 > 0$
$K'(2) = -0.6 < 0$

(a) K is increasing on $(0, 1)$.

(b) K is decreasing on $(1, \infty)$.

43. (a) $F(t) = -10.28 + 175.9te^{-t/1.3}$

$$F'(t) = (175.9)(e^{-t/1.3})$$

$$+ (175.9.9t)\left(-\frac{1}{1.3}e^{-t/1.3}\right)$$

$$= (175.9)(e^{-t/1.3})\left(1 - \frac{t}{1.3}\right)$$

$$\approx 175.9e^{-t/1.3}(1 - 0.769t)$$

(b) $F'(t)$ is equal to 0 at $t = 1.3$. Therefore, 1.3 is a critical number. Since the domain is $(0, \infty)$, test values in the intervals from $(0, 1.3)$ and $(1.3, \infty)$.

$$F'(1) \approx 18.83 > 0 \text{ and } F'(2) \approx -20.32 < 0$$

$F'(t)$ is increasing on $(0, 1.3)$ and decreasing on $(1.3, \infty)$.

45. (a) $W_2(t) = 532(1 - 0.911e^{-0.0021t})^{1.2466}$

$W_2'(t) \approx 663.1912(1 - 0.911e^{-0.0021t})(0.001913e^{-0.0021t})$

Since $W_2'(t) > 0$ when t is a positive number, $W_2(t)$ is increasing on $(0, \infty)$.

(b)

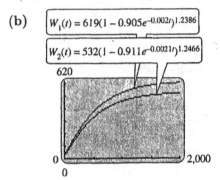

$W_1(t) = 619(1 - 0.905e^{-0.002t})^{1.2386}$

$W_2(t) = 532(1 - 0.911e^{-0.0021t})^{1.2466}$

620

0

0   2,000

47. $C(t) = 37.29 + 0.46 \cos\left(\dfrac{2\pi(t - 16.37)}{24} \right)$

$C'(t) = \left(\dfrac{2\pi}{24} \right)(0.46)\left(-\sin\left[\dfrac{2\pi(t - 16.37)}{24} \right] \right)$

$\quad = \dfrac{-0.46\pi}{12}\left(\sin\left[\dfrac{\pi(t - 16.37)}{12} \right] \right)$

$C'(t)$ is zero when $t = 4.37$, and $t = 16.37$.
Since $C'(t) > 0$ when x is in the interval $(4.37, 16.37)$, the function is increasing on $(4.37, 16.37)$.
Since $C'(t) < 0$ when x is in the interval $(0, 4.37)$ or $(16.37, 24)$, the function is decreasing on $(0, 4.37) \cup (16.37, 24)$.

49. As shown on the graph,

(a) horsepower increases with engine speed on $(1,000, 6,100)$;

(b) horsepower decreases with engine speed on $(6,100, 6,500)$;

(c) torque increases with engine speed on $(1,000, 3,000)$ and $(3,600, 4,200)$;

(d) torque decreases with engine speed on $(3,000, 3,600)$ and $(4,200, 6,500)$.

51. $H(r) = \dfrac{300}{1 + 0.03r^2} = 300(1 + 0.03r^2)^{-1}$

$H'(r) = 300[-1(1 + 0.03r^2)^{-2}(0.06r)]$

$\quad = \dfrac{-18r}{(1 + 0.03r^2)^2}$

Since r is a mortgage rate (in percent), it is always positive. Thus, $H'(r)$ is always negative.

(a) H is increasing on nowhere.

(b) H is decreasing on $(0, \infty)$.

5.2 Relative Extrema

1. As shown on the graph, the relative minimum of -4 occurs when $x = 1$.

3. As shown on the graph, the relative maximum of 3 occurs when $x = -2$.

5. As shown on the graph, the relative maximum of 3 occurs when $x = -4$ and the relative minimum of 1 occurs when $x = -2$.

7. As shown on the graph, the relative maximum of 3 occurs when $x = -4$; the relative minimum of -2 occurs when $x = -7$ and $x = -2$.

9. $f(x) = x^2 + 12x - 8$
$f'(x) = 2x + 12 = 2(x + 6)$
$f'(x)$ is zero when $x = -6$.

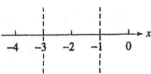

Test $f'(x)$ at -10 and 0.
$$f'(-10) = -8 < 0$$
$$f'(0) = 12 > 0$$

Thus, we see that f is decreasing on $(-\infty, -6)$ and increasing on $(-6, \infty)$, so $f(-6)$ is a relative minimum.

$$f(-6) = (-6)^2 + 12(-6) - 8 = 36 - 72 - 8 = -44$$

Relative minimum of -44 at -6

11. $f(x) = x^3 + 6x^2 + 9x - 8$
$f'(x) = 3x^2 + 12x + 9 = 3(x^2 + 4x + 3)$
$\quad = 3(x + 3)(x + 1)$

$f'(x)$ is zero when $x = -1$ or $x = -3$.

$$f'(-4) = 9 > 0$$
$$f'(-2) = -3 < 0$$
$$f'(0) = 9 > 0$$

Thus, f is increasing on $(-\infty, -3)$, decreasing on $(-3, -1)$, and increasing on $(-1, \infty)$.

f has a relative maximum at -3 and a relative minimum at -1.

$$f(-3) = -8$$
$$f(-1) = -12$$

Relative maximum of -8 at -3; relative minimum of -12 at -1

13. $f(x) = -\dfrac{4}{3}x^3 - \dfrac{21}{2}x^2 - 5x + 8$

$f'(x) = -4x^2 - 21x - 5$
$\quad\ = (-4x - 1)(x + 5)$

$f'(x)$ is zero when $x = -5$, or $x = -\frac{1}{4}$.

$$f'(-6) = -23 < 0$$
$$f'(-4) = 15 > 0$$
$$f'(0) = -5 < 0$$

f is decreasing on $(-\infty, -5)$, increasing on $\left(-5, -\frac{1}{4}\right)$, and decreasing on $\left(-\frac{1}{4}, \infty\right)$. f has a relative minimum at -5 and a relative maximum at $-\frac{1}{4}$.

$$f(-5) = -\frac{377}{6}$$
$$f\left(-\frac{1}{4}\right) = \frac{827}{96}$$

Relative maximum of $\frac{827}{96}$ at $-\frac{1}{4}$; relative minimum of $-\frac{377}{6}$ at -5

15. $f(x) = x^4 - 18x^2 - 4$
$f'(x) = 4x^3 - 36x$
$\quad\ = 4x(x^2 - 9)$
$\quad\ = 4x(x + 3)(x - 3)$

$f'(x)$ is zero when $x = 0$ or $x = -3$ or $x = 3$.

$$f'(-4) = 4(-4)^3 - 36(-4) = -112 < 0$$
$$f'(-1) = -4 + 36 = 32 > 0$$
$$f'(1) = 4 - 36 = -32 < 0$$
$$f'(4) = 4(4)^3 - 36(4) = 112 > 0.$$

f is decreasing on $(-\infty, -3)$ and $(0, 3)$; f is increasing on $(-3, 0)$ and $(3, \infty)$.

$$f(-3) = -85$$
$$f(0) = -4$$
$$f(3) = -85$$

Relative maximum of -4 at 0; relative minimum of -85 at 3 and -3

17. $f(x) = -(8 - 5x)^{2/3}$

$f'(x) = -\dfrac{2}{3}(8 - 5x)^{-1/3}(-5)$

$\qquad = \dfrac{10}{3(8 - 5x)^{1/3}}$

$f'(x)$ is never zero, but fails to exist if $8 - 5x = 0$, so $\frac{8}{5}$ is a critical number.

$$f'(0) = \frac{10}{3(8)^{1/3}} = \frac{5}{3} > 0$$

$$f'(2) = \frac{10}{3(8 - 10)^{1/3}} \approx -2.6 < 0$$

f is increasing on $\left(-\infty, \frac{8}{5}\right)$ and decreasing on $\left(\frac{8}{5}, \infty\right)$.

$$f\left(\frac{8}{5}\right) = -\left[8 - 5\left(\frac{8}{5}\right)\right]^{2/3} = 0$$

Relative maximum of 0 at $\frac{8}{5}$

19. $f(x) = 2x + 3x^{2/3}$
$f'(x) = 2 + 2x^{-1/3}$

$\qquad = 2 + \dfrac{2}{\sqrt[3]{x}}$

Find the critical numbers.

$f'(x) = 0$ when

$$2 + \frac{2}{\sqrt[3]{x}} = 0$$

$$\frac{2}{\sqrt[3]{x}} = -2$$

$$\frac{1}{\sqrt[3]{x}} = -1$$

$$x = (-1)^3$$
$$x = -1.$$

$f'(x)$ does not exist when

$$\sqrt[3]{x} = 0$$
$$x = 0.$$

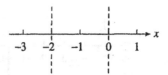

$$f'(-2) = 2 + \frac{2}{\sqrt[3]{-2}} \approx 0.41 > 0$$

$$f'\left(-\frac{1}{2}\right) = 2 + \frac{2}{\sqrt[3]{-\frac{1}{2}}}$$

$$= 2 + \frac{2\sqrt[3]{2}}{-1} \approx -0.52 < 0$$

$$f'(1) = 2 + \frac{2}{\sqrt[3]{1}} = 4 > 0$$

f is increasing on $(-\infty, -1)$ and $(0, \infty)$.
f is decreasing on $(-1, 0)$.

$$f(-1) = 2(-1) + 3(-1)^{2/3} = 1$$
$$f(0) = 0$$

Relative maximum of 1 at -1; relative minimum
of 0 at 0

21. $f(x) = x - \dfrac{1}{x}$

$f'(x) = 1 + \frac{1}{x^2}$ is never zero, but fails to exist at
$x = 0$.
Since $f(x)$ also fails to exist at $x = 0$, there are
no critical numbers and no relative extrema.

23. $f(x) = \dfrac{x^2 - 2x + 1}{x - 3}$

$$f'(x) = \frac{(x-3)(2x-2) - (x^2 - 2x + 1)(1)}{(x-3)^2}$$

$$= \frac{x^2 - 6x + 5}{(x-3)^2}$$

Find the critical numbers:

$$x^2 - 6x + 5 = 0$$
$$(x-5)(x-1) = 0$$
$$x = 5 \quad \text{or} \quad x = 1$$

Note that $f(x)$ and $f'(x)$ do not exist at $x = 3$, so
the only critical numbers are 1 and 5.

$$f'(0) = \frac{5}{9} > 0$$

$$f'(2) = -3 < 0$$

$$f'(6) = \frac{5}{9} > 0$$

$f(x)$ is increasing on $(-\infty, 1)$ and $(5, \infty)$.
$f(x)$ is decreasing on $(1, 5)$.

$$f(1) = 0$$
$$f(5) = 8$$

Relative maximum of 0 at 1; relative minimum of
8 at 5

25. $f(x) = x^2 e^x - 3$
$f'(x) = x^2 e^x + 2x e^x$
$\quad\;\; = x e^x (x + 2)$

$f'(x)$ is zero at $x = 0$ and $x = -2$.

$$f'(-3) = 3e^{-3} = \frac{3}{e^3} > 0$$

$$f'(-1) = -e^{-1} = \frac{-1}{e} < 0$$

$$f'(1) = 3e^1 > 0$$

f is increasing on $(-\infty, -2)$ and $(0, \infty)$.
f is decreasing on $(-2, 0)$.

$$f(0) = 0 \cdot e^0 - 3 = -3$$
$$f(-2) = (-2)^2 e^{-2} - 3$$

$$= \frac{4}{e^2} - 3$$

$$\approx -2.46$$

Relative minimum of -3 at 0; relative maximum
of -2.46 at -2

27. $f(x) = 2x + \ln x$

$$f'(x) = 2 + \frac{1}{x} = \frac{2x + 1}{x}$$

$f'(x)$ is zero at $x = -\frac{1}{2}$. The domain of $f(x)$ is
$(0, \infty)$. Therefore $f'(x)$ is never zero in the domain
of $f(x)$.
$f'(1) = 3 > 0$. Since $f(x)$ is always increasing, f
has no relative extrema.

29. $f(x) = \sin(\pi x)$

$f'(x) = \pi \cos(\pi x)$

The critical numbers are $x = \pm\dfrac{1}{2}, \pm\dfrac{3}{2}, \pm\dfrac{5}{2}, \ldots$

$f/(0) = \pi > 0$

$f'(1) = -\pi < 0$

$f'(2) = \pi > 0$

$\vdots$

$f(x)$ is increasing on the intervals

$$\cdots\left(-\frac{5}{2}, -\frac{3}{2}\right), \left(-\frac{1}{2}, \frac{1}{2}\right), \left(\frac{3}{2}, \frac{5}{2}\right)\cdots$$

$f(x)$ is decreasing on the intervals

$$\cdots\left(-\frac{3}{2}, -\frac{1}{2}\right), \left(\frac{1}{2}, \frac{3}{2}\right), \left(\frac{5}{2}, \frac{7}{2}\right)\cdots$$

Relative maximum of 1 at $x = \cdots-\dfrac{7}{2}, -\dfrac{3}{2}, \dfrac{1}{2}, \dfrac{5}{2}, \cdots$

Relative minimum of -1 at $x = \cdots-\dfrac{5}{2}, -\dfrac{1}{2}, \dfrac{3}{2}, \dfrac{7}{2}, \cdots$

31. $y = -2x^2 + 8x - 1$

$y' = -4x + 8$

$\quad = 4(2 - x)$

The vertex occurs when $y' = 0$ or when

$$2 - x = 0$$
$$x = 2.$$

When $x = 2$,

$$y = -2(2)^2 + 8(2) - 1 = 7.$$

The vertex is at $(2, 7)$.

33. $f(x) = x^5 - x^4 + 4x^3 - 30x^2 + 5x + 6$

$f'(x) = 5x^4 - 4x^3 + 12x^2 - 60x + 5$

Graph f' on a graphing calculator. A suitable choice for the viewing window is $[-4, 4]$ by $[-50, 50]$, Yscl = 10.

Use the calculator to estimate the x-intercepts of this graph. These numbers are the solutions of the equation $f'(x) = 0$ and thus the critical numbers for f. Rounded to three decimal places, these x-values are 0.85 and 2.161.

Examine the graph of f' near $x = 0.085$ and $x = 2.161$. Observe that $f'(x) > 0$ to the left of $x = 0.085$ and $f'(x) < 0$ to the right of $x = 0.085$.

Also observe that $f'(x) < 0$ to the left of $x = 2.161$ and $f'(x) > 0$ to the right of $x = 2.161$. The first derivative test allows us to conclude that f has a relative maximum at $x = 0.08$ and a relative minimum at $x = 2.161$.

$$f(0.085) \approx 6.211$$
$$f(2.161) \approx -57.607$$

Relative maximum of 6.211 at 0.085; relative minimum of -57.607 at 2.161.

35. $f(x) = 2\,|x + 1| + 4\,|x - 5| - 20$

Graph this function in the window $[-10, 10]$ by $[-15, 30]$, Yscl = 5.

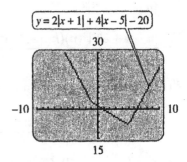

The graph shows that f has no relative maxima, but there is a relative minimum at $x = 5$.
(Note that the graph has a sharp point at $(5, -8)$, indicating that $f'(5)$ does not exist.)

37. $a(t) = 0.008t^3 - 0.288t^2 + 2.304t + 7$

$a'(t) = 0.024t^2 - 0.576t + 2.304$

Set $a' = 0$ and use the quadratic formula to solve for t.

$$0.024t^2 - 0.576t + 2.304 = 0$$
$$0.024(t^2 - 24t + 96) = 0$$
$$t^2 - 24t + 96 = 0$$
$$t \approx 5.07 \text{ or } t \approx 18.93$$

$t = 5.07 = 5$ hours $+ .07 \cdot 60$ minutes corresponds to 5:04 P.M.

$t = 18.93 = 18$ hours $+ .93 \cdot 60$ minutes corresponds to 6:56 A.M.

39. $M(t) = 369(0.93)^t(t)^{0.36}$

$M'(t) = (369)(0.93)^t \ln(0.93)(t^{0.36}) + 369(0.93)^t(0.36)(t)^{-0.64}$

$\qquad = (369t^{0.36})(0.93^t \ln 0.93) + \dfrac{132.84(0.93)^t}{t^{0.64}}$

$M'(t) = 0$ when $t \approx 4.96$.

Verify that $t \approx 4.96$ gives a maximum.

$$M'(4) > 0$$
$$M'(5) < 0$$

Find $M(4.96)$

$$M(4.96) = 369(0.93)^{4.96}(4.96)^{0.36} \approx 458.22$$

The female moose reaches a maximum weight of about 458.22 kilograms at about 4.96 years.

41. $f(t) = \dfrac{k(n-1)n^2 e^{nt}}{[n-1+e^{nt}]^2}$

$f'(t) = \dfrac{[n-1+e^{nt}]^2(n)(k(n-1)n^2 e^{nt}) - (2)(n-1+e^{nt})(ne^{nt})(k)(n-1)(n^2)e^{nt}}{[n-1+e^{nt}]^4}$

Set $f'(t)$ equal to zero and solve for t.

$$\dfrac{[n-1+e^{nt}]^2(n)(k(n-1)n^2 e^{nt}) - (2)(n-1+e^{nt})(ne^{nt})(k)(n-1)(n^2)e^{nt}}{[n-1+e^{nt}]^4} = 0$$

$$[n-1+e^{nt}]^2(n)(k(n-1)n^2 e^{nt}) - (2)(n-1+e^{nt})(ne^{nt})(k)(n-1)(n^2)e^{nt} = 0$$

$$(n-1+e^{nt}) - 2e^{nt} = 0$$
$$n-1-e^{nt} = 0$$
$$n-1 = e^{nt}$$
$$\ln(n-1) = \ln(e^{nt})$$
$$\ln(n-1) = nt$$
$$\dfrac{\ln(n-1)}{n} = t$$

The rate of change in the number of infected individuals reaches a maximum at $t = \dfrac{\ln(n-1)}{n}$.

43. $D(x) = -x^4 + 8x^3 + 80x^2$

$D'(x) = -4x^3 + 24x^2 + 160x = -4x(x^2 - 6x - 40) = -4x(x+4)(x-10)$

$D'(x) = 0$ when $x = 0$, $x = -4$, or $x = 10$.

Disregard the nonpositive values.

Verify that $x = 10$ gives a maximum.

$$D'(9) = 468 > 0$$
$$D'(11) = -660 < 0$$

The speaker should aim for a degree of discrepancy of 10.

45. $s(t) = -16t^2 + 64t + 3$

$s'(t) = -32t + 64$

When $s'(t) = 0$,

$$0 = -32t + 64$$
$$32t = 64$$
$$t = 2.$$

Verify that $t = 2$ gives a maximum.

$$s'(1) = 32 > 0$$
$$s'(3) = -32 < 0$$

Now find the height when $t = 2$.

$$s(2) = -16(2)^2 + 64(2) + 3$$
$$= 67$$

Therefore, the maximum height is 67 ft.

5.3   Higher Derivatives, Concavity, and the Second Derivative Test

1. $f(x) = 3x^3 - 4x + 5$

$f'(x) = 9x^2 - 4$

$f''(x) = 18x$

$f''(0) = 18(0) = 0$

$f''(2) = 18(2) = 36$

3. $f(x) = 3x^4 - 5x^3 + 2x^2$

$f'(x) = 12x^3 - 15x^2 + 4x$

$f''(x) = 36x^2 - 30x + 4$

$f''(0) = 36(0)^2 - 30(0) + 4 = 4$

$f''(2) = 36(2)^2 - 30(2) + 4 = 88$

5. $f(x) = 3x^2 - 4x + 8$

$f'(x) = 6x - 4$

$f''(x) = 6$

$f''(0) = 6$

$f''(2) = 6$

7. $f(x) = \dfrac{x^2}{1+x}$

$f'(x) = \dfrac{(1+x)(2x) - x^2(1)}{(1+x)^2} = \dfrac{2x + x^2}{(1+x)^2}$

$f''(x) = \dfrac{(1+x)^2(2 + 2x) - (2x + x^2)(2)(1+x)}{(1+x)^4}$

$\quad = \dfrac{(1+x)(2 + 2x) - (2x + x^2)(2)}{(1+x)^3} = \dfrac{2}{(1+x)^3}$

$f''(0) = 2$

$f''(2) = \dfrac{2}{27}$

9. $f(x) = \sqrt{x+4} = (x+4)^{1/2}$

$f'(x) = \dfrac{1}{2}(x+4)^{-1/2}$

$f''(x) = \dfrac{1}{2}\left(-\dfrac{1}{2}\right)(x+4)^{-3/2}$

$\quad = -\dfrac{1}{4}(x+4)^{-3/2} = \dfrac{-1}{4(x+4)^{3/2}}$

$f''(0) = \dfrac{-1}{4(0+4)^{3/2}} = \dfrac{-1}{4(4^{3/2})} = \dfrac{-1}{4(8)} = -\dfrac{1}{32}$

$f''(2) = \dfrac{-1}{4(2+4)^{3/2}} = \dfrac{-1}{4(6^{3/2})} \approx -0.0170$

11. $f(x) = 5x^{3/5}$

$f'(x) = 3x^{-2/5}$

$f''(x) = -\dfrac{6}{5}x^{-7/5} \quad$ or $\quad \dfrac{-6}{5x^{7/5}}$

$f''(0)$ does not exist.

$f''(2) = -\dfrac{6}{5}(2^{-7/5}) = \dfrac{-6}{5(2^{7/5})} \approx -0.4547$

13. $f(x) = 5e^{-x^2}$

$f'(x) = 5e^{-x^2}(-2x) = -10xe^{-x^2}$

$f''(x) = -10xe^{-x^2}(-2x) + e^{-x^2}(-10)$

$\quad = 20x^2e^{-x^2} - 10e^{-x^2}$

$f''(0) = 20(0^2)e^{-0^2} - 10e$

$\quad = 0 - 10 = -10$

$f''(2) = 20(2^2)e^{-(2^2)} - 10e^{-(2^2)}$

$\quad = 80e^{-4} - 10e^{-4} = 70e^{-4}$

$\quad \approx 1.282$

15. $f(x) = \dfrac{\ln x}{4x}$

$f'(x) = \dfrac{4x\left(\frac{1}{x}\right) - (\ln x)(4)}{(4x)^2}$

$\quad = \dfrac{4 - 4\ln x}{16x^2} = \dfrac{1 - \ln x}{4x^2}$

$f''(x) = \dfrac{4x^2\left(-\frac{1}{x}\right) - (1 - \ln x)8x}{16x^4}$

$\quad = \dfrac{-4x - 8x + 8x\ln x}{16x^4} = \dfrac{-12x + 8x\ln x}{16x^4}$

$\quad = \dfrac{4x(-3 + 2\ln x)}{16x^4} = \dfrac{-3 + 2\ln x}{4x^3}$

$f''(0)$ does not exist because $\ln 0$ is undefined.

$f''(2) = \dfrac{-3 + 2\ln 2}{4(2)^3} = \dfrac{-3 + 2\ln 2}{32} \approx 0.0504$

17. $f(x) = -x^4 + 2x^2 + 8$
$f'(x) = -4x^3 + 4x$
$f''(x) = -12x^2 + 4$
$f'''(x) = -24x$
$f^{(4)}(x) = -24$

19. $f(x) = 4x^5 + 6x^4 - x^2 + 2$
$f'(x) = 20x^4 + 24x^3 - 2x$
$f''(x) = 80x^3 + 72x^2 - 2$
$f'''(x) = 240x^2 + 144x$
$f^{(4)}(x) = 480x + 144$

21. $f(x) = \dfrac{x-1}{x+2}$

$f'(x) = \dfrac{(x+2)-(x-1)}{(x+2)^2}$

$\qquad = \dfrac{3}{(x+2)^2}$

$f''(x) = \dfrac{-3(2)(x+2)}{(x+2)^4}$

$\qquad = \dfrac{-6}{(x+2)^3}$

$f'''(x) = \dfrac{(-6)(-3)(x+2)^2}{(x+2)^6}$

$\qquad = 18(x+2)^{-4} \quad \text{or} \quad \dfrac{18}{(x+2)^4}$

$f^{(4)}(x) = \dfrac{-18(4)(x+2)^3}{(x+2)^8}$

$\qquad = -72(x+2)^{-5} \quad \text{or} \quad \dfrac{-72}{(x+2)^5}$

23. $f(x) = \dfrac{3x}{x-2}$

$f'(x) = \dfrac{(x-2)(3)-3x(1)}{(x-2)^2} = \dfrac{-6}{(x-2)^2}$

$f''(x) = \dfrac{-6(-2)(x-2)}{(x-2)^4} = \dfrac{12}{(x-2)^3}$

$f'''(x) = \dfrac{-12(3)(x-2)^2}{(x-2)^6} = -36(x-2)^{-4}$

$\qquad \text{or } \dfrac{-36}{(x-2)^4}$

$f^{(4)}(x) = \dfrac{-36(-4)(x-2)^3}{(x-2)^8}$

$\qquad = 144(x-2)^{-5}$

$\qquad \text{or } \dfrac{144}{(x-2)^5}$

25. $f(x) = \ln x$

(a) $f'(x) = \dfrac{1}{x} = x^{-1}$

$f''(x) = -x^{-2} = \dfrac{-1}{x^2}$

$f'''(x) = 2x^{-3} = \dfrac{2}{x^3}$

$f^{(4)}(x) = -6x^{-4} = \dfrac{-6}{x^4}$

$f^{(5)}(x) = 24x^{-5} = \dfrac{24}{x^5}$

(b) $f^{(n)}(x) = \dfrac{(-1)^{n-1}(n-1)!}{x^n}$

27. Concave upward on $(2, \infty)$
Concave downward on $(-\infty, 2)$
Point of inflection at $(2, 3)$

29. Concave upward on $(-\infty, -1)$ and $(8, \infty)$
Concave downward on $(-1, 8)$
Points of inflection at $(-1, 7)$ and $(8, 6)$

31. Concave upward on $(2, \infty)$
Concave downward on $(-\infty, 2)$
No points of inflection

33. $f(x) = x^2 + 10x - 9$
$f'(x) = 2x + 10$
$f''(x) = 2 > 0$ for all x.

Always concave upward
No points of inflection

35. $f(x) = -2x^3 + 9x^2 + 168x - 3$
$f'(x) = -6x^2 + 18x + 168$
$f''(x) = -12x + 18$
$f''(x) = -12x + 18 > 0 \quad$ when
$\qquad -6(2x - 3) > 0$
$\qquad\quad 2x - 3 < 0$
$\qquad\qquad x < \dfrac{3}{2}.$

Concave upward on $\left(-\infty, \frac{3}{2}\right)$

$f''(x) = -12x + 18 < 0$ when
$\qquad -6(2x - 3) < 0$
$\qquad\quad 2x - 3 > 0$
$\qquad\qquad x > \dfrac{3}{2}.$

Concave downward on $\left(\frac{3}{2}, \infty\right)$

$f''(x) = -12x + 18 = 0$ when
$$-6(2x + 3) = 0$$
$$2x + 3 = 0$$
$$x = \frac{3}{2}.$$
$$f\left(\frac{3}{2}\right) = \frac{525}{2}$$

Point of inflection at $\left(\frac{3}{2}, \frac{525}{2}\right)$

37. $f(x) = \dfrac{3}{x - 5}$

$$f'(x) = \frac{-3}{(x - 5)^2}$$

$$f''(x) = \frac{-3(-2)(x - 5)}{(x - 5)^4} = \frac{6}{(x - 5)^3}$$

$$f''(x) = \frac{6}{(x - 5)^3} > 0 \text{ when}$$
$$(x - 5)^3 > 0$$
$$x - 5 > 0$$
$$x > 5.$$

Concave upward on $(5, \infty)$

$$f''(x) = \frac{6}{(x - 5)^3} < 0 \text{ when}$$
$$(x - 5)^3 < 0$$
$$x - 5 < 0$$
$$x < 5.$$

Concave downward on $(-\infty, 5)$

$f''(x) \neq 0$ for any value for x; it does not exist when $x = 5$. There is a change of concavity there, but no point of inflection since $f(5)$ does not exist.

39. $f(x) = x(x + 5)^2$
$$f'(x) = x(2)(x + 5) + (x + 5)^2$$
$$= (x + 5)(2x + x + 5)$$
$$= (x + 5)(3x + 5)$$
$$f''(x) = (x + 5)(3) + (3x + 5)$$
$$= 3x + 15 + 3x + 5 = 6x + 20$$
$$f''(x) = 6x + 20 > 0 \text{ when}$$
$$2(3x + 10) > 0$$
$$3x > -10$$
$$x > -\frac{10}{3}.$$

Concave upward on $\left(-\frac{10}{3}, \infty\right)$

$$f''(x) = 6x + 20 < 0 \text{ when}$$
$$2(3x + 10) < 0$$
$$3x < -10$$
$$x < -\frac{10}{3}.$$

Concave downward on $\left(-\infty, -\frac{10}{3}\right)$

$$f\left(-\frac{10}{3}\right) = -\frac{10}{3}\left(-\frac{10}{3} + 5\right)^2$$
$$= \frac{-10}{3}\left(\frac{-10 + 15}{3}\right)^2$$
$$= -\frac{10}{3} \cdot \frac{25}{9} = -\frac{250}{27}$$

Point of inflection at $\left(-\frac{10}{3}, -\frac{250}{27}\right)$

41. $f(x) = 18x - 18e^{-x}$
$f'(x) = 18 - 18e^{-x}(-1) = 18 + 18e^{-x}$
$f''(x) = 18e^{-x}(-1) = -18e^{-x}$

$f''(x) = -18e^{-x} < 0$ for all x

$f(x)$ is never concave upward and always concave downward. There are no points of inflection since $-18e^{-x}$ is never equal to 0.

43. $f(x) = x^{8/3} - 4x^{5/3}$

$$f'(x) = \frac{8}{3}x^{5/3} - \frac{20}{3}x^{2/3}$$

$$f''(x) = \frac{40}{9}x^{2/3} - \frac{40}{9}x^{-1/3} = \frac{40(x - 1)}{9x^{1/3}}$$

$f''(x) = 0$ when $x = 1$
$f''(x)$ fails to exist when $x = 0$
Note that both $f(x)$ and $f'(x)$ exist at $x = 0$.
Check the sign of $f''(x)$ in the three intervals determined by $x = 0$ and $x = 1$ using test points.

$$f''(-1) = \frac{40(-2)}{9(-1)} = \frac{80}{9} > 0$$

$$f''\left(\frac{1}{8}\right) = \frac{40\left(-\frac{7}{8}\right)}{9\left(\frac{1}{2}\right)} = -\frac{70}{9} < 0$$

$$f''(8) = \frac{40(7)}{9(2)} = \frac{140}{9} > 0$$

Concave upward on $(-\infty, 0)$ and $(1, \infty)$; concave downward on $(0, 1)$

$$f(0) = (0)^{8/3} - 4(0)^{5/3} = 0$$
$$f(1) = (1)^{8/3} - 4(1)^{5/3} = -3$$

Points of inflection at $(0, 0)$ and $(1, -3)$

45. $f(x) = \sin(2x)$
$f'(x) = 2\cos(2x)$
$f''(x) = -4\sin(2x)$

$f''(x) = 0$ when $x = 0, \pm\dfrac{\pi}{2}, \pm\pi, \pm\dfrac{3\pi}{2}, \pm 2, \ldots$

Use test points from the intervals,

$$\left(-\frac{\pi}{2}, 0\right), \left(0, \frac{\pi}{2}\right), \left(\frac{\pi}{2}, \pi\right) \cdots$$

$f''(-1) \approx 3.64 > 0$
$f''(1) \approx -3.64 < 0$
$f''(2) \approx 3.03 > 0$

Concave upward on

$$\cdots \cup \left(-\frac{3\pi}{2}, -\pi\right) \cup \left(-\frac{\pi}{2}, 0\right) \cup \left(\frac{\pi}{2}, \pi\right) \cup \cdots$$

Concave downward on

$$\cdots \cup \left(-\pi, -\frac{\pi}{2}\right) \cup \left(0, \frac{\pi}{2}\right) \cup \left(\pi, \frac{3\pi}{2}\right) \cup \cdots$$

47. (a)

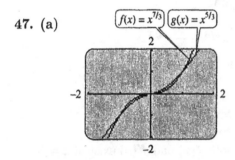

(b) Both $f(x)$ and $g(x)$ are concave down on $(-\infty, 0)$ and concave up on $(0, \infty)$. Thus, they both have a point of inflection at $(0, 0)$.

(c)

$f(x) = x^{7/3}$	$g(x) = x^{5/3}$
$f'(x) = \dfrac{7}{3}x^{4/3}$	$g'(x) = \dfrac{5}{3}x^{2/3}$
$f''(x) = \dfrac{28}{9}x^{1/3}$	$g''(x) = \dfrac{10}{9}x^{-1/3}$
$f''(0) = 0$	$g''(0)$ is undefined

(d) No.

49. The slope of the tangent line to $f(x) = \ln x$ as $x \to \infty$ is 0. The slope of the tangent line to the graph of $f(x) = \ln x$ as $x \to 0$ approaches ∞. For example, the slope of the tangent line at $x = 0.00001$ is $100,000$.

51. $f(x) = x^2 - 12x + 36$
$f'(x) = 2x - 12$

$f'(x) = 0$ when

$$2x - 12 = 0$$
$$x = 6.$$

Critical number: 6

$f''(x) = 2 > 0$ for all x.

The curve is concave upward, which means a relative minimum occurs at $x = 6$.

53. $f(x) = 2x^3 - 4x^2 + 2$
$f'(x) = 6x^2 - 8x$

$f'(x) = 0$ when

$$6x^2 - 8x = 0$$
$$2x(3x - 4) = 0$$
$$x = 0 \quad \text{or} \quad x = \frac{4}{3}.$$

Critical numbers: $0, \frac{4}{3}$

$f''(x) = 12x - 8$
$f''(0) = -8 < 0$, which means that a relative maximum occurs at $x = 0$.
$f''\left(\frac{4}{3}\right) = 8 > 0$, which means that a relative minimum occurs at $x = \frac{4}{3}$.

55. $f(x) = x^3$
$f'(x) = 3x^2$

$f'(x) = 0$ when

$$3x^2 = 0$$
$$x = 0.$$

Critical number: 0

$$f''(x) = 6x$$
$$f''(0) = 0$$

The second derivative test fails.
Use the first derivative test.

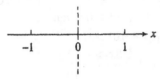

$$f'(-1) = 3(-1)^2 = 3 > 0$$

This indicates that f is increasing on $(-\infty, 0)$.

$$f'(1) = 3(1)^2 = 3 > 0$$

This indicates that f is increasing on $(0, \infty)$.
Neither a relative maximum nor relative minimum occurs at $x = 0$.

57. $f'(x) = x^3 - 6x^2 + 7x + 4$
$f''(x) = 3x^2 - 12x + 7$

Graph f' and f'' in the window $[-5, 5]$ by $[-5, 15]$,
Xscl $= 0.5$.
Graph of f':

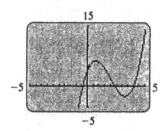

Graph of f'':

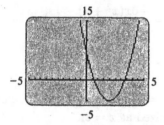

(a) f has relative extrema where $f'(x) = 0$. Use the graph to approximate the x-intercepts of the graph of f'. These numbers are the solutions of the equation $f'(x) = 0$. We find that the critical numbers of f are about $-0.4, 2.4$, and 4.0.
By either looking at the graph of f' and applying the first derivative test or by looking at the graph of f'' and applying the second derivative test, we see that f has relative minima at about -0.4 and 4.0 and a relative maximum at about 2.4.

(b) Examine the graph of f' to determine the intervals where the graph lies above and below the x-axis. We see that $f'(x) > 0$ on about $(-0.4, 2.4)$ and $(4.0, \infty)$, indicating that f is increasing on the same intervals. We also see that $f'(x) < 0$ on about $(-\infty, -0.4)$ and $(2.4, 4.0)$, indicating that f is decreasing on the same intervals.

(c) Examine the graph of f''. We see that this graph has two x-intercepts, so there are two x-values where $f''(x) = 0$. These x-values are about 0.7 and 3.3. Because the sign of f'' changes at these two values, we see that the x-values of the inflection points of the graph of f are about 0.7 and 3.3.

(d) We observe from the graph of f'' that $f''(x) > 0$ on about $(-\infty, 0.7)$ and $(3.3, \infty)$, so f is concave upward on the same intervals. Likewise, we observe that $f''(x) < 0$ on about $(0.7, 3.3)$, so f is concave downward on the same interval.

59. $f'(x) = \dfrac{1 - x^2}{(x^2 + 1)^2}$

$$f''(x) = \frac{(x^2+1)^2(-2x) - (1-x^2)(2)(x^2+1)2x}{(x^2+1)^4}$$

$$= \frac{-2x(x^2+1)[(x^2+1) + 2(1-x^2)]}{(x^2+1)^4}$$

$$= \frac{-2x(3 - x^2)}{(x^2+1)^3}$$

Graph f' and f'' in the window $[-3, 3]$ by $[-1.5, 1, 5]$,
Xscl $= 0.2$.
Graph of f':

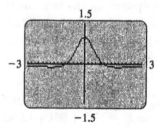

Graph of f'':

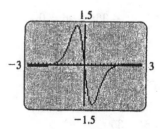

(a) The critical numbers of f are the x-intercepts of the graph of f'. (Note that there are no values where f' does not exist.) We see from the graph that these x-values are -1 and 1.

By either looking at the graph of f' and applying the first derivative test or by looking at the graph of f'' and applying the second derivative test, we see that f has a relative minimum at 1 and a relative maximum at -1.

(b) Examine the graph of f' to determine the intervals where the graph lies above and below the x-axis. We see that $f'(x) > 0$ on $(-1, 1)$, indicating that f is increasing on the same interval. We also see that $f'(x) < 0$ on $(-\infty, -1)$ and $(1, \infty)$, indicating that f is decreasing on the same intervals.

(c) Examine the graph of f''. We see that this graph has three x-intercepts, so there are three values where $f''(x) = 0$. These x-values are about -1.7, 0, and about 1.7. Because the sign of f'' and thus the concavity of f changes at these three values, we see that the x-values of the inflection points of the graph of f are about $-1.7, 0$, and about 1.7.

(d) We observe from the graph of f'' that $f''(x) > 0$ on about $(-1.7, 0)$ and $(1.7, \infty)$, so f is concave upward on the same intervals.

Likewise, we observe that $f''(x) < 0$ on about $(-\infty, -1.7)$ and $(0, 1.7)$, so f is concave downward on the same intervals.

61. (a) $R(t) = t^2(t - 18) + 96t + 1{,}000; \ 0 < t < 8$
$$= t^3 - 18t^2 + 96t + 1{,}000$$
$$R'(t) = 3t^2 - 36t + 96$$

Set $R'(t) = 0$.

$$3t^2 - 36t + 96 = 0$$
$$t^2 - 12t + 32 = 0$$
$$(t - 8)(t - 4) = 0$$
$$t = 8 \quad \text{or} \quad t = 4$$

8 is not in the domain of $R(t)$.
$R''(t) = 6t - 36$
$R''(4) = -12 < 0$ implies that $R(t)$ is maximized at $t = 4$, so the population is maximized at 4 hours.

(b) $R(4) = 16(-14) + 96(4) + 1{,}000$
$$= -224 + 384 + 1{,}000$$
$$= 1{,}160$$

The maximum population is 1,160 million.

63. $K(x) = \dfrac{3x}{x^2 + 4}$

(a) $K'(x) = \dfrac{3(x^2 + 4) - (2x)(3x)}{(x^2 + 4)^2}$
$$= \dfrac{-3x^2 + 12}{(x^2 + 4)^2} = 0$$
$$-3x^2 + 12 = 0$$
$$x^2 = 4$$
$$x = 2 \quad \text{or} \quad x = -2$$

For this application, the domain of K is $[0, \infty)$, so the only critical number is 2.

$$K''(x) = \dfrac{(x^2 + 4)^2(-6x) - (-3x^2 + 12)(2)(x^2 + 4)(2x)}{(x^2 + 4)^4}$$
$$= \dfrac{-6x(x^2 + 4) - 4x(-3x^2 + 12)}{(x^2 + 4)^3}$$
$$= \dfrac{6x^3 - 72x}{(x^2 + 4)^3}$$

$K''(2) = \dfrac{-96}{512} = -\dfrac{3}{16} < 0$ implies that $K(x)$ is maximized at $x = 2$.

Thus, the concentration is a maximum after 2 hours.

(b) $K(2) = \dfrac{3(2)}{(2)^2 + 4} = \dfrac{3}{4}$

The maximum concentration is $\frac{3}{4}\%$.

65. $v(x) = -35.98 + 12.09x - 0.4450x^2$
$v'(x) = 12.09 - 0.89x$
$v''(x) = -0.89$

Since $-0.89 < 0$, the function is always concave down.

67. $G(t) = \dfrac{10{,}000}{1 + 49e^{-0.1t}}$

$G'(t) = \dfrac{(1 + 49e^{-0.1t})(0) - (10{,}000)(-4.9e^{-0.1t})}{(1 + 49e^{-0.1t})^2}$
$$= \dfrac{49{,}000e^{-0.1t}}{(1 + 49e^{-0.1t})^2}$$

To find $G''(t)$, apply the quotient rule to find the derivative of $G'(t)$.

The numerator of $G''(t)$ will be

$$(1 + 49e^{-0.1t})^2(-4{,}900e^{-0.1t}) - (49{,}000e^{-0.1t})(2)(1 + 49e^{-0.1t})(-4.9e^{-0.1t})$$
$$= (1 + 49e^{-0.1t})(-4{,}900e^{-0.1t}) \cdot [(1 + 49e^{-0.1t}) - 20(4.9e^{-0.1t})]$$
$$= (-4{,}900e^{-0.1t})[1 + 49e^{-0.1t} - 98e^{-0.1t}]$$
$$= (-4{,}900e^{-0.1t})(1 - 49e^{-0.1t}).$$

Thus,

$$G''(t) = \frac{(-4{,}900e^{-0.1t})(1 - 49e^{-0.1t})}{(1 + 49e^{-0.1t})^4}.$$

$G''(t) = 0$ when $-4{,}900e^{-0.1t} = 0$ or $1 - 49e^{-0.1t} = 0$.
$-4{,}900e^{-0.1t} < 0$, and thus never equals zero.

$$1 - 49e^{-0.1t} = 0$$
$$1 = 49e^{-0.1t}$$
$$\frac{1}{49} = e^{-0.1t}$$
$$\ln\left(\frac{1}{49}\right) = -0.1t$$

$$\ln 1 - \ln 49 = -0.1t$$
$$-\ln 49 = -0.1t$$
$$\ln 49 = 0.1t$$
$$\ln 7^2 = 0.1t$$
$$2 \ln 7 = 0.1t$$
$$20 \ln 7 = t$$
$$38.9182 \approx t$$

The point of inflection is $(38.9182, 5{,}000)$.

69. $L(t) = Be^{-ce^{-kt}}$
$L'(t) = Be^{-ce^{-kt}}(-ce^{-kt})' = Be^{-ce^{-kt}}[-ce^{-kt}(-kt)'] = Bcke^{-ce^{-kt}-kt}$
$L''(t) = Bcke^{-ce^{-kt}-kt}(-ce^{-kt} - kt)' = Bcke^{-ce^{-kt}-kt}[-ce^{-kt}(-kt)' - k] = Bcke^{-ce^{-kt}-kt}(cke^{-kt} - k)$
$\quad = Bck^2e^{-ce^{-kt}-kt}(ce^{-kt} - 1)$

$L''(t) = 0$ when $ce^{-kt} - 1 = 0$

$$ce^{-kt} - 1 = 0$$
$$\frac{c}{e^{kt}} = 1$$
$$e^{kt} = c$$
$$kt = \ln c$$
$$t = \frac{\ln c}{k}$$

Letting $c = 7.267963$ and $k = 0.670840$

$$t = \frac{\ln 7.267963}{0.670840} \approx 2.96 \text{ years}$$

Verify that there is a point of inflection at $t = \frac{\ln c}{k} \approx 2.96$. For

$$L''(t) = Bck^2 e^{-ce^{-kt} - kt}(ce^{-kt} - 1),$$

we only need to test the factor $ce^{-kt} - 1$ on the intervals determined by $t \approx 2.96$ since the other factors are always positive.

$L''(1)$ has the same sign as

$$7.267963 e^{-0.670840(1)} - 1 \approx 2.72 > 0.$$

$L''(3)$ has the same sign as

$$7.267963 e^{-0.670840(3)} - 1 \approx -0.029 < 0.$$

Therefore L, is concave up on $\left(0, \frac{\ln c}{k} \approx 2.96\right)$ and concave down on $\left(\frac{\ln c}{k}, \infty\right)$, so there is a point of inflection at $t = \frac{\ln c}{k} \approx 2.96$ years.

$$L(2.96) = 14.3032 e^{-7.267963 e^{-0.6708040t}} \approx 5.27$$

The inflection point is at $(2.69, 5.27)$.

This signifies the time when the rate of growth begins to slow down since L changes from concave up to concave down at this inflection point.

71. $N(t) = 71.8 e^{-8.96 e^{-0.0685t}}$
$N'(t) = 71.8(-8.96)(-0.0685)e^{-0.0685t}(e^{-8.96 e^{-0.0685t}}) = (44.067968 e^{-0.0685t})(e^{-8.96 e^{-0.0685t}})$
$N''(t) = (44.067968)(0.61376 e^{-0.0685t} - 0.0685)e^{(-8.96 e^{-0.0685t} - 0.0685t)}$

$N''(t) = 0$ when $0.61376 e^{-0.0685t} - 0.0685 = 0$
$$0.61376 e^{-0.0685t} - 0.0685 = 0$$
$$e^{-0.0685t} \approx 0.11162$$
$$-0.0685t \approx \ln 0.11162$$
$$t \approx 32.01$$

$N(t)$ has an inflection point at $(32.01, N(32.01))$ or $(32.01, 26.41)$.

73. $V(x) = 12x(100 - x) = 1{,}200x - 12x^2$
$V'(x) = 1{,}200 - 24x$

Set $V'(x) = 0$.

$$1{,}200 - 24x = 0$$
$$x = 50$$
$$V''(x) = -24$$
$$V''(50) = -24$$

Thus, a value of 50 will produce a maximum rate of reaction.

75. $s(t) = -16t^2 + 140t + 37$
$s'(t) = -32t + 140$
$s'(t) = 0$ when $-32t + 140 = 0$
$$140 = 32t$$
$$4.375 = t.$$

$$s'(1) = -32(1) + 140 > 0$$
$$s'(5) = -32(5) + 140 < 0$$
$$s(4.375) = 343.25$$

(a) Maximum height of 343.25 ft is reached 4.375 sec after the ball is thrown.

(b) The ball hits ground when $s(t) = 0$.

$$\frac{-140 \pm \sqrt{140^2 - 4(-16)(37)}}{-32} \approx 9$$

$$s'(9) = -32(9) + 140$$
$$= -148$$

The ball hits ground in about 9 seconds with a speed of about -148 ft/sec.

77. $s(t) = 1.5t^2 + 4t$

(a) $s(10) = 1.5(10^2) + 4(10)$
$\qquad = 150 + 40$
$\qquad = 190$

The car will move 190 ft in 10 seconds.

(b) $v(t) = s'(t) = 3t + 4$

$$s'(5) = 3 \cdot 5 + 4 = 19$$
$$s'(10) = 3 \cdot 10 + 4 = 34$$

The velocity at 5 sec is 19 ft/sec, and the velocity at 10 sec is 34 ft/sec.

(c) The car stops when $v(t) = 0$, but $v(t) > 0$ for all $t \geq 0$.

(d) $a(t) = v'(t) = s''(t) = 3$

The acceleration at 5 sec is 3 ft/sec^2, and the acceleration at 10 sec is also 3 ft/sec^2.

(e) As t increases, the velocity increases, but the acceleration is constant.

79. (a) The left side of the graph changes from concave upward to concave downward at the point of inflection between cellular phones and digital video disc players. The rate of growth of sales begins to decline at the point of inflection.

(b) Food processors are closest to the right-hand point of inflection. This inflection point indicates that the rate of decline of sales is beginning to slow.

81. $R(x) = \frac{4}{27}(-x^3 + 66x^2 + 1,050x - 400)$

$0 \leq x \leq 25$

$$R'(x) = \frac{4}{27}(-3x^2 + 132x + 1,050)$$

$$R''(x) = \frac{4}{27}(-6x + 132)$$

A point of diminishing returns occurs at a point of inflection, or where $R''(x) = 0$.

$$\frac{4}{27}(-6x + 132) = 0$$

$$-6x + 132 = 0$$
$$6x = 132$$
$$x = 22$$

Test $R''(x)$ to determine whether concavity changes at $x = 22$.

$$R''(20) = \frac{4}{27}(-6 \cdot 20 + 132) = \frac{16}{9} > 0$$
$$R''(24) = \frac{4}{27}(-6 \cdot 24 + 132) = -\frac{16}{9} < 0$$

$R(x)$ is concave upward on $(0, 22)$ and concave downward on $(22, 25)$.

$$R(22) = \frac{4}{27}[-(22)^3 + 66(22)^2 + 1,060(22) - 400]$$

$$\approx 6,517.9$$

The point of diminishing returns is $(22, 6,517.9)$.

83. $R(x) = -0.6x^3 + 3.7x^2 + 5x, \ 0 \leq x \leq 6$
$R'(x) = -1.8x^2 + 7.4x + 5$
$R''(x) = -3.6x + 7.4$

A point of diminishing returns occurs at a point of inflection or where $R''(x) = 0$.

$$-3.6x + 7.4 = 0$$
$$-3.6x = -7.4$$

$$x = \frac{-7.4}{-3.6} \approx 2.06$$

Test $R''(x)$ to determine whether concavity changes at $x = 2.05$.

$$R''(2) = -3.6(2) + 7.4$$
$$= -7.2 + 7.4 = 0.2 > 0$$
$$R''(3) = -3.6(3) + 7.4$$
$$= -10.8 + 7.4 = -3.4 < 0$$

$R(x)$ is concave upward on $(0, 2.06)$ and concave downward on $(2.06, 6)$.

$$R(2.06) = -0.6(2.06)^3 + 3.7(2.06)^2 + 5(2.06)$$
$$\approx 20.8$$

The point of diminishing returns is $(2.06, 20.8)$.

5.4   Curve Sketching

1. Graph $y = x \ln |x|$ on a graphing calculator. A suitable choice for the viewing window is $[-1, 1]$ by $[-1, 1]$, Xscl = 0.1, Yscl = 0.1.

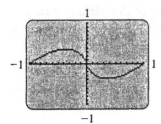

The calculator shows no y-value when $x = 0$ because 0 is not in the domain of this function. However, we see from the graph that

$$\lim_{x \to 0^-} x \ln |x| = 0$$

and

$$\lim_{x \to 0^+} x \ln |x| = 0.$$

Thus,

$$\lim_{x \to 0} x \ln |x| = 0.$$

3. $f(x) = -2x^3 - 9x^2 + 108x - 10$
$f'(x) = -6x^2 - 18x + 108$
$\quad = -6(x^2 + 3x - 18)$
$\quad = -6(x + 6)(x - 3)$

$f'(x) = 0$ when $x = -6$ or $x = 3$.
Critical numbers: -6 and 3
Critical points: $(-6, -550)$ and $(3, 179)$

$\quad f''(x) = -12x - 18$
$\quad f''(-6) = 54 > 0$
$\quad f''(3) = -54 < 0$

Relative maximum at 3, relative minimum at -6
Increasing on $(-6, 3)$
Decreasing on $(-\infty, -6)$ and $(3, \infty)$

$\quad f''(x) = -12x - 18 = 0$
$\quad -6(2x + 3) = 0$
$\quad\quad\quad x = -\dfrac{3}{2}$

Point of inflection at $(-1.5, -185.5)$
Concave upward on $(-\infty, -1.5)$
Concave downward on $(-1.5, \infty)$
y-intercept:

$y = -2(0)^3 - 9(0)^2 + 108(0) - 10 = -10$

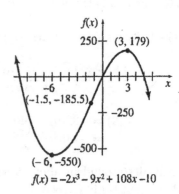

$$f(x) = -2x^3 - 9x^2 + 108x - 10$$

5. $f(x) = -3x^3 + 6x^2 - 4x - 1$
$f'(x) = -9x^2 + 12x - 4$
$\quad = -(3x - 2)^2$
$(3x - 2)^2 = 0$

$\quad\quad x = \dfrac{2}{3}$

Critical number: $\dfrac{2}{3}$

$$f\left(\frac{2}{3}\right) = -3\left(\frac{2}{3}\right)^3 + 6\left(\frac{2}{3}\right)^2 - 4\left(\frac{2}{3}\right) - 1 = -\frac{17}{9}$$

Critical point: $\left(\dfrac{2}{3}, -\dfrac{17}{9}\right)$

$f'(0) = -9(0)^2 + 12(0) - 4 = -4 < 0$
$f'(1) = -9(1)^2 + 12(1) - 4 = -1 < 0$

No relative extremum at $\left(\dfrac{2}{3}, -\dfrac{17}{9}\right)$

Decreasing on $(-\infty, \infty)$
$f''(x) = -18x + 12$
$\quad = -6(3x - 2)$

$3x - 2 = 0$

$\quad\quad x = \dfrac{2}{3}$

$f''(0) = -18(0) + 12 = 12 > 0$
$f''(1) = -18(1) + 12 = -6 < 0$

Concave upward on $\left(-\infty, \frac{2}{3}\right)$

Concave downward on $\left(\frac{2}{3}, \infty\right)$

Point of inflection at $\left(\frac{2}{3}, -\frac{17}{9}\right)$

y-intercept: $y = -3(0)^3 + 6(0)^2 - 4(0) - 1 = -1$

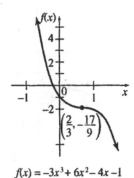

$$f(x) = -3x^3 + 6x^2 - 4x - 1$$

y-intercept: $y = 0^4 - 18(0)^2 + 5 = 5$

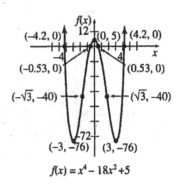

$$f(x) = x^4 - 18x^2 + 5$$

7. $f(x) = x^4 - 18x^2 + 5$

$$\begin{aligned}
f'(x) &= 4x^3 - 36x = 0 \\
4x(x^2 - 9) &= 0 \\
4(x)(x - 3)(x + 3) &= 0
\end{aligned}$$

Critical numbers 0, 3, and -3
Critical points: $(0, 5)$, $(3, -76)$ and $(-3, -76)$

$$\begin{aligned}
f''(x) &= 12x^2 - 36 \\
f''(-3) &= 72 > 0 \\
f''(0) &= -36 < 0 \\
f''(3) &= 72 > 0
\end{aligned}$$

Relative maximum at 0, relative minima at -3
and 3
Increasing on $(-3, 0)$ and $(3, \infty)$
Decreasing on $(-\infty, -3)$ and $(0, 3)$

$$\begin{aligned}
f''(x) &= 12x^2 - 36 = 0 \\
12(x^2 - 3) &= 0 \\
x &= \pm\sqrt{3}
\end{aligned}$$

Points of inflection at $(\sqrt{3}, -40)$, $(-\sqrt{3}, -40)$
Concave upward on $(-\infty, -\sqrt{3})$ and $(\sqrt{3}, \infty)$
Concave downward on $(-\sqrt{3}, \sqrt{3})$

x-intercepts: $0 = x^4 - 18x^2 + 5$
Let $u = x^2$.
$0 = u^2 - 18u + 5$

Use the quadratic formula.

$$\begin{aligned}
u &= 9 \pm 2\sqrt{19} \approx 17.72 \text{ or } 0.28 \\
x &= u^2 = \pm 4.2 \text{ or } \pm 0.53
\end{aligned}$$

9. $f(x) = x^4 - 2x^3$

$$\begin{aligned}
f'(x) &= 4x^3 - 6x^2 = 0 \\
2x^2(2x - 3) &= 0
\end{aligned}$$

Critical numbers: 0 and $\dfrac{3}{2}$

Critical points: $(0, 0)$ and $\left(\dfrac{3}{2}, -\dfrac{27}{16}\right)$

$$\begin{aligned}
f''(x) &= 12x^2 - 12x \\
f''(0) &= 0
\end{aligned}$$

$$f''\left(\frac{3}{2}\right) = 9 > 0$$

Relative minimum at $\dfrac{3}{2}$

No relative extremum at 0

Increasing on $\left(\dfrac{3}{2}, \infty\right)$

Decreasing on $\left(-\infty, \dfrac{3}{2}\right)$

$$\begin{aligned}
f''(x) &= 12x^2 - 12x = 0 \\
12x(x - 1) &= 0 \\
x &= 0 \text{ or } x = 1
\end{aligned}$$

Points of inflection at $(0, 0)$ and $(1, -1)$
Concave upward on $(-\infty, 0)$ and $(1, \infty)$
Concave downward on $(0, 1)$

x-intercepts: $0 = x^4 - 2x^3$
$0 = x^3(x - 2)$
$x = 0 \text{ or } x = 2$

y-intercept: $y = 0^4 - 2(0)^3 = 0$

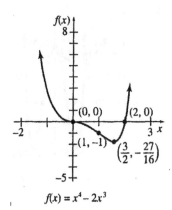

$$f(x) = x^4 - 2x^3$$

11. $f(x) = x + \dfrac{2}{x}$

$$= x + 2x^{-1}$$

Vertical asymptote at $x = 0$

$$f'(x) = 1 - 2x^{-2} = 1 - \frac{2}{x^2}$$

$$1 - \frac{2}{x^2} = 0$$

$$\frac{x^2 - 2}{x^2} = 0$$

$$x = \pm\sqrt{2}$$

Critical numbers: $\sqrt{2}$ and $-\sqrt{2}$
Critical points: $(\sqrt{2}, 2\sqrt{2})$ and $(-\sqrt{2}, -2\sqrt{2})$

$$f''(x) = 4x^{-3} = \frac{4}{x^3}$$

$$f''(\sqrt{2}) = \frac{2}{\sqrt{2}} > 0$$

$$f''(-\sqrt{2}) = -\frac{2}{\sqrt{2}} < 0$$

Relative maximum at $-\sqrt{2}$
Relative minimum at $\sqrt{2}$
Increasing on $(-\infty, -\sqrt{2})$ and $(\sqrt{2}, \infty)$
Decreasing on $(-\sqrt{2}, 0)$ and $(0, \sqrt{2})$
(Recall that $f(x)$ does not exist at $x = 0$.)

$f''(x) = \dfrac{4}{x^3}$ is never zero.

There are no points of inflection.
Concave upward on $(0, \infty)$
Concave downward on $(-\infty, 0)$
$f(x)$ is never zero, so there are no x-intercepts.
$f(x)$ does not exist for $x = 0$, so there is no y-intercept.

$y = x$ is an oblique asymptote.

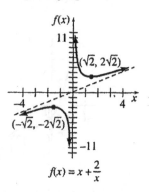

$$f(x) = x + \frac{2}{x}$$

13. $f(x) = \dfrac{x - 1}{x + 1}$

Vertical asymptote at $x = -1$
Horizontal asymptote at $y = 1$

$$f'(x) = \frac{(x + 1) - (x - 1)}{(x + 1)^2}$$

$$= \frac{2}{(x + 1)^2}$$

$f'(x)$ is never zero.
$f'(x)$ fails to exist for $x = -1$.

$$f''(x) = \frac{(x + 1)^2(0) - 2(2)(x + 1)}{(x + 1)^4}$$

$$= \frac{-4}{(x + 1)^3}$$

$f''(x)$ fails to exist for $x = -1$.
No critical values; no relative extrema
No points of inflection

$$f''(-2) = 4 > 0$$
$$f''(0) = -4 < 0$$

Concave upward on $(-\infty, -1)$
Concave downward on $(-1, \infty)$

x-intercept: $\dfrac{x - 1}{x + 1} = 0$
$$x = 1$$

y-intercept: $y = \dfrac{0 - 1}{0 + 1} = -1$

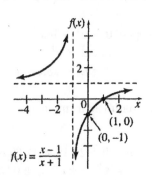

$$f(x) = \frac{x - 1}{x + 1}$$

15. $f(x) = \dfrac{1}{x^2 + x - 2}$

$= \dfrac{1}{(x+2)(x-1)}$

There are vertical asymptotes where $f(x)$ is undefined, namely, $x = -2$ and $x = 1$.

There is a horizontal asymptote at $y = 0$.

There is no x-intercept since $f(x)$ is never zero.

There is a y-intercept at

$$y = \dfrac{1}{0^2 + 0 - 2} = -\dfrac{1}{2}.$$

$$f'(x) = \dfrac{(x^2 + x - 2)(0) - (1)(2x + 1)}{(x^2 + x - 2)^2}$$

$$= \dfrac{-2x - 1}{(x^2 + x - 2)^2}$$

$$= \dfrac{-2x - 1}{[(x+2)(x-1)]^2}$$

There are critical numbers at $-2, 1$, and $-\frac{1}{2}$.

$$f'(-3) = \dfrac{-2(-3) - 1}{[(-3+2)(-3-1)]^2} = \dfrac{5}{16} > 0$$

$$f'(-1) = \dfrac{-2(-1) - 1}{[(-1+2)(-1-1)]^2} = \dfrac{1}{4} > 0$$

$$f'(0) = \dfrac{-2(0) - 1}{[(0+2)(0-1)]^2} = \dfrac{-1}{4} < 0$$

$$f'(2) = \dfrac{-2(2) - 1}{[(2+2)(2-1)]^2} = \dfrac{-5}{16} < 0$$

f is increasing on $(-\infty, -2)$ and $\left(-2, -\frac{1}{2}\right)$ and decreasing on $\left(-\frac{1}{2}, 1\right)$ and $(1, \infty)$.

$$f\left(-\dfrac{1}{2}\right) = \dfrac{1}{\left(-\frac{1}{2}\right)^2 + \left(-\frac{1}{2}\right) - 2}$$

$$= \dfrac{1}{\frac{1}{4} - \frac{2}{4} - \frac{8}{4}}$$

$$= \dfrac{1}{-\frac{9}{4}} = -\dfrac{4}{9}$$

There is a relative maximum at $\left(-\frac{1}{2}, -\frac{4}{9}\right)$.

$$f''(x) = \dfrac{(x^2 + x - 2)^2(-2) - (-2x - 1)2(x^2 + x - 2)(2x + 1)}{(x^2 + x - 2)^4}$$

$$= \dfrac{-2(x^2 + x - 2)[(x^2 + x - 2) + (-2x - 1)(2x + 1)]}{(x^2 + x - 2)^4}$$

$$= \dfrac{-2[x^2 + x - 2 - 4x^2 - 4x - 1]}{(x^2 + x - 2)^3}$$

$$= \dfrac{-2(-3x^2 - 3x - 3)}{(x^2 + x - 2)^3}$$

$$= \dfrac{6(x^2 + x + 1)}{(x^2 + x - 2)^3}$$

Since $x^2 + x + 1 = 0$ has no real solutions, there are no x-values where $f''(x) = 0$.

Since $x^2 + x - 2 = (x+2)(x-1)$, $f''(x)$ does not exist when $x = -2$ or $x = 1$. Since $f(x)$ does not exist at these x-values, there are no points of inflection.

$$f''(-3) = \dfrac{6[(-3)^2 + (-3) + 1]}{[(-3)^2 + (-3) - 2]^3} = \dfrac{21}{32} > 0$$

$$f''(0) = \dfrac{6(0^2 + 0 + 1)}{(0^2 + 0 - 2)^3} = -\dfrac{3}{4} < 0$$

$$f''(2) = \dfrac{6(2^2 + 2 + 1)}{(2^2 + 2 - 2)^3} = \dfrac{21}{32} > 0$$

f is concave upward on $(-\infty, -2)$ and $(1, \infty)$ and concave downward on $(-2, 1)$.

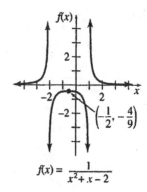

$$f(x) = \dfrac{1}{x^2 + x - 2}$$

17. $f(x) = \dfrac{x}{x^2 + 1}$

Horizontal asymptote at $y = 0$

$$f'(x) = \dfrac{(x^2 + 1)(1) - x(2x)}{(x^2 + 1)^2}$$

$$= \dfrac{1 - x^2}{(x^2 + 1)^2}$$

$$1 - x^2 = 0$$

Critical numbers: 1 and -1

Critical points: $\left(1, \frac{1}{2}\right)$ and $\left(-1, -\frac{1}{2}\right)$

$$f''(x) = \dfrac{(x^2 + 1)^2(-2x) - (1 - x^2)(2)(x^2 + 1)(2x)}{(x^2 + 1)^4}$$

$$= \dfrac{-2x^3 - 2x - 4x + 4x^3}{(x^2 + 1)^3}$$

$$= \dfrac{2x^3 - 6x}{(x^2 + 1)^3}$$

$$f''(1) = -\dfrac{1}{2} < 0$$

$$f''(-1) = \dfrac{1}{2} > 0$$

Relative maximum at 1
Relative minimum at -1
Increasing on $(-1, 1)$
Decreasing on $(-\infty, -1)$ and $(1, \infty)$

$$f''(x) = \frac{2x^3 - 6x}{(x^2 + 1)^3} = 0$$
$$2x^3 - 6x = 0$$
$$2x(x^2 - 3) = 0$$
$$x = 0, \ x = \pm\sqrt{3}$$

Points of inflection at $(0, 0)$, $\left(\sqrt{3}, \frac{\sqrt{3}}{4}\right)$ and
$\left(-\sqrt{3}, -\frac{\sqrt{3}}{4}\right)$

Concave upward on $(-\sqrt{3}, 0)$ and $(\sqrt{3}, \infty)$
Concave downward on $(-\infty, -\sqrt{3})$ and $(0, \sqrt{3})$

x-intercept: $0 = \dfrac{x}{x^2 + 1}$
$$0 = x$$

y-intercept: $y = \dfrac{0}{0^2 + 1} = 0$

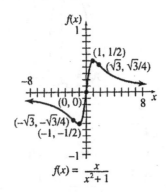

$$f(x) = \frac{x}{x^2 + 1}$$

19. $f(x) = \dfrac{1}{x^2 - 4}$

Vertical asymptotes at $x = 2$ and $x = -2$
Horizontal asymptote at $y = 0$

$$f'(x) = \frac{(x^2 - 4)(0) - 1(2x)}{(x^2 - 4)^2}$$
$$= \frac{-2x}{(x^2 - 4)^2} = 0$$

Critical number: 0

Critical point: $\left(0, -\frac{1}{4}\right)$

$$f''(x) = \frac{(x^2 - 4)^2(-2) - (-2x)(2)(x^2 - 4)(2x)}{(x^2 - 4)^4}$$
$$= \frac{6x^2 + 8}{(x^2 - 4)^3}$$

$$f''(0) = -\frac{1}{8} < 0$$

Relative maximum at 0
Increasing on $(-\infty, -2)$ and $(-2, 0)$
Decreasing on $(0, 2)$ and $(2, \infty)$
(Recall that $f(x)$ does not exist at $x = 2$ and $x = -2$.)

$f''(x) = \dfrac{6x^2 + 8}{(x^2 - 4)^3}$ can never be zero.
There are no points of inflection.

$$f''(-3) = \frac{62}{125} > 0$$
$$f''(3) = \frac{62}{125} > 0$$

Concave upward on $(-\infty, -2)$ and $(2, \infty)$
Concave downward on $(-2, 2)$
$f(x)$ is never zero so there is no x-intercept.

y-intercept: $y = \dfrac{1}{0^2 - 4} = -\dfrac{1}{4}$

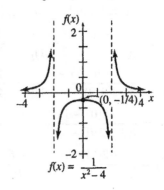

$$f(x) = \frac{1}{x^2 - 4}$$

21. $y = x \ln |x|$

Note that the domain of this function is $(-\infty, 0) \cup (0, \infty)$.
There is no y-intercept.

x-intercept: $0 = x \ln |x|$
$$x = 0 \quad \text{or} \quad \ln |x| = 0$$
$$|x| = e^0 = 1$$
$$x = \pm 1$$

Since 0 is not in the domain, the only x-intercepts are -1 and 1.

$$y' = x \cdot \frac{1}{x} + \ln |x|$$
$$= 1 + \ln |x|$$

$y' = 0$ when

$$0 = 1 + \ln |x|$$
$$-1 = \ln |x|$$
$$e^{-1} = |x|$$
$$x = \pm\frac{1}{e} \approx \pm 0.37.$$

Critical numbers: $\pm\frac{1}{e} \approx \pm 0.37$.

$$y'(-1) = 1 + \ln |-1| = 1 > 0$$
$$y'(-0.1) = 1 + \ln |-0.1| \approx -1.3 < 0$$
$$y'(0.1) = 1 + \ln |0.1| \approx -1.3 < 0$$
$$y'(1) = 1 + \ln |1| = 1 > 0$$

$$y\left(\frac{1}{e}\right) = \frac{1}{e} \ln \left|\frac{1}{e}\right| = -\frac{1}{e}$$

$$y\left(-\frac{1}{e}\right) = -\frac{1}{e} \ln \left|-\frac{1}{e}\right| = \frac{1}{e}$$

Increasing on $\left(-\infty, -\frac{1}{e}\right)$ and $\left(\frac{1}{e}, \infty\right)$ and decreasing on $\left(-\frac{1}{e}, 0\right)$ and $\left(0, \frac{1}{e}\right)$.

Relative maximum of $\left(-\frac{1}{e}, \frac{1}{e}\right)$; relative minimum of $\left(\frac{1}{e}, -\frac{1}{e}\right)$.

$$y'' = \frac{1}{x}$$

$$y''(-1) = \frac{1}{-1} = -1 < 0$$

$$y''(1) = \frac{1}{1} = 1 > 0$$

Concave downward on $(-\infty, 0)$;
Concave upward on $(0, \infty)$.

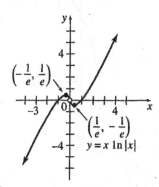

23. $y = \dfrac{\ln x}{x}$

Note that the domain of this function is $(0, \infty)$. Since $x \neq 0$, there is no y-intercept.

x-intercept: $y = 0$ when $\ln x = 0$
$$x = e^0 = 1$$

$$y' = \frac{x\left(\frac{1}{x}\right) - \ln x(1)}{x^2}$$

$$= \frac{1 - \ln x}{x^2}$$

Critical numbers:

$$1 - \ln x = 0$$
$$1 = \ln x$$
$$e^1 = x$$

$$y'(1) = \frac{1 - \ln 1}{1^2} = \frac{1}{1} = 1 > 0$$

$$y'(3) = \frac{1 - \ln 3}{3^2} = -0.01 < 0$$

The function is increasing on $(0, e)$ and decreasing on (e, ∞).

$$y(e) = \frac{\ln e}{e} = \frac{1}{e}$$

There is a relative maximum at $\left(e, \frac{1}{e}\right)$.

$$y'' = \frac{x^2\left(-\frac{1}{x}\right) - (1 - \ln x)2x}{x^4}$$

$$= \frac{-x - 2x(1 - \ln x)}{x^4}$$

$$= \frac{-x[1 + 2(1 - \ln x)]}{x^4}$$

$$= \frac{-(1 + 2 - 2\ln x)}{x^3}$$

$$= \frac{-3 + 2\ln x}{x^3}$$

$y'' = 0$ when $-3 + 2\ln x = 0$
$$2\ln x = 3$$

$$\ln x = \frac{3}{2} = 1.5$$

$$x = e^{1.5} \approx 4.48.$$

$$y''(1) = \frac{-3 + 2\ln 1}{1^3} = -3 < 0$$

$$y''(5) = \frac{-3 + 2\ln 5}{5^3} = 0.0018 > 0$$

Concave downward on $(0, e^{1.5})$; concave upward on $(e^{1.5}, \infty)$

$$y(e^{1.5}) = \frac{\ln e^{1.5}}{e^{1.5}} = \frac{1.5}{e^{1.5}}$$

$$= \frac{3}{2e^{1.5}} \approx 0.33$$

Point of inflection at $\left(e^{1.5}, \frac{1.5}{e^{1.5}}\right) \approx (4.48, 0.33)$

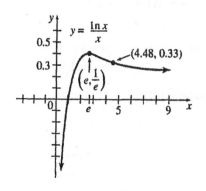

25. $y = xe^{-x}$
$$y' = -xe^{-x} + e^{-x}$$
$$= e^{-x}(1 - x)$$
$$y'' = e^{-x}(-1) + (1-x)(-e^{-x})$$
$$= -e^{-x}(1 + 1 - x)$$
$$= -e^{-x}(2 - x)$$

y-intercept: 0
$y' = 0$ when $e^{-x}(1 - x) = 0$
$$x = 1$$

$$y'(0) = e^{-0}(1 - 0) = 1 > 0$$

$$y'(2) = e^{-2}(1 - 2) = \frac{-1}{e^2} < 0$$

Increasing on $(-\infty, 1)$; decreasing on $(1, \infty)$
Relative maximum at $\left(1, \frac{1}{e}\right)$
$y'' = 0$ when $-e^{-x}(2 - x) = 0$
$$x = 2.$$

$$y''(0) = -e^{-0}(2 - 0) = -2 < 0$$

$$y''(3) = -e^{-3}(2 - 3) = \frac{1}{e^3} > 0$$

Concave downward on $(-\infty, 2)$, concave upward on $(2, \infty)$
Point of inflection at $\left(2, \frac{2}{e^2}\right)$

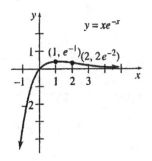

27. $y = (x - 1)e^{-x}$
$$y' = -(x - 1)e^{-x} + e^{-x}(1)$$
$$= e^{-x}[-(x - 1) + 1]$$
$$= e^{-x}(2 - x)$$
$$y'' = -e^{-x} + (2 - x)(-e^{-x})$$
$$= -e^{-x}[1 + (2 - x)]$$
$$= e^{-x}(3 - x)$$

y-intercept: $y = (0 - 1)e^{-0}$
$$= (-1)(1) = -1$$

x-intercept: $0 = (x - 1)e^{-x}$
$$x - 1 = 0$$
$$x = 1$$

$y' = 0$ when $e^{-x}(2 - x) = 0$
$$x = 2.$$

$$y''(2) = -e^{-2}(3 - 2) = \frac{-1}{e^2} < 0$$

Relative maximum at $\left(2, \frac{1}{e^2}\right)$

$$y'(0) = e^{-0}(2 - 0) = 2 > 0$$

$$y'(3) = e^{-3}(2 - 3) = \frac{-1}{e^3} < 0$$

Increasing $(-\infty, 2)$; decreasing on $(2, \infty)$.

$y'' = 0$ when $-e^{-x}(3 - x) = 0$
$$x = 3.$$

$$y''(0) = -e^{-0}(3 - 0) = -3 < 0$$

$$y''(4) = -e^{-4}(3 - 4) = \frac{1}{e^4} > 0$$

Concave downward on $(-\infty, 3)$; concave upward on $(3, \infty)$

$$f(3) = (3 - 1)e^{-3} = \frac{2}{e^3}$$

Point of inflection at $\left(3, \frac{2}{e^3}\right)$

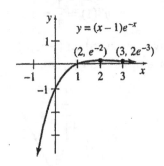

29. $y = x^{2/3} - x^{5/3}$
$$y' = \frac{2}{3}x^{-1/3} - \frac{5}{3}x^{2/3}$$
$$= \frac{2 - 5x}{3x^{1/3}}$$
$$y'' = \frac{3x^{1/3}(-5) - (2 - 5x)3\left(\frac{1}{3}\right)x^{-2/3}}{(3x^{1/3})^2}$$
$$= \frac{-15x^{1/3} - (2 - 5x)x^{-2/3}}{9x^{2/3}}$$
$$= \frac{-15x - (2 - 5x)}{9x^{4/3}}$$
$$= \frac{-10x - 2}{9x^{4/3}}$$

y-intercept: $y = 0^{2/3} - 0^{5/3} = 0$
x-intercept: $0 = x^{2/3} - x^{5/3}$
$$= x^{2/3}(1 - x)$$
$$x = 0 \text{ or } x = 1$$
$y' = 0$ when $2 - 5x = 0$
$$x = \frac{2}{5}$$

$$y'' \left(\frac{2}{5}\right) = \frac{-10\left(\frac{2}{5}\right) - 2}{9\left(\frac{2}{5}\right)^{4/3}}$$

$$\approx -2.262 < 0$$

$$y\left(\frac{2}{5}\right) = \left(\frac{2}{5}\right)^{2/3} - \left(\frac{2}{5}\right)^{5/3}$$

$$= \frac{3 \cdot 2^{2/3}}{5^{5/3}} \approx 0.326$$

Relative maximum at $\left(\dfrac{2}{5}, \dfrac{3 \cdot 2^{2/3}}{5^{5/3}}\right) \approx (0.4, .326)$

y' does not exist when $x = 0$
Since $y''(0)$ is undefined, use the first derivative test.

$$y'(-1) = \frac{2 - 5(-1)}{3(-1)^{1/3}} = \frac{7}{-3} < 0$$

$$y'\left(\frac{1}{8}\right) = \frac{2 - 5\left(\frac{1}{8}\right)}{3\left(\frac{1}{8}\right)^{1/3}} = \frac{11}{12} > 0$$

$$y'(1) = \frac{2 - 5}{3 \cdot 1^{1/3}} = -1 < 0$$

y increases on $\left(0, \frac{2}{5}\right)$.
y decreases on $(-\infty, 0)$ and $\left(\frac{2}{5}, \infty\right)$.
Relative minimum at $(0, 0)$

$$y'' = 0 \quad \text{when} \quad -10x - 2 = 0$$

$$x = -\frac{1}{5}$$

y'' undefined when $9x^{4/3} = 0$
$$x = 0$$

$$y''(-1) = \frac{-10(-1) - 2}{9(-1)^{4/3}} = \frac{8}{9} > 0$$

$$y''\left(-\frac{1}{8}\right) = \frac{-10\left(-\frac{1}{8}\right) - 2}{9\left(-\frac{1}{8}\right)^{4/3}} = -\frac{4}{3} < 0$$

$$y''(1) = \frac{-10(1) - 2}{9(1)^{4/3}} = -\frac{4}{3} < 0$$

Concave upward on $\left(-\infty, -\dfrac{1}{5}\right)$

Concave downward on $\left(-\dfrac{1}{5}, \infty\right)$

Point of inflection at $\left(-\dfrac{1}{5}, \dfrac{6}{5^{5/3}}\right) \approx (-0.2, 0.410)$

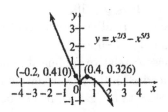

31. $f(x) = x + \cos x$
   $f'(x) = 1 - \sin x$
   $f''(x) = -\cos x$

y-intercept; $f(0) = 0 + \cos 0 = 1$
$f'(x) = 0$ when $1 - \sin x = 0$.
$$1 = \sin x$$
$$x = \frac{\pi}{2}, \frac{3\pi}{2}, \frac{5\pi}{2}, \cdots$$

$f'(1) = 1 - \sin 1 \approx 0.159 > 0$
$f'(3) = 1 - \sin 3 \approx 0.859 > 0$
$f'(6) = 1 - \sin 6 \approx 1.279 > 0$

Increasing $(-\infty, \infty)$.
$f''(x) = 0$ when $-\cos x = 0$

$$x = \pm\frac{\pi}{2}, \pm\frac{3\pi}{2}, \pm\frac{5\pi}{2}, \cdots$$

$f''(0) = -\cos 0 = -1 < 0$
$f''(2) = -\cos 2 \approx 0.416 > 0$
$f''(6) = -\cos 6 \approx -0.96 < 0$

Concave upward on

$$\cdots \left(-\frac{3\pi}{2}, -\frac{\pi}{2}\right), \left(\frac{\pi}{2}, \frac{3\pi}{2}\right), \left(\frac{5\pi}{2}, \frac{7\pi}{2}\right) \cdots$$

Concave downward on

$$\cdots \left(-\frac{\pi}{2}, \frac{\pi}{2}\right), \left(\frac{3\pi}{2}, \frac{5\pi}{2}\right), \left(\frac{7\pi}{2}, \frac{9\pi}{2}\right) \cdots$$

Points of inclection at

$$\cdots \left(-\frac{\pi}{2}, -\frac{\pi}{2}\right), \left(\frac{\pi}{2}, \frac{\pi}{2}\right), \left(\frac{3\pi}{2}, \frac{3\pi}{2}\right)$$

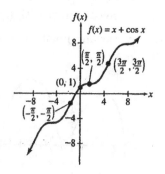

33. In exercises 3 and 7, either the relative maximum or relative minimum is outside of the normal window of $-10 \le y \le 10$.
Exercise 3: relative maximum at $(3, 179)$; relative minimum at $(-6, -550)$
Exercise 7: relative minima at $(-3, -76)$ and $(3, -76)$
In Exercise 15, the relative maximum at $\left(-\frac{1}{2}, -\frac{4}{9}\right)$ is so close to the x-axis, it may be hard to distinguish.

35. In the following exercises, either the relative maximum, relative minimum, or point of inflection is so small, it may be hard to distinguish: 17, 19, 23, 25, and 27.

37. **(a)** indicates a smooth, continuous curve except where there is a vertical asymptote.

(b) indicates that the function decreases on both sides of the asymptote, so there are no relative extrema.

(c) gives the horizontal asymptote $y = 2$.

(d) and **(e)** indicate that concavity does not change left of the asymptote, but that the right portion of the graph changes concavity at $x = 2$ and $x = 4$. There are points of inflection at 2 and 4.

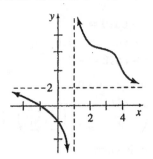

39. **(a)** indicates that there can be no asymptotes, sharp "corners", holes, or jumps. The graph must be one smooth curve.

(b) and **(c)** indicate relative maxima at -3 and 4 and a relative minimum at 1.

(d) and **(e)** are consistent with **(g)**.

(f) indicates turning points at the critical numbers -3 and 4.

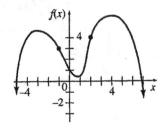

41. **(a)** indicates that the curve may not contain breaks.

(b) indicates that there is a sharp "corner" at 4.

(c) gives a point at $(1, 5)$.

(d) shows critical numbers.

(e) and **(f)** indicate (combined with **(c)** and **(d)**) a relative maximum at $(1, 5)$, and (combined with **(b)**) a relative minimum at 4.

(g) is consistent with **(b)**.

(h) indicates the curve is concave upward on $(2, 3)$.

(i) indicates the curve is concave downward on $(-\infty, 2), (3, 4)$ and $(4, \infty)$.

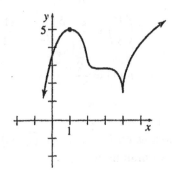

43. Sketch the curve for $\ell_1(v) = 0.08e^{0.33v}$

$$\ell_1'(v) = 0.0264e^{0.33v}$$
$$e^{0.33v} \ne 0$$

$\ell_1'(v)$ has no critical points.

$$\ell_1''(v) = 0.008712e^{0.33v}$$
$$e^{0.33v} \ne 0$$

$\ell_1''(v)$ has no inflection points.

Sketch the curve for $\ell_2 = -0.87v^2 + 28.17v - 211.41$

$$\ell_2'(v) = -1.74v + 28.17$$
$$-1.74v + 28.17 = 0$$
$$v \approx 16.19$$

Critical point: $(16.19, 16.62)$

$$\ell_2''(v) = -1.74$$

$\ell'(v)$ has no inflection points.
$\ell_2'(v)$ has a relative maximum at $(16.19, 16.62)$.

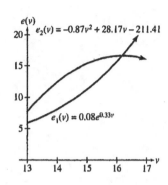

45. Sketch the graph of $I(n) = \dfrac{an^2}{b + n^2}$ when $a = 0.5$ and $b = 10$.

$$I(n) = \frac{0.5n^2}{10 + n^2}$$

$$I'(n) = \frac{(10 + n^2)n - (0.5n^2)(2n)}{(10 + n^2)^2} = \frac{n[(10 + n^2) - n^2]}{(10 + n^2)^2} = \frac{10n}{(10 + n^2)^2}$$

Set this derivative to zero and solve for n.

$$10n = 0$$
$$n = 0$$

There is one critical point at $(0, 0)$.

$$I''(n) = \frac{(10 + n^2)^2(10) - 40n^2(10 + n^2)}{(10 + n^2)^4} = \frac{100 - 30n^2}{(10 + n^2)^3}$$

Set this derivative to zero and solve for n.

$$100 - 30n^2 = 0$$

$$n = \sqrt{\frac{10}{3}} \approx 1.83$$

inflection point: $(1.83, 0.125)$

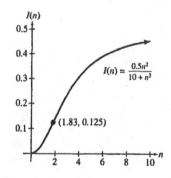

47. Sketch the curve for $L(C) = \dfrac{aC^n}{k + C^n}$ when $n = 4, k = 3,$ and $a = 100.$

$$L(C) = \frac{100C^4}{3 + C^4}$$

$$L'(C) = \frac{(3 + C^4)(400C^3) - (100C^4)(4C^3)}{(3 + C^4)^2}$$

$$= \frac{1{,}200C^3}{(3 + C^4)^2}$$

$$1{,}200C^3 = 0 \qquad 3 + C^4 = 0$$
$$C = 0 \qquad\qquad C^4 \neq -3$$

Critical point: $(0, 0)$

$$L''(C) = \frac{3{,}600C^2(3 + C^4)^2 - 9{,}600C^6(3 + C^4)}{(3 + C^4)^4}$$

$$= \frac{1{,}200C^2(9 - 5C^4)}{(3 + C^4)^3}$$

Set this derivative to zero and solve for C.

$$1200C^2(9 - 5C^4) = 0$$
$$-5C^4 = -9$$

$$C^4 = \frac{9}{5}$$

$$C = \left(\frac{9}{5}\right)^{1/4} \approx 1.16$$

inflection point: $(1.16, 37.5)$

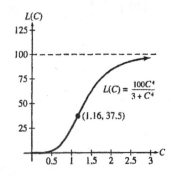

Chapter 5 Review Exercises

5. $f(x) = x^2 - 5x + 3$
$f'(x) = 2x - 5 = 0$

Critical number: $\frac{5}{2}$

$f''(x) = 2 > 0$ for all x.

$f(x)$ is a minimum at $x = \frac{5}{2}$.

f is increasing on $\left(\frac{5}{2}, \infty\right)$ and decreasing on $\left(-\infty, \frac{5}{2}\right)$.

7. $f(x) = -x^3 - 5x^2 + 8x - 6$
$f'(x) = -3x^2 - 10x + 8$
$$3x^2 + 10x - 8 = 0$$
$$(3x - 2)(x + 4) = 0$$

Critical numbers: $\frac{2}{3}$ and -4

$$f''(x) = -6x - 10$$

$$f''\left(\frac{2}{3}\right) = -14 < 0$$

$f(x)$ is a maximum at $x = \frac{2}{3}$.

$f''(-4) = 14 > 0$
$f(x)$ is a minimum at $x = -4$.
f is increasing on $\left(-4, \frac{2}{3}\right)$ and decreasing on $(-\infty, -4)$ and on $\left(\frac{2}{3}, \infty\right)$.

9. $f(x) = \dfrac{6}{x - 4}$

$f'(x) = \dfrac{-6}{(x-4)^2} < 0$ for all x, but not defined for $x = 4$.

f is never increasing; it is decreasing on $(-\infty, 4)$ and $(4, \infty)$.

11. $f(x) = \ln(x^2 - 1)$

$$f'(x) = \frac{1}{x^2 - 1} \cdot 2x$$

$$= \frac{2x}{x^2 - 1}$$

Note that the domain for the function is $(-\infty, -1) \cup (1, \infty)$.
$f'(x) = 0$ when $2x = 0$, which is when $x = 0$. But $x = 0$ is not in the domain of the function.

$$f'(-2) = \frac{2(-2)}{(-2)^2 - 1} = \frac{-4}{3} < 0$$

$$f'(2) = \frac{2(2)}{2^2 - 1} = \frac{4}{3} > 0$$

The function is increasing on $(1, \infty)$ and decreasing on $(-\infty, -1)$.

13. $f(x) = -x^2 + 4x - 8$
$f'(x) = -2x + 4 = 0$

Critical number: $x = 2$
$f''(x) = -2 < 0$ for all x, so $f(2)$ is a relative maximum.

$$f(2) = -4$$

Relative maximum of -4 at 2

15. $f(x) = 2x^2 - 8x + 1$
$f'(x) = 4x - 8 = 0$

Critical number: $x = 2$
$f''(x) = 4 > 0$ for all x, so $f(2)$ is a relative minimum.

$$f(2) = -7$$

Relative minimum of -7 at 2

17. $f(x) = 2x^3 + 3x^2 - 36x + 20$
$f'(x) = 6x^2 + 6x - 36 = 0$
$6(x^2 + x - 6) = 0$
$(x + 3)(x - 2) = 0$

Critical numbers: -3 and 2

$$f''(x) = \quad 12x + 6$$
$$f''(-3) = \quad -30 < 0, \text{ so a maximum occurs}$$
$$\text{at } x = -3.$$
$$f''(2) = \quad 30 > 0, \text{ so a minimum occurs}$$
$$\text{at } x = 2.$$
$$f(-3) = \quad 101$$
$$f(2) = \quad -24$$

Relative maximum of 101 at -3
Relative minimum of -24 at 2

19. $f(x) = \dfrac{xe^x}{x - 1}$

$$f'(x) = \frac{(x - 1)(xe^x + e^x) - xe^x(1)}{(x - 1)^2}$$

$$= \frac{x^2 e^x + xe^x - xe^x - e^x - xe^x}{(x - 1)^2}$$

$$= \frac{x^2 e^x - xe^x - e^x}{(x - 1)^2}$$

$$= \frac{e^x(x^2 - x - 1)}{(x - 1)^2}$$

$f'(x)$ is undefined at $x = 1$, but 1 is not in the domain of $f(x)$.
$f'(x) = 0$ when $x^2 - x - 1 = 0$

$$x = \frac{1 \pm \sqrt{1 - 4(1)(-1)}}{2}$$

$$= \frac{1 \pm \sqrt{5}}{2}$$

$$\frac{1 + \sqrt{5}}{2} \approx 1.618 \quad \text{or} \quad \frac{1 - \sqrt{5}}{2} = -0.618$$

Critical numbers are -0.618 and 1.618.

$$f'(1.4) = \frac{e^{1.4}(1.4^2 - 1.4 - 1)}{(1.4 - 1)^2} \approx -11.15 < 0$$

$$f'(2) = \frac{e^2(2^2 - 2 - 1)}{(2 - 1)^2} = e^2 \approx 7.39 > 0$$

$$f'(-1) = \frac{e^{-1}[(-1)^2 - (-1) - 1]}{(-1 - 1)^2} \approx 0.09 > 0$$

$$f'(0) = \frac{e^0(0^2 - 0 - 1)}{(0 - 1)^2} = -1 < 0$$

There is a relative maximum at $(-0.618, 0.206)$ and a relative minimum at $(1.618, 13.203)$.

21. $f(x) = 3x^4 - 5x^2 - 11x$
$f'(x) = 12x^3 - 10x - 11$
$f''(x) = 36x^2 - 10$
$f''(1) = 36(1)^2 - 10 = 26$
$f''(-3) = 36(-3)^2 - 10 = 314$

23. $f(x) = \dfrac{5x - 1}{2x + 3}$

$$f'(x) = \frac{(2x + 3)(5) - (5x - 1)(2)}{(2x + 3)^2} = \frac{17}{(2x + 3)^2}$$

$$f''(x) = \frac{(2x + 3)^2(0) - 17(2)(2x + 3)(2)}{(2x + 3)^4}$$

$$= \frac{-68}{(2x + 3)^3} \quad \text{or} \quad -68(2x + 3)^{-3}$$

$$f''(1) = \frac{-68}{[2(1) + 3]^3} = -\frac{68}{125}$$

$$f''(-3) = \frac{-68}{[2(-3) + 3]^3} = \frac{-68}{-27} = \frac{68}{27}$$

25. $f(t) = \sqrt{t^2 + 1} = (t^2 + 1)^{1/2}$

$$f'(t) = \frac{1}{2}(t^2 + 1)^{-1/2}(2t) = t(t^2 + 1)^{-1/2}$$

$$f''(t) = (t^2 + 1)^{-1/2}(1)$$

$$+ t\left[\left(-\frac{1}{2}\right)(t^2 + 1)^{-3/2}(2t)\right]$$

$$= (t^2 + 1)^{-1/2} - t^2(t^2 + 1)^{-3/2}$$

$$= \frac{1}{(t^2 + 1)^{1/2}} - \frac{t^2}{(t^2 + 1)^{3/2}} = \frac{t^2 + 1 - t^2}{(t^2 + 1)^{3/2}}$$

$$= (t^2 + 1)^{-3/2} \quad \text{or} \quad \frac{1}{(t^2 + 1)^{3/2}}$$

$$f''(1) = \frac{1}{(1 + 1)^{3/2}} = \frac{1}{2^{3/2}} \approx 0.354$$

$$f''(-3) = \frac{1}{(9 + 1)^{3/2}} = \frac{1}{10^{3/2}} \approx 0.032$$

27. $f(x) = -2x^3 - \frac{1}{2}x^2 + x - 3$

$f'(x) = -6x^2 - x + 1 = 0$

$(3x - 1)(2x + 1) = 0$

Critical numbers: $\frac{1}{3}$ and $-\frac{1}{2}$

Critical points: $\left(\frac{1}{3}, -2.80\right)$ and $\left(-\frac{1}{2}, -3.375\right)$

$$f''(x) = -12x - 1$$

$$f''\left(\frac{1}{3}\right) = -5 < 0$$

$$f''\left(-\frac{1}{2}\right) = 5 > 0$$

Relative maximum at $\frac{1}{3}$

Relative minimum at $-\frac{1}{2}$

Increasing on $\left(-\frac{1}{2}, \frac{1}{3}\right)$

Decreasing on $\left(-\infty, -\frac{1}{2}\right)$ and $\left(\frac{1}{3}, \infty\right)$

$$f''(x) = -12x - 1 = 0$$

$$x = -\frac{1}{12}$$

Point of inflection at $\left(-\frac{1}{12}, -3.09\right)$

Concave upward on $\left(-\infty, -\frac{1}{12}\right)$

Concave downward on $\left(-\frac{1}{12}, \infty\right)$

y-intercept:

$$y = -2(0)^3 - \frac{1}{2}(0)^2 + (0) - 3 = -3$$

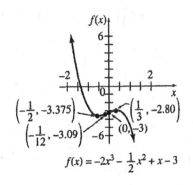

$f(x) = -2x^3 - \frac{1}{2}x^2 + x - 3$

29. $f(x) = x^4 - \frac{4}{3}x^3 - 4x^2 + 1$

$f'(x) = 4x^3 - 4x^2 - 8x = 0$

$4x(x^2 - x - 2) = 0$

$4x(x - 2)(x + 1) = 0$

Critical numbers: 0, 2, and -1

Critical points: $(0, 1)$, $\left(2, -\frac{29}{3}\right)$ and $\left(-1, -\frac{2}{3}\right)$

$$f''(x) = 12x^2 - 8x - 8$$

$$= 4(3x^2 - 2x - 2)$$

$$f''(-1) = 12 > 0$$

$$f''(0) = -8 < 0$$

$$f''(2) = 24 > 0$$

Relative maximum at 0

Relative minima at -1 and 2

Increasing on $(-1, 0)$ and $(2, \infty)$

Decreasing on $(-\infty, -1)$ and $(0, 2)$

$$f''(x) = 4(3x^2 - 2x - 2) = 0$$

$$x = \frac{2 \pm \sqrt{4 - (-24)}}{6}$$

$$= \frac{1 \pm \sqrt{7}}{3}$$

Points of inflection at $\left(\frac{1+\sqrt{7}}{3}, -5.12\right)$ and $\left(\frac{1-\sqrt{7}}{3}, 0.11\right)$

Concave upward on $\left(-\infty, \frac{1-\sqrt{7}}{3}\right)$ and $\left(\frac{1+\sqrt{7}}{3}, \infty\right)$

Concave downward on $\left(\frac{1-\sqrt{7}}{3}, \frac{1+\sqrt{7}}{3}\right)$

y-intercept:

$$y = (0)^4 - \frac{4}{3}(0)^3 - 4(0)^2 + 1 = 1$$

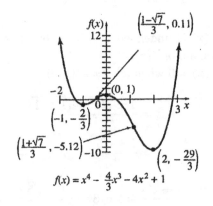

$f(x) = x^4 - \frac{4}{3}x^3 - 4x^2 + 1$

31. $f(x) = \dfrac{x-1}{2x+1}$

Vertical asymptote at $x = -\frac{1}{2}$

Horizontal asymptote at $y = \frac{1}{2}$

$$f'(x) = \frac{(2x+1)(1) - (x-1)(2)}{(2x+1)^2}$$

$$= \frac{3}{(2x+1)^2}$$

f' is never zero.
$f'(-\frac{1}{2})$ does not exist, but $-\frac{1}{2}$ is not a critical number because $-\frac{1}{2}$ is not in the domain of f. Thus, there are no critical numbers, so $f(x)$ has no relative extrema.

$$f''(x) = \frac{-12}{(2x+1)^3}$$

$$f''(0) = -12 < 0$$
$$f''(-1) = 12 > 0$$

Concave upward on $\left(-\infty, -\frac{1}{2}\right)$

Concave downward on $\left(-\frac{1}{2}, \infty\right)$

x-intercept: $\dfrac{x-1}{2x+1} = 0$

$$x = 1$$

y-intercept: $y = \dfrac{0-1}{2(0)+1} = -1$

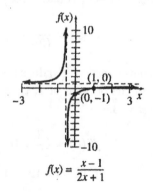

$f(x) = \dfrac{x-1}{2x+1}$

33. $f(x) = -4x^3 - x^2 + 4x + 5$
$\quad f'(x) = -12x^2 - 2x + 4$
$\qquad\quad = -2(6x^2 + x - 2) = 0$
$\qquad\quad (3x+2)(2x-1) = 0$

Critical numbers: $-\frac{2}{3}$ and $\frac{1}{2}$

Critical points: $\left(-\frac{2}{3}, 3.07\right)$ and $\left(\frac{1}{2}, 6.25\right)$

$$f''(x) = -24x - 2$$
$$= -2(12x+1)$$

$$f''\left(-\frac{2}{3}\right) = 14 > 0$$

$$f''\left(\frac{1}{2}\right) = -14 < 0$$

Relative maximum at $\frac{1}{2}$

Relative minimum at $-\frac{2}{3}$

Increasing on $\left(-\frac{2}{3}, \frac{1}{2}\right)$

Decreasing on $\left(-\infty, -\frac{2}{3}\right)$ and $\left(\frac{1}{2}, \infty\right)$

$$f''(x) = -2(12x+1) = 0$$
$$x = -\frac{1}{12}$$

Point of inflection at $\left(-\frac{1}{12}, 4.66\right)$

Concave upward on $\left(-\infty, -\frac{1}{12}\right)$

Concave downward on $\left(-\frac{1}{12}, \infty\right)$

y-intercept:

$$y = -4(0)^3 - (0)^2 + 4(0) + 5 = 5$$

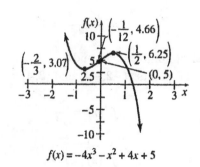

$f(x) = -4x^3 - x^2 + 4x + 5$

35. $f(x) = x^4 + 2x^2$

$f'(x) = 4x^3 + 4x$

$\qquad = 4x(x^2 + 1) = 0$

Critical number: 0

Critical point: $(0, 0)$

$$f''(x) = 12x^2 + 4$$
$$f''(0) = 4 > 0$$

Relative minimum at 0

Increasing on $(0, \infty)$

Decreasing on $(-\infty, 0)$

$$f''(x) = 12x^2 + 4$$
$$4(3x^2 + 1) \neq 0 \text{ for any } x$$

No points of inflection

$$f''(-1) = 16 > 0$$
$$f''(1) = 16 > 0$$

Concave upward on $(-\infty, \infty)$

x-intercept: 0; y-intercept: 0

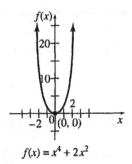

$$f(x) = x^4 + 2x^2$$

37. $f(x) = \dfrac{x^2 + 4}{x}$

Vertical asymptote at $x = 0$

Oblique asymptote at $y = x$

$$f'(x) = \frac{x(2x) - (x^2 + 4)}{x^2}$$
$$= \frac{x^2 - 4}{x^2} = 0$$

Critical numbers: -2 and 2

Critical points: $(-2, -4)$ and $(2, 4)$

$$f''(x) = \frac{8}{x^3}$$
$$f''(-2) = -1 < 0$$
$$f''(2) = 1 > 0$$

Relative maximum at -2

Relative minimum at 2

Increasing on $(-\infty, -2)$ and $(2, \infty)$

Decreasing on $(-2, 0)$ and $(0, 2)$

$f''(x) = \dfrac{8}{x^3} > 0$ for all x.

No inflection points

Concave upward on $(0, \infty)$

Concave downward on $(-\infty, 0)$

No x- or y-intercepts

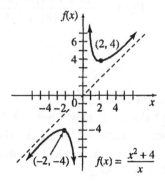

$$f(x) = \frac{x^2 + 4}{x}$$

39. $f(x) = \dfrac{2x}{3 - x}$

Vertical asymptote at $x = 3$

Horizontal asymptote at $y = -2$

$$f'(x) = \frac{(3 - x)(2) - (2x)(-1)}{(3 - x)^2}$$
$$= \frac{6}{(3 - x)^2}$$

$f'(x)$ is never zero. $f'(3)$ does not exist, but since 3 is not in the domain of f, it is not a critical number.

No critical numbers, so no relative extrema

$$f'(0) = \frac{2}{3} > 0$$
$$f'(4) = 6 > 0$$

Increasing on $(-\infty, 3)$ and $(3, \infty)$

$$f''(x) = \frac{12}{(3 - x)^3}$$

$f''(x)$ is never zero. $f''(3)$ does not exist, but since 3 is not in the domain of f, there is no inflection point at $x = 3$.

$$f''(0) = \frac{12}{27} > 0$$
$$f''(4) = -12 < 0$$

Concave upward on $(-\infty, 3)$

Concave downward on $(3, \infty)$

x-intercept: 0; y-intercept: 0

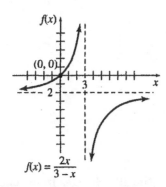

41. $f(x) = \dfrac{\ln x}{x^2}$

The domain is $(0, \infty)$. The function has an
x-intercept at $x = 1$.
Vertical asymptote at $x = 0$.
Horizontal asymptote at $y = 0$.

$$f'(x) = \frac{(x^2)\left(\frac{1}{x}\right) - (\ln x)(2x)}{x^4}$$

$$= \frac{1 - 2\ln x}{x^3}$$

$f'(x)$ is zero when

$$\frac{1 - 2\ln x}{x^3} = 0$$
$$1 - 2\ln x = 0$$
$$1 = 2\ln x$$
$$e^{1/2} = x$$

f is increasing on $(0, e^{1/2})$ and decreasing $(e^{1/2}, \infty)$.
Critical point: $(e^{1/2}, f(e^{1/2}))$ or $(1.65, 0.18)$.

$$f''(x) = \frac{(x^3)\left(-\frac{2}{x}\right) - (3x^2)(1 - 2\ln x)}{x^6}$$

$$= \frac{-5 + 6\ln x}{x^4}$$

$$-5 + 6\ln x = 0$$
$$6\ln x = 5$$
$$x = e^{5/6}$$

inflection point: $(e^{5/6}, f(e^{5/6}))$ or $(2.3, 0.16)$.

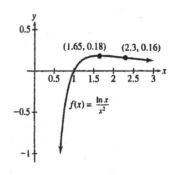

43. $f(x) = x - \sin x$

The domain is $(-\infty, \infty)$, the y-intercept is located
at $y = f(0) = 0$.
Solve $f(x) = 0$ to find the x-intercept.

$$x - \sin x = 0$$
$$x = 0$$

$f'(x) = 1 - \cos x$
$f'(x)$ is zero when

$$1 - \cos x = 0$$
$$x = 0, \pm 2\pi, \pm 4\pi$$

Critical numbers: $0, \pm 2\pi, \pm 4\pi, \ldots$
Critical points:
$$\ldots (-4\pi, -4\pi), (-2\pi, -2\pi), (0, 0), (2\pi, 2\pi), (4\pi, 4\ldots$$

$f'(\pi) = 2 > 0$
$f'(3\pi) = 2 > 0$

increasing on the intervals
$$\ldots (-4\pi, -2\pi), (-2\pi, 0), (0, 2\pi), (2\pi, 4\pi) \ldots$$

$f''(x) = \sin x$
$f''(x) = 0$ when $x = 0, \pm \pi, \pm 2\pi, \pm 3\pi, \ldots$
inflection points: $\ldots (-\pi, -\pi), (0, 0), (\pi, \pi) \ldots$

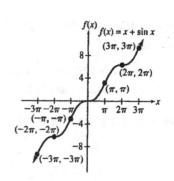

45.

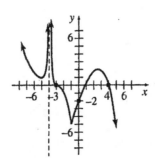

47. (a) Since the second derivative has many sign changes, the graph continually changes from concave upward to concave downward. Since there is a nonlinear decline, the graph must be one that declines, levels off, declines, levels off, etc. Therefore, the first derivative has many critical numbers where the first derivative is zero.

(b) The curve is always decreasing except at frequent points of inflection.

49. (a) Set the two formulas equal to each other.

$$1{,}486S^2 - 4{,}106S + 4{,}514 = 1{,}486S - 825$$
$$1{,}486S^2 - 5{,}592S + 5{,}339 = 0$$

Take the derivative.

$$2{,}972S - 5{,}592 = 0$$

Solve for S.

$$S \approx 1.88$$

For males with 1.88 square meters of surface area, the red cell volume increases approximately 1,486 ml for each additional square meter of surface area.

(b) Set the formulas equal to each other.

$$995e^{0.6085S} = 1{,}578S$$
$$995e^{0.6085S} - 1{,}578S = 0$$

Take the derivative.

$$605.4575e^{0.6085S} - 1{,}578 = 0$$
$$e^{0.6085S} \approx 2.6063$$
$$0.6085S \approx \ln 2.6063$$
$$S \approx 1.57 \text{ square meters}$$

By plugging the exact value of S into the two formulas given for PV, we get about 2,593 ml (Harley) and 2,484 for Pearson et al.??

(c) For males with 1.57 square meters of surface area, the red cell volume increases approximately 1,578 ml for each additional square meter of surface area.

(d) When f and g are closest together, their absolute difference is minimized.

$$\frac{d}{dx}\,|f(x_0) - g(x_0)| = 0$$
$$|f'(x_0) - g'(x_0)| = 0$$
$$f'(x_0) = g'(x_0)$$

51. (a) $z = (1-S)[1-(1+kS)^{-(f-1)}]$

$$\frac{dz}{dS} = (1-S)((f-1)(1+kS)^{-(f-1)-1}(k)) + (-1)[1-(1+kS)^{-(f-1)}]$$

$$= (1-S)((f-1)(1+kS)^{-f}(k)) - [1-(1+kS)]^{-(f-1)}$$

$$= \frac{k(f-1)(1-S)}{(1+kS)^f} - 1 + \frac{1}{(1+kS)^{f-1}}$$

$$= \frac{k(f-1)(1-S)}{(1+kS)^f} - 1 + \frac{(1+kS)}{(1+kS)^f}$$

$$= \frac{k(f-1)(1-S) + (1+kS)}{(1+kS)^f} - 1$$

$$= -1 + (1+kS)^{-f}[1+kS+k(f-1)(1-S)]$$

(d) Let $f = 2$ and $S = \dfrac{-1+\sqrt{1+k}}{k}$.

$$\frac{dz}{dS} = -1 + \left(1 + \frac{k(-1+\sqrt{1+k})}{k}\right)^{-2}\left[1 + \frac{k(-1+\sqrt{1+k})}{k} + k(2-1)\left(1 - \frac{-1+\sqrt{1+k}}{k}\right)\right]$$

$$= -1 + (1-1+\sqrt{1+k})^{-2}\left[1 - 1 + \sqrt{1+k} + k(1)\left(1 + \frac{1-\sqrt{1+k}}{k}\right)\right]$$

$$= -1 + \sqrt{1+k}^{-2}[\sqrt{1+k} + k + 1 - \sqrt{1+k}]$$

$$= -1 + (1+k)^{-1}(k+1)$$

$$= -1 + 1$$

$$= 0$$

53. (a) $R(c) = \dfrac{ac^2}{1+bc^2} - kc$

Let $a = 10, b = 0.08$, and $k = 7$.

$$R(c) = \frac{10c^2}{1+0.08c^2} - 7c$$

$$R'(c) = \frac{(1+0.08c^2)(20c) - (10c^2)(0.16c)}{(1+0.08c^2)^2} - 7$$

$$R'(c) = \frac{-0.0448c^4 - 1.12c^2 + 20c - 7}{(1+0.08c^2)^2}$$

Set $R'(c) = 0$.

$$-0.0448c^4 - 1.12c^2 + 20c - 7 = 0$$
$$c \approx 6.4$$

Critical point: $(6.4, 50.97)$

$f'(4) \approx 8.39 > 0$
$f'(7) \approx -1.22 < 0$

There is a maximum at $(6.4, 50.97)$.

Find the x-intercepts.

$$\frac{10c^2}{1 + 0.08c^2} - 7c = 0$$

$$10c^2 = 7c + 0.56c^3$$

$$0.56c^3 - 10c^2 + 7c = 0$$

$$0.56c^2 - 10c + 7 = 0$$

By the quadratic formula, $c \approx 0.73$ and $c \approx 17.13$

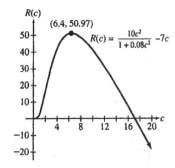

(b) Since R cannot be negative, the domain is $0.73 \le c \le 17.13$

55. $y(t) = A^{c^t}$

$$y'(t) = (\ln A)A^{c^t} \cdot \frac{d}{dt}c^t$$

$$= (\ln A)(\ln c)c^t A^{c^t}$$

$$y''(t) = (\ln A)(\ln c)$$
$$\cdot [(\ln c)c^t A^{c^t} + c^t(\ln A)(\ln c)c^t A^{c^t}]$$

$$= (\ln A)(\ln c)^2 c^t A^{c^t}[1 + (\ln A)c^t]$$

$y''(t) = 0$ when $1 + \ln(A)c^t = 0$

$$c^t = -\frac{1}{\ln A}$$

$$t \ln c = \ln\left(-\frac{1}{\ln A}\right)$$

$$t = -\frac{\ln(-\ln A)}{\ln c}$$

$$= -\frac{\ln(-\ln(0.3982 \cdot 10^{-291}))}{\ln 0.4152}$$

By properties of logarithms,

$$-\ln(0.3982 \cdot 10^{-291}) = -[\ln(0.3982) + \ln(10^{-291})]$$
$$= -[\ln(0.3982) - 291\ln(10)]$$
$$= -\ln 0(0.3982) + 291\ln(10)$$

So,

$$t = -\frac{\ln(-\ln(0.3982) + 291\ln(10))}{\ln(0.4152)}$$

$$\approx 7.405$$

At about 7.405 years the rate of learning to pass the test begins to slow down.

57. **(a)** $s(t) = 512t - 16t^2$
$$v(t) = s'(t) = 512 - 32t$$
$$a(t) = v'(t) = s''(t) = -32$$

(b) $v(t) = 0$ when $512 - 32t = 0$
$$t = 16.$$

$$v(0) = 512 > 0$$
$$v(20) = 512 - 640 = -128 < 0$$

The velocity reaches a maximum when $t = 16$.

$$s(16) = 512 \cdot 16 - 16(16^2) = 4{,}096$$

The maximum height is 4,096 ft.

(c) The projectile hits the ground when $s(t) = 0$.

$$512t - 16t^2 = 0$$
$$16t(32 - t) = 0$$
$$t = 0 \text{ or } t = 32$$

$$v(32) = 512 - 32(32) = -512$$

The projectile hits the ground after 32 seconds with a velocity of -512 ft/sec.

59. **(a)** When stock reaches its highest price of the day, $P'(t)$ is zero and $P''(t)$ is negative.

Chapter 5 Test

[5.1]

Find the largest open intervals where each of the following functions is (a) increasing or (b) decreasing.

1.

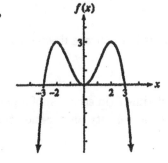

2. $f(x) = 3x^3 - 3x^2 - 3x + 5$

3. $f(x) = \dfrac{x-2}{x+3}$

4. $f(x) = \cos x$

5. What are critical numbers? Why are they significant?

[5.2]

Find the x-values of all points where the following functions have any relative extrema. Find the value(s) of any relative extrema.

6. $f(x) = 4x^3 - \dfrac{9}{2}x^2 - 3x - 1$ 7. $f(x) = 4x^{2/3}$ 8. $f(x) = \cos(\pi x)$

9. Use the derivative to find the vertex of the parabola $y = -2x^2 + 6x + 9$.

[5.3]

Find $f'''(x)$, the third derivative of f, and $f^{(4)}(x)$, the fourth derivative of f, for each of the following functions.

10. $f(x) = 2e^{3x}$

11. $f(x) = \dfrac{x}{3x+2}$

12. $f(x) = 4x^6 - 3x^4 + 2x^3 - 7x + 5$

13. $f(x) = \sin(3x)$

Find any points of inflection for the following functions.

14. $f(x) = 6x^3 - 18x^2 + 12x - 15$ **15.** $f(x) = (x-1)^3 + 2$ **16.** $f(x) = \dfrac{2}{1+x^2}$

Find the second derivative of each function; then find $f''(-3)$.

17. $f(x) = -5x^3 + 3x^2 - x + 1$ **18.** $f(t) = \sqrt{t^2 - 5}$ **19.** $f(x) = \tan\left(\dfrac{\pi x}{4}\right)$

20. Find the largest open intervals where $f(x) = x^3 - 3x^2 - 9x - 1$ is concave upward or concave downward. Find the location of any points of inflection. Graph the function.

21. The function $s(t) = t^3 - 9t^2 + 15t + 25$ gives the displacement in centimeters at time t (in seconds) of a particle moving along a line. Find the velocity and acceleration functions. Then find the velocity and acceleration at $t = 0$ and $t = 2$.

22. What is the difference between velocity and acceleration?

[5.4]

Find the horizontal asymptotes for the graphs of the following functions.

23. $f(x) = \dfrac{2x}{7x+1}$ **24.** $f(x) = \dfrac{3x^2 + 2x}{4x^3 - x}$

25. Find the vertical asymptotes for the function, $f(x) = \tan(\pi x)$.

26. Graph $f(x) = \frac{2}{3}x^3 - \frac{5}{2}x^2 - 3x + 1$. Give critical points, intervals where the function is increasing or decreasing, points of inflection, and intervals where the function is concave upward or concave downward.

27. Graph $f(x) = x + \frac{32}{x^2}$. Give relative extrema, regions where the function is increasing or decreasing, points of inflection, regions where the function is concave upward or downward, intercepts where possible, and asymptotes where applicable.

28. Sketch a graph of a single function that has all the properties listed.

(a) Continuous for all real numbers

(b) Increasing on $(-3, 2)$ and $(4, \infty)$

(c) Decreasing on $(-\infty, -3)$ and $(2, 4)$

(d) Concave upward on $(-\infty, 0)$

(e) Concave downward on $(0, 4)$ and $(4, \infty)$

(f) Differentiable everywhere except $x = 4$

(g) $f'(-3) = f'(2) = 0$

(h) An inflection point at $(0, 0)$

Chapter 5 Test Answers

1. **(a)** $(-\infty, -2)$ and $(0, 2)$ **(b)** $(-2, 0)$ and $(2, \infty)$

2. **(a)** $\left(-\infty, -\frac{1}{3}\right)$ and $(1, \infty)$ **(b)** $\left(-\frac{1}{3}, 1\right)$

3. **(a)** $(-\infty, -3)$ and $(-3, \infty)$
   (b) Nowhere decreasing

4. **(a)** $(n\pi, (n + 1)\pi)$ where n is an odd integer
   (b) $(n\pi, (n + 1)\pi)$ where n is an even integer

5. Critical numbers are x–values for which $f'(x) = 0$ or $f'(x)$ does not exist. They tell where the derivative may change signs, and are used to tell where a function is increasing or decreasing.

6. Relative maximum of $-\frac{19}{32}$ at $-\frac{1}{4}$; relative minimum of $-\frac{9}{2}$ at 1

7. Relative minimum of 0 at 0

8. Relative maximum of 1 at
   $x = \ldots -4, -2, 0, 2, 4, \ldots$;
   relative minimum of -1 at
   $x = \ldots -3, -1, 1, 3, \ldots$

9. $\left(\frac{3}{2}, \frac{27}{2}\right)$

10. $f'''(x) = 54e^{3x}$, $f^{(4)}(x) = 162e^{3x}$

11. $f'''(x) = \frac{108}{(3x+2)^4}$, $f^{(4)}(x) = \frac{-1,296}{(3x+2)^5}$

12. $f'''(x) = 480x^3 - 72x + 12$, $f^{(4)} = 1,440x^2 - 72$

13. $f'''(x) = -27\cos(3x)$, $f^{(4)} = 81\sin(3x)$

14. $(1, -15)$

15. $(1, 2)$

16. $\left(-\frac{\sqrt{3}}{3}, \frac{3}{2}\right), \left(\frac{\sqrt{3}}{3}, \frac{3}{2}\right)$

17. $f''(x) = -30x + 6$; $f''(-3) = 96$

18. $f''(t) = \frac{-5}{(t^2-5)^{3/2}}$; $f''(-3) = -\frac{5}{8}$

19. $f''(x) = \frac{1}{8}\pi^2 \sec^2\left(\frac{\pi x}{4}\right)\tan\left(\frac{\pi x}{4}\right)$; $f(-3) = \frac{x^2}{4}$

20. Concave upward on $(1, \infty)$;
    concave downward on $(-\infty, 1)$;
    point of inflection $(1, -12)$

21. $v(t) = 3t^2 - 18t + 15$; $a(t) = 6t - 18$;
    $v(0) = 15$ cm/sec; $a(0) = -18$ cm/sec^2;
    $v(2) = -9$ cm/sec; $a(2) = -6$ cm/sec^2

22. Velocity gives the rate of change of position relative to time. Acceleration gives the rate of change of velocity relative to time.

23. $y = \frac{2}{7}$

24. $y = 0$

25. $y = \ldots, -\frac{3}{2}, -\frac{1}{2}, \frac{1}{2}, \frac{3}{2}, \frac{5}{2}, \ldots$

26. Critical points: $\left(-\frac{1}{2}, \frac{43}{24}\right)$ or $(-0.5, 1.79)$ (relative maximum) and $\left(3, -\frac{25}{2}\right)$ or $(3, -12.5)$ (relative minimum); increasing on $\left(-\infty, -\frac{1}{2}\right)$ or $(-\infty, -0.5)$ and $(3, \infty)$; decreasing on $\left(-\frac{1}{2}, 3\right)$ or $(-0.5, 3)$; point of inflection: $\left(\frac{5}{4}, -\frac{257}{48}\right)$ or $(1.25, -5.35)$; concave up on $\left(\frac{5}{4}, \infty\right)$ or $(1.25, \infty)$; concave down on $\left(-\infty, \frac{5}{4}\right)$ or $(-\infty, 1.25)$

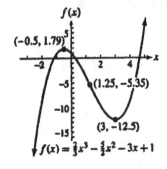

27. Relative extrema: Relative minimum at $(4, 6)$, no relative maxima; increasing on $(-\infty, 0)$ and $(4, \infty)$; decreasing on $(0, 4)$; points of inflection: none; concave upward on $(-\infty, 0)$ and $(0, \infty)$; concave downward nowhere; x-intercept: $-\sqrt[3]{32} \approx -3.17$; y-intercept: none; oblique asymptote: $y = x$

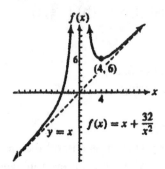

28. One such graph is shown. Other answers are possible.

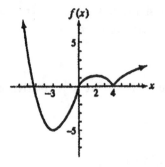

APPLICATIONS OF THE DERIVATIVE

6.1   Absolute Extrema

1. As shown on the graph, the absolute maximum occurs at x_3; there is no absolute minimum. (There is no functional value that is less than all others.)

3. As shown on the graph, there are no absolute extrema.

5. As shown on the graph, the absolute minimum occurs at x_1; there is no absolute maximum.

7. As shown on the graph, the absolute maximum occurs at x_1; the absolute minimum occurs at x_2.

11. $f(x) = x^3 - 6x^2 + 9x - 8;\ [0, 5]$

    Find critical numbers:

    $$f'(x) = 3x^2 - 12x + 9 = 0$$
    $$x^2 - 4x + 3 = 0$$
    $$(x-3)(x-1) = 0$$
    $$x = 1 \quad \text{or} \quad x = 3$$

x	$f(x)$	
0	-8	Absolute minimum
1	-4	
3	-8	Absolute minimum
5	-12	Absolute maximum

13. $f(x) = \frac{1}{3}x^3 + \frac{3}{2}x^2 - 4x + 1;\ [-5, 2]$

    Find critical numbers:

    $$f'(x) = x^2 + 3x - 4 = 0$$
    $$(x+4)(x-1) = 0$$
    $$x = -4 \quad \text{or} \quad x = 1$$

x	$f(x)$	
-4	$\frac{59}{3} \approx 19.67$	Absolute maximum
1	$-\frac{7}{6} \approx -1.17$	Absolute minimum
-5	$\frac{101}{6} \approx 16.83$	
2	$\frac{5}{3} \approx 1.67$	

15. $f(x) = x^4 - 18x^2 + 1;\ [-4, 4]$

    $$f'(x) = 4x^3 - 36x = 0$$
    $$4x(x^2 - 9) = 0$$
    $$4x(x+3)(x-3) = 0$$

    $$x = 0 \quad \text{or} \quad x = -3 \quad \text{or} \quad x = 3$$

x	$f(x)$	
-4	-31	
-3	-80	Absolute minimum
0	1	Absolute maximum
3	-80	Absolute minimum
4	-31	

17. $f(x) = \frac{1-x}{3+x};\ [0, 3]$

    $$f'(x) = \frac{-4}{(3+x)^2}$$

    No critical numbers

x	$f(x)$	
0	$\frac{1}{3}$	Absolute maximum
3	$-\frac{1}{3}$	Absolute minimum

19. $f(x) = \frac{x-1}{x^2+1};\ [1, 5]$

    $$f'(x) = \frac{-x^2 + 2x + 1}{(x^2+1)^2}$$

    $f'(x) = 0$ when

    $$-x^2 + 2x + 1 = 0$$
    $$x = 1 \pm \sqrt{2},$$

    but $1 - \sqrt{2}$ is not in $[1, 5]$.

x	$f(x)$	
1	0	Absolute minimum
5	$\frac{4}{27} \approx 0.15$	
$1 + \sqrt{2}$	≈ 0.21	Absolute maximum

21. $f(x) = (x^2 + 4)^{1/3}$; $[-2, 2]$

$$f'(x) = \frac{2x}{3(x^2 + 4)^{2/3}} = 0$$

$$x = 0$$

x	$f(x)$	
-2	2	Absolute maximum
0	1.59	Absolute minimum
2	2	Absolute maximum

23. $f(x) = (x - 3)(x - 1)^3$; $[-2, 3]$

$$f'(x) = (1)(x - 1)^3 + 3(x - 1)^2(x - 3)$$
$$= (x - 1)^2(4x - 10) = 0$$
$$= 2(x - 1)^2(2x - 5) = 0$$

$$x = 1 \quad \text{or} \quad \frac{5}{2}$$

x	$f(x)$	
-2	135	Absolute maximum
1	0	
$\frac{5}{2}$	-1.687	Absolute minimum
3	0	

25. $f(x) = x \ln x$; $\left[\frac{1}{4}, e\right]$

$$f'(x) = x\left(\frac{1}{x}\right) + (1)\ln x$$

$$= 1 + \ln x$$

$f'(x) = 0$ when

$$1 + \ln x = 0$$
$$\ln x = -1$$
$$x = e^{-1}$$

x	$f(x)$	
$\frac{1}{4}$	$\frac{1}{4}\ln\frac{1}{4} \approx -0.35$	
e^{-1}	$-e^{-1} \approx -0.37$	Absolute minimum
e	$e \approx 2.72$	Absolute maximum

27. $f(x) = \dfrac{-5x^4 + 2x^3 + 3x^2 + 9}{x^4 - x^3 + x^2 + 7}$; $[-1, 1]$

The indicated domain tells us the x-values to use for the viewing window, but we must experiment to find a suitable range for the y-values. In order to show the absolute extrema on $[-1, 1]$, we find that a suitable window is $[-1, 1]$ by $[0, 1.5]$ with Xscl = 0.1, Yscl = 0.1.

From the graph, we se that on $[-1, 1]$, f has an absolute maximum at about 0.6085 and an absolute minimum at -1.

29. $f(x) = x \cos\dfrac{\pi}{2}x$; $[0, 4]$

A suitable range for the y-values can be found by noting $-1 \leq \cos\dfrac{\pi}{2}x \leq 1$, for all x. On the interval $[0, 4]$, then, $-4 \leq x\cos\dfrac{\pi}{2}x \leq 4$. So a suitable window is $[0, 4]$ by $[-4, 4]$. From the graph, we see that on $[0, 4]$, f has an absolute maximum at 4 and an absolute minimum at about 2.18.

31. $f(x) = 2x + \dfrac{8}{x^2} + 1$, $x > 0$

$$f'(x) = 2 - \frac{16}{x^3}$$
$$= \frac{2x^3 - 16}{x^3}$$
$$= \frac{2(x - 2)(x^2 + 2x + 4)}{x^3}$$

Since the specified domain is $(0, \infty)$, a critical number is $x = 2$.

x	$f(x)$
2	7

There is an absolute minimum at $x = 2$; there is no absolute maximum, as can be seen by looking at the graph of f.

33. $f(x) = -3x^4 + 8x^3 + 18x^2 + 2$
$$f'(x) = -12x^3 + 24x^2 + 36x$$
$$= -12x(x^2 - 2x - 3)$$
$$= -12x(x - 3)(x + 1)$$

Critical numbers are 0, 3, and -1.

x	$f(x)$
-1	9
0	2
3	137

There is an absolute maximum at $x = 3$; there is no absolute minimum, as can be seen by looking at the graph of f.

35. $f(x) = \dfrac{x-1}{x^2+2x+6}$

$$f'(x) = \frac{(x^2+2x+6)(1)-(x-1)(2x+2)}{(x^2+2x+6)^2}$$

$$= \frac{x^2+2x+6-2x^2+2}{(x^2+2x+6)^2}$$

$$= \frac{-x^2+2x+8}{(x^2+2x+6)^2}$$

$$= \frac{-(x^2-2x-8)}{(x^2+2x+6)^2}$$

$$= \frac{-(x-4)(x+2)}{(x^2+2x+6)^2}$$

Critical numbers are 4 and -2.

x	$f(x)$
-2	$-\dfrac{1}{2}$
4	0.1

There is an absolute maximum at $x = 4$ and an absolute minimum at $x = -2$. This can be verified by looking at the graph of f.

37. $f(x) = \dfrac{x^2+36}{2x}, \; 1 \le x \le 12$

$$f'(x) = \frac{2x(2x)-(x^2+36)(2)}{(2x)^2} = \frac{4x^2-2x^2-72}{4x^2}$$

$$= \frac{2x^2-72}{4x^2} = \frac{2(x^2-36)}{4x^2}$$

$$= \frac{(x+6)(x-6)}{2x^2}$$

$f'(x) = 0$ when $x = 6$ and when $x = -6$. Only 6 is in the interval $1 \le x \le 12$.
Test for relative maximum or minimum.

$$f'(5) = \frac{(11)(-1)}{50} < 0$$

$$f'(7) = \frac{(13)(1)}{98} > 0$$

The minimum occurs at $x = 6$, or at 6 months. Since $f(6) = 6$, $f(1) = 18.5$, and $f(12) = 7.5$, the minimum percent is 6%.

39. Since we are only interested in the length during weeks 22 through 28, the domain of the function for this problem is $[22, 28]$. We now look for any critical numbers in this interval. We find

$$L'(t) = 0.788 - 0.02t$$

There is a critical number at $t = \dfrac{0.788}{0.02} = 39.4$, which is not in the interval. Thus, the maximum value will occur at one of the endpoints.

t	$L(t)$
22	5.4
28	7.2

The maximum length is about 7.2 millimeters.

41. (a)

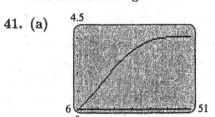

(b) The function is defined on the interval $[6, 48]$. We first find the derivative of the function.

$$M'(t) = 0.0229883 + 0.0216042t - 0.0009417t^2$$
$$+ 0.00001t^3$$

To determine where this function has a maximum, use a graphing calculator to graph the function, and we find one critical number in the interval at about 15.11. So the dentin is growing most rapidly on about day 15.

43. (a)

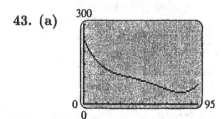

(b) To determine where the maximum and minimum numbers occur, use a graphing calculator to locate any extreme points on the graph. One critical number is found at about 81.51.

t	$P(t)$
0	241.75
81.51	43.60
95	71.16

The maximum number of polygons is about 242. The minimum number of polygons is about 44.

45. (a) The slope of the line tangent to the curve at the point corresponding to V_{mp} is approximately 0. The power expended is not changing at that point.

(b) The slope of the curve at the point corresponding to V_{mr} is about 0.1. The power expended is increasing 0.1 unit per unit increase in speed.

(c) The slope of the curve at the point corresponding to V_{opt} is about 0.12. The power expended is increasing 0.12 unit for each 1 unit increase in speed and is higher than the power required at V_{mr} and V_{opt}.

(d) As the speed increases, the power level at first decreases to the point corresponding to V_{mp}, which minimizes the power expended. As the speed continues to increase after V_{mp}, the power expended increases at an alarming rate.

(e) The slope of the line drawn from the origin to the corresponding points on the graph is smallest if it is drawn through V_{mr}. Since this is interpreted as the smallest ratio of energy per unit distance, it seems logical it should occur at the point representing minimal energy costs per unit of distance, or V_{mr}.

47. $M(x) = -.018x^2 + 1.24x + 6.2,\ 30 \le x \le 60$

$$M'(x) = -.036x + 1.24 = 0$$
$$x = 34.4$$

x	$M(x)$
30	27.2
34.4	27.6
60	15.8

The absolute minimum of 15.8 mpg occurs at 60 mph.
The absolute maximum of 27.6 mpg occurs at 34.4 mph.

49. Total area $= A(x)$

$$= \pi \left(\frac{x}{2\pi} \right)^2 + \left(\frac{12 - x}{4} \right)^2$$

$$= \frac{x^2}{4\pi} + \frac{(12 - x)^2}{16}$$

$$A'(x) = \frac{x}{2\pi} - \frac{12 - x}{8} = 0$$

$$\frac{4x - \pi(12 - x)}{8\pi} = 0$$

$$x = \frac{12\pi}{4 + \pi} \approx 5.28$$

x	Area
0	9
5.28	5.04
12	11.46

The total area is maximized when all 12 feet of wire are used to form the circle.

51. The value $x = 20$ minimizes $\frac{f(x)}{x}$ because this is the point where the line from the origin to the curve is tangent to the curve.
A production level of 20 units results in the minimum cost per unit.

53. The value $x = 300$ maximizes $\frac{f(x)}{x}$ because this is the point where the line from the origin to the curve is tangent to the curve.
A production level of 300 units results in the minimum cost per unit.

6.2   Applications of Extrema

1. $x + y = 100,\ P = xy$

(a) $y = 100 - x$

(b) $P = xy = x(100 - x)$
$\qquad = 100x - x^2$

(c) Since $y = 100 - x$ and x and y are nonnegative numbers, $x \ge 0$ and $100 - x \ge 0$ or $x \le 100$. The domain of P is $[0, 100]$.

(d) $P' = 100 - 2x$

$$100 - 2x = 0$$
$$2(50 - x) = 0$$
$$x = 50$$

(e)

x	P
0	0
50	2500
100	0

(f) From the chart, the maximum value of P is 2500; this occurs when $x = 50$ and $y = 50$.

3. $x + y = 150$

Maximize $x^2 y$.

(a) $y = 150 - x$

(b) Let $P = x^2 y = x^2(150 - x)$
$$= 150x^2 - x^3.$$

(c) Since $y = 150 - x$ and x and y are nonnegative, the domain of P is $[0, 150]$.

(d) $P' = 300x - 3x^2$

$$300x - 3x^2 = 0$$
$$3x(100 - x) = 0$$
$$x = 0 \quad \text{or} \quad x = 100$$

(e)

x	P
0	0
100	500,000
150	0

(f) The maximum value of $x^2 y$ occurs when $x = 100$ and $y = 50$. The maximum value is 500,000.

5. $p(t) = \dfrac{20t^3 - t^4}{1{,}000}$, $[0, 20]$

(a) $p'(t) = \dfrac{3}{50}t^2 - \dfrac{1}{250}t^3$

$$= \frac{1}{50}t^2 \left[3 - \frac{1}{5}t \right]$$

Critical numbers:

$$\frac{1}{50}t^2 = 0 \quad \text{or} \quad 3 - \frac{1}{5}t = 0$$
$$t = 0 \quad \text{or} \qquad t = 15$$

t	$p(t)$
0	0
15	16.875
20	0

The number of people infected reaches a maximum in 15 days.

(b) $P(15) = 16.875\%$

7. **(a)** $p(t) = 10te^{-t/8}$, $[0, 40]$

$$p'(t) = 10te^{-t/8}\left(-\frac{1}{8}\right) + e^{-t/8}(10)$$

$$= 10e^{-t/8}\left(-\frac{t}{8} + 1\right)$$

Critical numbers:

$p'(t) = 0$ when

$$-\frac{t}{8} + 1 = 0$$
$$t = 8.$$

t	$p(t)$
0	0
8	29.43
40	2.6952

The percent of the population infected reaches a maximum in 8 days.

(b) $P(8) = 29.43\%$

9. $H(S) = f(S) - S$

$$f(S) = \frac{25S}{S + 2}$$

$$H'(S) = \frac{(S + 2)(25) - 25S}{(S + 2)^2} - 1$$

$$= \frac{25S + 50 - 25S - (S + 2)^2}{(S + 2)^2}$$

$$= \frac{50 - (S^2 + 4S + 4)}{(S + 2)^2}$$

$$= \frac{-S^2 - 4S + 46}{(S + 2)^2}$$

$H'(S) = 0$ when

$$S^2 + 4S - 46 = 0$$
$$S = \frac{-4 \pm \sqrt{16 + 184}}{2} = 5.071.$$

(Discard the negative solution.)

The number of creatures needed to sustain the population is $S_0 = 5.071$ thousand.

$$H'' = \frac{(S+2)^2(-2S-4) - (S^2-4S+46)(2S+4)}{(S + 2)^4} < 0,$$

so H is a maximum at $S_0 = 5.071$.

$$H(S_0) = \frac{25(5.071)}{7.071} - 5.071 \approx 12.86$$

The maximum sustainable harvest is 12.86 thousand.

11. $r = 0.1$, $P = 100$

$$f(S) = Se^{r(1-S/P)}$$

$$f'(S) = -\frac{1}{1,000} \cdot Se^{0.1(1-S/100)} + e^{0.1(1-S/100)}$$

$$f'(S_0) = -0.001 S_0 e^{0.1(1-S_0/100)} + e^{0.1(1-S_0/100)}$$

Graph

$$Y_1 = -0.001 x e^{0.1(1-x/100)} + e^{0.1(1-x/100)}$$

and

$$Y_2 = 1$$

on the same screen. A suitable choice for the viewing window is $[0, 60]$ by $[0.5, 1.5]$ with Xscl $= 10$ and Yscl $= 0.5$. By zooming or using the "intersect" option, we find the graphs intersect when $x \approx 49.37$.

The maximum sustainable harvest is 49.37.

13. Let x = distance from P to A.

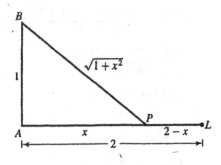

Energy used over land: 1 unit per mile

Energy used over water: $\frac{4}{3}$ units per mile

Distance over land: $(2 - x)$ mi
Distance over water: $\sqrt{1 + x^2}$ mi
Find the location of P to minimize energy used.

$$E(x) = 1(2 - x) + \frac{4}{3}\sqrt{1 + x^2}, \text{ where } 0 \le x \le 2.$$

$$E'(x) = -1 + \frac{4}{3}\left(\frac{1}{2}\right)(1 + x^2)^{-1/2}(2x)$$

If $E'(x) = 0$,

$$\frac{4}{3}x(1 + x^2)^{-1/2} = 1$$

$$\frac{4x}{3(1 + x^2)^{1/2}} = 1$$

$$\frac{4}{3}x = (1 + x^2)^{1/2}$$

$$\frac{16}{9}x^2 = 1 + x^2$$

$$\frac{7}{9}x^2 = 1$$

$$x^2 = \frac{9}{7}$$

$$x = \frac{3}{\sqrt{7}}$$

$$= \frac{3\sqrt{7}}{7}.$$

x	$E(x)$
0	3.3333
1.134	2.8819
2	2.9814

The absolute minimum occurs at $x \approx 1.134$.
Point P is $\frac{3\sqrt{7}}{7} \approx 1.134$ mi from Point A.

15. Graph $y = d(x)$ on a graphing calculator. (Be sure the angel mode is degrees.) A suitable choice for the viewing window is $[0, 90]$ by $[0, 15]$ with $X_{scl} = 10$ and $Y_{scl} = 3$. We see the graph has a maximum at about $x \approx 40$. Thus the release angle that maximizes the distance traveled is about $40°$. Using this value, we find

$$d(40) = 5.1 \sin(2 \cdot 40)\left[\sqrt{1 + \frac{0.41}{\sin^2(40)}}\right]$$

$$\approx 12.1118.$$

So, the maximum distance is about 12m.

17. (a) Solve the given equation for effective power for T, time.

$$\frac{kE}{T} = aSv^3 + I$$

$$\frac{kE}{aSv^3 + I} = T$$

Since distance is velocity, v, times time, T, we have

$$D(v) = v\frac{kE}{aSv^3 + I}$$

$$= \frac{kEv}{aSv^3 + I}.$$

(b) $D'(v) = \dfrac{(aSv^3 + I)kE - kEv(3aSv^2)}{(aSv^3 + I)^2}$

$\qquad\quad = \dfrac{kE(aSv^3 + I - 3aSv^3)}{(aSv^3 + I)^2}$

$\qquad\quad = \dfrac{kE(I - 2aSv^3)}{(aSv^3 + I)^2}$

Find the critical numbers by solving $D'(v) = 0$ for v.

$$I - 2aSv^3 = 0$$
$$2aSv^3 = I$$
$$v^3 = \dfrac{I}{2aS}$$
$$v = \left(\dfrac{I}{2aS}\right)^{1/3}$$

19. Let $\quad 8 - x =$ the distance the hunter will travel on the river.

Then $\sqrt{9 + x^2} =$ the distance he will travel on land.

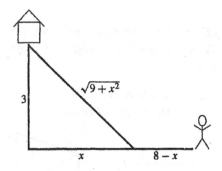

The rate on the river is 5 mph, the rate on land is 2 mph. Using $t = \dfrac{d}{r}$,

$\dfrac{8-x}{5} =$ the time on the river,

$\dfrac{\sqrt{9+x^2}}{2} =$ the time on land.

The total time is

$$T(x) = \dfrac{8-x}{5} + \dfrac{\sqrt{9+x^2}}{2}$$
$$= \dfrac{8}{5} - \dfrac{1}{5}x + \dfrac{1}{2}(9+x^2)^{1/2}.$$
$$T' = -\dfrac{1}{5} + \dfrac{1}{4}\cdot 2x(9+x^2)^{-1/2}$$

$$-\dfrac{1}{5} + \dfrac{x}{2(9+x^2)^{1/2}} = 0$$
$$\dfrac{1}{5} = \dfrac{x}{2(9+x^2)^{1/2}}$$
$$2(9+x^2)^{1/2} = 5x$$
$$4(9+x^2) = 25x^2$$
$$36 + 4x^2 = 25x^2$$
$$36 = 21x^2$$
$$\dfrac{6}{\sqrt{21}} = x$$
$$\dfrac{6\sqrt{21}}{21} = \dfrac{2\sqrt{21}}{7} = x$$
$$1.31 = x$$

x	$T(x)$
0	3.1
1.31	2.98
8	4.27

Since the minimum time is 2.98 hr, the hunter should travel $8 - 1.31$ or about 6.7 miles along the river.

21. $C(x) = \dfrac{1}{2}x^3 + 2x^2 - 3x + 35$

The average cost function is

$$A(x) = \overline{C}(x) = \dfrac{C(x)}{x}$$
$$= \dfrac{\frac{1}{2}x^3 + 2x^2 - 3x + 35}{x}$$
$$= \dfrac{1}{2}x^2 + 2x - 3 + \dfrac{35}{x}$$

or $\quad \dfrac{1}{2}x^2 + 2x - 3 + 35x^{-1}.$

Then

$$A'(x) = x + 2 - 35x^{-2}$$

or $\quad x + 2 - \dfrac{35}{x^2}.$

Graph $y = A'(x)$ on a graphing calculator. A suitable choice for the viewing window is $[0, 10]$ by $[-10, 10]$. (Negative values of x are not meaningful in this application.) Using the calculator, we see that the graph has an x-intercept or "zero" at $x \approx 2.722$. Thus, 2.722 is a critical number.

Now graph $y = A(x)$ and use this graph to confirm that a minimum occurs at $x \approx 2.722$.

Thus, the average cost is smallest at $x \approx 2.722$.

23. Let $x =$ the width
and $y =$ the length.

(a) The perimeter is

$$P = 2x + y$$
$$= 1,200,$$

so

$$y = 1,200 - 2x.$$

(b) Area $= xy = x(1,200 - 2x)$
$$A(x) = 1,200x - 2x^2$$

(c) $A' = 1,200 - 4x$
$$1,200 - 4x = 0$$
$$1,200 = 4x$$
$$300 = x$$

$A'' = -4$, which implies that $x = 300$ m leads to
the maximum area.

(d) If $x = 300$,

$$y = 1,200 - 2(300) = 600.$$

The maximum area is $(300)(600) = 180,000$ m^2.

25. Let $x =$ the width of the rectangle
and $y =$ the total length of the
rectangle.

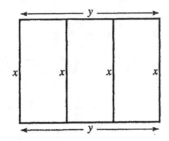

An equation for the fencing is

$$3,600 = 4x + 2y$$
$$2y = 3,600 - 4x$$
$$y = 1,800 - 2x.$$

Area $= xy = x(1,800 - 2x)$
$$A(x) = 1,800x - 2x^2$$

$A' = 1,800 - 4x$

$$1,800 - 4x = 0$$
$$1,800 = 4x$$
$$450 = x$$

$A'' = -4$, which implies that $x = 450$ is the location of a maximum.

If $x = 450$, $y = 1,800 - 2(450) = 900$.
The maximum area is

$$(450)(900) = 405,000 \text{ m}^2.$$

27. Let $x =$ the length at \$4 per meter
$y =$ the width at \$2 per meter.

$$xy = 15,625$$
$$y = \frac{15,625}{x}$$

Perimeter $= x + 2y = x + \dfrac{31,250}{x}$

Cost $= x(4) + \dfrac{31,250}{x}(2)$

$$= 4x + \frac{62,500}{x}$$

Minimize cost.

$$C' = 4 - \frac{62,500}{x^2}$$
$$4 - \frac{62,500}{x^2} = 0$$
$$4x^2 = 62,500$$
$$x^2 = 15,625$$
$$x = 125$$

$$y = \frac{15,625}{125} = 125$$

125 m at \$4 per meter will cost \$500. 250 m at
\$2 per meter will cost \$500. The total cost will be
\$1,000.

29. Let $x =$ a side of the base
$h =$ the height of the box.

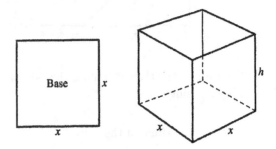

An equation for the volume of the box is

$$V = x^2 h,$$
so $32 = x^2 h$

$$h = \frac{32}{x^2}.$$

The box is open at the top so the area of the
surface material $m(x)$ in square inches is the area
of the base plus the area of the four sides.

$$m(x) = x^2 + 4xh$$

$$= x^2 + 4x\left(\frac{32}{x^2}\right)$$

$$= x^2 + \frac{128}{x}$$

$$m'(x) = 2x - \frac{128}{x^2}$$

$$\frac{2x^3 - 128}{x^2} = 0$$

$$2x^3 - 128 = 0$$
$$2(x^3 - 64) = 0$$
$$x = 4$$

$m''(x) = 2 + \frac{256}{x^3} > 0$ since $x > 0$.

So, $x = 4$ minimizes the surface material.

If $x = 4$,

$$h = \frac{32}{x^2} = \frac{32}{16} = 2.$$

The dimensions that will minimize the surface material are 4 in by 4 in by 2 in.

31. Let $\quad x =$ the width.

Then $2x =$ the length

and $\quad h =$ the height.

An equation for volume is

$$V = (2x)(x)h = 2x^2h$$
$$36 = 2x^2h.$$

So, $h = \frac{18}{x^2}$.

The surface area $S(x)$ is the sum of the areas of the base and the four sides.

$$S(x) = (2x)(x) + 2xh + 2(2x)h$$
$$= 2x^2 + 6xh$$

$$= 2x^2 + 6x\left(\frac{18}{x^2}\right)$$

$$= 2x^2 + \frac{108}{x}$$

$$S'(x) = 4x - \frac{108}{x^2}$$

$$\frac{4x^3 - 108}{x^2} = 0$$

$$4(x^3 - 27) = 0$$
$$x = 3$$

$$S''(x) = 4 + \frac{108(2)}{x^3}$$

$$= 4 + \frac{216}{x^3} > 0 \text{ since } x > 0.$$

So $x = 3$ minimizes the surface material.

If $x = 3$,

$$h = \frac{18}{x^2} = \frac{18}{9} = 2.$$

The dimensions are 3 ft by 6 ft by 2 ft.

33. 120 centimeters of ribbon are available; it will cover 4 heights and 8 radii.

$$4h + 8r = 120$$
$$h + 2r = 30$$
$$h = 30 - 2r$$

$$V = \pi r^2 h$$
$$V = \pi r^2(30 - 2r)$$
$$= 30\pi r^2 - 2\pi r^3$$

Maximize volume.

$$V' = 60\pi r - 6\pi r^2$$
$$60\pi r - 6\pi r^2 = 0$$
$$6\pi r(10 - r) = 0$$
$$r = 0 \quad \text{or} \quad r = 10$$

If $r = 0$, there is no box, so we discard this value.
$V'' = 6\pi - 12\pi r < 0$ for $r = 10$, which implies that $r = 10$ gives maximum volume.
When $r = 10$, $h = 30 - 2(10) = 10$.
The volume is maximized when the radius and height are both 10 cm.

35. $V = \pi r^2 h = 16$

$$h = \frac{16}{\pi r^2}$$

The total cost is the sum of the cost of the top and bottom and the cost of the sides or

$$C = 2(2)(\pi r^2) + 1(2\pi rh)$$

$$= 4(\pi r^2) + 1(2\pi r)\left(\frac{16}{\pi r^2}\right)$$

$$= 4\pi r^2 + \frac{32}{r}.$$

Minimize cost.

$$C' = 8\pi r - \frac{32}{r^2}$$

$$8\pi r - \frac{32}{r^2} = 0$$

$$8\pi r^3 = 32$$
$$\pi r^3 = 4$$

$$r = \sqrt[3]{\frac{4}{\pi}} \approx 1.08$$

$$h = \frac{16}{\pi (1.08)^2} \approx 4.34$$

The radius should be 1.08 ft and the height should be 4.34 ft. If these rounded values for the height and radius are used, the cost is

$$\begin{aligned} \$2(2)(\pi r^2) &+ \$1(2\pi rh) \\ &= 4\pi(1.08)^2 + 2\pi(1.08)(4.34) \\ &= \$44.11. \end{aligned}$$

37. (a) From Example 3, the area of the base is $(12 - 2x)(12 - 2x) = 4x^2 - 48x + 144$ and the total area of all four walls is $4x(12 - 2x) = -8x^2 + 48x$. Since the box has maximum volume when $x = 2$, the area of the base is $4(2)^2 - 48(2) + 144 = 64$ square inches and the total area of all four walls is $-8(2)^2 + 48(2) = 64$ square inches. So, both are 64 square inches.

(b) From Exercise 36, the area of the base is $(3 - 2x)(8 - 2x) = 4x^2 - 22x + 24$ and the total area of all four walls is $2x(3 - 2x) + 2x(8 - 2x) = -8x^2 + 22x$. Since the box has maximum volume when $x = \frac{2}{3}$, the area of the base is $4\left(\frac{2}{3}\right)^2 - 22\left(\frac{2}{3}\right) + 24 = \frac{100}{9}$ square feet and the total area of all four walls is $-8\left(\frac{2}{3}\right)^2 + 22\left(\frac{2}{3}\right) = \frac{100}{9}$ square feet. So, both are $\frac{100}{9}$ square feet.

(c) Based on the results from parts (a) and (b), it appears that the area of the base and the total area of the walls for the box with maximum volume are equal. (This conjecture is true.)

39. Distance on shore: $9 - x$ miles
Cost on shore: \$400 per mile
Distance underwater: $\sqrt{x^2 + 36}$
Cost underwater: \$500 per mile
Find the distance from A, that is, $(9 - x)$, to minimize cost, $C(x)$.

$$\begin{aligned} C(x) &= (9 - x)(400) + (\sqrt{x^2 + 36})(500) \\ &= 3{,}600 - 400x + 500(x^2 + 36)^{1/2} \end{aligned}$$

$$C'(x) = -400 + 500 \left(\frac{1}{2}\right)(x^2 + 36)^{-1/2}(2x)$$

$$= -400 + \frac{500x}{\sqrt{x^2 + 36}}$$

If $C'(x) = 0$,

$$\frac{500x}{\sqrt{x^2 + 36}} = 400$$

$$\frac{5x}{4} = \sqrt{x^2 + 36}$$

$$\frac{25}{16}x^2 = x^2 + 36$$

$$\frac{9}{16}x^2 = 36$$

$$x^2 = \frac{36 \cdot 16}{9}$$

$$x = \frac{6 \cdot 4}{3} = 8.$$

(Discard the negative solution.)
Then the distance should be

$$\begin{aligned} 9 - x &= 9 - 8 \\ &= 1 \text{ mile from point A.} \end{aligned}$$

41.   Let $x =$ the number of additional tables.
Then $90 - 0.25x =$ the cost per table and
and $300 + x =$ the number of tables ordered.

$$\begin{aligned} R &= (90 - 0.25x)(300 + x) \\ &= 27{,}000 + 15x - 0.25x^2 \\ R' &= 15 - 0.5x = 0 \\ &\qquad\qquad x = 30 \end{aligned}$$

$R'' = -0.5 < 0$, so when $300 + 30 = 330$ tables are ordered, revenue is maximum.

Thus, the maximum revenue is

$$\begin{aligned} R(30) &= 27{,}000 + 15(30) - 0.25(30)^2 \\ &= \$27{,}225 \end{aligned}$$

The maximum revenue is \$27,225.

Minimum revenue is found by letting $R = 0$.

$$(90 - 0.25x)(300 + x) = 0$$
$$90 - 0.25x = 0 \quad \text{or} \quad 300 + x = 0$$
$$x = 360 \quad \text{or} \qquad\qquad x = -300$$
$$\text{(impossible)}$$

So when $300 + 360 = 660$ tables are ordered, revenue is 0, that is, each table is free.
I would fire the assistant.

43. In Exercise 42, we found that the cost of the aluminum to make the can is

$$0.03\left(2\pi r^2 + \frac{2{,}000}{r}\right) = 0.06\pi r^2 + \frac{60}{r}.$$

The cost for the vertical seam is $0.01h$. From Example 4, we see that h and r are related by the equation

$$h = \frac{1{,}000}{\pi r^2},$$

so the sealing cost is

$$0.01h = 0.01\left(\frac{1{,}000}{\pi r^2}\right)$$

$$= \frac{10}{\pi r^2}.$$

Thus, the total cost is given by the function

$$C(r) = 0.06\pi r^2 + \frac{60}{r} + \frac{10}{\pi r^2}$$

$$\text{or}\quad 0.06\pi r^2 + 60r^{-1} + \frac{10}{\pi}r^{-2}.$$

Then

$$C'(x) = 0.12\pi r - 60r^{-2} - \frac{20}{\pi}r^{-3}$$

$$\text{or}\quad 0.12\pi r - \frac{60}{r^2} - \frac{20}{\pi r^3}.$$

Graph

$$y = 0.12\pi r - \frac{60}{r^2} - \frac{20}{\pi r^3}$$

on a graphing calculator. Since r must be positive, our window should not include negative values of x. A suitable choice for the viewing window is $[0, 10]$ by $[-10, 10]$. From the graph, we find that $C'(x) = 0$ when $x \approx 5.454$.
Thus, the cost is minimized when the radius is about 5.454 cm.
We can find the corresponding height by using the equation

$$h = \frac{1{,}000}{\pi r^2}$$

from Example 4.
If $r = 5.454$,

$$h = \frac{1{,}000}{\pi(5.454)^2} \approx 10.70.$$

To minimize cost, the can should have radius 5.454 cm and height 10.70 cm.

45. Let x be the length of the ladder and let y be the distance from the wall to the bottom of the ladder. Then

$$\cos\theta = \frac{y+2}{x} \quad\text{and}\quad \cot\theta = \frac{y}{9}$$

$$x\cos\theta = y + 2 \quad\text{and}\quad \cdot \quad y = 9\cot\theta.$$

Thus, $x\cos\theta = 9\cot\theta + 2$

$$x = \frac{9\cot\theta + 2}{\cos\theta}$$

$$x = 9\csc\theta + 2\sec\theta.$$

This expression gives the length of the ladder as a function of θ. Find the minimum value of this function.

$$\frac{dx}{d\theta} = -9\csc\theta\cot\theta + 2\sec\theta\tan\theta$$

$$= -9\left(\frac{1}{\sin\theta}\right)\left(\frac{\cos\theta}{\sin\theta}\right) + 2\left(\frac{1}{\cos\theta}\right)\left(\frac{\sin\theta}{\cos\theta}\right)$$

$$= \frac{-9\cos\theta}{\sin^2\theta} + \frac{2\sin\theta}{\cos^2\theta}$$

If $\frac{dx}{d\theta} = 0$, then

$$\frac{2\sin\theta}{\cos^2\theta} = \frac{9\cos\theta}{\sin^2\theta}$$

$$2\sin^3\theta = 9\cos^3\theta$$

$$\frac{\sin^3\theta}{\cos^3\theta} = \frac{9}{2}$$

$$\tan^3\theta = \frac{9}{2}$$

$$\tan\theta = \sqrt[3]{\frac{9}{2}}$$

$$\theta \approx 1.02619 \text{ radians.}$$

If $\theta < 1.02619$, $\frac{dx}{d\theta} < 0$.

If $\theta > 1.02619$, $\frac{dx}{d\theta} > 0$.

Therefore, there is a minimum when $\theta = 1.02619$.
If $\theta = 1.02619$,

$$x \approx 14.383.$$

The minimum length of the ladder is approximately 14.38 ft.

6.3   Implicit Differentiation

1. $4x^2 + 3y^2 = 6$

$$\frac{d}{dx}\left(4x^2 + 3y^2\right) = \frac{d}{dx}\left(6\right)$$

$$\frac{d}{dx}\left(4x^2\right) + \frac{d}{dx}\left(3y^2\right) = \frac{d}{dx}\left(6\right)$$

$$8x + 3 \cdot 2y\frac{dy}{dx} = 0$$

$$6y\frac{dy}{dx} = -8x$$

$$\frac{dy}{dx} = -\frac{4x}{3y}$$

3. $6x^2 + 8xy + y^2 = 6$

$$\frac{d}{dx}\left(6x^2 + 8xy + y^2\right) = \frac{d}{dx}\left(6\right)$$

$$12x + \frac{d}{dx}\left(8xy\right) + \frac{d}{dx}\left(y^2\right) = 0$$

$$12x + 8x\frac{dy}{dx} + y\frac{d}{dx}\left(8x\right) + 2y\frac{dy}{dx} = 0$$

$$12x + 8x\frac{dy}{dx} + 8y + 2y\frac{dy}{dx} = 0$$

$$\left(8x + 2y\right)\frac{dy}{dx} = -12x - 8y$$

$$\frac{dy}{dx} = \frac{-12x - 8y}{8x + 2y}$$

$$\frac{dy}{dx} = \frac{-6x - 4y}{4x + y}$$

5. $x^3 = y^2 + 4$

$$\frac{d}{dx}\left(x^3\right) = \frac{d}{dx}\left(y^2 + 4\right)$$

$$3x^2 = \frac{d}{dx}\left(y^2\right) + \frac{d}{dx}\left(4\right)$$

$$3x^2 = 2y\frac{dy}{dx} + 0$$

$$\frac{3x^2}{2y} = \frac{dy}{dx}$$

7. $3x^2 = \dfrac{2 - y}{2 + y}$

$$\frac{d}{dx}\left(3x^2\right) = \frac{d}{dx}\left(\frac{2 - y}{2 + y}\right)$$

$$6x = \frac{(2 + y)\frac{d}{dx}(2 - y) - (2 - y)\frac{d}{dx}(2 + y)}{(2 + y)^2}$$

$$6x = \frac{(2 + y)\left(-\frac{dy}{dx}\right) - (2 - y)\frac{dy}{dx}}{(2 + y)^2}$$

$$6x = \frac{-4\frac{dy}{dx}}{(2 + y)^2}$$

$$6x(2 + y)^2 = -4\frac{dy}{dx}$$

$$-\frac{3x(2 + y)^2}{2} = \frac{dy}{dx}$$

9. $\sqrt{x} + \sqrt{y} = 4$

$$\frac{d}{dx}\left(x^{1/2} + y^{1/2}\right) = \frac{d}{dx}\,4$$

$$\frac{1}{2}x^{-1/2} + \frac{1}{2}y^{-1/2}\frac{dy}{dx} = 0$$

$$\frac{1}{2}y^{-1/2}\frac{dy}{dx} = -\frac{1}{2}x^{-1/2}$$

$$\frac{dy}{dx} = 2y^{1/2}\left(-\frac{1}{2}x^{-1/2}\right)$$

$$\frac{dy}{dx} = -\frac{y^{1/2}}{x^{1/2}}$$

11. $x^4y^3 + 4x^{3/2} = 6y^{3/2} + 5$

$$\frac{d}{dx}\left(x^4y^3 + 4x^{3/2}\right) = \frac{d}{dx}\left(6y^{3/2} + 5\right)$$

$$\frac{d}{dx}\left(x^4y^3\right) + \frac{d}{dx}\left(4x^{3/2}\right) = \frac{d}{dx}\left(6y^{3/2}\right) + \frac{d}{dx}\left(5\right)$$

$$4x^3y^3 + x^4 \cdot 3y^2\frac{dy}{dx} + 6x^{1/2} = 9y^{1/2}\frac{dy}{dx} + 0$$

$$4x^3y^3 + 6x^{1/2} = 9y^{1/2}\frac{dy}{dx} - 3x^4y^2\frac{dy}{dx}$$

$$4x^3y^3 + 6x^{1/2} = \left(9y^{1/2} - 3x^4y^2\right)\frac{dy}{dx}$$

$$\frac{4x^3y^3 + 6x^{1/2}}{9y^{1/2} - 3x^4y^2} = \frac{dy}{dx}$$

13. $e^{x^2 y} = 5x + 4y + 2$

$$\frac{d}{dx}(e^{x^2 y}) = \frac{d}{dx}(5x + 4y + 2)$$

$$e^{x^2 y} \frac{d}{dx}(x^2 y) = \frac{d}{dx}(5x) + \frac{d}{dx}(4y) + \frac{d}{dx}(2)$$

$$e^{x^2 y}\left(2xy + x^2 \frac{dy}{dx}\right) = 5 + 4\frac{dy}{dx} + 0$$

$$2xye^{x^2 y} + x^2 e^{x^2 y}\frac{dy}{dx} = 5 + 4\frac{dy}{dx}$$

$$x^2 e^{x^2 y}\frac{dy}{dx} - 4\frac{dy}{dx} = 5 - 2xye^{x^2 y}$$

$$(x^2 e^{x^2 y} - 4)\frac{dy}{dx} = 5 - 2xye^{x^2 y}$$

$$\frac{dy}{dx} = \frac{5 - 2xye^{x^2 y}}{x^2 e^{x^2 y} - 4}$$

15. $x + \ln y = x^2 y^3$

$$\frac{d}{dx}(x + \ln y) = \frac{d}{dx}(x^2 y^3)$$

$$1 + \frac{1}{y}\frac{dy}{dx} = 2xy^3 + 3x^2 y^2 \frac{dy}{dx}$$

$$\frac{1}{y}\frac{dy}{dx} - 3x^2 y^2 \frac{dy}{dx} = 2xy^3 - 1$$

$$\left(\frac{1}{y} - 3x^2 y^2\right)\frac{dy}{dx} = 2xy^3 - 1$$

$$\frac{dy}{dx} = \frac{2xy^3 - 1}{\frac{1}{y} - 3x^2 y^2}$$

$$= \frac{y(2xy^3 - 1)}{1 - 3x^2 y^3}$$

17. $\sin(xy) = x$

$$\frac{d}{dx}[\sin(xy)] = \frac{d}{dx}(x)$$

$$\cos(xy)\left(x\frac{dy}{dx} + y\right) = 1$$

$$x\cos(xy)\frac{dy}{dx} + y\cos(xy) = 1$$

$$x\cos(xy)\frac{dy}{dx} = 1 - y\cos(xy)$$

$$\frac{dy}{dx} = \frac{1 - y\cos(xy)}{x\cos(xy)}$$

$$= \frac{1}{x\cos(xy)} - \frac{y\cos(xy)}{x\cos(xy)}$$

$$= \frac{\sec(xy)}{x} - \frac{y}{x}$$

19. $x^2 + y^2 = 25$; tangent at $(-3, 4)$

$$\frac{d}{dx}(x^2 + y^2) = \frac{d}{dx}(25)$$

$$2x + 2y\frac{dy}{dx} = 0$$

$$2y\frac{dy}{dx} = -2x$$

$$\frac{dy}{dx} = -\frac{x}{y}$$

$$m = -\frac{x}{y} = -\frac{-3}{4} = \frac{3}{4}$$

$$y - y_1 = m(x - x_1)$$

$$y - 4 = \frac{3}{4}[x - (-3)]$$

$$4y - 16 = 3x + 9$$

$$4y = 3x + 25$$

21. $x^2 y^2 = 1$; tangent at $(-1, 1)$

$$\frac{d}{dx}(x^2 y^2) = \frac{d}{dx}(1)$$

$$x^2 \frac{d}{dx}(y^2) + y^2 \frac{d}{dx}(x^2) = 0$$

$$x^2(2y)\frac{dy}{dx} + y^2(2x) = 0$$

$$2x^2 y\frac{dy}{dx} = -2xy^2$$

$$\frac{dy}{dx} = \frac{-2xy^2}{2x^2 y} = -\frac{y}{x}$$

$$m = -\frac{y}{x} = -\frac{1}{-1} = 1$$

$$y - 1 = 1[x - (-1)]$$

$$y = x + 1 + 1$$

$$y = x + 2$$

23. $2y^2 - \sqrt{x} = 4$; tangent at $(16, 2)$

$$\frac{d}{dx}(2y^2 - \sqrt{x}) = \frac{d}{dx}(4)$$

$$4y\frac{dy}{dx} - \frac{1}{2}x^{-1/2} = 0$$

$$4y\frac{dy}{dx} = \frac{1}{2x^{1/2}}$$

$$\frac{dy}{dx} = \frac{1}{8yx^{1/2}}$$

$$m = \frac{1}{8yx^{1/2}} = \frac{1}{8(2)(16)^{1/2}}$$

$$= \frac{1}{8(2)(4)} = \frac{1}{64}$$

$$y - 2 = \frac{1}{64}(x - 16)$$

$$64y - 128 = x - 16$$

$$64y = x + 112$$

25. $x - \sin(\pi y) = 1$; tangent at $(1, 0)$

$$\frac{d}{dx}[x - \sin(\pi y) = \frac{d}{dx}(1)$$

$$1 - \cos(\pi y)\frac{d}{dx}(\pi y) = 0$$

$$1 - \pi\cos(\pi y)\frac{dy}{dx} = 0$$

$$-\pi\cos(\pi y)\frac{dy}{dx} = -1$$

$$\frac{dy}{dx} = \frac{1}{\pi\cos(\pi y)}$$

$$m = \frac{1}{\pi\cos(\pi \cdot 0)} = \frac{1}{\pi}$$

$$y - 0 = \frac{1}{\pi}(x - 1)$$

$$y = \frac{x}{\pi} - \frac{1}{\pi}$$

27. $y^3 + xy - y = 8x^4$; $x = 1$

First, find the y-value of the point.

$$y^3 + (1)y - y = 8(1)^4$$
$$y^3 = 8$$
$$y = 2$$

The point is $(1, 2)$.

Find $\frac{dy}{dx}$.

$$3y^2\frac{dy}{dx} + x\frac{dy}{dx} + y - \frac{dy}{dx} = 32x^3$$

$$(3y^2 + x - 1)\frac{dy}{dx} = 32x^3 - y$$

$$\frac{dy}{dx} = \frac{32x^3 - y}{3y^2 + x - 1}$$

At $(1, 2)$,

$$\frac{dy}{dx} = \frac{32(1)^3 - 2}{3(2)^2 + 1 - 1}$$

$$= \frac{30}{12}$$

$$= \frac{5}{2}.$$

$$y - 2 = \frac{5}{2}(x - 1)$$

$$y - 2 = \frac{5}{2}x - \frac{5}{2}$$

$$y = \frac{5}{2}x - \frac{1}{2}$$

29. $y^3 + xy^2 + 1 = x + 2y^2$; $x = 2$

Find the y-value of the point.

$$y^3 + 2y^2 + 1 = 2 + 2y^2$$
$$y^3 + 1 = 2$$
$$y^3 = 1$$
$$y = 1$$

The point is $(2, 1)$.

Find $\frac{dy}{dx}$.

$$3y^2\frac{dy}{dx} + x\,2y\frac{dy}{dx} + y^2 = 1 + 4y\frac{dy}{dx}$$

$$3y^2\frac{dy}{dx} + 2xy\frac{dy}{dx} - 4y\frac{dy}{dx} = 1 - y^2$$

$$(3y^2 + 2xy - 4y)\frac{dy}{dx} = 1 - y^2$$

$$\frac{dy}{dx} = \frac{1 - y^2}{3y^2 + 2xy - 4y}$$

At $(2, 1)$,

$$\frac{dy}{dx} = \frac{1 - 1^2}{3(1)^2 + 2(2)(1) - 4(1)}$$

$$= 0.$$

$$y - 0 = 0(x - 2)$$
$$y = 1$$

31. $2y^3(x-3) + x\sqrt{y} = 3;\ x = 3$

Find the y-value of the point.

$$2y^3(3-3) + 3\sqrt{y} = 3$$
$$3\sqrt{y} = 3$$
$$\sqrt{y} = 1$$
$$y = 1$$

The point is $(3,1)$

Find $\frac{dy}{dx}$.

$$2y^3(1) + 6y^2(x-3)\frac{dy}{dx}$$
$$+ x\left(\frac{1}{2}\right)y^{-1/2}\frac{dy}{dx} + \sqrt{y} = 0$$
$$6y^2(x-3)\frac{dy}{dx} + \frac{x}{2\sqrt{y}}\frac{dy}{dx} = -2y^3 - \sqrt{y}$$
$$\left[6y^2(x-3) + \frac{x}{2\sqrt{y}}\right]\frac{dy}{dx} = -2y^3 - \sqrt{y}$$
$$\frac{dy}{dx} = \frac{-2y^3 - \sqrt{y}}{6y^2(x-3) + \frac{x}{2\sqrt{y}}}$$
$$= \frac{-4y^{7/2} - 2y}{12y^{5/2}(x-3) + x}$$

At $(3,1)$,

$$\frac{dy}{dx} = \frac{-4(1) - 2}{12(1)(3-3) + 3}$$
$$= \frac{-6}{3}$$
$$= -2.$$

$$y - 1 = -2(x-3)$$
$$y - 1 = -2x + 6$$
$$y = -2x + 7$$

33. $x^2 + y^2 = 100$

(a) Lines are tangent at points where $x = 6$. By substituting $x = 6$ in the equation, we find that the points are $(6,8)$ and $(6,-8)$.

$$\frac{d}{dx}(x^2 + y^2) = \frac{d}{dx}(100)$$
$$2x + 2y\frac{dy}{dx} = 0$$
$$2y\frac{dy}{dx} = -2x$$
$$dy = -\frac{x}{y}$$

$$m_1 = -\frac{x}{y} = -\frac{6}{8}$$
$$= -\frac{3}{4}$$
$$m_2 = -\frac{x}{y} = -\frac{6}{-8}$$
$$= \frac{3}{4}$$

First tangent:

$$y - 8 = -\frac{3}{4}(x-6)$$
$$y = -\frac{3}{4}x + \frac{50}{4}$$
$$3x + 4y = 50$$

Second tangent:

$$y - (-8) = \frac{3}{4}(x-6)$$
$$y + 8 = \frac{3}{4}x - \frac{18}{4}$$
$$y = \frac{3}{4}x - \frac{50}{4}$$
$$-3x + 4y = -50$$

(b)

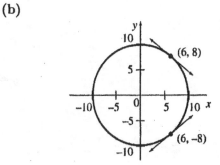

35. $3(x^2 + y^2)^2 = 25(x^2 - y^2)$; $(2, 1)$

Find $\dfrac{dy}{dx}$.

$$6(x^2 + y^2)\frac{d}{dx}(x^2 + y^2) = 25\frac{d}{dx}(x^2 - y^2)$$

$$6(x^2 + y^2)\left(2x + 2y\frac{dy}{dx}\right) = 25\left(2x - 2y\frac{dy}{dx}\right)$$

$$12x^3 + 12x^2y\frac{dy}{dx} + 12xy^2 + 12y^3\frac{dy}{dx} = 50x - 50y\frac{dy}{dx}$$

$$12x^2y\frac{dy}{dx} + 12y^3\frac{dy}{dx} + 50y\frac{dy}{dx} = -12x^3 - 12xy^2 + 50x$$

$$(12x^2y + 12y^3 + 50y)\frac{dy}{dx} = -12x^3 - 12xy^2 + 50x$$

$$\frac{dy}{dx} = \frac{-12x^3 - 12xy^2 + 50x}{12x^2y + 12y^3 + 50y}$$

At $(2, 1)$,

$$\frac{dy}{dx} = \frac{-12(2)^3 - 12(2)(1)^2 + 50(2)}{12(2)^2 + 12(1)^3 + 50(1)}$$

$$= \frac{-20}{110}$$

$$= -\frac{2}{11}$$

$$y - 1 = -\frac{2}{11}(x - 2)$$

$$y - 1 = -\frac{2}{11}x + \frac{4}{11}$$

$$y = -\frac{2}{11}x + \frac{15}{11}$$

37. $2(x^2 + y^2)^2 = 25xy^2$; $(2, 1)$

Find $\dfrac{dy}{dx}$.

$$4(x^2 + y^2)\frac{d}{dx}(x^2 + y^2) = 25\frac{d}{dx}(xy^2)$$

$$4(x^2 + y^2)\left(2x + 2y\frac{dy}{dx}\right) = 25\left(y^2 + 2xy\frac{dy}{dx}\right)$$

$$8x^3 + 8x^2y\frac{dy}{dx} + 8xy^2 + 8y^3\frac{dy}{dx} = 25y^2 + 50xy\frac{dy}{dx}$$

$$8x^2y\frac{dy}{dx} + 8y^3\frac{dy}{dx} - 50xy\frac{dy}{dx} = -8x^3 - 8xy^2 + 25y^2$$

$$(8x^2y + 8y^3 - 50xy)\left(\frac{dy}{dx}\right) = -8x^3 - 8xy^2 + 25y^2$$

$$\frac{dy}{dx} = \frac{-8x^3 - 8xy^2 + 25y^2}{8x^2y + 8y^3 - 50xy}$$

At $(2, 1)$,

$$\frac{dy}{dx} = \frac{-8(2)^3 - 8(2)(1)^2 + 25(1)^2}{8(2)^2(1) + 8(1)^3 - 50(2)(1)}$$

$$= \frac{-55}{-60} = \frac{11}{12}$$

$$y - 1 = \frac{11}{12}(x - 2)$$

$$y - 1 = \frac{11}{12}x - \frac{11}{6}$$

$$y = \frac{11}{12}x - \frac{5}{6}$$

39. $\sqrt{u} + \sqrt{2v + 1} = 5$

$$\frac{du}{dv}\left(\sqrt{u} + \sqrt{2v + 1}\right) = \frac{du}{dv}(5)$$

$$\frac{1}{2}u^{-1/2}\frac{du}{dv} + \frac{1}{2}(2v + 1)^{-1/2}(2) = 0$$

$$\frac{1}{2}u^{-1/2}\frac{du}{dv} = -\frac{1}{(2v + 1)^{1/2}}$$

$$\frac{du}{dv} = -\frac{2u^{1/2}}{(2v + 1)^{1/2}}$$

41. First note that

if $\log R(w) = 1.83 - 0.43\log(w)$

then $R(w) = 10^{1.83 - 0.43\log(w)}$

$\qquad\quad = 10^{1.83}10^{-0.43\log(w)}$

$\qquad\quad = 10^{1.83}\left[10^{\log(w)}\right]^{-0.43}$

$\qquad\quad = 10^{1.83}w^{-0.43}$

(a) $\dfrac{d}{dw}[\log R(w)] = \dfrac{d}{dw}[1.83 - 0.43\log(w)]$

$$\frac{1}{\ln 10}\frac{1}{R(w)}\frac{dR}{dw} = 0 - 0.43\frac{1}{\ln 10}\frac{1}{w}$$

$$\frac{dR}{dw} = -0.43\frac{R(w)}{w}$$

$$= -0.43\frac{10^{1.83}w^{-0.43}}{w}$$

$$\approx -29.07w^{-1.43}$$

(b) $\qquad R(w) = 10^{1.83}w^{-0.43}$

$$\frac{d}{dw}[R(w)] = \frac{d}{dw}[10^{1.83}w^{-0.43}]$$

$$\frac{dR}{dw} = 10^{1.83}(-0.43)w^{-1.43}$$

$$\approx -29.07w^{-1.43}$$

43. $\qquad\qquad xy^a = k$

$$\frac{d}{dx}(xy^a) = \frac{d}{dx}(k)$$

$$x\frac{d}{dx}(y^a) + y^a(1) = 0$$

$$x\left(ay^{a-1}\frac{dy}{dx}\right) + y^a = 0$$

$$axy^{a-1}\frac{dy}{dx} = -y^a$$

$$\frac{dy}{dx} = -\frac{y^a}{axy^{a-1}}$$

$$\frac{dy}{dx} = -\frac{y}{ax}$$

45. $2s^2 + \sqrt{st} - 4 = 3t$

$$4s\frac{ds}{dt} + \frac{1}{2}(st)^{-1/2}\left(s + t\frac{ds}{dt}\right) = 3$$

$$4s\frac{ds}{dt} + \frac{s + t\frac{ds}{dt}}{2\sqrt{st}} = 3$$

$$\frac{8s(\sqrt{st})\frac{ds}{dt} + s + t\frac{ds}{dt}}{2\sqrt{st}} = 3$$

$$\frac{(8s\sqrt{st} + t)\frac{ds}{dt} + s}{2\sqrt{st}} = 3$$

$$(8s\sqrt{st} + t)\frac{ds}{dt} = 6\sqrt{st} - s$$

$$\frac{ds}{dt} = \frac{-s + 6\sqrt{st}}{8s\sqrt{st} + t}$$

6.4   Related Rates

1. $y^2 - 5x^2 = -1$; $\dfrac{dx}{dt} = -3$, $x = 1$, $y = 2$

$$2y\frac{dy}{dt} - 10x\frac{dx}{dt} = 0$$

$$y\frac{dy}{dt} = 5x\frac{dx}{dt}$$

$$2\frac{dy}{dt} = 5(-3)$$

$$\frac{dy}{dt} = -\frac{15}{2}$$

3. $xy - 5x + 2y^3 = -70$; $\dfrac{dx}{dt} = -5$, $x = 2$, $y = -3$

$$x\frac{dy}{dt} + y\frac{dx}{dt} - 5\frac{dx}{dt} + 6y^2\frac{dy}{dt} = 0$$

$$(x + 6y^2)\frac{dy}{dt} + (y - 5)\frac{dx}{dt} = 0$$

$$(x + 6y^2)\frac{dy}{dt} = (5 - y)\frac{dx}{dt}$$

$$\frac{dy}{dt} = \frac{(5 - y)\frac{dx}{dt}}{x + 6y^2}$$

$$= \frac{[5 - (-3)](-5)}{2 + 6(-3)^2}$$

$$= \frac{-40}{56} = -\frac{5}{7}$$

5. $\dfrac{x^2 + y}{x - y} = 9; \ \dfrac{dx}{dt} = 2, \ x = 4, \ y = 2$

$$\frac{(x-y)\left(2x\frac{dx}{dt} + \frac{dy}{dt}\right) - (x^2 + y)\left(\frac{dx}{dt} - \frac{dy}{dt}\right)}{(x - y)^2} = 0$$

$$\frac{2x(x-y)\frac{dx}{dt} + (x-y)\frac{dy}{dt} - (x^2+y)\frac{dx}{dt} + (x^2+y)\frac{dy}{dt}}{(x - y)^2} = 0$$

$$[2x(x-y) - (x^2+y)]\frac{dx}{dt} + [(x-y) + (x^2+y)]\frac{dy}{dt} = 0$$

$$\frac{dy}{dt} = \frac{[(x^2 + y) - 2x(x - y)]\frac{dx}{dt}}{(x - y) + (x^2 + y)}$$

$$\frac{dy}{dt} = \frac{(-x^2 + y + 2xy)\frac{dx}{dt}}{x + x^2}$$

$$= \frac{[-(4)^2 + 2 + 2(4)(2)](2)}{4 + 4^2}$$

$$= \frac{4}{20} = \frac{1}{5}$$

7. $xe^y = 1 + \ln x; \ \dfrac{dx}{dt} = 6, \ x = 1, y = 0$

$$e^y\frac{dx}{dt} + xe^y\frac{dy}{dt} = 0 + \frac{1}{x}\frac{dx}{dt}$$

$$xe^y\frac{dy}{dt} = \left(\frac{1}{x} - e^y\right)\frac{dx}{dt}$$

$$\frac{dy}{dt} = \frac{\left(\frac{1}{x} - e^y\right)\frac{dx}{dt}}{xe^y}$$

$$= \frac{(1 - xe^y)\frac{dx}{dt}}{x^2 e^y}$$

$$= \frac{[1 - (1)e^0](6)}{1^2 e^0}$$

$$= \frac{0}{1} = 0$$

9. $V = k(R^2 - r^2); \ k = 555.6, \ R = 0.02$ mm, $\frac{dR}{dt} = 0.003$ mm per minute; r is constant.

$$V = k(R^2 - r^2)$$

$$V = 555.6(R^2 - r^2)$$

$$\frac{dV}{dt} = 555.6\left(2R\frac{dR}{dt} - 0\right)$$

$$= 555.6(2)(0.02)(0.003)$$

$$= 0.067 \text{ mm/min}$$

11. $b = 0.22w^{0.87}$

$$\frac{db}{dt} = 0.22(0.87)w^{-0.13}\frac{dw}{dt}$$

$$= 0.1914w^{-0.13}\frac{dw}{dt}$$

$$\frac{dw}{dt} = \frac{w^{0.13}}{0.1914}\frac{db}{dt}$$

$$= \frac{25^{0.13}}{0.1914}(0.25)$$

$$\approx 1.9849$$

The rate of change of the total weight is about 1.9849 g/day.

13. $m = 85.65w^{0.54}$

(a) $\dfrac{dm}{dt} = 85.65(0.54)w^{-0.46}\dfrac{dw}{dt}$

$$= 46.251w^{-0.46}\frac{dw}{dt}$$

(b) $\dfrac{dm}{dt} = 46.251(0.25)^{-0.46}(0.01)$

$$\approx 0.875$$

The rate of change of the average daily metabolic rate is about 0.875 kcal/day^2.

15. $E = 26.5w^{-0.34}$

$$\frac{dE}{dt} = 26.5(-0.34)w^{-1.34}\frac{dw}{dt}$$

$$= -9.01w^{-1.34}\frac{dw}{dt}$$

$$= -9.01(5)^{-1.34}(0.05)$$

$$\approx -0.0521$$

The rate of change of the energy expenditure is about -0.0521 kcal/kg/km/day.

17. $C = \dfrac{1}{10}(T - 60)^2 + 100$

$\dfrac{dC}{dt} = \dfrac{1}{5}(T - 60)\dfrac{dT}{dt}$

If $T = 76°$ and $\dfrac{dT}{dt} = 8$,

$\dfrac{dC}{dt} = \dfrac{1}{5}(76 - 60)(8) = \dfrac{1}{5}(16)(8)$

$= 25.6.$

The crime rate is rising at the rate of 25.6 crimes/month.

19. $W(t) = \dfrac{-0.02t^2 + t}{t + 1}$

$\dfrac{dW}{dt} = \dfrac{(-0.04t + 1)(t + 1) - (1)(-0.02t^2 + t)}{(t + 1)^2}$

If $t = 5$,

$\dfrac{dW}{dt} = \dfrac{(-0.2 + 1)(6) - (-0.5 + 5)}{6^2}$

$= \dfrac{4.8 - 4.5}{36}$

$= 0.008.$

21. (a) Let $x =$ the distance one car travels west;
$y =$ the distance the other car travels north;
$s =$ the distance between the two cars.

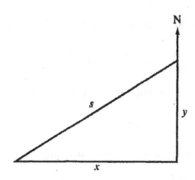

$s^2 = x^2 + y^2$

$2s\dfrac{ds}{dt} = 2x\dfrac{dx}{dt} + 2y\dfrac{dy}{dt}$

$s\dfrac{ds}{dt} = x\dfrac{dx}{dt} + y\dfrac{dy}{dt}$

Use $d = rt$ to find x and y.

$x = (40)(2) = 80$ mi
$y = (30)(2) = 60$ mi
$s = \sqrt{x^2 + y^2}$
$= \sqrt{(80)^2 + (60)^2}$
$= 100$

The distance between the cars after 2 hours is 100 mi.

$\dfrac{dx}{dt} = 40$ mph and $\dfrac{dy}{dt} = 30$ mph

$(100)\dfrac{ds}{dt} = (80)(40) + (60)(30)$

$\dfrac{ds}{dt} = \dfrac{5,000}{100} = 50$

The distance between the two cars is changing at the rate of 50 mph.

(b) From part (a), we have

$$s\dfrac{ds}{dt} = x\dfrac{dx}{dt} + y\dfrac{dy}{dt}$$

Use $d = rt$ to find x and y. When the second car has traveled 1 hour, the first car has traveled 2 hours.

$x = 40(1) = 40$ mi
$y = 30(2) = 60$ mi
$s = \sqrt{x^2 + y^2}$
$= \sqrt{(40)^2 + (60)^2}$
$= 72.11$

The distance between the cars after the second car has traveled 1 hour is about 72.11 mi.

$72.11\dfrac{ds}{dt} = (40)(40) + (60)(30)$

$\dfrac{ds}{dt} = \dfrac{3400}{72.11}$

≈ 47.15

The distance between the two cars is changing at a rate of about 47.15 mph.

23. $V = \frac{4}{3}\pi r^3$, $r = 4$ in, and $\dfrac{dr}{dt} = -\frac{1}{4}$ in/hr

$\dfrac{dV}{dt} = 4\pi r^2 \dfrac{dr}{dt}$

$= 4\pi(4)^2\left(-\dfrac{1}{4}\right)$

$= -16\pi$ in^3/hr

25. Let $y=$ the length of the man's shadow;

$x=$ the distance of the man from the lamp post;

$h=$ the height of the lamp post.

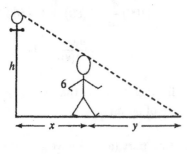

$$\frac{dx}{dt} = 50 \text{ ft/min}$$

Find $\frac{dy}{dt}$ when $x=25$ ft.

Now $\frac{h}{x+y} = \frac{6}{y}$, by similar triangles.

When $x=8$, $y=10$,

$$\frac{h}{18} = \frac{6}{10}$$
$$h = 10.8.$$

$$\frac{10.8}{x+y} = \frac{6}{y},$$
$$10.8y = 6x + 6y$$
$$4.8y = 6x$$
$$y = 1.25x$$

$$\frac{dy}{dt} = 1.25\frac{dx}{dt}$$
$$= 1.25(50)$$
$$\frac{dy}{dt} = 62.5$$

The length of the shadow is increasing at the rate of 62.5 ft/min.

27. Let $x=$ the distance from the docks

$s=$ the length of the rope.

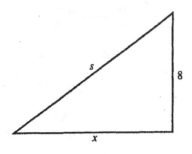

$$\frac{ds}{dt} = 1 \text{ ft/sec}$$
$$s^2 = x^2 + (8)^2$$
$$2s\frac{ds}{dt} = 2x\frac{dx}{dt} + 0$$
$$s\frac{ds}{dt} = x\frac{dx}{dt}$$

If $x=8$,

$$s = \sqrt{(8)^2 + (8)^2} = \sqrt{128} = 8\sqrt{2}.$$

Then,

$$8\sqrt{2}(1) = 8\frac{dx}{dt}$$
$$\frac{dx}{dt} = \sqrt{2} \approx 1.41$$

The boat is approaching the deck at $\sqrt{2} \approx 1.41$ ft/sec.

29. $C = 0.1x^2 + 10{,}000$; $x = 100$, $\frac{dx}{dt} = 10$

$$\frac{dC}{dt} = 0.1(2x)\frac{dx}{dt}$$
$$= 0.1(200)(10)$$
$$= 200$$

The cost is changing at a rate of \$200 per month.

31. $R = 50x - 0.4x^2$, $C = 5x + 15$; $x = 200$, $\frac{dx}{dt} = 50$

(a) $\frac{dR}{dt} = 50\frac{dx}{dt} - 0.8x\frac{dx}{dt}$

$$= 50(50) - 0.8(200)(50)$$
$$= 2{,}500 - 8{,}000$$
$$= -5{,}500$$

Revenue is decreasing at a rate of \$5,500 per day.

(b) $\frac{dC}{dt} = (5)\frac{dx}{dt}$

$$= (5)(50)$$
$$= 250$$

Cost is increasing at a rate of \$250 per day.

(c) $P = R - C$

$$\frac{dP}{dt} = \frac{dR}{dt} - \frac{dC}{dt}$$
$$= -5{,}500 - 250$$
$$= -5{,}750$$

Profit is decreasing at a rate of \$5,750 per day.

6.5 Differentials: Linear Approximation

1. $y = 2x^2 - 5x$; $x = -2$, $\Delta x = 0.2$

$dy = (4x - 5)\,dx$

$\quad \approx (4x - 5)\,\Delta x$

$\quad = [4(-2) - 5](0.2)$

$\quad = -2.6$

3. $y = x^3 - 2x^2 + 3$, $x = 1$, $\Delta x = -0.1$

$dy = (3x^2 - 4x)\,dx$

$\quad \approx (3x^2 - 4x)\,\Delta x$

$\quad = [3(1^2) - 4(1)](-0.1)$

$\quad = 0.1$

5. $y = \sqrt{3x}$, $x = 1$, $\Delta x = 0.15$

$dy = \sqrt{3}\left(\dfrac{1}{2}x^{-1/2}\right)dx$

$\quad \approx \dfrac{\sqrt{3}}{2\sqrt{x}}\,\Delta x$

$\quad = \dfrac{1.73}{2}(0.15)$

$\quad = 0.130$

7. $y = \dfrac{2x - 5}{x + 1}$; $x = 2$, $\Delta x = -0.03$

$dy = \dfrac{(x + 1)(2) - (2x - 5)(1)}{(x + 1)^2}\,dx$

$\quad = \dfrac{7}{(x + 1)^2}\,dx$

$\quad = \dfrac{7}{(x + 1)^2}\,\Delta x$

$\quad = \dfrac{7}{(2 + 1)^2}(-0.03)$

$\quad = -0.023$

9. $\sqrt{145}$

We know $\sqrt{144} = 12$, so $f(x) = \sqrt{x}$, $x = 144$, $dx = 1$.

$\dfrac{dy}{dx} = \dfrac{1}{2}x^{-1/2}$

$dy = \dfrac{1}{2\sqrt{x}}\,dx$

$dy = \dfrac{1}{2\sqrt{144}}(1) = \dfrac{1}{24}$

$\sqrt{145} \approx f(x) + dy = 12 + \dfrac{1}{24}$

$\quad \approx 12.0417$

By calculator, $\sqrt{145} \approx 12.0416$.
The difference is $|12.0417 - 12.0416| = 0.0001$.

11. $\sqrt{0.99}$

We know $\sqrt{1} = 1$, so $f(x) = \sqrt{x}$, $x = 1$, $dx = -0.01$.

$\dfrac{dy}{dx} = \dfrac{1}{2}x^{-1/2}$

$dy = \dfrac{1}{2\sqrt{x}}\,dx$

$dy = \dfrac{1}{2\sqrt{1}}(-0.01) = -0.005$

$\sqrt{0.99} \approx f(x) + dy = 1 - 0.005$

$\quad = 0.995$

By calculator, $\sqrt{0.99} \approx 0.9950$.
The difference is $|0.995 - 0.9950| = 0$.

13. $e^{0.01}$

We know $e^0 = 1$, so $f(x) = e^x$, $x = 0$, $dx = 0.01$.

$\dfrac{dy}{dx} = e^x$

$dy = e^x\,dx$

$dy = e^0(0.01) = 0.01$

$e^{0.01} \approx f(x) + dy = 1 + 0.01 = 1.01$

By calculator, $e^{0.01} \approx 1.0101$.
The difference is $|1.01 - 1.0101| = 0.0001$.

15. $\ln 1.05$

We know $\ln 1 = 0$, so $f(x) = \ln x$, $x = 1$, $dx = 0.05$.

$\dfrac{dy}{dx} = \dfrac{1}{x}$

$dy = \dfrac{1}{x}\,dx$

$dy = \dfrac{1}{1}(0.05) = 0.05$

$\ln 1.05 \approx f(x) + dy = 0 + 0.05 = 0.05$

By calculator, $\ln 1.05 \approx 0.0488$.
The difference is $|0.05 - 0.0488| = 0.0012$.

17. $\sin(0.03)$

We know $\sin(0) = 0$, so $f(x) = \sin(x)$, $x = 0$, $dx = 0.03$.

$\dfrac{dy}{dx} = \cos(x)$

$dy = \cos(x)\,dx$

$dy = \cos(0)(0.03) = 1(0.03) = 0.03$

$\sin(0.03) \approx f(x) + dy = 0 + 0.03 = 0.03$

By calculator, $\sin(0.03) \approx 0.0300$.

The difference is $|0.03 - 0.0300| = 0$.

19. (a) $A(x) = y = \dfrac{-7}{480}x^3 + \dfrac{127}{120}x$

Let $x = 3$, $dx = 0.5$.

$$\frac{dy}{dx} = \frac{-21}{480}x^2 + \frac{127}{120}$$

$$dy = \left(\frac{-21}{480}x^2 + \frac{127}{120}\right)dx = \left(\frac{-21}{480}(9) + \frac{127}{120}\right)(0.5) = 0.3323$$

The alcohol concentration increases by about 0.33 tenths of a percent.

(b) Let $x = 6$, $dx = 0.25$.

$$dy = \left(\frac{-21}{480}(36) + \frac{127}{120}\right)(0.25) = -0.1292$$

The alcohol concentration decreases by about 0.13 tenths of a percent.

21. $P(x) = \dfrac{25x}{8 + x^2}$

$$dP = \frac{(8 + x^2)(25) - 25x(2x)}{(8 + x^2)^2}\,dx = \frac{(8 + x^2)(25) - 25x(2x)}{(8 + x^2)^2}\,\Delta x$$

(a) $x = 2$, $\Delta x = 0.5$

$$dP = \frac{[(8 + 4)(25) - (25)(2)(4)](0.5)}{(8 + 4)^2} = 0.347 \text{ million}$$

(b) $x = 3$, $\Delta x = 0.25$

$$dP = \frac{[(8 + 9)(25) - 25(3)(6)]0.25}{(8 + 9)^2} \approx -0.022 \text{ million}$$

23. r changes from 14 mm to 16 mm, so $\Delta r = 2$.

$$V = \frac{4}{3}\pi r^3$$

$$dV = \frac{4}{3}(3)\pi r^2\,dr$$

$$\Delta V \approx 4\pi r^2\,\Delta r = 4\pi(14)^2(2) \doteq 1{,}568\pi \text{ mm}^3$$

25. r increases from 20 mm to 22 mm, so $\Delta r = 2$.

$$A = \pi r^2$$
$$dA = 2\pi r\,dr$$
$$\Delta A \approx 2\pi r\,\Delta r = 2\pi(20)(2) \doteq 80\pi \text{ mm}^2$$

27. $W(t) = -3.5 + 197.5e^{-e^{(-0.01394(t - 108.4))}}$

(a) $dW = 197.5e^{-e^{(-0.01394(t - 108.4))}}(-1)e^{(-0.01394(t - 108.4))}(-0.01394)dt$

$\qquad = 2.75315e^{-e^{(-0.01394(t - 108.4))}}e^{(-0.01394(t - 108.4))}dt$

We are given $t = 80$ and $dt = 90 - 80 = 10$.

$$dW \approx 9.258$$

The pig will gain about 9.3 kg.

(b) The actual weight gain is calculated as

$$W(90) - W(80) \approx 50.736 - 41.202 = 9.534$$

or about 9.5 kg.

29. $r = 3$ cm, $\Delta r = -0.2$ cm

$$V = \frac{4}{3}\pi r^3$$

$$dV = 4\pi r^2\, dr$$
$$\Delta V \approx 4\pi r^2\, \Delta r$$
$$= 4\pi(9)(-0.2)$$
$$= -7.2\pi \text{ cm}^3$$

31. $r = 4.87$ in, $\Delta r = \pm .040$

$$A = \pi r^2$$
$$dA = 2\pi r\, dr$$
$$\Delta A \approx 2\pi r\, \Delta r$$
$$= 2\pi(4.87)(\pm .040)$$
$$= \pm 1.224 \text{ in}^2$$

33. $h = 7.284$ in, $r = 1.09 \pm 0.007$ in

$$V = \frac{1}{3}\pi r^2 h$$

$$dV = \frac{2}{3}\pi r h\, dr$$

$$\Delta V \approx \frac{2}{3}\pi r h\, \Delta r$$

$$= \frac{2}{3}\pi(1.09)(7.284)(0.007)$$

$$= \pm 0.116 \text{ in}^3$$

35. If a cube is given a coating 0.1 in thick, each edge increases in length by twice that amount, or 0.2 in because there is a face at both ends of the edge.

$V = x^3$, $x = 4$, $\Delta x = 0.2$

$$dV = 3x^2\, dx$$
$$\Delta V \approx 3x^2\, \Delta x = 3(4^2)(0.2) = 9.6$$

For 1,000 cubes $9.6(1,000) = 9,600$ in^3 of coating should be ordered.

Chapter 6 Review Exercises

1. $f(x) = -x^2 + 5x + 1$; $[1, 4]$
$f'(x) = -2x + 5 = 0$ when $x = \frac{5}{2}$.

$$f(1) = 5$$
$$f\left(\frac{5}{2}\right) = \frac{29}{4}$$
$$f(4) = 5$$

Absolute maximum of $\frac{29}{4}$ at $\frac{5}{2}$; absolute minimum of 5 at 1 and 4

3. $f(x) = x^3 + 2x^2 - 15x + 3$; $[-4, 2]$
$f'(x) = 3x^2 + 4x - 15 = 0$ when
$$(3x - 5)(x + 3) = 0$$

$$x = \frac{5}{3} \quad \text{or} \quad x = -3.$$

$$f(-4) = 31$$
$$f(-3) = 39$$
$$f\left(\frac{5}{3}\right) = -\frac{319}{27}$$
$$f(2) = -11$$

Absolute maximum of 39 at -3; absolute minimum of $-\frac{319}{27}$ at $\frac{5}{3}$

9. $x^2y^3 + 4xy = 2$

$$\frac{d}{dx}\left(x^2y^3 + 4xy\right) = \frac{d}{dx}(2)$$

$$2xy^3 + 3y^2\left(\frac{dy}{dx}\right)x^2 + 4y + 4x\frac{dy}{dx} = 0$$

$$(3x^2y^2 + 4x)\frac{dy}{dx} = -2xy^3 - 4y$$

$$\frac{dy}{dx} = \frac{-2xy^2 - 4y}{3x^2y^2 + 4x}$$

11. $9\sqrt{x} + 4y^3 = \frac{2}{x}$

$$\frac{d}{dx}\left(9\sqrt{x} + 4y^3\right) = \frac{d}{dx}\left(\frac{2}{x}\right)$$

$$\frac{9}{2}x^{-1/2} + 12y^2\frac{dy}{dx} = \frac{-2}{x^2}$$

$$12y^2\frac{dy}{dx} = \frac{-2}{x^2} - \frac{9x^{-1/2}}{2}$$

$$12y^2\frac{dy}{dx} = \frac{-4 - 9x^{3/2}}{2x^2}$$

$$\frac{dy}{dx} = \frac{-4 - 9x^{3/2}}{24x^2y^2}$$

13. $\dfrac{x+2y}{x-3y} = y^{1/2}$

$$x + 2y = y^{1/2}(x - 3y)$$

$$\frac{d}{dx}(x + 2y) = \frac{d}{dx}[y^{1/2}(x - 3y)]$$

$$1 + 2\frac{dy}{dx} = y^{1/2}\left(1 - 3\frac{dy}{dx}\right)$$

$$+ \frac{1}{2}(x - 3y)y^{-1/2}\frac{dy}{dx}$$

$$1 + 2\frac{dy}{dx} = y^{1/2} - 3y^{1/2}\frac{dy}{dx} + \frac{1}{2}xy^{-1/2}\frac{dy}{dx}$$

$$- \frac{3}{2}y^{1/2}\frac{dy}{dx}$$

$$\left(2 + 3y^{1/2} - \frac{1}{2}xy^{-1/2} + \frac{3}{2}y^{1/2}\right)\frac{dy}{dx} = y^{1/2} - 1$$

$$\frac{2y^{1/2}\left(2 + \frac{9}{2}y^{1/2} - \frac{1}{2}xy^{-1/2}\right)}{2y^{1/2}}\frac{dy}{dx} = y^{1/2} - 1$$

$$\left(\frac{4y^{1/2} + 9y - x}{2y^{1/2}}\right)\frac{dy}{dx} = y^{1/2} - 1$$

$$\frac{dy}{dx} = \frac{2y - 2y^{1/2}}{4y^{1/2} + 9y - x}$$

15. $\sqrt{2x} - 4yx = -22$, tangent line at $(2,3)$

$$\frac{d}{dx}(\sqrt{2x} - 4yx) = \frac{d}{dx}(-22)$$

$$\frac{1}{2}(2x)^{-1/2}(2) - 4y - 4x\frac{dy}{dx} = 0$$

$$(2x)^{-1/2} - 4y = 4x\frac{dy}{dx}$$

$$\frac{dy}{dx} = \frac{(2x)^{-1/2} - 4y}{4x} = \frac{\frac{1}{\sqrt{2x}} - 4y}{4x}$$

At $(2,3)$,

$$m = \frac{\frac{1}{\sqrt{2\cdot2}} - 4\cdot3}{4\cdot2} = \frac{\frac{1}{2} - 12}{8}$$

$$= -\frac{23}{16}.$$

The equation of the tangent line is

$$y - y_1 = m(x - x_1)$$

$$y - 3 = -\frac{23}{16}(x - 2)$$

$$23x + 16y = 94.$$

19. $y = \dfrac{9 - 4x}{3 + 2x}$; $\dfrac{dx}{dt} = -1$, $x = -3$

$$\frac{dy}{dt} = \frac{(-4)(3 + 2x) - (2)(9 - 4x)}{(3 + 2x)^2}\frac{dx}{dt}$$

$$= \frac{-30}{(3 + 2x)^2}\frac{dx}{dt}$$

$$= \frac{-30}{[3 + 2(-3)]^2}(-1) = \frac{30}{9} = \frac{10}{3}$$

21. $\dfrac{x^2 + 5y}{x - 2y} = 2$; $\dfrac{dx}{dt} = 1$, $x = 2$, $y = 0$

$$x^2 + 5y = 2(x - 2y)$$

$$x^2 + 5y = 2x - 4y$$

$$9y = -x^2 + 2x$$

$$y = \frac{1}{9}(-x^2 + 2x)$$

$$= -\frac{1}{9}x^2 + \frac{2}{9}x$$

$$\frac{dy}{dt} = \left(-\frac{2}{9}x + \frac{2}{9}\right)\frac{dx}{dt}$$

$$= \left[\left(-\frac{2}{9}\right)(2) + \frac{2}{9}\right](1)$$

$$= -\frac{4}{9} + \frac{2}{9} = -\frac{2}{9}$$

23. $y = 8 - x^2 + x^3$, $x = -1$, $\Delta x = 0.02$

$$dy = (-2x + 3x^2)\,dx$$

$$\approx (-2x + 3x^2)\,\Delta x$$

$$= [-2(-1) + 3(-1)^2](0.02)$$

$$= 0.1$$

25. $y = \sin(x)$; $x = \dfrac{\pi}{3}$, $\Delta x = -0.01$

$$dy = \cos(x)\,dx$$

$$\approx \cos\left(\frac{\pi}{3}\right)(-0.01)$$

$$= \frac{1}{2}(-0.01)$$

$$= -0.005$$

27. $-12x + x^3 + y + y^2 = 4$

$$\frac{dy}{dx}(-12x + x^3 + y + y^2) = \frac{d}{dx}(4)$$

$$-12 + 3x^2 + \frac{dy}{dx} + 2y\frac{dy}{dx} = 0$$

$$(1 + 2y)\frac{dy}{dx} = 12 - 3x^2$$

$$\frac{dy}{dx} = \frac{12 - 3x^2}{1 + 2y}$$

(a) If $\dfrac{dy}{dx} = 0,$

$$12 - 3x^2 = 0$$
$$12 = 3x^2$$
$$\pm 2 = x.$$

$x = 2:$

$$-24 + 8 + y + y^2 = 4$$
$$y + y^2 = 20$$
$$y^2 + y - 20 = 0$$
$$(y + 5)(y - 4) = 0$$
$$y = -5 \text{ or } y = 4$$

$(2, -5)$ and $(2, 4)$ are critical points.

$x = -2:$

$$24 - 8 + y + y^2 = 4$$
$$y + y^2 = -12$$
$$y^2 + y + 12 = 0$$
$$y = \frac{-1 \pm \sqrt{1^2 - 48}}{2}$$

This leads to imaginary roots.
$x = -2$ does not produce critical points.

(b)

x	y_1	y_2
1.9	-4.99	3.99
2	-5	4
2.1	-4.99	3.99

The point $(2, -5)$ is a relative minimum.
The point $(2, 4)$ is a relative maximum.

(c) There is no absolute maximum or minimum for x or y.

29. $A = \pi r^2;\ \dfrac{dr}{dt} = 4$ ft/min, $r = 7$ ft

$$\frac{dA}{dt} = 2\pi r \frac{dr}{dt}$$
$$\frac{dA}{dt} = 2\pi(7)(4)$$
$$\frac{dA}{dt} = 56\pi$$

The rate of change of the area is 56π ft²/min.

31. To find the relative humidity is minimized, use a graphing calculator to locate any extreme points on the graph. A suitable window is $[15, 55]$ by $[70, 85]$. Critical numbers are found at about 19.82 and 41.09.

T	$R(T)$
15	76.76
19.82	75.82
41.09	81.11
55	71.35

The minimum relative humidity occurs at 55°C.

33. (a)

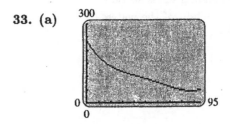

(b) To find where the maximum and minimum numbers occur, use a graphing calculator to locate any extreme points on the graph. One critical number is formed at about 87.78.

t	$P(t)$
0	237.09
87.78	43.56
95	48.66

The maximum number of polygons is about 237 at birth. The minimum number is about 44.

35. Let $x =$ the distance from the base of the ladder to the building;

$y =$ the height on the building at the top of the ladder.

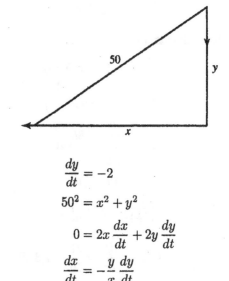

$$\frac{dy}{dt} = -2$$
$$50^2 = x^2 + y^2$$
$$0 = 2x\frac{dx}{dt} + 2y\frac{dy}{dt}$$
$$\frac{dx}{dt} = -\frac{y}{x}\frac{dy}{dt}$$

When $x = 30$, $y = \sqrt{2{,}500 - (30)^2} = 40.$
So

$$\frac{dx}{dt} = \frac{-40}{30}(-2) = \frac{80}{30} = \frac{8}{3}$$

The base of the ladder is slipping away from the building at a rate of $\frac{8}{3}$ ft/min.

37. Let $x =$ one-half the width of the
    triangular cross section;
    $h =$ the height of the water;
    $V =$ the volume of the water.

$$\frac{dV}{dt} = 3.5 \ ft^3/\text{min}$$

Find $\frac{dV}{dt}$ when $h = \frac{1}{3}$.

$$V = \begin{pmatrix} \text{Area of} \\ \text{triangular} \\ \text{side} \end{pmatrix} (\text{length})$$

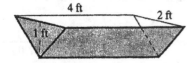

Area of triangular cross section

$$= \frac{1}{2}(\text{base})(\text{altitude})$$

$$= \frac{1}{2}(2x)(h) = xh$$

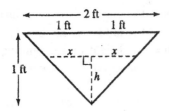

By similar triangles, $\frac{2x}{h} = \frac{2}{1}$, so $x = h$.

$$V = (xh)(4)$$
$$= h^2 \cdot 4$$
$$= 4h^2$$

$$\frac{dV}{dt} = 8h\frac{dh}{dt}$$

$$\frac{1}{8h} \cdot \frac{dV}{dt} = \frac{dh}{dt}$$

$$\frac{1}{8\left(\frac{1}{3}\right)}(3.5) = \frac{dh}{dt}$$

$$\frac{dh}{dt} = \frac{21}{16} = 1.3125$$

The depth of water is changing at the rate of 1.3125 ft/min.

39. $A = s^2$; $s = 9.2$, $\Delta s = \pm 0.04$
    $ds = 2s\,ds$
    $\Delta A \approx 2s\,\Delta s$
    $\quad = 2(9.2)(\pm 0.04)$
    $\quad = \pm 0.736 \text{ in}^2$

41. We need to minimize y. Note that $x > 0$.

$$\frac{dy}{dx} = \frac{x}{8} - \frac{2}{x}$$

Set the derivative equal to 0.

$$\frac{x}{8} - \frac{2}{x} = 0$$

$$\frac{x}{8} = \frac{2}{x}$$

$$x^2 = 16$$

$$x = 4$$

Since $\lim\limits_{x \to 0} y = \infty$, $\lim\limits_{x \to \infty} y = \infty$, and $x = 4$ is the only critical value in $(0, \infty)$, $x = 4$ produces a minimum value.

$$y = \frac{4^2}{16} - 2\ln 4 + \frac{1}{4} + 2\ln 6$$

$$= 1.25 + 2(\ln 6 - \ln 4)$$

$$= 1.25 + 2\ln 1.5$$

The y coordinate of the Southern most point of the second boat's path is $1.25 + 2\ln 1.5$.

43. Let $x =$ the length and width of a side
    of the base;
    $h =$ the height.

The volume is 32 m^3; the base is square and there is no top. Find the height, length, and width for minimum surface area.

$$\text{Volume} = x^2h$$
$$x^2h = 32$$

$$h = \frac{32}{x^2}$$

$$\text{Surface area} = x^2 + 4xh$$

$$A = x^2 + 4x\left(\frac{32}{x^2}\right)$$

$$= x^2 + 128x^{-1}$$
$$A' = 2x - 128x^{-2}$$

If $A' = 0$,

$$\frac{2x^3 - 128}{x^2} = 0$$

$$x^3 = 64$$
$$x = 4.$$

$$A''(x) = 2 + 2(128)x^{-3}$$
$$A''(4) = 6 > 0$$

The minimum is at $x = 4$, where

$$h = \frac{32}{4^2} = 2.$$

The dimensions are 2 m by 4 m by 4 m.

45. In Exercises 43 and 44, we found that the cost of the aluminum to make the can is $0.06\pi r^2 + \frac{60}{r}$, the cost to seal the top and bottom is $0.08\pi r$, and the cost to seal the vertical seam is $\frac{10}{\pi r^2}$.

Thus, the total cost is now given by the function

$$C(r) = 0.06\pi r^2 + \frac{60}{r} + 0.08\pi r + \frac{10}{\pi r^2}$$

$$\text{or} \quad 0.06\pi r^2 + 60r^{-1} + 0.08\pi r + \frac{10}{\pi}r^{-2}.$$

Then

$$C'(r) = 0.12\pi r - 60r^{-2} + 0.08\pi - \frac{20}{\pi}r^{-3}$$

$$\text{or} \quad 0.12\pi r - \frac{60}{r^2} + 0.08\pi - \frac{20}{\pi r^3}.$$

Graph

$$y = 0.12\pi r - \frac{60}{r^2} + 0.08\pi - \frac{20}{\pi r^3}$$

on a graphing calculator. A suitable choice for the viewing window is $[0, 10]$ by $[-10, 10]$. From the graph, we find that $C'(x) = 0$ when $x \approx 5.242$. Thus, the cost is minimized when the radius is about 5.242 cm.

To find the corresponding height, use the equation

$$h = \frac{1,000}{\pi r^2}$$

from Example 4.

If $r = 5.242$,

$$h = \frac{1,000}{\pi(5.242)^2} \approx 11.58.$$

To minimize cost, the can should have radius 5.242 cm and height 11.58 cm.

Chapter 6 Test

[6.1]

Find the locations of any absolute extrema for the functions with graphs as follows.

1.

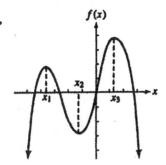

2.

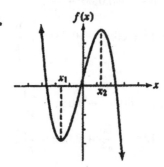

Find the locations of all absolute extrema for the functions defined as follows, with the specified domains.

3. $f(x) = x^3 - 12x$; $[0, 4]$ 4. $f(x) = \dfrac{x^2 + 4}{x}$; $[-3, -1]$ 5. $f(x) = x^2 \ln x$; $\left[\dfrac{1}{4}, e\right]$

6. Why is it important to check the endpoints of the domain when checking for absolute extrema?

[6.2]

7. Find two nonnegative numbers x and y such that $x + y = 50$ and $P = xy$ is maximized.

8. A travel agency offers a tour to the Bahamas for 12 people at \$800 each. For each person more than the original 12, up to a total of 30 people, who signs up for the cruise, the fare is reduced by \$20. What tour group size produces the greatest revenue for the travel agency?

9. \$320 is available for fencing for a rectangular garden. The fencing for the two sides parallel to the back of the house costs \$6 per linear foot, while the fencing for the other sides costs \$2 per linear foot. Find the dimensions that will maximize the area of the garden.

[6.3]

10. Find $\dfrac{dy}{dx}$ for $x^3 - x^2y + y^2 = 0$.   11. Find $\dfrac{dy}{dx}$ for $\cos(xy) = x$.

12. Find an equation for the tangent line to the graph of $2x^2 - y^2 = 2$ at $(3, 4)$.

13. Find an equation for the tangent line to the graph of $x - \cos(\pi y) = 1$ at $\left(1, \frac{1}{2}\right)$.

14. Suppose $3x^2 + 4y^2 + 6 = 0$. Use implicit differentiation to find $\dfrac{dy}{dx}$. Explain why your result is meaningless.

[6.4]

15. Find $\frac{dy}{dt}$ if $y = 3x^3 - 2x^2$, $\frac{dx}{dt} = -2$, and $x = 3$.

16. Find $\frac{dy}{dt}$ if $2xy - y^2 = x$, $x = -1$, $y = 3$, and $\frac{dx}{dt} = 2$.

17. When solving related rates problems, how should you interpret a negative derivative?

18. A 25-foot ladder is leaning against a vertical wall. If the bottom of the ladder is pulled horizontally away from the wall at 3 feet per second, how fast is the top of the ladder sliding down the wall when the bottom is 15 feet from the wall?

19. A real estate developer estimates that his monthly sales are given by

$$S = 30y - xy - \frac{y^2}{3,000},$$

where y is the average cost of a new house in the development and x percent is the current interest rate for home mortgages. If the current rate is 11% and is rising at a rate of $\frac{1}{4}$% per month and the current average price of a new home is \$90,000 and is increasing at a rate of \$500 per month, how fast are his expected sales changing.

20. **Brain Weight** The brain weight of a fetus can be estimated using the total weight of the fetus by the function

$$b = 0.22w^{0.87},$$

where w is the weight of the fetus (in g) and b is the brain weight (in g). Suppose the brain weight of a 30 g fetus is changing at a rate of 0.30 g/day. Use this to estimate the rate of change of the total weight of the fetus $\frac{dw}{dt}$.

[6.5]

21. If $y = \frac{2x-3}{x^2-7}$, find dy.

22. Find the value of dy if $y = \sqrt{5 - x^2}$, $x = 1$, $\triangle x = 0.01$.

23. Use the differential to approximate $\tan(0.07)$. Then use a calculator to approximate the quantity, and give the absolute value of the difference in the two results to four decimal places.

24. The radius of a sphere is claimed to be 5 cm with a possible error of 0.01 cm. Use differentials to estimate the possible error in the volume of the sphere.

Chapter 6 Test Answers

1. Absolute maximum at x_3

2. No absolute extrema

3. Absolute maximum of 16 at 4; absolute minimum of -16 at 2

4. Absolute maximum of -4 at -2; absolute minimum of -5 at -1

5. Absolute minimum at $\frac{1}{\sqrt{e}}$, or about 0.6065 absolute maximum at e, or about 2.7183.

6. The smallest or largest value in a closed interval may occur at an endpoint.

7. 25 and 25

8. 26 people

9. $13\frac{1}{2}$ ft on each side parallel to the back, 40 ft on each of the other sides

10. $\frac{dy}{dx} = \frac{2xy - 3x^2}{2y - x^2}$

11. $\frac{dy}{dx} = -\frac{1 + y\sin(xy)}{x\sin(xy)} = -\frac{y}{x} - \frac{\csc(xy)}{x}$

12. $3x - 2y = 1$ or $y = \frac{3}{2}x - \frac{1}{2}$

13. $y = -\frac{x}{\pi} + \frac{2 + \pi}{2\pi}$

14. $\frac{-3x}{4y}$; no function exists such that $3x^2 + 4y^2 + 6 = 0$.

15. -138

16. 1.25

17. A decrease in rate

18. $-\frac{9}{4}$ ft/sec

19. Decreasing at \$43,000/month

20. 2.4390 g/day

21. $dy = \frac{-2x^2 + 6x - 14}{(x^2 - 7)^2}dx$

22. -0.005

23. 0.07; 0.07115; 0.0012

24. 3.14 cm^3

INTEGRATION

7.1   Antiderivatives

1. If $F(x)$ and $G(x)$ are both antiderivatives of $f(x)$, then there is a constant C such that

$$F(x) - G(x) = C.$$

The two functions can differ only by a constant.

5. $\displaystyle\int 6\,dk = 6\int 1\,dk$

$$= 6\int k^0\,dy$$

$$= 6 \cdot \frac{1}{1}k^{0+1} + C$$

$$= 6k + C$$

7. $\displaystyle\int (2z+3)\,dz$

$$= 2\int z\,dz + 3\int z^0\,dz$$

$$= 2 \cdot \frac{1}{1+1}z^{1+1} + 3 \cdot \frac{1}{0+1}z^{0+1} + C$$

$$= z^2 + 3z + C$$

9. $\displaystyle\int (t^2 - 4t + 5)\,dt$

$$= \int t^2\,dt - 4\int t\,dt + 5\int t^0\,dt$$

$$= \frac{t^3}{3} - \frac{4t^2}{2} + 5t + C$$

$$= \frac{t^3}{3} - 2t^2 + 5t + C$$

11. $\displaystyle\int (4z^3 + 3z^2 + 2z - 6)\,dz$

$$= 4\int z^3\,dz + 3\int z^2\,dz + 2\int z\,dz$$

$$- 6\int z^0\,dz$$

$$= \frac{4z^4}{4} + \frac{3z^3}{3} + \frac{2z^2}{2} - 6z + C$$

$$= z^4 + z^3 + z^2 - 6z + C$$

13. $\displaystyle\int 5\sqrt{z}\,dz = 5\int z^{1/2}\,dz$

$$= \frac{5z^{3/2}}{\frac{3}{2}} + C$$

$$= 5\left(\frac{2}{3}\right)z^{3/2} + C$$

$$= \frac{10z^{3/2}}{3} + C$$

15. $\displaystyle\int x(x^2 - 3)\,dx = \int (x^3 - 3x)\,dx$

$$= \frac{x^4}{4} - \frac{3x^2}{2} + C$$

17. $\displaystyle\int (4\sqrt{v} - 3v^{3/2})\,dv$

$$= 4\int v^{1/2}\,dv - 3\int v^{3/2}\,dv$$

$$= \frac{4v^{3/2}}{\frac{3}{2}} - \frac{3v^{5/2}}{\frac{5}{2}} + C$$

$$= \frac{8v^{3/2}}{3} - \frac{6v^{5/2}}{5} + C$$

19. $\displaystyle\int (10u^{3/2} - 14u^{5/2})\,du$

$$= 10\int u^{3/2}\,du - 14\int u^{5/2}\,du$$

$$= \frac{10u^{5/2}}{\frac{5}{2}} - \frac{14u^{7/2}}{\frac{7}{2}} + C$$

$$= 10\left(\frac{2}{5}\right)u^{5/2} - 14\left(\frac{2}{7}\right)u^{7/2} + C$$

$$= 4u^{5/2} - 4u^{7/2} + C$$

21. $\displaystyle\int \left(\frac{1}{z^2}\right)dz = \int z^{-2}\,dz$

$$= \frac{z^{-2+1}}{-2+1} + C$$

$$= \frac{z^{-1}}{-1} + C$$

$$= -\frac{1}{z} + C$$

23. $\int \left(\dfrac{1}{y^3} - \dfrac{1}{\sqrt{y}} \right) dy = \int y^{-3}\, dy - \int y^{-1/2}\, dy$

$$= \dfrac{y^{-2}}{-2} - \dfrac{y^{1/2}}{\frac{1}{2}} + C$$

$$= -\dfrac{1}{2y^2} - 2y^{1/2} + C$$

25. $\int (-9t^{-2} - 2t^{-1})\, dt$

$$= -9 \int t^{-2}\, dt - 2 \int t^{-1}\, dt$$

$$= \dfrac{-9t^{-1}}{-1} - 2 \int \dfrac{dt}{t}$$

$$= \dfrac{9}{t} - 2 \ln |t| + C$$

27. $\int \dfrac{1}{3x^2}\, dx = \int \dfrac{1}{3} x^{-2}\, dx$

$$= \dfrac{1}{3} \int x^{-2}\, dx$$

$$= \dfrac{1}{3} \left(\dfrac{x^{-1}}{-1} \right) + C$$

$$= -\dfrac{1}{3} x^{-1} + C$$

$$= -\dfrac{1}{3x} + C$$

29. $\int 3e^{-0.2x}\, dx = 3 \int e^{-0.2x}\, dx$

$$= 3 \left(\dfrac{1}{-0.2} \right) e^{-0.2x} + C$$

$$= \dfrac{3(e^{-0.2x})}{-0.2} + C$$

$$= -15e^{-0.2x} + C$$

31. $\int \left(\dfrac{3}{x} + 4e^{-0.5x} \right) dx$

$$= 3 \int \dfrac{dx}{x} + 4 \int e^{-0.5x}\, dx$$

$$= 3 \ln |x| + \dfrac{4e^{-0.5x}}{-0.5} + C$$

$$= 3 \ln |x| - 8e^{-0.5x} + C$$

33. $\int \dfrac{1 + 2t^3}{t}\, dt = \int \left(\dfrac{1}{t} + 2t^2 \right) dt$

$$= \int \dfrac{1}{t}\, dt + 2 \int t^2\, dt$$

$$= \ln |t| + \dfrac{2t^3}{3} + C$$

35. $\int (e^{2u} + 4u)\, du = \dfrac{e^{2u}}{2} + \dfrac{4u^2}{2} + C$

$$= \dfrac{e^{2u}}{2} + 2u^2 + C$$

37. $\int (x + 1)^2\, dx = \int (x^2 + 2x + 1)\, dx$

$$= \dfrac{x^3}{3} + \dfrac{2x^2}{2} + x + C$$

$$= \dfrac{x^3}{3} + x^2 + x + C$$

39. $\int \dfrac{\sqrt{x} + 1}{\sqrt[3]{x}}\, dx = \int \left(\dfrac{\sqrt{x}}{\sqrt[3]{x}} + \dfrac{1}{\sqrt[3]{x}} \right) dx$

$$= \int (x^{(1/2 - 1/3)} + x^{-1/3})\, dx$$

$$= \int x^{1/6}\, dx + \int x^{-1/3}\, dx$$

$$= \dfrac{x^{7/6}}{\frac{7}{6}} + \dfrac{x^{2/3}}{\frac{2}{3}} + C$$

$$= \dfrac{6x^{7/6}}{7} + \dfrac{3x^{2/3}}{2} + C$$

41. $\int 10^x\, dx = \dfrac{10^x}{\ln 10} + C$

43. Find $f(x)$ such that $f'(x) = x^{2/3}$, and $\left(1, \frac{3}{5}\right)$ is on the curve.

$$\int x^{2/3}\, dx = \dfrac{x^{5/3}}{\frac{5}{3}} + C$$

$$f(x) = \dfrac{3x^{5/3}}{5} + C$$

Since $\left(1, \frac{3}{5}\right)$ is on the curve,

$$f(1) = \dfrac{3}{5}.$$

$$f(1) = \dfrac{3(1)^{5/3}}{5} + C = \dfrac{3}{5}$$

$$\dfrac{3}{5} + C = \dfrac{3}{5}$$

$$C = 0.$$

Thus,

$$f(x) = \dfrac{3x^{5/3}}{5}.$$

45. $\displaystyle \int \frac{g(x)}{x}\,dx = \int \frac{a-bx}{x}\,dx$

$\displaystyle \qquad = \int \left(\frac{a}{x}-b\right)dx$

$\displaystyle \qquad = a\int \frac{dx}{x} - b\int dx$

$\displaystyle \qquad = a\ln|x| - bx + c$

Since x represents a positive quantity, the absolute value sign can be dropped.

$$\int \frac{g(x)}{x}\,dx = a\ln x - bx + C$$

47. (a) $\displaystyle c(t) = (c_0 - C)e^{-kAt/V} + M$

$\displaystyle \quad c'(t) = (c_0 - C)\left(\frac{-kA}{V}\right)e^{-kAt/V}$

$\displaystyle \qquad = \frac{-kA}{V}(c_0 - C)e^{-kAt/V}$

(b) Since equation (1) states

$$c'(t) = \frac{kA}{V}[C - c(t)],$$

then from (a) and by substituting from equation (2), we obtain

$(c_0 - C)\left(\dfrac{-kA}{V}\right)e^{-kAt/V}$

$\displaystyle = \frac{kA}{V}C - \frac{kA}{V}[(c_0 - C)e^{-kAt/V} + M]$

$\dfrac{-kA}{V}(c_0 - C)e^{-kAt/V} + M$

$\displaystyle = \frac{-kA}{V}(c_0 - C)e^{-kAt/V} + \frac{kA}{V}C - \frac{kA}{V}M.$

If $t=0$, $c(t) = c_0$, so

$\begin{aligned}
c_0 &= (c_0 - C)e^0 + M \quad \textit{Equation (2)}\\
c_0 &= c_0 - C + M\\
\text{or}\quad & C = M.
\end{aligned}$

Thus,

$\dfrac{-kA}{V}(c_0 - C)e^{-kAt/V}$

$\displaystyle = \frac{-kA}{V}(c_0 - C)e^{-kAt/V} + \frac{kA}{V}M - \frac{kA}{V}M$

or

$\displaystyle \frac{-kA}{V}(c_0 - C)e^{-kAt/V} = \frac{-kA}{V}(c_0 - C)e^{-kAt/V}.$

49. $g'(x) = -0.00156x^3 + 0.0312x^2 - 0.264x + 0.137$

(a) $\displaystyle g(x) = \frac{-0.00156}{4}x^4 + \frac{0.0312}{3}x^3 - \frac{0.264}{2}x^2$

$\qquad + 0.137x + C$

$\qquad = -0.00039x^4 + 0.0104x^3 - 0.132x^2$

$\qquad + 0.137x + C$

Since $g(0) = 2.4$, $C = 2.4$. Therefore,

$g(x) = -0.00039x^4 + 0.0104x^3 - 0.132x^2$

$\qquad + 0.137x + 2.4.$

(b) 1986 corresponds to $x = 1$.

$g(1) = -0.00039(1)^4 + 0.0104(1)^3 - 0.132(1)^2$

$\qquad + 0.137(1) + 2.4$

$\quad \approx 2.4$

1990 corresponds to $x = 5$.

$g(5) = -0.00039(5)^4 + 0.0104(5)^3 - 0.132(5)^2$

$\qquad + 0.137(5) + 2.4$

$\quad \approx 0.8 \approx 1$

There were about 2.4 deaths per 100 million miles in 1986 and about 1 in 1990.

51. $a(t) = t^2 + 1$

$\displaystyle v(t) = \int (t^2 + 1)\,dt = \frac{t^3}{3} + t + C$

$\displaystyle v(0) = \frac{0^3}{3} + 0 + C$

Since $v(0) = 6$, $C = 6$.

$$v(t) = \frac{t^3}{3} + t + 6$$

53. $a(t) = -32$

$\displaystyle v(t) = \int -32\,dt = -32t + C_1$

$v(0) = -32(0) + C_1$

Since $v(0) = 0$, $C_1 = 0$.

$$v(t) = -32t$$

$$s(t) = \int -32t\,dt$$

$$\qquad = \frac{-32t^2}{2} + C_2$$

$$\qquad = -16t^2 + C_2$$

At $t = 0$, the plane is at 6,400 ft.

That is, $s(0) = 6,400$.

$$s(0) = -16(0)^2 + C_2$$
$$6,400 = 0 + C_2$$
$$C_2 = 6,400$$
$$s(t) = -16t^2 + 6,400$$

When the object hits the ground, $s(t) = 0$.

$$-16t^2 + 6,400 = 0$$
$$-16t^2 = -6,400$$
$$t^2 = 400$$
$$t = \pm 20$$

Discard -20 since time must be positive. The object hits the ground in 20 sec.

55. $a(t) = \dfrac{15}{2}\sqrt{t} = 3e^{-t}$

$$v(t) = \int \left(\frac{15}{2}\sqrt{t} + 3e^{-t}\right) dt$$

$$= \int \left(\frac{15}{2}t^{1/2} + 3e^{-t}\right) dt$$

$$= \frac{15}{2}\left(\frac{t^{3/2}}{\frac{3}{2}}\right) + 3\left(\frac{1}{-1}e^{-t}\right) + C_1$$

$$= 5t^{3/2} - 3e^{-t} + C_1$$

$$V(0) = 5(0)^{3/2} - 3e^{-0} + C_1 = -3 + C_1$$

Since $v(0) = -3$, $C_1 = 0$.

$$v(t) = 5t^{3/2} - 3e^{-t}$$

$$s(t) = \int (5t^{3/2} - 3e^{-t})\,dt$$

$$= 5\left(\frac{t^{5/2}}{\frac{5}{2}}\right) - 3\left(-\frac{1}{1}e^{-t}\right) + C_2$$

$$= 2t^{5/2} + 3e^{-t} + C_2$$
$$s(0) = 2(0)^{5/2} + 3e^{-0} + C_2$$
$$= 3 + C_2$$

Since $s(0) = 4$, $C_2 = 1$.
Thus,

$$s(t) = 2t^{5/2} + 3e^{-t} + 1.$$

7.2 Substitution

3. $\displaystyle\int 4(2x+3)^4\,dx = 2\int 2(2x+3)^4\,dx$

Let $u = 2x + 3$, so that $du = 2\,dx$.

$$= 2\int u^4\,du$$
$$= \frac{2 \cdot u^5}{5} + C$$
$$= \frac{2(2x+3)^5}{5} + C$$

5. $\displaystyle\int \frac{2\,dm}{(2m+1)^3} = \int 2(2m+1)^{-3}\,dm$

Let $u = 2m + 1$, so that $du = 2\,dm$.

$$= \int u^{-3}\,du$$
$$= \frac{u^{-2}}{-2} + C$$
$$= \frac{-(2m+1)^{-2}}{2} + C$$

7. $\displaystyle\int \frac{2x+2}{(x^2+2x-4)^4}\,dx$

$$= \int (2x+2)(x^2+2x-4)^{-4}\,dx$$

Let $u = x^2 + 2x - 4$, so that $du = (2x+2)\,dx$.

$$= \int u^{-4}\,du$$
$$= \frac{u^{-3}}{-3} + C$$
$$= \frac{-(x^2+2x-4)^{-3}}{3} + C$$

9. $\displaystyle\int z\sqrt{z^2-5}\,dz = \int z(z^2-5)^{1/2}\,dz$

$$= \frac{1}{2}\int 2z(z^2-5)^{1/2}\,dz$$

Let $u = z^2 - 5$, so that $du = 2z\,dz$.

$$= \frac{1}{2}\int u^{1/2}\,du$$
$$= \frac{1}{2} \cdot \frac{u^{3/2}}{3/2} + C$$
$$= \frac{1}{2}\left(\frac{2}{3}\right)u^{3/2} + C$$
$$= \frac{(z^2-5)^{3/2}}{3} + C$$

11. $\int (-4e^{2p})\, dp = -2 \int 2e^{2p}\, dp$

Let $u = 2p$, so that $du = 2\, dp$.

$$= -2 \int e^u\, du$$

$$= -2e^u + C$$
$$= -2e^{2p} + C$$

13. $\int 3x^2\, e^{2x^3}\, dx = \frac{1}{2} \int 2 \cdot 3x^2\, e^{2x^3}\, dx$

Let $u = 2x^3$, so that $du = 6x^2\, dx$.

$$= \frac{1}{2} \int e^u\, du$$

$$= \frac{1}{2} e^u + C$$

$$= \frac{e^{2x^3}}{2} + C$$

15. $\int (1-t)e^{2t-t^2}\, dt$

$$= \frac{1}{2} \int 2(1-t)e^{2t-t^2}\, dt$$

Let $u = 2t - t^2$, so that $du = (2 - 2t)\, dt$.

$$= \frac{1}{2} \int e^u\, du$$

$$= \frac{e^u}{2} + C$$

$$= \frac{e^{2t-t^2}}{2} + C$$

17. $\int \frac{e^{1/z}}{z^2}\, dz = -\int e^{1/z} \cdot \frac{-1}{z^2}\, dz$

Let $u = \frac{1}{z}$, so that $du = \frac{-1}{z^2}\, dx$.

$$= -\int e^u\, du$$

$$= -e^u + C$$
$$= -e^{1/z} + C$$

19. $\int (x^3 + 2x)(x^4 + 4x^2 + 7)^8 dx$

$$= \frac{1}{4} \int (x^4 + 4x^2 + 7)^8 (4x^3 + 8x) dx$$

Let $u = x^4 + 4x^2 + 7$, so that $du = (4x^3 + 8x)dx$

$$= \frac{1}{4} \int u^8 du = \frac{1}{4} \int \left(\frac{u^9}{9} \right) + C$$

$$= \frac{u^9}{36} + C = \frac{(x^4 + 4x^2 + 7)^9}{36} + C$$

21. $\int \frac{2x + 1}{(x^2 + x)^3}\, dx$

$$= \int (2x + 1)(x^2 + x)^{-3}\, dx$$

Let $u = x^2 + x$, so that $du = (2x + 1)\, dx$.

$$= \int u^{-3}\, du = \frac{u^{-2}}{-2} + C$$

$$= \frac{-1}{2u^2} + C = \frac{-1}{2(x^2 + x)^2} + C$$

23. $\int p(p+1)^5\, dp$

Let $u = p + 1$, so that $du = dp$; also, $p = u - 1$.

$$= \int (u-1)u^5\, du$$

$$= \int (u^6 - u^5)\, du$$

$$= \frac{u^7}{7} - \frac{u^6}{6} + C$$

$$= \frac{(p+1)^7}{7} - \frac{(p+1)^6}{6} + C$$

25. $\int \frac{u}{\sqrt{u-1}}\, du$

$$= \int u(u-1)^{-1/2}\, du$$

Let $w = u - 1$, so that $dw = du$ and $u = w + 1$.

$$= \int (w+1)w^{-1/2}\, dw$$

$$= \int (w^{1/2} + w^{-1/2})\, dw$$

$$= \frac{w^{3/2}}{\frac{3}{2}} + \frac{w^{1/2}}{\frac{1}{2}} + C$$

$$= \frac{2(u-1)^{3/2}}{3} + 2(u-1)^{1/2} + C$$

27. $\int (\sqrt{x^2 + 12x})(x + 6)\, dx$

$$= \int (x^2 + 12x)^{1/2}(x + 6)\, dx$$

Let $u = x^2 + 12x$, so that

$$du = (2x + 12)\, dx$$
$$du = 2(x + 6)\, dx.$$

$$= \frac{1}{2} \int u^{1/2}\, du = \frac{1}{2} \left(\frac{2}{3} \right) u^{3/2} + C$$

$$= \frac{(x^2 + 12x)^{3/2}}{3} + C$$

29. $\displaystyle\int \frac{t}{t^2+2}\,dt$

Let $u = t^2 + 2$, so that $du = 2t\,dt$.

$$= \frac{1}{2}\int \frac{du}{u}$$

$$= \frac{1}{2}\ln|u| + C$$

$$= \frac{\ln(t^2+2)}{2} + C$$

31. $\displaystyle\int \frac{(1+\ln x)^2}{x}\,dx$

Let $u = 1 + \ln x$, so that $du = \frac{1}{x}\,dx$.

$$= \int u^2\,du$$

$$= \frac{1}{3}u^3 + C$$

$$= \frac{(1+\ln x)^3}{3} + C$$

33. $\displaystyle\int \frac{e^{2x}}{e^{2x}+5}\,dx$

Let $u = e^{2x} + 5$, so that $du = 2e^{2x}\,dx$.

$$= \frac{1}{2}\int \frac{du}{u}$$

$$= \frac{1}{2}\ln|u| + C$$

$$= \frac{1}{2}\ln\left|e^{2x}+5\right| + C$$

$$= \frac{1}{2}\ln(e^{2x}+5) + C$$

37. $N'(t) = Ae^{kt}$

(a) $N(t) = \dfrac{A}{k}e^{kt} + C$

$A = 50$, $N(t) = 300$ when $t = 0$.

$$N(0) = \frac{50}{k}e^0 + C = 300$$

$$N'(5) = 250$$

Therefore,

$$N'(5) = 50e^{5k} = 250$$
$$e^{5k} = 5$$
$$5k = \ln 5$$
$$k = \frac{\ln 5}{5}.$$

$$N(0) = \frac{50}{\frac{\ln 5}{5}} + C = 300$$

$$\frac{250}{\ln 5} + C = 300$$

$$C = 300 - \frac{250}{\ln 5}$$

$$\approx 144.666$$

$$N(t) = \frac{50}{\frac{\ln 5}{5}}e^{(\ln 5/5)t} + 144.666$$

$$= 155.3337e^{0.321888t} + 144.666$$

(b) $N(12) = 155.3337e^{0.321888(12)} + 144.666$
$$\approx 7{,}537$$

39. (a) $S'(t) = 4.4e^{0.16t}$

$$S(t) = \int 4.4e^{0.16t}\,dt$$

$$= \frac{4.4}{0.16}e^{0.16t} + k$$

$$= 27.5e^{0.16t} + k$$

$$S(0) = 27.5e^{0.16(0)} + k$$

$$= 27.5 + k$$

Since $S(0) = 27.3$, $k = -0.2$.
Thus,

$$S(t) = 27.5e^{0.16t} - 0.2.$$

(b) $S(t) = 27.5e^{0.16t} - 0.2 = 2(27.3)$
$$27.5e^{0.16t} = 54.8$$
$$e^{0.16t} = 1.99\overline{27}$$
$$\ln e^{0.16t} = \ln 1.99\overline{27}$$
$$0.16t(\ln e) = \ln 1.99\overline{27}$$
$$t = \frac{\ln 1.99\overline{27}}{0.16}$$
$$t \approx 4.3$$

The number of Canadian cross-border shoppers will double in 4.3 yr.

7.3   Area and the Definite Integral

3. $f(x) = 2x + 1$, $x_1 = 0$, $x_2 = 2$, $x_3 = 4$, $x_4 = 6$, and $\Delta x = 2$

(a) $\displaystyle\sum_{i=1}^{4} f(x_i)\Delta x$

$$
\begin{aligned}
&= f(x_1)\Delta x + f(x_2)\Delta x + f(x_3)\Delta x \\
&\quad + f(x_4)\Delta x \\
&= f(0)(2) + f(2)(2) + f(4)(2) + f(6)(2) \\
&= [2(0) + 1](2) + [2(2) + 1](2) \\
&\quad + [2(4) + 1](2) + [2(6) + 1](2) \\
&= 2 + 5(2) + 9(2) + 13(2) \\
&= 56
\end{aligned}
$$

(b)

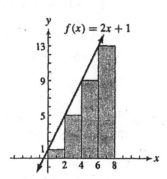

The sum of these rectangles approximates

$$\int_0^8 (2x + 1)\,dx.$$

7. $f(x) = x + 5$ from $x = 2$ to $x = 4$

For $n = 4$ rectangles:

$$\Delta x = \frac{4 - 2}{4} = \frac{1}{2}.$$

(a) Using the left endpoints:

i	x_i	$f(x_i)$
1	2	7
2	$\dfrac{5}{2}$	$\dfrac{15}{2}$
3	3	8
4	$\dfrac{7}{2}$	$\dfrac{17}{2}$

$$A = \sum_{i=1}^{4} f(x_i)\Delta x$$

$$= 7\left(\frac{1}{2}\right) + \frac{15}{2}\left(\frac{1}{2}\right) + 8\left(\frac{1}{2}\right) + \frac{17}{2}\left(\frac{1}{2}\right)$$

$$= \frac{31}{2} = 15.5$$

(b) Using the right endpoints:

i	x_i	$f(x_i)$
1	$\dfrac{5}{2}$	7.5
2	3	8
3	$\dfrac{7}{2}$	8.5
4	4	9

$$A = \frac{1}{2}(7.5) + \frac{1}{2}(8) + \frac{1}{2}(8.5) + \frac{1}{2}(9)$$

$$= \frac{33}{2} = 16.5$$

(c) Average $= \dfrac{15.5 + 16.5}{2} = 16$

(d) Using the midpoints:

i	x_i	$f(x_i)$
1	$\dfrac{9}{4}$	$\dfrac{29}{4}$
2	$\dfrac{11}{4}$	$\dfrac{31}{4}$
3	$\dfrac{13}{4}$	$\dfrac{33}{4}$
4	$\dfrac{15}{4}$	$\dfrac{35}{4}$

$$A = \sum_{i=1}^{4} f(x_i)\Delta x$$

$$= \frac{29}{4}\left(\frac{1}{2}\right) + \frac{31}{4}\left(\frac{1}{2}\right) + \frac{33}{4}\left(\frac{1}{2}\right) + \frac{35}{4}\left(\frac{1}{2}\right)$$

$$= 16$$

9. $f(x) = -x^2 + 4$ from $x = -2$ to $x = 2$

For $n = 4$ rectangles:

$$\Delta x = \frac{2 - (-2)}{4} = 1$$

(a) Using the left endpoints:

i	x_i	$f(x_i)$
1	-2	$-(-2)^2 + 4 = 0$
2	-1	$-(-1)^2 + 4 = 3$
3	0	$-(0)^2 + 4 = 4$
4	1	$-(1)^2 + 4 = 3$

$$A = \sum_{i=1}^{4} f(x_i)\Delta x$$

$$= (0)(1) + (3)(1) + (4)(1) + (3)(1)$$

$$= 10$$

(b) Using the right endpoints:

i	x_i	$f(x_i)$
1	-1	3
2	0	4
3	1	3
4	2	0

$$\text{Area} = 1(3) + 1(4) + 1(3) + 1(0)$$
$$= 10$$

(c) Average $= \dfrac{10 + 10}{2} = 10$

(d) Using the midpoints:

i	x_i	$f(x_i)$
1	$-\dfrac{3}{2}$	$\dfrac{7}{4}$
2	$-\dfrac{1}{2}$	$\dfrac{15}{4}$
3	$\dfrac{1}{2}$	$\dfrac{15}{4}$
4	$\dfrac{3}{2}$	$\dfrac{7}{4}$

$$A = \sum_{i=1}^{4} f(x_i)\Delta x$$
$$= \frac{7}{4}(1) + \frac{15}{4}(1) + \frac{15}{4}(1) + \frac{7}{4}(1)$$
$$= 11$$

11. $f(x) = e^x + 1$ from $x = -2$ to $x = 2$

For $n = 4$ rectangles:

$$\Delta x = \frac{2 - (-2)}{4} = 1$$

(a) Using the left endpoints:

i	x_i	$f(x_i)$
1	-2	$e^{-2} + 1$
2	-1	$e^{-1} + 1$
3	0	$e^0 + 1 = 2$
4	1	$e^1 + 1$

$$A = \sum_{i=1}^{4} f(x_i)\Delta x$$
$$= \sum_{i=1}^{4} f(x_i)(1)$$
$$= \sum_{i=1}^{4} f(x_i)$$
$$= (e^{-2} + 1) + (e^{-1} + 1) + 2 + e^1 + 1$$
$$\approx 8.2215 \approx 8.22$$

(b) Using the right endpoints:

i	x_i	$f(x_i)$
1	-1	$e^{-1} + 1$
2	0	2
3	1	$e + 1$
4	2	$e^2 + 1$

$$\text{Area} = 1(e^{-1} + 1) + 1(2) + 1(e + 1) + 1(e^2 + 1)$$
$$\approx 15.4752 \approx 15.48$$

(c) Average $= \dfrac{8.2215 + 15.4752}{2} = 11.84835 \approx 11.85$

(d) Using the midpoints:

i	x_i	$f(x_i)$
1	$-\dfrac{3}{2}$	$e^{-3/2} + 1$
2	$-\dfrac{1}{2}$	$e^{-1/2} + 1$
3	$\dfrac{1}{2}$	$e^{1/2} + 1$
4	$\dfrac{3}{2}$	$e^{3/2} + 1$

$$A = \sum_{i=1}^{4} f(x_i)\Delta x$$
$$= (e^{-3/2} + 1)(1) + (e^{-1/2} + 1)(1) + (e^{1/2} + 1)(1)$$
$$+ (e^{3/2} + 1)(1)$$
$$\approx 10.9601 \approx 10.96$$

13. $f(x) = \dfrac{2}{x}$ from $x = 1$ to $x = 9$

For $n = 4$ rectangles:

$$\Delta x = \frac{9-1}{4} = 2$$

(a) Using the left endpoints:

i	x_i	$f(x_i)$
1	1	$\dfrac{2}{1} = 2$
2	3	$\dfrac{2}{3}$
3	5	$\dfrac{2}{5} = 0.4$
4	7	$\dfrac{2}{7}$

$$A = \sum_{i=1}^{4} f(x_i)\Delta x$$

$$= (2)(2) + \frac{2}{3}(2) + (0.4)(2) + \left(\frac{2}{7}\right)(2)$$

$$\approx 6.7048 \approx 6.70$$

(b) Using the right endpoints:

i	x_i	$f(x_i)$
1	3	$\dfrac{2}{3}$
2	5	$\dfrac{2}{5}$
3	7	$\dfrac{2}{7}$
4	9	$\dfrac{2}{9}$

$$\text{Area} = 2\left(\frac{2}{3}\right) + 2\left(\frac{2}{5}\right) + 2\left(\frac{2}{7}\right) + 2\left(\frac{2}{9}\right)$$

$$= \frac{4}{3} + \frac{4}{5} + \frac{4}{7} + \frac{4}{9}$$

$$\approx 3.1492 \approx 3.15$$

(c) Average $= \dfrac{6.7 + 3.15}{2} = 4.93$

(d) Using the midpoints:

i	x_i	$f(x_i)$
1	2	1
2	4	$\dfrac{1}{2}$
3	6	$\dfrac{1}{3}$
4	8	$\dfrac{1}{4}$

$$A = \sum_{i=1}^{4} f(x_i)\Delta x$$

$$= 1(2) + \frac{1}{2}(2) + \frac{1}{3}(2) + \frac{1}{4}(2)$$

$$\approx 4.1667 \approx 4.17$$

15. $\displaystyle\int_0^5 (5-x)\,dx$

Graph $y = 5 - x$.

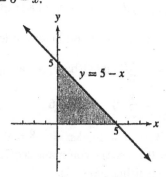

$\displaystyle\int_0^5 (5-x)\,dx$ is the area of a triangle with

base $= 5 - 0 = 5$ and altitude $= 5$.

$$\text{Area} = \frac{1}{2}(\text{altitude})(\text{base})$$

$$= \frac{1}{2}(5)(5) = 12.5$$

17. $\displaystyle\int_{-3}^{3} \sqrt{9 - x^2}\,dx$

Graph $y = \sqrt{9 - x^2}$.

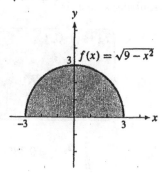

$\int_{-3}^{3} \sqrt{9 - x^2}\, dx$ is the area of a semicircle with radius 3 centered at the origin.

$$\text{Area} = \frac{1}{2}\pi r^2 = \frac{1}{2}\pi(3)^2 = \frac{9}{2}\pi$$

19. $\int_{1}^{3} (5 - x)\, dx$

Graph $y = 5 - x$.

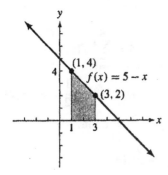

$\int_{1}^{3} (5 - x)\, dx$ is the area of a trapezoid with bases

of length 4 and 2 and height of length 2.

$$\text{Area} = \frac{1}{2}(\text{height})(\text{base}_1 + \text{base}_2)$$

$$= \frac{1}{2}(2)(4 + 2) = 6$$

21. **(a)** With $n = 10$, $\Delta x = \frac{1-0}{10} = 0.1$, and $x_1 = 0 + 0.1 = 0.1$, use the command seq(X^2,X,0.1,1,0.1) →L1. The resulting screen is:

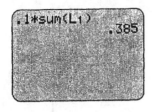

(b) Since $\sum_{i=1}^{n} f(x_i)\Delta x = \Delta x \left(\sum_{i=1}^{n} f(x_i) \right)$, use the

command 0.1*sum(L1) to approximate $\int_{0}^{1} x^2 dx$. The resulting screen is:

$\int\limits_{0}^{1} x^2 dx \approx 0.385$

(c) With $n = 100$, $\Delta x = \frac{1-0}{100} = 0.01$ and $x_1 = 0 + 0.01 = 0.01$, use the command seq(X^2,X,0.1,1,0.1) →L1. The resulting screen is:

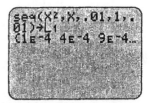

Use the command 0.01*sum(L1) to approximate $\int_{0}^{1} x^2 dx$. The resulting screen is:

$\int_{0}^{1} x^2 dx \approx 0.33835$

(d) With $n = 500$, $\Delta x = \frac{1-0}{500} = 0.002$, and $x_1 = 0 + 0.002 = 0.002$, use the command seq(X^2,X,0.1,1,0.1) →L1. The resulting screen is:

Use the command 0.002*sum(L1) to approximate $\int_{0}^{1} x^2 dx$. The resulting screen is:

$\int_{0}^{1} x^2 dx \approx 0.334334$

(e) As n gets larger the approximation for $\int_{0}^{1} x^2 dx$ seems to be approaching 0.333333 or $\frac{1}{3}$. We estimate $\int_{0}^{1} x^2 dx = \frac{1}{3}$.

For Exercises 23-31, readings on the graphs and answers may vary.

23. (a) Left endpoints:

Read values of the function from the graph for every hour from 8:15 to 2:15 and for the half hour beginning at 3:15. The values give the heights of 7 rectangles, each with width $\triangle x = 1$, and one rectangle with width $\triangle x = \frac{1}{2}$. We estimate the area under the curve as

$$\sum_{i=1}^{8} f(x_i)\,\triangle x_i = 0(1) + 16(1) + 23(1) + 29(1) + 30(1)$$
$$+ 28(1) + 20(1) + 12\left(\frac{1}{2}\right)$$
$$= 152.$$

Right endpoints:

Read values of the function from the graph for every hour from 9:15 to 3:15 and for the half hour ending at 3:45. The values give the heights of 7 rectangles, each of width $\triangle x = 1$, and one rectangle with width $\triangle x = \frac{1}{2}$. We estimate the area under the curve as

$$\sum_{i=1}^{8} f(x_i)\,\triangle x_i = 16(1) + 23(1) + 29(1) + 30(1) + 28(1)$$
$$+ 20(1) + 12(1) + 0\left(\frac{1}{2}\right)$$
$$= 158.$$

Average: $\dfrac{152 + 158}{2} = 155$

The total is about 155 cal/cm^2.

(b) Left endpoints:

Read values of the function from the graph for every hour from 7:30 to 3:30. The values give the heights of 9 rectangles. The width of each rectangle is $\triangle x = 1$. We estimate the rea under the curve as

$$\sum_{i=1}^{9} f(x_i)\,\triangle x_i = 0(1) + 42(1) + 70(1) + 80(1) + 90(1)$$
$$+ 90(1) + 80(1) + 70(1) + 42(1)$$
$$= 564.$$

Right endpoints:

Read values of the function from the graph for every hour from 9:30 to 4:30. The values give the heights of 9 rectangles. The width of each rectangle is $\triangle x = 1$. We estimate the rea under the curve as

$$\sum_{i=1}^{9} f(x_i)\,\triangle x_i = 42(1) + 70(1) + 80(1) + 90(1) + 90(1)$$
$$+ 80(1) + 70(1) + 42(1) + 0(1)$$
$$= 564.$$

Since both estimates are the same, the total is about 564 cal/cm^2.

25. Left endpoints:

Read values of the function on the graph every hour from 0 to 7. These values give us the heights of 8 rectangles. The width of each rectangle is $\triangle x = 1$. We estimate the area under the curve as

$$A = \sum_{i=1}^{8} f(x_i)\triangle x$$
$$= 0(1) + 1.2(1) + 2.1(1) + 2.9(1) + 3.5(1)$$
$$+ 3.7(1) + 3.3(1) + 2.4(1)$$
$$= 19.1.$$

Right endpoints:

Read values of the function on the graph every hour from 1 to 8. Now we estimate the area under the curve as

$$A = \sum_{i=1}^{8} f(x_i)\triangle x$$
$$= 1.2(1) + 2.1(1) + 2.9(1) + 3.5(1)$$
$$+ 3.7(1) + 3.3(1) + 2.4(1) + 1.0(1)$$
$$= 20.1.$$

Average:

$$\frac{19.1 + 20.1}{2} = 19.6 \approx 20$$

The area under the curve represents the total alcohol concentration. We estimate that this concentration is about 20 units.

27. (a) Left endpoints:

Read values for estimated production per day, current policy base, on the graph for every 2 yr from 1990 to 2008. These are the left sides of rectangles with width $\triangle x = 2$.

$$A = \sum_{i=1}^{10} f(x_i)\triangle x = 8.5(2) + 8.0(2) + 7.7(2) + 7.5(2)$$
$$+ 7.2(2) + 7.0(2) + 6.7(2)$$
$$+ 6.6(2) + 6.4(2) + 6.3(2)$$
$$\approx 143.8 \text{ million barrels}$$

This sum must be multiplied by 365 since there are 365 days in a year.

This gives an estimate of

$$365(143.8 \text{ million}) = 52,487 \text{ million}.$$

Right endpoints:

Read values on the graph for every 2 yr from 1992 to 2010.

$$A = \sum_{i=1}^{10} f(x_i)\Delta x = 8.0(2) + 7.7(2) + 7.5(2) + 7.2(2)$$
$$+ 7.0(2) + 6.7(2) + 6.6(2)$$
$$+ 6.4(2) + 6.3(2) + 6.3(2)$$
$$= 139.4$$

This gives an estimate of

$$365(139.4 \text{ million}) = 50,881 \text{ million}.$$

Average

$$\frac{52,487 + 50,881}{2} = 51,684 \text{ million}$$
$$\approx 52,000 \text{ million}$$
$$\text{or 52 billion}$$

We estimate the total oil production as about 52 billion barrels.

(b) Left endpoints:

Read values for estimated production, with strategy, on the graph for every 2 yr from 1990 to 2008. These are the left sides of rectangles with width $\Delta x = 2$.

$$\sum = 8.5(2) + 8(2) + 7.7(2) + 7.7(2)$$
$$+ 8.4(2) + 9(2) + 9.5(2)$$
$$+ 9.8(2) + 10.2(2) + 10(2)$$
$$= 177.6$$

Right endpoints:

Read values for estimated production, with strategy, for every 2 yr from 1992 to 2010.

$$\sum = 8(2) + 7.7(2) + 7.7(2) + 8.4(2)$$
$$+ 9(2) + 9.5(2) + 9.8(2)$$
$$+ 10.2(2) + 10(2) + 10.0(2)$$
$$= 180.6$$

Average:
$$\frac{177.6 + 180.6}{2} = 179.1$$

This gives an estimate of

$$365(179.1) = 65,372 \text{ million}$$
$$\approx 65,000 \text{ million}$$
$$\text{or 65 billion}$$

We estimate the total oil production as about 65 billion gallons.

29. Read the value of the function for every 5 sec from $x = 2.5$ to $x = 12.5$. These are the midpoints of rectangles with width $\Delta x = 5$. Then read the function for $x = 17$, which is the midpoint of a rectangle with width $\Delta x = 4$.

$$\sum_{i=1}^{4} f(x_i)\Delta x \approx 36(5) + 63(5) + 84(5) + 95(4)$$
$$\approx 1,295$$

$$\frac{1,295}{3,600}(5,280) \approx 1,900$$

The Porsche 928 traveled about 1,900 ft.

31. (a) Read values of the function on the plain glass graph every 2 hr from 6 to 6. These are at midpoints of the widths $\Delta x = 2$ and represent the heights of the rectangles.

$$f(x_i)\Delta x$$
$$= 132(2) + 215(2) + 150(2) + 44(2)$$
$$+ 34(2) + 26(2) + 12(2)$$
$$\approx 1,226$$

The total heat gain was about 1,230 BTUs per square foot.

(b) Read values on the ShadeScreen graph every 2 hr from 6 to 6.

$$\sum f(x_i)\Delta x$$
$$= 38(2) + 25(2) + 16(2) + 12(2)$$
$$+ 10(2) + 10(2) + 5(2)$$
$$\approx 232$$

The total heat gain was about 230 BTUs per square foot.

33. (a) The area of a trapezoid is

$$A = \frac{1}{2}h(b_1 + b_2) = \frac{1}{2}(6)(1 + 2) = 9.$$

Car A has traveled 9 ft.

(b) Car A is furthest ahead of car B at 2 sec. Notice that from $t = 0$ to $t = 2$, $v(t)$ is larger for car A than for car B. For $t > 2$, $v(t)$ is larger for car B than for car A.

(c) As seen in part (a), car A drove 9 ft after 2 sec. The distance of car B can be calculated as follows:

$$\frac{2-0}{4} = \frac{1}{2} = \text{width}$$

$$\text{Distance} = \frac{1}{2} \cdot v(0.25) + \frac{1}{2}v(0.75) + \frac{1}{2}v(1.25)$$

$$+ \frac{1}{2}v(1.75)$$

$$= \frac{1}{2}(0.2) + \frac{1}{2}(1) + \frac{1}{2}(2.6) + \frac{1}{2}(5)$$

$$= 4.4$$

$$9 - 4.4 = 4.6$$

The furthest car A can get ahead of car B is about 4.6 ft.

(d) At $t = 3$, car A travels $\frac{1}{2}(6)(2+3) = 15$ ft and car B travels approximately 13 ft.
At $t = 3.5$, car A travels $\frac{1}{2}(6)(2.5+3.5) = 18$ ft and car B travels approximately 18.25 ft. Therefore, car B catches up with car A between 3 and 3.5 sec.

35. Using the left endpoints:

$$\begin{aligned}
\text{Distance} &= v_0(1) + v_1(1) + v_2(1) \\
&= 10 + 6.5 + 6 \\
&= 22.5 \text{ ft}
\end{aligned}$$

Using the right endpoints:

$$\begin{aligned}
\text{Distance} &= v_1(1) + v_2(1) + v_3(1) \\
&= 6.5 + 6 + 5.5 \\
&= 18 \text{ ft}
\end{aligned}$$

7.4   The Fundamental Theorem of Calculus

1. $\displaystyle\int_{-2}^{4} (-1)\,dp = -1 \int_{-2}^{4} dp$

$$= -1 \cdot p \Big|_{-2}^{4}$$

$$= -1[4 - (-2)]$$

$$= -6$$

3. $\displaystyle\int_{-1}^{2} (3t - 1)\,dt$

$$= 3 \int_{-1}^{2} t\,dt - \int_{-1}^{2} dt$$

$$= \frac{3}{2}t^2 \Big|_{-1}^{2} - t \Big|_{-1}^{2}$$

$$= \frac{3}{2}[2^2 - (-1)^2] - [2 - (-1)]$$

$$= \frac{3}{2}(4 - 1) - (2 - 1)$$

$$= \frac{9}{2} - 3 = \frac{9 - 6}{2}$$

$$= \frac{3}{2}$$

5. $\displaystyle\int_{0}^{2} (5x^2 - 4x + 2)\,dx$

$$= 5 \int_{0}^{2} x^2\,dx - 4 \int_{0}^{2} x\,dx + 2 \int_{0}^{2} dx$$

$$= \frac{5x^3}{3} \Big|_{0}^{2} - 2x^2 \Big|_{0}^{2} + 2x \Big|_{0}^{2}$$

$$= \frac{5}{3}(2^3 - 0^3) - 2(2^2 - 0^2) + 2(2 - 0)$$

$$= \frac{5}{3}(8) - 2(4) + 2(2)$$

$$= \frac{40 - 24 + 12}{3} = \frac{28}{3}$$

7. $\displaystyle\int_{0}^{2} 3\sqrt{4u + 1}\,du$

Let $4u + 1 = x$, so that $4\,du = dx$.

When $u = 0$, $x = 4(0) + 1 = 1$.
When $u = 2$, $x = 4(2) + 1 = 9$.

$$\int_{0}^{2} 3\sqrt{4u + 1}\,du$$

$$= \frac{3}{4} \int_{0}^{2} \sqrt{4u + 1}\,(4\,du)$$

$$= \frac{3}{4} \int_{1}^{9} x^{1/2}\,dx$$

$$= \frac{3}{4} \cdot \frac{x^{3/2}}{3/2} \Big|_{1}^{9}$$

$$= \frac{3}{4} \cdot \frac{2}{3}(9^{3/2} - 1^{3/2})$$

$$= \frac{1}{2}(27 - 1) = \frac{26}{2}$$

$$= 13$$

9. $\displaystyle\int_0^1 2(t^{1/2} - t)\,dt$

$$= 2\int_0^1 t^{1/2} - 2\int_0^1 t\,dt$$

$$= 2\frac{t^{3/2}}{\frac{3}{2}}\,\Big|_0^1 - 2\frac{t^2}{2}\,\Big|_0^1$$

$$= \frac{4}{3}(1^{3/2} - 0^{3/2}) - (1^2 - 0^2)$$

$$= \frac{4}{3} - 1 = \frac{1}{3}$$

11. $\displaystyle\int_1^4 (5y\sqrt{y} + 3\sqrt{y})\,dy$

$$= 5\int_1^4 y^{3/2}\,dy + 3\int_1^4 y^{1/2}\,dy$$

$$= 5\left(\frac{y^{5/2}}{\frac{5}{2}}\right)\Big|_1^4 + 3\left(\frac{y^{3/2}}{\frac{3}{2}}\right)\Big|_1^4$$

$$= 2y^{5/2}\Big|_1^4 + 2y^{3/2}\Big|_1^4$$

$$= 2(4^{5/2} - 1) + 2(4^{3/2} - 1)$$
$$= 2(32 - 1) + 2(8 - 1)$$
$$= 62 + 14$$
$$= 76$$

13. $\displaystyle\int_4^6 \frac{2}{(x - 3)^2}\,dx$

Let $x - 3 = u$, so that

$$dx = du$$
$$x = u + 3.$$

When $x = 6$, $u = 6 - 3 = 3$.
When $x = 4$, $u = 4 - 3 = 1$.

$$\int_4^6 \frac{2}{(x - 3)^2}\,dx = 2\int_4^6 (x - 3)^{-2}\,dx$$

$$= 2\int_1^3 u^{-2}\,du$$

$$= 2\cdot\frac{u^{-1}}{-1}\,\Big|_1^3 = -2\cdot u^{-1}\,\Big|_1^3$$

$$= -2\left(\frac{1}{3} - 1\right) = -2\left(-\frac{2}{3}\right)$$

$$= \frac{4}{3}$$

15. $\displaystyle\int_1^5 (5n^{-2} + n^{-3})\,dn$

$$= \int_1^5 5n^{-2}\,dn + \int_1^5 n^{-3}\,dn$$

$$= \frac{5n^{-1}}{-1}\,\Big|_1^5 + \frac{n^{-2}}{-2}\,\Big|_1^5$$

$$= \frac{-5}{n}\,\Big|_1^5 + \frac{-1}{2n^2}\,\Big|_1^5$$

$$= \frac{-5}{5} - \left(\frac{-5}{1}\right) + \frac{-1}{2(25)} - \frac{-1}{2(1)}$$

$$= \frac{-5}{5} + \frac{5}{1} - \frac{1}{50} + \frac{1}{2}$$

$$= \frac{-50 + 250 - 1 + 25}{50} = \frac{224}{50}$$

$$= \frac{112}{25}$$

17. $\displaystyle\int_2^3 \left(2e^{-0.1A} + \frac{3}{A}\right)\,dA$

$$= 2\int_2^3 e^{-0.1A}\,dA + 3\int_2^3 \frac{1}{A}\,dA$$

$$= 2\frac{e^{-0.1A}}{-0.1}\,\Big|_2^3 + 3\,\ln|A|\,\Big|_2^3$$

$$= -20e^{-0.1A}\,\Big|_2^3 + 3\,\ln|A|\,\Big|_2^3$$

$$= 20e^{-0.2} - 20e^{-0.3} + 3\ln 3 - 3\ln 2$$
$$\approx 2.775$$

19. $\displaystyle\int_1^2 \left(e^{5u} - \frac{1}{u^2}\right)\,du$

$$= \int_1^2 e^{5u}\,du - \int_1^2 \frac{1}{u^2}\,du$$

$$= \frac{e^{5u}}{5}\,\Big|_1^2 + \frac{1}{u}\,\Big|_1^2$$

$$= \frac{e^{10}}{5} - \frac{e^5}{5} + \frac{1}{2} - 1$$

$$= \frac{e^{10}}{5} - \frac{e^5}{5} - \frac{1}{2}$$

$$\approx 4{,}375.1$$

21. $\displaystyle\int_{-1}^{0} y(2y^2 - 3)^5 \, dy$

Let $u = 2y^2 - 3$, so that

$$du = 4y \, dy \text{ and } \frac{1}{4} du = y \, dy.$$

When $y = -1$, $u = 2(-1)^2 - 3 = -1$.
When $y = 0$, $u = 2(0)^2 - 3 = -3$.

$$\frac{1}{4}\int_{-1}^{-3} u^5 \, du = \frac{1}{4} \cdot \frac{u^6}{6} \Big|_{-1}^{-3}$$

$$= \frac{1}{24} u^6 \Big|_{-1}^{-3}$$

$$= \frac{1}{24}(-3)^6 - \frac{1}{24}(-1)^6$$

$$= \frac{729}{24} - \frac{1}{24}$$

$$= \frac{728}{24}$$

$$= \frac{91}{3}$$

23. $\displaystyle\int_{1}^{64} \frac{\sqrt{z} - 2}{\sqrt[3]{z}} \, dz$

$$= \int_{1}^{64} \left(\frac{z^{1/2}}{z^{1/3}} - 2z^{-1/3} \right) dz$$

$$= \int_{1}^{64} z^{1/6} \, dz - 2\int_{1}^{64} z^{-1/3} \, dz$$

$$= \frac{z^{7/6}}{\frac{7}{6}} \Big|_{1}^{64} - 2\frac{z^{2/3}}{\frac{2}{3}}$$

$$= \frac{6z^{7/6}}{7} \Big|_{1}^{64} - 3z^{2/3} \Big|_{1}^{64}$$

$$= \frac{6(64)^{7/6}}{7} - \frac{6(1)^{7/6}}{7}$$

$$\qquad - 3(64^{2/3} - 1^{2/3})$$

$$= \frac{6(128)}{7} - \frac{6}{7} - 3(16 - 1)$$

$$= \frac{768 - 6 - 315}{7}$$

$$= \frac{447}{7}$$

$$\approx 63.857$$

25. $\displaystyle\int_{1}^{2} \frac{\ln x}{x} \, dx$

Let $u = \ln x$, so that

$$du = \frac{1}{x} \, dx.$$

When $x = 1$, $u = \ln 1 = 0$.
When $x = 2$, $u = \ln 2$.

$$\int_{0}^{\ln 2} u \, du = \frac{u^2}{2} \Big|_{0}^{\ln 2} = \frac{(\ln 2)^2}{2} - 0$$

$$= \frac{(\ln 2)^2}{2} \approx 0.24023$$

27. $\displaystyle\int_{0}^{8} x^{1/3}\sqrt{x^{4/3} + 9} \, dx$

Let $u = x^{4/3} + 9$, so that

$$du = \frac{4}{3}x^{1/3} \, dx \text{ and } \frac{3}{4} du = x^{1/3} \, dx.$$

When $x = 0$, $u = 0^{4/3} + 9 = 9$.
When $x = 8$, $u = 8^{4/3} + 9 = 25$.

$$\frac{3}{4}\int_{9}^{25} \sqrt{u} \, du = \frac{3}{4}\int_{9}^{25} u^{1/2} \, du$$

$$= \frac{3}{4} \cdot \frac{u^{3/2}}{\frac{3}{2}} \Big|_{9}^{25}$$

$$= \frac{1}{2} u^{3/2} \Big|_{9}^{25}$$

$$= \frac{1}{2}(25)^{3/2} - \frac{1}{2}(9)^{3/2}$$

$$= \frac{125}{2} - \frac{27}{2} = 49$$

29. $\displaystyle\int_{0}^{1} \frac{e^t}{(3 + e^t)^2} \, dt$

Let $u = 3 + e^t$, so that

$$du = e^t \, dt.$$

When $t = 0$, $u = 3 + e^0 = 4$.
When $t = 1$, $u = 3 + e$.

$$\int_{4}^{3+e} \frac{1}{u^2} \, du = \int_{4}^{3+e} u^{-2} \, du = \frac{u^{-1}}{-1} \Big|_{4}^{3+e}$$

$$= \frac{-1}{u} \Big|_{4}^{3+e} = -\frac{1}{3+e} + \frac{1}{4}$$

$$\approx 0.075122$$

31. $f(x) = 2x + 3;\ [8, 10]$

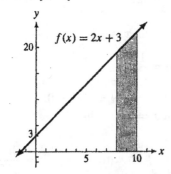

Graph does not cross the x-axis in given interval $[8, 10]$.

$$\int_8^{10} (2x + 3)\, dx = \left(\frac{2x^2}{2} + 3x\right)\bigg|_8^{10}$$
$$= (10^2 + 30) - (8^2 + 24) = 42$$

33. $f(x) = 2 - 2x^2;\ [0, 5]$

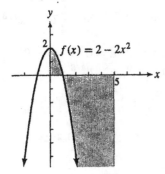

Find the points where the graph crosses the x-axis by solving $2 - 2x^2 = 0$.

$$2 - 2x^2 = 0$$
$$2x^2 = 2$$
$$x^2 = 1$$
$$x = \pm 1.$$

The only solution in the interval $[0, 5]$ is 1. The total area is

$$\int_0^1 (2 - 2x^2)\, dx + \left|\int_1^5 (2 - 2x^2)\, dx\right|$$
$$= \left(2x - \frac{2x^3}{3}\right)\bigg|_0^1 + \left|\left(2x - \frac{2x^3}{3}\right)\bigg|_1^5\right|$$
$$= 2 - \frac{2}{3} + \left|10 - \frac{2(5^3)}{3} - 2 + \frac{2}{3}\right|$$
$$= \frac{4}{3} + \left|\frac{-224}{3}\right|$$
$$= \frac{228}{3}$$
$$= 76.$$

35. $f(x) = x^3;\ [-1, 3]$

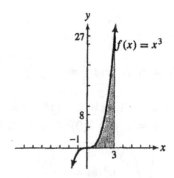

The solution

$$x^3 = 0$$
$$x = 0$$

indicates that the graph crosses the x-axis at 0 in the given interval $[-1, 3]$.

The total area is

$$\left|\int_{-1}^0 x^3\, dx\right| + \int_0^3 x^3\, dx$$
$$= \left|\frac{x^4}{4}\bigg|_{-1}^0\right| + \left|\frac{x^4}{4}\bigg|_0^3\right|$$
$$= \left|\left(0 - \frac{1}{4}\right)\right| + \left(\frac{3^4}{4} - 0\right)$$
$$= \frac{1}{4} + \frac{81}{4} = \frac{82}{4}$$
$$= \frac{41}{2}.$$

37. $f(x) = e^x - 1;\ [-1, 2]$

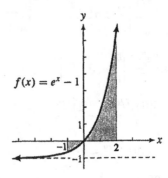

Solve

$$e^x - 1 = 0.$$
$$e^x = 1$$
$$x \ln e = \ln 1$$
$$x = 0$$

The graph crosses the x-axis at 0 in the given interval $[-1, 2]$.

The total area is

$$\left| \int_{-1}^{0} (e^x - 1)\,dx \right| + \int_{0}^{2} (e^x - 1)\,dx$$

$$= \left| (e^x - x) \Big|_{-1}^{0} \right| + (e^x - x) \Big|_{0}^{2}$$

$$= \left| (1 - 0) - (e^{-1} + 1) \right|$$
$$\quad + (e^2 - 2) - (1 - 0)$$

$$= \left| 1 - e^{-1} - 1 \right| + e^2 - 2 - 1$$

$$= \frac{1}{e} + e^2 - 3$$

$$\approx 4.757.$$

39. $f(x) = \dfrac{1}{x};\ [1, e]$

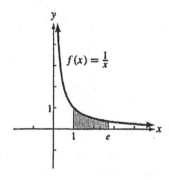

$\dfrac{1}{x} = 0$ has no solution so the graph does not cross the x-axis in the given interval $[1, e]$.

$$\int_{1}^{e} \frac{1}{x}\,dx = \ln x \Big|_{1}^{e}$$

$$= \ln e - \ln 1$$

$$= 1$$

41. $y = 4 - x^2;\ [0, 3]$

From the graph, we see that the total area is

$$\int_{0}^{2} (4 - x^2)\,dx + \left| \int_{2}^{3} (4 - x^2)\,dx \right|$$

$$= \left(4x - \frac{x^3}{3} \right) \Big|_{0}^{2} + \left| \left(4x - \frac{x^3}{3} \right) \Big|_{2}^{3} \right|$$

$$= \left[\left(8 - \frac{8}{3} \right) - 0 \right]$$

$$\quad + \left| \left[(12 - 9) - \left(8 - \frac{8}{3} \right) \right] \right|$$

$$= \frac{16}{3} + \left| 3 - \frac{16}{3} \right|$$

$$= \frac{16}{3} + \frac{7}{3}$$

$$= \frac{23}{3}$$

43. $y = e^x - e;\ [0, 2]$

From the graph, we see that the total area is

$$\left| \int_{0}^{1} (e^x - e)\,dx \right| + \int_{1}^{2} (e^x - e)\,dx$$

$$= \left| (e^x - xe) \Big|_{0}^{1} \right| + \left| (e^x - xe) \Big|_{1}^{2} \right|$$

$$= \left| (e^1 - e) - (e^0 + 0) \right|$$
$$\quad + (e^2 - 2e) - (e^1 - e)$$

$$= \left| -1 \right| + e^2 - 2e$$

$$= 1 + e^2 - 2e$$

$$\approx 2.9525.$$

45. The equation for Exercise 44 is

$$\int_{a}^{c} f(x)\,dx = \int_{a}^{b} f(x)\,dx + \int_{b}^{c} f(x)\,dx.$$

(a) If b is replaced by d, we get

$$\int_{a}^{c} f(x)\,dx = \int_{a}^{d} f(x)\,dx + \int_{d}^{c} f(x)\,dx.$$

This is a correct statement.

(b) If b is replaced by g, we get

$$\int_{a}^{c} f(x)\,dx = \int_{a}^{g} f(x)\,dx + \int_{g}^{c} f(x)\,dx.$$

This is a correct statement.

47. Prove: $\displaystyle\int_{a}^{b} k\, f(x)\,dx = k \int_{a}^{b} f(x)\,dx.$

Assume $F(x)$ is an antiderivative of $f(x)$. Then $kF(x)$ is an antiderivative of $k f(x)$.

$$\int_{a}^{b} k\, f(x)\,dx = kF(x) \Big|_{a}^{b}$$

$$= kF(b) - kF(a)$$

$$= k[F(b) - F(a)]$$

$$= k \int_{a}^{b} f(x)\,dx$$

49. Prove: $\displaystyle\int_{a}^{b} f(x)\,dx = - \int_{b}^{a} f(x)\,dx.$

Assume $F(x)$ is an antiderivative of $f(x)$.

$$\int_{a}^{b} f(x)\,dx = [F(x)] \Big|_{a}^{b}$$

$$= F(b) - F(a)$$

$$= (-1)[F(a) - F(b)]$$

$$= -1 \int_{b}^{a} f(x)\,dx$$

$$= - \int_{b}^{a} f(x)\,dx$$

51. $\int_0^1 e^{x^2}\, dx = 1.46265;\quad \int_0^2 e^{x^2}\, dx = 16.45263$

(a) Since e^{x^2} is symmetric about the y-axis,

$$\int_{-1}^1 e^{x^2}\, dx = 2\int_0^1 e^{x^2}\, dx$$
$$= 2(1.46265)$$
$$= 2.92530.$$

(b) $\int_1^2 e^{x^2}\, dx = \int_0^2 e^{x^2}\, dx - \int_0^1 e^{x^2}\, dx$

$$= 16.45263 - 1.46265$$
$$= 14.98998$$

53. $\int_3^9 (0.1762x^2 - 3.986x + 22.68)dx$

$$= \left(\frac{0.1762}{3}x^3 - \frac{3.986}{2}x^2 + 22.68x\right)\Big|_3^9$$
$$= 85.5036 - 51.6888$$
$$= 33.8148$$

The total increase in the length of a ram's horn during the period is about 33.8 cm.

55. $t = \int_0^n \frac{du}{K - au}$

(a) Let $x = K - au$, so that $dx = -a\, du$, or $du = -\frac{1}{a}dx$.

When $u = n, x = K - an$.
When $u = 0,\ x = K$.

$$t = \int_K^{K-an} \frac{1}{x}\left(-\frac{1}{a}\right)dx$$
$$= \left(-\frac{1}{a}\ln|x|\right)\Big|_K^{K-an}$$
$$= -\frac{1}{a}\ln|K - an| + \frac{1}{a}\ln|K|$$
$$= \frac{1}{a}\ln\frac{K}{K - an}$$

To find the number of neural circuits, solve this equation for n.

$$at = \ln\frac{K}{K - an}$$
$$e^{at} = \frac{K}{K - an}$$
$$K - an = \frac{K}{e^{at}}$$
$$-an = Ke^{-at} - K$$
$$n = \frac{K - Ke^{-at}}{a}$$

(b) As time progresses (i.e., as $t \to \infty$), the factor $e^{-at} = \frac{1}{e^{at}}$ approaches zero. So, n approaches $\frac{K}{a}$.

57. $C(t) = e^{-kt}\int_0^t e^{ky}D(y)dy$

(a) Assume $D(y) = D$, a constant, and let $u = ky$, so that $du = k\, dy$, or $dy = \frac{1}{k}du$.

When $y = t, u = kt$.
When $y = 0, u = 0$.

$$C(t) = e^{-kt}\int_0^{kt} e^u D\left(\frac{1}{k}\right)du$$
$$= \frac{D}{k}e^{-kt}e^u\Big|_0^{kt}$$
$$= \frac{D}{k}e^{-kt}(e^{kt} - e^0)$$
$$= \frac{D}{k}[e^{-kt}e^{kt} - e^{-kt}(1)]$$
$$= \frac{D}{k}(1 - e^{-kt})$$

(b) $A(t) = Ne^{-\int_0^t mC(z)dz}$

Evaluate the integral in the exponent using the result from part (a).

$$\int_0^t mC(z)dz = \int_0^t m\frac{D}{k}(1 - e^{-kz})dz$$
$$= \frac{mD}{k}\int_0^t (1 - e^{-kz})dz$$

Let $u = -kz$, so that $du = -k\, dz$, or $dz = -\frac{1}{k}du$.

When $z = t, u = -kt$.
When $z = 0, u = 0$.

$$\int_0^t mC(z)dz = \frac{mD}{k}\int_0^{-kt}(1 - e^u)\left(-\frac{1}{k}\right)du$$
$$= -\frac{mD}{k}\left(\frac{u - e^u}{k}\right)\Big|_0^{-kt}$$
$$= -\frac{mD}{k}\left(\frac{-kt - e^{-kt}}{k} - \frac{0 - e^0}{k}\right)$$
$$= -\frac{mD}{k}\left(-t + \frac{1 - e^{-kt}}{k}\right)$$
$$= \frac{mD}{k}\left(t + \frac{-1 + e^{-kt}}{k}\right)$$

Therefore,

$$A(t) = N\exp\left[\frac{-mD}{k}\left(t + \frac{-1 + e^{-kt}}{k}\right)\right].$$

59. The tanker is leaking oil at a rate in barrels per hour of

$$L'(t) = \frac{80 \ln (t + 1)}{t + 1}.$$

(a) $\int_0^{24} \frac{80 \ln (t + 1)}{t + 1} dt$

Let $u = \ln (t + 1)$, so that $du = \frac{1}{t+1} dt$.

When $t = 24$, $u = \ln 25$.
When $t = 0$, $u = \ln 1 = 0$.

$$80 \int_0^{\ln 25} u \, du = 80 \frac{u^2}{2} \Big|_0^{\ln 25}$$

$$= 40u^2 \Big|_0^{\ln 25}$$

$$= 40(\ln 25)^2 - 40(0)^2$$

$$\approx 414$$

About 414 barrels will leak on the first day.

(b) $\int_{24}^{48} \frac{80 \ln (t + 1)}{t + 1} dt$

Let $u = \ln (t + 1)$, so that the limits of integration with respect to u are $\ln 25$ and $\ln 49$.

$$80 \int_{\ln 25}^{\ln 49} u \, du = 40u^2 \Big|_{\ln 25}^{\ln 49}$$

$$= 40(\ln 49)^2 - 40(\ln 25)^2$$

$$\approx 191$$

About 191 barrels will leak on the second day.

(c) $\lim_{t \to \infty} L'(t) = \lim_{t \to \infty} \frac{80 \ln (t + 1)}{t + 1} = 0$

The number of barrels of oil leaking per day is decreasing to 0.

61. Total growth after 2.5 days is

$$\int_0^{2.5} R'(x) \, dx = \int_0^{2.5} 200e^{0.2x} \, dx$$

$$= 200 \frac{e^{0.2x}}{0.2} \Big|_0^{2.5}$$

$$= 1{,}000 e^{0.2x} \Big|_0^{2.5}$$

$$= 1{,}000 e^{0.5} - 1{,}000 e^0$$

$$\approx 648.72.$$

63. $F(T) = \int_0^T f(x) \, dx$

$$= \int_0^T kb^x \, dx$$

$$= \int_0^T ke^{(\ln b)x} \, dx$$

$$= k \int_0^T e^{(\ln b)x} \, dx$$

$$= \frac{k}{\ln b} \cdot e^{(\ln b)x} \Big|_0^T$$

$$= \frac{k}{\ln b} \left[e^{(\ln b)T} - 1 \right]$$

$$= \frac{k}{\ln b} \left[b^T - 1 \right]$$

65. $$w'(t) = (4t + 1)^{1/3}$$

$$w(t) = \int_0^3 (4t + 1)^{1/3} \, dt = \frac{1}{4} \cdot \frac{(4t + 1)^{4/3}}{\frac{4}{3}} \Big|_0^3$$

$$= \frac{3(4t + 1)^{4/3}}{16} \Big|_0^3$$

$$= \frac{3}{16} \left(13^{4/3} - 1^{4/3} \right)$$

$$= \frac{3}{16} \left(13^{4/3} - 1 \right)$$

$$= 5.5439$$

The change in weight is 5.5439 mg.

67. $\int_0^{100} 0.85 e^{0.0133x} \, dx$

Let $u = 0.0133x$, so that $du = 0.0133 dx$, or $dx = \frac{1}{0.0133} du$
When $x = 100, u = 1.33$.
When $x = 0, u = 0$.

$$\int_0^{1.33} 0.85 e^u \left(\frac{1}{0.0133} \right) du = \frac{0.85}{0.0133} e^u \Big|_0^{1.33}$$

$$= \frac{0.85}{0.0133} (e^{1.33} - e^0)$$

$$\approx 177.736$$

The total mass of the column is about 178 g.

69. $\int_{2.5}^{5} (15.6 - 0.02x - 0.154x^2)\, dx$

$$= \left(15.6x - \frac{0.02}{2}x^2 - \frac{0.154}{3}x^3\right)\Big|_{2.5}^{5}$$

$$= \left(15.6x - 0.01x^2 - \frac{0.154}{3}x^3\right)\Big|_{2.5}^{5}$$

$$= 71.333 - 38.1354$$

$$= 33.198$$

In 1990, approximately 33% of the population had an income between \$25,000 and \$50,000.

71. $c'(t) = 1.2e^{0.04t}$

$$c(T) = \int_0^T 1.2e^{0.04t}$$

$$= \frac{1.2}{.04}e^{0.04t}\Big|_0^T$$

$$= 30(e^{0.04T} - e^0)$$

$$= 30\left(e^{0.04T} - 1\right)$$

In 5 yr,

$$c(5) = 30(e^{0.04(5)} - 1)$$

$$= 30(e^{0.2} - 1)$$

$$\approx 6.64 \text{ billion barrels.}$$

7.5 Integrals of Trigonometric Functions

1. $\int \cos 5x\, dx$

Let $u = 5x$.
Then $du = 5\, dx$

$$\frac{1}{5}\, du = dx.$$

$$\int \cos 5x\, dx = \int \cos u \cdot \frac{1}{5}\, du$$

$$= \frac{1}{5}\int \cos u\, du$$

$$= \frac{1}{5}\sin u + C$$

$$= \frac{1}{5}\sin 5x + C$$

3. $\int (5\cos x + 2\sin x)\, dx$

$$= \int 5\cos x\, dx + \int 2\sin x\, dx$$

$$= 5\int \cos x\, dx + 2\int \sin x\, dx$$

$$= 5\sin x - 2\cos x + C$$

5. $\int x\sin x^2\, dx$

Let $u = x^2$.
Then $du = 2x\, dx$

$$\frac{1}{2}\, du = x\, dx.$$

$$\int x\sin x^2\, dx = \int \sin u \cdot \frac{1}{2}\, du$$

$$= \frac{1}{2}\int \sin u\, du$$

$$= -\frac{1}{2}\cos u + C$$

$$= -\frac{1}{2}\cos x^2 + C$$

7. $-\int 6\sec^2 2x\, dx$

Let $u = 2x$.
Then $du = 2\, dx$

$$\frac{1}{2}\, du = dx.$$

$$-\int 6\sec^2 2x\, dx = -\int 6\sec^2 u \cdot \frac{1}{2}\, du$$

$$= -3\int \sec^2 u\, du$$

$$= -3\tan u + C$$

$$= -3\tan 2x + C$$

9. $\int \sin^7 x\cos x\, dx$

Let $u = \sin x$.
Then $du = \cos x\, dx$.

$$\int \sin^7 x(\cos x)\, dx = \int u^7\, du$$

$$= \frac{1}{8}u^8 + C$$

$$= \frac{1}{8}\sin^8 x + C$$

11. $\int \sqrt{\sin x}\,(\cos x)\,dx$

Let $u = \sin x$.
Then $du = \cos x\,dx$.

$\int \sqrt{\sin x}\,(\cos x)\,dx$

$\displaystyle = \int \sqrt{u}\,du$

$\displaystyle = \int u^{1/2}\,du$

$\displaystyle = \frac{2}{3}u^{3/2} + C$

$\displaystyle = \frac{2}{3}(\sin x)^{3/2} + C$

13. $\displaystyle \int \frac{\sin x}{1 + \cos x}\,dx$

Let $u = 1 + \cos x$.
Then $\quad du = -\sin x$
$\qquad -du = \sin x\,dx$.

$\displaystyle \int \frac{\sin x}{1 + \cos x}\,dx = \int \frac{1}{u}(-du) = -\int \frac{1}{u}\,du$

$\displaystyle \qquad = -\ln|u| + C$
$\displaystyle \qquad = -\ln|1 + \cos x| + C$

15. $\int x^5 \cos x^6\,dx$

Let $u = x^6$.
Then $\quad du = 6x^5\,dx$
$\qquad \dfrac{1}{6}\,du = x^5\,dx$.

$\displaystyle \int x^5 \cos x^6\,dx = \int (\cos u)\cdot \frac{1}{6}\,du$

$\displaystyle \qquad = \frac{1}{6}\int \cos u\,du$

$\displaystyle \qquad = \frac{1}{6}\sin u + C$

$\displaystyle \qquad = \frac{1}{6}\sin x^6 + C$

17. $\int \tan \frac{1}{4}x\,dx$

Let $u = \dfrac{1}{4}x$.

Then $4\,du = dx$.

$\displaystyle \int \tan \frac{1}{4}x\,dx = \int (\tan u)(4\,du)$

$\displaystyle \qquad = 4\int \tan u\,du$

$\displaystyle \qquad = -4\ln|\cos u| + C$

$\displaystyle \qquad = -4\ln\left|\cos \frac{1}{4}x\right| + C$

19. $\int x^2 \cot x^3\,dx$

Let $u = x^3$.

Then $\dfrac{1}{3}\,du = x^2\,dx$.

$\displaystyle \int x^2 \cot x^3\,dx = \int (\cot u)\left(\frac{1}{3}\,du\right)$

$\displaystyle \qquad = \frac{1}{3}\int \cot u\,du$

$\displaystyle \qquad = \frac{1}{3}\ln|\sin u| + C$

$\displaystyle \qquad = \frac{1}{3}\ln|\sin x^3| + C$

21. $\int e^x \sin e^x\,dx$

Let $u = e^x$.
Then $du = e^x\,dx$.

$\displaystyle \int e^x \sin e^x\,dx = \int \sin u\,du$

$\displaystyle \qquad = -\cos u + C$
$\displaystyle \qquad = -\cos e^x + C$

23. $\int e^x \csc e^x \cot e^x\,dx$

Let $u = e^x$, so that $du = e^x\,dx$.

$\displaystyle \int e^x \csc e^x \cot e^x\,dx = \int \csc e^x \cot e^x (e^x\,dx)$

$\displaystyle \qquad = \int \csc u \cot u\,du$

$\displaystyle \qquad = -\csc u + C$
$\displaystyle \qquad = -\csc e^x + C$

25. $\displaystyle\int_0^{\pi/4} \sin x\,dx = -\cos x\,\Big|_0^{\pi/4}$

$$= -\cos\frac{\pi}{4} - (-\cos 0)$$

$$= -\frac{\sqrt{2}}{2} + 1$$

$$= 1 - \frac{\sqrt{2}}{2}$$

27. $\displaystyle\int_0^{\pi/3} \tan x\,dx = -\ln|\cos x|\,\Big|_0^{\pi/3}$

$$= -\ln\left|\cos\frac{\pi}{3}\right| - (-\ln|\cos 0|)$$

$$= -\ln\frac{1}{2} + \ln 1 = -\ln\frac{1}{2} + 0$$

$$= -\ln\frac{1}{2}$$

$$\text{or } \ln\left(\frac{1}{2}\right)^{-1} = \ln 2$$

29. $\displaystyle\int_{\pi/2}^{2\pi/3} \cos x\,dx = \sin x\,\Big|_{\pi/2}^{2\pi/3}$

$$= \sin\frac{2\pi}{3} - \sin\frac{\pi}{2}$$

$$= \frac{\sqrt{3}}{2} - 1$$

31. $\displaystyle\int_\theta^\phi (1 + I\cos 2x)^{-1}dx = \int_0^T dt$

Substitute $I = 1$ and $\cos 2x = 2\cos^2 x - 1$ and evaluate the integral on the left side of the equation.

$$\int_\theta^\phi (1 + I\cos 2x)^{-1}dx = \int_\theta^\phi (1 + 2\cos^2 x - 1)^{-1}dx$$

$$= \int_\theta^\phi (2\cos^2 x)^{-1}dx$$

$$= \frac{1}{2}\int_\theta^\phi \sec^2 x\,dx$$

$$= \frac{1}{2}\tan x\,\Big|_\theta^\phi$$

$$= \frac{1}{2}(\tan\phi - \tan\theta)$$

Now evaluate the integral on the right side of the equation.

$$\int_0^T dt = t\,\Big|_0^T = T$$

Therefore,

$$T = \frac{1}{2}(\tan\phi - \tan\theta).$$

33. (a)

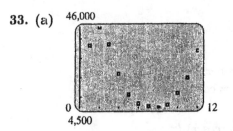

The data appears to be periodic, although there is a strong increase in February due to an extended period of cold weather.

(b) The function $C(x)$, derived by a $TI-83$ Plus calculator, is given by

$$C(x) = 20,277.8\sin(0.484742x + 1.02112) + 21,442.2.$$

(c) The estimate is given by

$$\int_0^{12} C(x)dx = \int_0^{12} [20,277.8\sin(0.484742x + 1.02112)$$

$$+ 21,442.2]dx$$

$$= \left[-\frac{20,277.8}{0.484742}\cos(0.484742x + 1.02112)\right.$$

$$\left.+ 21,442.2x\right]\Big|_0^{12}$$

$$\approx 243,603 \text{ million cubic feet.}$$

The actual value is 240,755 million cubic feet.

35. The total amount of daylight is given by

$$\int_0^{365} N(t)dt = \int_0^{365} [183.549\sin(0.0172t - 1.329)$$

$$+ 728.124]dt$$

$$= \left[-\frac{183.549}{0.0172}\cos(0.0172t - 1.329)\right.$$

$$\left.+ 728.124t\right]\Big|_0^{365}$$

$$\approx 265,819.0192 \text{ minutes}$$

$$\approx 4,430 \text{ hours.}$$

The result is relatively close to the actual value.

7.6   The Area Between Two Curves

1. $x = -2$, $x = 1$, $y = x^2 + 4$, $y = 0$

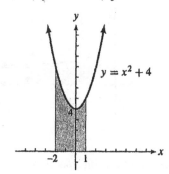

$$\int_{-2}^{1} [(x^2 + 4) - 0] \, dx$$

$$= \left(\frac{x^3}{3} + 4x \right) \Big|_{-2}^{1}$$

$$= \left(\frac{1}{3} + 4 \right) - \left(-\frac{8}{3} - 8 \right)$$

$$= 3 + 12 = 15$$

3. $x = -3$, $x = 1$, $y = x + 1$, $y = 0$

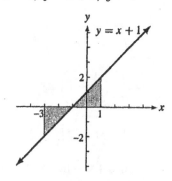

The region is composed of two separate regions because $y = x + 1$ intersects $y = 0$ at $x = -1$.
Let $f(x) = x + 1$, $g(x) = 0$.
In the interval $[-3, -1]$, $g(x) \geq f(x)$.
In the interval $[-1, 1]$, $f(x) \geq g(x)$.

$$\int_{-3}^{-1} [0 - (x + 1)] \, dx + \int_{-1}^{1} [(x + 1) - 0] \, dx$$

$$= \left(\frac{-x^2}{2} - x \right) \Big|_{-3}^{-1} + \left(\frac{x^2}{2} + x \right) \Big|_{-1}^{1}$$

$$= \left(-\frac{1}{2} + 1 \right) - \left(-\frac{9}{2} + 3 \right)$$

$$+ \left(\frac{1}{2} + 1 \right) - \left(\frac{1}{2} - 1 \right)$$

$$= 4$$

5. $x = -2$, $x = 1$, $y = 2x$, $y = x^2 - 3$

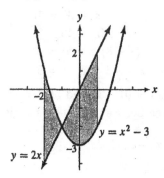

Find the points of intersection of the graphs of $y = 2x$ and $y = x^2 - 3$ by substituting for y.

$$2x = x^2 - 3$$
$$0 = x^2 - 2x - 3$$
$$0 = (x - 3)(x + 1)$$

The only intersection in $[-2, 1]$ is at $x = -1$.
In the interval $[-2, -1]$, $(x^2 - 3) \geq 2x$.
In the interval $[-1, 1]$, $2x \geq (x^2 - 3)$.

$$\int_{-2}^{-1} [(x^2 - 3) - (2x)] \, dx + \int_{-1}^{1} [(2x) - (x^2 - 3)] \, dx$$

$$= \int_{-2}^{-1} (x^2 - 3 - 2x) \, dx + \int_{-1}^{1} (2x - x^2 + 3) \, dx$$

$$= \left(\frac{x^3}{3} - 3x - x^2 \right) \Big|_{-2}^{-1} + \left(x^2 - \frac{x^3}{3} + 3x \right) \Big|_{-1}^{1}$$

$$= -\frac{1}{3} + 3 - 1 - \left(-\frac{8}{3} + 6 - 4 \right) + 1 - \frac{1}{3} + 3$$

$$- \left(1 + \frac{1}{3} - 3 \right)$$

$$= \frac{5}{3} + 6 = \frac{23}{3}$$

7. $y = x^2 - 30$
$y = 10 - 3x$

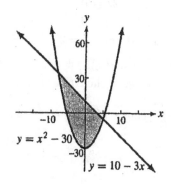

Find the points of intersection.

$$x^2 - 30 = 10 - 3x$$
$$x^2 + 3x - 40 = 0$$
$$(x+8)(x-5) = 0$$
$$x = -8 \quad \text{or} \quad x = 5$$

Let $f(x) = 10 - 3x$ and $g(x) = x^2 - 30$.
The area between the curves is given by

$$\int_{-8}^{5} [f(x) - g(x)]\,dx$$

$$= \int_{-8}^{5} [(10 - 3x) - (x^2 - 30)]\,dx$$

$$= \int_{-8}^{5} (-x^2 - 3x + 40)\,dx$$

$$= \left(\frac{-x^3}{3} - \frac{3x^2}{2} + 40x \right) \Big|_{-8}^{5}$$

$$= \frac{-5^3}{3} - \frac{3(5)^2}{2} + 40(5)$$
$$- \left[\frac{-(-8)^3}{3} - \frac{3(-8)^2}{2} + 40(-8) \right]$$

$$= \frac{-125}{3} - \frac{75}{2} + 200 - \frac{512}{3} + \frac{192}{2} + 320$$

$$\approx 366.1667.$$

9. $y = x^2$, $y = 2x$

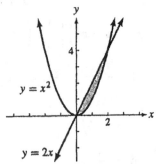

Find the points of intersection.

$$x^2 = 2x$$
$$x^2 - 2x = 0$$
$$x(x - 2) = 0$$
$$x = 0 \quad \text{or} \quad x = 2$$

Let $f(x) = 2x$ and $g(x) = x^2$.
The area between the curves is given by

$$\int_{0}^{2} [f(x) - g(x)]\,dx = \int_{0}^{2} (2x - x^2)\,dx$$

$$= \left(\frac{2x^2}{2} - \frac{x^3}{3} \right) \Big|_{0}^{2}$$

$$= 4 - \frac{8}{3} = \frac{4}{3}.$$

11. $x = 1$, $x = 6$, $y = \dfrac{1}{x}$, $y = -1$

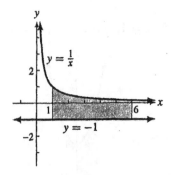

Find the points of intersection.

$$-1 = \frac{1}{x}$$
$$x = -1$$

The curves do not intersect in the interval $[1, 6]$.
If $f(x) = \frac{1}{x}$ and $g(x) = -1$, the area between the curves is

$$\int_{1}^{6} [f(x) - g(x)]\,dx$$

$$= \int_{1}^{6} \left[\frac{1}{x} - (-1) \right] dx$$

$$= \int_{1}^{6} \left(\frac{1}{x} + 1 \right) dx$$

$$= (\ln |x| + x) \Big|_{1}^{6}$$

$$= \ln 6 + 6 - \ln 1 - 1$$
$$= 5 + \ln 6$$
$$\approx 6.792.$$

13. $x = -1$, $x = 1$, $y = e^x$, $y = 3 - e^x$

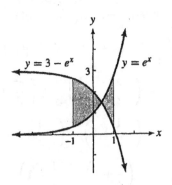

To find the point of intersection, set $e^x = 3 - e^x$ and solve for x.

$$e^x = 3 - e^x$$
$$2e^x = 3$$
$$e^x = \frac{3}{2}$$
$$\ln e^x = \ln \frac{3}{2}$$
$$x \ln e = \ln \frac{3}{2}$$
$$x = \ln \frac{3}{2}$$

The area of the region between the curves from $x = -1$ to $x = 1$ is

$$\int_{-1}^{\ln 3/2} [(3 - e^x) - e^x] \, dx + \int_{\ln 3/2}^{1} [e^x - (3 - e^x)] \, dx$$

$$= \int_{-1}^{\ln 3/2} (3 - 2e^x) \, dx + \int_{\ln 3/2}^{1} (2e^x - 3) \, dx$$

$$= (3x - 2e^x) \Big|_{-1}^{\ln 3/2} + (2e^x - 3x) \Big|_{\ln 3/2}^{1}$$

$$= \left[\left(3 \ln \frac{3}{2} - 2e^{\ln 3/2} \right) - [3(-1) - 2e^{-1}] \right]$$

$$\quad + \left[2e^1 - 3(1) - \left(2e^{\ln 3/2} - 3 \ln \frac{3}{2} \right) \right]$$

$$= \left[\left(3 \ln \frac{3}{2} - 3 \right) - \left(-3 - \frac{2}{e} \right) \right]$$

$$\quad + \left[2e - 3 - \left(3 - 3 \ln \frac{3}{2} \right) \right]$$

$$= 6 \ln \frac{3}{2} + \frac{2}{e} + 2e - 6 \approx 2.6051.$$

15. $x = 1, \ x = 2, \ y = e^x, \ y = \dfrac{1}{x}$

Sketch the graphs of $y = e^x$ and $y = \frac{1}{x}$.

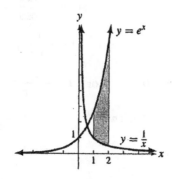

Let $f(x) = e^x$ and $g(x) = \frac{1}{x}$.
For all x in $[1, 2]$, $f(x) \geq g(x)$.
The area between the curves is

$$\int_{1}^{2} [f(x) - g(x)] \, dx = \int_{1}^{2} \left(e^x - \frac{1}{x} \right) \, dx$$

$$= (e^x - \ln |x|) \Big|_{1}^{2}$$

$$= e^2 - \ln 2 - e + \ln 1$$
$$= e^2 - e - \ln 2$$
$$\approx 3.978.$$

17. $y = x^3 - x^2 + x + 1, \ y = 2x^2 - x + 1$

Find the points of intersection.

$$x^3 - x^2 + x + 1 = 2x^2 - x + 1$$
$$x^3 - 3x^2 + 2x = 0$$
$$x(x^2 - 3x + 2) = 0$$
$$x(x - 2)(x - 1) = 0$$

The points of intersection are at $x = 0$, $x = 1$, and $x = 2$.

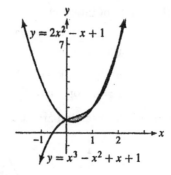

Area between the curves is

$$\int_{0}^{1} [(x^3 - x^2 + x + 1) - (2x^2 - x + 1)] \, dx$$

$$+ \int_{1}^{2} [(2x^2 - x + 1) - (x^3 - x^2 + x + 1)] \, dx$$

$$= \int_{0}^{1} (x^3 - 3x^2 + 2x) \, dx + \int_{0}^{1} (-x^3 + 3x^2 - 2x) \, dx$$

$$= \left(\frac{x^4}{4} - x^3 + x^2 \right) \Big|_{0}^{1} + \left(\frac{-x^4}{4} + x^3 - x^2 \right) \Big|_{1}^{2}$$

$$= \left[\left(\frac{1}{4} - 1 + 1 \right) - (0) \right]$$

$$\quad + \left[(-4 + 8 - 4) - \left(-\frac{1}{4} + 1 - 1 \right) \right]$$

$$= \frac{1}{4} + \frac{1}{4}$$

$$= \frac{1}{2}.$$

19. $y = x^4 + \ln(x + 10),$
$y = x^3 + \ln(x + 10)$

Find the points of intersection.

$$x^4 + \ln(x + 10) = x^3 + \ln(x + 10)$$
$$x^4 - x^3 = 0$$
$$x^3(x - 1) = 0$$
$$x = 0 \quad \text{or} \quad x = 1$$

The points of intersection are at $x = 0$ and $x = 1$.
The area between the curves is

$$\int_0^1 [(x^3 + \ln(x + 10)) - (x^4 + \ln(x + 10))]\, dx$$

$$= \int_0^1 (x^3 - x^4)\, dx$$

$$= \left(\frac{x^4}{4} - \frac{x^5}{5}\right)\Big|_0^1$$

$$= \left(\frac{1}{4} - \frac{1}{5}\right) - (0) = \frac{1}{20}.$$

21. $y = x^{4/3},\ y = 2x^{1/3}$

Find the points of intersection.

$$x^{4/3} = 2x^{1/3}$$
$$x^{4/3} - 2x^{1/3} = 0$$
$$x^{1/3}(x - 2) = 0$$
$$x = 0 \quad \text{or} \quad x = 2$$

The points of intersection are at $x = 0$ and $x = 2$.

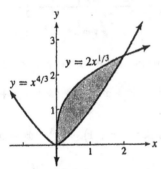

The area between the curves is

$$\int_0^2 (2x^{1/3} - x^{4/3})\, dx = 2\frac{x^{4/3}}{\frac{4}{3}} - \frac{x^{7/3}}{\frac{7}{3}}\Big|_0^2$$

$$= \frac{3}{2}x^{4/3} - \frac{3}{7}x^{7/3}\Big|_0^2$$

$$= \left[\frac{3}{2}(2)^{4/3} - \frac{3}{7}(2)^{7/3}\right] - 0$$

$$= \frac{3(2^{4/3})}{2} - \frac{3(2^{7/3})}{7}$$

$$\approx 1.6199.$$

23. $x = 0,\ x = \frac{\pi}{4},\ y = \cos x,\ y = \sin x$

Find the points of intersection.

$$\sin x = \cos x$$
$$\frac{\sin x}{\cos x} = 1$$
$$\tan x = 1$$
$$x = \frac{\pi}{4} \pm \pi n,\, n \text{ an integer}$$

The curves intersect at the right endpoint of the given interval.

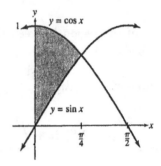

The area between the curves is

$$\int_0^{\pi/4} [\cos(x) - \sin(x)]\, dx$$

$$= [\sin(x) + \cos(x)]\Big|_0^{\pi/4}$$

$$= \left[\sin\left(\frac{\pi}{4}\right) + \cos\left(\frac{\pi}{4}\right)\right] - [\sin(0) + \cos(0)]$$

$$= \left(\frac{\sqrt{2}}{2} + \frac{\sqrt{2}}{2}\right) - (0 + 1)$$

$$= \sqrt{2} - 1$$

25. $x = \frac{\pi}{4},\ y = \tan x,\ y = \sin x$

Find the points of intersection.

$$\sin x = \tan x$$
$$\sin x - \tan x = 0$$
$$\sin x - \frac{\sin x}{\cos x} = 0$$
$$\sin x \left(1 - \frac{1}{\cos x}\right) = 0$$
$$\sin x = 0 \quad \text{or} \quad 1 - \frac{1}{\cos x} = 0$$
$$x = 0 \pm \pi n \qquad\qquad 1 = \frac{1}{\cos x}$$
$$\cos x = 1$$
$$x = 0 \pm \pi n,\, n \text{ an integer}$$

The curves intersect at $x = 0$.

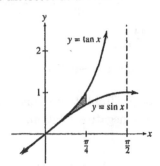

$$\int_0^{\pi/4} [\tan(x) - \sin(x)]dx$$

$$= [-\ln|\cos(x)| + \cos(x)]\Big|_0^{\pi/4}$$

$$= \left[-\ln\left|\cos\left(\frac{\pi}{4}\right)\right| + \cos\left(\frac{\pi}{4}\right)\right]$$
$$\quad - [-\ln|\cos(0)| + \cos(0)]$$

$$= \left[-\ln\left(\frac{\sqrt{2}}{2}\right) + \frac{\sqrt{2}}{2}\right] - [-\ln(1) + 1]$$

$$= \frac{\sqrt{2}}{2} - 1 + \ln\sqrt{2}$$

$$= \frac{\sqrt{2}}{2} - 1 + \frac{\ln 2}{2}$$

27. Graph $y_1 = e^x$ and $y_2 = -x^2 - 2x$ on your graphing calculator. Use the intersect command to find the two intersection points. The resulting screens are:

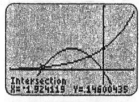

These screens show that $e^x = -x^2 - 2x$ when $x \approx -1.9241$ and $x \approx -0.4164$.
In the interval $[-1.9241, -0.4164]$,

$$e^x < -x^2 - 2x.$$

The area between the curves is given by

$$\int_{-1.9241}^{-0.4164} [(-x^2 - 2x) - e^x]dx.$$

Use the fnInt command to approximate this definite integral.
The resulting screen is:

The last screen shows that the area is approximately 0.6650.

29. (a) The pollution level in the lake is changing at the rate $f(t) - g(t)$ at any time t. We find the amount of pollution by integrating.

$$\int_0^{12} [f(t) - g(t)]dt$$

$$= \int_0^{12} [10(1 - e^{-0.5t}) - 0.4t]dt$$

$$= \left(10t - 10 \cdot \frac{1}{-0.5}e^{-0.5t} - 0.4 \cdot \frac{1}{2}t^2\right)\Big|_0^{12}$$

$$= (20e^{-0.5t} + 10t - 0.2t^2)\Big|_0^{12}$$

$$= [20e^{-0.5(12)} + 10(12) - 0.2(12)^2]$$
$$\quad - [20e^{-0.5(0)} + 10(0) - 0.2(0)^2]$$
$$= (20e^{-6} + 91.2) - (20)$$
$$= 20e^{-6} + 71.2 \approx 71.25$$

After 12 hours, there are about 71.25 gallons.

(b) The graphs of the functions intersect at about 25.00. So the rate that pollution enters the lake equals the rate the pollution is removed at about 25 hours.

(c) $\int_0^{25} [f(t) - g(t)]dt$

$$= (20e^{-0.5t} + 10t - 0.2t^2)\Big|_0^{25}$$
$$= [20e^{-0.5(25)} + 10(25) - 0.2(25)^2] - 20$$
$$= 20e^{-12.5} + 105$$
$$\approx 105$$

After 25 hours, there are about 105 gallons.

(d) For $t > 25$, $g(t) > f(t)$, and pollution is being removed at the rate $g(t) - f(t)$. So, we want to solve for c, where

$$\int_0^c [f(t) - g(t)]dt = 0.$$

(Altternatively, we could solve for c in

$$\int_{25}^c [g(t) - f(t)dt = 105.$$

One way to do this with a graphing calculator is to graph the function

$$y = \int_0^x [f(t) - g(t)]dt$$

and determine the values of x for which $y = 0$. The first window shows how the function can be defined.

A suitable window for the graph is $[0, 50]$ by $[0, 110]$.

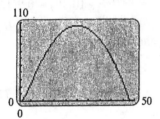

Use the calculator's features to approximate where the graph intersects the x-axis. These are at 0 and about 47.91. Therefore, the pollution will be removed from the lake after about 47.91 hours.

31. **(a)** It is profitable to use the machine until $S(x) = C(x)$.

$$150 - x^2 = x^2 + \frac{11}{4}x$$

$$2x^2 + \frac{11}{4}x - 150 = 0$$

$$8x^2 + 11x - 600 = 0$$

$$x = \frac{-11 \pm \sqrt{121 - 4(8)(-600)}}{16}$$

$$= \frac{-11 \pm 139}{16}$$

$$x = 8 \quad \text{or} \quad x = -9.375$$

It will be profitable to use this machine for 8 years. Reject the negative solution.

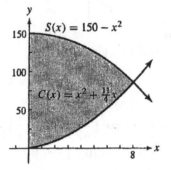

(b) Since $150 - x^2 > x^2 + \frac{11}{4}x$, in the interval $[0, 8]$, the net total savings in the first year are

$$\int_0^1 \left[(150 - x^2) - \left(x^2 + \frac{11}{4}x \right) \right] dx$$

$$= \int_0^1 \left(-2x^2 - \frac{11}{4}x + 150 \right) dx$$

$$= \left(\frac{-2x^3}{3} - \frac{11x^2}{8} + 150x \right) \Big|_0^1$$

$$= -\frac{2}{3} - \frac{11}{8} + 150$$

$$\approx \$148.$$

(c) The net total savings over the entire period of use are

$$\int_0^3 \left[(150 - x^2) - \left(x^2 + \frac{11}{4}x \right) \right] dx$$

$$= \left(\frac{-2x^3}{3} - \frac{11x^2}{8} + 150x \right) \Big|_0^8$$

$$= \frac{-2(8^3)}{3} - \frac{11(8^2)}{8} + 150(8)$$

$$= \frac{-1,024}{3} - \frac{704}{8} + 1,200$$

$$\approx \$771.$$

33. (a) $E(x) = e^{0.1x}$ and $I(x) = 98.8 - e^{0.1x}$

To find the point of intersection, where profit will be maximized, set the functions equal to each other and solve for x.

$$e^{0.1x} = 98.8 - e^{0.1x}$$
$$2e^{0.1x} = 98.8$$
$$e^{0.1x} = 49.4$$
$$0.1x = \ln 49.4$$
$$x = \frac{\ln 49.4}{0.1}$$
$$x \approx 39$$

The optimum number of days for the job to last is 39.

(b) The total income for 39 days is

$$\int_0^{39} (98.8 - e^{0.1x})\, dx$$

$$= \left(98.8x - \frac{e^{0.1x}}{0.1}\right)\Big|_0^{39}$$

$$= \left(98.8x - 10e^{0.1x}\right)\Big|_0^{39}$$

$$= [98.8(39) - 10e^{3.9}] - (0 - 10)$$
$$= \$3{,}369.18.$$

(c) The total expenditure for 39 days is

$$\int_0^{39} e^{0.1x}\, dx = \frac{e^{0.1x}}{0.1}\Big|_0^{39}$$

$$= 10e^{0.1x}\Big|_0^{39}$$

$$= 10e^{3.9} - 10$$
$$= \$484.02.$$

(d) Profit = Income − Expense
$$= 3{,}369.18 - 484.02$$
$$= \$2{,}885.16$$

Chapter 7 Review Exercises

5. $\displaystyle\int (2x + 3)\, dx = \frac{2x^2}{2} + 3x + C$

$$= x^2 + 3x + C$$

7. $\displaystyle\int (x^2 - 3x + 2)\, dx$

$$= \frac{x^3}{3} - \frac{3x^2}{2} + 2x + C$$

9. $\displaystyle\int 3\sqrt{x}\, dx = 3\int x^{1/2}\, dx$

$$= \frac{3x^{3/2}}{\frac{3}{2}} + C$$

$$= 2x^{3/2} + C$$

11. $\displaystyle\int (x^{1/2} + 3x^{-2/3})\, dx$

$$= \frac{x^{3/2}}{\frac{3}{2}} + \frac{3x^{1/3}}{\frac{1}{3}} + C$$

$$= \frac{2x^{3/2}}{3} + 9x^{1/3} + C$$

13. $\displaystyle\int \frac{-4}{x^3}\, dx = \int -4x^{-3}\, dx$

$$= \frac{-4x^{-2}}{-2} + C$$

$$= 2x^{-2} + C$$

15. $\displaystyle\int -3e^{2x}\, dx = \frac{-3e^{2x}}{2} + C$

17. $\displaystyle\int xe^{3x^2}\, dx = \frac{1}{6}\int 6xe^{3x^2}\, dx$

Let $u = 3x^2$, so that $du = 6x\, dx$.

$$= \frac{1}{6}\int e^u\, du$$

$$= \frac{1}{6}e^u + C$$

$$= \frac{e^{3x^2}}{6} + C$$

19. $\displaystyle\int \frac{3x}{x^2 - 1}\, dx = 3\left(\frac{1}{2}\right)\int \frac{2x\, dx}{x^2 - 1}$

Let $u = x^2 - 1$, so that $du = 2x\, dx$.

$$= \frac{3}{2}\int \frac{du}{u}$$

$$= \frac{3}{2}\ln|u| + C$$

$$= \frac{3\ln|x^2 - 1|}{2} + C$$

21. $\displaystyle\int \frac{x^2\,dx}{(x^3+5)^4} = \frac{1}{3}\int \frac{3x^2\,dx}{(x^3+5)^4}$

Let $u = x^3 + 5$, so that

$$du = 3x^2\,dx.$$

$$= \frac{1}{3}\int \frac{du}{u^4}$$

$$= \frac{1}{3}\int u^{-4}\,du$$

$$= \frac{1}{3}\left(\frac{u^{-3}}{-3}\right) + C$$

$$= \frac{-(x^3+5)^{-3}}{9} + C$$

23. $\displaystyle\int \frac{x^3}{e^{3x^4}}\,dx = \int x^3 e^{-3x^4}$

$$= -\frac{1}{12}\int -12x^3 e^{-3x^4}\,dx$$

Let $u = -3x^4$, so that $du = -12x^3\,dx$.

$$= -\frac{1}{12}\int e^u\,du$$

$$= -\frac{1}{12}e^u + C$$

$$= \frac{-e^{-3x^4}}{12} + C$$

25. $\displaystyle\int \sin 2x\,dx$

Let $u = 2x$.
Then $du = 2\,dx$.

$$\frac{1}{2}\,du = dx.$$

$$\int \sin 2x\,dx = \int (\sin u)\left(\frac{1}{2}\,du\right)$$

$$= \frac{1}{2}\int \sin u\,du$$

$$= \frac{1}{2}(-\cos u) + C$$

$$= -\frac{1}{2}\cos 2x + C$$

27. $\displaystyle\int \tan 9x\,dx$

Let $u = 9x$. Then $\frac{1}{9}\,du = dx$.

$$\int \tan 9x\,dx = \int (\tan u)\left(\frac{1}{9}\,du\right)$$

$$= \frac{1}{9}\int \tan u\,du$$

$$= \frac{1}{9}(-\ln|\cos u|) + C$$

$$= -\frac{1}{9}\ln|\cos 9x| + C$$

29. $\displaystyle\int 5\sec^2 x\,dx = 5\int \sec^2 x\,dx$

$$= 5\tan x + C$$

31. $\displaystyle\int x\sin 3x^2\,dx$

Let $u = 3x^2$. Then $\frac{1}{6}\,du = x\,dx$.

$$\int x\sin 3x^2\,dx = \int \sin u\left(\frac{1}{6}\,du\right)$$

$$= \frac{1}{6}\int \sin u\,du$$

$$= \frac{1}{6}(-\cos u) + C$$

$$= -\frac{1}{6}\cos 3x^2 + C$$

33. $\displaystyle\int \sqrt{\cos x}\,\sin x\,dx$

Let $u = \cos x$.
Then $du = -\sin x\,dx$
$-du = \sin x\,dx$.

$$\int \sqrt{\cos x}\,\sin x\,dx = \int \sqrt{u}(-du)$$

$$= -\int u^{1/2}\,du$$

$$= -\frac{2}{3}u^{3/2} + C$$

$$= -\frac{2}{3}(\cos x)^{3/2} + C$$

35. $\displaystyle\int x \tan 11x^2 \, dx$

Let $u = 11x^2$. Then $du = 22x \, dx$.

$\displaystyle\int x \tan 11x^2 \, dx$

$\displaystyle = \frac{1}{22} \int (\tan 11x^2) \cdot (22x \, dx)$

$\displaystyle = \frac{1}{22} \int \tan u \, du$

$\displaystyle = \frac{1}{22}(-\ln|\cos u|) + C$

$\displaystyle = -\frac{1}{22} \ln|\cos 11x^2| + C$

37. $\displaystyle\int (\sin x)^{5/2} \cos x \, dx$

Let $u = \sin x$. Then $du = \cos x \, dx$.

$\displaystyle\int (\sin x)^{5/2} \cos x \, dx$

$\displaystyle = \int u^{5/2} \, du$

$\displaystyle = \frac{u^{5/2+1}}{\frac{5}{2}+1} + C$

$\displaystyle = \frac{u^{7/2}}{\frac{7}{2}} + C$

$\displaystyle = \frac{2}{7} u^{7/2} + C$

$\displaystyle = \frac{2}{7}(\sin x)^{7/2} + C$

39. $\displaystyle\int \sec^2 3x \tan 3x \, dx$

Let $u = \tan 3x$. Then $du = 3 \sec^2 3x \, dx$.

$\displaystyle\int \sec^2 3x \tan 3x \, dx$

$\displaystyle = \frac{1}{3} \int u \, du$

$\displaystyle = \frac{1}{3} \cdot \frac{u^2}{2} + C$

$\displaystyle = \frac{1}{6} u^2 + C$

$\displaystyle = \frac{1}{6} \tan^2 3x + C$

41. (a) $\displaystyle\int_0^4 f(x) \, dx = 0$, since the area above the x-axis from 0 to 2 is identical to the area below the x-axis from 2 to 4.

(b) $\displaystyle\int_0^4 f(x) \, dx$ can be computed by calculating the area of the rectangle and triangle that make up the region shown in graph.

Area of rectangle = (length)(width)
$$= (3)(1) = 3$$

Area of triangle $= \dfrac{1}{2}$(base)(height)

$$= \frac{1}{2}(1)(3) = \frac{3}{2}$$

$$\int_0^4 f(x) \, dx = 3 + \frac{3}{2} = \frac{9}{2} = 4.5$$

43. $\displaystyle\int_0^4 (2x + 3) \, dx$

Graph $y = 2x + 3$.

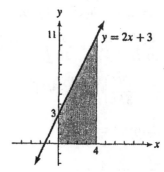

$\displaystyle\int_0^4 (2x+3) \, dx$ is the area of a trapezoid with $B = 11$, $b = 3$, $h = 4$. The formula for the area is

$$A = \frac{1}{2}(B + b)h.$$

$$A = \frac{1}{2}(11 + 3)(4)$$

$$A = 28,$$

so

$$\int_0^4 (2x + 3) \, dx = 28.$$

45. The Fundamental Theorem of Calculus states that

$$\int_a^b f(x) \, dx = F(x) \Big|_a^b = F(b) - F(a),$$

where f is continuous on $[a, b]$ and F is any anti-derivative of f.

47. $\displaystyle\int_1^6 (2x^2 + x)\, dx$

$$= \left(\frac{2x^3}{3} + \frac{x^2}{2}\right)\Bigg|_1^6$$

$$= \left[\frac{2(6)^3}{3} + \frac{(6)^2}{2}\right] - \left[\frac{2(1)^3}{3} + \frac{(1)^2}{2}\right]$$

$$= 144 + 18 - \frac{2}{3} - \frac{1}{2}$$

$$= 162 - \frac{2}{3} - \frac{1}{2}$$

$$= \frac{965}{6}$$

$$\approx 160.83$$

49. $\displaystyle\int_0^1 x\sqrt{5x^2 + 4}\, dx$

Let $u = 5x^2 + 4$, so that

$$du = 10x\, dx \text{ and } \frac{1}{10}\, du = x\, dx.$$

When $x = 0$, $u = 5(0^2) + 4 = 4$.
When $x = 1$, $u = 5(1^2) + 4 = 9$.

$$= \frac{1}{10}\int_4^9 \sqrt{u}\, du = \frac{1}{10}\int_4^9 u^{1/2}\, du$$

$$= \frac{1}{10}\cdot\frac{u^{3/2}}{3/2}\Bigg|_4^9 = \frac{1}{15}u^{3/2}\Bigg|_4^9$$

$$= \frac{1}{15}(9)^{3/2} - \frac{1}{15}(4)^{3/2}$$

$$= \frac{27}{15} - \frac{8}{15}$$

$$= \frac{19}{15}$$

51. $\displaystyle\int_1^6 8x^{-1}\, dx = \int_1^6 \frac{8}{x}\, dx$

$$= 8(\ln x)\Bigg|_1^6$$

$$= 8(\ln 6 - \ln 1)$$

$$= 8\ln 6$$

$$\approx 14.334$$

53. $\displaystyle\int_1^6 \frac{5}{2}e^{4x}\, dx = \frac{1}{4}\cdot\frac{5}{2}\int_1^6 4e^{4x}\, dx$

$$= \frac{5e^{4x}}{8}\Bigg|_1^6$$

$$= \frac{5(e^{24} - e^4)}{8}$$

$$\approx 1.656 \times 10^{10}$$

55. $\displaystyle\int_{\pi/2}^{\pi} \sin x\, dx$

$$= -\cos x \Bigg|_{\pi/2}^{\pi}$$

$$= -\cos \pi + \cos \frac{\pi}{2}$$

$$= 1 - 0 = 1$$

57. $\displaystyle\int_0^{2\pi} (5 + 5\sin x)\, dx$

$$= \int_0^{2\pi} 5\, dx + \int_0^{2\pi} 5\sin x\, dx$$

$$= 5x \Bigg|_0^{2\pi} + (-5\cos x)\Bigg|_0^{2\pi}$$

$$= (10\pi - 0) + [-5 - (-5)]$$

$$= 10\pi$$

59. $f(x) = \sqrt{x - 1};\ [1, 10]$

$$\text{Area} = \int_1^{10} \sqrt{x - 1}\, dx$$

$$= \int_1^{10} (x - 1)^{1/2}\, dx$$

$$= \frac{2}{3}(x - 1)^{3/2}\Bigg|_1^{10}$$

$$= \frac{2}{3}(9)^{3/2} - \frac{2}{3}(0)^{3/2}$$

$$= \frac{2}{3}(27)$$

$$= 18$$

61. $\displaystyle\int_0^2 e^x\, dx = e^x \Bigg|_0^2$

$$= e^2 - e^0$$

$$= e^2 - 1$$

$$\approx 6.3891$$

63. $f(x) = 5 - x^2$, $g(x) = x^2 - 3$

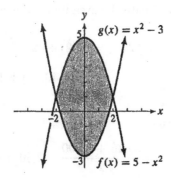

Points of intersection:

$$5 - x^2 = x^2 - 3$$
$$2x^2 - 8 = 0$$
$$2(x^2 - 4) = 0$$
$$x = \pm 2$$

Since $f(x) \geq g(x)$ in $[-2, 2]$, the area between the graphs is

$$\int_{-2}^{2} [f(x) - g(x)]\, dx$$

$$= \int_{-2}^{2} [(5 - x^2) - (x^2 - 3)]\, dx$$

$$= \int_{-2}^{2} (-2x^2 + 8)\, dx$$

$$= \left(\frac{-2x^3}{3} + 8x \right) \Big|_{-2}^{2}$$

$$= -\frac{2}{3}(8) + 16 + \frac{2}{3}(-8) - 8(-2)$$

$$= \frac{-32}{3} + 32$$

$$= \frac{64}{3}.$$

65. $f(x) = x^2 - 4x$, $g(x) = x + 1$, $x = 2$, $x = 4$

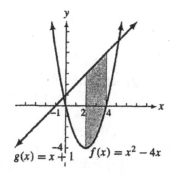

$g(x) > f(x)$ in the interval $[2, 4]$.

$$\int_{2}^{4} [(x + 1) - (x^2 - 4x)]\, dx$$

$$= \int_{2}^{4} (x + 1 - x^2 + 4x)\, dx$$

$$= \int_{2}^{4} (5x + 1 - x^2)\, dx$$

$$= \left(\frac{5x^2}{2} + x - \frac{x^3}{3} \right) \Big|_{2}^{4}$$

$$= \left(\frac{5}{2}(4)^2 + 4 - \frac{(4)^3}{3} \right)$$

$$- \left(\frac{5}{2}(2)^2 + 2 - \frac{(2)^3}{3} \right)$$

$$= \left(40 + 4 - \frac{64}{3} \right) - \left(10 + 2 - \frac{8}{3} \right)$$

$$= \frac{40}{3}$$

67. (a) The total area is the area of the triangle on $[0, 12]$ with height 0.024 plus the area of the rectangle on $[12, 17.6]$ with height 0.024.

$$A = \frac{1}{2}(12 - 0)(0.024) + (17.6 - 12)(0.024)$$

$$= 0.144 + 0.1344$$

$$= 0.2784$$

(b) On $[0, 12]$ we defined the function $f(x)$ with slope $\frac{0.024 - 0}{12 - 0} = 0.002$ and y-intercept 0.

$$f(x) = 0.002x$$

On $[12, 17.6]$, define $g(x)$ as the constant value.

$$g(x) = 0.024.$$

The area is the sum of the integrals of these two functions.

$$A = \int_{0}^{12} 0.002x\, dx + \int_{12}^{17.6} 0.024\, dx$$

$$= 0.001x^2 \Big|_{0}^{12} + 0.024x \Big|_{12}^{17.6}$$

$$= 0.001(12^2 - 0^2) + 0.024(17.6 - 12)$$

$$= 0.144 + 0.1344$$

$$= 0.2784$$

69. Since answers are found by estimating values on the graph, exact answers may vary slightly; however when rounded to the nearest hundred, all answers should be the same. Sample solution:

(a) Left endpoints:

Read the values of the function from the graph, using the open circles for the functional values. The values of x and $f(x)$ are listed in the table.

x	0	2	5	15	30	45	60
$f(x)$	30	50	60	105	85	70	55

The values give the heights of 6 rectangles. The width of each rectangle is found by subtracting subsequent values of x. We estimate the area under the curve as

$$\sum_{i=1}^{6} f(x_i) \triangle x_i = 30(2) + 50(3) + 60(10) + 105(15)$$
$$+ 85(15) + 70(15)$$
$$= 4,710.$$

Right endpoints:

We estimate the area under the curve as

$$\sum_{i=1}^{6} f(x_i) \triangle x_i = 50(2) + 60(3) + 105(10) + 85(15)$$
$$+ 70(15) + 55(15)$$
$$= 4,480.$$

Average:

$$\frac{4,710 + 4,480}{2} = 4,595 \approx 4,600$$

(b) Read the values of the function from the graph, using the closed circles for the functional values. The values of x and $g(x)$ are listed in the table.

x	0	2	5	15	30	45	60
$g(x)$	20	42	42	70	52	40	20

The values give the heights of 6 rectangles. The width of each rectangle is found by subtracting subsequent values of x. We estimate the area under the curve as

$$\sum_{i=1}^{6} g(x_i) \triangle x_i = 20(2) + 42(3) + 42(10) + 70(15)$$
$$+ 52(15) + 40(15)$$
$$= 3,016.$$

Right endpoints:

We estimate the area under the curve as

$$\sum_{i=1}^{6} g(x_i) \triangle x_i = 42(2) + 42(3) + 70(10) + 52(15)$$
$$+ 40(15) + 20(15)$$
$$= 2,590.$$

Average:

$$\frac{3,016 + 2,590}{2} = 2,803 \approx 2,800$$

(c) $\dfrac{4,600 - 2,800}{2,800} \approx 0.6428$

The area under the curve is about 64% more for the fasting sheep.

73. Total amount

$$= \int_0^t 100,000e^{0.03t}\, dt$$
$$= \frac{100,000e^{0.03t}}{0.03}\Big|_0^t$$
$$= \frac{10,000,000}{3}(e^{0.03t} - 1)$$

Set this expression equal to 4,000,000.

$$\frac{10,000,000}{3}(e^{0.03t} - 1) = 4,000,000$$
$$e^{0.03t} - 1 = 1.2$$
$$0.03t = \ln 2.2$$
$$t \approx 26.3$$

It will take him 26.3 years to use up the supply.

75. **(a)** $f(x) = \displaystyle\int 2.158e^{0.0198x}\, dx$

$$= \frac{2.158}{0.0198}e^{0.0198x} + C$$
$$= 109.0e^{0.0198x} + C$$

Since the productivity in 1992 ($x = 2$) was 115,

$$f(2) = 115.$$

Therefore,

$$f(2) = 109.0e^{0.0198(2)} + C = 115$$
$$113.4 + C = 115$$
$$C = 1.6.$$

Thus,

$$f(x) = 109.0e^{0.0198x} + 1.6.$$

(b) 1994 corresponds to $x = 4$.

$$f(x) = 109.0e^{0.0198(4)} + 1.6$$
$$= 119.6$$

If the actual productivity was 118.6, the functional value differs from the actual value by 1.0.

77. The total number of homicides during the 12 year period is

$$\int_0^{12} (5.45x^3 - 105x^2 + 391x + 1{,}798)dx,$$

where x is the time in years since 1988.

$$\int_0^{12} (5.45x^3 - 105x^2 + 391x + 1{,}798)dx$$

$$= \left(\frac{5.45}{4}x^4 - \frac{105}{3}x^3 + \frac{391}{2}x^2 + 1{,}798x\right)\Big|_0^{12}$$

$$= \frac{5.45}{4}(12)^4 - \frac{105}{3}(12)^3 + \frac{391}{2}(12)^2$$

$$+ 1{,}798(12) - 0$$

$$= 17{,}500.8$$

There were about 17,500 homicides during the time from the beginning of 1988 to the end of 1999.

79. For each month, subtract the average temperature from 65° (if it falls below 65°F), then multiply this number times the number of days in the month. The sum is the total number of heating degree days. Readings may vary, but the sum is approximately 4,800 degree-days. (The actual value is 4,868 degree-days.)

81. Let $x = $ base
$y = $ height

From the Pythagorean theorem

$$x^2 + y^2 = 6^2$$
$$y^2 = \sqrt{36 - x^2}$$

Use the area formula for a triangle.

$$A = \frac{1}{2} \cdot \text{base} \cdot \text{height}$$

$$= \frac{1}{2}x\sqrt{36 - x^2}$$

$$\frac{dA}{dx} = -\frac{x^2}{2\sqrt{36 - x^2}} + \frac{\sqrt{36 - x^2}}{2}$$

Set $\frac{dA}{dx}$ to zero and solve for x.

$$-\frac{x^2}{2\sqrt{36 - x^2}} + \frac{\sqrt{36 - x^2}}{2} = 0$$

$$\frac{x^2}{2\sqrt{36 - x^2}} = \frac{\sqrt{36 - x^2}}{2}$$

$$2x^2 = 2(36 - x^2)$$
$$4x^2 = 72$$
$$x = \sqrt{18}$$
$$x = 3\sqrt{2}$$

$$\theta = \cos^{-1}\left(\frac{3\sqrt{2}}{6}\right) = \cos^{-1}\left(\frac{\sqrt{2}}{2}\right) = \frac{\pi}{4}$$

The maximum area occurs when θ equals $\frac{\pi}{4}$ or 45°.

Chapter 7 Test

[7.1]

1. Explain what is meant by an antiderivative of a function $f(x)$.

Find each indefinite integral.

2. $\int \left(3x^3 - 5x^2 + x + 1\right)\, dx$ 3. $\int \left(\dfrac{5}{x} + e^{0.5x}\right)\, dx$ 4. $\int \dfrac{2x^3 - 3x^2}{\sqrt{x}}\, dx$

5. Find the cost function $C(x)$ if the marginal cost function is given by $C'(x) = 200 + 2x^{-1/4}$ and 16 units cost $4,000.

6. A ball is thrown upward at time $t = 0$ with initial velocity of 64 feet per second from a height of 100 feet. Assume that $a(t) = -32$ feet per second per second. Find $v(t)$ and $s(t)$.

7. **Biochemical Excretion** If the rate of excretion of a biochemical compound is given by
$$f'(t) = 0.02e^{-0.02t},$$
the total amount excreted by time t (in minutes) is $f(t)$.

   (a) Find an expression for $f(t)$.

   (b) If 0 units are excreted at time $t = 0$, how many units are excreted in 10 minutes?

[7.2]

Use substitution to find each indefinite integral.

8. $\int 6x^2 \left(3x^3 - 5\right)^8\, dx$ 10. $\int \sqrt[3]{2x^2 - 8x}\,(x - 2)\, dx$

9. $\int \dfrac{6x + 5}{3x^2 + 5x}\, dx$ 11. $\int 4x^3 e^{-x^4}\, dx$

12. A city's population is predicted to grow at a rate of
$$P'(t) = \dfrac{400e^{10t}}{1 + e^{10t}}$$
people per year where t is the time in years from the present. Find the total population 3 years from now if $P(0) = 100,000$.

13. **Infection Rate** The rate of infection of a disease (in people per month) is given by the function
$$I'(t) = \dfrac{100t}{t^2 + 1},$$
where t is the time in months since the disease broke out. Find the total number of infected people over the first three months of the disease.

[7.3]

14. Evaluate $\displaystyle\sum_{i=1}^{4} \frac{2}{i^2}$.

15. Approximate the area under the graph of $f(x) = x^2 + x$ and above the x-axis from $x = 0$ to $x = 2$ using four rectangles of equal width. Let the height of each rectangle be the function value at the left endpoint.

16. Approximate the value of $\int_1^5 x^2 \, dx$ by summing the areas of rectangles. Use four rectangles of equal width. Use the left endpoints, then the right endpoints; then give the average of these answers.

[7.4]

Evaluate the following definite integrals.

17. $\displaystyle\int_1^5 \left(4x^3 - 5x\right) dx$

18. $\displaystyle\int_1^5 \left(\frac{3}{x^2} + \frac{2}{x}\right) dx$

19. $\displaystyle\int_1^2 \sqrt{3r - 2} \, dr$

20. $\displaystyle\int_0^1 4xe^{x^2+1} \, dx$

21. $\displaystyle\int_{-2}^1 3x \left(x^2 - 4\right)^5 dx$

22. **Rams' Horns** The average annual increment in the horn length (in cm) of bighorn rams born since 1986 can be approximated by

$$y = 0.1762x^2 - 3.986x + 22.68,$$

where x is the ram's age in years, for x between 3 and 9. Integrate to find the total increase in the length of a ram's horn from year 4 to year 7.

23. Find the area of the region between the x-axis and $f(x) = e^{x/2}$ on the interval $[0, 2]$.

24. The rate at which a substance grows is given by $R(x) = 500e^{0.5x}$, where x is the time in days. What is the total accumulated growth after 4 days?

[7.5]

Find each integral.

25. $\displaystyle\int 3\sin 2x \, dx$

26. $\displaystyle\int 4\sec^2 x \, dx$

27. $\displaystyle\int x\cos\left(2x^2\right) dx$

28. $\displaystyle\int \tan 3x \, dx$

29. $\displaystyle\int \left(\cos x\right)^2 \sin x \, dx$

30. $\displaystyle\int_{\pi/6}^{\pi/3} \cos x \, dx$

31. $\displaystyle\int_0^{\pi/4} \sec^2 x \, dx$

32. $\displaystyle\int_0^{\pi/4} x \tan x^2 \, dx$

33. If b is a positive real number, explain why $\displaystyle\int_{-b}^{b} \sin x \, dx$ will always be zero.

34. Migratory Animals The number of migratory animals (in hundreds) counted at a certain checkpoint is given by
$$T(t) = 50 + 50 \cos\left(\frac{\pi}{6} t\right),$$
where t is time in months, with $t = 0$ corresponding to July. Use a definite integral to find the number of animals passing the checkpoint from July through December.

[7.6]

35. Find the area of the region enclosed by $f(x) = -x + 4$, $g(x) = -x^2 + 6x - 6$, $x = 2$, and $x = 4$.

36. Find the area of the region enclosed by $f(x) = 5x$ and $g(x) = x^3 - 4x$.

37. A company has determined that the use of a new process would produce a savings rate (in thousands of dollars) of
$$S(x) = 2x + 7,$$
where x is the number of years the process is used. However, the use of this process also creates additional costs (in thousands of dollars) according to the rate-of-cost function
$$C(x) = x^2 + 2x + 3.$$

(a) For how many years does the new process save the company money?

(b) Find the net savings in thousands of dollars over this period.

38. Pollution Pollution begins to enter a lake at time $t = 0$ at a rate (in gallons per hour) given by the formula
$$f(t) = 10\left(1 - e^{-0.5t}\right),$$
where t is the time in hours. At the same time, a pollution filter begins to remove the pollution at a rate
$$g(t) = 0.4t$$
as long as $g(t) \le f(t)$. If $g(t) > f(t)$, the pollution is removed at the same rate that it enters the lake. How much pollution is in the lake after 6 hours?

Chapter 7 Test Answers

1. An antiderivative of a function $f(x)$ is a function $F(x)$ such that $F'(x) = f(x)$.

2. $\frac{3}{4}x^4 - \frac{5}{3}x^3 + \frac{x^2}{2} + x + C$

3. $5\ln|x| + 2e^{0.5x} + C$

4. $\frac{4}{7}x^{7/2} - \frac{6}{5}x^{5/2} + C$

5. $C(x) = 200x + \frac{8}{3}x^{3/4} + 778.67$

6. $v(t) = -32t + 64$; $s(t) = -16t^2 + 64t + 100$

7. (a) $f(t) = -e^{-0.02t} + k$ (b) 0.181 unit

8. $\frac{2}{27}(3x^3 - 5)^9 + C$

9. $\ln|3x^2 + 5x| + C$

10. $\frac{3}{16}(2x^2 - 8x)^{4/3} + C$

11. $-e^{-x^4} + C$

12. 101,172

13. $50\ln 10 \approx 115.13$; about 115 people

14. $2\frac{61}{72}$ or 2.85

15. 3.25

16. Left endpoints: 30; right endpoints: 54; average: 42

17. 564

18. $2\ln 5 + \frac{12}{5}$ or 5.62

19. $\frac{14}{9}$ or 1.56

20. $2e^2 - 2e$ or 9.34

21. $\frac{729}{4}$ or 182.25

22. 18.7 cm

23. $2e - 2$ or 3.44

24. 6,389

25. $-\frac{3}{2}\cos 2x + C$

26. $4\tan x + C$

27. $\frac{1}{4}\sin(2x^2) + C$

28. $-\frac{1}{3}\ln|\cos 3x| + C$

29. $-\frac{1}{3}\cos^3 x + C$

30. $\frac{\sqrt{3}-1}{2}$ or 0.366

31. 1

32. $-\frac{1}{2}\ln\left(\cos\frac{\pi^2}{16}\right)$ or 0.101852

33. $\displaystyle\int_{-b}^{b} \sin x \, dx = -\cos x \Big|_{-b}^{b}$
$$= -\cos b - [-\cos(-b)]$$
$$= -\cos b + \cos(-b)$$

For the cosine function, $\cos(-x) = \cos x$ for all values of x. Thus, $-\cos b + \cos(-b) = 0$ for all values of b, and $\displaystyle\int_{-b}^{b} \sin x \, dx = 0$ for all real numbers b.

34. 30,000

35. $\frac{10}{3}$ or 3.33

36. 40.5

37. (a) 2 years   (b) $5.33 thousand

38. 33.80 gal

FURTHER TECHNIQUES AND APPLICATIONS OF INTEGRATION

8.1 Numerical Integration

1. $\int_0^2 x^2\,dx$

$n = 4,\ b = 2,\ a = 0,\ f(x) = x^2$

i	x_i	$f(x_i)$
0	0	0
1	$\frac{1}{2}$	0.25
2	1	1
3	$\frac{3}{2}$	2.25
4	2	4

(a) Trapezoidal rule:

$\int_0^2 x^2\,dx$

$\approx \dfrac{2-0}{4}\left[\dfrac{1}{2}(0) + 0.25 + 1 + 2.25 + \dfrac{1}{2}(4)\right]$

$= 0.5(5.5)$

$= 2.7500$

(b) Simpson's rule:

$\int_0^2 x^2\,dx$

$\approx \dfrac{2}{3(4)}[0 + 4(0.25) + 2(1) + 4(2.25) + 4]$

$= \dfrac{2}{12}(16)$

≈ 2.6667

(c) Exact value:

$$\int_0^2 x^2\,dx = \left.\dfrac{x^3}{3}\right|_0^2$$

$$= \dfrac{8}{3} \approx 2.6667$$

3. $\int_{-1}^3 \dfrac{1}{4-x}\,dx$

$n = 4,\ b = 3,\ a = -1,\ f(x) = \dfrac{1}{4-x}$

i	x_i	$f(x_i)$
0	-1	0.2
1	0	0.25
2	1	0.3333
3	2	0.5
4	3	1

(a) Trapezoidal rule:

$\int_{-1}^3 \dfrac{1}{4-x}\,dx$

$\approx \dfrac{3-(-1)}{4}\cdot\left[\dfrac{1}{2}(0.2) + 0.25 + 0.3333 + 0.5 + \dfrac{1}{2}(1)\right]$

≈ 1.6833

(b) Simpson's rule:

$\int_{-1}^3 \dfrac{1}{4-x}\,dx$

$\approx \dfrac{3-(-1)}{12}\cdot[0.2+4(0.25)+2(0.3333)+4(0.5)+1]$

≈ 1.6222

(c) Exact value:

$$\int_{-1}^3 \dfrac{1}{4-x}\,dx = -\int_{-1}^3 \dfrac{-dx}{4-x} = -(\ln|4-x|)\Big|_{-1}^3$$

$$= -\ln 1 + \ln 5 = \ln 5 \approx 1.6094$$

5. $\int_{-2}^2 (2x^2 + 1)\,dx$

$n = 4,\ b = 2,\ a = -2,\ f(x) = 2x^2 + 1$

i	x_i	$f(x)$
0	-2	9
1	-1	3
2	0	1
3	1	3
4	2	9

(a) Trapezoidal rule:

$$\int_{-2}^{2} (2x^2 + 1)\, dx$$

$$\approx \frac{2 - (-2)}{4} \cdot \left[\frac{1}{2}(9) + 3 + 1 + 3 + \frac{1}{2}(9)\right]$$

$$\approx 16$$

(b) Simpson's rule:

$$\int_{-2}^{2} (2x^2 + 1)\, dx$$

$$\approx \frac{2 - (-2)}{12}$$

$$\cdot \left[9 + 4(3) + 2(1) + 4(3) + 9\right]$$
$$\approx 14.6667$$

(c) Exact value:

$$\int_{-2}^{2} (2x^2 + 1)\, dx$$

$$= \left(\frac{2x^3}{3} + x\right)\Big|_{-2}^{2}$$

$$= \left(\frac{16}{3} + 2\right) - \left(-\frac{16}{3} - 2\right)$$

$$= \frac{32}{3} + 4$$

$$= \frac{44}{3} \approx 14.667$$

7. $\displaystyle\int_{1}^{5} \frac{1}{x^2}\, dx$

$n = 4,\ b = 5,\ a = 1,\ f(x) = \dfrac{1}{x^2}$

i	x_i	$f(x_i)$
0	1	1
1	2	0.25
2	3	0.1111
3	4	0.0625
4	5	0.04

(a) Trapezoidal rule:

$$\int_{1}^{5} \frac{1}{x^2}\, dx$$

$$\approx \frac{5 - 1}{4} \cdot \left[\frac{1}{2}(1) + 0.25 + 0.1111 + 0.0625 + \frac{1}{2}(0.04)\right]$$

$$\approx 0.9436$$

(b) Simpson's rule:

$$\int_{1}^{5} \frac{1}{x^2}\, dx$$

$$\approx \frac{5 - 1}{12} \cdot [1 + 4(0.25) + 2(0.1111) + 4(0.0625) + 0.04]$$

$$\approx 0.8374$$

(c) Exact value:

$$\int_{1}^{5} x^{-2}\, dx = -x^{-1}\Big|_{1}^{5}$$

$$= -\frac{1}{5} + 1$$

$$= \frac{4}{5} = 0.8$$

9. $\displaystyle\int_{0}^{1} xe^{x^2}\, dx$

$n = 4,\ b = 1,\ a = 0,\ f(x) = xe^{x^2}$

i	x_i	$f(x_i)$
0	0	0
1	$\frac{1}{4}$	0.26612
2	$\frac{1}{2}$	0.64201
3	$\frac{3}{4}$	1.3163
4	1	2.7183

(a) Trapezoidal rule:

$$\int_{0}^{1} xe^{x^2}\, dx$$

$$\approx \frac{1 - 0}{4} \left[\frac{1}{2}(0) + 0.26612 + 0.64201\right.$$

$$\left. + 1.3163 + \frac{1}{2}(2.7183)\right]$$

$$\approx \frac{1}{4}[3.58358] \approx 0.895895$$

(b) Simpson's rule:

$$\int_{0}^{1} xe^{x^2}\, dx$$

$$\approx \frac{1 - 0}{3(4)}[0 + 4(0.26612) + 2(0.64201)$$

$$+ 4(1.3163) + 2.7183]$$

$$\approx \frac{1}{12}[10.332] \approx 0.861$$

(c) Exact value:

$$\int_0^1 xe^{x^2}\,dx = \frac{1}{2}e^{x^2}\Big|_0^1$$

$$= \frac{e}{2} - \frac{1}{2}$$

$$= 0.859141$$

11. $\int_0^\pi \sin x\,dx$

$n = 4,\ b = \pi,\ a = 0,\ f(x) = \sin x$

i	x_i	$f(x_i)$
0	0	0
1	$\frac{\pi}{4}$	0.70711
2	$\frac{\pi}{2}$	1
3	$\frac{3\pi}{4}$	0.70711
4	π	0

(a) Trapezoidal rule:

$$\int_0^\pi \sin x\,dx$$

$$\approx \frac{\pi - 0}{4}\left[\frac{1}{2}(0) + 0.70711 + 1 + 0.70711 + \frac{1}{2}(0)\right]$$

$$\approx 1.8961$$

(b) Simpson's rule:

$$\int_0^\pi \sin x\,dx$$

$$\approx \frac{\pi - 0}{3(4)}[0 + 4(0.70711) + 2(1) + 4(0.70711) + 0]$$

$$\approx 2.0046$$

(c) Exact value:

$$\int_0^\pi \sin x\,dx = -\cos x\Big|_0^\pi$$

$$= -\cos(\pi) + \cos(0)$$

$$= -(-1) + 1$$

$$= 2$$

13. $y = \sqrt{4 - x^2}$

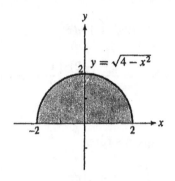

$n = 8,\ b = 2,\ a = -2,\ f(x) = \sqrt{4 - x^2}$

i	x_i	y
0	-2.0	0
1	-1.5	1.32289
2	-1.0	1.73205
3	-0.5	1.93649
4	0	2
5	0.5	1.93649
6	1.0	1.73205
7	1.5	1.32289
8	2.0	0

(a) Trapezoidal rule:

$$\int_{-2}^2 \sqrt{4 - x^2}\,dx$$

$$\approx \frac{2 - (-2)}{8}$$

$$\cdot \left[\frac{1}{2}(0) + 1.32289 + 1.73205 + \cdots + \frac{1}{2}(0)\right]$$

$$\approx 5.9914$$

(b) Simpson's rule:

$$\int_{-2}^2 \sqrt{4 - x^2}\,dx$$

$$\approx \frac{2 - (-2)}{3(8)}$$

$$\cdot [0 + 4(1.32289) + 2(1.73205) + 4(1.93649) + 2(2)$$
$$+ 4(1.93649) + 2(1.73205) + 4(1.32289) + 0]$$

$$\approx 6.1672$$

(c) Area of semicircle $= \frac{1}{2}\pi r^2$

$$= \frac{1}{2}\pi(2)^2$$

$$\approx 6.2832$$

Simpson's rule is more accurate.

15. Since $f(x) > 0$ and $f''(x) > 0$ for all x between a and b, we know the graph of $f(x)$ on the interval from a to b is concave upward. Thus, the trapezoid that approximates the area will have an area greater than the actual area. Thus,

$$T > \int_a^b f(x)\,dx.$$

The correct choice is (b).

17. (a) $\displaystyle\int_0^1 x^4\,dx = \left(\frac{1}{5}\right)x^5\Big|_0^1$

$$= \frac{1}{5}$$

$$= 0.2$$

(b) $n = 4$, $b = 1$, $a = 0$, $f(x) = x^4$

$$\int_0^1 x^4\,dx \approx \frac{1-0}{4}\left[\frac{1}{2}(0) + \frac{1}{256} + \frac{1}{16} + \frac{81}{256} + \frac{1}{2}(1)\right]$$

$$= \frac{1}{4}\left(\frac{226}{256}\right)$$

$$\approx 0.220703$$

$n = 8$, $b = 1$, $a = 0$, $f(x) = x^4$

$$\int_0^1 x^4\,dx \approx \frac{1-0}{8}\left[\frac{1}{2}(0) + \frac{1}{4,096} + \frac{1}{256} + \frac{81}{4,096}\right.$$

$$\left. + \frac{1}{16} + \frac{625}{4,096} + \frac{81}{256} + \frac{2,401}{4,096} + \frac{1}{2}(1)\right]$$

$$= \frac{1}{8}\left(\frac{6,724}{4,096}\right)$$

$$\approx 0.205200$$

$n = 16$, $b = 1$, $a = 0$, $f(x) = x^4$

$$\int_0^1 x^4\,dx \approx \frac{1-0}{16}\left[\frac{1}{2}(0) + \frac{1}{65,536} + \frac{1}{4,096}\right.$$

$$+ \frac{81}{65,536} + \frac{1}{256} + \frac{625}{65,536}$$

$$+ \frac{81}{4,096} + \frac{2,401}{65,536} + \frac{1}{16}$$

$$+ \frac{6,561}{65,536} + \frac{625}{4,096} + \frac{14,641}{65,536}$$

$$+ \frac{81}{256} + \frac{28,561}{65,536} + \frac{2,401}{4,096}$$

$$\left. + \frac{50,625}{65,536} + \frac{1}{2}(1)\right]$$

$$\approx \frac{1}{16}\left(\frac{211,080}{65,536}\right)$$

$$\approx 0.201302$$

$n = 32$, $b = 1$, $a = 0$, $f(x) = x^4$

$$\int_0^1 x^4\,dx$$

$$\approx \frac{1-0}{32}\left[\frac{1}{2}(0) + \frac{1}{1,048,576} + \frac{1}{65,536}\right.$$

$$+ \frac{81}{1,048,576} + \frac{1}{4,096} + \frac{625}{1,048,576}$$

$$+ \frac{81}{65,536} + \frac{2,401}{1,048,576} + \frac{1}{256} + \frac{6,561}{1,048,576}$$

$$+ \frac{625}{65,536} + \frac{14,641}{1,048,576} + \frac{81}{4,096} + \frac{28,561}{1,048,576}$$

$$+ \frac{2,401}{65,536} + \frac{50,625}{1,048,576} + \frac{1}{16} + \frac{83,521}{1,048,576}$$

$$+ \frac{6,561}{65,536} + \frac{130,321}{1,048,576} + \frac{625}{4,096} + \frac{194,481}{1,048,576}$$

$$+ \frac{14,641}{65,536} + \frac{279,841}{1,048,576} + \frac{81}{256} + \frac{390,625}{1,048,576}$$

$$+ \frac{28,561}{65,536} + \frac{531,441}{1,048,576} + \frac{2,401}{4,096} + \frac{707,281}{1,048,576}$$

$$\left. + \frac{50,625}{65,536} + \frac{923,521}{1,048,576} + \frac{1}{2}(1)\right]$$

$$\approx \frac{1}{32}\left(\frac{6,721,808}{1,048,576}\right) \approx 0.200325$$

To find error for each value of n, subtract as indicated.

$n = 4$: $(0.220703 - 0.2) = 0.020703$
$n = 8$: $(0.205200 - 0.2) = 0.005200$
$n = 16$: $(0.201302 - 0.2) = 0.001302$
$n = 32$: $(0.200325 - 0.2) = 0.000325$

(c) $p = 1$

$$4^1(0.020703) = 4(0.020703)$$
$$= 0.082812$$
$$8^1(0.005200) = 8(0.005200)$$
$$= 0.0416$$

Since these are not the same, try $p = 2$.

$p = 2$:

$$4^2(0.020703) = 16(0.020703)$$
$$= 0.331248$$
$$8^2(0.005200) = 64(0.005200) = 0.3328$$
$$16^2(0.001302) = 256(0.001302)$$
$$= 0.333312$$
$$32^2(0.000325) = 1,024(0.000325)$$
$$= 0.3328$$

Since these values are all approximately the same, the correct choice is $p = 2$.

19. (a) $\int_0^1 x^4\, dx = \frac{1}{5}x^5\Big|_0^1$

$$= \frac{1}{5}$$

$$= 0.2$$

(b) $n = 4$, $b = 1$, $a = 0$, $f(x) = x^4$

$$\int_0^1 x^4\, dx \approx \frac{1-0}{3(4)}\left[0 + 4\left(\frac{1}{256}\right) + 2\left(\frac{1}{16}\right) \right.$$
$$\left. + 4\left(\frac{81}{256}\right) + 1\right]$$

$$= \frac{1}{12}\left(\frac{77}{32}\right)$$

$$\approx 0.2005208$$

$n = 8$, $b = 1$, $a = 0$, $f(x) = x^4$

$$\int_0^1 x^4\, dx \approx \frac{1-0}{3(8)}\left[0 + 4\left(\frac{1}{4,096}\right) + 2\left(\frac{1}{256}\right)\right.$$
$$+ 4\left(\frac{81}{4,096}\right) + 2\left(\frac{1}{16}\right) + 4\left(\frac{625}{4,096}\right)$$
$$\left. + 2\left(\frac{81}{256}\right) + 4\left(\frac{2,401}{4,096}\right) + 1\right]$$

$$= \frac{1}{24}\left(\frac{4,916}{1,024}\right)$$

$$\approx 0.2000326$$

$n = 16$, $b = 1$, $a = 0$, $f(x) = x^4$

$$\int_0^1 x^4\, dx$$

$$\approx \frac{1-0}{3(16)}\left[0 + 4\left(\frac{1}{65,536}\right) + 2\left(\frac{1}{4,096}\right)\right.$$
$$+ 4\left(\frac{81}{65,536}\right) + 2\left(\frac{1}{256}\right) + 4\left(\frac{625}{65,536}\right)$$
$$+ 2\left(\frac{81}{4,096}\right) + 4\left(\frac{2,401}{65,536}\right) + 2\left(\frac{1}{16}\right)$$
$$+ 4\left(\frac{6,561}{65,536}\right) + 2\left(\frac{625}{4,096}\right) + 4\left(\frac{14,641}{65,536}\right)$$
$$+ 2\left(\frac{81}{256}\right) + 4\left(\frac{28,561}{65,536}\right) + 2\left(\frac{2,401}{4,096}\right)$$
$$\left. + 4\left(\frac{50,625}{65,536} + 1\right)\right]$$

$$= \frac{1}{48}\left(\frac{157,288}{16,384}\right) \approx 0.2000020$$

$n = 32$, $b = 1$, $a = 0$, $f(x) = x^4$

$$\int_0^1 x^4\, dx$$

$$\approx \frac{1-0}{3(32)}\left[0 + 4\left(\frac{1}{1,048,576}\right) + 2\left(\frac{1}{65,536}\right)\right.$$
$$+ 4\left(\frac{81}{1,048,576}\right) + 2\left(\frac{1}{4,096}\right) + 4\left(\frac{625}{1,048,576}\right)$$
$$+ 2\left(\frac{625}{65,536}\right) + 4\left(\frac{14,641}{1,048,576}\right) + 2\left(\frac{81}{4,096}\right)$$
$$+ 4\left(\frac{28,561}{1,048,576}\right) + 2\left(\frac{2,401}{65,536}\right) + 4\left(\frac{50,625}{1,048,576}\right)$$
$$+ 2\left(\frac{1}{16}\right) + 4\left(\frac{83,521}{1,048,576}\right) + 2\left(\frac{6,561}{65,536}\right)$$
$$+ 4\left(\frac{130,321}{1,048,576}\right) + 2\left(\frac{625}{4,096}\right) + 4\left(\frac{194,481}{1,048,576}\right)$$
$$+ 2\left(\frac{14,641}{65,536}\right) + 4\left(\frac{279,841}{1,048,576}\right) + 2\left(\frac{81}{256}\right)$$
$$+ 4\left(\frac{390,625}{1,048,576}\right) + 2\left(\frac{28,561}{65,536}\right) + 4\left(\frac{531,441}{1,048,576}\right)$$
$$+ 2\left(\frac{2,401}{4,096}\right) + 4\left(\frac{707,281}{1,048,576}\right) + 2\left(\frac{50,625}{65,536}\right)$$
$$\left. + 4\left(\frac{923,521}{1,048,576}\right) + 1\right]$$

$$= \frac{1}{96}\left(\frac{50,033,168}{262,144}\right) \approx 0.2000001$$

To find error for each value of n, subtract as indicated.

$n = 4$: $(0.2005208 - 0.2) = 0.0005208$

$n = 8$: $(0.2000326 - 0.2) = 0.0000326$

$n = 16$: $(0.2000020 - 0.2) = 0.0000020$

$n = 32$: $(0.2000001 - 0.2) = 0.0000001$

(c) $p = 1$:

$$4^1(0.0005208) = 4(0.0005208) = 0.0020832$$
$$8^1(0.0000326) = 8(0.0000326) = 0.0002608$$

Try $p = 2$:

$$4^2(0.0005208) = 16(0.0005208) = 0.0083328$$
$$8^2(0.0000326) = 64(0.0000326) = 0.0020864$$

Try $p = 3$:

$$4^3(0.0005208) = 64(0.0005208) = 0.0333312$$
$$8^3(0.0000326) = 512(0.0000326) = 0.0166912$$

Try $p = 4$:

$$4^4(0.0005208) = 256(0.0005208) = 0.1333248$$
$$8^4(0.0000326) = 4,096(0.0000326) = 0.1335296$$
$$16^4(0.0000020) = 65,536(0.0000020) = 0.131072$$
$$32^4(0.0000001) = 1,048,576(0.0000001) = 0.1048576$$

These are the closest values we can get; thus, $p = 4$.

21. Midpoint rule:

$n = 4$, $b = 5$, $a = 1$, $f(x) = \dfrac{1}{x^2}$, $\Delta x = 1$

i	x_i	$f(x_i)$
1	$\dfrac{3}{2}$	$\dfrac{4}{9}$
2	$\dfrac{5}{2}$	$\dfrac{4}{25}$
3	$\dfrac{7}{2}$	$\dfrac{4}{49}$
4	$\dfrac{9}{2}$	$\dfrac{4}{81}$

$$\int_1^5 \frac{1}{x^2}\,dx \approx \sum_{i=1}^{4} f(x_i)\Delta x$$

$$= \frac{4}{9}(1) + \frac{4}{25}(1) + \frac{4}{49}(1) + \frac{4}{81}(1)$$

$$\approx 0.7355$$

Simpson's rule:

$m = 8$, $b = 5$, $a = 1$, $f(x) = \dfrac{1}{x^2}$

i	x_i	$f(x_i)$
0	1	1
1	$\dfrac{3}{2}$	$\dfrac{4}{9}$
2	2	$\dfrac{1}{4}$
3	$\dfrac{5}{2}$	$\dfrac{4}{25}$
4	3	$\dfrac{1}{9}$
5	$\dfrac{7}{2}$	$\dfrac{4}{49}$
6	4	$\dfrac{1}{16}$
7	$\dfrac{9}{2}$	$\dfrac{4}{81}$
8	5	$\dfrac{1}{25}$

$$\int_1^5 \frac{1}{x^2}\,dx$$

$$\approx \frac{5-1}{3(8)}\left[1 + 4\left(\frac{4}{9}\right) + 2\left(\frac{1}{4}\right) + 4\left(\frac{4}{25}\right)\right.$$

$$+ 2\left(\frac{1}{9}\right) + 4\left(\frac{4}{49}\right) + 2\left(\frac{1}{16}\right)$$

$$\left. + 4\left(\frac{4}{81}\right) + \frac{1}{25}\right]$$

$$\approx \frac{1}{6}(4.82906)$$

$$\approx 0.8048$$

From #7 part a, $T \approx 0.9436$, when $n = 4$. To verify the formula evaluate $\frac{2M+T}{3}$.

$$\frac{2M+T}{3} \approx \frac{2(0.7355) + 0.9436}{3}$$

$$\approx 0.8048$$

23. $y = b_0 w^{b_1} e^{-b_2 w}$

(a) If $t = 7w$ then $w = \dfrac{t}{7}$.

$$y = b_0 \left(\frac{t}{7}\right)^{b_1} e^{-b_2 t/7}$$

(b) Replacing the constants with the given values, we have

$$y = 5.955 \left(\frac{t}{7}\right)^{0.233} e^{-0.027t/7}\,dt$$

In 25 weeks, there are 175 days.

$$\int_0^{175} 5.955 \left(\frac{t}{7}\right)^{0.233} e^{-0.027t/7}\,dt$$

$n = 10$, $b = 175$, $a = 0$,

$$f(t) = 5.955 \left(\frac{t}{7}\right)^{0.233} e^{-0.027t/7}$$

i	t_i	$f(t_i)$
0	0	0
1	17.5	6.89
2	35	7.57
3	52.5	7.78
4	70	7.77
5	87.5	7.65
6	105	7.46
7	122.5	7.23
8	140	6:97
9	157.5	6.70
10	175	6.42

Trapezoidal rule:

$$\int_0^{175} 5.955 \left(\frac{t}{7}\right)^{0.233} e^{-0.027t/7} dt$$

$$\approx \frac{175 - 0}{10}\left[\frac{1}{2}(0) + 6.89 + 7.57 + 7.78 + 7.77\right.$$

$$\left. + 7.65 + 7.46 + 7.23 + 6.97 + 6.70 + \frac{1}{2}(6.42)\right]$$

$$= 17.5(69.23)$$
$$= 1,211.525$$

The total milk consumed is about 1,212 kg.

Simpson's rule:

$$\int_0^{175} 5.955 \left(\frac{t}{7}\right)^{0.233} e^{-0.027t/7} dt$$

$$\approx \frac{175 - 0}{3(10)}[0 + 4(6.89) + 2(7.57) + 4(7.78)$$

$$+ 2(7.77) + 4(7.65) + 2(7.46) + 4(7.23)$$
$$+ 2(6.97) + 4(6.70) + 6.42]$$

The total milk consumed is about 1,231 kg.

(c) Replacing the constants with the given values, we have

$$y = 8.409 \left(\frac{t}{7}\right)^{0.143} e^{-0.037t/7}.$$

In 25 weeks, there are 175 days.

$$\int_0^{175} 8.409 \left(\frac{t}{7}\right)^{0.143} e^{-0.037t/7} dt$$

$n = 10$, $b = 175$, $a = 0$,

$$f(t) = 8.409 \left(\frac{t}{7}\right)^{0.143} e^{-0.037t/7}$$

i	t_i	$f(t_i)$
0	0	0
1	17.5	8.74
2	35	8.80
3	52.5	8.50
4	70	8.07
5	87.5	7.60
6	105	7.11
7	122.5	6.63
8	140	6.16
9	157.5	5.71
10	175	5.28

Trapezoidal rule:

$$\int_0^{175} 8.409 \left(\frac{t}{7}\right)^{0.143} e^{-0.037t/7} dt$$

$$\approx \frac{175 - 0}{10}\left[\frac{1}{2}(0) + 8.74 + 8.80 + 8.50 + 8.07\right.$$

$$\left. + 7.60 + 7.11 + 6.63 + 6.16 + 5.71 + \frac{1}{2}(5.28)\right]$$

$$= 17.5(69.96)$$
$$= 1,224.30$$

The total milk consumed is about 1,224 kg.

Simpson's rule:

$$\int_0^{175} 8.409 \left(\frac{t}{7}\right)^{0.143} e^{-0.037t/7} dt$$

$$\approx \frac{175 - 0}{3(10)}[0 + 4(8.74) + 2(8.80) + 4(8.50)$$

$$+ 2(8.07) + 4(7.60) + 2(7.11) + 4(6.63)$$
$$+ 2(6.16) + 4(5.71) + 5.28]$$

$$= \frac{35}{6}(214.28)$$

$$= 1,249.97$$

The total milk consumed is about 1,250 kg.

25. $y = e^{-t^2} + \frac{1}{t}$

The total reaction is

$$\int_1^9 \left(e^{-t^2} + \frac{1}{t}\right) dt.$$

$n = 8$, $b = 9$, $a = 1$, $f(t) = e^{-t^2} + \frac{1}{t}$

i	x_i	$f(x_i)$
0	1	1.3679
1	2	0.5183
2	3	0.3335
3	4	0.2500
4	5	0.2000
5	6	0.1667
6	7	0.1429
7	8	0.1250
8	9	0.1111

(a) Trapezoidal rule:

$$\int_1^9 \left(e^{-t^2} + \frac{1}{t} \right) dt$$

$$\approx \frac{9-1}{8} \left[\frac{1}{2}(1.3679) + 0.5183 + 0.3335 \right.$$

$$\left. + \cdots + \frac{1}{2}(0.1111) \right]$$

$$\approx 2.4759$$

(b) Simpson's rule:

$$\int_1^9 \left(e^{-t^2} + \frac{1}{t} \right) dt$$

$$\approx \frac{9-1}{3(8)} [1.3679 + 4(0.5183) + 2(0.3335)$$

$$+ 4(0.2500) + 2(0.2000) + 4(0.1667)$$
$$+ 2(0.1429) + 4(0.1250) + 0.1111]$$
$$\approx 2.3572$$

27. Note that heights may differ depending on the readings of the graph. Thus, answers may vary.
$n = 10$, $b = 20$, $a = 0$

i	x_i	$f(x_i)$
0	0	0
1	2	5
2	4	3
3	6	2
4	8	1.5
5	10	1.2
6	12	1
7	14	0.5
8	16	0.3
9	18	0.2
10	20	0.2

Area under curve for Formulation A

$$= \frac{20-0}{10} \left[\frac{1}{2}(0) + 5 + 3 + 2 + 1.5 + 1.2 \right.$$

$$\left. + 1 + 0.5 + 0.3 + 0.2 + \frac{1}{2}(0.2) \right]$$

$$= 2(14.8)$$

$$\approx 30 \text{ mcg/ml}$$

This represents the total amount of drug available to the patient.

29. As in Exercise 25, readings on the graph may vary, so answers may vary. The area both under the curve for Formulation A and above the minimum effective concentration line in on the interval $\left[\frac{1}{2}, 6 \right]$.

Area under curve for Formulation A on $\left[\frac{1}{2}, 1 \right]$, with $n = 1$

$$= \frac{1 - \frac{1}{2}}{1} \left[\frac{1}{2}(2 + 6) \right]$$

$$= \frac{1}{2}(4) = 2$$

Area under curve for Formulation A on $[1, 6]$, with $n = 5$

$$= \frac{6-1}{5} \left[\frac{1}{2}(6) + 5 + 4 + 3 + 2.4 + \frac{1}{2}(2) \right]$$

$$= 18.4$$

Area under minimum effective concentration line $\left[\frac{1}{2}, 6 \right]$

$$= 5.5(2) = 11.0$$

Area under the curve for Formulation A and above minimum effective concentration line

$$= 2 + 18.4 - 11.0$$

$$\approx 9 \text{ mcg/ml}$$

This represents the total effective amount of drug available to the patient.

31. (a)

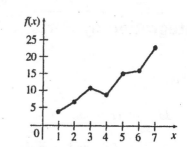

(b) $\dfrac{7-1}{6} \left[\dfrac{1}{2}(4) + 7 + 11 + 9 + 15 \right.$

$$\left. + 16 + \frac{1}{2}(23) \right]$$

$$= 71.5$$

(c) $\dfrac{7-1}{3(6)} [4 + 4(7) + 2(11) + 4(9)$

$$+ 2(15) + 4(16) + 23]$$
$$= 69.0$$

33. We need to evaluate

$$\int_{12}^{36} (105e^{0.01x} + 32)\, dx.$$

Using a calculator program for Simpson's rule with $n = 100$, we obtain 3,979.24 as the value of this integral. This indicates that the total revenue between the twelfth and thirty-sixth months is about 3,979.24.

35. Use a calculator program for Simpson's rule with $n = 100$ to evaluate each of the integrals in this exercise.

(a) $\int_{-1}^{1} \left(\frac{1}{\sqrt{2\pi}} e^{-x^2/2} \right) dx \approx 0.682689$

The probability that a normal random variable is within 1 standard deviation of the mean is about 0.682689.

(b) $\int_{-2}^{2} \left(\frac{1}{\sqrt{2\pi}} e^{-x^2/2} \right) dx \approx 0.954500$

The probability that a normal random variable is within 2 standard deviation of the mean is about 0.954500.

(c) $\int_{-3}^{3} \left(\frac{1}{\sqrt{2\pi}} e^{-x^2/2} \right) dx \approx 0.997300$

The probability that a normal random variable is within 3 standard deviation of the mean is about 0.997300.

8.2   Integration by Parts

1. $\int xe^x\, dx$

Let $dv = e^x\, dx$ and $u = x$.

Then $v = \int e^x\, dx$ and $du = dx$.

$$v = e^x + C$$

Use the formula

$$\int u\, dv = uv - \int v\, du.$$

$$\int xe^x\, dx = xe^x - \int e^x\, dx$$

$$= xe^x - e^x + C$$

3. $\int (5x - 9)e^{-3x}\, dx$

Let $dv = e^{-3x}\, dx$ and $u = 5x - 9$.

Then $v = \int e^{-3x}\, dx$ and $du = 5\, dx$.

$$v = \frac{e^{-3x}}{-3}$$

$$\int (5x - 9)e^{-3x}\, dx$$

$$= \frac{-(5x - 9)e^{-3x}}{3} + \int \frac{5}{3} e^{-3x}\, dx$$

$$= \frac{-5xe^{-3x}}{3} + 3e^{-3x} + \frac{5e^{-3x}}{-9} + C$$

$$= \frac{-5xe^{-3x}}{3} - \frac{5e^{-3x}}{9} + 3e^{-3x} + C$$

or $\dfrac{-5xe^{-3x}}{3} + \dfrac{22e^{-3x}}{9} + C$

5. $\int_0^1 \frac{2x + 1}{e^x}\, dx$

$$= \int_0^1 (2x + 1)e^{-x}\, dx$$

Let $dv = e^{-x}\, dx$ and $u = 2x + 1$.

Then $v = \int e^{-x}\, dx$ and $du = 2\, dx$.

$$v = -e^{-x}$$

$$\int \frac{2x + 1}{e^x}\, dx$$

$$= -(2x + 1)e^{-x} + \int 2e^{-x}\, dx$$

$$= -(2x + 1)e^{-x} - 2e^{-x}$$

$$\int_0^1 \frac{2x + 1}{e^x}\, dx$$

$$= \left[-(2x + 1)e^{-x} - 2e^{-x} \right] \Big|_0^1$$

$$= [-(3)e^{-1} - 2e^{-1}] - (-1 - 2)$$
$$= -5e^{-1} + 3$$
$$\approx 1.1606$$

7. $\int_1^4 \ln 2x\, dx$

Let $dv = dx$ and $u = \ln 2x$.

Then $v = x$ and $du = \frac{1}{x}$.

$$\int \ln 2x\, dx = x \ln 2x - \int dx$$

$$\int_1^4 \ln 2x\, dx$$

$$= (x \ln 2x - x)\Big|_1^4$$

$$= (4 \ln 8 - 4) - (\ln 2 - 1)$$
$$= 4 \ln 2^3 - 4 - \ln 2 + 1$$
$$= 12 \ln 2 - \ln 2 - 3$$
$$= 11 \ln 2 - 3$$
$$\approx 4.6246$$

9. $\int x \ln dx$

Let $dv = x\, dx$ and $u = \ln x$.

Then $v = \frac{x^2}{2}$ and $du = \frac{1}{x}\, dx$.

$$\int x \ln dx = \frac{x^2}{2} \ln x - \int \frac{x}{2}\, dx$$

$$= \frac{x^2 \ln x}{2} - \frac{x^2}{4} + C$$

11. $\int -6x \cos 5x\, dx$

Let $u = -6x$ and $dv = \cos 5x\, dx$.

Then $du = -6\, dx$ and $v = \frac{1}{5} \sin 5x$.

$$\int -6x \cos 5x\, dx$$

$$= (-6x)\left(\frac{1}{5} \sin 5x\right) - \int \left(\frac{1}{5} \sin 5x\right)(-6\, dx)$$

$$= -\frac{6}{5} \sin 5x + \frac{6}{5} \int \sin 5x\, dx$$

$$= -\frac{6}{5}x \sin 5x + \frac{6}{5} \cdot \frac{1}{5}(-\cos 5x) + C$$

$$= -\frac{6}{5}x \sin 5x - \frac{6}{25} \cos 5x + C$$

13. $\int 8x \sin x\, dx$

Let $u = 8x$ and $\frac{dv}{dx} = \sin x$.

Then $du = 8\, dx$ and $v = -\cos x$.

$$\int 8x \sin x\, dx$$

$$= (8x)(-\cos x) - \int (-\cos x)(8\, dx)$$

$$= -8x \cos x + 8 \int \cos x\, dx$$

$$= -8x \cos x + 8 \sin x + C$$

15. $\int -6x^2 \cos 8x\, dx$

Let $u = -6x^2$ and $dv = \cos 8x\, dx$.

Then $du = -12x\, dx$ and

$$v = \frac{1}{8} \sin 8x.$$

$$\int -6x^2 \cos 8x\, dx$$

$$= (-6x^2)\left(\frac{1}{8} \sin 8x\right) - \int \left(\frac{1}{8} \sin 8x\right)(-12x\, dx)$$

$$= -\frac{3}{4}x^2 \sin 8x + \frac{3}{2} \int x \sin 8x\, dx$$

In $\int x \sin 8x\, dx$, let

$$u = x \quad \text{and} \quad dv = \sin 8x\, dx.$$

Then $du = dx$ and $v = -\frac{1}{8} \cos 8x.$

$$\int -6x^2 \cos 8x\, dx$$

$$= -\frac{3}{4}x^2 \sin 8x + \frac{3}{2}\left[-\frac{1}{8}x \cos 8x - \int \left(-\frac{1}{8} \cos 8x\right) dx\right]$$

$$= -\frac{3}{4}x^2 \sin 8x - \frac{3}{16}x \cos 8x + \frac{3}{16} \int \cos 8x\, dx$$

$$= -\frac{3}{4}x^2 \sin 8x - \frac{3}{16}x \cos 8x + \frac{3}{16} \cdot \frac{1}{8} \sin 8x + C$$

$$= -\frac{3}{4}x^2 \sin 8x - \frac{3}{16}x \cos 8x + \frac{3}{128} \sin 8x + C$$

17. The area is $\int_2^4 (x-2)e^x \, dx$.

Let $dv = e^x \, dx$ and $u = x - 2$.
Then $v = e^x$ and $du = dx$.

$$\int (x-2)e^x \, dx = (x-2)e^x - \int e^x \, dx$$

$$\int_1^4 (x-2)e^x \, dx = \left[(x-2)e^x - e^x\right]\Big|_2^4$$

$$= (2e^4 - e^4) - (0 - e^2)$$
$$= e^4 + e^2 \approx 61.9872$$

19. $\int x^2 e^{2x} \, dx$

Let $u = x^2$ and $dv = e^{2x} \, dx$.
Use column integration.

D		I
x^2	$+$	e^{2x}
$2x$	$-$	$\dfrac{e^{2x}}{2}$
2	$+$	$\dfrac{e^{2x}}{4}$
0		$\dfrac{e^{2x}}{8}$

$$\int x^2 e^{2x} \, dx = x^2\left(\frac{e^{2x}}{2}\right) - 2x\left(\frac{e^{2x}}{4}\right) + \frac{2e^{2x}}{8} + C$$

$$= \frac{x^2 e^{2x}}{2} - \frac{x e^{2x}}{2} + \frac{e^{2x}}{4} + C$$

21. $\int_0^5 x\sqrt[3]{x^2 + 2} \, dx$

$$= \int_0^5 x(x^2 + 2)^{1/2} \, dx$$

$$= \frac{1}{2}\int_0^5 2x(x^2 + 2)^{1/3} \, dx$$

Let $u = x^2 + 2$. Then $du = 2x \, dx$.
If $x = 5$, $u = 27$. If $x = 0$, $u = 2$.

$$= \frac{1}{2}\int_2^{27} u^{1/3} \, du$$

$$= \frac{1}{2}\left(\frac{u^{4/3}}{1}\right)\left(\frac{3}{4}\right)\Big|_2^{27}$$

$$= \frac{3}{8}(27)^{4/3} - \frac{3}{8}(2)^{4/3}$$

$$= \frac{243}{8} - \frac{3(2^{4/3})}{8}$$

$$= \frac{243}{8} - \frac{3\sqrt[3]{2}}{4} \approx 29.4301$$

23. $\int (8x + 7) \ln(5x) \, dx$

Let $dv = (8x + 7)\, dx$ and $u = \ln 5x$.

Then $v = \dfrac{8x^2}{2} + 7x$ and $du = \dfrac{1}{5x}(5)\, dx$.

$\qquad\qquad = 4x^2 + 7x \qquad\qquad = \dfrac{1}{x}\, dx$.

$$\int (8x + 7)\ln(5x)\, dx$$

$$= (4x^2 + 7x)\ln 5x - \int (4x + 7)\, dx$$

$$= (4x^2 + 7x)\ln 5x - \left(\frac{4x^2}{2} + 7x\right) + C$$

$$= 4x^2 \ln(5x) + 7x\ln(5x) - 2x^2 - 7x + C$$

25. $\int x^2 \sqrt{x + 2} \, dx$

Let $u = x^2$ and $dv = (x + 2)^{1/2}\, dx$.
Use column integration.

D		I
x^2	$+$	$(x+2)^{1/2}$
$2x$	$-$	$(x+2)^{3/2}\left(\frac{2}{3}\right)$
2	$+$	$(x+2)^{5/2}\left(\frac{2}{3}\right)\left(\frac{2}{5}\right)$
0		$(x+2)^{7/2}\left(\frac{2}{3}\right)\left(\frac{2}{5}\right)\left(\frac{2}{7}\right)$

$$\int x^2 \sqrt{x + 2} \, dx$$

$$= x^2(x+2)^{3/2}\left(\frac{2}{3}\right) - 2x(x+2)^{5/2}\left(\frac{2}{3}\right)\left(\frac{2}{5}\right)$$

$$+ 2(x+2)^{7/2}\left(\frac{2}{3}\right)\left(\frac{2}{5}\right)\left(\frac{2}{7}\right) + C$$

$$= \frac{2x^2(x+2)^{3/2}}{3} - \frac{8x(x+2)^{5/2}}{15}$$

$$+ \frac{16(x+2)^{7/2}}{105} + C$$

27. $\displaystyle\int_0^1 \frac{x^3\,dx}{\sqrt{3+x^2}} = \int_0^1 x^3(3x+x^2)^{-1/2}\,dx$

Let $dv = x(3+x^2)^{-1/2}\,dx$ and $u = x^2$.

Then $v = \dfrac{2(3+x^2)^{1/2}}{2}$

$\qquad v = (3+x^2)^{1/2}$ and $du = 2x\,dx$.

$\displaystyle\int \frac{x^3\,dx}{\sqrt{3+x^2}}$

$\displaystyle = x^2(3+x^2)^{1/2} - \int 2x(3+x^2)^{1/2}\,dx$

$\displaystyle = x^2(3+x^2)^{1/2} - \frac{2}{3}(3+x^2)^{3/2}$

$\displaystyle\int_0^1 \frac{x^3\,dx}{\sqrt{3+x^2}}$

$\displaystyle = \left[x^2(3+x^2)^{1/2} - \frac{2}{3}(3+x^2)^{3/2}\right]\Big|_0^1$

$\displaystyle = 4^{1/2} - \frac{2}{3}(4^{3/2}) - 0 + \frac{2}{3}(3^{3/2})$

$\displaystyle = 2 - \frac{2}{3}(8) + \frac{2}{3}(3^{3/2})$

≈ 0.13077

29. $\displaystyle\int \frac{9}{\sqrt{x^2+9}}\,dx = 9\int \frac{dx}{\sqrt{x^2+9}}$

If $a = 3$, this integral matches entry 5 in the table.

$\displaystyle = 9\ln\left|x + \sqrt{x^2+9}\right| + C$

31. $\displaystyle\int \frac{3}{x\sqrt{121-x^2}}\,dx$

$\displaystyle = 3\int \frac{dx}{x\sqrt{11^2-x^2}}$

If $a = 11$, this integral matches entry 9 in the table.

$\displaystyle = 3\left(-\frac{1}{11}\ln\left|\frac{11+\sqrt{121-x^2}}{x}\right|\right) + C$

$\displaystyle = -\frac{3}{11}\ln\left|\frac{11+\sqrt{121-x^2}}{x}\right| + C$

33. $\displaystyle\int \frac{-3}{x(4x+3)^2}\,dx$

$\displaystyle = -3\int \frac{dx}{x(4x+3)^2}$

This matches entry 14 in the table with $a = 4$ and $b = 3$.

$\displaystyle = -3\left[\frac{1}{3(4x+3)} + \frac{1}{3^2}\ln\left|\frac{x}{4x+3}\right|\right] + C$

$\displaystyle = \frac{-1}{4x+3} - \frac{1}{3}\ln\left|\frac{x}{4x+3}\right| + C$

37. $\displaystyle\int x^n \cdot \ln|x|\,dx, n \neq -1$

Let $u = \ln|x|$ and $dv = x^n\,dx$.

Use column integration.

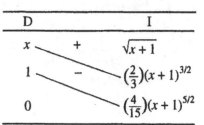

D		I
$\ln|x|$	$+$	x^n
$\dfrac{1}{x}$	$-$	$\dfrac{1}{n+1}x^{n+1}$

$\displaystyle\int x^n \cdot \ln|x|\,dx$

$\displaystyle = \frac{1}{n+1}x^{n+1}\ln|x| - \int\left[\frac{1}{x}\cdot\frac{1}{n+1}x^{n+1}\right]dx$

$\displaystyle = \frac{1}{n+1}x^{n+1}\ln|x| - \int \frac{1}{n+1}x^n\,dx$

$\displaystyle = \frac{1}{n+1}x^{n+1}\ln|x| - \frac{1}{(n+1)^2}x^{n+1} + C$

$\displaystyle = x^{n+1}\left[\frac{\ln|x|}{n+1} - \frac{1}{(n+1)^2}\right] + C$

39. $\displaystyle\int x\sqrt{x+1}\,dx$

(a) Let $u = x$ and $dv = \sqrt{x+1}\,dx$.
Use column integration.

D		I
x	$+$	$\sqrt{x+1}$
1	$-$	$\left(\frac{2}{3}\right)(x+1)^{3/2}$
0		$\left(\frac{4}{15}\right)(x+1)^{5/2}$

$\displaystyle\int x\sqrt{x+1}\,dx$

$\displaystyle = \left(\frac{2}{3}\right)x(x+1)^{3/2} - \left(\frac{4}{15}\right)(x+1)^{5/2} + C$

(b) Let $u = x+1$; then $u - 1 = x$ and $du = dx$.

$\displaystyle\int x\sqrt{x+1}\,dx$

$\displaystyle = \int (u-1)u^{1/2}\,du = \int (u^{3/2} - u^{1/2})\,du$

$\displaystyle = \frac{2}{5}u^{5/2} - \frac{2}{3}u^{3/2} + C$

$\displaystyle = \frac{2}{5}(x+1)^{5/2} - \frac{2}{3}(x+1)^{3/2} + C$

(c) Both results factor as $\frac{2}{15}(x+1)^{3/2}(3x-2) + C$, so they are equivalent.

41. $\displaystyle\int_0^1 ke^{-kt}(1-t)dt$

Let $u = 1-t$ and $dv = e^{-kt}dt$.

Then $du = -dt$ and $v = -\dfrac{1}{k}e^{-kt}$

$\displaystyle\int_0^1 ke^{-kt}(1-t)dt$

$= k\left[(1-t)\left(-\dfrac{1}{k}e^{-kt}\right) - \int\left(-\dfrac{1}{k}e^{-kt}\right)(-dt)\right]\Big|_0^1$

$= k\left[-\dfrac{1}{k}e^{-kt}(1-t) - \dfrac{1}{k}\int e^{-kt}dt\right]\Big|_0^1$

$= -e^{-kt}(1-t) + \dfrac{1}{k}e^{-kt}\Big|_0^1$

$= -e^{-kt}\left(1 - t - \dfrac{1}{k}\right)\Big|_0^1$

$= \left[-e^{-k}\left(-\dfrac{1}{k}\right)\right] - \left[-e^0\left(1 - \dfrac{1}{k}\right)\right]$

$= 1 - \dfrac{1}{k} + \dfrac{1}{k}e^{-k}$

(a) $k = \dfrac{1}{12}$:

$\displaystyle\int_0^1 \dfrac{1}{12}e^{-t/12}(1-t)dt$

$= 1 - \dfrac{1}{1/12} + \dfrac{1}{1/12}e^{-1/12}$

$= 12e^{-1/12} - 11$

≈ 0.0405

$k = \dfrac{1}{24}$:

$\displaystyle\int_0^1 \dfrac{1}{24}e^{-t/24}(1-t)dt$

$= 1 - \dfrac{1}{1/24} + \dfrac{1}{1/24}e^{-1/24}$

$24e^{-1/24} - 23$

≈ 0.0205

$k = \dfrac{1}{48}$:

$\displaystyle\int_0^1 \dfrac{1}{48}e^{-t/48}(1-t)dt$

$= 1 - \dfrac{1}{1/48} + \dfrac{1}{1/48}e^{-1/48}$

$= 48e^{-1/48} - 47$

≈ 0.0103

(b) $\displaystyle\int_1^6 ke^{-kt}\dfrac{6-t}{5}dt$

The integral, easily found by comparing it to the integral in part (a), is

$-e^{-kt}\left(\dfrac{6}{5} - \dfrac{t}{5} - \dfrac{1}{5k}\right)\Big|_1^6$

$= \left[-e^{-6k}\left(\dfrac{6}{5} - \dfrac{6}{5} - \dfrac{1}{5k}\right)\right]$

$\qquad - \left[-e^{-k}\left(\dfrac{6}{5} - \dfrac{1}{5} - \dfrac{1}{5k}\right)\right]$

$= \dfrac{1}{5k}e^{-6k} + \left(1 - \dfrac{1}{5k}\right)e^{-k}$

$k = \dfrac{1}{12}$:

$\displaystyle\int_1^6 \dfrac{1}{12}e^{-t/12}\dfrac{6-t}{5}dt$

$= \dfrac{1}{5(1/12)}e^{-6(1/12)} + \left[1 - \dfrac{1}{5(1/12)}\right]e^{-1/12}$

$= \dfrac{12}{5}e^{-1/2} - \dfrac{7}{5}e^{-1/12}$

≈ 0.1676

$k = \dfrac{1}{24}$:

$\displaystyle\int_1^6 \dfrac{1}{24}e^{-t/24}\dfrac{6-t}{5}dt$

$= \dfrac{1}{5(1/24)}e^{-6(1/24)} + \left[1 - \dfrac{1}{5(1/24)}\right]e^{-1/24}$

$= \dfrac{24}{5}e^{-1/4} - \dfrac{19}{5}e^{-1/24}$

≈ 0.0933

$k = \dfrac{1}{48}$:

$\displaystyle\int_1^6 \dfrac{1}{48}e^{-t/48}\dfrac{6-t}{5}dt$

$= \dfrac{1}{5(1/48)}e^{-6(1/48)} + \left[1 - \dfrac{1}{5(1/48)}\right]e^{-1/48}$

$= \dfrac{48}{5}e^{-1/8} - \dfrac{43}{5}e^{-1/48}$

≈ 0.0493

43. $\displaystyle\int_0^3 30xe^{2x}\,dx$

Let $dv = e^{2x}\,dx$ and $u = 30x$.

Then $v = \dfrac{e^{2x}}{2}$ and $du = 30\,dx$.

$$\int 30xe^{2x}\,dx = 15xe^{2x} - \int 15e^{2x}\,dx$$

$$= 15xe^{2x} - \frac{15}{2}e^{2x}$$

$$\int_0^3 30xe^{2x}\,dx = \left(15xe^{2x} - \frac{15}{2}e^{2x}\right)\Bigg|_0^3$$

$$= 45e^6 - \frac{15}{2}e^6 - 0 + \frac{15}{2}$$

$$= \frac{75}{2}e^6 + \frac{15}{2} = \frac{15}{2}(5e^6 + 1)$$

$$\approx 15,136$$

The total accumulated growth during the first 3 days was 15,136.

45. $R = \displaystyle\int_0^{12} (x+1)\ln(x+1)\,dx$

Let $u = \ln(x+1)$ and $dv = (x+1)\,dx$.

Then $du = \dfrac{1}{x+1}\,dx$ and $v = \dfrac{1}{2}(x+1)^2$.

$$\int (x+1)\ln(x+1)\,dx$$

$$= \frac{1}{2}(x+1)^2\ln(x+1)$$

$$\quad - \int \left[\frac{1}{2}(x+1)^2 \cdot \frac{1}{x+1}\right]dx$$

$$= \frac{1}{2}(x+1)^2\ln(x+1) - \int \frac{1}{2}(x+1)\,dx$$

$$= \frac{1}{2}(x+1)^2\ln(x+1) - \frac{1}{4}(x+1)^2 + C$$

$$\int_0^{12} (x+1)\ln(x+1)\,dx$$

$$= \left[\frac{1}{2}(x+1)^2\ln(x+1) - \frac{1}{4}(x+1)^2\right]\Bigg|_0^{12}$$

$$= \frac{169}{2}\ln 13 - 42 \approx \$174.74$$

8.3   Volume and Average Value

1. $f(x) = x,\ y = 0,\ x = 0,\ x = 2$

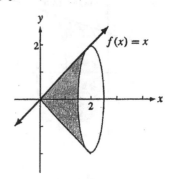

$$V = \pi \int_0^2 x^2\,dx = \frac{\pi x^3}{2}\Bigg|_0^2$$

$$= \frac{\pi(8)}{3} - 0$$

$$= \frac{8\pi}{3}$$

3. $f(x) = 2x + 1,\ y = 0,\ x = 0,\ x = 4$

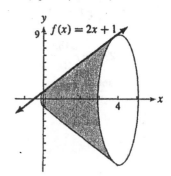

$$V = \pi \int_0^4 (2x+1)^2\,dx$$

Let $u = 2x + 1$. Then $du = 2\,dx$.
If $x = 4$, $u = 9$. If $x = 0$, $u = 1$.

$$V = \frac{1}{2}\pi \int_0^4 2(2x+1)^2\,dx$$

$$= \frac{1}{2}\pi \int_1^9 u^2\,du$$

$$= \frac{\pi}{2}\left(\frac{u^3}{3}\right)\Bigg|_1^9$$

$$= \frac{729\pi}{6} - \frac{\pi}{6}$$

$$= \frac{728\pi}{6}$$

$$= \frac{364\pi}{3}$$

5. $f(x) = \frac{1}{3}x + 2$, $y = 0$, $x = 1$, $x = 3$

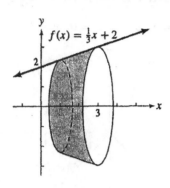

$$V = \pi \int_1^3 \left(\frac{1}{3}x + 2\right)^2 dx$$

$$= 3\pi \int_1^3 \frac{1}{3}\left(\frac{1}{3}x + 2\right)^2 dx$$

$$= 3\pi \frac{\left(\frac{1}{3}x + 2\right)^3}{3}\bigg|_1^3$$

$$= \pi \left(\frac{1}{3}x + 2\right)^3 \bigg|_1^3$$

$$= 27\pi - \frac{343\pi}{27}$$

$$= \frac{386\pi}{27}$$

7. $f(x) = \sqrt{x}$, $y = 0$, $x = 1$, $x = 2$

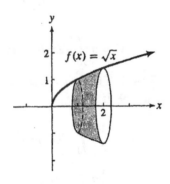

$$V = \pi \int_1^2 (\sqrt{x})^2 dx = \pi \int_1^2 x\, dx$$

$$= \frac{\pi x^2}{2}\bigg|_1^2$$

$$= 2\pi - \frac{\pi}{2}$$

$$= \frac{3\pi}{2}$$

9. $f(x) = \sqrt{2x + 1}$, $y = 0$, $x = 1$, $x = 4$

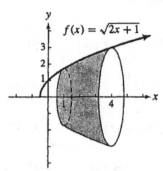

$$V = \pi \int_1^4 (\sqrt{2x + 1})^2 dx$$

$$= \pi \int_1^4 (2x + 1)\, dx$$

$$= \pi \left(\frac{2x^2}{2} + x\right)\bigg|_1^4$$

$$= \pi[(16 + 4) - 2]$$

$$= 18\pi$$

11. $f(x) = e^x$; $y = 0$, $x = 0$, $x = 2$

$$V = \pi \int_0^2 e^{2x}\, dx = \frac{\pi e^{2x}}{2}\bigg|_0^2$$

$$= \frac{\pi e^4}{2} - \frac{\pi}{2}$$

$$= \frac{\pi}{2}(e^4 - 1)$$

$$\approx 84.19$$

13. $f(x) = \frac{1}{\sqrt{x}}$, $y = 0$, $x = 1$, $x = 4$

$$V = \pi \int_1^4 \frac{1}{(\sqrt{x})^2}\, dx$$

$$= \pi \int_1^4 \frac{1}{x}\, dx$$

$$= \pi \ln |x| \bigg|_1^4$$

$$= \pi \ln 4 - \pi \ln 1$$

$$= \pi \ln 4 \approx 4.36$$

15. $f(x) = x^2$, $y = 0$, $x = 1$, $x = 5$

$$V = \pi \int_1^5 x^4\, dx = \frac{\pi x^5}{5}\bigg|_1^5$$

$$= 625\pi - \frac{\pi}{5}$$

$$= \frac{3{,}124\pi}{5}$$

17. $f(x) = 1 - x^2$, $y = 0$

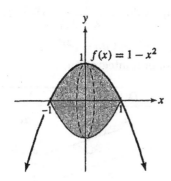

Since $f(x) = 1 - x^2$ intersects $y = 0$ where

$$1 - x^2 = 0$$
$$x = \pm 1,$$
$$a = -1 \quad \text{and} \quad b = 1.$$

$$V = \pi \int_{-1}^{1} (1 - x^2)^2 \, dx$$

$$= \pi \int_{-1}^{1} (1 - 2x^2 + x^4) \, dx$$

$$= \pi \left(x - \frac{2x^3}{3} + \frac{x^5}{5} \right) \Big|_{-1}^{1}$$

$$= \pi \left(1 - \frac{2}{3} + \frac{1}{5} \right) - \pi \left(-1 + \frac{2}{3} - \frac{1}{5} \right)$$

$$= 2\pi - \frac{4\pi}{3} + \frac{2\pi}{5}$$

$$\doteq \frac{16\pi}{15}$$

19. $f(x) = \sec x$, $y = 0$, $x = 0$, $x = \dfrac{\pi}{4}$

$$V = \pi \int_{0}^{\pi/4} (\sec x)^2 \, dx$$

$$= \pi \int_{0}^{\pi/4} \sec^2 x \, dx$$

$$= \pi (\tan x) \Big|_{0}^{\pi/4}$$

$$= \pi \left(\tan \frac{\pi}{4} - \tan 0 \right)$$

$$= \pi (1 - 0)$$

$$= \pi$$

21. $f(x) = \sqrt{1 - x^2}$
     $r = \sqrt{1} = 1$

$$V = \pi \int_{-1}^{1} (\sqrt{1 - x^2})^2 \, dx$$

$$= \pi \int_{-1}^{1} (1 - x^2) \, dx$$

$$= \pi \left(x - \frac{x^3}{3} \right) \Big|_{-1}^{1}$$

$$= \pi \left(1 - \frac{1}{3} \right) - \pi \left(-1 + \frac{1}{3} \right)$$

$$= 2\pi - \frac{2}{3}\pi$$

$$= \frac{4\pi}{3}$$

23. $f(x) = \sqrt{r^2 - x^2}$

$$V = \pi \int_{-r}^{r} (\sqrt{r^2 - x^2})^2 \, dx$$

$$= \pi \int_{-r}^{r} (r^2 - x^2) \, dx$$

$$= \pi \left(r^2 x - \frac{x^3}{3} \right) \Big|_{-r}^{r}$$

$$= \pi \left(r^3 - \frac{r^3}{3} \right) - \pi \left(-r^3 + \frac{r^3}{3} \right)$$

$$= 2r^3 \pi - \left(\frac{2r^3 \pi}{3} \right)$$

$$= \frac{4r^3 \pi}{3}$$

25. $f(x) = r$, $x = 0$, $x = h$

Graph $f(x) = r$; then show the solid of revolution formed by rotating about the x-axis the region bounded by $f(x)$, $x = 0$, $x = h$.

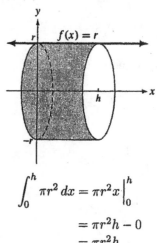

$$\int_{0}^{h} \pi r^2 \, dx = \pi r^2 x \Big|_{0}^{h}$$

$$= \pi r^2 h - 0$$

$$= \pi r^2 h$$

27. $f(x) = x^2 - 2;\ [0,5]$

Average value

$$= \frac{1}{5-0} \int_0^5 (x^2 - 2)\, dx$$

$$= \frac{1}{5} \left(\frac{x^3}{3} - 2x \right) \Big|_0^5$$

$$= \frac{1}{5} \left(\frac{125}{3} - 10 \right)$$

$$= \frac{1}{5} \left(\frac{95}{3} \right)$$

$$= \frac{19}{3}$$

29. $f(x) = \sqrt{x+1};\ [3,8]$

Average value

$$= \frac{1}{8-3} \int_3^8 \sqrt{x+1}\, dx$$

$$= \frac{1}{5} \int_3^8 (x+1)^{1/2}\, dx$$

$$= \frac{1}{5} \cdot \frac{2}{3} (x+1)^{3/2} \Big|_3^8$$

$$= \frac{2}{15} (9^{3/2} - 4^{3/2})$$

$$= \frac{2}{15} (27 - 8) = \frac{38}{15}$$

31. $f(x) = e^{x/5};\ [0,5]$

Average value

$$= \frac{1}{5-0} \int_0^5 e^{x/5}\, dx$$

$$= \int_0^5 \frac{1}{5} e^{x/5}\, dx$$

$$= e^{x/5} \Big|_0^5 = e^1 - e^0$$

$$= e - 1 \approx 1.718$$

33. $f(x) = x^2 e^{2x};\ [0,2]$

Average value $= \dfrac{1}{2-0} \displaystyle\int_0^2 x^2 e^{2x}\, dx$

Let $u = x^2$ and $dv = e^{2x}\, dx$.
Use column integration.

D	I
x^2 $+$	e^{2x}
$2x$ $-$	$\frac{1}{2} e^{2x}$
2 $+$	$\frac{1}{4} e^{2x}$
0	$\frac{1}{8} e^{2x}$

$$\frac{1}{2-0} \int_0^2 x^2 e^{2x}\, dx$$

$$= \frac{1}{2} \left[(x^2) \left(\frac{1}{2} \right) e^{2x} - (2x) \left(\frac{1}{4} \right) e^{2x} \right.$$

$$\left. + 2 \left(\frac{1}{8} \right) e^{2x} \right] \Big|_0^2$$

$$= \frac{1}{2} \left(2e^4 - e^4 + \frac{1}{4} e^4 - \frac{1}{4} \right)$$

$$= \frac{5e^4 - 1}{8} \approx 33.999$$

35. $f(x) = \sec^2 x;\ \left[0, \dfrac{\pi}{4} \right]$

Average value $= \dfrac{1}{\frac{\pi}{4} - 0} \displaystyle\int_0^{\pi/4} \sec^2 x\, dx$

$$= \frac{4}{\pi} (\tan x) \Big|_0^{\pi/4}$$

$$= \frac{4}{\pi} \left(\tan \frac{\pi}{4} - \tan 0 \right)$$

$$= \frac{4}{\pi}$$

37. $f(x) = e^{-x^2},\ y = 0,\ x = -2,\ x = 2$

$$V = \pi \int_{-2}^2 (e^{-x^2})^2\, dx$$

$$= \pi \int_{-2}^2 e^{-2x^2}\, dx$$

Using a graphing calculator with the *fn Int* feature to evaluate the integral, we get $3.937153082 \approx 3.9372$.

39. (a) $\int_0^R 2\pi r k(R^2 - r^2)\, dr$

$= 2\pi k \int_0^R r(R^2 - r^2)\, dr$

(b) $= 2\pi k \int_0^R (rR^2 - r^3)\, dr$

$= 2\pi k \left[\dfrac{r^2 R^2}{2} - \dfrac{r^4}{4} \right]\Big|_0^R$

$= 2\pi k \left[\dfrac{R^4}{2} - \dfrac{R^4}{4} - 0 \right]$

$= 2\pi k \left(\dfrac{R^4}{4} \right) = \dfrac{\pi k R^4}{2}$

41. For (a), (b), and (c), use substitution and column integration to find $\int \ln(t+1)\, dt$.

Let $y = t+1$; then $dy = dt$.

So $\int \ln(t+1)\, dt = \int \ln y\, dy$.

Let $u = \ln y$ and $dv = dy$.

D	I
$\ln y$ $\quad +$	1
$\frac{1}{y}$ $\quad -$	y

$\int \ln(t+1)\, dt$

$= \int \ln y\, dy = y \ln y - \int \left(\dfrac{1}{y} \cdot y \right) dy$

$= y \ln y - \int dy = y \ln y - y + C$

$= (t+1) \ln(t+1) - (t+1) + C$

(a) The average number of items produced daily after 5 days is given by

$\dfrac{1}{5-0} \int_0^5 35 \ln(t+1)\, dt$.

$\dfrac{1}{5} \int_0^5 35 \ln(t+1)\, dt$

$= \dfrac{35}{5} \int_0^5 \ln(t+1)\, dt$

$= 7([(t+1)\ln(t+1) - (t+1)]\Big|_0^5)$

$= 7(6 \ln 6 - 6 + 1)$

$= 7(6 \ln 6 - 5) \approx 40.254$

(b) After 10 days, the average number of items produced daily is

$\dfrac{1}{10-0} \int_0^{10} 35 \ln(t+1)\, dt$

$= \dfrac{35}{10}([(t+1)\ln(t+1) - (t+1)]\Big|_0^{10})$

$= \dfrac{7}{2}(11 \ln -11 + 1)$

$= \dfrac{7(11 \ln 11 - 10)}{2} \approx 57.319.$

(c) After 15 days, the average number of items produced daily is

$\dfrac{1}{15-0} \int_0^{15} 35 \ln(t+1)\, dt$

$= \dfrac{35}{15}([(t+1)\ln(t+1) - (t+1)]\Big|_0^{15})$

$= \dfrac{7}{3}(16 \ln 16 - 16 + 1)$

$= \dfrac{7(16 \ln 16 - 15)}{3} \approx 68.510.$

8.4 Improper Integrals

1. $\int_2^\infty \dfrac{1}{x^2}\, dx = \lim_{b \to \infty} \int_2^b x^{-2}\, dx = \lim_{b \to \infty} \int (-x^{-1})\Big|_2^b$

$= \lim_{b \to \infty} \left(-\dfrac{1}{b} + \dfrac{1}{2} \right)$

$= \lim_{b \to \infty} \left(-\dfrac{1}{b} \right) + \lim_{b \to \infty} \dfrac{1}{2}$

As $b \to \infty$, $-\dfrac{1}{b} \to 0$. The integral is convergent.

$\int_2^\infty \dfrac{1}{x^2}\, dx = 0 + \dfrac{1}{2} = \dfrac{1}{2}$

3. $\int_1^\infty \dfrac{1}{\sqrt{x}}\, dx = \lim_{b \to \infty} \int_1^b x^{-1/2}\, dx = \lim_{b \to \infty} (2x^{1/2})\Big|_1^b$

$= \lim_{b \to \infty} (2\sqrt{b} - 2)$

$= \lim_{b \to \infty} 2\sqrt{b} - \lim_{b \to \infty} 2$

As $b \to \infty$, $2\sqrt{b} \to \infty$.
The integral is divergent.

5. $\displaystyle\int_{-\infty}^{-1} \frac{2}{x^3}\, dx = \int_{-\infty}^{-1} 2x^{-3}\, dx = \lim_{a \to -\infty} \int_{a}^{-1} 2x^{-3}\, dx$

$\displaystyle = \lim_{a \to -\infty} \left. \left(\frac{2x^{-2}}{-2} \right) \right|_{a}^{-1}$

$\displaystyle = \lim_{a \to -\infty} \left(-1 + \frac{1}{a^2} \right)$

As $a \to -\infty$, $\frac{1}{a^2} \to 0$. The integral is convergent.

$\displaystyle \int_{-\infty}^{-1} \frac{2}{x^3}\, dx = -1 + 0 = -1$

7. $\displaystyle\int_{1}^{\infty} \frac{1}{x^{1.0001}}\, dx$

$\displaystyle = \int_{1}^{\infty} x^{-1.0001}\, dx$

$\displaystyle = \lim_{b \to \infty} \int_{1}^{b} x^{-1.0001}\, dx$

$\displaystyle = \lim_{b \to \infty} \left. \left(\frac{x^{-0.0001}}{-0.0001} \right) \right|_{1}^{b}$

$\displaystyle = \lim_{b \to \infty} \left(-\frac{1}{(0.0001) b^{0.0001}} + \frac{1}{0.0001} \right)$

As $b \to \infty$, $-\frac{1}{0.0001 b^{0.0001}} \to 0$.

The integral is convergent.

$\displaystyle \int_{1}^{\infty} \frac{1}{x^{1.0001}}\, dx = 0 + \frac{1}{0.0001} = 10{,}000$

9. $\displaystyle\int_{-\infty}^{-1} x^{-2}\, dx = \lim_{a \to -\infty} \int_{a}^{-1} x^{-2}\, dx$

$\displaystyle = \lim_{a \to -\infty} \left. (-x^{-1}) \right|_{a}^{-1}$

$\displaystyle = \lim_{a \to -\infty} \left(1 + \frac{1}{a} \right)$

$= 1 + 0$

$= 1$

The integral is convergent.

11. $\displaystyle\int_{-\infty}^{-1} x^{-8/3}\, dx = \lim_{a \to -\infty} \int_{a}^{-1} x^{-8/3}\, dx$

$\displaystyle = \lim_{a \to -\infty} \left. \left(-\frac{3}{5} x^{-5/3} \right) \right|_{a}^{-1}$

$\displaystyle = \lim_{a \to -\infty} \left(\frac{3}{5} + \frac{3}{5a^{5/3}} \right)$

$\displaystyle = \frac{3}{5} + 0$

$\displaystyle = \frac{3}{5}$

The integral is convergent.

13. $\displaystyle\int_{0}^{\infty} 4e^{-4x}\, dx = \lim_{b \to \infty} \int_{0}^{b} 4e^{-4x}\, dx$

$\displaystyle = \lim_{b \to \infty} \left. \left(\frac{4e^{-4x}}{-4} \right) \right|_{0}^{b}$

$\displaystyle = \lim_{b \to \infty} (-e^{-4b} + 1)$

$\displaystyle = \lim_{b \to \infty} \left(-\frac{1}{e^{4b}} + 1 \right)$

$= 0 + 1 = 1$

The integral is convergent.

15. $\displaystyle\int_{-\infty}^{0} 4e^{x}\, dx = \lim_{a \to -\infty} \int_{a}^{0} 4e^{x}\, dx$

$\displaystyle = \lim_{a \to -\infty} \left. (4e^{x}) \right|_{a}^{0}$

$\displaystyle = \lim_{a \to -\infty} (4 - 4e^{a})$

Since a approaches $-\infty$, e^{a} is in the denominator of a fraction.

As $a \to \infty$, $-4e^{a} \to 0$. The integral is convergent.

$\displaystyle \int_{-\infty}^{0} 4e^{x}\, dx = 4 - 0 = 4$

17. $\displaystyle\int_{-\infty}^{-1} \ln |x|\, dx = \lim_{a \to -\infty} \int_{a}^{-1} \ln |x|\, dx$

Let $u = \ln |x|$ and $dv = dx$.

Then $du = \frac{1}{x}\, dx$ and $v = x$.

$\displaystyle \int \ln |x|\, dx = x \ln |x| - \int \frac{x}{x}\, dx$

$\displaystyle = x \ln |x| - x + C$

$\displaystyle \int_{-\infty}^{-1} \ln |x|\, dx = \lim_{a \to -\infty} \left. (x \ln |x| - x) \right|_{a}^{-1}$

$\displaystyle = \lim_{a \to -\infty} (-\ln 1 + 1 - a \ln |a| + a)$

$\displaystyle = \lim_{a \to -\infty} (1 + a - a \ln |a|)$

The integral is divergent, since as $a \to -\infty$.

$\displaystyle (a - a \ln |a|) = -a(-1 + \ln |a|) \to \infty.$

19. $\displaystyle\int_{0}^{\infty} \frac{dx}{(x+1)^2} = \lim_{b \to \infty} \int_{0}^{b} \frac{dx}{(x+1)^2}$ *Use substitution*

$\displaystyle = \lim_{b \to \infty} \left. -(x+1)^{-1} \right|_{0}^{b}$

$\displaystyle = \lim_{b \to \infty} \left(\frac{-1}{b+1} + 1 \right)$

As $b \to \infty$, $-\frac{1}{b+1} \to 0$. The integral is convergent.

$$\int_0^\infty \frac{dx}{(x+1)^2} = 0 + 1 = 1$$

21. $\displaystyle\int_{-\infty}^{-1} \frac{2x-1}{x^2-x}\, dx$

$$= \lim_{a \to -\infty} \int_a^{-1} \frac{2x-1}{x^2-x}\, dx \quad \textit{Use substitution}$$

$$= \lim_{a \to -\infty} \ln\left|x^2-x\right|\Big|_a^{-1}$$

$$= \lim_{a \to -\infty} \left(\ln 2 - \ln\left|a^2-a\right|\right)$$

As $a \to -\infty$, $\ln\left|a^2-a\right| \to \infty$. The integral is divergent.

23. $\displaystyle\int_2^\infty \frac{1}{x \ln x}\, dx$

$$= \lim_{b \to \infty} \int_2^b \frac{1}{x \ln x}\, dx \quad \textit{Use substitution}$$

$$= \lim_{b \to \infty} \left[\ln(\ln x)\Big|_2^b\right]$$

$$= \lim_{b \to \infty} \left[\ln(\ln b) - \ln(\ln 2)\right]$$

As $b \to \infty$, $\ln(\ln b) \to \infty$. The integral is divergent.

25. $\displaystyle\int_0^\infty xe^{2x}\, dx = \lim_{b \to \infty} \int_0^b xe^{2x}\, dx$

Let $\quad dv = e^{2x}\, dx \quad$ and $\quad u = x$.

Then $\quad v = \frac{1}{2}e^{2x} \quad$ and $du = dx$.

$$\int xe^{2x}\, dx = \frac{x}{2}e^{2x} - \int \frac{1}{2}e^{2x}\, dx$$

$$= \frac{x}{2}e^{2x} - \frac{1}{4}e^{2x} + C$$

$$= \frac{1}{4}(2x-1)e^{2x} + C$$

$$\int_0^\infty xe^{2x}\, dx$$

$$= \lim_{b \to \infty} \left[\frac{1}{4}(2x-1)e^{2x}\right]\Big|_0^b$$

$$= \lim_{b \to \infty} \left[\frac{1}{4}(2b-1)e^{2b} - \frac{1}{4}(-1)(1)\right]$$

$$= \lim_{b \to \infty} \left[\frac{1}{4}(2b-1)e^{2b} + \frac{1}{4}\right]$$

As $b \to \infty$, $\frac{1}{4}(2b-1)e^{2b} \to \infty$. The integral is divergent.

27. $\displaystyle\int_{-\infty}^\infty x^3 e^{-x^4}\, dx = \int_{-\infty}^0 x^3 e^{-x^4}\, dx + \int_0^\infty x^3 e^{-x^4}\, dx$

We evaluate each of the two improper integrals on the right.

$$\int_{-\infty}^0 x^3 e^{-x^4}\, dx = \lim_{b \to -\infty} \int_b^0 x^3 e^{-x^4}\, dx \quad \textit{Use substitution}$$

$$= \lim_{b \to -\infty} \left[-\frac{1}{4}e^{-x^4}\Big|_b^0\right]$$

$$= \lim_{b \to -\infty} \left[-\frac{1}{4} + \frac{1}{4e^{b^4}}\right]$$

As $b \to -\infty$, $\frac{1}{4e^{b^4}} \to 0$. The integral is convergent.

$$\int_{-\infty}^0 x^3 e^{-x^4}\, dx = -\frac{1}{4} + 0 = -\frac{1}{4}$$

$$\int_0^\infty x^3 e^{-x^4}\, dx = \lim_{b \to \infty} \int_0^b x^3 e^{-x^4}\, dx \quad \textit{Use substitution}$$

$$= \lim_{b \to \infty} \left[-\frac{1}{4}e^{-x^4}\Big|_0^b\right]$$

$$= \lim_{b \to \infty} \left[-\frac{1}{4e^{b^4}} + \frac{1}{4}\right]$$

As $b \to \infty$, $-\frac{1}{4e^{b^4}} \to 0$. The integral is convergent.

$$\int_0^\infty x^3 e^{-x^4}\, dx = 0 + \frac{1}{4} = \frac{1}{4}$$

Since each of the improper integrals converges, the original improper integral converges.

$$\int_{-\infty}^\infty x^3 e^{-x^4}\, dx = -\frac{1}{4} + \frac{1}{4} = 0$$

29. $\displaystyle\int_{-\infty}^\infty \frac{x}{x^2+1}\, dx = \int_{-\infty}^0 \frac{x}{x^2+1}\, dx + \int_0^\infty \frac{x}{x^2+1}\, dx$

We evaluate the first improper integral on the right.

$$\int_{-\infty}^0 \frac{x}{x^2+1}\, dx = \lim_{b \to -\infty} \int_b^0 \frac{x}{x^2+1}\, dx \quad \textit{Use substitution}$$

$$= \lim_{b \to -\infty} \left[\frac{1}{2}\ln(x^2+1)\Big|_b^0\right]$$

$$= \lim_{b \to -\infty} \left[0 - \frac{1}{2}\ln(b^2+1)\right]$$

As $b \to -\infty$, $\ln(b^2+1) \to \infty$. The integral is divergent. Since one of the two improper integrals on the right diverges, the original improper integral diverges.

31. $f(x) = \dfrac{1}{x-1}$ for $(-\infty, 0]$

$$\int_{-\infty}^{0} \frac{1}{x-1}\,dx = \lim_{a \to -\infty} \int_{a}^{0} \frac{dx}{x-1}$$

$$= \lim_{a \to -\infty} \left(\ln|x-1| \Big|_{a}^{0} \right)$$

$$= \lim_{a \to -\infty} \left(\ln|-1| - \ln|a-1| \right)$$

But $\lim\limits_{a \to -\infty} \left(\ln|a-1| \right) = \infty$.

The integral is divergent, so the area cannot be found.

33. $f(x) = \dfrac{1}{(x-1)^2}$ for $(-\infty, 0]$

$$\int_{-\infty}^{0} \frac{1}{(x-1)^2}$$

$$= \lim_{a \to -\infty} \int_{a}^{0} \frac{1}{(x-1)^2} \quad \textit{Use substitution}$$

$$= \lim_{a \to -\infty} -(x-1)^{-1} \Big|_{a}^{0}$$

$$= \lim_{a \to -\infty} \left(-\frac{1}{-1} + \frac{1}{a-1} \right)$$

As $a \to -\infty$, $\dfrac{1}{a-1} \to 0$. The integral is convergent.

$$= 1 + 0 = 1$$

Therefore, the area is 1.

35. $\displaystyle\int_{-\infty}^{\infty} xe^{-x^2}\,dx$

Let $u = -x^2$, so that $du = -2x\,dx$.

$$= \lim_{a \to -\infty} \left(-\frac{1}{2} \int_{a}^{0} -2xe^{-x^2}\,dx \right)$$

$$+ \lim_{b \to \infty} \left(-\frac{1}{2} \int_{0}^{b} -2xe^{-x^2}\,dx \right)$$

$$= \lim_{a \to -\infty} \left(-\frac{1}{2} e^{-x^2} \right) \Big|_{a}^{0}$$

$$+ \lim_{b \to \infty} \left(-\frac{1}{2} e^{-x^2} \right) \Big|_{0}^{b}$$

$$= \lim_{a \to -\infty} \left(-\frac{1}{2} + \frac{1}{2e^{-a^2}} \right)$$

$$+ \lim_{b \to \infty} \left(-\frac{1}{2e^{b^2}} + \frac{1}{2} \right)$$

$$= -\frac{1}{2} + \frac{1}{2} = 0$$

37. $\displaystyle\int_{1}^{\infty} \frac{1}{x^p}\,dx$

Case 1a $\quad p < 1$:

$$\int_{1}^{\infty} \frac{1}{x^p}\,dx = \int_{1}^{\infty} x^{-p}\,dx$$

$$= \lim_{a \to \infty} \int_{1}^{a} x^{-p}\,dx$$

$$= \lim_{a \to \infty} \left[\frac{x^{-p+1}}{(-p+1)} \Big|_{1}^{a} \right]$$

$$= \lim_{a \to \infty} \left[\frac{1}{(-p+1)} (a^{-p+1} - 1) \right]$$

$$= \lim_{a \to \infty} \left[\frac{1}{(-p+1)} a^{1-p} - \frac{1}{(-p+1)} \right]$$

Since $p < 1$, $1 - p$ is positive and, as $a \to \infty$, $a^{1-p} \to \infty$. The integral diverges.

Case 1b $\quad p = 1$:

$$\int_{1}^{\infty} \frac{1}{x^p}\,dx = \int_{1}^{\infty} \frac{1}{x}\,dx$$

$$= \lim_{a \to \infty} \int_{1}^{a} \frac{1}{x}\,dx$$

$$= \lim_{a \to \infty} \left(\ln|x| \Big|_{1}^{a} \right)$$

$$= \lim_{a \to \infty} \left(\ln|a| - \ln 1 \right)$$

$$= \lim_{a \to \infty} \ln|a|$$

As $a \to \infty$, $\ln|a| \to \infty$. The integral diverges.

Therefore, $\displaystyle\int_{1}^{\infty} \frac{1}{x^p}$ diverges when $p \leq 1$.

Case 2 $\quad p > 1$:

$$\int_{1}^{\infty} \frac{1}{x^p}\,dx = \lim_{a \to \infty} \int_{1}^{a} x^{-p}\,dx$$

$$= \lim_{a \to \infty} \left(\frac{x^{-p+1}}{-p+1} \Big|_{1}^{a} \right)$$

$$= \lim_{a \to \infty} \left[\frac{a^{-p+1}}{(-p+1)} - \frac{1}{(-p+1)} \right]$$

Since $p > 1$, $-p + 1 < 0$; thus as $a \to \infty$, $\dfrac{a^{-p+1}}{(-p+1)} \to 0$.

Hence,

$$\lim_{a \to \infty} \left[\frac{a^{-p+1}}{(-p+1)} - \frac{1}{(-p+1)} \right] = 0 - \frac{1}{(-p+1)}$$

$$= \frac{-1}{-p+1} = \frac{1}{p-1}.$$

The integral converges.

39. (a) Use the *fnInt* feature on a graphing utility to obtain

$$\int_1^{20} \frac{1}{\sqrt{1+x^2}}\, dx \approx 2.8081;$$

$$\int_1^{50} \frac{1}{\sqrt{1+x^2}}\, dx \approx 3.7239;$$

$$\int_1^{100} \frac{1}{\sqrt{1+x^2}}\, dx \approx 4.4170;$$

$$\int_1^{1,000} \frac{1}{\sqrt{1+x^2}}\, dx \approx 6.7195;$$

$$\int_1^{10,000} \frac{1}{\sqrt{1+x^2}}\, dx \approx 9.0221.$$

(b) Since the values of the integrals in part a do not appear to be approaching some fixed finite number but get bigger, the integral $\int_1^{\infty} \frac{1}{\sqrt{1+x^2}}\, dx$ appears to be divergent.

(c) Use the *fnInt* feature on a graphing utility to obtain

$$\int_1^{20} \frac{1}{\sqrt{1+x^4}}\, dx \approx 0.8770;$$

$$\int_1^{50} \frac{1}{\sqrt{1+x^4}}\, dx \approx 0.9070;$$

$$\int_1^{100} \frac{1}{\sqrt{1+x^4}}\, dx \approx 0.9170;$$

$$\int_1^{1,000} \frac{1}{\sqrt{1+x^4}}\, dx \approx 0.9260;$$

$$\int_1^{10,000} \frac{1}{\sqrt{1+x^4}}\, dx \approx 0.9269.$$

(d) Since the values of the integrals in part c appear to be approaching some fixed finite number, the integral $\int_1^{\infty} \frac{1}{\sqrt{1+x^4}}\, dx$ appears to be convergent.

(e) For large x, we may consider $1+x^2 \approx x^2$ and $1+x^4 \approx x^4$.
Thus,

$$\frac{1}{\sqrt{1+x^4}} \approx \frac{1}{\sqrt{x^2}} = \frac{1}{x} \text{ and}$$

$$\frac{1}{\sqrt{1+x^4}} \approx \frac{1}{\sqrt{x^4}} = \frac{1}{x^2}.$$

In Example 1(a) on page 455, we showed that $\int_1^{\infty} \frac{1}{x}\, dx$ diverges. Thus, we might guess that $\int_1^{\infty} \frac{1}{\sqrt{1+x^2}}\, dx$ diverges as well. In Exercise 1, we saw that $\int_2^{\infty} \frac{1}{x^2}\, dx$ converges. Thus, we might guess that $\int_1^{\infty} \frac{1}{\sqrt{1+x^4}}\, dx$ converges as well.

41. (a) Use the *fnInt* feature on a graphing utility to obtain

$$\int_0^{10} e^{-0.00001x}\, dx \approx 9.9995;$$

$$\int_0^{50} e^{-0.00001x}\, dx \approx 49.9875;$$

$$\int_0^{100} e^{-0.00001x}\, dx \approx 99.9500;$$

$$\int_0^{1,000} e^{-0.00001x}\, dx \approx 995.0166.$$

(b) Since the values of the integrals in part a do not appear to be approaching some fixed finite number, the integral $\int_0^{\infty} e^{-0.00001x}\, dx$ appears to be divergent.

(c) $\int_0^{\infty} e^{-0.00001x}\, dx$

$$= \lim_{b \to \infty} \int_0^b e^{-0.00001x}\, dx$$

$$= \lim_{b \to \infty} \left[\frac{e^{-0.00001x}}{-0.00001} \Big|_0^b \right]$$

$$= \lim_{b \to \infty} \left[-\frac{1}{0.00001 e^{0.00001b}} + \frac{1}{0.00001} \right]$$

$$= 0 + 100,000 = 100,000$$

43. Use the *fnInt* feature on a graphing calculator to obtain:

$$\int_0^1 e^{-x} \cos x\, dx \approx 0.555396883$$

$$\int_0^{10} e^{-x} \cos x\, dx \approx 0.500006698$$

$$\int_0^{100} e^{-x} \cos x\, dx \approx 0.500000000.$$

Based on these results, it appears the integral is convergent and $\int_0^{\infty} e^{-x} \cos x\, dx = \frac{1}{2}$.

45. $P = \int_0^{\infty} e^{-rt}(at+b)K\, dt$

$$= K \lim_{c \to \infty} \int_0^c (at+b)e^{-rt}\, dt$$

Evaluate $\int_0^c (at+b)e^{-rt}\, dt$ using integration by parts.

Let $u = at+b$ and $dv = e^{-rt}\, dt$.
Then $du = a\, dt$ and $v = -\frac{1}{r}e^{-rt}$.

$$\int_0^c (at+b)e^{-rt}\,dt$$

$$= \left[(at+b)\left(-\frac{1}{r}e^{-rt}\right) - \int\left(-\frac{1}{r}e^{-rt}\right)a\,dt\right]\Bigg|_0^c$$

$$= \left[-\frac{at+b}{r}e^{-rt} + \frac{a}{r}\int e^{-rt}dt\right]\Bigg|_0^c$$

$$= \left[-\frac{at+b}{r}e^{-rt} - \frac{a}{r^2}e^{-rt}\right]\Bigg|_0^c$$

$$= \left(-\frac{ac+b}{r}e^{-rc} - \frac{a}{r^2}e^{-rc}\right) - \left(-\frac{b}{r}e^0 - \frac{a}{r^2}e^0\right)$$

Therefore,

$$K\lim_{c\to\infty}\int_0^c (at+b)e^{-rt}dt$$

$$= K\lim_{c\to\infty}\left(-\frac{ac+b}{r}e^{-rc} - \frac{a}{r^2}e^{-rc} + \frac{b}{r} + \frac{a}{r^2}\right)$$

$$= K\left(0 - 0 + \frac{b}{r} + \frac{a}{r^2}\right)$$

$$= \frac{K(a+br)}{r^2}$$

47. $\displaystyle\int_0^\infty 50e^{-0.06t}\,dt = 50\lim_{b\to\infty}\int_0^b e^{-0.06t}\,dt$

$$= 50\lim_{b\to\infty}\frac{e^{-0.06t}}{-0.06}\Bigg|_0^b$$

$$= \frac{50}{-0.06}\lim_{b\to\infty}(e^{-0.06b} - e^0)$$

$$= -\frac{50}{0.06}(0-1) = \frac{50}{0.06}$$

$$\approx 833.33$$

Chapter 8 Review Exercises

5. $\displaystyle\int_1^3 \frac{\ln x}{x}\,dx$

Trapezoidal Rule:

$n = 4,\ b = 3,\ a = 1,\ f(x) = \frac{\ln x}{x}$

i	x_1	$f(x_i)$
0	1	0
1	1.5	0.27031
2	2	0.34657
3	2.5	0.36652
4	3	0.3662

$$\int_1^3 \frac{\ln x}{x}\,dx \approx \frac{3-1}{4}\left[\frac{1}{2}(0) + 0.27031 + 0.34657\right.$$

$$\left. + 0.36652 + \frac{1}{2}(0.3662)\right]$$

$$= 0.58325$$

Exact value:

$$\int_1^3 \frac{\ln x}{x}\,dx$$

$$= \frac{1}{2}(\ln x)^2\Bigg|_1^3$$

$$= \frac{1}{2}(\ln 3)^2 - \frac{1}{2}(\ln 1)^2$$

$$\approx 0.60347$$

7. $\displaystyle\int_0^1 e^x\sqrt{e^x + 1}\,dx$

Trapezoidal Rule:

$n = 4,\ b = 1,\ a = 0,\ f(x) = e^x\sqrt{e^x+1}$

i	x_i	$f(x_i)$
0	0	1.4142
1	0.25	1.9405
2	0.5	2.6833
3	0.75	3.7376
4	1	5.2416

$$\int_0^1 e^x\sqrt{e^x+1}\,dx$$

$$= \frac{1-0}{4}\left[\frac{1}{2}(1.4142) + 1.905 + 2.6833\right.$$

$$\left. + 3.7376 + \frac{1}{2}(5.2416)\right]$$

$$\approx 2.9223$$

Exact Value:

$$\int_0^1 e^x\sqrt{e^x+1}\,dx$$

$$= \int_0^1 e^x(e^x+1)^{1/2}\,dx$$

$$= \frac{2}{3}(e^x+1)^{3/2}\Bigg|_0^1$$

$$= \frac{2}{3}(e+1)^{3/2} - \frac{2}{3}(2)^{3/2}$$

$$\approx 2.8943$$

9. $\displaystyle\int_2^{10} \frac{x\,dx}{x-1}$

Simpson's Rule:

i	x_i	$f(x_1)$
0	2	2
1	4	$\frac{4}{3}$
2	6	$\frac{6}{5}$
3	8	$\frac{8}{7}$
4	10	$\frac{10}{9}$

$$\int_2^{10} \frac{x}{x-1}\,dx$$

$$\approx \frac{10-2}{3(4)}\left[2 + 4\left(\frac{4}{3}\right) + 2\left(\frac{6}{5}\right) + 4\left(\frac{8}{7}\right) + \frac{10}{9}\right]$$

$$\approx 10.28$$

This answer is close to the answer of 10.197 obtained from the exact integral in Exercise 49.

11. (a) $\displaystyle\int_1^5 \left[\sqrt{x-1} - \left(\frac{x-1}{2}\right)\right]\,dx$

$$= \int_1^5 \left(\sqrt{x-1} - \frac{x}{2} + \frac{1}{2}\,dx\right)$$

$$= \left(\frac{2}{3}(x-1)^{3/2} - \frac{x^2}{4} + \frac{x}{2}\right)\Big|_1^5$$

$$= \left(\frac{16}{3} - \frac{25}{4} + \frac{5}{2}\right) - \left(0 - \frac{1}{4} + \frac{1}{2}\right)$$

$$= \frac{16}{3} - 6 + 2 = \frac{4}{3}$$

(b) $n=4$, $b=5$, $a=1$, $f(x) = \sqrt{x-1} - \frac{x}{2} + \frac{1}{2}$

i	x_i	$f(x_i)$
0	1	0
1	2	0.5
2	3	0.41421
3	4	0.23205
4	5	0

$$\int_1^5 \left(\sqrt{x-1} - \frac{x}{2} + \frac{1}{2}\right)\,dx$$

$$= \left(\frac{5-1}{4}\right)\left[\frac{1}{2}(0) + 0.5 + 0.41421\right.$$

$$\left. + 0.23205 + \frac{1}{2}(0)\right]$$

$$= 1.14626$$

(c) $\displaystyle\int_1^5 \left(\sqrt{x-1} - \frac{x}{2} + \frac{1}{2}\right)\,dx$

$$= \left(\frac{5-1}{3(4)}\right)[0 + 4(0.5) + 2(0.41421)$$

$$+ 4(0.23205) + 0]$$

$$= \left(\frac{1}{3}\right)(3.75662)$$

$$= 1.2522$$

13. $\displaystyle\int x(8-x)^{3/2}\,dx$

Let $u = x$ and $dv = (8-x)^{3/2}$.

Then $du = dx$ and $v = -\frac{2}{5}(8-x)^{5/2}$.

$$\int x(8-x)^{3/2}\,dx$$

$$= -\frac{2}{5}x(8-x)^{5/2} + \int \frac{2}{5}(8-x)^{5/2}\,dx$$

$$= -\frac{2}{5}x(8-x)^{5/2} - \frac{2}{5}\left(\frac{2}{7}\right)(8-x)^{7/2} + C$$

$$= -\frac{2x}{5}(8-x)^{5/2} - \frac{4}{35}(8-x)^{7/2} + C$$

15. $\displaystyle\int xe^x\,dx$

Let $u = x$ and $dv = e^x\,dx$.

Then $du = dx$ and $v = e^x$.

$$\int xe^x\,dx = xe^x - \int e^x\,dx$$

$$= xe^x - e^x + C$$

17. $\displaystyle\int \ln|2x+3|\,dx$

First, use substitution.

Let $a = 2x+3$. Then $da = 2\,dx$.

$$\int \ln|2x+3|\,dx = \frac{1}{2}\int \ln|a|\,da$$

Second, integrate by parts.

Let $u = \ln|a|$ and $dv = da$.

Then $du = \frac{1}{a}\,da$ and $v = a$.

$$\frac{1}{2}\int \ln|a|\,da$$

$$= \frac{1}{2}\left(a\ln|a| - \int da\right)$$

$$= \frac{1}{2}(a\ln|a| - a) + C$$

Finally, substitute $2x + 3$ for a.

$$\int \ln |2x + 3| \, dx$$

$$= \frac{1}{2}[(2x + 3) \ln |2x + 3| - (2x + 3) + C]$$

$$= \frac{1}{2}(2x + 3)[\ln |2x + 3| - 1] + C$$

19. $\int \dfrac{x}{9 - 4x^2} \, dx$

Use substitution.
Let $u = 9 - 4x^2$. Then $du = -8x \, dx$.

$$\int \frac{x}{9 - 4x^2} \, dx = -\frac{1}{8} \int \frac{-8x \, dx}{9 - 4x^2}$$

$$= -\frac{1}{8} \int \frac{du}{u}$$

$$= -\frac{1}{8} \ln |u| + C$$

$$= -\frac{1}{8} \ln |9 - 4x^2| + C$$

21. $\int (x + 2) \sin x \, dx$

Let $u = x + 2$ and $dv = \sin x \, dx$.
Then $du = dx$ and $v = -\cos x$.

$$\int (x + 2) \sin x \, dx = (x + 2)(-\cos x) - \int (-\cos x) dx$$

$$= -(x + 2) \cos x + \int \cos x \, dx$$

$$= -(x + 2) \cos x + \sin x + C$$

23. $\int_1^e x^3 \ln x \, dx$

Let $u = \ln x$ and $dv = x^3 \, dx$.
Use column integration.

D		I
$\ln x$	$+$	x^3
$\dfrac{1}{x}$	$-$	$\dfrac{1}{4}x^4$

$$\int x^3 \ln x \, dx$$

$$= \frac{1}{4}x^4 \ln x - \int \left(\frac{1}{4}x^4 \cdot \frac{1}{x} \right) dx$$

$$= \frac{1}{4}x^4 \ln x - \frac{1}{4} \int x^3 \, dx$$

$$= \frac{1}{4}x^4 \ln x - \frac{1}{16}x^4 + C$$

$$\int_1^e x^3 \ln x \, dx$$

$$= \left(\frac{1}{4}x^4 \ln x - \frac{1}{16}x^4 \right)\Big|_1^e$$

$$= \left(\frac{e^4}{4} \right) (1) - \frac{e^4}{16} - 0 + \frac{1}{16}$$

$$= \frac{e^4}{4} - \frac{e^4}{16} + \frac{1}{16}$$

$$= \frac{3e^4 + 1}{16}$$

$$\approx 10.300$$

25. $A = \displaystyle\int_0^1 (3 + x^2)e^{2x} \, dx$

$$= \int_0^1 3e^{2x} \, dx + \int_0^1 x^2 e^{2x} \, dx$$

$$\int 3e^{2x} \, dx = \frac{3}{2}e^{2x} + C$$

For the second integral, $\int x^2 e^{2x} \, dx$, let $u = x^2$ and $dv = e^{2x} \, dx$.
Use column integration.

D		I
x^2	$+$	e^{2x}
$2x$	$-$	$\dfrac{e^{2x}}{2}$
2	$+$	$\dfrac{e^{2x}}{4}$
0		$\dfrac{e^{2x}}{8}$

$$\int x^2 e^{2x} \, dx = x^2 \frac{e^{2x}}{2} - 2x \frac{e^{2x}}{4} + 2 \frac{e^{2x}}{8}$$

$$= \frac{x^2 e^{2x}}{2} - \frac{xe^{2x}}{2} + \frac{e^{2x}}{4}$$

$$A = \left(\frac{3}{2}e^{2x} + \frac{x^2 e^{2x}}{2} - \frac{xe^{2x}}{2} + \frac{e^{2x}}{4} \right)\Big|_0^1$$

$$= \frac{3}{2}e^2 + \frac{e^2}{2} - \frac{e^2}{2} + \frac{e^2}{4} - \left(\frac{3}{2} + \frac{1}{4} \right)$$

$$= \left(\frac{6 + 2 - 2 + 1}{4} \right) e^2 - \left(\frac{7}{4} \right)$$

$$= \frac{7}{4}(e^2 - 1) \approx 11.181$$

27. $f(x) = 2x - 1$, $y = 0$, $x = 3$

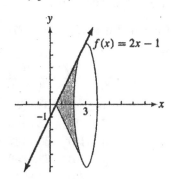

Since $f(x) = 2x - 1$ intersects $y = 0$ at $x = \frac{1}{2}$, the integral has a lower bound $a = \frac{1}{2}$.

$$V = \pi \int_{1/2}^{3} (2x - 1)^2 \, dx$$

$$= \pi \int_{1/2}^{3} (4x^2 - 4x + 1) \, dx$$

$$= \pi \left(\frac{4x^3}{3} - \frac{4x^2}{2} + x \right) \Big|_{1/2}^{3}$$

$$= \pi \left(36 - 18 + 3 - \frac{1}{6} + \frac{1}{2} - \frac{1}{2} \right)$$

$$= \pi \left(21 - \frac{1}{6} \right) = \frac{125\pi}{6} \pi \approx 65.45$$

29. $f(x) = e^{-x}$, $y = 0$, $x = -2$, $x = 1$

$$V = \pi \int_{-2}^{1} e^{-2x} \, dx = \frac{\pi e^{-2x}}{-2} \Big|_{-2}^{1}$$

$$= \frac{\pi e^{-2}}{-2} + \frac{\pi e^4}{2} = \frac{\pi(e^4 - e^{-2})}{2}$$

$$\approx 85.55$$

31. $f(x) = 4 - x^2$, $y = 0$, $x = -1$, $x = 1$

$$V = \pi \int_{-1}^{1} (4 - x^2)^2 \, dx$$

$$= \pi \int_{-1}^{1} (16 - 8x^2 + x^4) \, dx$$

$$= \pi \left(16x - \frac{8x^3}{3} + \frac{x^5}{5} \right) \Big|_{-1}^{1}$$

$$= \pi \left(16 - \frac{8}{3} + \frac{1}{5} + 16 - \frac{8}{3} + \frac{1}{5} \right)$$

$$= \pi \left(32 - \frac{16}{3} + \frac{2}{5} \right)$$

$$= \frac{406\pi}{15}$$

$$\approx 85.03$$

33. The frustum may be shown as follows.

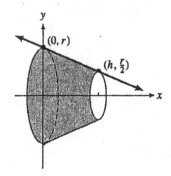

Use the two points given to find

$$f(x) = -\frac{r}{2h}x + r.$$

$$V = \pi \int_{0}^{h} \left(-\frac{r}{2h}x + r \right)^2 \, dx$$

$$= -\frac{2\pi h}{3r} \left(-\frac{r}{2h}x + r \right)^3 \Big|_{0}^{h}$$

$$= -\frac{2\pi h}{3r} \left[\left(-\frac{r}{2} + r \right)^3 - (0 + r)^3 \right]$$

$$= -\frac{2\pi h}{3r} \left[\left(\frac{r}{2} \right)^3 - r^3 \right]$$

$$= -\frac{2\pi h}{3r} \left(\frac{r^3}{8} - r^3 \right)$$

$$= -\frac{2\pi h}{3r} \left(-\frac{7r^3}{8} \right)$$

$$= \frac{7\pi r^2 h}{12}$$

35. $f(x) = \sqrt{x + 1}$

$$\frac{1}{b - a} \int_{a}^{b} f(x) \, dx$$

$$= \frac{1}{8 - 0} \int_{0}^{8} \sqrt{x + 1} \, dx$$

$$= \frac{1}{8} \int_{0}^{8} (x + 1)^{1/2} \, dx$$

$$= \frac{1}{8} \left(\frac{2}{3} \right) (x + 1)^{3/2} \Big|_{0}^{8}$$

$$= \frac{1}{12} (9)^{3/2} - \frac{1}{12} (1)$$

$$= \frac{27}{12} - \frac{1}{12} = \frac{26}{12}$$

$$= \frac{13}{6}$$

37. $\displaystyle\int_1^\infty x^{-1}\,dx = \lim_{b\to\infty}\int_1^b \frac{dx}{x}$

$\displaystyle\qquad = \lim_{b\to\infty}\ln x \Big|_1^b$

$\displaystyle\qquad = \lim_{b\to\infty}\ln b$

As $b\to\infty$, $\ln b\to\infty$. The integral is divergent.

39. $\displaystyle\int_0^\infty \frac{dx}{(5x+2)^2}$

$\displaystyle\qquad = \lim_{b\to\infty}\int_0^b (5x+2)^{-2}\,dx$

$\displaystyle\qquad = \lim_{b\to\infty}\left[-\frac{1}{5}(5x+2)^{-1}\right]\Big|_0^b$

$\displaystyle\qquad = \lim_{b\to\infty}\left[\frac{-1}{5(5x+2)}\right]\Big|_0^b$

$\displaystyle\qquad = \lim_{b\to\infty}\left[\frac{-1}{5(5b+2)}+\frac{1}{10}\right]$

$\displaystyle\qquad = \frac{1}{10}$

41. $\displaystyle\int_{-\infty}^0 \frac{x}{x^2+3}\,dx$

$\displaystyle\qquad = \lim_{a\to-\infty}\frac{1}{2}\int_a^0 \frac{2x\,dx}{x^2+3}$

$\displaystyle\qquad = \lim_{a\to-\infty}\frac{1}{2}(\ln|x^2+3|)\Big|_a^0$

$\displaystyle\qquad = \lim_{a\to-\infty}\frac{1}{2}(\ln 3 - \ln|a^2+3|)$

As $a\to-\infty$, $\frac{1}{2}(\ln 3 - \ln|a^2+3|)\to\infty$.
The integral is divergent.

43. $\displaystyle A = \int_{-\infty}^1 \frac{3}{(x-2)^2}\,dx$

$\displaystyle\qquad = \lim_{a\to-\infty}\int_a^1 3(x-2)^{-2}\,dx$

$\displaystyle\qquad = \lim_{a\to-\infty}\frac{3(x-2)^{-1}}{-1}\Big|_a^1$

$\displaystyle\qquad = \lim_{a\to-\infty}\left(-\frac{3}{x-2}\right)\Big|_a^1$

$\displaystyle\qquad = \lim_{a\to-\infty}\left(-\frac{3}{-1}+\frac{3}{a-2}\right)$

$\displaystyle\qquad = 3$

45. $\displaystyle\int_0^{320} 1.87t^{1.49}e^{-0.189(\ln t)^2}\,dt$

$n=8, b=320, a=1, f(t)=1.87t^{1.49}e^{-0.189(\ln t)^2}$

i	t_i	$f(t_i)$
0	1	1.8700
1	41	34.9086
2	81	33.9149
3	121	30.7147
4	161	27.5809
5	201	24.8344
6	241	22.4794
7	281	20.4622
8	321	18.7255

(a) Trapezoidal rule:

$\displaystyle\int_1^{321} 1.87t^{1.49}e^{-0.189(\ln t)^2}\,dt$

$\displaystyle\approx \frac{321-1}{8}\left[\frac{1}{2}(1.87)+34.9086+33.9149+30.7147\right.$

$\displaystyle\qquad +27.5809+24.8344+22.4794+20.4622$

$\displaystyle\qquad \left.+\frac{1}{2}(18.7255)\right]$

$\displaystyle\approx 8{,}208$

The total amount of milk produced is about 8,208 kg.

(b) $\displaystyle\int_1^{321} 1.87t^{1.49}e^{-0.189(\ln t)^2}\,dt$

$\displaystyle\approx \frac{321-1}{3(8)}[1.87+4(34.9086)+2(33.9149)$
$\displaystyle\qquad +4(30.7147)+2(27.5809)+4(24.8344)$
$\displaystyle\qquad +2(22.4794)+4(20.4622)+18.7255]$
$\displaystyle\approx 8{,}430$

The total amount of milk produced is about 8,430 kg.

(c) Using a graphing calculator's *fnInt* feature:

$\displaystyle\int_1^{321} 1.87t^{1.49}e^{-0.189(\ln t)^2}\,dt \approx 8{,}558.$

The total amount of milk produced is about 8,558 kg.

47. $\displaystyle\int_0^5 0.5xe^{-x}\,dx$

$$= 0.5\int_0^5 xe^{-x}\,dx$$

Let $\quad u = x$ and $dv = e^{-x}\,dx$.

Then $du = dx$ and $v = \frac{e^{-x}}{-1}$.

$$\int xe^{-x}\,dx = \frac{xe^{-x}}{-1} + \int e^{-x}\,dx$$

$$= -xe^{-x} + \frac{e^{-x}}{-1}$$

$$0.5\int_0^5 xe^{-x}\,dx = 0.5(-xe^{-x} - e^{-x})\Big|_0^5$$

$$= 0.5(-5e^{-5} - e^{-5} + e^0)$$
$$= 0.480$$

The total reaction over the first 5 hr is 0.480.

49. (a) $\displaystyle \overline{T} = \frac{1}{10-0}\int_0^{10}(400 - 0.25x^2)\,dx$

$$= \frac{1}{10}\left(400x - \frac{0.25x^3}{3}\right)\Big|_0^{10}$$

$$= \frac{1}{10}\left[400(10) - \frac{0.25}{3}(10)^3\right]$$

$$= \frac{1}{10}(3{,}916.7)$$

$$= 391.7$$

(b) $\displaystyle \overline{T} = \frac{1}{40-10}\int_{10}^{40}(400 - 0.25x^2)\,dx$

$$= \frac{1}{30}\left(400x - \frac{0.25x^3}{3}\right)\Big|_{10}^{40}$$

$$= \frac{1}{30}\left[\left(400(40) - \frac{0.25(40)^3}{3}\right)\right.$$

$$\left. - \left(400(10) - \frac{0.025(10)^3}{3}\right)\right]$$

$$= \frac{1}{30}(10{,}666.7 - 3{,}916.67)$$

$$= 225$$

(c) $\displaystyle \overline{T} = \frac{1}{40-0}\int_0^{40}(400 - 0.25x^2)\,dx$

$$= \frac{1}{40}\left(400x - \frac{0.25x^3}{3}\right)\Big|_0^{40}$$

$$= \frac{1}{40}\left[(400)(40) - \frac{(0.25)(40)^3}{3}\right]$$

$$= \frac{1}{40}[10{,}666.71]$$

$$= 266.7$$

Chapter 8 Test

[8.1]

1. Use $n = 4$ to approximate the value of the given integral by the following methods: (a) the trapezoidal rule and (b) Simpson's rule. (c) Find the exact value by integration.

$$\int_0^2 x\sqrt{x^2 + 1}\, dx$$

2. Use $n = 4$ to approximate the value of the given integral by the following methods: (a) the trapezoidal rule and (b) Simpson's rule.

$$\int_0^2 \frac{1}{\sqrt{1 + x^3}}\, dx$$

3. Use Simpson's rule with $n = 6$ to approximate the value of $\int_0^3 \frac{1}{x^2 + 1}\, dx$.

4. Use Simpson's rule with $n = 6$ to approximate the value of $\int_0^{\pi/3} \tan x^2\, dx$.

5. Find the area between the curve $y = e^{-x^2}$ and the x-axis from $x = 0$ to $x = 3$, using the trapezoidal rule with $n = 6$.

6. Find the area between the curve $y = \frac{x}{x^2 + 1}$ and the x-axis from $x = 1$ to $x = 4$, using Simpson's rule with $n = 6$.

7. **Milk Production**   Researchers report that the average amount of milk produced (in kg/day) by a 4- to 5-year-old cow weighing 700 kg can be approximated by

$$y = 1.87t^{1.49}e^{-0.189(\ln t)^2},$$

where t is the number of days into lactation.*

(a) Approximate the total amount of milk produced during days 1 to 121 using the trapezoidal rule with $n = 6$.

(b) Repeat part (a) using Simpson's rule with $n = 6$.

(c) Repeat part (a) using the integration feature of a graphing calculator, and compare your answer with the answers to parts (a) and (b).

---

* Freeze, Brian S., and Timothy J. Richards, "Lactation Curve Estimation for Use in Economic Optimization Models in the Dairy Industry," *Journal of Dairy Science*, Vol. 75, 1992, pp. 2984-2989.

[8.2]

Use integration by parts to find the following integrals.

8. $\int x \ln|5x| \, dx$

11. $\int \frac{x+3}{(3x-1)^4} \, dx$

9. $\int x\sqrt{2x-1} \, dx$

12. $\int_4^{11} x\sqrt[3]{x-3} \, dx$

10. $\int (2x-1) e^x \, dx$

13. Given that the marginal profit in dollars earned from the sale of x computers is

$$P'(x) = xe^{0.001x} - 100,$$

find the total profit from the sale of the first 500 computers.

14. Use integration by parts to find $\int x^2 \sin(x) \, dx$.

15. Thermic Effect of Food A person's metabolic rate tends to go up after eating a meal and then, after some time has passed, it returns to a resting metabolic rate. As we saw in the section on Increasing and Decreasing Functions, this phenomenon is known as the thermic effect of food, and the effect (in kJ/hr) for one individual is

$$F(t) = -10.28 + 175.9te^{-t/1.3},$$

where t is the number of hours that have elapsed since eating a meal.[*] Find the total thermic energy of a meal for the next eight hours after a meal by integrating the thermic effect function between $t = 0$ and $t = 8$.

[8.3]

Find the volume of the solid of revolution formed by rotating each of the following bounded regions about the x-axis.

16. $f(x) = 3x - 1$, $y = 0$, $x = 4$

18. $f(x) = e^x$, $y = 0$, $x = -1$, $x = 2$

17. $f(x) = \frac{1}{\sqrt{3x+1}}$, $y = 0$, $x = 0$, $x = 5$

19. $f(x) = x^2$, $y = 0$, $x = 2$

20. Find the average value of $f(x) = x^3 - x^2$ on the interval $[-1, 1]$.

21. The rate of depreciation t years after purchase of a certain machine is

$$D = 10{,}000(t - 7), \ 0 \le t \le 6.$$

What is the average depreciation over the first 3 years?

---

[*] Reed, George, and James Hill, "Measuring the Thermic Effect of Food," *American Journal of Clinical Nutrition*, Vol. 63, 1996, pp. 164-169.

22. Find the volume of the solid of revolution formed by rotating $f(x)\sec(x)$ about the x-axis from $x = \frac{\pi}{4}$ to $x = \frac{\pi}{3}$.

23. Find the average value of $f(x) = \frac{\pi}{x^2}\sin\frac{\pi}{x}$ on the interval $[2, 6]$.

24. Determine whether the improper integral $\displaystyle\int_0^{\pi/2} \frac{\cos x}{1 - \sin x}\, dx$ converges or diverges. If it converges, find the value.

[8.4]

Find the value of each integral that converges.

25. $\displaystyle\int_2^\infty \frac{14}{(x+1)^2}\, dx$

26. $\displaystyle\int_{-\infty}^2 \frac{4}{x+1}\, dx$

27. $\displaystyle\int_1^\infty e^{-3x}\, dx$

28. $\displaystyle\int_1^\infty \frac{1}{x}\, dx$

29. $\displaystyle\int_1^\infty x^{-1/3}\, dx$

30. $\displaystyle\int_{100}^\infty x^{-3/2}\, dx$

Find the area between the graph of the function and the x-axis over the given interval, if possible.

31. $f(x) = 8e^{-x}$ on $[1, \infty)$

32. $f(x) = \dfrac{1}{(2x-1)^2}$ on $[2, \infty)$

33. Find the capital value of an asset that generates income at an annual rate of \$4,000 if the interest rate is 8% compounded continuously.

Chapter 8 Test Answers

1. (a) 3.457 (b) 3.392 (c) 3.393

2. (a) 1.397 (b) 1.405

3. 1.25

4. 0.4837

5. 0.89

6. 1.07

7. (a) 3,656 (b) 3,740 (c) 3,761

8. $\frac{1}{2}x^2 \ln|5x| - \frac{1}{4}x^2 + C$

9. $\frac{x}{3}(2x-1)^{3/2} - \frac{1}{15}(2x-1)^{5/2} + C$

10. $(2x-3)e^x + C$

11. $-\frac{1}{9}\left[\frac{x+3}{(3x-1)^3}\right] - \frac{1}{54(3x-1)^2} + C$ or $-\frac{9x+17}{54(3x-1)^3} + C$

12. 88.18

13. $125,639.36

14. $-x^2\cos(x) + 2x\sin(x) + 2\cos(x) + C$

15. 211 kJ

16. $\frac{1331}{9}\pi$ or 464.61

17. 2.90

18. 85.55

19. $\frac{32\pi}{5}$ or 20.11

20. $-\frac{1}{3}$

21. 55,000 per year

22. $\pi\left(\sqrt{3}-1\right)$

23. $\frac{\sqrt{3}}{8}$

24. Diverges

25. $\frac{14}{3}$

26. Divergent

27. 0.017

28. Divergent

29. Divergent

30. $\frac{1}{5}$

31. $\frac{8}{e}$ or 2.943

32. $\frac{1}{6}$

33. $50,000

MULTIVARIABLE CALCULUS

9.1 Functions of Several Variables

1. $f(x, y) = 4x + 5y + 3$

    (a) $f(2, -1) = 4(2) + 5(-1) + 3 = 6$

    (b) $f(-4, 1) = 4(-4) + 5(1) + 3 = -8$

    (c) $f(-2, -3) = 4(-2) + 5(-3) + 3 = -20$

    (d) $f(0, 8) = 4(0) + 5(8) + 3 = 43$

3. $h(x, y) = \sqrt{x^2 + 2y^2}$

    (a) $h(5, 3) = \sqrt{25 + 2(9)} = \sqrt{43}$

    (b) $h(2, 4) = \sqrt{4 + 32} = 6$

    (c) $h(-1, -3) = \sqrt{1 + 18} = \sqrt{19}$

    (d) $h(-3, -1) = \sqrt{9 + 2} = \sqrt{11}$

5. $f(x, y) = x \sin(x^2 y)$

    (a) $f\left(1, \dfrac{\pi}{2}\right) = (1) \sin\left[(1)^2 \left(\dfrac{\pi}{2}\right)\right]$

$$= \sin \frac{\pi}{2} = 1$$

    (b) $f\left(\dfrac{1}{2}, \pi\right) = \left(\dfrac{1}{2}\right) \sin\left[\left(\dfrac{1}{2}\right)^2 (\pi)\right]$

$$= \frac{1}{2} \sin \frac{\pi}{4}$$

$$= \frac{1}{2\sqrt{2}}$$

    (c) $f\left(\sqrt{\pi}, \dfrac{1}{2}\right) = (\sqrt{\pi}) \sin\left[(\sqrt{\pi})^2 \left(\dfrac{1}{2}\right)\right]$

$$= \sqrt{\pi} \sin \frac{\pi}{2} = \sqrt{\pi}$$

    (d) $f\left(-1, -\dfrac{\pi}{2}\right) = (-1) \sin\left[(-1)^2 \left(-\dfrac{\pi}{2}\right)\right]$

$$= -\sin\left(-\frac{\pi}{2}\right) = 1$$

7. $x + y + z = 12$

    To find x-intercept, let $y = 0$, $z = 0$.

$$x + 0 + 0 = 12$$
$$x = 12$$

    To find y-intercept, let $x = 0$, $z = 0$.

$$y = 12$$

    To find z-intercept, let $x = 0$, $y = 0$.

$$z = 12$$

Sketch the portion of the plane in the first octant that contains these intercepts.

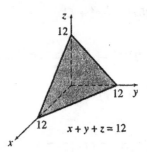

9. $4x + 2y + 3z = 24$

    x-intercept: $y = 0$, $z = 0$

$$4x = 24$$
$$x = 6$$

    y-intercept: $x = 0$, $z = 0$

$$2y = 24$$
$$y = 12$$

    z-intercept: $x = 0$, $y = 0$

$$3z = 24$$
$$z = 8$$

Sketch the portion of the plane in the first octant that contains these intercepts.

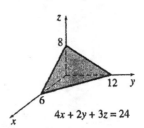

11. $y + z = 5$

x-intercept: $y = 0$, $z = 0$

$$0 = 5$$

Impossible, so no x-intercept

y-intercept: $x = 0$, $z = 0$

$$y = 5$$

z-intercept: $x = 0$, $y = 0$

$$z = 5$$

Sketch the portion of the plane in the first octant that contains these intercepts and its parallel to the x-axis.

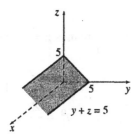

13. $z = 3$

No x-intercept, no y-intercept
Sketch the portion of the plane in the first octant that passes through $(0, 0, 3)$ parallel to the xy-plane.

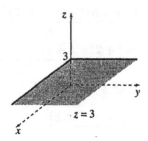

15. $x + 3y + 2z = 8$

If $z = 0$, we have $x + 3y = 8$. The level curve is a line in the xy-plane with an x-intercept of 8 and a y-intercept of $\frac{8}{3}$.
If $z = 2$, we have $x + 3y = 4$. The level curve is a line in the plane $z = 2$ passing through the points $(4, 0, 2)$ and $\left(0, \frac{4}{3}, 2\right)$.
If $z = 4$, we have $x + 3y = 0$. The level curve is a line in the plane $z = 4$ passing through the point $(0, 0, 4)$ on the z-axis.

Sketch segments of these lines in the first octant.

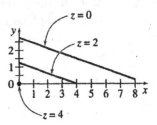

17. $2y - \dfrac{x^2}{3} = z$

If $z = 0$, we have $y = \frac{1}{6}x^2$. The level curve is a parabola in the xy-plane with vertex at the origin.
If $z = 2$, we have $y = \frac{1}{6}x^2 + 1$. The level curve is a parabola in the plane $z = 2$ with vertex at the point $(0, 1, 2)$.
If $z = 4$, we have $y = \frac{1}{6}x^2 + 2$. The level curve is a parabola in the plane $z = 4$ with vertex at the point $(0, 2, 4)$.

Sketch portions of these curves in the first octant.

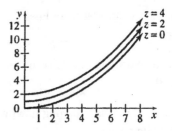

19. x-intercept: $y = 0, z = 0 \quad (a \neq 0)$

$$ax = d$$

$$x = \frac{d}{a}$$

y-intercept: $x = 0, z = 0 \quad (b \neq 0)$

$$by = d$$

$$y = \frac{d}{b}$$

z-intercept: $x = 0, y = 0 \quad (c \neq 0)$

$$cz = d$$

$$z = \frac{d}{c}$$

If $d \neq 0$, then for the plane to have a portion in the first octant, one of $\frac{d}{a}, \frac{d}{b}$, or $\frac{d}{c}$ must be positive, so at least one of a, b, or c must have the same sign as d.

If $d = 0$, then the trace in the xy-plane is the line $x = -\frac{b}{a}y$, the trace in the yz-plane is $y = -\frac{c}{b}z$, and the trace in the xz-plane is $x = -\frac{c}{a}z$. $-\frac{c}{a}$, $-\frac{c}{b}$, or $-\frac{b}{a}$ must be positive, so $a, b,$ and c cannot all have the same sign.

23. $z^2 - y^2 - x^2 = 1$

If $z = 0$,
$$-(y^2 + x^2) = 1.$$

This is impossible so there is no xy-trace.
If $x = 0$,
$$z^2 - y^2 = 1.$$

yz-trace: hyperbola
If $y = 0$,
$$z^2 - x^2 = 1.$$

xz-trace: hyperbola
The equation is represented by a hyperboloid of two sheets, as shown in (f).

25. $z = y^2 - x^2$

If $z = 0$,
$$x^2 = y^2$$
$$x = \pm y.$$

xy-trace: two intersecting lines.
If $x = 0$,
$$z = y^2.$$

yz-trace: parabola, opening upward
If $y = 0$,
$$z = -x^2.$$

xz-trace: parabola, opening downward
Hyperbolic paraboloid

Both (a) and (e) are hyperbolic paraboloids, but only (a) has traces described by this function.

27. $z = 5(x^2 + y^2)^{-1/2} = \dfrac{5}{\sqrt{x^2 + y^2}}$

Note that $z > 0$ for all values of x and y.
xz-trace: $y = 0$
$$z = \frac{5}{\sqrt{x^2}} = \frac{5}{|x|}$$

This gives one branch of the hyperbola $xz = 5$, where $z > 0$ and $x > 0$, and one branch of the hyperbola $xz = -5$, where $z > 0$ and $x < 0$.

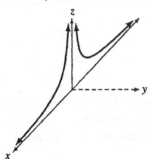

yz-trace: $x = 0$
$$z = \frac{5}{\sqrt{y^2}} = \frac{5}{|y|}$$

This gives one branch of the hyperbola $yz = 5$, where $z > 0$ and $y > 0$, and one branch of the hyperbola $yz = -5$, where $z > 0$ and $y < 0$.

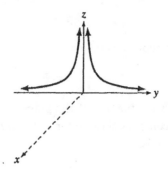

Level curves on planes $z = k$, where $k > 0$, are
$$k = \frac{5}{\sqrt{x^2 + y^2}}$$
$$x^2 + y^2 = \frac{25}{k^2}.$$

These are circles with centers $(0, 0, k)$ and radii $\frac{5}{k}$. The graph of $z = 5(x^2 + y^2)^{-1/2}$ is (d).

29. $f(x, y) = 7x^3 + 8y^2$

(a) $\dfrac{f(x + h, y) - f(x, y)}{h}$

$= \dfrac{[7(x + h)^3 + 8y^2] - (7x^3 + 8y^2)}{h}$

$= \dfrac{7x^3 + 21x^2 h + 21xh^2 + 7h^3 - 7x^3}{h}$

$= 21x^2 + 21xh + 7h^2$

(b) $\dfrac{f(x, y + h) - f(x, y)}{h}$

$= \dfrac{[7x^3 + 8(y + h)^2] - (7x^3 + 8y^2)}{h}$

$= \dfrac{8y^2 + 16yh + 8h^2 - 8y^2}{h}$

$= 16y + 8h$

(c) $\displaystyle\lim_{h \to 0} \dfrac{f(x + h, y) - f(x, y)}{h}$

$= \displaystyle\lim_{h \to 0} (21x^2 + 21xh + 7h^2)$

$= 21x^2 + 21x(0) + 7(0)^2$

$= 21x^2$

(d) $\lim\limits_{h \to 0} \dfrac{f(x, y+h) - f(x,y)}{h}$

$= \lim\limits_{h \to 0} (16y + 8h)$

$= 16y + 8(0)$

$= 16y$

31. A linear regression on the $y = 0$ column gives
$f_0(x) = 4.019 + 1.989x$.
The $y = 1$ column gives

$$f_1(x) = 7.045 + 1.98x.$$

The $y = 2$ column gives

$$f_2(x) = 9.978 + 2.013x.$$

The $y = 3$ column gives

$$f_3(x) = 12.989 + 2.019x.$$

Use the nearest integer values.

$$f_0(x) = f(x, 0) = 4 + 2x$$
$$= a + bx + c(0)$$
$$= a + bx$$

$$f_1(x) = f(x, 1) = 7 + 2x$$
$$= a + bx + c(1)$$
$$= a + bx + c$$

$$f_2(x) = f(x, 2) = 10 + 2x$$
$$= a + bx + c(2)$$
$$= a + bx + 2c$$

$$f_3(x) = f(x, 3) = 13 + 2x$$
$$= a + bx + c(3)$$
$$= a + bx + 3c$$

Thus, $b = 2$ and we have

$$4 = a$$
$$7 = a + c$$
$$10 = a + 2c$$
$$13 = a + 3c$$

so $a = 4$ and $c = 3$.

$$f(x, y) = 4 + 2x + 3y$$

33. $F = \dfrac{v^2}{gl}$

(a) $2.56 = \dfrac{v^2}{(9.81)(0.09)}$

$v^2 = (2.56)(9.81)(0.09)$

$v^2 \approx 2.260$

$v \approx 1.5$

For a ferret, this change occurs at 1.5 m/sec.

$2.56 = \dfrac{v^2}{(9.81)(1.2)}$

$v^2 = (2.56)(9.81)(1.2)$

$v^2 \approx 30.136$

$v \approx 5.5$

For a rhinoceros, this change occurs at 5.5 m/sec.

(b) $.025 = \dfrac{v^2}{(9.81)(4)}$

$v^2 = (0.025)(9.81)(4)$

$v^2 = 0.981$

$v \approx 0.99$

The sauropods were traveling at 1 m/sec.

35. (a) $L(F, H, T) = F - (H + 0.16T)$

(b) $L(2{,}000, 1{,}300, 2{,}500)$
$= 2{,}000 - [1{,}300 + 0.16(2{,}500)]$
$= 2{,}000 - 1{,}700$
$= 300$ mg/ℓ

37. (a) $\ln(D) = 4.890179 + 0.005163R - 0.004345L$
$\qquad\qquad - 0.000019F$

$D = e^{4.890179 + 0.005163R - 0.004345L - 0.000019F}$

$= e^{4.890179} e^{0.005163R - 0.004345L - 0.000019F}$

$= 132.977375 e^{0.005163R - 0.004345L - 0.000019F}$

(b) Replace R with 150, L with 75, and F with 3,500 in the formula from part (a).

$D = 132.977375 e^{0.005163(150) - 0.004345(75) - 0.000019(3{,}500)}$

$\approx 132.977375(1.465322)$

≈ 194.854670

The day of passage is about day 194. This corresponds to July 13.

39. (a) $P = \dfrac{\lambda_1 E_1}{1 + \lambda_1 h_1}$

$= \dfrac{0.2(162)}{1 + 0.2(3.6)}$

≈ 18.84

The profitability of worms only is about 18.8 kcal/min.

(b) $P = \dfrac{\lambda_1 E_1 + \lambda_2 E_2}{1 + \lambda_1 h_1 + \lambda_2 h_2}$

$= \dfrac{0.2(162) + 3.0(24)}{1 + 0.2(3.6) + 3.0(0.6)}$

≈ 29.66

The profitability of worms and moths is about 29.7 kcal/min.

(c) $P(\text{worms}) \approx 18.84$

$$P(\text{moths}) = \frac{3.0(24)}{1 + 3.0(0.6)} \approx 25.71$$

$$P(\text{grubs}) = \frac{3.0(40)}{1 + 3.0(1.6)} \approx 20.69$$

$P(\text{worms, moths}) \approx 29.66$

$$P(\text{worms, grubs}) = \frac{0.2(162) + 3.0(40)}{1 + 0.2(3.6) + 3.0(1.6)}$$

$$\approx 23.37$$

$$P(\text{moths, grubs}) = \frac{3.0(24) + 3.0(40)}{1 + 3.0(0.6) + 3.0(1.6)}$$

$$\approx 25.26$$

$P(\text{worms, moths, grubs})$

$$= \frac{0.2(162) + 3.0(24) + 3.0(40)}{1 + 0.2(3.6) + 3.0(0.6) + 3.0(1.6)}$$

$$\approx 26.97$$

The combination that provides maximum profitability is worms and moths.

41. $N(R, C) = 329.32 + 0.0377R - 0.0171C$

(a) $N(141,319, 37,960)$
$$= 329.32 + 0.0377(141,319)$$
$$- 0.0171(37,960)$$
$$\approx 5,008$$

Approximately 5008 deer were harvested in Tuscarawas County.

(b)

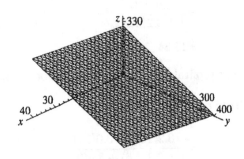

43. Let the area be given by $g(L, W, H)$.
Then,

$$g(L, W, H) = 2LW + 2WH + 2LH \text{ ft}^2.$$

45. $P(x, y) = 100 \left[\dfrac{3}{5} x^{-2/5} + \dfrac{2}{5} y^{-2/5} \right]^{-5}$

(a) $P(32, 1)$

$$= 100 \left[\frac{3}{5}(32)^{-2/5} + \frac{2}{5}(1)^{-2/5} \right]^{-5}$$

$$= 100 \left[\frac{3}{5}\left(\frac{1}{4}\right) + \frac{2}{5}(1) \right]^{-5}$$

$$= 100 \left(\frac{11}{20} \right)^{-5}$$

$$= 100 \left(\frac{20}{11} \right)^{5}$$

$$\approx 1,986.95$$

The production is approximately 1987 cameras.

(b) $P(1, 32)$

$$= 100 \left[\frac{3}{5}(1)^{-2/5} + \frac{2}{5}(32)^{-2/5} \right]^{-5}$$

$$= 100 \left[\frac{3}{5}(1) + \frac{2}{5}\left(\frac{1}{4}\right) \right]^{-5}$$

$$= 100 \left(\frac{7}{10} \right)^{-5}$$

$$= 100 \left(\frac{10}{7} \right)^{5}$$

$$\approx 595$$

The production is approximately 595 cameras.

(c) 32 work hours means that $x = 32$. 243 units of capital means that $y = 243$.

$P(32, 243)$

$$= 100 \left[\frac{3}{5}(32)^{-2/5} + \frac{2}{5}(243)^{-2/5} \right]^{-5}$$

$$= 100 \left[\frac{3}{5}\left(\frac{1}{4}\right) + \frac{2}{5}\left(\frac{1}{9}\right) \right]^{-5}$$

$$= 100 \left(\frac{7}{36} \right)^{-5}$$

$$= 100 \left(\frac{36}{7} \right)^{5}$$

$$\approx 359,767.81$$

The production is approximately 359,768 cameras.

47. $M = f(n, i, t) = \dfrac{(1+i)^n(1-t)+t}{[1+(1-t)i]^n}$

Therefore,

$f(25, 0.06, 0.33)$

$= \dfrac{(1+0.06)^{25}(1-0.33)+0.33}{[1+(1-0.33)(0.06)]^{25}}$

$= \dfrac{(1.06)^{25}(0.67)+0.33}{[1+(0.67)(0.06)]^{25}}$

≈ 1.197

Since the value of $M \approx 1.197$, which is greater than 1, the IRA account grows faster.

9.2   Partial Derivatives

1. $z = f(x, y) = 12x^2 - 8xy + 3y^2$

(a) $\dfrac{\partial z}{\partial x} = 24x - 8y$

(b) $\dfrac{\partial z}{\partial y} = -8x + 6y$

(c) $\dfrac{\partial f}{\partial x}(2, 3) = 24(2) - 8(3) = 24$

(d) $f_y(1, -2) = -8(1) + 6(-2)$
$= -20$

3. $f(x, y) = -2xy + 6y^3 + 2$

$f_x(x, y) = -2y$
$f_y(x, y) = 2x + 18y^2$
$f_x(2, -1) = -2(-1) = 2$
$f_y(-4, 3) = -2(-4) + 18(3)^2$
$= 8 + 18(9)$
$= 170$

5. $f(x, y) = 3x^3y^2$
$f_x(x, y) = 9x^2y^2$
$f_y(x, y) = 6x^3y$
$f_x(2, -1) = 9(4)(1) = 36$
$f_y(-4, 3) = 6(-64)(3)$
$= -1,152$

7. $f(x, y) = e^{x+y}$
$f_x(x, y) = e^{x+y}$
$f_y(x, y) = e^{x+y}$
$f_x(2, -1) = e^{2-1}$
$= e^1 = e$

$f_y(-4, 3) = e^{-4+3}$
$= e^{-1}$
$= \dfrac{1}{e}$

9. $f(x, y) = -5e^{3x-4y}$
$f_x(x, y) = -15e^{3x-4y}$
$f_y(x, y) = 20e^{3x-4y}$
$f_x(2, -1) = -15e^{10}$
$f_y(-4, 3) = 20e^{-12-12}$
$= 20e^{-24}$

11. $f(x, y) = \dfrac{x^2 + y^3}{x^3 - y^2}$

$f_x(x, y) = \dfrac{2x(x^3 - y^2) - 3x^2(x^2 + y^3)}{(x^3 - y^2)^2}$

$= \dfrac{2x^4 - 2xy^2 - 3x^4 - 3x^2y^3}{(x^3 - y^2)^2}$

$= \dfrac{-x^4 - 2xy^2 - 3x^2y^3}{(x^3 - y^2)^2}$

$f_y(x, y) = \dfrac{3y^2(x^3 - y^2) - (-2y)(x^2 + y^3)}{(x^3 - y^2)^2}$

$= \dfrac{3x^3y^2 - 3y^4 + 2x^2y + 2y^4}{(x^3 - y^2)^2}$

$= \dfrac{3x^3y^2 - y^4 + 2x^2y}{(x^3 - y^2)^2}$

$f_x(2, -1) = \dfrac{-2^4 - 2(2)(-1)^2 - 3(2^2)(-1)^3}{[2^3 - (-1)^2]^2}$

$= -\dfrac{8}{49}$

$f_y(-4, 3) = \dfrac{3(-4)^3(3)^2 - 3^4 + 2(-4)^2(3)}{[(-4)^3 - 3^2]^2}$

$= -\dfrac{1,713}{5,329}$

13. $f(x, y) = \ln \left| 1 + 3x^2y^3 \right|$

$f_x(x, y) = \dfrac{1}{1 + 3x^2y^3} \cdot 6xy^3$

$= \dfrac{6xy^3}{1 + 3x^2y^3}$

$f_y(x, y) = \dfrac{9x^2y^2}{1 + 3x^2y^3}$

$f_x(2, -1) = \dfrac{6(2)(-1)}{1 + 3(4)(-1)}$

$= \dfrac{-12}{1 - 12} = \dfrac{12}{11}$

$f_y(-4, 3) = \dfrac{9(16)(9)}{1 + 3(16)(27)}$

$= \dfrac{1,296}{1,297}$

15. $f(x,y) = xe^{x^2 y}$

$$f_x(x,y) = e^{x^2 y} \cdot 1 + x(2xy)(e^{x^2 y})$$
$$= e^{x^2 y}(1 + 2x^2 y)$$
$$f_y(x,y) = x^3 e^{x^2 y}$$
$$f_x(2,-1) = e^{-4}(1 - 8) = -7e^{-4}$$
$$f_y(-4,3) = -64e^{48}$$

17. $f(x,y) = \sqrt{x^4 + 3xy + y^4 + 10}$

$$f_x(x,y) = \frac{4x^3 + 3y}{2\sqrt{x^4 + 3xy + y^4 + 10}}$$

$$f_y(x,y) = \frac{3x + 4y^3}{2\sqrt{x^4 + 3xy + y^4 + 10}}$$

$$f_x(2,-1) = \frac{4(2)^3 + 3(-1)}{2\sqrt{2^4 + 3(2)(-1) + (-1)^4 + 10}}$$

$$= \frac{29}{2\sqrt{21}}$$

$$f_y(-4,3) = \frac{3(-4) + 4(3)^3}{2\sqrt{(-4)^4 + 3(-4)(3) + 3^4 + 10}}$$

$$= \frac{48}{\sqrt{311}}$$

19. $f(x,y) = \dfrac{3x^2 y}{e^{xy} + 2}$

$$f_x(x,y) = \frac{6xy(e^{xy} + 2) - ye^{xy}(3x^2 y)}{(e^{xy} + 2)^2}$$

$$= \frac{6xy(e^{xy} + 2) - 3x^2 y^2 e^{xy}}{(e^{xy} + 2)^2}$$

$$f_y(x,y) = \frac{3x^2(e^{xy} + 2) - xe^{xy}(3x^2 y)}{(e^{xy} + 2)^2}$$

$$= \frac{3x^2(e^{xy} + 2) - 3x^3 y e^{xy}}{(e^{xy} + 2)^2}$$

$$f_x(2,-1) = \frac{6(2)(-1)(e^{2(-1)} + 2) - 3(2)^2(-1)^2 e^{2(-1)}}{(e^{2(-1)} + 2)^2}$$

$$= \frac{-12e^{-2} - 24 - 12e^{-2}}{(e^{-2} + 2)^2}$$

$$= \frac{-24(e^{-2} + 1)}{(e^{-2} + 2)^2}$$

$$f_y(-4,3) = \frac{3(-4)^2(e^{(-4)(3)} + 2) - 3(-4)^3(3)e^{(-4)(3)}}{(e^{(-4)(3)} + 2)^2}$$

$$= \frac{48e^{-12} + 96 + 576e^{-12}}{(e^{-12} + 2)^2}$$

$$= \frac{624e^{-12} + 96}{(e^{-12} + 2)^2}$$

21. $f(x,y) = x\sin(\pi y)$

$$f_x(x,y) = \sin(\pi y)$$
$$f_y(x,y) = \pi x \cos(\pi y)$$
$$f_x(2,-1) = \sin(-\pi) = 0$$
$$f_y(-4,3) = \pi(-4)\cos(3\pi) = 4\pi$$

23. $f(x,y) = 6x^3 y - 9y^2 + 2x$

$$f_x(x,y) = 18x^2 y + 2$$
$$f_y(x,y) = 6x^3 - 18y$$
$$f_{xx}(x,y) = 36xy$$
$$f_{yy}(x,y) = -18$$
$$f_{xy}(x,y) = 18x^2 = f_{yx}(x,y)$$

25. $R(x,y) = 4x^2 - 5xy^3 + 12y^2 x^2$

$$R_x(x,y) = 8x - 5y^3 + 24y^2 x$$
$$R_y(x,y) = -15xy^2 + 24yx^2$$
$$R_{xx}(x,y) = 8 + 24y^2$$
$$R_{yy}(x,y) = -30xy + 24x^2$$
$$R_{xy}(x,y) = -15y^2 + 48xy$$
$$= R_{yx}(x,y)$$

27. $r(x,y) = \dfrac{4x}{x+y}$

$$r_x(x,y) = \frac{4(x+y) - 4x}{(x+y)^2}$$

$$= 4y(x+y)^{-2}$$

$$r_y(x,y) = \frac{-4x}{(x+y)^2}$$

$$r_{xx}(x,y) = -8y(x+y)^{-3}$$

$$= \frac{-8y}{(x+y)^3}$$

$$r_{yy}(x,y) = 8x(x+y)^{-3}$$

$$= \frac{8x}{(x+y)^3}$$

$$r_{xy}(x,y) = 4(x-y)(x+y)^{-3}$$

$$= \frac{4x - 4y}{(x+y)^3}$$

$$= r_{yx}(x,y)$$

29. $z = 4xe^y$

$$z_x(x,y) = 4e^y$$
$$z_y(x,y) = 4xe^y$$
$$z_{xx}(x,y) = 0$$
$$z_{yy}(x,y) = 4xe^y$$
$$z_{xy}(x,y) = 4e^y = z_{yx}(x,y)$$

31. $r = \ln|x+y|$

$$r_x(x,y) = \frac{1}{x+y}$$

$$r_y(x,y) = \frac{1}{x+y}$$

$$r_{xx}(x,y) = \frac{-1}{(x+y)^2}$$

$$r_{yy}(x,y) = \frac{-1}{(x+y)^2}$$

$$r_{xy}(x,y) = \frac{-1}{(x+y)^2}$$

$$= r_{yx}(x,y)$$

33. $z = x\ln|xy|$

$$z_x(x,y) = \ln|xy| + 1$$

$$z_y(x,y) = \frac{x}{y}$$

$$z_{xx}(x,y) = \frac{1}{x}$$

$$z_{yy}(x,y) = -xy^{-2} = \frac{-x}{y^2}$$

$$z_{xy}(x,y) = \frac{1}{y} = z_{yz}(x,y)$$

35. $f(x,y) = 6x^2 + 6y^2 + 6xy + 36x - 5$

First, $f_x = 12x + 6y + 36$ and $f_y = 12y + 6x$.
We must solve the system

$$12x + 6y + 36 = 0$$
$$12y + 6x = 0.$$

Multiply both sides of the first equation by -2 and add.

$$\begin{array}{r} -24x - 12y - 72 = 0 \\ \underline{6x + 12y \qquad = 0} \\ -18x \qquad - 72 = 0 \\ x = -4 \end{array}$$

Substitute into either equation to get $y = 2$.
The solution is $x = -4, y = 2$.

37. $f(x,y) = 9xy - x^3 - y^3 - 6$

First, $f_x = 9y - 3x^2$ and $f_y = 9x - 3y^2$.
We must solve the system

$$9y - 3x^2 = 0$$
$$9x - 3y^2 = 0.$$

From the first equation, $y = \frac{1}{3}x^2$.
Substitute into the second equation to get

$$9x - 3\left(\frac{1}{3}x^2\right)^2 = 0$$

$$9x - 3\left(\frac{1}{9}x^4\right) = 0$$

$$9x - \frac{1}{3}x^4 = 0.$$

Multiply by 3 to get

$$27x - x^4 = 0.$$

Now factor.

$$x(27 - x^3) = 0$$

Set each factor equal to 0.

$$\begin{array}{cc} x = 0 & \text{or} \quad 27 - x^3 = 0 \\ & x = 3 \end{array}$$

Substitute into $y = \frac{x^2}{3}$.

$$y = 0 \quad \text{or} \quad y = 3$$

The solutions are $x = 0$, $y = 0$ and $x = 3$, $y = 3$.

39. $f(x,y,z) = x^2 + yz + z^4$

$$f_x(x,y,z) = 2x$$
$$f_y(x,y,z) = z$$
$$f_z(x,y,z) = y + 4z^3$$
$$f_{yz}(x,y,z) = 1$$

41. $f(x,y,z) = \dfrac{6x - 5y}{4z + 5}$

$$f_x(x,y,z) = \frac{6}{4z + 5}$$

$$f_y(x,y,z) = \frac{-5}{4z + 5}$$

$$f_z(x,y,z) = \frac{-4(6x - 5y)}{(4z + 5)^2}$$

$$f_{yz}(x,y,z) = \frac{20}{(4z + 5)^2}$$

43. $f(x, y, z) = \ln \left| x^2 - 5xz^2 + y^4 \right|$

$$f_x(x, y, z) = \frac{2x - 5z^2}{x^2 - 5xz^2 + y^4}$$

$$f_y(x, y, z) = \frac{4y^3}{x^2 - 5xz^2 + y^4}$$

$$f_z(x, y, z) = \frac{-10xz}{x^2 - 5xz^2 + y^4}$$

$$f_{yz}(x, y, z) = \frac{4y^3(10zx)}{(x^2 - 5xz^2 + y^4)^2}$$

$$= \frac{40xy^3z}{(x^2 - 5xz^2 + y^4)^2}$$

45. $f(x, y) = \left(x + \dfrac{y}{2} \right)^{x + y/2}$

(a) $f_x(1, 2) = \lim\limits_{h \to 0} \dfrac{f(1 + h, 2) - f(1, 2)}{h}$

We will use a small value for h. Let $h = 0.00001$.

$$f_x(1, 2) \approx \frac{f(1.00001, 2) - f(1, 2)}{0.00001}$$

$$\approx \frac{\left(1.00001 + \frac{2}{2}\right)^{1.00001 + 2/2} - \left(1 + \frac{2}{2}\right)^{1 + 2/2}}{0.00001}$$

$$\approx \frac{2.00001^{2.00001} - 2^2}{0.00001}$$

$$\approx 6.773$$

(b) $f_y(1, 2) = \lim\limits_{h \to 0} \dfrac{f(1, 2 + h) - f(1, 2)}{h}$

Again, let $h = 0.00001$.

$$f_y(1, 2) \approx \frac{f(1, 2.00001) - f(1, 2)}{0.00001}$$

$$\approx \frac{\left(1 + \frac{2.00001}{2}\right)^{1 + 2.00001/2} - \left(1 + \frac{2}{2}\right)^{1 + 2/2}}{0.0001}$$

$$\approx \frac{2.000005^{2.000005} - 2^2}{0.00001}$$

$$\approx 3.386$$

47. $M(x, y) = 2xy + 10xy^2 + 30y^2 + 20$

(a) $\dfrac{\partial M}{\partial x} = 2y + 10y^2$

$\dfrac{\partial M}{\partial x}(20, 4) = 168$

(b) $\dfrac{\partial M}{\partial y} = 2x + 20yx + 60y$

$\dfrac{\partial M}{\partial y}(24, 10) = 5,448$

(c) An increase in days since rain causes more of an increase in matings.

49. $C(a, b, v) = \dfrac{b}{a - v} = b(a - v)^{-1}$

(a) $C(160, 200, 125) = \dfrac{200}{160 - 125} = \dfrac{200}{35} \approx 5.71$

(b) $C_a(a, b, v) = -b(a - v)^{-2} \cdot 1 = -\dfrac{b}{(a - v)^2}$

$$C_a(160, 200, 125) = -\frac{200}{(160 - 125)^2}$$

$$= -\frac{200}{35^2} \approx -0.163$$

(c) $C_b(a, b, v) = (a - v)^{-1}$

$$C_b(160, 200, 125) = (160 - 125)^{-1} = \frac{1}{35}$$

$$\approx 0.0286$$

(d) $C_v(a, b, v) = -b(a - v)^{-2} \cdot (-1) = \dfrac{b}{(a - v)^2}$

$$C_v(160, 200, 125) = \frac{200}{(160 - 125)^2}$$

$$= \frac{200}{35^2}$$

$$\approx 0.163$$

(e) Changing a by 1 unit produces the greatest decrease in the liters of blood pumped, while changing v by 1 unit produces the same amount of increase in the liters of blood pumped.

51. $B(w, h) = \dfrac{703w}{h^2}$

(a) $B(220, 74) = \dfrac{703(220)}{74^2} \approx 28$

(b) $\dfrac{\partial B}{\partial w} = \dfrac{703}{h^2}$

$\dfrac{\partial B}{\partial h} = \dfrac{-2(703)w}{h^3} = -\dfrac{1,406w}{h^3}$

Since $\dfrac{\partial B}{\partial w}$ is positive, as w increases, so does B. Since $\dfrac{\partial B}{\partial h}$ is negative, B decreases as h increases.

(c) If w_m and h_m represent a person's weight and height, in kilograms and meters, respectively, then $w_m = 0.4555w$, where w is weight in pounds, and $h_m = 0.0254h$, where h is height in inches. To transform the formula $B = \dfrac{703w}{h^2}$ to handle metric inputs, make the substitutions $w = \dfrac{w_m}{0.4555}$ and $h = \dfrac{h_m}{0.0254}$.

$$B_m = \frac{703 \frac{w_m}{0.4555}}{\left(\frac{h_m}{0.0254}\right)^2} = \frac{703(0.0254)^2}{0.4555} \cdot \frac{w_m}{h_m^2}$$

Since $\dfrac{703(0.0254)^2}{0.4555} \approx 0.996 \approx 1$, use $B_m = \dfrac{w_m}{h_m^2}$.

53. $p(l, t) = 1 + \dfrac{l}{33}(1 - 2^{-t/5})$

(a) $p(33, 10) = 1 + \dfrac{33}{33}(1 - 2^{-10/5}) = 1 + (1)(1 - 2^{-2}) = 1 + \dfrac{3}{4} = 1.75$

The pressure at 33 feet for a 10-minute dive is 1.75 atmospheres.

(b) $p_l(l, t) = \dfrac{1}{33}(1 - 2^{-t/5})$

$p_t(l, t) = \dfrac{l}{33}(-(\ln 2)\left(-\dfrac{1}{5}2^{-t/5}\right)) = \dfrac{l}{165}(\ln 2)(2^{-t/5})$

$p_l(33, 10) = \dfrac{1}{33}(1 - 2^{-10/5}) = \dfrac{1}{33}\left(\dfrac{3}{4}\right) \approx 0.023 \text{ atm/ft}$

Increasing the depth of the dive by 1 foot (to 34 feet), while keeping the dive length at 10 minutes, increases the pressure by approximately 0.023 atmospheres.

$$p_t(33, 10) = \dfrac{33}{165}(\ln 2)(2^{-10/5}) = \dfrac{1}{20}\ln 2 \approx 0.035 \text{ atm/min}$$

Increasing the dive time by 1 minute (to 11 minutes), while keeping the depth of the dive to 33 feet increases the pressure by approximately 0.035 atmospheres.

(c) Solve $p(66, t) = 2.15$

$$1 + \dfrac{66}{33}(1 - 2^{-t/5}) = 2.15$$

$$2(1 - 2^{-t/5}) = 1.15$$
$$1 - 2^{-t/5} = 0.575$$
$$2^{-t/5} = 0.425$$

$$-\dfrac{t}{5}\ln 2 = \ln 0.425$$

$$t = -\dfrac{5\ln 0.425}{\ln 2}$$

$$t \approx 6.17$$

The maximum dive length is 6.17 minutes.

55. (a) $f(90, 30) \approx 90$

(b) $f(90, 75) \approx 109$

(c) $f(80, 75) \approx 86$

(d) $f_T(90, 30) \approx \dfrac{f(95, 30) - f(90, 30)}{5} \approx \dfrac{95 - 90}{5} = 1$

(e) $f_H(90, 30) \approx \dfrac{f(90, 35) - f(90, 30)}{5} \approx \dfrac{92 - 90}{5} = 0.4$

(f) $f_T(90, 75) \approx \dfrac{f(95, 75) - f(90, 75)}{5} \approx \dfrac{130 - 109}{5} = 4.2$

(g) $f_H(90, 75) \approx \dfrac{f(90, 80) - f(90, 75)}{5} \approx \dfrac{114 - 109}{5} = 1$

57. $\ln(D) = 4.890179 + 0.005163R - 0.004345L$
$$- 0.000019F$$

(a) From part a of Exercises 37 of the previous section:

$$D(R, L, F) = 132.977375e^{0.005163R-0.004345L-0.000019F}$$
$$D_F(R, L, F) = 132.977375(-0.000019)e^{0.005163R-0.004345L-0.000019F}$$
$$= -0.002527e^{0.005163R-0.004345L-0.000019F}$$

(b)$D_F(150, 75, 3,500) = -0.002527e^{0.005163(150)-0.004345(75)-0.000019(3,500)}$
$$\approx -0.00370$$

For a $1,000\text{m}^3/\text{s}$ increase of water flow, the date of passage will be about

$$1,000D_F(150,\ 75,\ 3,500) \approx 1,000(-0.00370) \approx -3.7$$

or about 3.7 days later.

59. $w = -\dfrac{1}{a}\ln\left[1 - \dfrac{(N-1)b(1-e^{-kn})}{N}\right]$

(a) $\dfrac{\partial w}{\partial n} = -\dfrac{1}{a}\cdot\dfrac{1}{1-\frac{(N-1)b(1-e^{-kn})}{N}}\cdot\dfrac{(N-1)b(-e^{-kn})}{N}\cdot(-k) == -\dfrac{(N-1)bke^{-kn}}{aN\left[1-\frac{(N-1)b(1-e^{-kn})}{N}\right]}$

This represents the rate of change of the number of wrong attempts with respect to the number of trials.

(b) $\dfrac{\partial w}{\partial N} = -\dfrac{1}{a}\cdot\dfrac{1}{1-\frac{(N-1)b(1-e^{-kn})}{N}}\cdot\left[\dfrac{Nb(1-e^{-kn})-(N-1)b(1-e^{-kn})}{N^2}\right]$

$$= \dfrac{b(1-e^{-kn})[N-(N-1)]}{aN^2\left[1-\frac{(N-1)b(1-e^{-kn})}{N}\right]}$$

This represents the rate of change of the number of wrong attempts with respect to the number of stimuli.

61. $p = f(s, n, a) = 0.003a + 0.1(sn)^{1/2}$

(a) $f(8, 6, 450) = 0.003(450) + 0.1[(8)(6)]^{1/2} = 1.35 + 0.1(48)^{1/2} = 2.0428$
$$p \approx 2.04\%$$

(b) $f(3, 3, 320) = 0.003(320) + 0.1[(3)(3)]^{1/2} = 0.96 + 0.1(9)^{1/2} = 1.26$
$$p = 1.26\%$$

(c) $f_n(s, n, a) = 0.003(0) + 0.1\left[\dfrac{1}{2}(sn)^{-1/2}(s)\right]$

$$f_n(3, 3, 320) = 0.1\left(\dfrac{1}{2}\right)[(3)(3)]^{-1/2}(3) = (0.1)\left(\dfrac{1}{2}\right)\left(\dfrac{1}{3}\right)(3) = 0.05$$

$$p_n = 0.05\%$$

$f_a(s, n, a) = 0.003$ for all ordered triples (s, n, a). Therefore, $p_a = 0.003\%$. $p_n = 0.05\%$ is the rate of change of the probability for an additional semester of high school math.

$p_a = 0.003\%$ is the rate of change of the probability per unit of change in an SAT score.

63. $w(x, y) = \dfrac{x + y}{1 + \frac{xy}{c^2}}$

(a) $w(50{,}000, 150{,}000)$

$$= \dfrac{50{,}000 + 150{,}000}{1 + \frac{(50{,}000)(150{,}000)}{(186{,}282)^2}}$$

$$\approx 164{,}456$$

The rocket is traveling 164,456 m/sec relative to the stationary observer.

(b) $w_x = \dfrac{\left(\left(1 + \frac{xy}{c^2}\right) - \frac{y}{c^2}(x + y)\right)}{\left(1 + \frac{xy}{c^2}\right)^2}$

$$= \dfrac{1 + \frac{xy}{c^2} - \frac{xy}{c^2} - \frac{y^2}{c^2}}{\left(1 + \frac{xy}{c^2}\right)^2}$$

$$= \dfrac{1 - \left(\frac{y}{c}\right)^2}{\left(1 + \frac{xy}{c^2}\right)^2}$$

$w_x(50{,}000, 150{,}000)$

$$= \dfrac{1 - \left(\frac{150{,}000}{186{,}282}\right)^2}{\left(1 + \frac{(50{,}000)(150{,}000)}{(186{,}282)^2}\right)^2} \approx 0.238$$

The instantaneous rate of change is 0.238 m/sec per m/sec.

(c) $w(c, c) = \dfrac{c + c}{1 + \frac{(c)(c)}{c^2}} = \dfrac{2c}{2} = c$

The speed is the speed of light, c.

65. $R(x, y) = 5x^2 + 9y^2 - 4xy$

(a) $R_x(x, y) = 10x - 4y$
$R_x(9, 5) = 10(9) - 4(5) = 70$

R would increase by about \$70.

(b) $R_y(x, y) = 18y - 4x$
$R_y(9, 5) = 18(5) - 4(9) = 54$

R would increase by about \$54.

9.3   Maxima and Minima

1. $f(x, y) = xy + x - y$

$f_x(x, y) = y + 1, \ f_y(x, y) = x - 1$

If $f_x(x, y) = 0, \ y = -1$.
If $f_y(x, y) = 0, \ x = 1$.

Therefore, $(1, -1)$ is the critical point.

$$f_{xx}(x, y) = 0$$
$$f_{yy}(x, y) = 0$$
$$f_{xy}(x, y) = 1$$

For $(1, -1)$,

$$D = 0 \cdot 0 - 1^2 = -1 < 0.$$

A saddle point is at $(1, -1)$.

3. $f(x, y) = x^2 - 2xy + 2y^2 + x - 5$

$f_x(x, y) = 2x - 2y + 1, \ f_y(x, y) = -2x + 4y$

Solve the system $f_x(x, y) = 0, \ f_y(x, y) = 0$.

$$\begin{array}{r} 2x - 2y + 1 = 0 \\ -2x + 4y \quad\ = 0 \\ \hline 2y + 1 = 0 \end{array}$$

$$y = -\dfrac{1}{2}$$

$$-2x + 4\left(-\dfrac{1}{2}\right) = 0$$

$$-2x = 2$$

$$x = -1$$

Therefore, $\left(-1, -\frac{1}{2}\right)$ is the critical point.

$$f_{xx}(x, y) = 2$$
$$f_{yy}(x, y) = 4$$
$$f_{xy}(x, y) = -2$$

For $\left(-1, -\frac{1}{2}\right)$,

$$D = 2 \cdot 4 - (-2)^2 = 4 > 0.$$

Since $f_{xx}(x, y) = 2 > 0$, then a relative minimum is at $\left(-1, -\frac{1}{2}\right)$.

5. $f(x, y) = x^2 - xy + y^2 + 2x + 2y + 6$

$f_x(x, y) = 2x - y + 2, \ f_y(x, y) = -x + 2y + 2$

Solve the system $f_x(x, y) = 0, \ f_y(x, y) = 0.$

$$2x - y + 2 = 0$$
$$\underline{-x + 2y + 2 = 0}$$

$$2x - y + 2 = 0$$
$$\underline{-2x + 4y + 4 = 0}$$
$$3y + 6 = 0$$
$$y = -2$$

$$-x + 2(-2) + 2 = 0$$
$$x = -2$$

$(-2, -2)$ is the critical point.

$$f_{xx}(x, y) = 2$$
$$f_{yy}(x, y) = 2$$
$$f_{xy}(x, y) = -1$$

For $(-2, -2)$,

$$D = (2)(2) - (-1)^2 = 3 > 0.$$

Since $f_{xx}(x, y) > 0$, a relative minimum is at $(-2, -2)$.

7. $f(x, y) = x^2 + 3xy + 3y^2 - 6x + 3y$

$f_x(x, y) = 2x + 3y - 6, \ f_y(x, y) = 3x + 6y + 3$

Solve the system $f_x(x, y) = 0, \ f_y(x, y) = 0.$

$$2x + 3y - 6 = 0$$
$$3x + 6y + 3 = 0$$

$$-4x - 6y + 12 = 0$$
$$\underline{3x + 6y + 3 = 0}$$
$$-x + 15 = 0$$
$$x = 15$$

$$3(15) + 6y + 3 = 0$$
$$6y = -48$$
$$y = -8$$

$(15, -8)$ is the critical point.

$$f_{xx}(x, y) = 2$$
$$f_{yy}(x, y) = 6$$
$$f_{xy}(x, y) = 3$$

For $(15, -8)$,

$$D = 2 \cdot 6 - 9 = 3 > 0.$$

Since $f_{xx}(x, y) > 0$, a relative minimum is at $(15, -8)$.

9. $f(x, y) = 4xy - 10x^2 - 4y^2 + 8x + 8y + 9$

$f_x(x, y) = 4y - 20x + 8, \ f_y(x, y) = 4x - 8y + 8$

$$4y - 20x + 8 = 0$$
$$4x - 8y + 8 = 0$$

$$4y - 20x + 8 = 0$$
$$\underline{-4y + 2x + 4 = 0}$$
$$-18x + 12 = 0$$
$$x = \frac{2}{3}$$

$$4y - 20\left(\frac{2}{3}\right) + 8 = 0$$

The critical point is $\left(\frac{2}{3}, \frac{4}{3}\right)$.

$$f_{xx}(x, y) = -20$$
$$f_{yy}(x, y) = -8$$
$$f_{xy}(x, y) = 4$$

For $\left(\frac{2}{3}, \frac{4}{3}\right)$,

$$D = (-20)(-8) - 16 = 144 > 0.$$

Since $f_{xx}(x, y) < 0$, a relative maximum is at $\left(\frac{2}{3}, \frac{4}{3}\right)$.

11. $f(x, y) = x^2 + xy - 2x - 2y + 2$

$f_x(x, y) = 2x + y - 2, \ f_y(x, y) = x - 2$

$$2x + y - 2 = 0$$
$$x \qquad - 2 = 0$$
$$x = 2$$
$$2(2) + y - 2 = 0$$
$$y = -2$$

The critical point is $(2, -2)$.

$$f_{xx}(x, y) = 2$$
$$f_{yy}(x, y) = 0$$
$$f_{xy}(x, y) = 1$$

For $(2, -2)$,

$$D = 2 \cdot 0 - 1^2 = -1 < 0.$$

A saddle point is at $(2, -2)$.

13. $f(x, y) = 2x^3 + 3y^2 - 12xy + 4$

$f_x(x, y) = 6x^2 - 12y,\ f_y(x, y) = 6y - 12x$

$$6x^2 - 12y = 0$$
$$6y - 12x = 0$$

If $6y - 12x = 0,\ y = 2x.$
Substitute for y in the first equation.

$$6x^2 - 12(2x) = 0$$
$$6x(x - 4) = 0$$
$$x = 0 \quad \text{or} \quad x = 4$$
Then $\qquad y = 0 \quad \text{or} \quad y = 8.$

The critical points are $(0, 0)$ and $(4, 8)$.

$$f_{xx}(x, y) = 12x$$
$$f_{yy}(x, y) = 6$$
$$f_{xy}(x, y) = -12$$

For $(0, 0)$,

$$D = 12(0)6 - (-12)^2$$
$$= -144 < 0.$$

A saddle point is at $(0, 0)$.
For $(4, 8)$,

$$D = 12(4)6 - (-12)^2$$
$$= 144 > 0.$$

Since $f_{xx}(x, y) = 12(4) = 48 > 0$, a relative minimum is at $(4, 8)$.

15. $f(x, y) = x^2 + 4y^3 - 6xy - 1$

$f_x(x, y) = 2x - 6y,\ f_y(x, y) = 12y^2 - 6x$

Solve $f_x(x, y) = 0$ for x.

$$2x + 6y = 0$$
$$x = 3y$$

Substitute for x in $12y^2 - 6x = 0$.

$$12y^2 - 6(3y) = 0$$
$$6y(2y - 3) = 0$$
$$y = 0 \quad \text{or} \quad y = \frac{3}{2}$$
Then $\qquad x = 0 \quad \text{or} \quad x = \frac{9}{2}.$

The critical points are $(0, 0)$ and $\left(\frac{9}{2}, \frac{3}{2}\right)$.

$$f_{xx}(x, y) = 2$$
$$f_{yy}(x, y) = 24y$$
$$f_{xy}(x, y) = -6$$

For $(0, 0)$,

$$D = 2 \cdot 24(0) - (-6)^2$$
$$= -36 < 0.$$

A saddle point is at $(0, 0)$.
For $\left(\frac{9}{2}, \frac{3}{2}\right)$,

$$D = 2 \cdot 24\left(\frac{3}{2}\right) - (-6)^2$$
$$= 36 > 0.$$

Since $f_{xx}(x, y) > 0$, a relative minimum is at $\left(\frac{9}{2}, \frac{3}{2}\right)$.

17. $f(x, y) = e^{xy}$
$\qquad f_x = ye^{xy}$
$\qquad f_y = xe^{xy}$
$\qquad ye^{xy} = 0$
$\qquad xe^{xy} = 0$
$\qquad x = y = 0$

The critical point is $(0, 0)$.

$$f_{xx} = y^2 e^{xy},$$
$$f_{yy} = x^2 e^{xy},$$
$$f_{xy} = e^{xy} + xye^{xy}$$

For $(0, 0)$,

$$D = 0 \cdot 0 - (e^0)^2 = -1 < 0.$$

A saddle point is at $(0, 0)$.

21. $z = -3xy + x^3 - y^3 + \dfrac{1}{8}$

$f_x(x, y) = -3y + 3x^2,\ f_y(x, y) = -3x - 3y^2$

Solve the system $f_x = 0,\ f_y = 0$.

$$-3y + 3x^2 = 0$$
$$-3x - 3y^2 = 0$$

$$-y + x^2 = 0$$
$$\underline{-x - y^2 = 0}$$

Solve the first equation for y, substitute into the second, and solve for x.

$$y = x^2$$
$$-x - x^4 = 0$$
$$x(1 + x^3) = 0$$
$$x = 0 \quad \text{or} \quad x = -1$$
Then $\qquad y = 0 \quad \text{or} \quad y = 1.$

The critical points are $(0, 0)$ and $(-1, 1)$.

$$f_{xx}(x, y) = 6x$$
$$f_{yy}(x, y) = -6y$$
$$f_{xy}(x, y) = -3$$

For $(0,0)$,

$$D = 0 \cdot 0 - (-3)^2 = -9 < 0.$$

A saddle point is at $(0,0)$.
For $(-1,1)$,

$$D = -6(-6) - (-3)^2 = 27 > 0.$$

$$f_{xx}(x,y) = 6(-1) = -6 < 0.$$

$$f(-1,1) = -3(-1)(1) + (-1)^3 - 1^3 + \frac{1}{8}$$

$$= 1\frac{1}{8}$$

A relative maximum of $1\frac{1}{8}$ is at $(-1,1)$.
The equation matches graph (a).

23. $z = y^4 - 2y^2 + x^2 - \frac{17}{16}$

$f_x(x,y) = 2x, \ f_y(x,y) = 4y^3 - 4y$

Solve the system $f_x = 0, \ f_y = 0$.

$$\begin{aligned} 2x &= 0 \quad (1) \\ 4y^3 - 4y &= 0 \quad (2) \\ 4y(y^2 - 1) &= 0 \\ 4y(y+1)(y-1) &= 0 \end{aligned}$$

Equation (1) gives $x = 0$ and equation (2) gives $y = 0$, $y = -1$, or $y = 1$.
The critical points are $(0,0)$, $(0,-1)$, and $(0,1)$.

$$\begin{aligned} f_{xx}(x,y) &= 2, \\ f_{yy}(x,y) &= 12y^2 - 4, \\ f_{xy}(x,y) &= 0 \end{aligned}$$

For $(0,0)$,

$$D = 2(12 \cdot 0^2 - 4) - 0 = -8 < 0.$$

A saddle point is at $(0,0)$.
For $(0,-1)$,

$$D = 2[12(-1)^2 - 4] - 0 = 16 > 0.$$

$$f_{xx}(x,y) = 2 > 0$$

$$f(0,-1) = (-1)^4 - 2(-1)^2 + 0^2 - \frac{17}{16}$$

$$= -2\frac{1}{16}$$

A relative minimum of $-2\frac{1}{16}$ is at $(0,-1)$.
For $(0,1)$,

$$D = 2(12 \cdot 1^2 - 4) - 0 = 16 > 0$$

$$f_{xx}(x,y) = 2 > 0$$

$$f(0,1) = 1^4 - 2 \cdot 1^2 + 0^2 - \frac{17}{16}$$

$$= -2\frac{1}{16}$$

A relative minimum of $-2\frac{1}{16}$ is at $(0,-1)$.
The equation matches graph (b).

25. $z = -x^4 + y^4 + 2x^2 - 2y^2 + \frac{1}{16}$

$f_x(x,y) = -4x^3 + 4x, \ f_y(x,y) = 4y^3 - 4y$

Solve $f_x(x,y) = 0, \ f_y(x,y) = 0$.

$$\begin{aligned} -4x^3 + 4x &= 0 \quad (1) \\ 4y^3 - 4y &= 0 \quad (2) \\ -4x(x^2 - 1) &= 0 \quad (1) \\ -4x(x+1)(x-1) &= 0 \\ 4y(y^2 - 1) &= 0 \quad (2) \\ 4y(y+1)(y-1) &= 0 \end{aligned}$$

Equation (1) gives $x = 0$, -1, or 1.
Equation (2) gives $y = 0$, -1, or 1.
Critical points are $(0,0)$, $(0,-1)$, $(0,1)$, $(-1,0)$, $(-1,-1)$, $(-1,1)$, $(1,0)$, $(1,-1)$, $(1,1)$.

$$\begin{aligned} f_{xx}(x,y) &= -12x^2 + 4, \\ f_{yy}(x,y) &= 12y^2 - 4, \\ f_{xy}(x,y) &= 0 \end{aligned}$$

For $(0,0)$,

$$D = 4(-4) - 0 = -16 < 0.$$

For $(0,-1)$,

$$D = 4(8) - 0 = 32 > 0,$$

and $f_{xx}(x,y) = 4 > 0$.

$$f(0,-1) = -\frac{15}{16}$$

For $(0,1)$,

$$D = 4(8) - 0 = 32 > 0,$$

and $f_{xx}(x,y) = 4 > 0$.

$$f(0,1) = -\frac{15}{16}$$

For $(-1, 0)$,

$$D = -8(-4) - 0 = 32 > 0,$$

and $f_{xx}(x, y) = -8 < 0$.

$$f(-1, 0) = 1\tfrac{1}{16}$$

For $(-1, -1)$,

$$D = -8(8) - 0 = -64 < 0.$$

For $(-1, 1)$,

$$D = -8(8) - 0 = -64 < 0.$$

For $(1, 0)$,

$$D = -8(-4) = 32 > 0,$$

and $f_{xx}(x, y) = -8 < 0$.

$$f(1, 0) = 1\tfrac{1}{16}$$

For $(1, -1)$,

$$D = -8(8) - 0 = -64 < 0.$$

For $(1, 1)$,

$$D = -8(8) - 0 = -64 < 0.$$

Saddle points are at $(0, 0)$, $(-1, -1)$, $(-1, 1)$, $(1, -1)$, and $(1, 1)$.
Relative maximum of $1\tfrac{1}{16}$ is at $(-1, 0)$ and $(1, 0)$.
Relative minimum of $-\tfrac{15}{16}$ is at $(0, -1)$ and $(0, 1)$.
The equation matches graph (e).

27. $f(x, y) = 1 - x^4 - y^4$

$$f_x(x, y) = -4x^3, \quad f_y(x, y) = -4y^3$$

The system

$$f_x(x, y) = -4x^3 = 0, f_y(x, y) = -4y^3 = 0$$

gives the critical point $(0, 0)$.

$$f_{xx}(x, y) = -12x^2$$
$$f_{yy}(x, y) = -12y^3$$
$$f_{xy}(x, y) = 0$$

For $(0, 0)$,

$$D = 0 \cdot 0 - 0^2 = 0.$$

Therefore, the test gives no information. Examine a graph of the function drawn by using level curves.

If $f(x, y) = 1$, then $x^4 + y^4 = 0$. The level curve is the point $(0, 0, 1)$.
If $f(x, y) = 0$, then $x^4 + y^4 = 1$. The level curve is the circle with center $(0, 0, 0)$ and radius 1.
If $f(x, y) = -15$, then $x^4 + y^4 = 16$. The level curve is the curve with center $(0, 0, -15)$ and radius 2.
The xz-trace is

$$z = 1 - x^4.$$

This curve has a maximum at $(0, 0, 1)$ and opens downward.
The yz-trace is

$$z = 1 - y^4.$$

This curve also has a maximum at $(0, 0, 1)$ and opens downward.

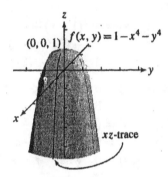

If $f(x, y) > 1$, then $x^4 + y^4 < 0$, which is impossible, so the function does not exist. Thus, the function has a relative maximum of 1 at $(0, 0)$.

31. $f(x, y) = ax^2e^y + y^2e^x$

(a) $f_x(x, y) = 2axe^y + y^2e^x$
$f_y(x, y) = ax^2e^y + 2ye^x$
$f_x(0, 0) = 2a(0)e^0 + 0^2e^0 = 0$
$f_y(0, 0) = a(0)^2e^0 + 2(0)e^0 = 0$

Thus, $(0, 0)$ is a critical point for all values of a.

(b) $f_{xx}(x, y) = 2ae^y + y^2e^x$
$f_{yy}(x, y) = ax^2e^y + 2e^x$
$f_{xy}(x, y) = 2axe^y + 2ye^x$
$f_{xx}(0, 0) = 2ae^0 + 0^2e^0 = 2a$
$f_{yy}(0, 0) = a(0)^2e^0 + 2e^0 = 2$
$f_{xy}(0, 0) = 2a(0)e^0 + 2(0)e^0 = 0$
$\quad D = (2a)(2) - 0^2 = 4a$

$(0, 0)$ is a relative minimum when $4a > 0$ and $f_{xx}(0, 0) = 2a > 0$, hence when $a > 0$.

33. $P(\alpha, r, s) = \alpha(3r^2(1-r) + r^3) + (1-\alpha)$
$(3s^2(1-s) + s^3)$

(a) $P(0.9, 0.5, 0.6)$
$= 0.9(3(0.5)^2(1-0.5) + (0.5)^3)$
$+ (1-0.9)(3(0.6)^2(1-0.6) + (0.6)^3)$
$= 0.5148$

$P(0.1, 0.8, 0.4)$
$= 0.1(3(0.8)^2(1-0.8) + (0.8)^3)$
$+ (1-0.1)(3(0.4)^2(1-0.4) + (0.4)^3)$
$= 0.4064$

The jury is less likely to make the correct decision in the second situation.

(b) If $r = s = 1$ then $P(\alpha, 1, 1) = 1$, so the jury always makes a correct decision. These values do not depend on α, but in a real-life situation α is likely to influence r and s.

(c) When P reaches a maximum, $P_\alpha, P_r,$ and P_s equal 0.

$P_\alpha(\alpha, r, s) = 3r^2(1-r) + r^3$
$\qquad\qquad - (3s^2(1-s) + s^3)$
$\qquad = 3r^2 - 2r^3 - (3s^2 - 2s^3)$

$P_r(\alpha, r, s) = \alpha(6r(1-r) - 3r^2 + 3r^2)$
$\qquad = 6\alpha r(1-r)$

$P_s(\alpha, r, s) = (1-\alpha)(6s(1-s) - 3s^2 + 3s^2)$
$\qquad = 6s(1-\alpha)(1-s)$

$P_\alpha(\alpha, r, s) = 0$ when $r = s$.

Since $P_r(\alpha, r, s) = 6\alpha r(1-r)$, and $P_s(\alpha, r, s) = 6(1-\alpha)s(1-s)$, then $P_\alpha, P_r,$ and P_s are simultaneously 0 at the points $(\alpha, 1, 1)$ and $(\alpha, 0, 0)$. So $(\alpha, 1, 1)$ and $(\alpha, 0, 0)$ are critical points.
$P(\alpha, 0, 0) = 0$ while $P(\alpha, 1, 1) = 1$
Since $P(\alpha, r, s)$ represents a probability, $0 \le P(\alpha, r, s) \le 1$. Thus, $P(\alpha, 1, 1) = 1$ is a maximum value of the function.

35. $P(x, y) = 1,000 + 24x - x^2 + 80y - y^2$

$P_x(x, y) = 24 - 2x$
$P_y(x, y) = 80 - 2y$
$24 - 2x = 0$
$80 - 2y = 0$
$12 = x$
$40 = y$

A critical point is $(12, 40)$.

$P_{xx}(x, y) = -2,\ P_{yy}(x, y) = -2,\ P_{xy}(x, y) = 0$

For $(12, 40)$,

$$D = -2(-2) - 0^2 = 4 > 0.$$

Since $P_{xx}(x, y) < 0$,

$$P(12, 40) = 1,000 + 24(12) - (12)^2$$
$$+ 80(40) - (40)^2$$
$$= 2,744.$$

The maximum profit is $274,400.

37. $C(x, y) = 2x^2 + 3y^2 - 2xy$
$\qquad\qquad + 2x - 126y + 3,800$

$C_x = 4x - 2y + 2$
$C_y = 6y - 2x - 126$

$0 = 4x - 2y + 2$
$0 = 6y - 2x - 126$

$\begin{aligned}0 &= 2x - y + 1\\ 0 &= -2x + 6y - 126 \\ \hline 0 &= \qquad 5y - 125\end{aligned}$

$y = 25$
$0 = 2x - 25 + 1$

If $y = 25$, $x = 12$.
$(12, 25)$ is a critical point.

$C_{xx}(x, y) = 4$
$C_{yy}(x, y) = 6$
$C_{xy}(x, y) = -2$

For $(12, 25)$,

$$D = 4 \cdot 6 - 4 = 20 > 0.$$

Since $C_{xx}(x, y) > 0$, 12 units of electrical tape and 25 units of packing tape should be produced to yield a minimum cost.

$C(12, 25)$
$= 2(12)^2 + 3(25)^2 - 2(12 \cdot 25)$
$\quad + 2(12) - 126(25) + 3,800$
$= 2,237$

The minimum cost is $2,237.

9.4   Total Differentials and Approximations

1. Let $z = f(x, y) = \sqrt{x^2 + y^2}$.

Then

$$dz = f_x(x, y)\, dx + f_y(x, y)\, dy$$

$$dz = \frac{1}{2}(x^2 + y^2)^{-1/2}(2x)\, dx$$

$$+ \frac{1}{2}(x^2 + y^2)^{-1/2}(2y)\, dy$$

$$= \frac{x}{\sqrt{x^2 + y^2}}\, dx + \frac{y}{\sqrt{x^2 + y^2}}\, dy$$

To approximate $\sqrt{3.04^2 + 4.06^2}$, we let $x = 3$, $dx = 0.04$, $y = 4$ and $dy = 0.06$.

$$dz = \frac{3}{\sqrt{3^2 + 4^2}}(0.04) + \frac{4}{\sqrt{3^2 + 4^2}}(0.06)$$

$$= \frac{3}{5}(0.04) + \frac{4}{5}(0.06) = 0.024 + 0.048 = 0.072$$

$$f(3.04, 4.06) = f(3, 4) + \Delta z$$
$$\approx f(3, 4) + dz$$
$$= \sqrt{3^2 + 4^2} + 0.072$$
$$f(3.04, 406) \approx 5.072$$

Using a calculator, $\sqrt{3.04^2 + 4.06^2} \approx 5.0720$.
The absolute value of the difference of the two results is $|5.072 - 5.0720| = 0$.

3. Let $z = f(x, y) = (x^2 + y^2)^{1/3}$.

Then

$$dz = f_x(x, y)\, dx + f_y(x, y)\, dy$$

$$dz = \frac{1}{3}(x^2 + y^2)^{-2/3}(2x)\, dx$$

$$+ \frac{1}{3}(x^2 + y^2)^{-2/3}(2y)\, dy$$

$$= \frac{2x}{3(x^2 + y^2)^{2/3}}\, dx + \frac{2y}{3(x^2 + y^2)^{2/3}}\, dy$$

To approximate $(1.92^2 + 2.1^2)^{1/3}$, we let $x = 2$, $dx = -0.08$, $y = 2$, and $dy = 0.1$.

$$dz = \frac{2(2)}{3[(2)^2 + (2)^2]^{2/3}}(-0.08)$$

$$+ \frac{2(2)}{3[(2)^2 + (2)^2]^{2/3}}(0.1)$$

$$= \frac{4}{12}(-0.08) + \frac{4}{12}(0.1)$$

$$= 0.00\bar{6}$$

$$f(1.92, 2.1) = f(2, 2) + \Delta z$$
$$\approx f(2, 2) + dz$$
$$= 2 + 0.00\bar{6}$$
$$f(1.92, 2.1) \approx 2.0067$$

Using a calculator, $(1.92^2 + 2.1^2)^{1/3} \approx 2.0080$.
The absolute value of the difference of the two results is $|2.0067 - 2.0080| = 0.0013$.

5. Let $z = f(x, y) = xe^y$.

Then

$$dz = f_x(x, y)\, dx + f_y(x, y)\, dy$$
$$= e^y\, dx + xe^y\, dy.$$

To approximate $0.97e^{.02}$ we let $x = 1$, $dx = -0.03$, $y = 0$, and $dy = 0.02$.

$$dz = e^0(-0.03) + (1)e^0(0.02)$$
$$dz = -0.01$$

$$f(0.97, 0.02) = f(1, 0) + \Delta z$$
$$\approx f(1.0) + dz$$
$$= (1)e^0 + (-0.01)$$
$$f(0.97, 0.02) \approx 0.99$$

Using a calculator, $0.97e^{0.02} \approx 0.9896$.
The absolute value of the difference of the two results is $|0.99 - 0.9896| = 0.0004$.

7. Let $z = x \ln y$.

Then

$$dz = f_x(x, y)\, dx + f_y(x, y)\, dy$$

$$= (\ln y)\, dx + \frac{x}{y}\, dy.$$

To approximate $1.04 \ln 0.95$ we let $x = 1$, $dx = 0.04$, $y = 1$, and $dy = -0.05$.

$$dz = (\ln 1)(0.04) + \frac{1}{1}(-0.05)$$

$$dz = -0.05$$

$$f(1.04, 0.95) = f(1, 1) + \Delta z$$
$$\approx f(1, 1) + dz$$
$$= (1)(\ln 1) + (-0.05)$$
$$f(1.04, 0.95) \approx -0.05$$

Using a calculator, $1.04 \ln 0.95 \approx -0.0533$.
The absolute value of the difference of the two results is $|-0.05 - (-0.0533)| = 0.0033$.

9. $z = x^2 + 3xy + y^2$

$x = 4$, $y = -2$, $dx = 0.02$, $dy = -0.03$

$$f_x(x, y) = 2x + 3y$$
$$f_y(x, y) = 3x + 2y$$
$$dz = (2x + 3y)\,dx + (3x + 2y)\,dy$$

Substitute the given values to get

$$dz = [2(4) + 3(-2)](0.02) + [3(4) + 2(-2)](-0.03)$$
$$= 0.04 + (-0.24) = -0.2.$$

11. $z = \dfrac{y^2 + 3x}{y^2 - x}$, $x = 4$, $y = -4$,

$dx = 0.01$, $dy = 0.03$

$$dz = \frac{(y^2 - x) \cdot 3 - (y^2 + 3x) \cdot (-1)}{(y^2 - x)^2}\,dx$$
$$+ \frac{(y^2 - x) \cdot 2y - (y^2 + 3x) \cdot 2y}{(y^2 - x)^2}\,dx$$
$$= \frac{4y^2}{(y^2 - x)^2}\,dx - \frac{8xy}{(y^2 - x)^2}\,dy$$
$$= \frac{4(-4)^2}{[(-4)^2 - 4]^2}(0.01) - \frac{8(4)(-4)}{[(-4)^2 - 4]^2}(0.03)$$
$$\approx 0.0311$$

13. $w = \dfrac{5x^2 + y^2}{z + 1}$

$x = -2$, $y = 1$, $z = 1$

$dx = 0.02$, $dy = -0.03$, $dz = 0.02$

$$f_x(x, y) = \frac{(z + 1)10x - (5x^2 + y^2)(0)}{(z + 1)^2}$$
$$= \frac{10x}{z + 1}$$
$$f_y(x, y) = \frac{(z + 1)(2y) - (5x^2 + y^2)(0)}{(z + 1)^2}$$
$$= \frac{2y}{z + 1}$$
$$f_z(x, y) = \frac{(z + 1)(0) - (5x^2 + y^2)(1)}{(z + 1)^2}$$
$$= \frac{-5x^2 - y^2}{(z + 1)^2}$$
$$dw = \frac{10x}{z + 1}\,dx + \frac{2y}{z + 1}\,dy + \frac{-5x^2 - y^2}{(z + 1)^2}\,dz$$

Substitute the given values.

$$dw = \frac{-20}{2}(0.02) + \frac{2}{2}(-0.03) + \frac{[-5(4) - 1](0.02)}{(2)^2}$$
$$= -0.2 - 0.03 - \frac{21}{4}(0.02)$$
$$= -0.355$$

15. The volume of the bone is

$$V = \pi r^2 h,$$

with $h = 7$, $r = 1.4$, $dr = 0.09$, $dh = 2(0.09) = 0.18$

$$dV = 2\pi r h\,dr + \pi r^2\,dh$$
$$= 2\pi(1.4)(7)(0.09) + \pi(1.4)^2(0.18)$$
$$= 6.65$$

6.65 cm^3 of preservative are used.

17. $C = \dfrac{b}{a - v} = b(a - v)^{-1}$

$a = 160$,
$b = 200$, $v = 125$
$da = 145 - 160 = -15$
$db = 190 - 200 = -10$
$dv = 130 - 125 = 5$

$$dC = -b(a - v)^{-2}\,da$$
$$+ \frac{1}{a - v}\,db + b(a - v)^{-2}\,dv$$
$$= \frac{-b}{(a - v)^2}\,da + \frac{1}{a - v}\,db + \frac{b}{(a - v)^2}\,dv$$
$$= \frac{-200}{(160 - 125)^2}(-15) + \frac{1}{160 - 125}(-10)$$
$$+ \frac{200}{(160 - 125)^2}(5)$$
$$\approx 2.98 \text{ liters}$$

19. $C(t, g) = 0.6(0.96)^{(210t/1{,}500)-1}$

$$+ \frac{gt}{126t - 900}\left(1 - (0.96)^{(210t/1{,}500)-1}\right)$$

(a) $C(180, 8) = 0.6(0.96)^{(210(180)/1{,}500)-1}$

$$+ \frac{(8)(180)}{126(180) - 900}$$
$$\cdot \left(1 - (0.96)^{(210(180)/1{,}500)-1}\right)$$
$$\approx 0.265$$

(b) $C_t(t,g)$

$$= 0.6(\ln 0.96)\left(\frac{210}{1,500}\right)(0.96)^{(210t/1,500)-1}$$

$$+ \frac{g(126t-900)-126(gt)}{(126t-900)^2}$$

$$\cdot(1-(0.96)^{(210t/1,500)-1})$$

$$-\frac{gt}{126t-900}(\ln 0.96)\left(\frac{210}{1,500}\right)(0.96)^{(210t/1,500)-1}$$

$C_g(t,g)$

$$= \frac{t}{126t-900}(1-(0.96)^{(210t/1,500)-1})$$

$C(180-10,8+1)$

$$\approx C(180,8)+C_t(180,8)\cdot(-10)$$
$$+C_g(180,8)\cdot(1)$$
$$\approx 0.265+(-0.00115)(-10)+0.00519(1)$$
$$\approx 0.282$$

$C(170,9)\approx 0.282$

The approximation is very good.

21. $t(x,y,p,C)=\dfrac{\sqrt{x^2+(y-p)^2}}{331.45+0.6C}$

(a) $t(5,-2,20,20)=\dfrac{\sqrt{5^2+(-2-20)^2}}{331.45+0.6(20)}$

$$=\frac{\sqrt{509}}{343.45}$$

$$\approx 0.066$$

$t(5,-2,10,20)=\dfrac{\sqrt{5^2+(-2-10)^2}}{331.45+0.6(20)}$

$$=\frac{\sqrt{169}}{343.45}$$

$$\approx 0.038$$

In a close race, this difference could certainly affect the outcome.

(b) Since the starter remains stationary, $dx=dy=0$, so $t_x(x,y,p,C)$ and $t_y(x,y,p,C)$ do not need to be computed.

$t_p(x,y,p,C)$

$$=\frac{-2(y-p)}{2(331.45+0.6C)\sqrt{x^2+(y-p)^2}}$$

$$=\frac{p-y}{(331.45+0.6C)\sqrt{x^2+(y-p)^2}}$$

$t_C(x,y,p,C)=-\dfrac{0.6\sqrt{x^2+(y-p)^2}}{(331.45+0.6C)^2}$

$dx=0, dy=0, dp=0.5, dC=-5$

$dt = t_x(5,-2,20,20)\cdot 0+t_y(5,-2,20,20)\cdot 0$
$$+t_p(5,-2,20,20)\cdot 0.5$$
$$+t_C(5,-2,20,20)\cdot(-5)$$

$$=\frac{20-(-2)}{(331.45+0.6(20))\sqrt{5^2+(-2-20)^2}}\cdot 0.5$$

$$-\frac{0.6\sqrt{5^2+(-2-20)^2}}{(331.45+0.6(20))^2}\cdot(-5)$$

$$\approx 0.00199 \text{ sec}$$

This is the approximate change in the time when the swimmer stands 0.5 m farther from the starter and the temperature decreases by 5°C.

23. $V=\dfrac{1}{3}\pi r^2 h$, $r=2.9$, $h=8.4$,

$dr=dh=\pm 0.1$

$dV=\dfrac{2}{3}\pi rh\,dr+\dfrac{1}{3}\pi r^2\,dh$

$$=\frac{2}{3}\pi(2.9)(8.4)(\pm 0.1)$$

$$+\frac{1}{3}\pi(2.9)^2(\pm 0.1)$$

$$\approx \pm 5.98$$

The maximum possible error is 5.98 cm³.

25. Let $z=f(r,h)=\dfrac{1}{3}\pi r^2 h$.

Then

$$dz=f_r(r,h)\,dr+f_h(r,h)\,dh$$

$$dz=\frac{2}{3}\pi rh\,dr+\frac{1}{3}\pi r^2\,dh.$$

Since there is a maximum error of $a\%$ in measuring the radius, the maximum value of dr is $\frac{a}{100}r$. Similarly, the maximum value of dh is $\frac{b}{100}h$. Therefore, the maximum value of dz is

$$dz=\frac{2}{3}\pi rh\left(\frac{a}{100}r\right)+\frac{1}{3}\pi r^2\left(\frac{b}{100}h\right)$$

$$=\frac{2a}{100}\left(\frac{1}{3}\pi r^2 h\right)+\frac{b}{100}\left(\frac{1}{3}\pi r^2 h\right)$$

$$=\left(\frac{2a+b}{100}\right)\left(\frac{1}{3}\pi r^2 h\right).$$

The maximum percent error in calculating the volume is $(2a+b)\%$.

Since the volume of a cylinder is $\pi r^2 h$, the maximum percent error in calculating the volume of the cylinder is the same.

27. Let r be the radius inside the tumbler and h be the height inside.

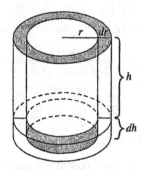

$$V = \pi r^2 h, \; r = 1.5, \; h = 9, \; dr = dh = 0.2$$

$$\begin{aligned} dV &= 2\pi r h \, dr + \pi r^2 \, dh \\ &= 2\pi(1.5)(9)(0.2) + \pi(1.5)^2(0.2) \\ &\approx 18.4 \end{aligned}$$

Approximately 18.4 cm^3 of material is needed.

9.5 Double Integrals

1. $\displaystyle\int_0^3 (x^3 y + y) \, dx = \left(\frac{x^4}{4} y + yx \right) \Big|_0^3$

$$= \frac{81y}{4} + 3y = \frac{93y}{4}$$

3. $\displaystyle\int_4^8 \sqrt{6x + y} \, dx = \int_4^8 (6x + y)^{1/2} \, dx$

Let $u = 6x + y$. Then $du = 6 \, dx$.
When $x = 8$, $u = 48 + y$.
When $x = 4$, $u = 24 + y$.

$$\int_{y+24}^{y+48} u^{1/2} \cdot \frac{1}{6} \, du$$

$$= \frac{1}{6} \cdot \frac{2}{3} u^{3/2} \Big|_{y+24}^{y+48}$$

$$= \frac{1}{9} u^{3/2} \Big|_{y+24}^{y+48}$$

$$= \frac{1}{9} [(48 + y)^{3/2} - (24 + y)^{3/2}]$$

5. $\displaystyle\int_4^5 x\sqrt{x^2 + 3y} \, dy$

$$= \int_4^5 x(x^2 + 3y)^{1/2} \, dy$$

$$= \frac{2x}{9} (x^2 + 3y)^{3/2} \Big|_4^5$$

$$= \frac{2x}{9} [(x^2 + 15)^{3/2} - (x^2 + 12)^{3/2}]$$

7. $\displaystyle\int_4^9 \frac{3 + 5y}{\sqrt{x}} \, dx = (3 + 5y) \int_4^9 x^{-1/2} \, dx$

$$= (3 + 5y) 2x^{1/2} \Big|_4^9$$

$$= (3 + 5y) 2[\sqrt{9} - \sqrt{4}]$$

$$= 6 + 10y$$

9. $\displaystyle\int_{-1}^1 e^{x+4y} \, dy$

Let $u = x + 4y$. Then $du = 4 \, dy$.
When $y = 1$, $u = x + 4$.
When $y = -1$, $u = x - 4$.

$$\frac{1}{4} \int_{x-4}^{x+4} e^u \, du = \frac{1}{4} e^u \Big|_{x-4}^{x+4}$$

$$= \frac{1}{4} (e^{x+4} - e^{x-4})$$

$$= \frac{1}{4} e^{x+4} - \frac{1}{4} e^{x-4}$$

11. $\displaystyle\int_0^5 xe^{x^2 + 9y} \, dx$

Let $u = x^2 + 9y$. Then $du = 2x \, dx$.
When $x = 5$, $u = 25 + 9y$.
When $x = 0$, $u = 9y$.

$$\frac{1}{2} \int_{9y}^{25+9y} e^u \, du = \frac{1}{2} (e^{25+9y} - e^{9y})$$

$$= \frac{1}{2} e^{25+9y} - \frac{1}{2} e^{9y}$$

13. $\displaystyle\int_1^2 \left[\int_0^3 (x^3 y + y) \, dx \right] dy$

From Exercise 1,

$$\int_0^3 (x^3 y + y) \, dx = \frac{93y}{4}.$$

$$\int_1^2 \left[\int_0^3 (x^3 y + y) \, dx \right] dy = \int_1^2 \frac{93y}{4} \, dy$$

$$= \frac{93}{8} y^2 \Big|_1^2$$

$$= \frac{93}{8} (4 - 1)$$

$$= \frac{279}{8}$$

15. $\int_0^1 \left[\int_3^6 x\sqrt{x^2+3y}\,dx \right] dy$

From Exercise 6,

$\int_3^6 x\sqrt{x^2+3y}\,dx$

$= \frac{1}{3}[(36+3y)^{3/2} - (9+3y)^3].$

$\int_0^1 \left[\int_3^6 x\sqrt{x^2+3y}\,dx \right] dy$

$= \int_0^1 \frac{1}{3}[(36+3y)^{3/2} - (9+3y)^{3/2}]\,dy$

Let $u = 36 + 3y$. Then $du = 3\,dy$.
When $y=0$, $u=36$.
When $y=1$, $u=39$.
Let $z = 9 + 3y$. Then $dz = 3\,dy$.
When $y=0$, $z=9$.
When $y=1$, $z=12$.

$\frac{1}{9}\left[\int_{36}^{39} u^{3/2}\,du - \int_9^{12} z^{3/2}\,dz \right]$

$= \frac{1}{9} \cdot \frac{2}{5}[(39)^{5/2} - (36)^{5/2}$

$\qquad - (12)^{5/2} + (9)^{5/2}]$

$= \frac{2}{45}[(39)^{5/2} - (12)^{5/2} - 6^5 + 3^5]$

$= \frac{2}{45}(39^{5/2} - 12^{5/2} - 7{,}533)$

17. $\int_1^2 \left[\int_4^9 \frac{3+5y}{\sqrt{x}}\,dx \right] dy$

From Exercise 7,

$\int_4^9 \frac{3+5y}{\sqrt{x}}\,dx = 6 + 10y.$

$\int_1^2 \left[\int_4^9 \frac{3+5y}{\sqrt{x}}\,dx \right] dy$

$= \int_1^2 (6 + 10y)\,dy$

$= 6y \Big|_1^2 + 5y^2 \Big|_1^2$

$= 6(2-1) + 5(4-1)$

$= 6 + 15$

$= 21$

19. $\int_1^2 \int_1^2 \frac{dx\,dy}{xy}$

$= \int_1^2 \left(\frac{\ln|x|}{y} \right)\Big|_1^2 dy$

$= \int_1^2 \frac{1}{y}(\ln 2 - \ln 1)\,dy$

$= \ln 2 \int_1^2 \frac{1}{y}\,dy$

$= \ln 2 (\ln|y|) \Big|_1^2$

$= \ln 2 (\ln 2 - \ln 1)$

$= (\ln 2)^2$

21. $\int_2^4 \int_3^5 \left(\frac{x}{y} + \frac{y}{3} \right) dx\,dy$

$= \int_2^4 \left(\frac{x^2}{2y} + \frac{yx}{3} \right)\Big|_3^5 dy$

$= \int_2^4 \left[\frac{25}{2y} + \frac{5y}{3} - \left(\frac{9}{2y} + \frac{3y}{3} \right) \right] dy$

$= \int_2^4 \left(\frac{16}{2y} + \frac{2y}{3} \right) dy$

$= \left(8\ln|y| + \frac{y^2}{3} \right)\Big|_2^4$

$= 8(\ln 4 - \ln 2) + \frac{16}{3} - \frac{4}{3}$

$= 8\ln\frac{4}{2} + \frac{12}{3}$

$= 8\ln 2 + 4$

23. $\int_R \int (x + 3y^2)\,dx\,dy;\ 0 \le x \le 2,\ 1 \le y \le 5$

$\int_R \int (x + 3y^2)\,dx\,dy$

$= \int_1^5 \int_0^2 (x + 3y^2)\,dx\,dy$

$= \int_1^5 \left(\frac{x^2}{2} + 3y^2 x \right)\Big|_0^2 dy$

$= \int_1^5 \left(\frac{4}{2} + 6y^2 - 0 \right) dy$

$= (2y + 2y^3) \Big|_1^5$

$= 2 \cdot 5 + 2 \cdot 125 - (2 \cdot 1 + 2 \cdot 1)$

$= 256$

25. $\int_R \int \sqrt{x+y}\,dy\,dx; \ 1 \leq x \leq 3, 0 \leq y \leq 1$

$\int_R \int \sqrt{x+y}\,dy\,dx$

$$= \int_1^3 \int_0^1 (x+y)^{1/2}\,dy\,dx$$

$$= \int_1^3 \left[\frac{2}{3}(x+y)^{3/2}\right]\Big|_0^1 dx$$

$$= \int_1^3 \frac{2}{3}[(x+1)^{3/2} - x^{3/2}]\,dx$$

$$= \frac{2}{3} \cdot \frac{2}{5}[(x+1)^{5/2} - x^{5/2}]\Big|_1^3$$

$$= \frac{4}{15}(4^{5/2} - 3^{5/2} - 2^{5/2} + 1^{5/2})$$

$$= \frac{4}{15}(32 - 3^{5/2} - 2^{5/2} + 1)$$

$$= \frac{4}{15}(33 - 3^{5/2} - 2^{5/2})$$

27. $\int_R \int \frac{2}{(x+y)^2}\,dy\,dx; \ 2 \leq x \leq 3, 1 \leq y \leq 5$

$\int_R \int \frac{2}{(x+y)^2}\,dy\,dx$

$$= \int_2^3 \int_1^5 2(x+y)^{-2}\,dy\,dx$$

$$= \int_2^3 -2(x+y)^{-1}\Big|_1^5 dx$$

$$= -2\int_2^3 \left[\frac{1}{(5+x)} - \frac{1}{(1+x)}\right]dx$$

$$= -2(\ln|5+x| - \ln|1+x|)\Big|_2^3$$

$$= -2\left(\ln\left|\frac{5+x}{1+x}\right|\right)\Big|_2^3$$

$$= -2\left(\ln 2 - \ln\frac{7}{3}\right)$$

$$= -2\ln\frac{6}{7} \quad \text{or} \quad 2\ln\frac{7}{6}$$

29. $\int_R \int ye^{(x+y^2)}\,dx\,dy; \ 2 \leq x \leq 3, 0 \leq y \leq 2$

$\int_R \int ye^{(x+y^2)}\,dx\,dy$

$$= \int_0^2 \int_2^3 ye^{x+y^2}\,dx\,dy$$

$$= \int_0^2 ye^{x+y^2}\Big|_2^3 dy$$

$$= \int_0^2 (ye^{3+y^2} - ye^{2+y^2})\,dy$$

$$= e^3 \int_0^2 ye^{y^2}\,dy - e^2 \int_0^2 ye^{y^2}\,dy$$

$$= \frac{e^3}{2}(e^{y^2})\Big|_0^2 - \frac{e^2}{2}(e^{y^2})\Big|_0^2$$

$$= \frac{e^3}{2}(e^4 - e^0) - \frac{e^2}{2}(e^4 - e^0)$$

$$= \frac{1}{2}(e^7 - e^6 - e^3 + e^2)$$

31. $\int \int_R x\cos(xy)\,dy\,dx; \ \frac{\pi}{2} \leq x \leq \pi, 0 \leq y \leq 1$

$\int \int_R x\cos(xy)\,dy\,dx$

$$= \int_{\pi/2}^\pi \int_0^1 x\cos(xy)\,dy\,dx$$

$$= \int_{\pi/2}^\pi \frac{1}{x}x\sin(xy)\Big|_0^1 dx$$

$$= \int_{\pi/2}^\pi [\sin(x) - \sin(0)]\,dx$$

$$= \int_{\pi/2}^\pi \sin(x)\,dx$$

$$= -\cos(x)\Big|_{\pi/2}^\pi$$

$$= -\cos(\pi) + \cos\left(\frac{\pi}{2}\right)$$

$$= 1$$

33. $z = 6x + 2y + 5; \; -1 \le x \le 1, \; 0 \le y \le 3$

$$V = \int_{-1}^{1} \int_{0}^{3} (6x + 2y + 5) \, dy \, dx$$

$$= \int_{-1}^{1} (6xy + 5y + y^2) \Big|_{0}^{3} \, dx$$

$$= \int_{-1}^{1} (18x + 15 + 9) \, dx$$

$$= (9x^2 + 24x) \Big|_{-1}^{1}$$

$$= 9 + 24 - (9 - 24) = 48$$

35. $z = x^2; \; 0 \le x \le 1, \; 0 \le y \le 4$

$$V = \int_{0}^{1} \int_{0}^{4} x^2 \, dy \, dx$$

$$= \int_{0}^{1} (x^2 y) \Big|_{0}^{4} \, dx$$

$$= \int_{0}^{1} 4x^2 \, dx$$

$$= \frac{4}{3} x^3 \Big|_{0}^{1} = \frac{4}{3}$$

37. $z = x\sqrt{x^2 + y}; \; 0 \le x \le 1, \; 0 \le y \le 1$

$$V = \int_{0}^{1} \int_{0}^{1} x\sqrt{x^2 + y} \, dx \, dy$$

Let $u = x^2 + y$. Then $du = 2x \, dx$.
When $x = 0$, $u = y$.
Wen $x = 1$, $u = 1 + y$.

$$= \int_{0}^{1} \left[\int_{y}^{1+y} u^{1/2} \, du \right] dy$$

$$= \int_{0}^{1} \frac{1}{2} \left(\frac{2}{3} u^{3/2} \right) \Big|_{y}^{1+y} \, dy$$

$$= \int_{0}^{1} \frac{1}{3} [(1+y)^{3/2} - y^{3/2}] \, dy$$

$$= \frac{1}{3} \cdot \frac{2}{5} [(1+y)^{5/2} - y^{5/2}] \Big|_{0}^{1}$$

$$= \frac{2}{15} (2^{5/2} - 1 - 1)$$

$$= \frac{2}{15} (2^{5/2} - 2)$$

39. $z = \dfrac{xy}{(x^2 + y^2)^2}; \; 1 \le x \le 2, \; 1 \le y \le 4$

$$V = \int_{1}^{2} \int_{1}^{4} \frac{xy}{(x^2 + y^2)^2} \, dy \, dx$$

$$= \int_{1}^{2} \left[\int_{1}^{4} xy(x^2 + y^2)^{-2} \, dy \right] dx$$

$$= \int_{1}^{2} \left[\int_{1}^{4} \frac{1}{2} x(x^2 + y^2)^{-2}(2y) \, dy \right] dx$$

$$= \int_{1}^{2} \left[-\frac{1}{2} x(x^2 + y^2)^{-1} \right] \Big|_{1}^{4} \, dx$$

$$= \int_{1}^{2} \left[-\frac{1}{2} x(x^2 + 16)^{-1} + \frac{1}{2} x(x^2 + 1)^{-1} \right] dx$$

$$= -\frac{1}{2} \int_{1}^{2} \frac{1}{2} (x^2 + 16)^{-1}(2x) \, dx$$

$$\quad + \frac{1}{2} \int_{1}^{2} \frac{1}{2} (x^2 + 1)^{-1}(2x) \, dx$$

$$= -\frac{1}{2} \cdot \frac{1}{2} \ln |x^2 + 16| \Big|_{1}^{2} + \frac{1}{2} \cdot \frac{1}{2} \ln |x^2 + 1| \Big|_{1}^{2}$$

$$= -\frac{1}{4} \cdot \ln 20 + \frac{1}{4} \ln 17 + \frac{1}{4} \ln 5 - \frac{1}{4} \ln 2$$

$$= \frac{1}{4} (-\ln 20 + \ln 17 + \ln 5 - \ln 2)$$

$$= \frac{1}{4} \ln \frac{(17)(5)}{(20)(2)}$$

$$= \frac{1}{4} \ln \frac{17}{8}$$

41. $\displaystyle\int_{R}\int xe^{xy} \, dx \, dy; \; 0 \le x \le 2; \; 0 \le y \le 1$

$$\int_{R}\int xe^{xy} \, dx \, dy$$

$$= \int_{0}^{2} \int_{0}^{1} xe^{xy} \, dy \, dx$$

$$= \int_{0}^{2} \frac{x}{x} e^{xy} \Big|_{0}^{1} \, dx$$

$$= \int_{0}^{2} (e^x - e^0) \, dx$$

$$= (e^x - x) \Big|_{0}^{2}$$

$$= e^2 - 2 - e^0 + 0$$

$$= e^2 - 3$$

43. $\displaystyle\int_2^4 \int_2^{x^2} (x^2 + y^2)\, dy\, dx$

$\displaystyle = \int_2^4 \left(x^2 y + \frac{y^3}{3}\right)\Bigg|_2^{x^2} dx$

$\displaystyle = \int_2^4 \left(x^4 + \frac{x^6}{3} - 2x^2 - \frac{8}{3}\right) dx$

$\displaystyle = \left(\frac{x^5}{5} + \frac{x^7}{21} - \frac{2}{3}x^3 - \frac{8}{3}x\right)\Bigg|_2^4$

$\displaystyle = \frac{1{,}024}{5} + \frac{16{,}384}{21} - \frac{2}{3}(64) - \frac{8}{3}(4)$

$\displaystyle \quad - \left(\frac{32}{5} + \frac{128}{21} - \frac{16}{3} - \frac{16}{3}\right)$

$\displaystyle = \frac{1{,}024}{5} - \frac{32}{5} + \frac{16{,}384 - 128}{21}$

$\displaystyle \quad - \frac{128}{3} - \frac{32}{3} - \left(\frac{-32}{3}\right)$

$\displaystyle = \frac{992}{5} + \frac{16{,}256}{21} - \frac{128}{3}$

$\displaystyle = \frac{20{,}832}{105} + \frac{81{,}280}{105} - \frac{4{,}480}{105}$

$\displaystyle = \frac{97{,}632}{105}$

$\displaystyle \approx 929.83$

45. $\displaystyle\int_0^4 \int_0^x \sqrt{xy}\, dy\, dx$

$\displaystyle = \int_0^4 \int_0^x (xy)^{1/2}\, dy\, dx$

$\displaystyle = \int_0^4 \left[\frac{2(xy)^{3/2}}{3x}\right]\Bigg|_0^x dx$

$\displaystyle = \frac{2}{3}\int_0^4 \left[\frac{(\sqrt{x^2})^3}{x} - \frac{0}{x}\right] dx$

$\displaystyle = \frac{2}{3}\int_0^4 x^2\, dx = \frac{2}{3}\cdot\frac{x^3}{3}\Bigg|_0^4 = \frac{2}{9}(64)$

$\displaystyle = \frac{128}{9}$

47. $\displaystyle\int_1^2 \int_y^{3y} \frac{1}{x}\, dx\, dy = \int_1^2 (\ln|x|)\Bigg|_y^{3y} dy$

$\displaystyle = \int_1^2 (\ln|3y| - \ln|y|)\, dy$

$\displaystyle = \int_1^2 \ln\left|\frac{3y}{y}\right| dy$

$\displaystyle = (\ln 3)(y)\Bigg|_1^2 = \ln 3$

49. $\displaystyle\int_0^4 \int_1^{e^x} \frac{x}{y}\, dy\, dx$

$\displaystyle = \int_0^4 (x \ln|y|)\Bigg|_1^{e^x} dx$

$\displaystyle = \int_0^4 (x \ln e^x - x \ln 1)\, dx$

$\displaystyle = \int_0^4 x^2\, dx = \frac{x^3}{3}\Bigg|_0^4 = \frac{64}{3}$

51. $\displaystyle\int_0^{\ln 2} \int_{e^y}^2 \frac{1}{\ln x}\, dx\, dy$

Changing the order of integration,

$\displaystyle\int_0^{\ln 2} \int_{e^y}^2 \frac{1}{\ln x}\, dx\, dy = \int_1^2 \int_0^{\ln x} \frac{1}{\ln x}\, dy\, dx$

$\displaystyle = \int_1^2 \left[\frac{1}{\ln x} y\Big|_0^{\ln x}\right] dx$

$\displaystyle = \int_1^2 (1 - 0)\, dx$

$\displaystyle = x\Big|_1^2$

$\displaystyle = 2 - 1 = 1$

53. $\displaystyle\int_R \int (4x + 7y)\, dy\, dx;\ 1 \le x \le 3;$

$0 \le y \le x + 1$

$\displaystyle\int_R \int (4x + 7y)\, dy\, dx$

$\displaystyle\int_1^3 \int_0^{x+1} (4x + 7y)\, dy\, dx$

$\displaystyle = \int_1^3 \left(4xy + \frac{7y^2}{2}\right)\Bigg|_0^{x+1} dx$

$\displaystyle = \int_1^3 \left[4x^2 + 4x + \frac{7}{2}(x^2 + 2x + 1)\right] dx$

$\displaystyle = \int_1^3 \left(\frac{15}{2}x^2 + 11x + \frac{7}{2}\right) dx$

$\displaystyle = \left(\frac{15}{2}\cdot\frac{x^3}{3} + \frac{11x^2}{2} + \frac{7x}{2}\right)\Bigg|_1^3$

$\displaystyle = \frac{5}{2}\cdot 27 + \frac{11(9)}{2} + \frac{7(3)}{2}$

$\displaystyle \quad - \left(\frac{5}{2} + \frac{11}{2} + \frac{7}{2}\right)$

$\displaystyle = \frac{130}{2} + \frac{88}{2} + \frac{14}{2}$

$\displaystyle = \frac{232}{2} = 116$

55. $\int_R \int (4 - 4x^2)\, dy\, dx;\ 0 \le x \le 1,$

$0 \le y \le 2 - 2x$

$\int_R \int (4 - 4x^2)\, dy\, dx$

$= \int_0^2 \int_0^{2-2x} 4(1 - x^2)\, dy\, dx$

$= \int_0^1 [4(1 - x^2)y]\Big|_0^{2(1-x)}\, dx$

$= \int_0^1 4(1 - x^2)(2)(1 - x)\, dx$

$= 8 \int_0^1 (1 - x - x^2 + x^3)\, dx$

$= 8 \left(x - \frac{x^2}{2} - \frac{x^3}{3} + \frac{x^4}{4} \right)\Big|_0^1$

$= 8 \left(1 - \frac{1}{2} - \frac{1}{3} + \frac{1}{4} \right)$

$= 8 \left(\frac{1}{2} - \frac{1}{12} \right)$

$= 8 \cdot \frac{5}{12} = \frac{10}{3}$

57. $\int_R \int e^{x/y^2}\, dx\, dy;\ 1 \le y \le 2,\ 0 \le x \le y^2$

$\int_R \int e^{x/y^2}\, dx\, dy$

$= \int_1^2 \int_0^{y^2} e^{x/y^2}\, dx\, dy$

$= \int_1^2 [y^2 e^{x/y^2}]\Big|_0^{y^2}\, dy$

$= \int_1^2 (y^2 e^{y^2/y^2} - y^2 e^0)\, dy$

$= \int_1^2 (ey^2 - y^2)\, dy$

$= (e - 1) \frac{y^3}{3}\Big|_1^2$

$= (e - 1) \left(\frac{8}{3} - \frac{1}{3} \right)$

$= \frac{7(e - 1)}{3}$

59. $\int_R \int x^3 y\, dy\, dx;\ R$ bounded by $y = x^2,\ y = 2x$

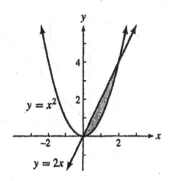

The points of intersection can be determined by solving the following system for x.

$$y = x^2$$
$$y = 2x$$

$$x^2 = 2x$$
$$x(x - 2) = 0$$
$$x = 0 \quad \text{or} \quad x = 2$$

Therefore,

$\int_R \int x^3 y\, dx\, dy$

$= \int_0^2 \int_{x^2}^{2x} x^3 y\, dy\, dx$

$= \int_0^2 \left(x^3 \frac{y^2}{2} \right)\Big|_{x^2}^{2x}\, dx$

$= \int_0^2 \left[x^3 \frac{(4x^2)}{2} - x^3 \frac{(x^4)}{2} \right]\, dx$

$= \int_0^2 \left(2x^5 - \frac{x^7}{2} \right)\, dx$

$= \left(\frac{1}{3} x^6 - \frac{1}{16} x^8 \right)\Big|_0^2$

$= \frac{1}{3} \cdot 2^6 - \frac{1}{16} \cdot 2^8$

$= \frac{64}{3} - 16$

$= \frac{16}{3}.$

61. $\displaystyle\int_R\!\!\int \frac{dy\,dx}{y}$; R bounded by $y = x$, $y = \dfrac{1}{x}$, $x = 2$.

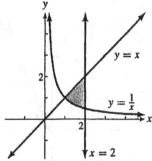

The graphs of $y = x$ and $y = \frac{1}{x}$ intersect at $(1,1)$.

$$\int_1^2 \int_{1/x}^x \frac{dy}{y}\,dx = \int_1^2 \ln y \Big|_{1/x}^x\,dx$$

$$= \int_1^2 \left(\ln x - \ln\frac{1}{x}\right)dx$$

$$= \int_1^2 2\ln x\,dx$$

$$= 2(x\ln x - x)\Big|_1^2$$

$$= 2[(2\ln 2 - 2) - (\ln 1 - 1)]$$

$$= 4\ln 2 - 2$$

65. $f(x,y) = x^2 + y^2$; $0 \le x \le 2$, $0 \le y \le 3$

The area of region R is

$$A = (2 - 0)(3 - 0) = 6.$$

The average value of

$$f(x,y) = x^2 + y^2$$

over R is

$$\frac{1}{A}\int_R\!\!\int (x^2 + y^2)\,dy\,dx$$

$$= \frac{1}{6}\int_0^2 \int_0^3 (x^2 + y^2)\,dy\,dx$$

$$= \frac{1}{6}\int_0^2 \left(x^2 y + \frac{y^3}{3}\right)\Big|_0^3\,dx$$

$$= \frac{1}{6}\int_0^2 (3x^2 + 9)\,dx$$

$$= \frac{1}{6}(x^3 + 9x)\Big|_0^2$$

$$= \frac{1}{6}(8 + 18 - 0) = \frac{1}{6}\cdot 26$$

$$= \frac{13}{3}.$$

67. $f(x,y) = e^{2x+y}$; $1 \le x \le 2$, $2 \le y \le 3$

The area of region R is

$$A = (2 - 1)(3 - 2) = 1.$$

The average value of f over R is

$$\frac{1}{A}\int_R\!\!\int e^{2x+y}\,dy\,dx = \int_R\!\!\int e^{2x+y}\,dy\,dx$$

$$= \int_1^2 \int_2^3 e^{2x+y}\,dy\,dx$$

$$= \int_1^2 \left(e^{2x+y}\right)\Big|_2^3\,dx$$

$$= \int_1^2 \left(e^{2x+3} - e^{2x+2}\right)dx$$

$$= \frac{1}{2}(e^{2x+3} - e^{2x+2})\Big|_1^2$$

$$= \frac{1}{2}(e^{4+3} - e^{2+3} - e^{4+2} + e^4)$$

$$= \frac{e^7 - e^6 - e^5 + e^4}{2}.$$

69. $C(x,y) = \dfrac{1}{9}x^2 + 2x + y^2 + 5y + 100$

$$\text{Area} = (75 - 48)(60 - 20) = 27(40) = 1,080$$

The average cost is

$$\frac{1}{1,080}\int_R\!\!\int \left(\frac{1}{9}x^2 + 2x + y^2 + 5y + 100\right)dx\,dy$$

$$= \frac{1}{1,080}\int_{48}^{75}\int_{20}^{60}\left[\frac{1}{9}x^2 + 2x + 100 + y^2 + 5y\right]dy\,dx$$

$$= \frac{1}{1,080}\int_{48}^{75}\left[\left(\frac{1}{9}x^2 + 2x + 100\right)y + \frac{y^3}{3} + \frac{5}{2}y^2\right]\Big|_{20}^{60}\,dx$$

$$= \frac{1}{1,080}\int_{48}^{75}\left[\left(\frac{1}{9}x^2 + 2x + 100\right)(60 - 20)\right.$$

$$\left. + \frac{1}{3}(60^3 - 20^3) + \frac{5}{2}(60^2 - 20^2)\right]dx$$

$$= \frac{40}{1,080}\left(\frac{1}{27}x^3 + x^2 + 100x\right)\Big|_{48}^{75}$$

$$+ \frac{1}{1,080}\left[\frac{208,000}{3} + \frac{5}{2}(3,200)\right]x\Big|_{48}^{75}$$

$$= \frac{40}{1,080}\cdot\left[\frac{1}{27}(75^3 - 48^3) + 75^2 - 48^2 + 100(75 - 48)\right]$$

$$+ \frac{1}{1,080}\left(\frac{208,000}{3} + 8,000\right)(75 - 48)$$

$$= \frac{40}{1,080}\left(\frac{311,283}{27} + 3,321 + 2,700\right) + \frac{1}{1,080}\left(\frac{232,000}{3}\right) 27$$

$$= \frac{40}{1,080}(11,529 + 6,021) + \frac{23,200}{12} \approx \$2,583.$$

71. $R = q_1 p_1 + q_2 p_2$ where $q_1 = 300 - 2p_1$,
$q_2 = 500 - 1.2p_2$, $25 \le p_1 \le 50$, and
$50 \le p_2 \le 75$.

$$A = 25 \cdot 25 = 625$$
$$R = (300 - 2p_1)p_1 + (500 - 1.2p_2)p_2$$
$$R = 300p_1 - 2p_1^2 + 500p_2 - 1.2p_2^2$$

Average Revenue:

$$\frac{1}{625}\int_{25}^{50}\int_{50}^{75}(300p_1 - 2p_1^2 + 500p_2 - 1.2p_2^2)\,dp_2\,dp_1$$

$$= \frac{1}{625}\int_{25}^{50}(300p_1p_2 - 2p_1^2p_2 + 250p_2^2 - 0.4p_2^3)\Big|_{50}^{75}\,dp_1$$

$$= \frac{1}{625}\int_{25}^{50}(22,500p_1 - 150p_1^2 + 1,406,250 + 50,000)dp$$

$$-168,750 - 15,000p_1 + 100p_1^2 - 625,000$$

$$= \frac{1}{625}\int_{25}^{50}(662,500 + 7,500p_1 - 50p_1^2)\,dp_1$$

$$= \frac{1}{625}\left(662,500p_1 + 3,750p_1^2 - \frac{50p_1^3}{3}\right)\Big|_{25}^{50}$$

$$= \frac{1}{625}\left(33,125,000 + 9,375,000 - \frac{6,250,000}{3}\right.$$

$$\left. - 16,562,500 - 2,343,750 + \frac{781,250}{3}\right)$$

$$\approx \$34,833$$

Chapter 9 Review Exercises

5. $f(x,y) = 3x^2y^2 - 5x + 2y$

$$f(-1,2) = 12 + 5 + 4 = 21$$
$$f(6,-3) = 972 - 30 - 6 = 936$$

7. $f(x,y) = \dfrac{\sqrt{x^2 + y^2}}{x - y}$

$$f(-1,2) = \frac{\sqrt{1+4}}{-1-2} = -\frac{\sqrt{5}}{3}$$

$$f(6,-3) = \frac{\sqrt{36+9}}{6+3} = \frac{\sqrt{45}}{9} = \frac{\sqrt{5}}{3}$$

For Exercies 9-13, see the answer graphs in the back of the textbook.

9. $x + y + 4z = 8$

x-intercept: $y = 0$, $z = 0$
$$x = 8$$

y-intercept: $x = 0$, $z = 0$
$$y = 8$$

z-intercept: $x = 0$, $y = 0$
$$4z = 8$$
$$z = 2$$

11. $3x + 5z = 15$

No y-intercept
x-intercept: $y = 0$, $z = 0$
$$3x = 15$$
$$x = 5$$

z-intercept: $x = 0$, $y = 0$
$$5z = 15$$
$$z = 3$$

13. $y = 2$

No x-intercept, no z-intercept
The graph is a plane parallel to the xz-plane.

15. $z = f(x,y) = \dfrac{x + y^2}{x - y^2}$

(a) $\dfrac{\partial z}{\partial y} = \dfrac{(x - y^2) \cdot 2y - (x + y^2)(-2y)}{(x - y^2)^2}$

$$= \frac{4xy}{(x - y^2)^2}$$

(b) $\dfrac{\partial z}{\partial x} = \dfrac{(x - y^2) \cdot 1 - (x + y^2) \cdot 1}{(x - y^2)^2}$

$$= \frac{-2y^2}{(x - y^2)^2}$$

$$= -2y^2(x - y^2)^{-2}$$

$$\left(\frac{\partial z}{\partial x}\right)(0,2) = \frac{-8}{(-4)^2} = -\frac{1}{2}$$

(c) $f_{xx}(x,y) = 4y^2(x - y^2)^{-3}$

$$= \frac{4y^2}{(x - y^2)^3}$$

$$f_{xx}(-1,0) = \frac{0}{1} = 0$$

17. $f(x,y) = 6x^5y - 8xy^9$

$f_x(x,y) = 30x^4y - 8y^9$
$f_y(x,y) = 6x^5 - 72xy^8$

19. $f(x,y) = \dfrac{2x + 5y^2}{3x^2 + y^2}$

$f_x(x,y) = \dfrac{(3x^2 + y^2) \cdot 2 - (2x + 5y^2) \cdot 6x}{(3x^2 + y^2)^2}$

$\qquad = \dfrac{2y^2 - 6x^2 - 30xy^2}{(3x^2 + y^2)^2}$

$f_y(x,y) = \dfrac{(3x^2 + y^2) \cdot 10y - (2x + 5y^2) \cdot 2y}{(3x^2 + y^2)^2}$

$\qquad = \dfrac{30x^2y - 4xy}{(3x^2 + y^2)^2}$

21. $f(x,y) = (y - 2)^2 e^{x+2y}$

$f_x(x,y) = (y - 2)^2 e^{x+2y}$
$f_y(x,y) = e^{x+2y} \cdot 2(y - 2) + (y - 2)^2 \cdot 2e^{x+2y}$
$\qquad = 2(y - 2)[1 + (y - 2)]e^{x+2y}$
$\qquad = 2(y - 2)(y - 1)e^{x+2y}$

23. $f(x,y) = \ln|2 - x^2y^3|$

$f_x(x,y) = \dfrac{1}{2 - x^2y^3} \cdot (-2xy^3)$

$\qquad = \dfrac{-2xy^3}{2 - x^2y^3}$

$f_y(x,y) = \dfrac{1}{2 - x^2y^3} \cdot (-3x^2y^2)$

$\qquad = \dfrac{-3x^2y^2}{2 - x^2y^3}$

25. $f(x,y) = x\tan(7x^2 + 4y^2)$

$f_x(x,y) = \tan(7x^2 + 4y^2) + x \cdot \sec^2(7x^2 + 4y^2) \cdot 14x$
$\qquad = \tan(7x^2 + 4y^2) + 14x^2\sec^2(7x^2 + 4y^2)$
$f_y(x,y) = x \cdot \sec^2(7x^2 + 4y^2) \cdot 8y$
$\qquad = 8xy\sec^2(7x^2 + 4y^2)$

27. $f(x,y) = -6xy^4 + x^2y$

$f_x(x,y) = -6y^4 + 2xy$
$f_{xx}(x,y) = 2y$
$f_{xy}(x,y) = -24y^3 + 2x$

29. $f(x,y) = \dfrac{3x + y}{x - 1}$

$f_x(x,y) = \dfrac{(x - 1) \cdot 3 - (3x + y) \cdot 1}{(x - 1)^2}$

$\qquad = \dfrac{-3 - y}{(x - 1)^2} = (-3 - y)(x - 1)^{-2}$

$f_{xx}(x,y) = -2(-3 - y)(x - 1)^{-3}$

$\qquad = \dfrac{2(3 + y)}{(x - 1)^3}$

$f_{xy}(x,y) = \dfrac{-1}{(x - 1)^2}$

31. $f(x,y) = ye^{x^2}$

$f_x(x,y) = 2xye^{x^2}$
$f_{xx}(x,y) = 2xy \cdot 2xe^{x^2} + e^{x^2} \cdot 2y$
$\qquad = 2ye^{x^2}(2x^2 + 1)$
$f_{xy}(x,y) = 2xe^{x^2}$

33. $f(x,y) = \ln|1 + 3xy^2|$

$f_x(x,y) = \dfrac{1}{1 + 3xy^2} \cdot 3y^2$

$\qquad = \dfrac{3y^2}{1 + 3xy^2}$

$\qquad = 3y^2(1 + 3xy^2)^{-1}$

$f_{xx}(x,y) = 3y^2 \cdot (-3y^2)(1 + 3xy^2)^{-2}$

$\qquad = \dfrac{-9y^4}{(1 + 3xy^2)^2}$

$f_{xy}(x,y) = \dfrac{(1 + 3xy^2) \cdot 6y - 3y^2(6xy)}{(1 + 3xy^2)^2}$

$\qquad = \dfrac{6y}{(1 + 3xy^2)^2}$

35. $z = x^2 + y^2 + 9x - 8y + 1$

$z_x(x,y) = 2x + 9, \ z_y(x,y) = 2y - 8$

$2x + 9 = 0$

$x = -\dfrac{9}{2}$

$2y - 8 = 0$
$y = 4$

$z_{xx}(x,y) = 2, \ z_{yy}(x,y) = 2, \ z_{xy}(x,y) = 0$

$D = 2(2) - (0)^2 = 4 > 0$ and $z_{xx}(x,y) > 0$.

Relative minimum at $\left(-\dfrac{9}{2}, 4\right)$

37. $z = x^3 - 8y^2 + 6xy + 4$

$z_x(x, y) = 3x^2 + 6y, \; z_y(x, y) = -16y + 6x$

$$3x^2 + 6y = 0$$
$$x^2 + 2y = 0$$

$$y = -\frac{x^2}{2}$$

$$-16y + 6x = 0$$
$$-8y + 3x = 0$$

Substituting, we have

$$-8\left(-\frac{x^2}{2}\right) + 3x = 0$$

$$4x^2 + 3x = 0$$
$$x(4x + 3) = 0$$

$$x = 0 \quad \text{or} \quad x = -\frac{3}{4}$$

$$y = 0 \quad \text{or} \quad y = -\frac{9}{32}.$$

$z_{xx}(x, y) = 6x, \; z_{yy}(x, y) = -16, \; z_{xy}(x, y) = 6$

$D = 6x(-16) - (6)^2 = -96x - 36$

At $(0, 0)$, $D = -36 < 0$.
Saddle point at $(0, 0)$.

At $\left(-\frac{3}{4}, -\frac{9}{32}\right)$, $D = 36 > 0$ and $z_{xx}(x, y) = -\frac{9}{2} < 0$.

Relative maximum at $\left(-\frac{3}{4}, -\frac{9}{32}\right)$

39. $f(x, y) = 3x^2 + 2xy + 2y^2 - 3x + 2y - 9$

$f_x(x, y) = 6x + 2y - 3, \; f_y(x, y) = 2x + 4y + 2$

$$6x + 2y - 3 = 0$$
$$2x + 4y + 2 = 0$$

$$-12x - 4y + 6 = 0$$
$$2x + 4y + 2 = 0$$
$$\overline{-10x \qquad + 8 = 0}$$

$$x = \frac{4}{5}$$

$$2\left(\frac{4}{5}\right) + 4y + 2 = 0$$

$$4y = -\frac{18}{5}$$

$$y = -\frac{9}{10}$$

$f_{xx}(x, y) = 6, \; f_{yy}(x, y) = 4, \; f_{xy}(x, y) = 2$

$D = 6(4) - (2)^2 = 2 > 0$ and $f_{xx}(x, y) > 0$.

Relative minimum at $\left(\frac{4}{5}, -\frac{9}{10}\right)$

41. $f(x, y) = 7x^2 + y^2 - 3x + 6y - 5xy$

$f_x(x, y) = 14x - 3 - 5y, \; f_y(x, y) = 2y + 6 - 5x$

$$14x - 5y - 3 = 0$$
$$-5x + 2y + 6 = 0$$

$$28x - 10y - 6 = 0$$
$$-25x + 10y + 30 = 0$$
$$\overline{3x \qquad + 24 = 0}$$

$$x = -8$$

$$-5(-8) + 2y + 6 = 0$$
$$2y = -46$$
$$y = -23$$

$f_{xx}(x, y) = 14, \; f_{yy}(x, y) = 2, \; f_{xy}(x, y) = -5$

$D = 14(2) - (-5)^2 = 3 > 0$ and $f_{xx}(x, y) > 0$.

Relative minimum at $(-8, -23)$

43. $z = 2x^2 - 4y^2 + 6xy; \; x = 2, \; y = -3,$
$dx = 0.01, \; dy = 0.05$

$$f_x(x, y) = 4x + 6y$$
$$f_y(x, y) = -8y + 6x$$
$$dz = (4x + 6y)\,dx + (-8y + 6x)\,dy$$

Substitute.

$$dz = [4(2) + 6(-3)](0.01)$$
$$\quad + [-8(-3) + 6(2)](0.05)$$
$$= -0.1 + 1.8 = 1.7$$

45. Let $z = f(x, y) = \sqrt{x^2 + y^2}$.

Then

$$dz = f_x(x, y)\,dx + f_y(x, y)\,dy.$$

$$dz = \frac{1}{2}(x^2 + y^2)^{-1/2}(2x)\,dx$$

$$\quad + \frac{1}{2}(x^2 + y^2)(2y)\,dx$$

$$= \frac{x}{\sqrt{x^2 + y^2}}\,dx + \frac{y}{\sqrt{x^2 + y^2}}\,dy$$

To approximate $\sqrt{5.1^2 + 12.05^2}$, we let $x = 5$, $dx = 0.1$, $y = 12$, and $dy = 0.05$.

Then,

$$dz = \frac{5}{\sqrt{5^2 + 12^2}}(0.1) + \frac{12}{\sqrt{5^2 + 12^2}}(0.05)$$

$$= \frac{5}{13}(0.1) + \frac{12}{13}(0.05)$$

$$\approx 0.0846.$$

Therefore,

$$f(5.1, 12.05) = f(5, 12) + \Delta z$$
$$\approx f(5, 12) + dz$$
$$= \sqrt{5^2 + 12^2} + 0.0846$$
$$f(5.1, 12.05) \approx 13.0846$$

Using a calculator, $\sqrt{5.1^2 + 12.05^2} \approx 13.0848$.
The absolute value of the difference of the two results is $|13.0846 - 13.0848| = 0.0002$.

47. $\int_4^9 \dfrac{6y - 8}{\sqrt{x}}\, dx$

$$= (6y - 8)(2\sqrt{x})\Big|_4^9$$
$$= 2(3y - 4)(2)(\sqrt{9} - \sqrt{4})$$
$$= 4(3y - 4)$$
$$= 12y - 16$$

49. $\int_0^5 \dfrac{6x}{\sqrt{4x^2 + 2y^2}}\, dx$

Let $u = 4x^2 + 2y^2$; then $du = 8x\, dx$.
When $x = 0$, $u = 2y^2$.
When $x = 5$, $u = 100 + 2y^2$.

$$= \frac{3}{4} \int_{2y^2}^{100+2y^2} u^{-1/2}\, du$$
$$= \frac{3}{4}(2u^{1/2})\Big|_{2y^2}^{100+2y^2}$$
$$= \frac{3}{4} \cdot 2[(100 + 2y^2)^{1/2} - (2y^2)^{1/2}]$$
$$= \frac{3}{2}[(100 + 2y^2)^{1/2} - (2y^2)^{1/2}]$$

51. $\int_0^2 \left[\int_0^4 (x^2 y^2 + 5x)\, dx \right] dy$

$$= \int_0^2 \left(\frac{1}{3}x^3 y^2 + \frac{5}{2}x^2 \right)\Big|_0^4 dx$$
$$= \int_0^2 \left(\frac{64}{3}y^2 + 40 \right) dy$$
$$= \left(\frac{64y^3}{9} + 40y \right)\Big|_0^2$$
$$= \frac{64}{9}(8) + 40(2)$$
$$= \frac{512}{9} + \frac{720}{9}$$
$$= \frac{1{,}232}{9}$$

53. $\int_3^4 \left[\int_2^5 \sqrt{6x + 3y}\, dx \right] dy$

$$= \int_3^4 \frac{1}{9}(6x + 3y)^{3/2}\Big|_2^5 dx$$
$$= \int_3^4 \frac{1}{9}[(30 + 3y)^{3/2} - (12 + 3y)^{3/2}]\, dy$$
$$= \frac{1}{3} \cdot \frac{1}{9} \cdot \frac{2}{5}$$
$$\cdot [(30 + 3y)^{5/2} - (12 + 3y)^{5/2}]\Big|_3^4$$
$$= \frac{2}{135}[(42)^{5/2} - (24)^{5/2} - (39)^{5/2} + (21)^{5/2}]$$

55. $\int_2^4 \int_2^4 \dfrac{dx\, dy}{y}$

$$= \int_2^4 \left(\frac{1}{y}x \right)\Big|_2^4 dy$$
$$= \int_2^4 \left[\frac{1}{y}(4 - 2) \right] dy$$
$$= 2 \ln |y| \Big|_2^4$$
$$= 2 \ln \left| \frac{4}{2} \right|$$
$$= 2 \ln 2 \text{ or } \ln 4$$

57. $\displaystyle\int\int_R (x^2 + y^2)\, dx\, dy;\ 0 \le x \le 2,\ 0 \le y \le 3$

$$\int\int_R (x^2 + y^2)\, dx\, dy$$
$$= \int_0^3 \int_0^2 (x^2 + y^2)\, dx\, dy$$
$$= \int_0^3 \left(\frac{x^3}{3} + y^2 x \right)\Big|_0^2 dy$$
$$= \int_0^3 \left(\frac{8}{3} + 2y^2 \right) dy$$
$$= \left(\frac{8}{3}y + \frac{2}{3}y^3 \right)\Big|_0^3$$
$$= 8 + 18 = 26$$

59. $\iint_R \sqrt{y+x}\,dx\,dy;\ 0 \le x \le 7,\ 1 \le y \le 9$

$$\iint_R \sqrt{y+x}\,dx\,dy$$

$$= \int_0^7 \int_1^9 \sqrt{y+x}\,dy\,dx$$

$$= \int_0^7 \left[\frac{2}{3}(y+x)^{3/2}\right]\Big|_1^9 dx$$

$$= \int_0^7 \frac{2}{3}[(9+x)^{3/2} - (1+x)^{3/2}]\,dx$$

$$= \frac{2}{3}\cdot\frac{2}{5}[(9+x)^{5/2} - (1+x)^{5/2}]\Big|_0^7$$

$$= \frac{4}{15}[(16)^{5/2} - (8)^{5/2} - (9)^{5/2} + (1)^{5/2}]$$

$$= \frac{4}{15}[4^5 - (2\sqrt{2})^5 - 3^5 + 1]$$

$$= \frac{4}{15}(1{,}024 - 32(4\sqrt{2}) - 243 + 1)$$

$$= \frac{4}{15}(782 - 128\sqrt{2})$$

$$= \frac{4}{15}(782 - 8^{5/2})$$

61. $\iint_R xy\sin(xy^2)\,dy\,dx;\ 0 \le x \le \frac{\pi}{2},$

$\quad 0 \le y \le 1$

$$\iint_R xy\sin(xy^2)\,dy\,dx$$

$$= \int_0^{\pi/2} \int_0^1 xy\sin(xy^2)\,dy\,dx$$

$$= \int_0^{\pi/2} \frac{1}{2}[-\cos(xy^2)]\Big|_0^1 dx$$

$$= -\frac{1}{2}\int_0^{\pi/2} (\cos x - 1)\,dx$$

$$= -\frac{1}{2}(\sin x - x)\Big|_0^{\pi/2}$$

$$= -\frac{1}{2}\left[\left(1 - \frac{\pi}{2}\right) - 0\right]$$

$$= -\frac{1}{2} + \frac{\pi}{4}$$

63. $z = x + 9y + 8;\ 1 \le x \le 6,\ 0 \le y \le 8$

$$V = \iint_R (x+9y+8)\,dx\,dy$$

$$= \int_0^8 \int_1^6 (x+9y+8)\,dx\,dy$$

$$= \int_0^8 \left[\frac{x^2}{2} + (9y+8)x\,dx\right]\Big|_1^6 dy$$

$$= \int_0^8 \left[\frac{36-1}{2} + (9y+8)(6-1)\right] dy$$

$$= \left[\frac{35}{2}y + 5\left(\frac{9}{2}y^2 + 8y\right)\right]\Big|_0^8$$

$$= \frac{35}{2}(8) + 5\left[\left(\frac{9}{2}\right)(64) + 8\cdot 8\right]$$

$$= \frac{35(8)}{2} + 5\left(\frac{9(64)}{2} + 8\cdot 8\right)$$

$$= 140 + 5(352)$$

$$= 1{,}900$$

65. $\int_0^1 \int_0^{2x} xy\,dy\,dx$

$$= \int_0^1 \left(\frac{xy^2}{2}\right)\Big|_0^{2x} dx$$

$$= \int_0^1 \frac{x}{2}(4x^2 - 0)\,dx$$

$$= \int_0^1 2x^3\,dx$$

$$= \left(\frac{1}{2}x^4\right)\Big|_0^1 = \frac{1}{2}$$

67. $\int_0^1 \int_{x^2}^x x^3 y\,dy\,dx$

$$\int_0^1 \left(\frac{x^3}{2}y^2\right)\Big|_{x^2}^x dx$$

$$= \int_0^1 \frac{x^3}{2}(x^2 - x^4)\,dx$$

$$= \frac{1}{2}\int_0^1 (x^5 - x^7)\,dx$$

$$= \frac{1}{2}\left(\frac{x^6}{6} - \frac{x^8}{8}\right)\Big|_0^1$$

$$= \frac{1}{2}\left(\frac{1}{6} - \frac{1}{8}\right) = \frac{1}{2}\cdot\frac{1}{24} = \frac{1}{48}$$

69. $\displaystyle\int_0^2 \int_{x/2}^1 \frac{1}{y^2+1}\,dy\,dx$

Change the order of integration.

$$\int_0^2 \int_{x/2}^1 \frac{1}{y^2+1}\,dy\,dx$$

$$= \int_0^1 \int_0^{2y} \frac{1}{y^2+1}\,dx\,dy$$

$$= \int_0^1 \frac{x}{y^2+1}\bigg|_0^{2y}\,dy$$

$$= \int_0^1 \left[\frac{1}{y^2+1}(2y) - \frac{1}{y^2+1}(0)\right]\,dy$$

$$= \int_0^1 \frac{2y}{y^2+1}\,dy$$

$$= \ln(y^2+1)\bigg|_0^1$$

$$= \ln 2 - \ln 1$$

$$= \ln 2 - 0 = \ln 2$$

71. $\displaystyle\iint_R (2x+3y)\,dx\,dy;\ 0 \le y \le 1,$

$y \le x \le 2-y$

$$\int_0^1 \int_y^{2-y} (2x+3y)\,dx\,dy$$

$$= \int_0^1 (x^2+3xy)\bigg|_y^{2-y}\,dy$$

$$= \int_0^1 [(2-y)^2 - y^2 + 3y(2-y-y)]\,dy$$

$$= \int_0^1 (4-4y+y^2-y^2+6y-6y^2)\,dy$$

$$= \int_0^1 (4+2y-6y^2)\,dy$$

$$= (4y+y^2-2y^3)\bigg|_0^1$$

$$= 4+1-2 = 3$$

73. Assume that blood vessels are cylindrical.

$V = \pi r^2 h$

$r = 0.7,\ h = 2.7$

$dr = dh = \pm 0.1$

$dV = 2\pi rh\,dr + \pi r^2\,dh$

$\quad = 2\pi(0.7)(2.7)(\pm 0.1) + \pi(0.7)^2(\pm 0.1)$

$\quad \approx \pm 1.3$

The possible error is 1.3 cm^3.

75. $L(w,t) = (0.00082t + 0.0955)e^{(\ln w + 10.49)/2.842}$

(a) $L(450,4) = [0.00082(4) + 0.0955]e^{(\ln(450)+10.49)/2.84}$

$\quad\quad \approx 33.982$

The length is about 33.98 cm.

(b) $L_w(w,t) = (0.00082t + 0.0955)e^{(\ln w + 10.49)/2.842}$

$\quad\quad\quad \cdot \dfrac{1}{2.842w}$

$L_w(450,7) \approx 0.027$

The approximate change in the length of a trout if its weight increases from 450 to 451 g while age is held constant at 7 yr is 0.027 cm.

$$L_t(w,t) = 0.00082e^{(\ln w + 10.49)/2.842}$$
$$L_t(450,7) \approx 0.28$$

The approximate change in the length of a trout if its age increases from 7 to 8 yr while weight is held constant at 450 g is 0.28 cm.

77. $f(a,b) = \dfrac{1}{4}b\sqrt{4a^2-b^2}$

(a) $f(3,2) = \dfrac{1}{4}(2)\sqrt{4(3)^2-2^2}$

$\quad\quad = \dfrac{1}{2}\sqrt{32}$

$\quad\quad = 2\sqrt{2}$

$\quad\quad \approx 2.83$

The area of the bottom of the planter is approximately 2.83 ft^2.

(b) $A = \dfrac{1}{4}b\sqrt{4a^2-b^2}$

$dA = \dfrac{1}{4}b \cdot \dfrac{1}{2}(4a^2-b^2)^{-1/2}(8a)da$

$\quad\quad + \left[\dfrac{1}{4}b \cdot \dfrac{1}{2}(4a^2-b^2)^{-1/2}(-2b)\right.$

$\quad\quad \left. + \dfrac{1}{4}(4a^2-b^2)^{1/2}\right]db$

$dA = \dfrac{ab}{\sqrt{4a^2-b^2}}da + \dfrac{1}{4}\left(\dfrac{-b^2}{\sqrt{4a^2-b^2}} + \sqrt{4a^2-b^2}\right)$

If $a=3$, $b=2$, $da=0$, and $db=0.5$,

$$dA = \dfrac{1}{4}\left(\dfrac{-2^2}{\sqrt{4(3)^2-2^2}} + \sqrt{4(3)^2-2^2}\right)(0.5)$$

$dA \approx 0.6187.$

The approximate effect on the area is an increase of 0.6187 ft^2.

79. Maximize $f(x,y) = xy$, subject to $2x + y = 400$.

1. $g(x,y) = 2x + y - 400$

2. $F(x,y,\lambda) = xy + \lambda(2x + y - 400)$

3. $F_x = y + 2\lambda$
$F_y = x + \lambda$
$F_\lambda = 2x + y - 400$

4. $\quad y + 2\lambda = 0$
$\quad\quad x + \lambda = 0$
$2x + y - 400 = 0$

5. $\lambda = -\dfrac{y}{2}, \; \lambda = -x$

$$-\dfrac{y}{2} = -x$$

$$y = 2x$$

Substituting into $2x + y - 400$, we have

$$2x + 2x - 400 = 0,$$

so $\quad x = 100, \; y = 200.$

Dimensions are 100 feet by 200 feet for maximum area of 20,000 ft^2.

81. $c(x,y) = 2x + y^2 + 4xy + 25$

(a) $c_x = 2 + 4y$
$c_x(640, 6) = 2 + 4(6)$
$\quad\quad\quad\quad = 26$

For an additional 1 Mb of memory, the approximate change in cost is $26.

(b) $c_y = 2y + 4x$
$c_y(640, 6) = 2(6) + 4(640)$
$\quad\quad\quad\quad = 2,572$

For an additional hour of labor, the approximate change in cost is $2,572.

83. $C(x,y) = \ln(x^2 + y) + e^{xy/20}$
$x = 15, \; y = 9, \; dx = 1, \; dy = -1$

$$dC = \left(\frac{2x}{x^2 + y} + \frac{y}{20}e^{xy/20}\right)dx$$

$$+ \left(\frac{1}{x^2 + y} + \frac{x}{20}e^{xy/20}\right)dy$$

$dC(15, 9)$

$$= \left(\frac{2(15)}{15^2 + 9} + \frac{9}{20}e^{(15)(9)/20}\right)(1)$$

$$+ \left(\frac{1}{15^2 + 9} + \frac{15}{20}e^{(9)(15)/20}\right)(-1)$$

$$= \frac{29}{234} - \frac{3}{10}e^{27/4}$$

$$= -256.10$$

Costs decrease by $256.10.

85. $V = \dfrac{4}{3}\pi r^3, \; r = 2$ ft,

$$dr = 1 \text{ in} = \frac{1}{12} \text{ ft}$$

$$dV = 4\pi r^2 \, dr = 4\pi(2)^2 \left(\frac{1}{12}\right) \approx 4.19 \text{ ft}^3$$

87. $P(x,y) = 0.01(-x^2 + 3xy + 160x - 5y^2$
$\quad\quad\quad\quad + 200y + 2,600)$

with $x + y = 280$.

(a) $y = 280 - x$

$P(x) = 0.01[-x^2 + 3x(280 - x) + 160x$
$\quad\quad - 5(280 - x)^2 + 200(280 - x)$
$\quad\quad + 2,600]$
$\quad = 0.01(-x^2 + 840x - 3x^2 + 160x$
$\quad\quad - 392,000 + 2,800x - 5x^2$
$\quad\quad + 56,000 - 200x + 2,600)$
$P(x) = 0.01(-9x^2 + 3,600x - 333,400)$

$P'(x) = 0.01(-18x + 3,600)$
$\quad 0.01(-18x + 3,600) = 0$
$\quad\quad\quad\quad -18x = -3,600$
$\quad\quad\quad\quad\quad\quad x = 200$

If $x < 200$, $P'(x) > 0$, and if $x > 200$, $P'(x) < 0$. Therefore, P is maximum when $x = 200$. If $x = 200$, $y = 80$.

$P(200, 80)$
$\quad = 0.01[-200^2 + 3(200)(80) + 160(200)$
$\quad\quad - 5(80)^2 + 200(80) + 2,600]$
$\quad = 0.01(26,600)$
$\quad = 266$

Thus, $200 spent on fertilizer and $80 spent on seed will produce a maximum profit of $266 per acre.

(b) $P(x, y) = 0.01(-x^2 + 3xy + 160x - 5y^2$
$$+ 200y + 2,600)$$
$$P_x(x, y) = 0.01(-2x + 3y + 160)$$
$$P_y(x, y) = 0.01(3x - 10y + 200)$$
$$0.01(-2x + 3y + 160) = 0$$
$$0.01(3x - 10y + 200) = 0$$

These equations simplify to

$$-2x + 3y = -160$$
$$3x - 10y = -200.$$

Solve this system.

$$-6x + 9y = -480$$
$$\underline{6x - 20y = -400}$$
$$-11y = -880$$
$$y = 80$$

If $y = 80$,

$$3x - 10(80) = -200$$
$$3x = 600$$
$$x = 200.$$

$$P_{xx}(x, y) = 0.01(-2) = -0.02$$
$$P_{yy}(x, y) = 0.01(-10) = -0.1$$
$$P_{xy}(x, y) = 0$$

For $(200, 80)$,

$$D = (-0.02)(-0.1) - 0^2 = 0.002 > 0, \text{ and}$$

$P_{xx} < 0$, so there is a relative maximum at $(200, 80)$.

$P(200, 80) = 266$, as in part (a). Thus, \$200 spent on fertilizer and \$80 spent on seed will produce a maximum profit of \$266 per acre.

(c) Maximize $P(x, y)$
$$= 0.01(-x^2 + 3xy + 160x - 5y^2$$
$$+ 200y + 2,600)$$
subject to $x + y = 280$.

1. $g(x, y) = x + y - 280$

2. $F(x, y, \lambda)$
$$= 0.01(-x^2 + 3xy + 160x - 5y^2$$
$$+ 200y + 2,600) + \lambda(x + y - 280)$$

3. $F_x = 0.01(-2x + 3y + 160) + \lambda$
$F_y = 0.01(3x - 10y + 200) + \lambda$
$F_\lambda = x + y - 280$

4. $0.01(-2x + 3y + 160) + \lambda = 0$ (1)
$0.01(3x - 10y + 200) + \lambda = 0$ (2)
$x + y - 280 = 0$ (3)

5. Equations (1) and (2) give

$$0.01(-2x + 3y + 160) = 0.01(3x - 10y + 200)$$
$$-2x + 3y + 160 = 3x - 10y + 200$$
$$-5x + 13y = 40.$$

Multiplying equation (3) by 5 gives

$$5x + 5y - 1,400 = 0.$$

$$-5x + 13y = 40$$
$$\underline{5x + 5y = 1,400}$$
$$18y = 1,440$$
$$y = 80$$

If $y = 80$,

$$5x + 5(80) = 1,400$$
$$5x = 1,000$$
$$x = 200.$$

Thus, $P(200, 80)$ is a maximum. As before, $P(200, 80) = 266$.

Thus, \$200 spent on fertilizer and \$80 spent on seed will produce a maximum profit of \$266 per acre.

Chapter 9 Test

[9.1]

1. Find $f(-2, 1)$ for $f(x, y) = 2x^2 - 4xy + 7y$.

2. Find $g(-1, 3)$ for $g(x, y) = \sqrt{2x^2 - xy^2}$.

3. Complete the ordered triples $(0, 0, \quad)$, $(0, \quad, 0)$ and $(\quad, 0, 0)$ for the plane $2x - 4y + 8z = 8$.

4. Graph the first octant portion of the plane $4x + 2y + 8z = 8$.

5. Let $f(x, y) = x^2 + 2y^2$. Find $\dfrac{f(x + h, y) - f(x, y)}{h}$.

6. Let $f(x, y) = xy \sin\left(\dfrac{\pi}{y^2}\right)$. Find the following.

   (a) $f(1, 2)$ (b) $f(\sqrt{2}, \sqrt{2})$ (c) $f(2, -1)$

7. **Snake River Salmon** Researchers have determined that the date D (day of the year) in which a Chinook Salmon passes the Lower Granite Dam is related to the day of release R (day of the year) of the tagged salmon, the length of the salmon L (in mm), and the mean daily flow of the water F (in m³/s) such that

$$\ln(D) = 4.890179 + 0.005163R - 0.004345L - 0.000019F.$$

   (a) Solve this expression for D.

   (b) Estimate the date of passage for a 68 mm salmon that was released on the day 125 with an average flow rate of 4,200 m³/s.

[9.2]

8. Let $z = f(x, y) = 3x^3 - 5x^2y + 4y^2$. Find each of the following.

   (a) $\dfrac{\partial z}{\partial x}$ (b) $\dfrac{\partial z}{\partial y}(1, 1)$ (c) f_{xx}

9. Let $f(x, y) = \dfrac{x}{2x + y^2}$. Find each of the following.

   (a) $f_x(1, -1)$ (b) $f_y(2, 1)$

10. Let $f(x, y) = \sqrt{2x^2 - y^2}$. Find each of the following.

   (a) f_x (b) f_y (c) f_{xy}

---

* Conner, W., R. Steinhorst, and H. Burge, "Forecasting Survival and Passage of Migratory Juvenile Salmonids," *North American Journal of Fisheries Management*, Vol. 20, 2000, pp. 651-660.

11. Let $f(x, y) = xe^{y^2}$. Find each of the following.

   (a) f_x **(b)** f_y **(c)** f_{yx}

12. Let $f(x, y) = \ln(2x^2y^2 + 1)$. Find each of the following.

   (a) f_{xx} **(b)** f_{xy} **(c)** f_{yy}

13. The production function for a certain country is

$$z = 2x^5y^4,$$

where x represents the amount of labor and y the amount of capital. Find the marginal productivity of

   (a) labor;   **(b)** capital.

14. Let $f(x, y) = x^2 \sin(x + y^2)$. Find the following.

   (a) f_x **(b)** f_y **(c)** f_{xx} **(d)** f_{xy}

15. Total Body Water   Accurate prediction of total body water is critical in determining adequate dialysis doses for patients with renal disease. For white males, total body water can be estimated by the function

$$T(A, W, B) = 23.04 - 0.03A + 0.50W - 0.62B,$$

where T is the total body water (in liters), A is age (in yr), W is weight (in kg), and B is the body mass index.* Find T_A and T_W and interpret each.

[9.3]

Find all points where the functions defined below have any relative extrema. Find any saddle points.

16. $z = 4x^2 + 2y^2 - 8x$ **17.** $z = 2x^2 - 4xy + y^3$ **18.** $z = 1 - 2y - x^2 - 2xy - 2y^2$

19. Find all points where $f(x, y) = x \sin y$ has any relative extrema. Identify any saddle points.

20. A company manufactures two calculator models. The total revenue from x thousand solar calculators and y thousand battery-operated calculators is

$$R(x, y) = 4{,}000 - 5y^2 - 8x^2 - 2xy + 42y + 102x.$$

Find x and y so that revenue is maximized.

---

* Churnlea, W., S. Guo, C. Zellar, et. al., "Total Body Water Reference Values and Prediction Equations for Adults," *Kidney International*, Vol. 59, 2001, pp. 2250-2258.

[9.4]

21. Use the total differential to approximate the quantity $\sqrt{3.05^2 + 4.02^2}$. Then use a calculator to approximate the quantity and give the absolute value of the difference in the two results to four decimal places.

22. (a) Find dz for $z = e^{x+y} \ln xy$.

    (b) Evaluate dz when $x = 1$, $y = 1$, $dx = 0.02$ and $dy = 0.01$.

23. A sphere of radius 3 feet is to receive an insulating coating $\frac{1}{2}$ inch thick. Use differentials to approximate the volume of the coating needed.

24. Evaluate dz for $z = \cos(x+y) - \sin(x-y)$, $x = \dfrac{\pi}{4}$, $y = \dfrac{\pi}{12}$, $dx = 0.05$, and $dy = 0.02$.

25. **Horn Volume** The volume of the horns from bighorn sheep were estimated by researchers using the equation
$$V = \frac{h\pi}{3}\left(r_1^2 + r_1 r_2 + r_2^2\right),$$
where h is the length of a horn segment (in cm) and r_1 and r_2 are the radii of the two ends of the horn segment (in cm).*

    (a) Determine the volume of a segment of horn that is 50 cm long with radii of 6 and 4 cm respectively.

    (b) Use the total differential to estimate the volume of the segment of horn if the horn segment from part a was actually 52 cm long with radii of 5.9 cm and 4.2 cm, respectively. Compare this with the actual volume.

[9.5]

26. Evaluate $\displaystyle\int_0^4 \int_1^4 \sqrt{x + 3y}\, dx\, dy$.

27. Find the volume under the surface $z = 2xy$ and above the rectangle with boundaries $0 \le x \le 1$, $0 \le y \le 4$.

28. Evaluate $\displaystyle\int_0^2 \int_{y/2}^3 (x+y)\, dx\, dy$.

29. Use the region R with boundaries $0 \le y \le 2$ and $0 \le x \le y$ to evaluate
$$\iint_R (6 - x - y)\, dx\, dy.$$

30. Evaluate the integral $\displaystyle\int_0^{\pi/4} \int_0^{\pi/4} \sin(x+y)\, dx\, dy$.

---

* Fitzsimmons, N., S. Buskirk, and M. Smith, "Population History, Genetic Variability, and Horn Growth in Bighorn Sheep," *Conservation Biology*, No. 9, No. 2, April 1995, pp. 314-323.

31. Evaluate the integral

$$\iint_R (\cos x) e^{y + \sin x} \, dx \, dy$$

on the region R with boundaries $0 \le x \le \dfrac{\pi}{2}$, $0 \le y \le 1$.

Chapter 9 Test Answers

1. 23

2. $\sqrt{11}$

3. $(0,0,1)$, $(0,-2,0)$, $(4,0,0)$

4.

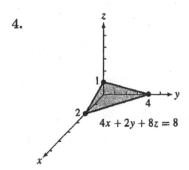

$4x + 2y + 8z = 8$

5. $2x + h$

6. (a) $\sqrt{2}$ (b) 2 (c) 0

7. (a) $D \approx 132.977375e^{0.005163R - 0.004345L - 0.000019F}$

(b) About day 174.

8. (a) $9x^2 - 10xy$ (b) 3 (c) $18x - 10y$

9. (a) $\frac{1}{9}$ (b) $\frac{-4}{25}$

10. (a) $\frac{2x}{\sqrt{2x^2-y^2}}$ (b) $\frac{-y}{\sqrt{2x^2-y^2}}$ (c) $\frac{2xy}{(2x^2-y^2)^{3/2}}$

11. (a) e^{y^2} (b) $2xye^{y^2}$ (c) $2ye^{y^2}$

12. (a) $\frac{-8x^2y^4+4y^2}{(2x^2y^2+1)^2}$ (b) $\frac{8xy}{(2x^2y^2+1)^2}$ (c) $\frac{4x^2-8x^4y^2}{(2x^2y^2+1)^2}$

13. (a) $10x^4y^4$ (b) $8x^5y^3$

14. (a) $f_x(x,y) = x^2\cos(x+y^2) + 2x\sin(x+y^2)$

(b) $f_y(x,y) = 2x^2y\cos(x+y^2)$

(c) $f_{xx}(x,y) = -x^2\sin(x+y^2) + 4x\cos(x+y^2) + 2\sin(x+y^2)$

(d) $f_{xy}(x,y) = -2x^2y\sin(x+y^2) + 4xy\cos(x+y^2)$

15. $T_A = -0.03$; the approximate change in total body water if age is increased by one year while weight and the body mass index are held constant is a decrease of approximately 0.03 liters.

$T_W = 0.50$; the approximate change in total body water if weight is increased by one kg while age and the body mass index are held constant is an increase of approximately 0.50 liters.

16. Relative minimum of -4 at $(1,0)$

17. Relative minimum of $-\frac{32}{27}$ at $\left(\frac{4}{3}, \frac{4}{3}\right)$, saddle point at $(0,0)$

18. Relative maximum of 2 at $(1,-1)$

19. No relative extrema; saddle points at $(0,0)$, $(0,\pm\pi)$, $(0,\pm 2\pi)$, etc.

20. 6 thousand solar calculators and 3 thousand battery-operated calculators

21. 5.046; 5.0461; 0.0001

22. (a) $\left(e^{x+y}\ln(xy) + \frac{e^{x+y}}{x}\right)dx + \left(e^{x+y}\ln xy + \frac{e^{x+y}}{y}\right)dy$ (b) $0.03e^2 \approx 0.222$

23. 4.71 cu ft

24. $-\frac{\sqrt{3}}{20} \approx -0.0866$

25. (a) 3,979.4 cm^3 (b) 4,201.4 cm^3; 4,205.5 cm^3

26. $\frac{4}{45}\left(993 - 13^{5/2}\right)$ **27.** 8   **28.** $\frac{40}{3}$

29. 8   **30.** $\sqrt{2} - 1$ **31.** $e^2 - 2e + 1 = (e-1)^2$

MATRICES

10.1 Solution of Linear Systems

1. $2x + 3y = 11$
$\quad x + 2y = 8$

The equations are already in proper form. The augmented matrix obtained from the coefficients and the constants is

$$\begin{bmatrix} 2 & 3 & | & 11 \\ 1 & 2 & | & 8 \end{bmatrix}.$$

3. $2x + y + z = 3$
$\quad 3x - 4y + 2z = -7$
$\quadx + y + z = 2$

leads to the augmented matrix

$$\begin{bmatrix} 2 & 1 & 1 & | & 3 \\ 3 & -4 & 2 & | & -7 \\ 1 & 1 & 1 & | & 2 \end{bmatrix}.$$

5. We are given the augmented matrix

$$\begin{bmatrix} 1 & 0 & | & 2 \\ 0 & 1 & | & 3 \end{bmatrix}.$$

This is equivalent to the system of equations

$$x = 2$$
$$ y = 3,$$

or $x = 2$, $y = 3$.

7. $\begin{bmatrix} 1 & 0 & 0 & | & 2 \\ 0 & 1 & 0 & | & 3 \\ 0 & 0 & 1 & | & -2 \end{bmatrix}$

The system associated with this matrix is

$$x = 2$$
$$ y = 3$$
$$ z = -2,$$

or $x = 2$, $y = 3$, $z = -2$.

9. *Row operations* on a matrix correspond to transformations of a system of equations.

11. $\begin{bmatrix} 2 & 3 & 8 & | & 20 \\ 1 & 4 & 6 & | & 12 \\ 0 & 3 & 5 & | & 10 \end{bmatrix}$

Find $R_1 + (-2)R_2$. In row 2, column 1,

$$2 + (-2)1 = 0.$$

In row 2, column 2,

$$3 + (-2)4 = -5.$$

In row 2, column 3,

$$8 + (-2)6 = -4.$$

In row 2, column 4,

$$20 + (-2)12 = -4$$

Replace R_2 with these values. The new matrix is

$$\begin{bmatrix} 2 & 3 & 8 & | & 20 \\ 0 & -5 & -4 & | & -4 \\ 0 & 3 & 5 & | & 10 \end{bmatrix}.$$

13. $\begin{bmatrix} 1 & 4 & 2 & | & 9 \\ 0 & 1 & 5 & | & 14 \\ 0 & 3 & 8 & | & 16 \end{bmatrix}$

$-4R_2 + R_1 \rightarrow R_1$

$$\begin{bmatrix} -4(0)+1 & -4(1)+4 & -4(5)+2 & | & -4(14)+9 \\ 0 & 1 & 5 & | & 14 \\ 0 & 3 & 8 & | & 16 \end{bmatrix}$$

$$= \begin{bmatrix} 1 & 0 & -18 & | & -47 \\ 0 & 1 & 5 & | & 14 \\ 0 & 3 & 8 & | & 16 \end{bmatrix}$$

15. $\begin{bmatrix} 3 & 0 & 0 & | & 18 \\ 0 & 5 & 0 & | & 9 \\ 0 & 0 & 4 & | & 8 \end{bmatrix}$

$\frac{1}{3}R_1 \to R_1$

$\begin{bmatrix} \frac{1}{3}(3) & \frac{1}{3}(0) & \frac{1}{3}(0) & | & \frac{1}{3}(18) \\ 0 & 5 & 0 & | & 9 \\ 0 & 0 & 4 & | & 8 \end{bmatrix} = \begin{bmatrix} 1 & 0 & 0 & | & 6 \\ 0 & 5 & 0 & | & 9 \\ 0 & 0 & 4 & | & 8 \end{bmatrix}$

17. $x + y = 5$
$x - y = -1$

has augmented matrix

$\begin{bmatrix} 1 & 1 & | & 5 \\ 1 & -1 & | & -1 \end{bmatrix}.$

Use row operations as follows.

$-1R_1 + R_2 \to R_2 \quad \begin{bmatrix} 1 & 1 & | & 5 \\ 0 & -2 & | & -6 \end{bmatrix}$

$-\frac{1}{2}R_2 \to R_2 \quad \begin{bmatrix} 1 & 1 & | & 5 \\ 0 & 1 & | & 3 \end{bmatrix}$

$-1R_2 + R_1 \to R_1 \quad \begin{bmatrix} 1 & 0 & | & 2 \\ 0 & 1 & | & 3 \end{bmatrix}$

Read the solution from the last column of the matrix. The solution is $(2, 3)$.

19. $2x - 5y = 10$
$4x - 5y = 15$

Write the augmented matrix and use row operations.

$\begin{bmatrix} 2 & -5 & | & 10 \\ 4 & -5 & | & 15 \end{bmatrix}$

$-2R_1 + R_2 \to R_2 \quad \begin{bmatrix} 2 & -5 & | & 10 \\ 0 & 5 & | & -5 \end{bmatrix}$

$R_1 + R_2 \to R_1 \quad \begin{bmatrix} 2 & 0 & | & 5 \\ 0 & 5 & | & -5 \end{bmatrix}$

$\frac{1}{2}R_1 \to R_1 \quad \begin{bmatrix} 1 & 0 & | & \frac{5}{2} \\ 0 & 1 & | & -1 \end{bmatrix}$
$\frac{1}{5}R_2 \to R_2$

The solution is $\left(\frac{5}{2}, -1\right)$.

21. $2x - 3y = 2$
$4x - 6y = 1$

Write the augmented matrix and use row operations.

$\begin{bmatrix} 2 & -3 & | & 2 \\ 4 & -6 & | & 1 \end{bmatrix}$

$-2R_1 + R_2 \to R_2 \quad \begin{bmatrix} 2 & -3 & | & 2 \\ 0 & 0 & | & -3 \end{bmatrix}$

The system associated with the last matrix is

$2x - 3y = 2$
$0x + 0y = -3.$

Since the second equation, $0 = -3$, is false, the system is inconsistent and therefore has no solution.

23. $6x - 3y = 1$
$-12x + 6y = -2$

Write the augmented matrix of the system and use row operations.

$\begin{bmatrix} 6 & -3 & | & 1 \\ -12 & 6 & | & -2 \end{bmatrix}$

$2R_1 + R_2 \to R_2 \quad \begin{bmatrix} 6 & -3 & | & 1 \\ 0 & 0 & | & 0 \end{bmatrix}$

$\frac{1}{6}R_1 \to R_1 \quad \begin{bmatrix} 1 & -\frac{1}{2} & | & \frac{1}{6} \\ 0 & 0 & | & 0 \end{bmatrix}$

This is as far as we can go with the Gauss-Jordan method. To complete the solution, write the equation that corresponds to the first row of the matrix.

$x - \frac{1}{2}y = \frac{1}{6}$

Solve this equation for x in terms of y.

$x = \frac{1}{2}y + \frac{1}{6} = \frac{3y + 1}{6}$

The solution is $\left(\frac{3y+1}{6}, y\right)$, where y is any real number.

25. $y = x - 1$
$y = 6 + z$
$z = -1 - x$

First write the system in proper form.

$$\begin{aligned} -x + y \quad\;\; &= -1 \\ y - z &= \;\;6 \\ x \quad\;\; + z &= -1 \end{aligned}$$

Write the augmented matrix and use row operations.

$$\begin{bmatrix} -1 & 1 & 0 & | & -1 \\ 0 & 1 & -1 & | & 6 \\ 1 & 0 & 1 & | & -1 \end{bmatrix}$$

$$R_1 + R_3 \rightarrow R_3 \quad \begin{bmatrix} -1 & 1 & 0 & | & -1 \\ 0 & 1 & -1 & | & 6 \\ 0 & 1 & 1 & | & -2 \end{bmatrix}$$

$$\begin{aligned} -1R_2 + R_1 \rightarrow R_1 \\ -1R_2 + R_3 \rightarrow R_3 \end{aligned} \quad \begin{bmatrix} -1 & 0 & 1 & | & -7 \\ 0 & 1 & -1 & | & 6 \\ 0 & 0 & 2 & | & -8 \end{bmatrix}$$

$$\begin{aligned} R_3 + (-2)R_1 \rightarrow R_1 \\ 2R_2 + R_3 \rightarrow R_2 \end{aligned} \quad \begin{bmatrix} 2 & 0 & 0 & | & 6 \\ 0 & 2 & 0 & | & 4 \\ 0 & 0 & 2 & | & -8 \end{bmatrix}$$

$$\begin{aligned} \tfrac{1}{2}R_1 \rightarrow R_1 \\ \tfrac{1}{2}R_2 \rightarrow R_2 \\ \tfrac{1}{2}R_3 \rightarrow R_3 \end{aligned} \quad \begin{bmatrix} 1 & 0 & 0 & | & 3 \\ 0 & 1 & 0 & | & 2 \\ 0 & 0 & 1 & | & -4 \end{bmatrix}$$

The solution is $(3, 2, -4)$.

27. $2x - 2y \quad\;\; = -2$
$y + z = \;\;4$
$x \quad\;\; + z = \;\;1$

Write the augmented matrix and use row operations.

$$\begin{bmatrix} 2 & -2 & 0 & | & -2 \\ 0 & 1 & 1 & | & 4 \\ 1 & 0 & 1 & | & 1 \end{bmatrix}$$

$$R_1 + (-2)R_3 \rightarrow R_3 \quad \begin{bmatrix} 2 & -2 & 0 & | & -2 \\ 0 & 1 & 1 & | & 4 \\ 0 & -2 & -2 & | & -4 \end{bmatrix}$$

$$\begin{aligned} 2R_2 + R_1 \rightarrow R_1 \\ 2R_2 + R_3 \rightarrow R_3 \end{aligned} \quad \begin{bmatrix} 2 & 0 & 2 & | & 6 \\ 0 & 1 & 1 & | & 4 \\ 0 & 0 & 0 & | & 4 \end{bmatrix}$$

This matrix corresponds to the system

$$\begin{aligned} 2x + 2z &= 6 \\ y + \;z &= 4 \\ 0 &= 4. \end{aligned}$$

The false statement $0 = 4$ indicates that the system is inconsistent and therefore has no solution.

29. $4x + 4y - 4z = 24$
$2x - \;\;y + \;\;z = -9$
$x - 2y + 3z = \;\;1$

Write the augmented matrix and use row operations.

$$\begin{bmatrix} 4 & 4 & -4 & | & 24 \\ 2 & -1 & 1 & | & -9 \\ 1 & -2 & 3 & | & 1 \end{bmatrix}$$

$$\begin{aligned} R_1 + (-2)R_2 \rightarrow R_2 \\ R_1 + (-4)R_3 \rightarrow R_3 \end{aligned} \quad \begin{bmatrix} 4 & 4 & -4 & | & 24 \\ 0 & 6 & -6 & | & 42 \\ 0 & 12 & -16 & | & 20 \end{bmatrix}$$

$$\begin{aligned} 2R_2 + (-3)R_1 \rightarrow R_1 \\ -2R_2 + R_3 \rightarrow R_3 \end{aligned} \quad \begin{bmatrix} -12 & 0 & 0 & | & 12 \\ 0 & 6 & -6 & | & 42 \\ 0 & 0 & -4 & | & -64 \end{bmatrix}$$

$$-3R_3 + 2R_2 \rightarrow R_2 \quad \begin{bmatrix} -12 & 0 & 0 & | & 12 \\ 0 & 12 & 0 & | & 276 \\ 0 & 0 & -4 & | & -64 \end{bmatrix}$$

$$\begin{aligned} -\tfrac{1}{12}R_1 \rightarrow R_1 \\ -\tfrac{1}{12}R_2 \rightarrow R_2 \\ -\tfrac{1}{4}R_3 \rightarrow R_3 \end{aligned} \quad \begin{bmatrix} 1 & 0 & 0 & | & -1 \\ 0 & 1 & 0 & | & 23 \\ 0 & 0 & 1 & | & 16 \end{bmatrix}$$

The solution is $(-1, 23, 16)$.

31. $3x + \;\;5y - \;\;z = 0$
$4x - \;\;y + 2z = 1$
$-6x - 10y + 2z = 0$

Write the augmented matrix and use row operations.

$$\begin{bmatrix} 3 & 5 & -1 & | & 0 \\ 4 & -1 & 2 & | & 1 \\ -6 & -10 & 2 & | & 0 \end{bmatrix}$$

$$\begin{aligned} 4R_1 + (-3)R_2 \rightarrow R_2 \\ 2R_1 + R_3 \rightarrow R_3 \end{aligned} \quad \begin{bmatrix} 3 & 5 & -1 & | & 0 \\ 0 & 23 & -10 & | & -3 \\ 0 & 0 & 0 & | & 0 \end{bmatrix}$$

$$23R_1 + (-5)R_2 \rightarrow R_1 \quad \begin{bmatrix} 69 & 0 & 27 & | & 15 \\ 0 & 23 & -10 & | & -3 \\ 0 & 0 & 0 & | & 0 \end{bmatrix}$$

$$\begin{aligned} \tfrac{1}{69}R_1 \rightarrow R_1 \\ \tfrac{1}{23}R_2 \rightarrow R_2 \end{aligned} \quad \begin{bmatrix} 1 & 0 & \tfrac{9}{23} & | & \tfrac{5}{23} \\ 0 & 1 & -\tfrac{10}{23} & | & -\tfrac{3}{23} \\ 0 & 0 & 0 & | & 0 \end{bmatrix}$$

The row of zeros indicates dependent equations. Solve the first two equations respectively for x and y in terms of z to obtain

$$x = -\frac{9}{23}z + \frac{5}{23} = \frac{-9z+5}{23}$$

and

$$y = \frac{10}{23}z - \frac{3}{23} = \frac{10z-3}{23}.$$

The solution is $\left(\frac{-9z+5}{23}, \frac{10z-3}{23}, z\right)$, where z is any real number.

33. $5x - 4y + 2z = 4$
$\quad\ 5x + 3y - z = 17$
$\quad 15x - 5y + 3z = 25$

Write the augmented matrix and use row operations.

$$\begin{bmatrix} 5 & -4 & 2 & | & 4 \\ 5 & 3 & -1 & | & 17 \\ 15 & -5 & 3 & | & 25 \end{bmatrix}$$

$\begin{array}{l} -1R_1 + R_2 \to R_2 \\ -3R_1 + R_3 \to R_3 \end{array}$ $\begin{bmatrix} 5 & -4 & 2 & | & 4 \\ 0 & 7 & -3 & | & 13 \\ 0 & 7 & -3 & | & 13 \end{bmatrix}$

$\begin{array}{l} 4R_2 + 7R_1 \to R_1 \\[2pt] -1R_2 + R_3 \to R_3 \end{array}$ $\begin{bmatrix} 35 & 0 & 2 & | & 80 \\ 0 & 7 & -3 & | & 13 \\ 0 & 0 & 0 & | & 0 \end{bmatrix}$

$\begin{array}{l} \frac{1}{35}R_1 \to R_1 \\[4pt] \frac{1}{7}R_2 \to R_2 \end{array}$ $\begin{bmatrix} 1 & 0 & \frac{2}{35} & | & \frac{16}{7} \\ 0 & 1 & -\frac{3}{7} & | & \frac{13}{7} \\ 0 & 0 & 0 & | & 0 \end{bmatrix}$

The row of zeros indicates dependent equations. Solve the first two equations respectively for x and y in terms of z to obtain

$$x = -\frac{2}{35}z + \frac{16}{7} = \frac{-2z+80}{35}$$

and

$$y = \frac{3}{7}z + \frac{13}{7} = \frac{3z+13}{7}.$$

The solution is $\left(\frac{-2z+80}{35}, \frac{3z+13}{7}, z\right)$.

35. $2x + 3y + z = 9$
$\quad 4x + 6y + 2z = 18$
$\quad -\dfrac{1}{2}x - \dfrac{3}{4}y - \dfrac{1}{4}z = -\dfrac{9}{4}$

Write the augmented matrix and use row operations.

$$\begin{bmatrix} 2 & 3 & 1 & | & 9 \\ 4 & 6 & 2 & | & 18 \\ -\frac{1}{2} & -\frac{3}{4} & -\frac{1}{4} & | & -\frac{9}{4} \end{bmatrix}$$

$\begin{array}{l} -2R_1 + R_2 \to R_2 \\[2pt] \frac{1}{4}R_1 + R_3 \to R_3 \end{array}$ $\begin{bmatrix} 2 & 3 & 1 & | & 9 \\ 0 & 0 & 0 & | & 0 \\ 0 & 0 & 0 & | & 0 \end{bmatrix}$

The rows of zeros indicate dependent equations. Since the equation involves x, y, and z, let y and z be parameters. Solve the equation for x to obtain $x = \frac{9-3y-z}{2}$.

The solution is $\left(\frac{9-3y-z}{2}, y, z\right)$, where y and z are any real numbers.

37. $x + 2y \quad\ - w = 3$
$\quad 2x \quad\ + 4z + 2w = -6$
$\quad x + 2y - z \quad\ = 6$
$\quad 2x - y + z + w = -3$

Write the augmented matrix and use row operations.

$$\begin{bmatrix} 1 & 2 & 0 & -1 & | & 3 \\ 2 & 0 & 4 & 2 & | & -6 \\ 1 & 2 & -1 & 0 & | & 6 \\ 2 & -1 & 1 & 1 & | & -3 \end{bmatrix}$$

$\begin{array}{l} -2R_1 + R_2 \to R_2 \\ -1R_1 + R_3 \to R_3 \\ -2R_1 + R_4 \to R_4 \end{array}$ $\begin{bmatrix} 1 & 2 & 0 & -1 & | & 3 \\ 0 & -4 & 4 & 4 & | & -12 \\ 0 & 0 & -1 & 1 & | & 3 \\ 0 & -5 & 1 & 3 & | & -9 \end{bmatrix}$

$\begin{array}{l} R_2 + 2R_1 \to R_1 \\[10pt] -5R_2 + 4R_4 \to R_4 \end{array}$ $\begin{bmatrix} 2 & 0 & 4 & 2 & | & -6 \\ 0 & -4 & 4 & 4 & | & -12 \\ 0 & 0 & -1 & 1 & | & 3 \\ 0 & 0 & -16 & -8 & | & 24 \end{bmatrix}$

$\begin{array}{l} 4R_3 + R_1 \to R_1 \\ 4R_3 + R_2 \to R_2 \\[6pt] 16R_3 + (-1)R_4 \to R_4 \end{array}$ $\begin{bmatrix} 2 & 0 & 0 & 6 & | & 6 \\ 0 & -4 & 0 & 8 & | & 0 \\ 0 & 0 & -1 & 1 & | & 3 \\ 0 & 0 & 0 & 24 & | & 24 \end{bmatrix}$

$R_4 + (-4R_1) \rightarrow R_1$
$R_4 + (-3R_2) \rightarrow R_2$
$R_4 + (-24R_3) \rightarrow R_3$

$$\begin{bmatrix} -8 & 0 & 0 & 0 & 0 \\ 0 & 12 & 0 & 0 & 24 \\ 0 & 0 & 24 & 0 & -48 \\ 0 & 0 & 0 & 24 & 24 \end{bmatrix}$$

$-\frac{1}{8}R_1 \rightarrow R_1$
$\frac{1}{12}R_2 \rightarrow R_2$
$\frac{1}{24}R_3 \rightarrow R_3$
$\frac{1}{24}R_4 \rightarrow R_4$

$$\begin{bmatrix} 1 & 0 & 0 & 0 & 0 \\ 0 & 1 & 0 & 0 & 2 \\ 0 & 0 & 1 & 0 & -2 \\ 0 & 0 & 0 & 1 & 1 \end{bmatrix}$$

The solution is $x = 0$, $y = 2$, $z = -2$, $w = 1$, or $(0, 2, -2, 1)$.

39. $\begin{aligned} x + y - z + 2w &= -20 \\ 2x - y + z + w &= 11 \\ 3x - 2y + z - 2w &= 27 \end{aligned}$

$$\begin{bmatrix} 1 & 1 & -1 & 2 & -20 \\ 2 & -1 & 1 & 1 & 11 \\ 3 & -2 & 1 & -2 & 27 \end{bmatrix}$$

$-2R_1 + R_2 \rightarrow R_2$
$-3R_1 + R_3 \rightarrow R_3$

$$\begin{bmatrix} 1 & 1 & -1 & 2 & -20 \\ 0 & -3 & 3 & -3 & 51 \\ 0 & -5 & 4 & -8 & 87 \end{bmatrix}$$

$-\frac{1}{3}R_2 \rightarrow R_2$

$$\begin{bmatrix} 1 & 1 & -1 & 2 & -20 \\ 0 & 1 & -1 & 1 & -17 \\ 0 & -5 & 4 & -8 & 87 \end{bmatrix}$$

$-1R_2 + R_1 \rightarrow R_1$
$5R_2 + R_3 \rightarrow R_3$

$$\begin{bmatrix} 1 & 0 & 0 & 1 & -3 \\ 0 & 1 & -1 & 1 & -17 \\ 0 & 0 & -1 & -3 & 2 \end{bmatrix}$$

$-1R_3 \rightarrow R_3$

$$\begin{bmatrix} 1 & 0 & 0 & 1 & -3 \\ 0 & 1 & -1 & 1 & -17 \\ 0 & 0 & 1 & 3 & -2 \end{bmatrix}$$

$R_3 + R_2 \rightarrow R_2$

$$\begin{bmatrix} 1 & 0 & 0 & 1 & -3 \\ 0 & 1 & 0 & 4 & -19 \\ 0 & 0 & 1 & 3 & -2 \end{bmatrix}$$

This is as far as we can go using row operations. To complete the solution, write the equations that correspond to the matrix.

$$\begin{aligned} x + w &= -3 \\ y + 4w &= -19 \\ z + 3w &= -2 \end{aligned}$$

Let w be the parameter and express x, y, and z in terms of w. From the equations above, $x = -w - 3$, $y = -4w - 19$, and $z = -3w - 2$.

The solution is $(-w - 3, -4w - 19, -3w - 2, w)$, where w is any real number.

41. $\begin{aligned} 10.47x + 3.52y + 2.58z - 6.42w &= 218.65 \\ 8.62x - 4.93y - 1.75z + 2.83w &= 157.03 \\ 4.92x + 6.83y - 2.97z + 2.65w &= 462.3 \\ 2.86x + 19.10y - 6.24z - 8.73w &= 398.4 \end{aligned}$

Write the augmented matrix of the system.

$$\begin{bmatrix} 10.47 & 3.52 & 2.58 & -6.42 & 218.65 \\ 8.62 & -4.93 & -1.75 & 2.83 & 157.03 \\ 4.92 & 6.83 & -2.97 & 2.65 & 462.3 \\ 2.86 & 19.10 & -6.24 & -8.73 & 398.4 \end{bmatrix}$$

This exercise should be solved by graphing calculator or computer methods. The solution, which may vary slightly, is $x \approx 28.9436$, $y \approx 36.6326$, $z \approx 9.6390$, and $w \approx 37.1036$, or

$$(28.9436, 36.6326, 9.6390, 37.1036).$$

43. (a) For the first equation, the first sighting in 1980 was on day $y = 759 - 0.338(1980) = 89.76$, or during the eighty-ninth day of the year. Since 1980 was a leap year, the eighty-ninth day fell on March 29.

For the second equation, the first sighting in 1980 was on day $y = 1,637 - 0.779(1980) = 94.58$, or during the ninety-fourth day of the year. Since 1980 was a leap year, the ninety-fourth day fell on April 3.

(b) $y = 759 - 0.338x$
$y = 1,637 - 0.779x$

Put the equations in proper form to obtain the system

$0.338x + y = 759$
$0.779x + y = 1,637$

Use graphing calculator or computer methods to solve this system. The solution is approximately $(1991, 86)$. The two estimates agree that the first sighting occurred 86 days into the year in the year 1991.

45. Let $x =$ required units of corn,
$y =$ required units of soybeans, and
$z =$ required units of cottonseed.

The system to be solved is

$$\begin{aligned} 0.25x + 0.4y + 0.2z &= 22 \\ 0.4x + 0.2y + 0.3z &= 28 \\ 0.3x + 0.2y + 0.1z &= 18 \end{aligned}$$

The augmented matrix for this system is

$$\begin{bmatrix} 0.25 & 0.4 & 0.2 & | & 22 \\ 0.4 & 0.2 & 0.3 & | & 28 \\ 0.3 & 0.2 & 0.1 & | & 18 \end{bmatrix}$$

Use graphing calculator or computer methods to solve this system. The solution is 40 units of corn, 15 units of soybeans, and 30 units of cottonseed.

47. Let $x =$ the number of kilograms of the first chemical,

$y =$ the number of kilograms of the second chemical, and

$z =$ the number of kilograms of the third chemical.

The system to be solved is

$$x + y + z = 750$$
$$x = 0.108(750)$$
$$\frac{y}{z} = \frac{4}{3}.$$

Rewrite this system as

$$x + y + z = 750$$
$$x \qquad\quad = 81$$
$$3y - 4z = \quad 0.$$

Use graphing calculator or computer methods to solve this system. The solution, which may vary slightly, is to use 81 kg of the first chemical, about 382.286 kg of the second chemical, and about 286.714 kg of the third chemical.

49. Let $x =$ the number of species A,

$y =$ the number of species B, and

$z =$ the number of species C.

Use a chart to organize the information.

		Species A	B	C	Totals
	I	1.32	2.1	0.86	490
Food	II	2.9	0.95	1.52	897
	III	1.75	0.6	2.01	653

The system to be solved is

$$1.32x + 2.1y + 0.86z = 490$$
$$2.9x + 0.95y + 1.52z = 897$$
$$1.75x + 0.6y + 2.01z = 653.$$

Use graphing calculator or computer methods to solve this system. The solution, which may vary slightly, is to stock about 244 fish of species A, 39 fish of species B, and 101 fish of species C.

51. (a) Bulls:

The number of white ones was one half plus one third the number of black greater than the brown.

$$X = \left(\frac{1}{2} + \frac{1}{3}\right) Y + T$$
$$X = \frac{5}{6}Y + T$$
$$6X = 5Y + 6T$$
$$6X - 5Y = 6T$$

The number of the black, one quarter plus one fifth the number of the spotted greater than the brown.

$$Y = \left(\frac{1}{4} + \frac{1}{5}\right) Z + T$$
$$Y = \frac{9}{20}Z + T$$
$$20Y = 9Z + 20T$$
$$20Y - 9Z = 20T$$

The number of the spotted, one sixth and one seventh the number of the white greater than the brown.

$$Z = \left(\frac{1}{6} + \frac{1}{7}\right) X + T$$
$$Z = \frac{13}{42}X + T$$
$$42Z = 13X + 42T$$
$$42Z - 13X = 42T$$

So the system of equations for the bulls is

$$6X - 5Y = 6T$$
$$20Y - 9Z = 20T$$
$$42Z - 13X = 42T.$$

Cows:

The number of white ones was one third plus one quarter of the total black cattle.

$$x = \left(\frac{1}{3} + \frac{1}{4}\right)(Y + y)$$
$$x = \frac{7}{12}(Y + y)$$
$$12x = 7Y + 7y$$
$$12x - 7y = 7Y$$

The number of the black, one quarter plus one fifth the total of the spotted cattle.

$$y = \left(\frac{1}{4} + \frac{1}{5}\right)(Z + z)$$

$$y = \frac{9}{20}(Z + z)$$

$$20y = 9Z + 9z$$

$$20y - 9z = 9Z$$

The number of the spotted, one fifth plus one sixth the total of the brown cattle.

$$z = \left(\frac{1}{5} + \frac{1}{6}\right)(T + t)$$

$$z = \frac{11}{30}(T + t)$$

$$30z = 11T + 11t$$

$$30z - 11t = 11T$$

The number of the brown, one sixth plus one seventh the total of the white cattle.

$$t = \left(\frac{1}{6} + \frac{1}{7}\right)(X + x)$$

$$t = \frac{13}{42}(X + x)$$

$$42t = 13X + 13x$$

$$42t - 13x = 13X$$

So the system of equations for the cows is

$$12x - 7y = 7Y$$
$$20y - 9z = 9Z$$
$$30z - 11t = 11T$$
$$-13x + 42t = 13X$$

(b) For $T = 4{,}149{,}387$, the 3×3 system to be solved is

$$6X - 5Y \qquad = 24{,}896{,}322$$
$$20Y - 9Z = 82{,}987{,}740$$
$$-13X \qquad + 42Z = 174{,}274{,}254$$

Write the augmented matrix of the system.

$$\begin{bmatrix} 6 & -5 & 0 & \vline & 24{,}896{,}322 \\ 0 & 20 & -9 & \vline & 82{,}987{,}740 \\ -13 & 0 & 42 & \vline & 174{,}274{,}254 \end{bmatrix}$$

This exercise should be solved by graphing calculator or computer methods. The solution is $X = 10{,}366{,}482$ white bulls, $Y = 7{,}460{,}514$ black bulls, and $Z = 7{,}358{,}060$ spotted bulls.

For $X = 10{,}366{,}482$, $Y = 7{,}460{,}514$, and $Z = 7{,}358{,}060$, the 4×4 system to be solved is

$$12x - 7y = 52{,}223{,}598$$
$$20y - 9z = 66{,}222{,}540$$
$$30z - 11t = 45{,}643{,}257$$
$$-13x + 42t = 134{,}764{,}266$$

Write the augmented matrix of the system.

$$\begin{bmatrix} 12 & -7 & 0 & 0 & \vline & 52{,}223{,}598 \\ 0 & 20 & -9 & 0 & \vline & 66{,}222{,}540 \\ 0 & 0 & 30 & -11 & \vline & 45{,}643{,}257 \\ -13 & 0 & 0 & 42 & \vline & 134{,}764{,}266 \end{bmatrix}$$

This exercise should be solved by graphing calculator or computer methods. The solution is $x = 7{,}206{,}360$ white cows, $y = 4{,}893{,}246$ black cows, $z = 3{,}515{,}820$ spotted cows, and $t = 5{,}439{,}213$ brown cows.

53. Let $x =$ the number of chairs produced each week,
  $y =$ the number of cabinets produced each week, and
  $z =$ the number of buffets produced each week.

Make a table to organize the information.

	Chair	Cabinet	Buffet	Totals
Cutting	0.2	0.5	0.3	1,950
Assembly	0.3	0.4	0.1	1,490
Finishing	0.1	0.6	0.4	2,160

The system to be solved is

$$0.2x + 0.5y + 0.3z = 1{,}950$$
$$0.3x + 0.4y + 0.1z = 1{,}490$$
$$0.1x + 0.6y + 0.4z = 2{,}160.$$

Write the augmented matrix of the system.

$$\begin{bmatrix} 0.2 & 0.5 & 0.3 & | & 1,950 \\ 0.3 & 0.4 & 0.1 & | & 1,490 \\ 0.1 & 0.6 & 0.4 & | & 2,160 \end{bmatrix}$$

$$\begin{matrix} 10R_1 \to R_1 \\ 10R_2 \to R_2 \\ 10R_3 \to R_3 \end{matrix} \begin{bmatrix} 2 & 5 & 3 & | & 19,500 \\ 3 & 4 & 1 & | & 14,900 \\ 1 & 6 & 4 & | & 21,600 \end{bmatrix}$$

Interchange rows 1 and 3.

$$\begin{bmatrix} 1 & 6 & 4 & | & 21,600 \\ 3 & 4 & 1 & | & 14,900 \\ 2 & 5 & 3 & | & 19,500 \end{bmatrix}$$

$$\begin{matrix} -3R_1 + R_2 \to R_2 \\ -2R_1 + R_3 \to R_3 \end{matrix} \begin{bmatrix} 1 & 6 & 4 & | & 21,600 \\ 0 & -14 & -11 & | & -49,900 \\ 0 & -7 & -5 & | & -23,700 \end{bmatrix}$$

$$-\tfrac{1}{14}R_2 \to R_2 \begin{bmatrix} 1 & 6 & 4 & | & 21,600 \\ 0 & 1 & \frac{11}{14} & | & \frac{24,950}{7} \\ 0 & -7 & -5 & | & -23,700 \end{bmatrix}$$

$$\begin{matrix} -6R_2 + R_1 \to R_1 \\ \\ 7R_2 + R_3 \to R_3 \end{matrix} \begin{bmatrix} 1 & 0 & -\frac{5}{7} & | & \frac{1,500}{7} \\ 0 & 1 & \frac{11}{14} & | & \frac{24,950}{7} \\ 0 & 0 & \frac{1}{2} & | & 1,250 \end{bmatrix}$$

$$2R_3 \to R_3 \begin{bmatrix} 1 & 0 & -\frac{5}{7} & | & \frac{1,500}{7} \\ 0 & 1 & \frac{11}{14} & | & \frac{24,950}{7} \\ 0 & 0 & 1 & | & 2,500 \end{bmatrix}$$

$$\begin{matrix} \frac{5}{7}R_3 + R_1 \to R_1 \\ -\frac{11}{14}R_3 + R_2 \to R_2 \end{matrix} \begin{bmatrix} 1 & 0 & 0 & | & 2,000 \\ 0 & 1 & 0 & | & 1,600 \\ 0 & 0 & 1 & | & 2,500 \end{bmatrix}$$

The solution is (2,000, 1,600, 2,500). Therefore, 2,000 chairs, 1,600 cabinets, and 2,500 buffets should be produced.

55. Let $x =$ the amount borrowed at 13%,

$y =$ the amount borrowed at 14%,

and $z =$ the amount borrowed at 12%.

(a) The system to be solved is

$$\begin{aligned} x + \quad y + \quad z &= 25,000 & (1) \\ 0.13x + 0.14y + 0.12z &= 3,240 & (2) \\ y &= \tfrac{1}{2}x + 2,000. & (3) \end{aligned}$$

Multiply equation (2) by 100 and equation (3) by 2. Then rewrite the system.

$$\begin{aligned} x + \quad y + \quad z &= 25,000 & (1) \\ 13x + 14y + 12z &= 324,000 & (4) \\ -x + 2y \quad\quad &= 4,000 & (5) \end{aligned}$$

Write the augmented matrix of the system.

$$\begin{bmatrix} 1 & 1 & 1 & | & 25,000 \\ 13 & 14 & 12 & | & 324,000 \\ -1 & 2 & 0 & | & 4,000 \end{bmatrix}$$

$$\begin{matrix} -13R_1 + R_2 \to R_2 \\ R_1 + R_3 \to R_3 \end{matrix} \begin{bmatrix} 1 & 1 & 1 & | & 25,000 \\ 0 & 1 & -1 & | & -1,000 \\ 0 & 3 & 1 & | & 29,000 \end{bmatrix}$$

$$\begin{matrix} -1R_2 + R_1 \to R_1 \\ \\ -3R_2 + R_3 \to R_3 \end{matrix} \begin{bmatrix} 1 & 0 & 2 & | & 26,000 \\ 0 & 1 & -1 & | & -1,000 \\ 0 & 0 & 4 & | & 32,000 \end{bmatrix}$$

$$\tfrac{1}{4}R_3 \to R_3 \begin{bmatrix} 1 & 0 & 2 & | & 26,000 \\ 0 & 1 & -1 & | & -1,000 \\ 0 & 0 & 1 & | & 8,000 \end{bmatrix}$$

$$\begin{matrix} -2R_3 + R_1 \to R_1 \\ R_3 + R_2 \to R_2 \end{matrix} \begin{bmatrix} 1 & 0 & 0 & | & 10,000 \\ 0 & 1 & 0 & | & 7,000 \\ 0 & 0 & 1 & | & 8,000 \end{bmatrix}$$

The solution is (10,000, 7,000, 8,000). Borrow $10,000 at 13%, $7,000 at 14%, and $8,000 at 12%.

(b) If the condition is dropped, refer to the first two rows of the fourth augmented matrix of part (a).

$$\begin{bmatrix} 1 & 0 & 2 & | & 26,000 \\ 0 & 1 & -1 & | & -1,000 \end{bmatrix}$$

This gives

$$\begin{aligned} x &= 26,000 - 2z \\ y &= z - 1,000. \end{aligned}$$

Since all values must be nonnegative,

$$26,000 - 2z \geq 0 \quad \text{and} \quad z - 1,000 \geq 0$$
$$z \leq 13,000 \quad \text{and} \quad z \geq 1,000.$$

Therefore, the amount borrowed at 12% must be between $1,000 and $13,000. If $z = 5,000$, then

$$\begin{aligned} x &= 26,000 - 2(5,000) = 16,000 \text{ and} \\ y &= 5,000 - 1,000 = 4,000. \end{aligned}$$

Therefore, $16,000 is borrowed at 13% and $4,000 at 14%.

(c) Substitute $z = 6,000$ into equations (1), (4), and (5) from part (a) to obtain the system

$$\begin{aligned} x + \quad y + \quad 6,000 &= 25,000 \quad (6) \\ 13x + 14y + 12(6,000) &= 324,000 \quad (7) \\ -x + \quad 2y \qquad\quad &= \quad 4,000. \quad (5) \end{aligned}$$

This gives the system

$$\begin{aligned} x + \quad y &= 19,000 \quad (8) \\ 13x + 14y &= 252,000 \quad (9) \\ -x + \quad 2y &= \quad 4,000. \quad (5) \end{aligned}$$

The augmented matrix for this system is

$$\begin{bmatrix} 1 & 1 & | & 19,000 \\ 13 & 14 & | & 252,000 \\ -1 & 2 & | & 4,000 \end{bmatrix}.$$

$$\begin{matrix} -13R_1 + R_2 \rightarrow R_2 \\ R_1 + R_3 \rightarrow R_3 \end{matrix} \begin{bmatrix} 1 & 1 & | & 19,000 \\ 0 & 1 & | & 5,000 \\ 0 & 3 & | & 23,000 \end{bmatrix}.$$

$$-3R_2 + R_3 \rightarrow R_3 \begin{bmatrix} 1 & 1 & | & 19,000 \\ 0 & 1 & | & 5,000 \\ 0 & 0 & | & 8,000 \end{bmatrix}.$$

The equation that corresponds to the last row of this matrix is $0x + 0y = 8,000$, so there is no solution if \$6,000 is borrowed at 12%.

57. (a) $43,500x - y = 1,295,000$
$\qquad 27,000x - y = \quad 440,000$

Write the augmented matrix of the system.

$$\begin{bmatrix} 43,500 & -1 & | & 1,295,000 \\ 27,000 & -1 & | & 440,000 \end{bmatrix}.$$

$$-\frac{27,000}{43,500}R_1 + R_2 \rightarrow R_2 \begin{bmatrix} 43,500 & -1 & | & 1,295,000 \\ 0 & -\frac{11}{29} & | & -\frac{10,550,000}{29} \end{bmatrix}$$

$$29R_2 \rightarrow R_2 \begin{bmatrix} 43,500 & -1 & | & 1,295,000 \\ 0 & -11 & | & -10,550,000 \end{bmatrix}$$

$$-11R_1 + R_2 \rightarrow R_1 \begin{bmatrix} -478,500 & 0 & | & -24,795,000 \\ 0 & -11 & | & -10,550,000 \end{bmatrix}$$

$$-\frac{1}{478,500}R_1 \rightarrow R_1 \begin{bmatrix} 1 & 0 & | & \frac{570}{11} \\ 0 & 1 & | & \frac{10,550,000}{11} \end{bmatrix}$$
$$-\frac{1}{11}R_2 \rightarrow R_2$$

The solution is $\left(\frac{570}{11}, \frac{10,550,000}{11}\right)$, or approximately $(51.8, 959,091)$. The profit/loss from the show will be equal for each venue at about 51.8 weeks. The profit will be about \$959,091.

59. (a) The system to be solved is

$$\begin{aligned} 0 &= 200,000 - 0.5r - 0.3b \\ 0 &= 350,000 - 0.5r - 0.7b. \end{aligned}$$

First, write the system in proper form.

$$\begin{aligned} 0.5r + 0.3b &= 200,000 \\ 0.5r + 0.7b &= 350,000 \end{aligned}$$

Write the augmented matrix and use row operations.

$$\begin{bmatrix} 0.5 & 0.3 & | & 200,000 \\ 0.5 & 0.7 & | & 350,000 \end{bmatrix}.$$

$$\begin{matrix} 10R_1 \rightarrow R_1 \\ 10R_2 \rightarrow R_2 \end{matrix} \begin{bmatrix} 5 & 3 & | & 2,000,000 \\ 5 & 7 & | & 3,500,000 \end{bmatrix}$$

$$-1R_1 + R_2 \rightarrow R_2 \begin{bmatrix} 5 & 3 & | & 2,000,000 \\ 0 & 4 & | & 1,500,000 \end{bmatrix}$$

$$-\frac{3}{4}R_2 + R_1 \rightarrow R_1 \begin{bmatrix} 5 & 0 & | & 875,000 \\ 0 & 4 & | & 1,500,000 \end{bmatrix}$$

$$\begin{matrix} \frac{1}{5}R_1 \rightarrow R_1 \\ \frac{1}{4}R_2 \rightarrow R_2 \end{matrix} \begin{bmatrix} 1 & 0 & | & 175,000 \\ 0 & 1 & | & 375,000 \end{bmatrix}$$

The solution is $(175,000, 375,000)$. When the rate of increase for each is zero, there are 175,000 soldiers in the Red Army and 375,000 soldiers in the Blue Army.

61. (a) $5.4 = a(8)^2 + b(8) + c$
$\qquad 5.4 = 64a + 8b + c$
$\qquad 6.3 = a(13)^2 + b(13) + c$
$\qquad 6.3 = 169a + 13b + c$
$\qquad 5.6 = a(18)^2 + b(18) + c$
$\qquad 5.6 = 324a + 18b + c$

The linear system to be solved is

$$\begin{aligned} 64a + \quad 8b + c &= 5.4 \\ 169a + 13b + c &= 6.3 \\ 324a + 18b + c &= 5.6. \end{aligned}$$

Use a graphing calculator or computer methods to solve this system. The solution is $a = -0.032, b = 0.852$, and $c = 0.632$. Thus, the equation is

$$y = -0.032x^2 + 0.852x + 0.632.$$

(b)

The answer obtained using Gauss–Jordan elimination is the same as the answer obtained using the quadratic regression feature on a graphing calculator.

10.2 Addition and Subtraction of Matrices

1. $\begin{bmatrix} 1 & 3 \\ 5 & 7 \end{bmatrix} = \begin{bmatrix} 1 & 5 \\ 3 & 7 \end{bmatrix}$

   This statement is false, since not all corresponding elements are equal.

3. $\begin{bmatrix} x \\ y \end{bmatrix} = \begin{bmatrix} 3 \\ 5 \end{bmatrix}$ if $x = 3$ and $y = 5$.

   This statement is true. The matrices are the same size and corresponding elements are equal.

5. $\begin{bmatrix} 1 & 9 & -4 \\ 3 & 7 & 2 \\ -1 & 1 & 0 \end{bmatrix}$ is a square matrix.

   This statement is true. The matrix has 3 rows and 3 columns.

7. $\begin{bmatrix} -4 & 8 \\ 2 & 3 \end{bmatrix}$ is a 2×2 square matrix.

   Its additive inverse is $\begin{bmatrix} 4 & -8 \\ -2 & -3 \end{bmatrix}$.

9. $\begin{bmatrix} -6 & 8 & 0 & 0 \\ 4 & 1 & 9 & 2 \\ 3 & -5 & 7 & 1 \end{bmatrix}$ is a 3×4 matrix.

   Its additive inverse is
   $$\begin{bmatrix} 6 & -8 & 0 & 0 \\ -4 & -1 & -9 & -2 \\ -3 & 5 & -7 & -1 \end{bmatrix}.$$

11. $\begin{bmatrix} 2 \\ 4 \end{bmatrix}$ is a 2×1 column matrix.

    Its additive inverse is
    $$\begin{bmatrix} -2 \\ -4 \end{bmatrix}.$$

13. The sum of an $n \times m$ matrix and its additive inverse is the $n \times m$ zero matrix.

15. $\begin{bmatrix} 2 & 1 \\ 4 & 8 \end{bmatrix} = \begin{bmatrix} x & 1 \\ y & z \end{bmatrix}$

    Corresponding elements must be equal for the matrices to be equal. Therefore, $x = 2$, $y = 4$, and $z = 8$.

17. $\begin{bmatrix} x+6 & y+2 \\ 8 & 3 \end{bmatrix} = \begin{bmatrix} -9 & 7 \\ 8 & k \end{bmatrix}$

Corresponding elements must be equal.

$$x + 6 = -9 \qquad y + 2 = 7 \qquad k = 3$$
$$x = -15 \qquad\quad y = 5$$

Thus, $x = -15$, $y = 5$, and $k = 3$.

19. $\begin{bmatrix} -7+z & 4r & 8s \\ 6p & 2 & 5 \end{bmatrix} + \begin{bmatrix} -9 & 8r & 3 \\ 2 & 5 & 4 \end{bmatrix}$

   $$= \begin{bmatrix} 2 & 36 & 27 \\ 20 & 7 & 12a \end{bmatrix}$$

Add the two matrices on the left side of this equation to obtain

$$\begin{bmatrix} -7+z & 4r & 8s \\ 6p & 2 & 5 \end{bmatrix} + \begin{bmatrix} -9 & 8r & 3 \\ 2 & 5 & 4 \end{bmatrix}$$

$$= \begin{bmatrix} (-7+z)+(-9) & 4r+8r & 8s+3 \\ 6p+2 & 7 & 9 \end{bmatrix}$$

$$= \begin{bmatrix} -16+z & 12r & 8s+3 \\ 6p+2 & 7 & 9 \end{bmatrix}.$$

Corresponding elements of this matrix and the matrix on the right side of the original equation must be equal.

$$-16+z = 2 \qquad 12r = 36 \qquad 8s+3 = 27$$
$$z = 18 \qquad\quad r = 3 \qquad\quad s = 3$$

$$6p+2 = 20 \qquad 9 = 12a$$
$$p = 3 \qquad\qquad a = \frac{3}{4}$$

Thus, $z = 18$, $r = 3$, $s = 3$, $p = 3$, and $a = \frac{3}{4}$.

21. $\begin{bmatrix} 1 & 2 & 5 & -1 \\ 3 & 0 & 2 & -4 \end{bmatrix} + \begin{bmatrix} 8 & 10 & -5 & 3 \\ -2 & -1 & 0 & 0 \end{bmatrix}$

$= \begin{bmatrix} 1+8 & 2+10 & 5+(-5) & -1+3 \\ 3+(-2) & 0+(-1) & 2+0 & -4+0 \end{bmatrix}$

$= \begin{bmatrix} 9 & 12 & 0 & 2 \\ 1 & -1 & 2 & -4 \end{bmatrix}$

23. $\begin{bmatrix} 1 & 3 & -2 \\ 4 & 7 & 1 \end{bmatrix} + \begin{bmatrix} 3 & 0 \\ 6 & 4 \\ -5 & 2 \end{bmatrix}$

These matrices cannot be added since the first matrix has size 2×3, while the second has size 3×2. Only matrices that are the same size can be added.

25. The matrices have the same size, so the subtraction can be done. Let A and B represent the given matrices. Using the definition of subtraction, we have

$A - B = A + (-B)$

$= \begin{bmatrix} 2 & 8 & 12 & 0 \\ 7 & 4 & -1 & 5 \\ 1 & 2 & 0 & 10 \end{bmatrix} + \begin{bmatrix} -1 & -3 & -6 & -9 \\ -2 & 3 & 3 & -4 \\ -8 & 0 & 2 & -17 \end{bmatrix}$

$= \begin{bmatrix} 1 & 5 & 6 & -9 \\ 5 & 7 & 2 & 1 \\ -7 & 2 & 2 & -7 \end{bmatrix}.$

27. $\begin{bmatrix} 2 & 3 \\ -2 & 4 \end{bmatrix} + \begin{bmatrix} 4 & 3 \\ 7 & 8 \end{bmatrix} - \begin{bmatrix} 3 & 2 \\ 1 & 4 \end{bmatrix}$

$= \begin{bmatrix} 2+4-3 & 3+3-2 \\ -2+7-1 & 4+8-4 \end{bmatrix} = \begin{bmatrix} 3 & 4 \\ 4 & 8 \end{bmatrix}$

29. $\begin{bmatrix} 1 & 5 \\ -3 & 7 \end{bmatrix} - \begin{bmatrix} 6 & 3 \\ 2 & 4 \end{bmatrix} + \begin{bmatrix} 8 & 10 \\ -1 & 0 \end{bmatrix}$

$= \begin{bmatrix} 1-6+8 & 5-3+10 \\ -3-2+(-1) & 7-4+0 \end{bmatrix}$

$= \begin{bmatrix} 3 & 12 \\ -6 & 3 \end{bmatrix}$

31. $\begin{bmatrix} -4x+2y & -3x+y \\ 6x-3y & 2x-5y \end{bmatrix} + \begin{bmatrix} -8x+6y & 2x \\ 3y-5x & 6x+4y \end{bmatrix}$

$= \begin{bmatrix} (-4x+2y)+(-8x+6y) & (-3x+y)+2x \\ (6x-3y)+(3y-5x) & (2x-5y)+(6x+4y) \end{bmatrix}$

$= \begin{bmatrix} -12x+8y & -x+y \\ x & 8x-y \end{bmatrix}$

33. The additive inverse of

$$X = \begin{bmatrix} x & y \\ z & w \end{bmatrix}$$

is

$$-X = \begin{bmatrix} -x & -y \\ -z & -w \end{bmatrix}.$$

35. Show that $X + (T + P) = (X + T) + P$.

On the left side, the sum $T + P$ is obtained first, and then

$$X + (T + P).$$

This gives the matrix

$$\begin{bmatrix} x+(r+m) & y+(s+n) \\ z+(t+p) & w+(u+q) \end{bmatrix}.$$

For the right side, first the sum $X+T$ is obtained, and then

$$(X + T) + P.$$

This gives the matrix

$$\begin{bmatrix} (x+r)+m & (y+s)+n \\ (z+t)+p & (w+u)+q \end{bmatrix}.$$

Comparing corresponding elements, we see that they are equal by the associative property of addition of real numbers. Thus,

$$X + (T + P) = (X + T) + P.$$

37. Show that $P + O = P$.

$$P + O = \begin{bmatrix} m & n \\ p & q \end{bmatrix} + \begin{bmatrix} 0 & 0 \\ 0 & 0 \end{bmatrix}$$

$$= \begin{bmatrix} m+0 & n+0 \\ p+0 & q+0 \end{bmatrix}$$

$$= \begin{bmatrix} m & n \\ p & q \end{bmatrix}$$

$$= P$$

Thus, $P + O = P$.

39. (a) There are four food groups and three meals. To represent the data by a 3×4 matrix, we must use the rows to correspond to the meals, breakfast, lunch, and dinner, and the columns to correspond to the four food groups. Thus, we obtain the matrix

$$\begin{bmatrix} 2 & 1 & 2 & 1 \\ 3 & 2 & 2 & 1 \\ 4 & 3 & 2 & 1 \end{bmatrix}.$$

(b) There are four food groups. These will correspond to the four rows. There are three components in each food group: fat, carbohydrates, and protein. These will correspond to the three columns. The matrix is

$$\begin{bmatrix} 5 & 0 & 7 \\ 0 & 10 & 1 \\ 0 & 15 & 2 \\ 10 & 12 & 8 \end{bmatrix}.$$

(c) The matrix is

$$\begin{bmatrix} 8 \\ 4 \\ 5 \end{bmatrix}.$$

41.

Obtained Pain Relief

	Yes	No
Painfree	22	3
Placebo	8	17

(a) Of the 25 patients who took the placebo, 8 got relief.

(b) Of the 25 patients who took Painfree, 3 got no relief.

(c) $\begin{bmatrix} 22 & 3 \\ 8 & 17 \end{bmatrix} + \begin{bmatrix} 21 & 4 \\ 6 & 19 \end{bmatrix} + \begin{bmatrix} 19 & 6 \\ 10 & 15 \end{bmatrix} + \begin{bmatrix} 23 & 2 \\ 3 & 22 \end{bmatrix}$

$= \begin{bmatrix} 85 & 15 \\ 27 & 73 \end{bmatrix}$

(d) Yes, it appears that Painfree is effective. Of the 100 patients who took the medication, 85% got relief.

43. (a) The matrix for the life expectancy of African Americans is

	M	F
1970	60.0	68.3
1980	63.8	72.5
1990	64.5	73.6
1997	67.2	74.7

(b) The matrix for the life expectancy of White Americans is

	M	F
1970	68.0	75.6
1980	70.7	78.1
1990	72.7	79.4
1997	74.3	79.9

(c) The matrix showing the difference between the life expectancy between the two groups is

$$\begin{bmatrix} 60.0 & 68.3 \\ 63.8 & 72.5 \\ 64.5 & 73.6 \\ 67.2 & 74.7 \end{bmatrix} - \begin{bmatrix} 68.0 & 75.6 \\ 70.7 & 78.1 \\ 72.7 & 79.4 \\ 74.3 & 79.9 \end{bmatrix}$$

$$= \begin{bmatrix} 60.0 & 68.3 \\ 63.8 & 72.5 \\ 64.5 & 73.6 \\ 67.2 & 74.7 \end{bmatrix} + \begin{bmatrix} -68.0 & -75.6 \\ -70.7 & -78.1 \\ -72.7 & -79.4 \\ -74.3 & -79.9 \end{bmatrix}$$

$$= \begin{bmatrix} -8.0 & -7.3 \\ -6.9 & -5.6 \\ -8.2 & -5.8 \\ -7.1 & -5.2 \end{bmatrix}$$

45. (a) The matrix for the educational attainment of African American males is

	Four Years of High School or More	Four Years of College or More
1960	18.2	2.8
1970	30.1	4.2
1980	50.8	8.4
1990	65.8	11.9
1995	73.4	13.6
1998	75.2	13.9

(b) The matrix for the educational attainment of African American females is

	Four Years of High School or More	Four Years of College or More
1960	21.8	3.3
1970	32.5	4.6
1980	51.5	8.3
1990	66.5	10.8
1995	74.1	12.9
1998	76.7	15.4

(c) The matrix showing how much more (or less) education African American males have attained than African American females is

$$\begin{bmatrix} 18.2 & 2.8 \\ 30.1 & 4.2 \\ 50.8 & 8.4 \\ 65.8 & 11.9 \\ 73.4 & 13.6 \\ 75.2 & 13.9 \end{bmatrix} - \begin{bmatrix} 21.8 & 3.3 \\ 32.5 & 4.6 \\ 51.5 & 8.3 \\ 66.5 & 10.8 \\ 74.1 & 12.9 \\ 76.7 & 15.4 \end{bmatrix}$$

$$= \begin{bmatrix} 18.2 & 2.8 \\ 30.1 & 4.2 \\ 50.8 & 8.4 \\ 65.8 & 11.9 \\ 73.4 & 13.6 \\ 75.2 & 13.9 \end{bmatrix} + \begin{bmatrix} -21.8 & -3.3 \\ -32.5 & -4.6 \\ -51.5 & -8.3 \\ -66.5 & -10.8 \\ -74.1 & -12.9 \\ -76.7 & -15.4 \end{bmatrix}$$

$$= \begin{bmatrix} -3.6 & -0.5 \\ -2.4 & -0.4 \\ -0.7 & 0.1 \\ -0.7 & 1.1 \\ -0.7 & 0.7 \\ -1.5 & -1.5 \end{bmatrix}$$

47. (a) The production cost matrix for Chicago is

$$\begin{array}{c} \\ \text{Material} \\ \text{Labor} \end{array} \begin{array}{cc} \text{Phones} & \text{Calculators} \\ \begin{bmatrix} 4.05 & 7.01 \\ 3.27 & 3.51 \end{bmatrix} \end{array}.$$

The production cost matrix for Seattle is

$$\begin{array}{c} \\ \text{Material} \\ \text{Labor} \end{array} \begin{array}{cc} \text{Phones} & \text{Calculators} \\ \begin{bmatrix} 4.40 & 6.90 \\ 3.54 & 3.76 \end{bmatrix} \end{array}.$$

(b) The new production cost matrix for Chicago is

$$\begin{array}{c} \\ \text{Material} \\ \text{Labor} \end{array} \begin{array}{cc} \text{Phones} & \text{Calculators} \\ \begin{bmatrix} 4.05 + 0.37 & 7.01 + 0.42 \\ 3.27 + 0.11 & 3.51 + 0.11 \end{bmatrix} \end{array}$$

or $\begin{bmatrix} 4.42 & 7.43 \\ 3.38 & 3.62 \end{bmatrix}.$

10.3   Multiplication of Matrices

In Exercises 1-5, let

$$A = \begin{bmatrix} -2 & 4 \\ 0 & 3 \end{bmatrix} \text{ and } B = \begin{bmatrix} -6 & 2 \\ 4 & 0 \end{bmatrix}.$$

1. $2A = 2 \begin{bmatrix} -2 & 4 \\ 0 & 3 \end{bmatrix} = \begin{bmatrix} -4 & 8 \\ 0 & 6 \end{bmatrix}$

3. $-4B = -4 \begin{bmatrix} -6 & 2 \\ 4 & 0 \end{bmatrix} = \begin{bmatrix} 24 & -8 \\ -16 & 0 \end{bmatrix}$

5. $-4A + 5B = -4 \begin{bmatrix} -2 & 4 \\ 0 & 3 \end{bmatrix} + 5 \begin{bmatrix} -6 & 2 \\ 4 & 0 \end{bmatrix}$

$$= \begin{bmatrix} 8 & -16 \\ 0 & -12 \end{bmatrix} + \begin{bmatrix} -30 & 10 \\ 20 & 0 \end{bmatrix}$$

$$= \begin{bmatrix} -22 & -6 \\ 20 & -12 \end{bmatrix}$$

7. Matrix A size   Matrix B size

   $2 \times \underline{\mathbf{2}}$ $\underline{\mathbf{2}} \times 2$

   The number of columns of A is the same as the number of rows of B, so the product AB exists. The size of the matrix AB is 2×2.

   Matrix B size   Matrix A size

   $2 \times \underline{\mathbf{2}}$ $\underline{\mathbf{2}} \times 2$

   Since the number of columns of B is the same as the number of rows of A, the product BA also exists and has size 2×2.

9. Matrix A size   Matrix B size

   $3 \times \underline{\mathbf{5}}$ $\underline{\mathbf{5}} \times 2$

   Since matrix A has 5 columns and matrix B has 5 rows, the product AB exists and has size 3×2.

   Matrix B size   Matrix A size

   $5 \times \underline{\mathbf{2}}$ $\underline{\mathbf{3}} \times 5$

   Since B has 2 columns and A has 3 rows, the product BA does not exist.

11. Matrix A size   Matrix B size

    $4 \times \underline{\mathbf{2}}$ $\underline{\mathbf{3}} \times 4$

    The number of columns of A is not the same as the number of rows of B, so the product AB does not exist.

    Matrix B size   Matrix A size

    $3 \times \underline{\mathbf{4}}$ $\underline{\mathbf{4}} \times 2$

    The number of columns of B is the same as the number of rows of A, so the product BA exists and has size 3×2.

13. To find the product matrix AB, the number of *columns* of A must be the same as the number of *rows* of B.

15. Call the first matrix A and the second matrix B. The product matrix AB will have size 2×1.

Step 1: Multiply the elements of the first row of A by the corresponding elements of the column of B and add.

$$\begin{bmatrix} 1 & 2 \\ 3 & 4 \end{bmatrix} \begin{bmatrix} -1 \\ 7 \end{bmatrix} \qquad 1(-1) + 2(7) = 13$$

Therefore, 13 is the first row entry of the product matrix AB.

Step 2: Multiply the elements of the second row of A by the corresponding elements of the column of B and add.

$$\begin{bmatrix} 1 & 2 \\ 3 & 4 \end{bmatrix} \begin{bmatrix} -1 \\ 7 \end{bmatrix} \qquad 3(-1) + 4(7) = 25$$

The second row entry of the product is 25.

Step 3: Write the product using the two entries found above.

$$AB = \begin{bmatrix} 1 & 2 \\ 3 & 4 \end{bmatrix} \begin{bmatrix} -1 \\ 7 \end{bmatrix} = \begin{bmatrix} 13 \\ 25 \end{bmatrix}$$

17. $\begin{bmatrix} 1 & 5 & 3 \\ -1 & 2 & 7 \end{bmatrix} \begin{bmatrix} 4 \\ 2 \\ -3 \end{bmatrix}$

$$= \begin{bmatrix} 1 \cdot 4 + 5 \cdot 2 + 3(-3) \\ -1(4) + 2 \cdot 2 + 7(-3) \end{bmatrix}$$

$$= \begin{bmatrix} 5 \\ -21 \end{bmatrix}$$

19. $\begin{bmatrix} 5 & 1 \\ 2 & 3 \end{bmatrix} \begin{bmatrix} 3 & -1 & 0 \\ 1 & 0 & 2 \end{bmatrix}$

$$= \begin{bmatrix} 5 \cdot 3 + 1 \cdot 1 & 5(-1) + 1 \cdot 0 & 5 \cdot 0 + 1 \cdot 2 \\ 2 \cdot 3 + 3 \cdot 1 & 2(-1) + 3 \cdot 0 & 2 \cdot 0 + 3 \cdot 2 \end{bmatrix}$$

$$= \begin{bmatrix} 16 & -5 & 2 \\ 9 & -2 & 6 \end{bmatrix}$$

21. $\begin{bmatrix} 2 & 2 & -1 \\ 3 & 0 & 1 \end{bmatrix} \begin{bmatrix} 0 & 2 \\ -1 & 4 \\ 0 & 2 \end{bmatrix}$

$$= \begin{bmatrix} 2 \cdot 0 + 2(-1) + (-1)0 & 2 \cdot 2 + 2 \cdot 4 + (-1)2 \\ 3 \cdot 0 + 0(-1) + 1(0) & 3 \cdot 2 + 0 \cdot 4 + 1 \cdot 2 \end{bmatrix}$$

$$= \begin{bmatrix} -2 & 10 \\ 0 & 8 \end{bmatrix}$$

23. $\begin{bmatrix} 1 & 2 \\ 3 & 4 \end{bmatrix} \begin{bmatrix} -1 & 5 \\ 7 & 0 \end{bmatrix} = \begin{bmatrix} 1(-1) + 2 \cdot 7 & 1 \cdot 5 + 2 \cdot 0 \\ 3(-1) + 4 \cdot 7 & 3 \cdot 5 + 4 \cdot 0 \end{bmatrix}$

$$= \begin{bmatrix} 13 & 5 \\ 25 & 15 \end{bmatrix}$$

25. $\begin{bmatrix} -2 & -3 & 7 \\ 1 & 5 & 6 \end{bmatrix} \begin{bmatrix} 1 \\ 2 \\ 3 \end{bmatrix} = \begin{bmatrix} -2(1) + (-3)2 + 7 \cdot 3 \\ 1 \cdot 1 + 5 \cdot 2 + 6 \cdot 3 \end{bmatrix}$

$$= \begin{bmatrix} 13 \\ 29 \end{bmatrix}$$

27. $\left(\begin{bmatrix} 4 & 3 \\ 1 & 2 \\ 0 & -5 \end{bmatrix} \begin{bmatrix} 2 & -2 \\ 1 & -1 \end{bmatrix} \right) \begin{bmatrix} 10 \\ 0 \end{bmatrix}$

$$= \begin{bmatrix} 11 & -11 \\ 4 & -4 \\ -5 & 5 \end{bmatrix} \begin{bmatrix} 10 \\ 0 \end{bmatrix} = \begin{bmatrix} 110 \\ 40 \\ -50 \end{bmatrix}$$

29. $\begin{bmatrix} 2 & -2 \\ 1 & -1 \end{bmatrix} \left(\begin{bmatrix} 4 & 3 \\ 1 & 2 \end{bmatrix} + \begin{bmatrix} 7 & 0 \\ -1 & 5 \end{bmatrix} \right)$

$$= \begin{bmatrix} 2 & -2 \\ 1 & -1 \end{bmatrix} \begin{bmatrix} 11 & 3 \\ 0 & 7 \end{bmatrix}$$

$$= \begin{bmatrix} 22 & -8 \\ 11 & -4 \end{bmatrix}$$

31. (a) $AB = \begin{bmatrix} -2 & 4 \\ 1 & 3 \end{bmatrix} \begin{bmatrix} -2 & 1 \\ 3 & 6 \end{bmatrix} = \begin{bmatrix} 16 & 22 \\ 7 & 19 \end{bmatrix}$

(b) $BA = \begin{bmatrix} -2 & 1 \\ 3 & 6 \end{bmatrix} \begin{bmatrix} -2 & 4 \\ 1 & 3 \end{bmatrix} = \begin{bmatrix} 5 & -5 \\ 0 & 30 \end{bmatrix}$

(c) No, AB and BA are not equal here.

(d) No, AB does not always equal BA.

33. Verify that $P(X + T) = PX + PT$.
Find $P(X + T)$ and $PX + PT$ separately and compare their values to see if they are the same.

$P(X + T)$

$$= \begin{bmatrix} m & n \\ p & q \end{bmatrix} \left(\begin{bmatrix} x & y \\ z & w \end{bmatrix} + \begin{bmatrix} r & s \\ t & u \end{bmatrix} \right)$$

$$= \begin{bmatrix} m & n \\ p & q \end{bmatrix} \left(\begin{bmatrix} x+r & y+s \\ z+t & w+u \end{bmatrix} \right)$$

$$= \begin{bmatrix} m(x+r) + n(z+t) & m(y+s) + n(w+u) \\ p(x+r) + q(z+t) & p(y+s) + q(w+u) \end{bmatrix}$$

$$= \begin{bmatrix} mx + mr + nz + nt & my + ms + nw + nu \\ px + pr + qz + qt & py + ps + qw + qu \end{bmatrix}$$

$PX + PT$

$$= \begin{bmatrix} m & n \\ p & q \end{bmatrix} \begin{bmatrix} x & y \\ z & w \end{bmatrix} + \begin{bmatrix} m & n \\ p & q \end{bmatrix} \begin{bmatrix} r & s \\ t & u \end{bmatrix}$$

$$= \begin{bmatrix} mx + nz & my + nw \\ px + qz & py + qw \end{bmatrix} + \begin{bmatrix} mr + nt & ms + nu \\ pr + qt & ps + qu \end{bmatrix}$$

$$= \begin{bmatrix} (mx + nz) + (mr + nt) & (my + nw) + (ms + nu) \\ (px + qz) + (pr + qt) & (py + qw) + (ps + qu) \end{bmatrix}$$

$$= \begin{bmatrix} mx + nz + mr + nt & my + nw + ms + nu \\ px + qz + pr + qt & py + qw + ps + qu \end{bmatrix}$$

$$= \begin{bmatrix} mx + mr + nz + nt & my + ms + nw + nu \\ px + pr + qz + qt & py + ps + qw + qu \end{bmatrix}$$

Observe that the two results are identical. Thus, $P(X + T) = PX + PT$.

35. Verify that $(k+h)P = kP + hP$ for any real numbers k and h.

$$(k + h)P = (k + h)\begin{bmatrix} m & n \\ p & q \end{bmatrix}$$

$$= \begin{bmatrix} (k + h)m & (k + h)n \\ (k + h)p & (k + h)q \end{bmatrix}$$

$$= \begin{bmatrix} km + hm & kn + hn \\ kp + hp & kq + hq \end{bmatrix}$$

$$= \begin{bmatrix} km & kn \\ kp & kq \end{bmatrix} + \begin{bmatrix} hm & hn \\ hp & hq \end{bmatrix}$$

$$= k\begin{bmatrix} m & n \\ p & q \end{bmatrix} + h\begin{bmatrix} m & n \\ p & q \end{bmatrix}$$

$$= kP + hP$$

Thus, $(k + h)P = kP + hP$ for any real numbers k and h.

37. $\begin{bmatrix} 2 & 3 & 1 \\ 1 & -4 & 5 \end{bmatrix} \begin{bmatrix} x_1 \\ x_2 \\ x_3 \end{bmatrix} = \begin{bmatrix} 2x_1 + 3x_2 + x_3 \\ x_1 - 4x_2 + 5x_3 \end{bmatrix}$,

and $\begin{bmatrix} 2x_1 + 3x_2 + x_3 \\ x_1 - 4x_2 + 5x_3 \end{bmatrix} = \begin{bmatrix} 5 \\ 8 \end{bmatrix}$.

This is equivalent to

$$2x_1 + 3x_2 + x_3 = 5$$
$$x_1 - 4x_2 + 5x_3 = 8$$

since corresponding elements of equal matrices must be equal. Reversing this, observe that the given system of linear equations can be written as the matrix equation

$$\begin{bmatrix} 2 & 3 & 1 \\ 1 & -4 & 5 \end{bmatrix} \begin{bmatrix} x_1 \\ x_2 \\ x_3 \end{bmatrix} = \begin{bmatrix} 5 \\ 8 \end{bmatrix}.$$

39. (a) Using a graphing calculator or a computer to find the product matrix. The answer is

$$AC = \begin{bmatrix} 6 & 106 & 158 & 222 & 28 \\ 120 & 139 & 64 & 75 & 115 \\ -146 & -2 & 184 & 144 & -129 \\ 106 & 94 & 24 & 116 & 110 \end{bmatrix}.$$

(b) CA does not exist.

(c) AC and CA are clearly not equal, since CA does not even exist.

41. Use a graphing calculator or computer to find the matrix products and sums. The answers are as follows.

(a) $C + D = \begin{bmatrix} -1 & 5 & 9 & 13 & -1 \\ 7 & 17 & 2 & -10 & 6 \\ 18 & 9 & -12 & 12 & 22 \\ 9 & 4 & 18 & 10 & -3 \\ 1 & 6 & 10 & 28 & 5 \end{bmatrix}$

(b) $(C + D)B = \begin{bmatrix} -2 & -9 & 90 & 77 \\ -42 & -63 & 127 & 62 \\ 413 & 76 & 180 & -56 \\ -29 & -44 & 198 & 85 \\ 137 & 20 & 162 & 103 \end{bmatrix}$

(c) $CB = \begin{bmatrix} -56 & -1 & 1 & 45 \\ -156 & -119 & 76 & 122 \\ 315 & 86 & 118 & -91 \\ -17 & -17 & 116 & 51 \\ 118 & 19 & 125 & 77 \end{bmatrix}$

(d) $DB = \begin{bmatrix} 54 & -8 & 89 & 32 \\ 114 & 56 & 51 & -60 \\ 98 & -10 & 62 & 35 \\ -12 & -27 & 82 & 34 \\ 19 & 1 & 37 & 26 \end{bmatrix}$

(e) $CB + DB = \begin{bmatrix} -2 & -9 & 90 & 77 \\ -42 & -63 & 127 & 62 \\ 413 & 76 & 180 & -56 \\ -29 & -44 & 198 & 85 \\ 137 & 20 & 162 & 103 \end{bmatrix}$

(f) Yes, $(C+D)B$ and $CB+DB$ are equal, as can be seen by observing that the answers to parts (b) and (e) are identical.

43. (a) $XY = \begin{bmatrix} 2 & 1 & 2 & 1 \\ 3 & 2 & 2 & 1 \\ 4 & 3 & 2 & 1 \end{bmatrix} \begin{bmatrix} 5 & 0 & 7 \\ 0 & 10 & 1 \\ 0 & 15 & 2 \\ 10 & 12 & 8 \end{bmatrix}$

$= \begin{bmatrix} 20 & 52 & 27 \\ 25 & 62 & 35 \\ 30 & 72 & 43 \end{bmatrix}$

The rows give the amounts of fat, carbohydrates, and protein, respectively, in each of the daily meals.

(b) $YZ = \begin{bmatrix} 5 & 0 & 7 \\ 0 & 10 & 1 \\ 0 & 15 & 2 \\ 10 & 12 & 8 \end{bmatrix} \begin{bmatrix} 8 \\ 4 \\ 5 \end{bmatrix} = \begin{bmatrix} 75 \\ 45 \\ 70 \\ 168 \end{bmatrix}$

The rows give the number of calories in one exchange of each of the food groups.

(c) Use the matrices found for XY and YZ from parts (a) and (b).

$(XY)Z = \begin{bmatrix} 20 & 52 & 27 \\ 25 & 62 & 35 \\ 30 & 72 & 43 \end{bmatrix} \begin{bmatrix} 8 \\ 4 \\ 5 \end{bmatrix} = \begin{bmatrix} 503 \\ 623 \\ 743 \end{bmatrix}$

$X(YZ) = \begin{bmatrix} 2 & 1 & 2 & 1 \\ 3 & 2 & 2 & 1 \\ 4 & 3 & 2 & 1 \end{bmatrix} \begin{bmatrix} 75 \\ 45 \\ 70 \\ 168 \end{bmatrix} = \begin{bmatrix} 503 \\ 623 \\ 743 \end{bmatrix}$

The rows give the number of calories in each meal.

45. $\dfrac{1}{2}\left(\begin{bmatrix} 60.0 & 68.3 \\ 63.8 & 72.5 \\ 64.5 & 73.6 \\ 67.2 & 74.7 \end{bmatrix} + \begin{bmatrix} 68.0 & 75.6 \\ 70.7 & 78.1 \\ 72.7 & 79.4 \\ 74.3 & 79.9 \end{bmatrix} \right)$

$= \dfrac{1}{2} \begin{bmatrix} 128 & 143.9 \\ 134.5 & 150.6 \\ 137.2 & 153 \\ 141.5 & 154.6 \end{bmatrix}$

$\approx \begin{bmatrix} 64 & 72.0 \\ 67.3 & 75.3 \\ 68.6 & 76.5 \\ 70.8 & 77.3 \end{bmatrix}$

47. (a) The matrices are

$S = \begin{bmatrix} 0.038 & 0.014 \\ 0.022 & 0.008 \\ 0.023 & 0.007 \\ 0.014 & 0.009 \\ 0.010 & 0.011 \end{bmatrix}$ and

$P = \begin{bmatrix} 283 & 1{,}627 & 218 & 199 & 425 \\ 360 & 2{,}038 & 286 & 227 & 460 \\ 468 & 2{,}498 & 362 & 252 & 484 \\ 621 & 2{,}987 & 443 & 278 & 498 \\ 798 & 3{,}451 & 523 & 306 & 508 \end{bmatrix}.$

(b)

	Births	Deaths
1960	58.598	24.97
1970	72.872	30.449
$PS = \quad$ 1980	89.434	36.662
1990	108.373	43.671
2000	127.639	50.783

This product matrix gives the total number of births and deaths (in millions) in each year.

49. (a) $\begin{bmatrix} 10 & 4 & 3 & 5 & 6 \\ 7 & 2 & 2 & 3 & 8 \\ 4 & 5 & 1 & 0 & 10 \\ 0 & 3 & 4 & 5 & 5 \end{bmatrix} \begin{bmatrix} 2 & 3 \\ 1 & 1 \\ 4 & 3 \\ 3 & 3 \\ 1 & 2 \end{bmatrix}$

	A	B
Dept. 1	57	70
Dept. 2	41	54
Dept. 3	27	40
Dept. 4	39	40

with $=$ before Dept. 2.

(b) The total cost to buy from supplier A is $57 + 41 + 27 + 39 = \$164$, and the total cost to buy from supplier B is $70 + 54 + 40 + 40 = \$204$. The company should make the purchase from supplier A, since \$164 is a lower total cost than \$204.

10.4　Matrix Inverses

1. $\begin{bmatrix} 2 & 3 \\ 1 & 1 \end{bmatrix} \begin{bmatrix} -1 & 3 \\ 1 & -2 \end{bmatrix} = \begin{bmatrix} 1 & 0 \\ 0 & 1 \end{bmatrix} = I$

$\begin{bmatrix} -1 & 3 \\ 1 & -2 \end{bmatrix} \begin{bmatrix} 2 & 3 \\ 1 & 1 \end{bmatrix} = \begin{bmatrix} 1 & 0 \\ 0 & 1 \end{bmatrix} = I$

Yes, these matrices are inverses of each other since their product matrix (both ways) is I.

3. $\begin{bmatrix} 2 & 1 \\ 3 & 2 \end{bmatrix} \begin{bmatrix} 2 & 1 \\ -3 & 2 \end{bmatrix} = \begin{bmatrix} 1 & 4 \\ 0 & 7 \end{bmatrix} \neq I$

No, these matrices are not inverses of each other since their product matrix is not I.

5. $\begin{bmatrix} 1 & 2 & 0 \\ 0 & 1 & 0 \\ 0 & 1 & 0 \end{bmatrix} \begin{bmatrix} 1 & -2 & 0 \\ 0 & 1 & 0 \\ 0 & -1 & 1 \end{bmatrix} = \begin{bmatrix} 1 & 0 & 0 \\ 0 & 1 & 0 \\ 0 & 1 & 0 \end{bmatrix} \neq I$

No, these matrices are not inverses of each other.

7. $\begin{bmatrix} 1 & 3 & 3 \\ 1 & 4 & 3 \\ 1 & 3 & 4 \end{bmatrix} \begin{bmatrix} 7 & -3 & -3 \\ -1 & 1 & 0 \\ -1 & 0 & 1 \end{bmatrix} = \begin{bmatrix} 1 & 0 & 0 \\ 0 & 1 & 0 \\ 0 & 0 & 1 \end{bmatrix} = I$

$\begin{bmatrix} 7 & -3 & -3 \\ -1 & 1 & 0 \\ -1 & 0 & 1 \end{bmatrix} \begin{bmatrix} 1 & 3 & 3 \\ 1 & 4 & 3 \\ 1 & 3 & 4 \end{bmatrix} = \begin{bmatrix} 1 & 0 & 0 \\ 0 & 1 & 0 \\ 0 & 0 & 1 \end{bmatrix} = I$

Yes, these matrices are inverses of each other.

9. No, a matrix with a row of all zeros does not have an inverse; the row of all zeros makes it impossible to get all the 1's in the main diagonal of the identity matrix.

11. Let $A = \begin{bmatrix} 1 & -1 \\ 2 & 0 \end{bmatrix}$.

Form the augmented matrix $[A|I]$.

$[A|I] = \begin{bmatrix} 1 & -1 & 1 & 0 \\ 2 & 0 & 0 & 1 \end{bmatrix}$

Perform row operations on $[A|I]$ to get a matrix of the form $[I|B]$.

$\begin{bmatrix} 1 & -1 & 1 & 0 \\ 2 & 0 & 0 & 1 \end{bmatrix}$

$-2R_1 + R_2 \rightarrow R_2 \quad \begin{bmatrix} 1 & -1 & 1 & 0 \\ 0 & 2 & -2 & 1 \end{bmatrix}$

$2R_1 + R_2 \rightarrow R_1 \quad \begin{bmatrix} 2 & 0 & 0 & 1 \\ 0 & 2 & -2 & 1 \end{bmatrix}$

$\frac{1}{2}R_1 \rightarrow R_1 \quad \begin{bmatrix} 1 & 0 & 0 & \frac{1}{2} \\ 0 & 1 & -1 & \frac{1}{2} \end{bmatrix} = [I|B]$
$\frac{1}{2}R_2 \rightarrow R_2$

The matrix B in the last transformation is the desired multiplicative inverse.

$A^{-1} = \begin{bmatrix} 0 & \frac{1}{2} \\ -1 & \frac{1}{2} \end{bmatrix}$

This answer may be checked by showing that $AA^{-1} = I$ and $A^{-1}A = I$.

13. Let $A = \begin{bmatrix} 3 & -1 \\ -5 & 2 \end{bmatrix}$.

$[A|I] = \begin{bmatrix} 3 & -1 & 1 & 0 \\ -5 & 2 & 0 & 1 \end{bmatrix}$

$5R_1 + 3R_2 \rightarrow R_2 \quad \begin{bmatrix} 3 & -1 & 1 & 0 \\ 0 & 1 & 5 & 3 \end{bmatrix}$

$R_1 + R_2 \rightarrow R_1 \quad \begin{bmatrix} 3 & 0 & 6 & 3 \\ 0 & 1 & 5 & 3 \end{bmatrix}$

$\frac{1}{3}R_1 \rightarrow R_1 \quad \begin{bmatrix} 1 & 0 & 2 & 1 \\ 0 & 1 & 5 & 3 \end{bmatrix} = [I|B]$

The desired inverse is

$A^{-1} = \begin{bmatrix} 2 & 1 \\ 5 & 3 \end{bmatrix}.$

15. Let $A = \begin{bmatrix} -6 & 4 \\ -3 & 2 \end{bmatrix}.$

$[A|I] = \begin{bmatrix} -6 & 4 & 1 & 0 \\ -3 & 2 & 0 & 1 \end{bmatrix}$

$R_1 + (-2)R_2 \rightarrow R_2 \quad \begin{bmatrix} -6 & 4 & 1 & 0 \\ 0 & 0 & 1 & -2 \end{bmatrix}$

Because the last row has all zeros to the left of the vertical bar, there is no way to complete the desired transformation. A has no inverse.

17. Let $A = \begin{bmatrix} 1 & 0 & 0 \\ 0 & -1 & 0 \\ 1 & 0 & 1 \end{bmatrix}.$

$[A|I] = \begin{bmatrix} 1 & 0 & 0 & 1 & 0 & 0 \\ 0 & -1 & 0 & 0 & 1 & 0 \\ 1 & 0 & 1 & 0 & 0 & 1 \end{bmatrix}$

$-1R_1 + R_3 \rightarrow R_3 \quad \begin{bmatrix} 1 & 0 & 0 & 1 & 0 & 0 \\ 0 & -1 & 0 & 0 & 1 & 0 \\ 0 & 0 & 1 & -1 & 0 & 1 \end{bmatrix}$

$-1R_2 \rightarrow R_2 \quad \begin{bmatrix} 1 & 0 & 0 & 1 & 0 & 0 \\ 0 & 1 & 0 & 0 & -1 & 0 \\ 0 & 0 & 1 & -1 & 0 & 1 \end{bmatrix}$

$A^{-1} = \begin{bmatrix} 1 & 0 & 0 \\ 0 & -1 & 0 \\ -1 & 0 & 1 \end{bmatrix}$

19.   Let $A = \begin{bmatrix} -1 & -1 & -1 \\ 4 & 5 & 0 \\ 0 & 1 & -3 \end{bmatrix}.$

$[A|I] = \left[\begin{array}{ccc|ccc} -1 & -1 & -1 & 1 & 0 & 0 \\ 4 & 5 & 0 & 0 & 1 & 0 \\ 0 & 1 & -3 & 0 & 0 & 1 \end{array}\right]$

$4R_1 + R_2 \rightarrow R_2$ $\left[\begin{array}{ccc|ccc} -1 & -1 & -1 & 1 & 0 & 0 \\ 0 & 1 & -4 & 4 & 1 & 0 \\ 0 & 1 & -3 & 0 & 0 & 1 \end{array}\right]$

$R_2 + R_1 \rightarrow R_1$ $\left[\begin{array}{ccc|ccc} -1 & 0 & -5 & 5 & 1 & 0 \\ 0 & 1 & -4 & 4 & 1 & 0 \\ 0 & 0 & 1 & -4 & -1 & 1 \end{array}\right]$
$-1R_2 + R_3 \rightarrow R_3$

$5R_3 + R_1 \rightarrow R_1$ $\left[\begin{array}{ccc|ccc} -1 & 0 & 0 & -15 & -4 & 5 \\ 0 & 1 & 0 & -12 & -3 & 4 \\ 0 & 0 & 1 & -4 & -1 & 1 \end{array}\right]$
$4R_3 + R_2 \rightarrow R_2$

$-1R_1 \rightarrow R_1$ $\left[\begin{array}{ccc|ccc} 1 & 0 & 0 & 15 & 4 & -5 \\ 0 & 1 & 0 & -12 & -3 & 4 \\ 0 & 0 & 1 & -4 & -1 & 1 \end{array}\right]$

$A^{-1} = \begin{bmatrix} 15 & 4 & -5 \\ -12 & -3 & 4 \\ -4 & -1 & 1 \end{bmatrix}$

21. Let $A = \begin{bmatrix} 1 & 2 & 3 \\ -3 & -2 & -1 \\ -1 & 0 & 1 \end{bmatrix}.$

$[A|I] = \left[\begin{array}{ccc|ccc} 1 & 2 & 3 & 1 & 0 & 0 \\ -3 & -2 & -1 & 0 & 1 & 0 \\ -1 & 0 & 1 & 0 & 0 & 1 \end{array}\right]$

$3R_1 + R_2 \rightarrow R_2$ $\left[\begin{array}{ccc|ccc} 1 & 2 & 3 & 1 & 0 & 0 \\ 0 & 4 & 8 & 3 & 1 & 0 \\ 0 & 2 & 4 & 1 & 0 & 1 \end{array}\right]$
$R_1 + R_3 \rightarrow R_3$

$R_2 + (-2R_1) \rightarrow R_1$ $\left[\begin{array}{ccc|ccc} -2 & 0 & 2 & 1 & 1 & 0 \\ 0 & 4 & 8 & 3 & 1 & 0 \\ 0 & 0 & 0 & 1 & 1 & -2 \end{array}\right]$
$R_2 + (-2R_3) \rightarrow R_3$

Because the last row has all zeros to the left of the vertical bar, there is no way to complete the desired transformation. A has no inverse.

23. Let $A = \begin{bmatrix} 2 & 4 & 6 \\ -1 & -4 & -3 \\ 0 & 1 & -1 \end{bmatrix}.$

$[A|I] = \left[\begin{array}{ccc|ccc} 2 & 4 & 6 & 1 & 0 & 0 \\ -1 & -4 & -3 & 0 & 1 & 0 \\ 0 & 1 & -1 & 0 & 0 & 1 \end{array}\right]$

$R_1 + 2R_2 \rightarrow R_2$ $\left[\begin{array}{ccc|ccc} 2 & 4 & 6 & 1 & 0 & 0 \\ 0 & -4 & 0 & 1 & 2 & 0 \\ 0 & 1 & -1 & 0 & 0 & 1 \end{array}\right]$

$R_2 + R_1 \rightarrow R_1$ $\left[\begin{array}{ccc|ccc} 2 & 0 & 6 & 2 & 2 & 0 \\ 0 & -4 & 0 & 1 & 2 & 0 \\ 0 & 0 & -4 & 1 & 2 & 4 \end{array}\right]$
$R_2 + 4R_3 \rightarrow R_3$

$6R_3 + 4R_1 \rightarrow R_1$ $\left[\begin{array}{ccc|ccc} 8 & 0 & 0 & 14 & 20 & 24 \\ 0 & -4 & 0 & 1 & 2 & 0 \\ 0 & 0 & -4 & 1 & 2 & 4 \end{array}\right]$

$\frac{1}{8}R_1 \rightarrow R_1$ $\left[\begin{array}{ccc|ccc} 1 & 0 & 0 & \frac{7}{4} & \frac{5}{2} & 3 \\ 0 & 1 & 0 & -\frac{1}{4} & -\frac{1}{2} & 0 \\ 0 & 0 & 1 & -\frac{1}{4} & -\frac{1}{2} & -1 \end{array}\right]$
$-\frac{1}{4}R_2 \rightarrow R_2$
$-\frac{1}{4}R_3 \rightarrow R_3$

$A^{-1} = \begin{bmatrix} \frac{7}{4} & \frac{5}{2} & 3 \\ -\frac{1}{4} & -\frac{1}{2} & 0 \\ -\frac{1}{4} & -\frac{1}{2} & -1 \end{bmatrix}$

25.   Let $A = \begin{bmatrix} 1 & -2 & 3 & 0 \\ 0 & 1 & -1 & 1 \\ -2 & 2 & -2 & 4 \\ 0 & 2 & -3 & 1 \end{bmatrix}.$

$[A|I] = \left[\begin{array}{cccc|cccc} 1 & -2 & 3 & 0 & 1 & 0 & 0 & 0 \\ 0 & 1 & -1 & 1 & 0 & 1 & 0 & 0 \\ -2 & 2 & -2 & 4 & 0 & 0 & 1 & 0 \\ 0 & 2 & -3 & 1 & 0 & 0 & 0 & 1 \end{array}\right]$

$2R_1 + R_3 \rightarrow R_3$ $\left[\begin{array}{cccc|cccc} 1 & -2 & 3 & 0 & 1 & 0 & 0 & 0 \\ 0 & 1 & -1 & 1 & 0 & 1 & 0 & 0 \\ 0 & -2 & 4 & 4 & 2 & 0 & 1 & 0 \\ 0 & 2 & -3 & 1 & 0 & 0 & 0 & 1 \end{array}\right]$

$2R_2 + R_1 \rightarrow R_1$ $\left[\begin{array}{cccc|cccc} 1 & 0 & 1 & 2 & 1 & 2 & 0 & 0 \\ 0 & 1 & -1 & 1 & 0 & 1 & 0 & 0 \\ 0 & 0 & 2 & 6 & 2 & 2 & 1 & 0 \\ 0 & 0 & -1 & -1 & 0 & -2 & 0 & 1 \end{array}\right]$
$2R_2 + R_3 \rightarrow R_3$
$-2R_2 + R_4 \rightarrow R_4$

$R_3 + (-2)R_1 \rightarrow R_1$ $\left[\begin{array}{cccc|cccc} -2 & 0 & 0 & 2 & 0 & -2 & 1 & 0 \\ 0 & 2 & 0 & 8 & 2 & 4 & 1 & 0 \\ 0 & 0 & 2 & 6 & 2 & 2 & 1 & 0 \\ 0 & 0 & 0 & 4 & 2 & -2 & 1 & 2 \end{array}\right]$
$R_3 + 2R_2 \rightarrow R_2$

$R_3 + 2R_4 \rightarrow R_4$

$$
\begin{matrix}
-2R_1 + R_4 \rightarrow R_1 \\
R_2 + (-2)R_4 \rightarrow R_2 \\
2R_3 + (-3)R_4 \rightarrow R_3
\end{matrix}
\left[\begin{array}{cccc|cccc}
4 & 0 & 0 & 0 & 2 & 2 & -1 & 2 \\
0 & 2 & 0 & 0 & -2 & 8 & -1 & -4 \\
0 & 0 & 4 & 0 & -2 & 10 & -1 & -6 \\
0 & 0 & 0 & 4 & 2 & -2 & 1 & 2
\end{array}\right]
$$

$$
\begin{matrix}
\frac{1}{4}R_1 \rightarrow R_1 \\
\frac{1}{2}R_2 \rightarrow R_2 \\
\frac{1}{4}R_3 \rightarrow R_3 \\
\frac{1}{4}R_4 \rightarrow R_4
\end{matrix}
\left[\begin{array}{cccc|cccc}
1 & 0 & 0 & 0 & \frac{1}{2} & \frac{1}{2} & -\frac{1}{4} & \frac{1}{2} \\
0 & 1 & 0 & 0 & -1 & 4 & -\frac{1}{2} & -2 \\
0 & 0 & 1 & 0 & -\frac{1}{2} & \frac{5}{2} & -\frac{1}{4} & -\frac{3}{2} \\
0 & 0 & 0 & 1 & \frac{1}{2} & -\frac{1}{2} & \frac{1}{4} & \frac{1}{2}
\end{array}\right]
$$

$$
A^{-1} = \begin{bmatrix}
\frac{1}{2} & \frac{1}{2} & -\frac{1}{4} & \frac{1}{2} \\
-1 & 4 & -\frac{1}{2} & -2 \\
-\frac{1}{2} & \frac{5}{2} & -\frac{1}{4} & -\frac{3}{2} \\
\frac{1}{2} & -\frac{1}{2} & \frac{1}{4} & \frac{1}{2}
\end{bmatrix}
$$

27. $2x + 3y = 10$
$\qquad x - y = -5$

First write the system in matrix form.

$$
\begin{bmatrix} 2 & 3 \\ 1 & -1 \end{bmatrix}
\begin{bmatrix} x \\ y \end{bmatrix} =
\begin{bmatrix} 10 \\ -5 \end{bmatrix}
$$

Let $A = \begin{bmatrix} 2 & 3 \\ 1 & -1 \end{bmatrix}$, $X = \begin{bmatrix} x \\ y \end{bmatrix}$, $B = \begin{bmatrix} 10 \\ -5 \end{bmatrix}$.

The system is (in matrix form) $AX = B$.
Now use row operations to find A^{-1}.

$$
[A|I] = \begin{bmatrix} 2 & 3 & | & 1 & 0 \\ 1 & -1 & | & 0 & 1 \end{bmatrix}
$$

$$
R_1 + (-2)R_2 \rightarrow R_2 \quad \begin{bmatrix} 2 & 3 & | & 1 & 0 \\ 0 & 5 & | & 1 & -2 \end{bmatrix}
$$

$$
-3R_2 + 5R_1 \rightarrow R_1 \quad \begin{bmatrix} 10 & 0 & | & 2 & 6 \\ 0 & 5 & | & 1 & -2 \end{bmatrix}
$$

$$
\begin{matrix} \frac{1}{10}R_1 \rightarrow R_1 \\ \frac{1}{5}R_2 \rightarrow R_2 \end{matrix} \quad \begin{bmatrix} 1 & 0 & | & \frac{1}{5} & \frac{3}{5} \\ 0 & 1 & | & \frac{1}{5} & -\frac{2}{5} \end{bmatrix}
$$

$$
A^{-1} = \begin{bmatrix} \frac{1}{5} & \frac{3}{5} \\ \frac{1}{5} & -\frac{2}{5} \end{bmatrix}.
$$

Next, find the product $A^{-1}B$.

$$
A^{-1}B = \begin{bmatrix} \frac{1}{5} & \frac{3}{5} \\ \frac{1}{5} & -\frac{2}{5} \end{bmatrix} \begin{bmatrix} 10 \\ -5 \end{bmatrix} = \begin{bmatrix} -1 \\ 4 \end{bmatrix}
$$

Since $X = A^{-1}B$,

$$
X = \begin{bmatrix} x \\ y \end{bmatrix} = \begin{bmatrix} -1 \\ 4 \end{bmatrix}.
$$

Thus, the solution is $(-1, 4)$.

29. $2x + y = 5$
$\qquad 5x + 3y = 13$

Let $A = \begin{bmatrix} 2 & 1 \\ 5 & 3 \end{bmatrix}$, $X = \begin{bmatrix} x \\ y \end{bmatrix}$, $B = \begin{bmatrix} 5 \\ 13 \end{bmatrix}$.

Use row operations to obtain

$$
A^{-1} = \begin{bmatrix} 3 & -1 \\ -5 & 2 \end{bmatrix}.
$$

$$
X = A^{-1}B = \begin{bmatrix} 3 & -1 \\ -5 & 2 \end{bmatrix} \begin{bmatrix} 5 \\ 13 \end{bmatrix} = \begin{bmatrix} 2 \\ 1 \end{bmatrix}
$$

The solution is $(2, 1)$.

31. $-x + y = 1$
$\qquad 2x - y = 1$

Let $A = \begin{bmatrix} -1 & 1 \\ 2 & -1 \end{bmatrix}$, $X = \begin{bmatrix} x \\ y \end{bmatrix}$, $B = \begin{bmatrix} 1 \\ 1 \end{bmatrix}$.

Use row operations to obtain

$$
A^{-1} = \begin{bmatrix} 1 & 1 \\ 2 & 1 \end{bmatrix}.
$$

$$
X = A^{-1}B = \begin{bmatrix} 1 & 1 \\ 2 & 1 \end{bmatrix} \begin{bmatrix} 1 \\ 1 \end{bmatrix} = \begin{bmatrix} 2 \\ 3 \end{bmatrix}
$$

The solution is $(2, 3)$.

33. $-x - 8y = 12$
$\qquad 3x + 24y = -36$

Let $A = \begin{bmatrix} -1 & -8 \\ 3 & 24 \end{bmatrix}$, $X = \begin{bmatrix} x \\ y \end{bmatrix}$, $B = \begin{bmatrix} 12 \\ -36 \end{bmatrix}$.

Using row operations on $[A|I]$ leads to the matrix

$$
\begin{bmatrix} 1 & 8 & | & -1 & 0 \\ 0 & 0 & | & 3 & 1 \end{bmatrix},
$$

but the zeros in the second row indicate that matrix A does not have an inverse. We cannot complete the solution by this method.

Since the second equation is a multiple of the first, the equations are dependent. Solve the first equation of the system for x.

$$
\begin{aligned}
-x - 8y &= 12 \\
-x &= 8y + 12 \\
x &= -8y - 12
\end{aligned}
$$

The solution is $(-8y - 12, y)$, where y is any real number.

35. $-x - y - z = 1$
$4x + 5y \qquad = -2$
$\qquad y - 3z = 3$

has coefficient matrix

$$A = \begin{bmatrix} -1 & -1 & -1 \\ 4 & 5 & 0 \\ 0 & 1 & -3 \end{bmatrix}.$$

In Exercise 19, it was found that

$$A^{-1} = \begin{bmatrix} -1 & -1 & -1 \\ 4 & 5 & 0 \\ 0 & 1 & 3 \end{bmatrix}^{-1}$$

$$= \begin{bmatrix} 15 & 4 & -5 \\ -12 & -3 & 4 \\ -4 & -1 & 1 \end{bmatrix}.$$

Since $X = A^{-1}B$,

$$\begin{bmatrix} x \\ y \\ z \end{bmatrix} = \begin{bmatrix} 15 & 4 & -5 \\ -12 & -3 & 4 \\ -4 & -1 & 1 \end{bmatrix} \begin{bmatrix} 1 \\ -2 \\ 3 \end{bmatrix} = \begin{bmatrix} -8 \\ 6 \\ 1 \end{bmatrix}.$$

The solution is $(-8, 6, 1)$.

37. $2x + 4y + 6z = 4$
$-x - 4y - 3z = 8$
$\qquad y - z = -4$

has coefficient matrix

$$A = \begin{bmatrix} 2 & 4 & 6 \\ -1 & -4 & -3 \\ 0 & 1 & -1 \end{bmatrix}.$$

In Exercise 23, it was found that

$$A^{-1} = \begin{bmatrix} 2 & 4 & 6 \\ -1 & -4 & -3 \\ 0 & 1 & -1 \end{bmatrix}^{-1}$$

$$= \begin{bmatrix} \frac{7}{4} & \frac{5}{2} & 3 \\ -\frac{1}{4} & -\frac{1}{2} & 0 \\ -\frac{1}{4} & -\frac{1}{2} & -1 \end{bmatrix}.$$

Since $X = A^{-1}B$,

$$\begin{bmatrix} x \\ y \\ z \end{bmatrix} = \begin{bmatrix} \frac{7}{4} & \frac{5}{2} & 3 \\ -\frac{1}{4} & -\frac{1}{2} & 0 \\ -\frac{1}{4} & -\frac{1}{2} & -1 \end{bmatrix} \begin{bmatrix} 4 \\ 8 \\ -4 \end{bmatrix} = \begin{bmatrix} 15 \\ -5 \\ -1 \end{bmatrix}.$$

The solution is $(15, -5, -1)$.

39. $2x - 2y \qquad = 5$
$\qquad 4y + 8z = 7$
$x \qquad + 2z = 1$

has coefficient matrix

$$A = \begin{bmatrix} 2 & -2 & 0 \\ 0 & 4 & 8 \\ 1 & 0 & 2 \end{bmatrix}.$$

However, using row operations on $[A|I]$ shows that A does not have an inverse, so another method must be used.

Try the Gauss-Jordan method. The augmented matrix is

$$\begin{bmatrix} 2 & -2 & 0 & | & 5 \\ 0 & 4 & 8 & | & 7 \\ 1 & 0 & 2 & | & 1 \end{bmatrix}.$$

After several row operations, we obtain the matrix

$$\begin{bmatrix} 1 & 0 & 2 & | & \frac{17}{4} \\ 0 & 1 & 2 & | & \frac{7}{4} \\ 0 & 0 & 0 & | & 13 \end{bmatrix}.$$

The bottom row of this matrix shows that the system has no solution, since $0 = 13$ is a false statement.

41. $x - 2y + 3z \qquad = 4$
$\qquad y - z + w = -8$
$-2x + 2y - 2z + 4w = 12$
$\qquad 2y - 3z + w = -4$

has coefficient matrix

$$A = \begin{bmatrix} 1 & -2 & 3 & 0 \\ 0 & 1 & -1 & 1 \\ -2 & 2 & -2 & 4 \\ 0 & 2 & -3 & 1 \end{bmatrix}.$$

In Exercise 25, it was found that

$$A^{-1} = \begin{bmatrix} \frac{1}{2} & \frac{1}{2} & -\frac{1}{4} & \frac{1}{2} \\ -1 & 4 & -\frac{1}{2} & -2 \\ -\frac{1}{2} & \frac{5}{2} & -\frac{1}{4} & -\frac{3}{2} \\ \frac{1}{2} & -\frac{1}{2} & \frac{1}{4} & \frac{1}{2} \end{bmatrix}.$$

Since $X = A^{-1}B$,

$$\begin{bmatrix} x \\ y \\ z \\ w \end{bmatrix} = \begin{bmatrix} \frac{1}{2} & \frac{1}{2} & -\frac{1}{4} & \frac{1}{2} \\ -1 & 4 & -\frac{1}{2} & -2 \\ -\frac{1}{2} & \frac{5}{2} & -\frac{1}{4} & -\frac{3}{2} \\ \frac{1}{2} & -\frac{1}{2} & \frac{1}{4} & \frac{1}{2} \end{bmatrix} \begin{bmatrix} 4 \\ -8 \\ 12 \\ -4 \end{bmatrix} = \begin{bmatrix} -7 \\ -34 \\ -19 \\ 7 \end{bmatrix}.$$

The solution is $(-7, -34, -19, 7)$.

In Exercises 43-47, let $A = \begin{bmatrix} a & b \\ c & d \end{bmatrix}$.

43. $IA = \begin{bmatrix} 1 & 0 \\ 0 & 1 \end{bmatrix} \begin{bmatrix} a & b \\ c & d \end{bmatrix} = \begin{bmatrix} a & b \\ c & d \end{bmatrix} = A$

Thus, $IA = A$.

45. $A \cdot 0 = \begin{bmatrix} a & b \\ c & d \end{bmatrix} \begin{bmatrix} 0 & 0 \\ 0 & 0 \end{bmatrix} = \begin{bmatrix} 0 & 0 \\ 0 & 0 \end{bmatrix} = 0$

Thus, $A \cdot 0 = 0$.

47. In Exercise 46, it was found that

$$A^{-1} = \frac{1}{ad - bc} \begin{bmatrix} d & -b \\ -c & a \end{bmatrix}.$$

$$A^{-1}A = \left(\frac{1}{ad - bc} \begin{bmatrix} d & -b \\ -c & a \end{bmatrix} \right) \begin{bmatrix} a & b \\ c & d \end{bmatrix} = \frac{1}{ad - bc} \left(\begin{bmatrix} d & -b \\ -c & a \end{bmatrix} \begin{bmatrix} a & b \\ c & d \end{bmatrix} \right)$$

$$= \frac{1}{ad - bc} \begin{bmatrix} ad - bc & 0 \\ 0 & ad - bc \end{bmatrix} = \begin{bmatrix} 1 & 0 \\ 0 & 1 \end{bmatrix} = I$$

Thus, $A^{-1}A = I$.

49.
$$AB = O$$
$$A^{-1}(AB) = A^{-1} \cdot O$$
$$(A^{-1}A)B = O$$
$$I \cdot B = O$$
$$B = O$$

Thus, if $AB = O$ and A^{-1} exists, then $B = O$.

51. This exercise should be solved by graphing calculator or computer methods. The solution, which may vary slightly, is

$$C^{-1} = \begin{bmatrix} -0.0477 & -0.0230 & 0.0292 & 0.0895 & -0.0402 \\ 0.0921 & 0.0150 & 0.0321 & 0.0209 & -0.0276 \\ -0.0678 & 0.0315 & -0.0404 & 0.0326 & 0.0373 \\ 0.0171 & -0.0248 & 0.0069 & -0.0003 & 0.0246 \\ -0.0208 & 0.0740 & 0.0096 & -0.1018 & 0.0646 \end{bmatrix}.$$

(Entries are rounded to 4 places.)

53. This exercise should be solved by graphing calculator or computer methods. The solution, which may vary slightly, is

$$D^{-1} = \begin{bmatrix} 0.0394 & 0.0880 & 0.0033 & 0.0530 & -0.1499 \\ -0.1492 & 0.0289 & 0.0187 & 0.1033 & 0.1668 \\ -0.1330 & -0.0543 & 0.0356 & 0.1768 & 0.1055 \\ 0.1407 & 0.0175 & -0.0453 & -0.1344 & 0.0655 \\ 0.0102 & -0.0653 & 0.0993 & 0.0085 & -0.0388 \end{bmatrix}.$$

(Entries are rounded to 4 places.)

55. This exercise should be solved by graphing calculator or computer methods. The solution may vary slightly.

The answer is, yes, $D^{-1}C^{-1} = (CD)^{-1}$.

57. This exercise should be solved by graphing calculator or computer methods. The solution, which may vary slightly, is

$$\begin{bmatrix} 1.51482 \\ 0.053479 \\ -0.637242 \\ 0.462629 \end{bmatrix}.$$

59. Let $x =$ the number of Super Vim tablets,
$y =$ the number of Multitab tablets, and
$z =$ the number of Mighty Mix tablets.

The total number of vitamins is

$$x + y + z.$$

The total amount of niacin is

$$15x + 20y + 25z.$$

The total amount of Vitamin E is

$$12x + 15y + 35z.$$

(a) The system to be solved is

$$\begin{array}{rcrcrcr} x & + & y & + & z & = & 225 \\ 15x & + & 20y & + & 25z & = & 4{,}750 \\ 12x & + & 15y & + & 35z & = & 5{,}225. \end{array}$$

Let $A = \begin{bmatrix} 1 & 1 & 1 \\ 15 & 20 & 25 \\ 12 & 15 & 35 \end{bmatrix}$, $X = \begin{bmatrix} x \\ y \\ z \end{bmatrix}$, $B = \begin{bmatrix} 225 \\ 4{,}750 \\ 5{,}225 \end{bmatrix}$.

Thus, $AX = B$ and

$$\begin{bmatrix} 1 & 1 & 1 \\ 15 & 20 & 25 \\ 12 & 15 & 35 \end{bmatrix} \begin{bmatrix} x \\ y \\ z \end{bmatrix} = \begin{bmatrix} 225 \\ 4{,}750 \\ 5{,}225 \end{bmatrix}.$$

Use row operations to obtain the inverse of the coefficient matrix.

$$A^{-1} = \begin{bmatrix} \frac{65}{17} & -\frac{4}{17} & \frac{1}{17} \\ -\frac{45}{17} & \frac{23}{85} & -\frac{2}{17} \\ -\frac{3}{17} & -\frac{3}{85} & \frac{1}{17} \end{bmatrix}$$

Since $X = A^{-1}B$,

$$\begin{bmatrix} x \\ y \\ z \end{bmatrix} = \begin{bmatrix} \frac{65}{17} & -\frac{4}{17} & \frac{1}{17} \\ -\frac{45}{17} & \frac{23}{85} & -\frac{2}{17} \\ -\frac{3}{17} & -\frac{3}{85} & \frac{1}{17} \end{bmatrix} \begin{bmatrix} 225 \\ 4{,}750 \\ 5{,}225 \end{bmatrix} = \begin{bmatrix} 50 \\ 75 \\ 100 \end{bmatrix}.$$

There are 50 Super Vim tablets, 50 Multitab tablets, and 100 Mighty Mix tablets.

(b) The matrix of constants is changed to

$$B = \begin{bmatrix} 185 \\ 3{,}625 \\ 3{,}750 \end{bmatrix}.$$

$$\begin{bmatrix} x \\ y \\ z \end{bmatrix} = \begin{bmatrix} \frac{65}{17} & -\frac{4}{17} & \frac{1}{17} \\ -\frac{45}{17} & \frac{23}{85} & -\frac{2}{17} \\ -\frac{3}{17} & -\frac{3}{85} & \frac{1}{17} \end{bmatrix} \begin{bmatrix} 185 \\ 3{,}625 \\ 3{,}750 \end{bmatrix} = \begin{bmatrix} 75 \\ 50 \\ 60 \end{bmatrix}$$

There are 75 Super Vim tablets, 50 Multitab tablets, and 60 Mighty Mix tablets.

(c) The matrix of constants is changed to

$$B = \begin{bmatrix} 230 \\ 4{,}450 \\ 4{,}210 \end{bmatrix}.$$

$$\begin{bmatrix} x \\ y \\ z \end{bmatrix} = \begin{bmatrix} \frac{65}{17} & -\frac{4}{17} & \frac{1}{17} \\ -\frac{45}{17} & \frac{23}{85} & -\frac{2}{17} \\ -\frac{3}{17} & -\frac{3}{85} & \frac{1}{17} \end{bmatrix} \begin{bmatrix} 230 \\ 4{,}450 \\ 4{,}210 \end{bmatrix} = \begin{bmatrix} 80 \\ 100 \\ 50 \end{bmatrix}$$

There are 80 Super Vim tablets, 100 Multitab tablets, and 50 Mighty Mix tablets.

61. Let $x =$ the number of transistors,
$y =$ the number of resistors, and
$z =$ the number of computer chips.

Solve the following system:

$$\begin{array}{rcl} 3x + 3y + 2z & = & \text{amount of copper available;} \\ x + 2y + z & = & \text{amount of zinc available;} \\ 2x + y + 2z & = & \text{amount of glass available.} \end{array}$$

First, find the inverse of the coefficient matrix

$$\begin{bmatrix} 3 & 3 & 2 \\ 1 & 2 & 1 \\ 2 & 1 & 2 \end{bmatrix}.$$

$$\left[\begin{array}{ccc|ccc} 3 & 3 & 2 & 1 & 0 & 0 \\ 1 & 2 & 1 & 0 & 1 & 0 \\ 2 & 1 & 2 & 0 & 0 & 1 \end{array}\right]$$

$\begin{array}{c} R_1 + (-3)R_2 \to R_2 \\ 2R_1 + (-3)R_3 \to R_3 \end{array}$ $\left[\begin{array}{ccc|ccc} 3 & 3 & 2 & 1 & 0 & 0 \\ 0 & -3 & -1 & 1 & -3 & 0 \\ 0 & 3 & -2 & 2 & 0 & -3 \end{array}\right]$

$\begin{array}{c} R_2 + R_1 \to R_1 \\ \\ R_2 + R_3 \to R_3 \end{array}$ $\left[\begin{array}{ccc|ccc} 3 & 0 & 1 & 2 & -3 & 0 \\ 0 & -3 & -1 & 1 & -3 & 0 \\ 0 & 0 & -3 & 3 & -3 & -3 \end{array}\right]$

$\begin{array}{c} R_3 + 3R_1 \to R_1 \\ R_3 + (-3)R_2 \to R_2 \end{array}$ $\left[\begin{array}{ccc|ccc} 9 & 0 & 0 & 9 & -12 & -3 \\ 0 & 9 & 0 & 0 & 6 & -3 \\ 0 & 0 & -3 & 3 & -3 & -3 \end{array}\right]$

$\begin{array}{c} \frac{1}{9}R_1 \to R_1 \\ \frac{1}{9}R_2 \to R_2 \\ -\frac{1}{3}R_3 \to R_3 \end{array}$ $\left[\begin{array}{ccc|ccc} 1 & 0 & 0 & 1 & -\frac{4}{3} & -\frac{1}{3} \\ 0 & 1 & 0 & 0 & \frac{2}{3} & -\frac{1}{3} \\ 0 & 0 & 1 & -1 & 1 & 1 \end{array}\right]$

$$A^{-1} = \begin{bmatrix} 1 & -\frac{4}{3} & -\frac{1}{3} \\ 0 & \frac{2}{3} & -\frac{1}{3} \\ -1 & 1 & 1 \end{bmatrix}$$

(a) 810 units of copper, 410 units of zinc, and 490 units of glass

$$\begin{bmatrix} x \\ y \\ z \end{bmatrix} = \begin{bmatrix} 1 & -\frac{4}{3} & -\frac{1}{3} \\ 0 & \frac{2}{3} & -\frac{1}{3} \\ -1 & 1 & 1 \end{bmatrix} \begin{bmatrix} 810 \\ 410 \\ 490 \end{bmatrix} = \begin{bmatrix} 100 \\ 110 \\ 90 \end{bmatrix}$$

100 transistors, 110 resistors, and 90 computer chips can be made.

(b) 765 units of copper, 385 units of zinc, and 470 units of glass

$$\begin{bmatrix} x \\ y \\ z \end{bmatrix} = \begin{bmatrix} 1 & -\frac{4}{3} & -\frac{1}{3} \\ 0 & \frac{2}{3} & -\frac{1}{3} \\ -1 & 1 & 1 \end{bmatrix} \begin{bmatrix} 765 \\ 385 \\ 470 \end{bmatrix} = \begin{bmatrix} 95 \\ 100 \\ 90 \end{bmatrix}$$

95 transistors, 100 resistors, and 90 computer chips can be made.

(c) 1,010 units of copper, 500 units of zinc, and 610 units of glass

$$\begin{bmatrix} x \\ y \\ z \end{bmatrix} = \begin{bmatrix} 1 & -\frac{4}{3} & -\frac{1}{3} \\ 0 & \frac{2}{3} & -\frac{1}{3} \\ -1 & 1 & 1 \end{bmatrix} \begin{bmatrix} 1,010 \\ 500 \\ 610 \end{bmatrix} = \begin{bmatrix} 140 \\ 130 \\ 100 \end{bmatrix}$$

140 transistors, 130 resistors, and 100 computer chips can be made.

63. Let $x =$ the number of pounds of pretzels,
$\quad\quad y =$ the number of pounds of dried
$\quad\quad\quad$ fruit, and
$\quad\quad z =$ the number of pounds of nuts.

The total amount of trail mix is

$$x + y + z.$$

The cost of the trail mix is

$$3x + 4y + 8z.$$

Twice as many pretzels as dried fruit means

$$x = 2y.$$

(a) The system to be solved is

$$\begin{array}{rcl} x + y + z &=& 140 \\ 3x + 4y + 8z &=& 140(6) \\ x &=& 2y, \end{array}$$

which can be rewritten as

$$\begin{array}{rcl} x + y + z &=& 140 \\ 3x + 4y + 8z &=& 840 \\ x - 2y &=& 0. \end{array}$$

Let $A = \begin{bmatrix} 1 & 1 & 1 \\ 3 & 4 & 8 \\ 1 & -2 & 0 \end{bmatrix}$, $X = \begin{bmatrix} x \\ y \\ z \end{bmatrix}$,

$$B = \begin{bmatrix} 140 \\ 840 \\ 0 \end{bmatrix}.$$

Thus, $AX = B$, and

$$\begin{bmatrix} 1 & 1 & 1 \\ 3 & 4 & 8 \\ 1 & -2 & 0 \end{bmatrix} \begin{bmatrix} x \\ y \\ z \end{bmatrix} = \begin{bmatrix} 140 \\ 840 \\ 0 \end{bmatrix}.$$

Use row operations to obtain the inverse of the coefficient matrix.

$$A^{-1} = \begin{bmatrix} \frac{8}{7} & -\frac{1}{7} & \frac{2}{7} \\ \frac{4}{7} & -\frac{1}{14} & -\frac{5}{14} \\ -\frac{5}{7} & \frac{3}{14} & \frac{1}{14} \end{bmatrix}$$

Since $X = A^{-1}B$,

$$\begin{bmatrix} x \\ y \\ z \end{bmatrix} = \begin{bmatrix} \frac{8}{7} & -\frac{1}{7} & \frac{2}{7} \\ \frac{4}{7} & -\frac{1}{14} & -\frac{5}{14} \\ -\frac{5}{7} & \frac{3}{14} & \frac{1}{14} \end{bmatrix} \begin{bmatrix} 140 \\ 840 \\ 0 \end{bmatrix} = \begin{bmatrix} 40 \\ 20 \\ 80 \end{bmatrix}.$$

Use 40 lb of pretzels, 20 lb of dried fruit, and 80 lb of nuts.

(b) For 112 lb at \$6.50/lb, the matrix of constants is changed to

$$B = \begin{bmatrix} 112 \\ 112(6.50) \\ 0 \end{bmatrix} = \begin{bmatrix} 112 \\ 728 \\ 0 \end{bmatrix}.$$

$$\begin{bmatrix} x \\ y \\ z \end{bmatrix} = \begin{bmatrix} \frac{8}{7} & -\frac{1}{7} & \frac{2}{7} \\ \frac{4}{7} & -\frac{1}{14} & -\frac{5}{14} \\ -\frac{5}{7} & \frac{3}{14} & \frac{1}{14} \end{bmatrix} \begin{bmatrix} 112 \\ 728 \\ 0 \end{bmatrix} = \begin{bmatrix} 24 \\ 12 \\ 76 \end{bmatrix}.$$

Use 24 lb of pretzels, 12 lb of dried fruit, and 76 lb of nuts.

(c) For 126 lb at \$5/lb, the matrix of constants is changed to

$$B = \begin{bmatrix} 126 \\ 126(5) \\ 0 \end{bmatrix} = \begin{bmatrix} 126 \\ 630 \\ 0 \end{bmatrix}.$$

$$\begin{bmatrix} x \\ y \\ z \end{bmatrix} = \begin{bmatrix} \frac{8}{7} & -\frac{1}{7} & \frac{2}{7} \\ \frac{4}{7} & -\frac{1}{14} & -\frac{5}{14} \\ -\frac{5}{7} & \frac{3}{14} & \frac{1}{14} \end{bmatrix} \begin{bmatrix} 126 \\ 630 \\ 0 \end{bmatrix} = \begin{bmatrix} 54 \\ 27 \\ 45 \end{bmatrix}.$$

Use 54 lb of pretzels, 27 lb of dried fruit, and 45 lb of nuts.

10.5 Eigenvalues and Eigenvectors

1. $M = \begin{bmatrix} 5 & 0 \\ 2 & 1 \end{bmatrix}$

$$0 = \det(M - \lambda I)$$

$$= \det \begin{bmatrix} 5 - \lambda & 0 \\ 2 & 1 - \lambda \end{bmatrix}$$

$$= (5 - \lambda)(1 - \lambda) - 0(2)$$

$$= (5 - \lambda)(1 - \lambda)$$

$$0 = 5 - \lambda \quad \text{or} \quad 0 = 1 - \lambda$$
$$\lambda = 5 \quad \quad \text{or} \quad \lambda = 1$$

The eigenvalues are 1 and 5.

Find the eigenvector corresponding to $\lambda = 1$.

$$0 = (M - \lambda I)X$$

$$\begin{bmatrix} 0 \\ 0 \end{bmatrix} = \begin{bmatrix} 5 - 1 & 0 \\ 2 & 1 - 1 \end{bmatrix} \begin{bmatrix} x_1 \\ x_2 \end{bmatrix}$$

$$= \begin{bmatrix} 4 & 0 \\ 2 & 0 \end{bmatrix} \begin{bmatrix} x_1 \\ x_2 \end{bmatrix}$$

The augmented matrix for this system is

$$\begin{bmatrix} 4 & 0 & | & 0 \\ 2 & 0. & | & 0 \end{bmatrix}.$$

A solution to this system is

$$\begin{bmatrix} x_1 \\ x_2 \end{bmatrix} = \begin{bmatrix} 0 \\ 1 \end{bmatrix}.$$

Now find the eigenvector corresponding to $\lambda = 5$.

$$0 = (M - \lambda I)X$$

$$\begin{bmatrix} 0 \\ 0 \end{bmatrix} = \begin{bmatrix} 5 - 5 & 0 \\ 2 & 1 - 5 \end{bmatrix} \begin{bmatrix} x_1 \\ x_2 \end{bmatrix}$$

$$= \begin{bmatrix} 0 & 0 \\ 2 & -4 \end{bmatrix} \begin{bmatrix} x_1 \\ x_2 \end{bmatrix}$$

The augmented matrix for this system is

$$\begin{bmatrix} 0 & 0 & | & 0 \\ 2 & -4 & | & 0 \end{bmatrix}.$$

A solution to this system is

$$\begin{bmatrix} x_1 \\ x_2 \end{bmatrix} = \begin{bmatrix} 2 \\ 1 \end{bmatrix}.$$

The given matrix has the following eigenvalues and corresponding eigenvectors.

$$\lambda = 1, \begin{bmatrix} 0 \\ 1 \end{bmatrix}; \lambda = 5, \begin{bmatrix} 2 \\ 1 \end{bmatrix}$$

3. $M = \begin{bmatrix} 3 & 2 \\ 3 & 8 \end{bmatrix}$

$$0 = \det(M - \lambda I)$$

$$= \det \begin{bmatrix} 3 - \lambda & 2 \\ 3 & 8 - \lambda \end{bmatrix}$$

$$= (3 - \lambda)(8 - \lambda) - 3(2)$$

$$= \lambda^2 - 11\lambda + 18$$

$$= (\lambda - 9)(\lambda - 2)$$

$0 = \lambda - 9$ or $0 = \lambda - 2$

$\lambda = 9$ or $\lambda = 2$

The eigenvalues are 2 and 9.

Find the eigenvector corresponding to $\lambda = 2$.

$$0 = (M - \lambda I)X$$

$$\begin{bmatrix} 0 \\ 0 \end{bmatrix} = \begin{bmatrix} 3-2 & 2 \\ 3 & 8-2 \end{bmatrix} \begin{bmatrix} x_1 \\ x_2 \end{bmatrix}$$

$$= \begin{bmatrix} 1 & 2 \\ 3 & 6 \end{bmatrix} \begin{bmatrix} x_1 \\ x_2 \end{bmatrix}$$

The augmented matrix for this system is

$$\begin{bmatrix} 1 & 2 & | & 0 \\ 3 & 6 & | & 0 \end{bmatrix}.$$

A solution to this system is

$$\begin{bmatrix} x_1 \\ x_2 \end{bmatrix} = \begin{bmatrix} 2 \\ -1 \end{bmatrix}.$$

Now find the eigenvector corresponding to $\lambda = 9$.

$$0 = (M - \lambda I)X$$

$$\begin{bmatrix} 0 \\ 0 \end{bmatrix} = \begin{bmatrix} 3-9 & 2 \\ 3 & 8-9 \end{bmatrix} \begin{bmatrix} x_1 \\ x_2 \end{bmatrix}$$

$$= \begin{bmatrix} -6 & 2 \\ 3 & -1 \end{bmatrix} \begin{bmatrix} x_1 \\ x_2 \end{bmatrix}$$

The augmented matrix for this system is

$$\begin{bmatrix} -6 & 2 & | & 0 \\ 3 & -1 & | & 0 \end{bmatrix}.$$

A solution to this system is

$$\begin{bmatrix} x_1 \\ x_2 \end{bmatrix} = \begin{bmatrix} 1 \\ 3 \end{bmatrix}.$$

The given matrix has the following eigenvalues and corresponding eigenvectors.

$$\lambda = 2, \begin{bmatrix} 2 \\ -1 \end{bmatrix}, \lambda = 9, \begin{bmatrix} 1 \\ 3 \end{bmatrix}$$

5. $M = \begin{bmatrix} 4 & -3 \\ 2 & -1 \end{bmatrix}$

$0 = \det(M - \lambda I)$

$$= \det \begin{bmatrix} 4-\lambda & -3 \\ 2 & -1-\lambda \end{bmatrix}$$

$= (4-\lambda)(-1-\lambda) - (-3)(2)$

$= \lambda^2 - 3\lambda + 2$

$= (\lambda - 2)(\lambda - 1)$

$0 = \lambda - 2$ or $0 = \lambda - 1$

$\lambda = 2$ or $\lambda = 1$

The eigenvalues are 1 and 2.

Find the eigenvector corresponding to $\lambda = 1$.

$$0 = (M - \lambda I)X$$

$$\begin{bmatrix} 0 \\ 0 \end{bmatrix} = \begin{bmatrix} 4-1 & -3 \\ 2 & -1-1 \end{bmatrix} \begin{bmatrix} x_1 \\ x_2 \end{bmatrix}$$

$$= \begin{bmatrix} 3 & -3 \\ 2 & -2 \end{bmatrix} \begin{bmatrix} x_1 \\ x_2 \end{bmatrix}$$

The augmented matrix for this system is

$$\begin{bmatrix} 3 & -3 & | & 0 \\ 2 & -2 & | & 0 \end{bmatrix}.$$

A solution to this system is

$$\begin{bmatrix} x_1 \\ x_2 \end{bmatrix} = \begin{bmatrix} 1 \\ 1 \end{bmatrix}.$$

Now find the eigenvector corresponding to $\lambda = 2$.

$$0 = (M - \lambda I)X$$

$$\begin{bmatrix} 0 \\ 0 \end{bmatrix} = \begin{bmatrix} 4-2 & -3 \\ 2 & -1-2 \end{bmatrix} \begin{bmatrix} x_1 \\ x_2 \end{bmatrix}$$

$$= \begin{bmatrix} 2 & -3 \\ 2 & -3 \end{bmatrix} \begin{bmatrix} x_1 \\ x_2 \end{bmatrix}$$

The augmented matrix for this system is

$$\begin{bmatrix} 2 & -3 & | & 0 \\ 2 & -3 & | & 0 \end{bmatrix}.$$

A solution to this system is

$$\begin{bmatrix} x_1 \\ x_2 \end{bmatrix} = \begin{bmatrix} 3 \\ 2 \end{bmatrix}.$$

The given matrix has the following eigenvalues and corresponding eigenvectors.

$$\lambda = 1, \begin{bmatrix} 1 \\ 1 \end{bmatrix}, \lambda = 2, \begin{bmatrix} 3 \\ 2 \end{bmatrix}$$

7. $M = \begin{bmatrix} 4 & 0 & 0 \\ 3 & -1 & 0 \\ 2 & 5 & -3 \end{bmatrix}$

$0 = \det(M - \lambda I)$

$$= \det \begin{bmatrix} 4-\lambda & 0 & 0 \\ 3 & -1-\lambda & 0 \\ 2 & 5 & -3-\lambda \end{bmatrix}$$

$= (4-\lambda)[(-1-\lambda)(-3-\lambda) - 0(5)] - 0 + 0$

$= (4-\lambda)(-1-\lambda)(-3-\lambda)$

$0 = 4 - \lambda$ or $0 = -1 - \lambda$ or $0 = -3 - \lambda$
$\lambda = 4$ or $\lambda = -1$ or $\lambda = -3$

The eigenvalues are $-3, -1,$ and 4.
Find the eigenvector corresponding to $\lambda = -3$.

$$0 = (M - \lambda I)X$$

$$\begin{bmatrix} 0 \\ 0 \\ 0 \end{bmatrix} = \begin{bmatrix} 4-(-3) & 0 & 0 \\ 3 & -1-(-3) & 0 \\ -2 & 5 & -3-(-3) \end{bmatrix} \begin{bmatrix} x_1 \\ x_2 \\ x_3 \end{bmatrix}$$

$$= \begin{bmatrix} 7 & 0 & 0 \\ 3 & 2 & 0 \\ 2 & 5 & 0 \end{bmatrix} \begin{bmatrix} x_1 \\ x_2 \\ x_3 \end{bmatrix}$$

The augmented matrix for this system is

$$\begin{bmatrix} 7 & 0 & 0 & | & 0 \\ 3 & 2 & 0 & | & 0 \\ 2 & 5 & 0 & | & 0 \end{bmatrix}.$$

A solution to this system is

$$\begin{bmatrix} x_1 \\ x_2 \\ x_3 \end{bmatrix} = \begin{bmatrix} 0 \\ 0 \\ 1 \end{bmatrix}.$$

Now find the eigenvector corresponding to $\lambda = -1$.

$$0 = (M - \lambda I)X$$

$$\begin{bmatrix} 0 \\ 0 \\ 0 \end{bmatrix} = \begin{bmatrix} 4-(-1) & 0 & 0 \\ 3 & -1-(-1) & 0 \\ 2 & 5 & -3-(-1) \end{bmatrix} \begin{bmatrix} x_1 \\ x_2 \\ x_3 \end{bmatrix}$$

$$= \begin{bmatrix} 5 & 0 & 0 \\ 3 & 0 & 0 \\ 2 & 5 & -2 \end{bmatrix} \begin{bmatrix} x_1 \\ x_2 \\ x_3 \end{bmatrix}$$

The augmented matrix for this system is

$$\begin{bmatrix} 5 & 0 & 0 & | & x_1 \\ 3 & 0 & 0 & | & x_2 \\ 2 & 5 & -2 & | & x_3 \end{bmatrix}.$$

A solution to this system is

$$\begin{bmatrix} x_1 \\ x_2 \\ x_3 \end{bmatrix} = \begin{bmatrix} 0 \\ 2 \\ 5 \end{bmatrix}.$$

Now find the eigenvector corresponding to $\lambda = 4$.

$$0 = (M - \lambda I)X$$

$$\begin{bmatrix} 0 \\ 0 \\ 0 \end{bmatrix} = \begin{bmatrix} 4-4 & 0 & 0 \\ 3 & -1-4 & 0 \\ 2 & 5 & -3-4 \end{bmatrix} \begin{bmatrix} x_1 \\ x_2 \\ x_3 \end{bmatrix}$$

$$= \begin{bmatrix} 0 & 0 & 0 \\ 3 & -5 & 0 \\ 2 & 5 & -7 \end{bmatrix} \begin{bmatrix} x_1 \\ x_2 \\ x_3 \end{bmatrix}$$

The augmented matrix for this system is

$$\begin{bmatrix} 0 & 0 & 0 & | & 0 \\ 3 & -5 & 0 & | & 0 \\ 2 & 5 & -7 & | & 0 \end{bmatrix}.$$

A solution to this system is

$$\begin{bmatrix} x_1 \\ x_2 \\ x_3 \end{bmatrix} = \begin{bmatrix} 35 \\ 21 \\ 25 \end{bmatrix}.$$

The given matrix has the following eigenvalues and corresponding eigenvectors.

$$\lambda = -3, \begin{bmatrix} 0 \\ 0 \\ 1 \end{bmatrix}; \lambda = -1, \begin{bmatrix} 0 \\ 2 \\ 5 \end{bmatrix}; \lambda = 4, \begin{bmatrix} 35 \\ 21 \\ 25 \end{bmatrix}$$

9. $ax + by = 0$
$cx + dy = 0$

The augmented matrix for this system is

$$\begin{bmatrix} a & b & | & 0 \\ c & d & | & 0 \end{bmatrix}.$$

Now

$$ad - bc = 0$$
$$ad = bc$$
$$a = \frac{bc}{d}$$

Now substitute $\frac{bc}{d}$ for a in the augmented matrix.

$$\begin{bmatrix} \frac{bc}{d} & b & | & 0 \\ c & d & | & 0 \end{bmatrix}$$

$R_1/b \rightarrow R_1$ $\begin{bmatrix} \frac{c}{d} & 1 & | & 0 \\ c & d & | & 0 \end{bmatrix}$

$dR_1 \rightarrow R_1$ $\begin{bmatrix} c & d & | & 0 \\ c & d & | & 0 \end{bmatrix}$

Since R_1 is a multiple of R_2, the system of equations represented by the matrix is dependent.

11. $M = \begin{bmatrix} 10 & -9 \\ 4 & -2 \end{bmatrix}$

$0 = \det(M - \lambda I)$

$= \det \begin{bmatrix} 10 - \lambda & -9 \\ 4 & -2 - \lambda \end{bmatrix}$

$= (10 - \lambda)(-2 - \lambda) - (-9)(4)$

$= \lambda^2 - 8\lambda + 16$

$= (\lambda - 4)^2$

$0 = \lambda - 4$

$\lambda = 4$

The eigenvalue is 4.

Find the eigenvector corresponding to $\lambda = 4$.

$0 = (M - \lambda I)X$

$\begin{bmatrix} 0 \\ 0 \end{bmatrix} = \begin{bmatrix} 10 - 4 & -9 \\ 4 & -2 - 4 \end{bmatrix} \begin{bmatrix} x_1 \\ x_2 \end{bmatrix}$

$= \begin{bmatrix} 6 & -9 \\ 4 & -6 \end{bmatrix} \begin{bmatrix} x_1 \\ x_2 \end{bmatrix}$

The augmented matrix for this system is

$\begin{bmatrix} 6 & -9 & | & 0 \\ 4 & -6 & | & 0 \end{bmatrix}$

A solution to this system is

$\begin{bmatrix} x_1 \\ x_2 \end{bmatrix} = \begin{bmatrix} 3 \\ 2 \end{bmatrix}.$

The eigenvalue and eigenvector of the given system are 4, $\begin{bmatrix} 3 \\ 2 \end{bmatrix}$.

13. $M = \begin{bmatrix} 1 & 5 \\ -2 & 3 \end{bmatrix}$

$0 = \det(M - \lambda I)$

$= \begin{bmatrix} 1 - \lambda & 5 \\ -2 & 3 - \lambda \end{bmatrix}$

$= (1 - \lambda)(3 - \lambda) - 5(-2)$

$= \lambda^2 - 4\lambda + 3 + 10$

$= \lambda^2 - 4\lambda + 13$

Solve using the quadratic formula.

$\lambda = \dfrac{-(-4) \pm \sqrt{(-4)^2 - 4(13)}}{2}$

$= \dfrac{4 \pm \sqrt{-36}}{2}$

$= \dfrac{4 \pm 6i}{2}$

$= 2 \pm 3i$

The eigenvalues are $2 - 3i$ and $2 + 3i$.
Find the eigenvector corresponding to $2 - 3i$.

$0 = (M - \lambda I)X$

$\begin{bmatrix} 0 \\ 0 \end{bmatrix} = \begin{bmatrix} 1 - (2 - 3i) & 5 \\ -2 & 3 - (2 - 3i) \end{bmatrix} \begin{bmatrix} x_1 \\ x_2 \end{bmatrix}$

$= \begin{bmatrix} -1 + 3i & 5 \\ -2 & 1 + 3i \end{bmatrix} \begin{bmatrix} x_1 \\ x_2 \end{bmatrix}$

The augmented matrix for this system is

$\begin{bmatrix} -1 + 3i & 5 & | & 0 \\ -2 & 1 + 3i & | & 0 \end{bmatrix}$

$(1 + 3i)R_1 \rightarrow R_1 \quad \begin{bmatrix} -10 & 5(1 + 3i) & | & 0 \\ -2 & 1 + 3i & | & 0 \end{bmatrix}$

$\dfrac{R_1}{5} \rightarrow R_1 \quad \begin{bmatrix} -2 & 1 + 3i & | & 0 \\ -2 & 1 + 3i & | & 0 \end{bmatrix}$

A solution to this system is

$\begin{bmatrix} x_1 \\ x_2 \end{bmatrix} = \begin{bmatrix} 1 + 3i \\ 2 \end{bmatrix}.$

Now find the eigenvector corresponding to $\lambda = 2 + 3i$.

$0 = (M - \lambda I)X$

$\begin{bmatrix} 0 \\ 0 \end{bmatrix} = \begin{bmatrix} 1 - (2 + 3i) & 5 \\ -2 & 3 - (2 + 3i) \end{bmatrix} \begin{bmatrix} x_1 \\ x_2 \end{bmatrix}$

$= \begin{bmatrix} -1 - 3i & 5 \\ -2 & 1 - 3i \end{bmatrix} \begin{bmatrix} x_1 \\ x_2 \end{bmatrix}$

The augmented matrix for this system is

$\begin{bmatrix} -1 - 3i & 5 & | & 0 \\ -2 & 1 - 3i & | & 0 \end{bmatrix}$

A solution to this system is

$\begin{bmatrix} x_1 \\ x_2 \end{bmatrix} = \begin{bmatrix} 1 - 3i \\ 2 \end{bmatrix}.$

The given matrix has the following eigenvalues and corresponding eigenvectors.

$\lambda = 2 - 3i, \begin{bmatrix} 1 - 3i \\ 2 \end{bmatrix}; \lambda = 2 + 3i, \begin{bmatrix} 1 - 3i \\ 2 \end{bmatrix}$

15. Find the eigenvalues and eigenvectors.

$$M = \begin{bmatrix} 0.5 & 0.9 \\ 1.4 & 0 \end{bmatrix}$$

$$0 = \det(M - \lambda I)$$

$$= \det \begin{bmatrix} 0.5 - \lambda & 0.9 \\ 1.4 & 0 - \lambda \end{bmatrix}$$

$$= (0.5 - \lambda)(-\lambda) - 0.9(1.4)$$
$$= \lambda^2 - 0.5\lambda - 1.26$$

Solve using the quadratic formula.

$$\lambda = \frac{-(-0.5) \pm \sqrt{(-0.5)^2 - 4(-1.26)}}{2}$$

$$= \frac{0.5 \pm \sqrt{5.29}}{2}$$

$$= \frac{0.5 \pm 2.3}{2}$$

$$= 1.4, -0.9$$

Find the eigenvector corresponding to $\lambda = 1.4$.

$$0 = (M - \lambda I)X$$

$$\begin{bmatrix} 0 \\ 0 \end{bmatrix} = \begin{bmatrix} 0.5 - 1.4 & 0.9 \\ 1.4 & 0 - 1.4 \end{bmatrix} \begin{bmatrix} x_1 \\ x_2 \end{bmatrix}$$

$$= \begin{bmatrix} -0.9 & 0.9 \\ 1.4 & -1.4 \end{bmatrix} \begin{bmatrix} x_1 \\ x_2 \end{bmatrix}$$

The augmented matrix for this system is

$$\begin{bmatrix} -0.9 & 0.9 & | & 0 \\ 1.4 & -1.4 & | & 0 \end{bmatrix}$$

A solution to this system is

$$\begin{bmatrix} x_1 \\ x_2 \end{bmatrix} = \begin{bmatrix} 1 \\ 1 \end{bmatrix}.$$

The eigenvector corresponding to $\lambda = -0.9$ contains a negative value so it cannot be considered as a population.

(a) Since the initial population is 10,000 a multiple of the eigenvector is used.

$$\begin{bmatrix} x_1 \\ x_2 \end{bmatrix} = k \begin{bmatrix} 1 \\ 1 \end{bmatrix} = \begin{bmatrix} k \\ k \end{bmatrix}$$

$$k + k = 10,000$$
$$2k = 10,000$$
$$k = 5,000$$

(b) Therefore, the initial population is

$$\begin{bmatrix} 5,000 \\ 5,000 \end{bmatrix}$$

with a growth factor of 1.4.

17. Find the eigenvalues and eigenvectors.

$$M = \begin{bmatrix} 0.2 & 0.7 \\ 0.7 & 0.2 \end{bmatrix}$$

$$0 = \det(M - \lambda I)$$

$$= \det \begin{bmatrix} 0.2 - \lambda & 0.7 \\ 0.7 & 0.2 - \lambda \end{bmatrix}$$

$$= (0.2 - \lambda)(0.2 - \lambda) - 0.7(0.7)$$
$$= \lambda^2 - 0.4\lambda - 0.45$$

Solve using the quadratic formula.

$$\lambda = \frac{-(0.4) \pm \sqrt{(-0.4)^2 - 4(-0.45)}}{2}$$

$$= \frac{0.4 \pm \sqrt{1.96}}{2}$$

$$= \frac{0.4 \pm 1.4}{2}$$

$$= 0.9, -0.5$$

Find the eigenvector corresponding to $\lambda = 0.9$.

$$0 = (M - \lambda I)X$$

$$\begin{bmatrix} 0 \\ 0 \end{bmatrix} = \begin{bmatrix} 0.2 - 0.9 & 0.7 \\ 0.7 & 0.2 - 0.9 \end{bmatrix} \begin{bmatrix} x_1 \\ x_2 \end{bmatrix}$$

$$= \begin{bmatrix} -0.7 & 0.7 \\ 0.7 & -0.7 \end{bmatrix} \begin{bmatrix} x_1 \\ x_2 \end{bmatrix}$$

The augmented matrix for this system is

$$\begin{bmatrix} -0.7 & 0.7 & | & 0 \\ 0.7 & -0.7 & | & 0 \end{bmatrix}$$

A solution to this system is

$$\begin{bmatrix} x_1 \\ x_2 \end{bmatrix} = \begin{bmatrix} 1 \\ 1 \end{bmatrix}.$$

The eigenvector corresponding to $\lambda = -0.5$ contains a negative value so it cannot be considered as a population.

(a) Since the initial population is 10,000 a multiple of the eigenvector is used.

$$\begin{bmatrix} x_1 \\ x_2 \end{bmatrix} = k \begin{bmatrix} 1 \\ 1 \end{bmatrix} = \begin{bmatrix} k \\ k \end{bmatrix}$$

$$k + k = 10,000$$
$$2k = 10,000$$
$$k = 5,000$$

(b) Therefore, the initial population is

$$\begin{bmatrix} 5,000 \\ 5,000 \end{bmatrix}$$

with a growth factor of 0.9.

19. Find the eigenvalues and eigenvectors.

$$M = \begin{bmatrix} 0.1 & 1.2 \\ 0.4 & 0.3 \end{bmatrix}$$

$$0 = \det(M - \lambda I)$$

$$= \det \begin{bmatrix} 0.1 - \lambda & 1.2 \\ 0.4 & 0.3 - \lambda \end{bmatrix}$$

$$= (0.1 - \lambda)(0.3 - \lambda) - 1.2(0.4)$$

$$= \lambda^2 - 0.4\lambda - 0.45$$

Solve using the quadratic formula.

$$\lambda = \frac{-(0.4) \pm \sqrt{(-0.4)^2 - 4(-0.45)}}{2}$$

$$= \frac{0.4 \pm \sqrt{1.96}}{2}$$

$$= \frac{0.4 \pm 1.4}{2}$$

$$= 0.9, -0.5$$

Find the eigenvector corresponding to $\lambda = 0.9$.

$$0 = (M - \lambda I)X$$

$$\begin{bmatrix} 0 \\ 0 \end{bmatrix} = \begin{bmatrix} 0.1 - 0.9 & 1.2 \\ 0.4 & 0.3 - 0.9 \end{bmatrix} \begin{bmatrix} x_1 \\ x_2 \end{bmatrix}$$

$$= \begin{bmatrix} -0.8 & 1.2 \\ 0.4 & -0.6 \end{bmatrix} \begin{bmatrix} x_1 \\ x_2 \end{bmatrix}$$

The augmented matrix for this system is

$$\begin{bmatrix} -0.8 & 1.2 & 0 \\ 0.4 & -0.6 & 0 \end{bmatrix}$$

A solution to this system is

$$\begin{bmatrix} x_1 \\ x_2 \end{bmatrix} = \begin{bmatrix} 3 \\ 2 \end{bmatrix}.$$

The eigenvector corresponding to $\lambda = -0.5$ contains a negative value so it cannot be considered as a population.

(a) Since the initial population is 10,000 a multiple of the eigenvector is used.

$$\begin{bmatrix} x_1 \\ x_2 \end{bmatrix} = k \begin{bmatrix} 3 \\ 2 \end{bmatrix} = \begin{bmatrix} 3k \\ 2k \end{bmatrix}$$

$$3k + 2k = 10,000$$

$$5k = 10,000$$

$$k = 2,000$$

(b) Therefore, the initial population is

$$\begin{bmatrix} 3(2,000) \\ 2(2,000) \end{bmatrix} \text{ or } \begin{bmatrix} 6,000 \\ 4,000 \end{bmatrix}$$

with a growth factor of 0.9.

21. The Leslie matrix, $M = \begin{bmatrix} 0.5 & 0.9 \\ 1.4 & 0 \end{bmatrix}$ has

eigenvalues $\lambda_1 = -0.9$ and $\lambda_2 = 1.4$ with

corresponding eigenvectors $V_1 = \begin{bmatrix} -9 \\ 14 \end{bmatrix}$ and

$V_2 = \begin{bmatrix} 1 \\ 1 \end{bmatrix}$, respectively. The initial population

is $X(0) = \begin{bmatrix} 3,000 \\ 1,000 \end{bmatrix}$.

Find values a and b such that $aV_1 + bV_2 = X(0)$.

The equation

$$a \begin{bmatrix} -9 \\ 14 \end{bmatrix} + b \begin{bmatrix} 1 \\ 1 \end{bmatrix} = \begin{bmatrix} 3,000 \\ 1,000 \end{bmatrix}$$

corresponds to the augmented matrix

$$\begin{bmatrix} -9 & 1 & 3,000 \\ 14 & 1 & 1,000 \end{bmatrix}.$$

A solution to this system is

$$\begin{bmatrix} a \\ b \end{bmatrix} = \begin{bmatrix} \frac{-2,000}{23} \\ \frac{51,000}{23} \end{bmatrix}.$$

$$X_n = M^n X(0) = M^n \left(\frac{-2,000}{3} V_1 + \frac{51,000}{23} V_2 \right)$$

$$= \frac{-2,000}{23} M^n V_1 + \frac{51,000}{23} M^n V_2$$

$$= \frac{-2,000}{23} \lambda_1^n V_1 + \frac{51,000}{23} \lambda_1^n V_2$$

$$= \frac{-2,000}{23} (-0.9)^n \begin{bmatrix} 9 \\ 14 \end{bmatrix}$$

$$+ \frac{51,000}{23} (1.4)^n \begin{bmatrix} 1 \\ 1 \end{bmatrix}$$

Since $\lim_{n \to \infty} (-0.9)^n = 0$, the first term approaches zero as time passes. Therefore, when n is large,

$$X(n) \approx \frac{51,000}{23} (1.4)^n \begin{bmatrix} 1 \\ 1 \end{bmatrix}.$$

Chapter 10 Review Exercises

3. $2x + 3y = 10$ (1)
$-3x + y = 18$ (2)

Eliminate x in equation (2).

$$
\begin{array}{ll}
 & 2x + 3y = 10 \quad (1) \\
3R_1 + 2R_2 \rightarrow R_2 & 11y = 66 \quad (3)
\end{array}
$$

Make each leading coefficient equal 1.

$$
\begin{array}{ll}
\frac{1}{2}R_1 \rightarrow R_1 & x + \frac{3}{2}y = 5 \quad (4) \\
\frac{1}{11}R_2 \rightarrow R_2 & \phantom{x + \frac{3}{2}}y = 6 \quad (5)
\end{array}
$$

Substitute 6 for y in equation (4) to get $x = -4$.

The solution is $(-4, 6)$.

5. $2x - 3y + z = -5$ (1)
$x + 4y + 2z = 13$ (2)
$5x + 5y + 3z = 14$ (3)

Eliminate x in equations (2) and (3).

$$
\begin{array}{ll}
 & 2x - 3y + z = -5 \quad (1) \\
-2R_2 + R_1 \rightarrow R_2 & -11y - 3z = -31 \quad (4) \\
5R_1 + (-2)R_3 \rightarrow R_3 & -25y - z = -53 \quad (5)
\end{array}
$$

Eliminate y in equation (5).

$$
\begin{array}{ll}
 & 2x - 3y + z = -5 \quad (1) \\
 & -11y - 3z = -31 \quad (4) \\
-25R_2 + 11R_3 \rightarrow R_3 & 64z = 192 \quad (6)
\end{array}
$$

Make each leading coefficient equal 1.

$$
\begin{array}{ll}
\frac{1}{2}R_1 \rightarrow R_1 & x - \frac{3}{2}y + \frac{1}{2}z = -\frac{5}{2} \quad (7) \\
-\frac{1}{11}R_2 \rightarrow R_2 & \phantom{x - \frac{3}{2}}y + \frac{3}{11}z = \frac{31}{11} \quad (8) \\
\frac{1}{64}R_3 \rightarrow R_3 & \phantom{x - \frac{3}{2}y + \frac{3}{11}}z = 3 \quad (9)
\end{array}
$$

Substitute 3 for z in equation (8) to get $y = 2$.
Substitute 3 for z and 2 for y in equation (7) to get $x = -1$.

The solution is $(-1, 2, 3)$.

7. $2x + 4y = -6$
$-3x - 5y = 12$

Write the augmented matrix and use row operations.

$$
\begin{bmatrix} 2 & 4 & | & -6 \\ -3 & -5 & | & 12 \end{bmatrix}
$$

$$
3R_1 + 2R_2 \rightarrow R_2 \quad \begin{bmatrix} 2 & 4 & | & -6 \\ 0 & 2 & | & 6 \end{bmatrix}
$$

$$
-2R_2 + R_1 \rightarrow R_1 \quad \begin{bmatrix} 2 & 0 & | & -18 \\ 0 & 2 & | & 6 \end{bmatrix}
$$

$$
\begin{array}{l} \frac{1}{2}R_1 \rightarrow R_1 \\ \frac{1}{2}R_2 \rightarrow R_2 \end{array} \quad \begin{bmatrix} 1 & 0 & | & -9 \\ 0 & 1 & | & 3 \end{bmatrix}
$$

The solution is $(-9, 3)$.

9. $x - y + 3z = 13$
$4x + y + 2z = 17$
$3x + 2y + 2z = 1$

Write the augmented matrix and use row operations.

$$
\begin{bmatrix} 1 & -1 & 3 & | & 13 \\ 4 & 1 & 2 & | & 17 \\ 3 & 2 & 2 & | & 1 \end{bmatrix}
$$

$$
\begin{array}{l} -4R_1 + R_2 \rightarrow R_2 \\ -3R_1 + R_3 \rightarrow R_3 \end{array} \quad \begin{bmatrix} 1 & -1 & 3 & | & 13 \\ 0 & 5 & -10 & | & -35 \\ 0 & 5 & -7 & | & -38 \end{bmatrix}
$$

$$
\begin{array}{l} R_2 + 5R_1 \rightarrow R_1 \\ \\ -1R_2 + R_3 \rightarrow R_3 \end{array} \quad \begin{bmatrix} 5 & 0 & 5 & | & 30 \\ 0 & 5 & -10 & | & -35 \\ 0 & 0 & 3 & | & -3 \end{bmatrix}
$$

$$
\begin{array}{l} 5R_3 + (-3R_1) \rightarrow R_1 \\ 10R_3 + 3R_2 \rightarrow R_2 \end{array} \quad \begin{bmatrix} -15 & 0 & 0 & | & -105 \\ 0 & 15 & 0 & | & -135 \\ 0 & 0 & 3 & | & -3 \end{bmatrix}
$$

$$
\begin{array}{l} -\frac{1}{15}R_1 \rightarrow R_1 \\ \frac{1}{15}R_2 \rightarrow R_2 \\ \frac{1}{3}R_3 \rightarrow R_3 \end{array} \quad \begin{bmatrix} 1 & 0 & 0 & | & 7 \\ 0 & 1 & 0 & | & -9 \\ 0 & 0 & 1 & | & -1 \end{bmatrix}
$$

The solution is $(7, -9, -1)$.

11. $3x - 6y + 9z = 12$
$-x + 2y - 3z = -4$
$x + y + 2z = 7$

Write the augmented matrix and use row operations.

$$
\begin{bmatrix} 3 & -6 & 9 & | & 12 \\ -1 & 2 & -3 & | & -4 \\ 1 & 1 & 2 & | & 7 \end{bmatrix}
$$

$$
\begin{array}{l} R_1 + 3R_2 \rightarrow R_2 \\ -1R_1 + 3R_3 \rightarrow R_3 \end{array} \quad \begin{bmatrix} 3 & -6 & 9 & | & 12 \\ 0 & 0 & 0 & | & 0 \\ 0 & 9 & -3 & | & 9 \end{bmatrix}
$$

The zero in row 2, column 2 is an obstacle. To proceed, interchange the second and third rows.

$$\begin{bmatrix} 3 & -6 & 9 & | & 12 \\ 0 & 9 & -3 & | & 9 \\ 0 & 0 & 0 & | & 0 \end{bmatrix}$$

$$3R_1 + 2R_2 \to R_1 \quad \begin{bmatrix} 9 & 0 & 21 & | & 54 \\ 0 & 9 & -3 & | & 9 \\ 0 & 0 & 0 & | & 0 \end{bmatrix}$$

$$\begin{matrix} \frac{1}{9}R_1 \to R_1 \\ \frac{1}{9}R_2 \to R_2 \end{matrix} \quad \begin{bmatrix} 1 & 0 & \frac{7}{3} & | & 6 \\ 0 & 1 & -\frac{1}{3} & | & 1 \\ 0 & 0 & 0 & | & 0 \end{bmatrix}$$

The row of zeros indicates dependent equations. Solve the first two equations respectively for x and y in terms of z to obtain

$$x = 6 - \frac{7}{3}z \quad \text{and} \quad y = 1 + \frac{1}{3}z.$$

The solution of the system is

$$\left(6 - \frac{7}{3}z, 1 + \frac{1}{3}z, z \right),$$

where z is any real number.

13. $\begin{bmatrix} 2 & x \\ y & 6 \\ 5 & z \end{bmatrix} = \begin{bmatrix} a & -1 \\ 4 & 6 \\ p & 7 \end{bmatrix}$

The size of these matrices is 3×2. For matrices to be equal, corresponding elements must be equal, so $a = 2$, $x = -1$, $y = 4$, $p = 5$, and $z = 7$.

15. $\begin{bmatrix} a+5 & 3b & 6 \\ 4c & 2+d & -3 \\ -1 & 4p & q-1 \end{bmatrix} = \begin{bmatrix} -7 & b+2 & 2k-3 \\ 3 & 2d-1 & 4\ell \\ m & 12 & 8 \end{bmatrix}$

These are 3×3 square matrices. Since corresponding elements must be equal,

$a + 5 = -7$, so $\quad a = -12$;

$3b = b + 2$, so $\quad b = 1$;

$6 = 2k - 3$, so $\quad k = \frac{9}{2}$;

$4c = 3$, so $\quad c = \frac{3}{4}$;

$2 + d = 2d - 1$, so $\quad d = 3$;

$-3 = 4\ell$, so $\quad \ell = -\frac{3}{4}$;

$\quad\quad\quad\quad\quad\quad m = -1$;

$4p = 12$, so $\quad p = 3$; and

$q - 1 = 8$, so $\quad q = 9$.

17. $2G - 4F = 2\begin{bmatrix} 2 & 5 \\ 1 & 6 \end{bmatrix} - 4\begin{bmatrix} -1 & 4 \\ 3 & 7 \end{bmatrix}$

$\quad\quad = \begin{bmatrix} 4 & 10 \\ 2 & 12 \end{bmatrix} + \begin{bmatrix} 4 & -16 \\ -12 & -28 \end{bmatrix}$

$\quad\quad = \begin{bmatrix} 8 & -6 \\ -10 & -16 \end{bmatrix}$

19. Since B is a 3×3 matrix, and A is a 3×2 matrix, the calculation of $B - A$ is not possible.

21. A has size 3×2 and F has 2×2, so AF will have size 3×2.

$$AF = \begin{bmatrix} 4 & 10 \\ -2 & -3 \\ 6 & 9 \end{bmatrix} \begin{bmatrix} -1 & 4 \\ 3 & 7 \end{bmatrix} = \begin{bmatrix} 26 & 86 \\ -7 & -29 \\ 21 & 87 \end{bmatrix}$$

23. D has size 3×1 and E has size 1×3, so DE will have size 3×3.

$$DE = \begin{bmatrix} 6 \\ 1 \\ 0 \end{bmatrix} \begin{bmatrix} 1 & 3 & -4 \end{bmatrix} = \begin{bmatrix} 6 & 18 & -24 \\ 1 & 3 & -4 \\ 0 & 0 & 0 \end{bmatrix}$$

25. B has size 3×3 and D has size 3×1, so BD will have size 3×1.

$$BD = \begin{bmatrix} 2 & 3 & -2 \\ 2 & 4 & 0 \\ 0 & 1 & 2 \end{bmatrix} \begin{bmatrix} 6 \\ 1 \\ 0 \end{bmatrix} = \begin{bmatrix} 15 \\ 16 \\ 1 \end{bmatrix}$$

27. $\quad\quad\quad F = \begin{bmatrix} -1 & 4 \\ 3 & 7 \end{bmatrix}$

$[F | I] = \begin{bmatrix} -1 & 4 & | & 1 & 0 \\ 3 & 7 & | & 0 & 1 \end{bmatrix}$

$3R_1 + R_2 \to R_2 \quad \begin{bmatrix} -1 & 4 & | & 1 & 0 \\ 0 & 19 & | & 3 & 1 \end{bmatrix}$

$4R_2 + (-19R_1) \to R_1 \quad \begin{bmatrix} 19 & 0 & | & -7 & 4 \\ 0 & 19 & | & 3 & 1 \end{bmatrix}$

$\begin{matrix} \frac{1}{19}R_1 \to R_1 \\ \frac{1}{19}R_2 \to R_2 \end{matrix} \quad \begin{bmatrix} 1 & 0 & | & -\frac{7}{19} & \frac{4}{19} \\ 0 & 1 & | & \frac{3}{19} & \frac{1}{19} \end{bmatrix}$

$F^{-1} = \begin{bmatrix} -\frac{7}{19} & \frac{4}{19} \\ \frac{3}{19} & \frac{1}{19} \end{bmatrix}$

29. A and C are 3×2 matrices, so their sum $A + C$ is a 3×2 matrix. Only square matrices have inverses. Therefore, $(A + C)^{-1}$ does not exist.

31. Let
$$A = \begin{bmatrix} -4 & 2 \\ 0 & 3 \end{bmatrix}.$$

$$[A|I] = \begin{bmatrix} -4 & 2 & | & 1 & 0 \\ 0 & 3 & | & 0 & 1 \end{bmatrix}$$

$$2R_2 + (-3R_1) \rightarrow R_1 \quad \begin{bmatrix} 12 & 0 & | & -3 & 2 \\ 0 & 3 & | & 0 & 1 \end{bmatrix}$$

$$\begin{array}{l} \frac{1}{12}R_1 \rightarrow R_1 \\ \frac{1}{3}R_2 \rightarrow R_2 \end{array} \quad \begin{bmatrix} 1 & 0 & | & -\frac{1}{4} & \frac{1}{6} \\ 0 & 1 & | & 0 & \frac{1}{3} \end{bmatrix}$$

$$A^{-1} = \begin{bmatrix} -\frac{1}{4} & \frac{1}{6} \\ 0 & \frac{1}{3} \end{bmatrix}$$

33. Let
$$A = \begin{bmatrix} 6 & 4 \\ 3 & 2 \end{bmatrix}.$$

$$[A|I] = \begin{bmatrix} 6 & 4 & | & 1 & 0 \\ 3 & 2 & | & 0 & 1 \end{bmatrix}$$

$$R_1 + (-2)R_2 \rightarrow R_2 \quad \begin{bmatrix} 6 & 4 & | & 1 & 0 \\ 0 & 0 & | & 1 & -2 \end{bmatrix}$$

The zeros in the second row indicate that the original matrix has no inverse.

35. Let
$$A = \begin{bmatrix} 2 & 0 & 4 \\ 1 & -1 & 0 \\ 0 & 1 & -2 \end{bmatrix}.$$

$$[A|I] = \begin{bmatrix} 2 & 0 & 4 & | & 1 & 0 & 0 \\ 1 & -1 & 0 & | & 0 & 1 & 0 \\ 0 & 1 & -2 & | & 0 & 0 & 1 \end{bmatrix}$$

$$-2R_2 + R_1 \rightarrow R_2 \quad \begin{bmatrix} 2 & 0 & 4 & | & 1 & 0 & 0 \\ 0 & 2 & 4 & | & 1 & -2 & 0 \\ 0 & 1 & -2 & | & 0 & 0 & 1 \end{bmatrix}$$

$$-2R_3 + R_2 \rightarrow R_3 \quad \begin{bmatrix} 2 & 0 & 4 & | & 1 & 0 & 0 \\ 0 & 2 & 4 & | & 1 & -2 & 0 \\ 0 & 0 & 8 & | & 1 & -2 & -2 \end{bmatrix}$$

$$\begin{array}{l} -1R_3 + 2R_1 \rightarrow R_1 \\ -1R_3 + 2R_2 \rightarrow R_2 \end{array} \quad \begin{bmatrix} 4 & 0 & 0 & | & 1 & 2 & 2 \\ 0 & 4 & 0 & | & 1 & -2 & 2 \\ 0 & 0 & 8 & | & 1 & -2 & -2 \end{bmatrix}$$

$$\begin{array}{l} \frac{1}{4}R_1 \rightarrow R_1 \\ \frac{1}{4}R_2 \rightarrow R_2 \\ \frac{1}{8}R_3 \rightarrow R_3 \end{array} \quad \begin{bmatrix} 1 & 0 & 0 & | & \frac{1}{4} & \frac{1}{2} & \frac{1}{2} \\ 0 & 1 & 0 & | & \frac{1}{4} & -\frac{1}{2} & \frac{1}{2} \\ 0 & 0 & 1 & | & \frac{1}{8} & -\frac{1}{4} & -\frac{1}{4} \end{bmatrix}$$

$$A^{-1} = \begin{bmatrix} \frac{1}{4} & \frac{1}{2} & \frac{1}{2} \\ \frac{1}{4} & -\frac{1}{2} & \frac{1}{2} \\ \frac{1}{8} & -\frac{1}{4} & -\frac{1}{4} \end{bmatrix}.$$

37. Let
$$A = \begin{bmatrix} 2 & 3 & 5 \\ -2 & -3 & -5 \\ 1 & 4 & 2 \end{bmatrix}.$$

$$[A|I] = \begin{bmatrix} 2 & 3 & 5 & | & 1 & 0 & 0 \\ -2 & -3 & -5 & | & 0 & 1 & 0 \\ 1 & 4 & 2 & | & 0 & 0 & 1 \end{bmatrix}$$

$$\begin{array}{l} R_1 + R_2 \rightarrow R_2 \\ R_1 + (-2R_3) \rightarrow R_3 \end{array} \quad \begin{bmatrix} 2 & 3 & 5 & | & 1 & 0 & 0 \\ 0 & 0 & 0 & | & 1 & 1 & 0 \\ 0 & -5 & 1 & | & 1 & 0 & -2 \end{bmatrix}$$

The zeros in the second row to the left of the vertical bar indicate that the original matrix has no inverse.

39. $A = \begin{bmatrix} 1 & 2 \\ 2 & 4 \end{bmatrix}, \ B = \begin{bmatrix} 5 \\ 10 \end{bmatrix}$

Row operations may be used to see that matrix A has no inverse. The matrix equation $AX = B$ may be written as the system of equations

$$\begin{array}{ll} x + 2y = 5 & (1) \\ 2x + 4y = 10. & (2) \end{array}$$

Use the elimination method to solve this system. Begin by eliminating x in equation (2).

$$-2R_1 + R_2 \rightarrow R_2 \quad \begin{array}{ll} x + 2y = 5 & (1) \\ \quad\quad 0 = 0 & (3) \end{array}$$

The true statement in equation (3) indicates that the equations are dependent. Solve equation (1) for x in terms of y.

$$x = -2y + 5$$

The solution is $(-2y + 5, y)$, where y is any real number.

41. $A = \begin{bmatrix} 2 & 4 & 0 \\ 1 & -2 & 0 \\ 0 & 0 & 3 \end{bmatrix}, \ B = \begin{bmatrix} 72 \\ -24 \\ 48 \end{bmatrix}$

Use row operations to find the inverse of A, which is

$$A^{-1} = \begin{bmatrix} \frac{1}{4} & \frac{1}{2} & 0 \\ \frac{1}{8} & -\frac{1}{4} & 0 \\ 0 & 0 & \frac{1}{3} \end{bmatrix}.$$

Since $X = A^{-1}B$,

$$X = \begin{bmatrix} \frac{1}{4} & \frac{1}{2} & 0 \\ \frac{1}{8} & -\frac{1}{4} & 0 \\ 0 & 0 & \frac{1}{3} \end{bmatrix} \begin{bmatrix} 72 \\ -24 \\ 48 \end{bmatrix} = \begin{bmatrix} 6 \\ 15 \\ 16 \end{bmatrix}.$$

43. $5x + 10y = 80$
$3x - 2y = 120$

Let $A = \begin{bmatrix} 5 & 10 \\ 3 & -2 \end{bmatrix}$, $X = \begin{bmatrix} x \\ y \end{bmatrix}$, $B = \begin{bmatrix} 80 \\ 120 \end{bmatrix}$.

Use row operations to find the inverse of A, which is

$$A^{-1} = \begin{bmatrix} \frac{1}{20} & \frac{1}{4} \\ \frac{3}{40} & -\frac{1}{8} \end{bmatrix}.$$

Since $X = A^{-1}B$,

$$\begin{bmatrix} x \\ y \end{bmatrix} = \begin{bmatrix} \frac{1}{20} & \frac{1}{4} \\ \frac{3}{40} & -\frac{1}{8} \end{bmatrix} \begin{bmatrix} 80 \\ 120 \end{bmatrix} = \begin{bmatrix} 34 \\ -9 \end{bmatrix}.$$

The solution is $(34, -9)$.

45. $x + 4y - z = 6$
$2x - y + z = 3$
$3x + 2y + 3z = 16$

Let $A = \begin{bmatrix} 1 & 4 & -1 \\ 2 & -1 & 1 \\ 3 & 2 & 3 \end{bmatrix}$, $X = \begin{bmatrix} x \\ y \\ z \end{bmatrix}$, $B = \begin{bmatrix} 6 \\ 3 \\ 16 \end{bmatrix}$.

Use row operations to find the inverse of A, which is

$$A^{-1} = \begin{bmatrix} \frac{5}{24} & \frac{7}{12} & -\frac{1}{8} \\ \frac{1}{8} & -\frac{1}{4} & \frac{1}{8} \\ -\frac{7}{24} & -\frac{5}{12} & \frac{3}{8} \end{bmatrix}.$$

Since $X = A^{-1}B$,

$$\begin{bmatrix} x \\ y \\ z \end{bmatrix} = \begin{bmatrix} \frac{5}{24} & \frac{7}{12} & -\frac{1}{8} \\ \frac{1}{8} & -\frac{1}{4} & \frac{1}{8} \\ -\frac{7}{24} & -\frac{5}{12} & \frac{3}{8} \end{bmatrix} \begin{bmatrix} 6 \\ 3 \\ 16 \end{bmatrix} = \begin{bmatrix} 1 \\ 2 \\ 3 \end{bmatrix}.$$

The solution is $(1, 2, 3)$.

47. $M = \begin{bmatrix} 5 & 3 \\ 3 & 5 \end{bmatrix}$

$0 = \det(M - \lambda I)$

$= \det \begin{bmatrix} 5 - \lambda & 3 \\ 3 & 5 - \lambda \end{bmatrix}$

$= (5 - \lambda)(5 - \lambda) - 3(3)$
$= \lambda^2 - 10\lambda + 16$
$= (\lambda - 2)(\lambda - 8)$

$0 = \lambda - 2$ or $0 = \lambda - 8$
$\lambda = 2$ or $\lambda = 8$

The eigenvalues are 2 and 8.

Find the eigenvector corresponding to $\lambda = 2$.

$0 = (M - \lambda I)X$

$\begin{bmatrix} 0 \\ 0 \end{bmatrix} = \begin{bmatrix} 5 - 2 & 3 \\ 3 & 5 - 2 \end{bmatrix} \begin{bmatrix} x_1 \\ x_2 \end{bmatrix}$

$= \begin{bmatrix} 3 & 3 \\ 3 & 3 \end{bmatrix} \begin{bmatrix} x_1 \\ x_2 \end{bmatrix}$

The augmented matrix for this system is

$$\begin{bmatrix} 3 & 3 & | & 0 \\ 3 & 3 & | & 0 \end{bmatrix}.$$

A solution to this system is

$$\begin{bmatrix} x_1 \\ x_2 \end{bmatrix} = \begin{bmatrix} 1 \\ -1 \end{bmatrix}.$$

Now find the eigenvector corresponding to $\lambda = 8$.

$0 = (M - \lambda I)X$

$\begin{bmatrix} 0 \\ 0 \end{bmatrix} = \begin{bmatrix} 5 - 8 & 3 \\ 3 & 5 - 8 \end{bmatrix} \begin{bmatrix} x_1 \\ x_2 \end{bmatrix}$

$= \begin{bmatrix} -3 & 3 \\ 3 & -3 \end{bmatrix} \begin{bmatrix} x_1 \\ x_2 \end{bmatrix}$

The augmented matrix for this system is

$$\begin{bmatrix} -3 & 3 & | & 0 \\ 3 & -3 & | & 0 \end{bmatrix}.$$

A solution to this system is

$$\begin{bmatrix} x_1 \\ x_2 \end{bmatrix} = \begin{bmatrix} 1 \\ 1 \end{bmatrix}.$$

The given matrix has the following eigenvalues and corresponding eigenvectors.

$$\lambda = 2, \begin{bmatrix} 1 \\ -1 \end{bmatrix}; \lambda = 8, \begin{bmatrix} 1 \\ 1 \end{bmatrix}$$

49. $M = \begin{bmatrix} -2 & 0 & 0 \\ 1 & 3 & 0 \\ -1 & 1 & 2 \end{bmatrix}$

$0 = \det(M - \lambda I)$

$= \det \begin{bmatrix} -2 - \lambda & 0 & 0 \\ 1 & 3 - \lambda & 0 \\ -1 & 1 & 2 - \lambda \end{bmatrix}$

$= (-2 - \lambda)[(3 - \lambda)(2 - \lambda) - 0(1)]$
$= (-2 - \lambda)(3 - \lambda)(2 - \lambda)$

$0 = -2 - \lambda$ or $0 = 3 - \lambda$ or $0 = 2 - \lambda$
$\lambda = -2$ or $\lambda = 3$ or $\lambda = 2$

The eigenvalues are $-2, 3$, and 2.

Find the eigenvector corresponding to $\lambda = -2$.

$$0 = (M - \lambda I)X$$

$$\begin{bmatrix} 0 \\ 0 \\ 0 \end{bmatrix} = \begin{bmatrix} -2-(-2) & 0 & 0 \\ 1 & 3-(-2) & 0 \\ -1 & 1 & 2-(-2) \end{bmatrix} \begin{bmatrix} x_1 \\ x_2 \\ x_3 \end{bmatrix}$$

$$= \begin{bmatrix} 0 & 0 & 0 \\ 1 & 5 & 0 \\ -1 & 1 & 4 \end{bmatrix} \begin{bmatrix} x_1 \\ x_2 \\ x_3 \end{bmatrix}$$

The augmented matrix for this system is

$$\begin{bmatrix} 0 & 0 & 0 & | & 0 \\ 1 & 5 & 0 & | & 0 \\ -1 & 1 & 4 & | & 0 \end{bmatrix}.$$

A solution to this system is

$$\begin{bmatrix} x_1 \\ x_2 \\ x_3 \end{bmatrix} = \begin{bmatrix} 10 \\ -2 \\ 3 \end{bmatrix}.$$

Now find the eigenvector corresponding to $\lambda = 3$.

$$0 = (M - \lambda I)X$$

$$\begin{bmatrix} 0 \\ 0 \\ 0 \end{bmatrix} = \begin{bmatrix} -2-3 & 0 & 0 \\ 1 & 3-3 & 0 \\ -1 & 1 & 2-3 \end{bmatrix} \begin{bmatrix} x_1 \\ x_2 \\ x_3 \end{bmatrix}$$

$$= \begin{bmatrix} -5 & 0 & 0 \\ 1 & 0 & 0 \\ -1 & 1 & -1 \end{bmatrix} \begin{bmatrix} x_1 \\ x_2 \\ x_3 \end{bmatrix}$$

The augmented matrix for this system is

$$\begin{bmatrix} -5 & 0 & 0 & | & 0 \\ 1 & 0 & 0 & | & 0 \\ -1 & 1 & -1 & | & 0 \end{bmatrix}.$$

A solution to this system is

$$\begin{bmatrix} x_1 \\ x_2 \\ x_3 \end{bmatrix} = \begin{bmatrix} 0 \\ 1 \\ 1 \end{bmatrix}.$$

Now find the eigenvector corresponding to $\lambda = 2$.

$$0 = (M - \lambda I)X$$

$$\begin{bmatrix} 0 \\ 0 \\ 0 \end{bmatrix} = \begin{bmatrix} -2-2 & 0 & 0 \\ 1 & 3-2 & 0 \\ -1 & 1 & 2-2 \end{bmatrix} \begin{bmatrix} x_1 \\ x_2 \\ x_3 \end{bmatrix}$$

$$= \begin{bmatrix} -4 & 0 & 0 \\ 1 & 1 & 0 \\ -1 & 1 & 0 \end{bmatrix} \begin{bmatrix} x_1 \\ x_2 \\ x_3 \end{bmatrix}$$

The augmented matrix for this system is

$$\begin{bmatrix} -4 & 0 & 0 & | & 0 \\ 1 & 1 & 0 & | & 0 \\ -1 & 1 & 0 & | & 0 \end{bmatrix}.$$

A solution to this system is

$$\begin{bmatrix} x_1 \\ x_2 \\ x_3 \end{bmatrix} = \begin{bmatrix} 0 \\ 0 \\ 1 \end{bmatrix}.$$

The given matrix has the following eigenvalues and corresponding eigenvectors.

$$\lambda = -2, \begin{bmatrix} 10 \\ -2 \\ 3 \end{bmatrix}; \lambda = 3, \begin{bmatrix} 0 \\ 1 \\ 1 \end{bmatrix}; \lambda = 2, \begin{bmatrix} 0 \\ 0 \\ 1 \end{bmatrix}$$

51. The given information can be written as the following 4×3 matrix.

$$\begin{bmatrix} 8 & 8 & 8 \\ 10 & 5 & 9 \\ 7 & 10 & 7 \\ 8 & 9 & 7 \end{bmatrix}.$$

53. (a) $a + b = 0.60$ (1)
  $c + d = 0.75$ (2)
  $a + c = 0.65$ (3)
  $b + d = 0.70$ (4)

The augmented matrix of the system is

$$\begin{bmatrix} 1 & 1 & 0 & 0 & | & 0.60 \\ 0 & 0 & 1 & 1 & | & 0.75 \\ 1 & 0 & 1 & 0 & | & 0.65 \\ 0 & 1 & 0 & 1 & | & 0.70 \end{bmatrix}.$$

$-1R_1 + R_3 \rightarrow R_3$

$$\begin{bmatrix} 1 & 1 & 0 & 0 & | & 0.60 \\ 0 & 0 & 1 & 1 & | & 0.75 \\ 0 & -1 & 1 & 0 & | & 0.05 \\ 0 & 1 & 0 & 1 & | & 0.70 \end{bmatrix}$$

Interchange rows 2 and 4.

$$\begin{bmatrix} 1 & 1 & 0 & 0 & | & 0.60 \\ 0 & 1 & 0 & 1 & | & 0.70 \\ 0 & -1 & 1 & 0 & | & 0.05 \\ 0 & 0 & 1 & 1 & | & 0.75 \end{bmatrix}$$

$-1R_2 + R_1 \rightarrow R_1$

$R_2 + R_3 \rightarrow R_3$

$$\begin{bmatrix} 1 & 0 & 0 & -1 & | & -0.10 \\ 0 & 1 & 0 & 1 & | & 0.70 \\ 0 & 0 & 1 & 1 & | & 0.75 \\ 0 & 0 & 1 & 1 & | & 0.75 \end{bmatrix}$$

Since R_3 and R_4 are identical, there will be infinitely many solutions. We do not have enough information to determine the values of a, b, c, and d.

(b) i. If $d = 0.33$, the system of equations in part (a) becomes

$$
\begin{aligned}
a + b &= 0.60 \quad (1) \\
c + 0.33 &= 0.75 \quad (2) \\
a + c &= 0.65 \quad (3) \\
b + 0.33 &= 0.70. \quad (4)
\end{aligned}
$$

Equation (2) gives $c = 0.42$, and equation (4) gives $b = 0.37$. Substituting $c = 0.42$ into equation (3) gives $a = 0.23$. Therefore, $a = 0.23$, $b = 0.37$, $c = 0.42$, and $d = 0.33$.

Thus, A is healthy, B and D are tumorous, and C is bone.

ii. If $d = 0.43$, the system of equations in part (a) becomes

$$
\begin{aligned}
a + b &= 0.60 \quad (1) \\
c + 0.43 &= 0.75 \quad (2) \\
a + c &= 0.65 \quad (3) \\
b + 0.43 &= 0.70. \quad (4)
\end{aligned}
$$

Equation (2) gives $c = 0.32$, and equation (4) gives $b = 0.27$. Substituting $c = 0.32$ into equation (3) gives $a = 0.33$. Therefore, $a = 0.33$, $b = 0.27$, $c = 0.32$, and $d = 0.43$.

Thus, A and C are tumorous, B could be healthy or tumorous, and D is bone.

(c) The original system now has two additional equations.

$$
\begin{aligned}
a + b &= 0.60 \quad (1) \\
c + d &= 0.75 \quad (2) \\
a + c &= 0.65 \quad (3) \\
b + d &= 0.70 \quad (4) \\
b + c &= 0.85 \quad (5) \\
a + d &= 0.50 \quad (6)
\end{aligned}
$$

The augmented matrix of this system is

$$
\left[\begin{array}{cccc|c}
1 & 1 & 0 & 0 & 0.60 \\
0 & 0 & 1 & 1 & 0.75 \\
1 & 0 & 1 & 0 & 0.65 \\
0 & 1 & 0 & 1 & 0.70 \\
0 & 1 & 1 & 0 & 0.85 \\
1 & 0 & 0 & 1 & 0.50
\end{array}\right].
$$

Using the Gauss-Jordan method we obtain

$$
\left[\begin{array}{cccc|c}
1 & 0 & 0 & 0 & 0.20 \\
0 & 1 & 0 & 0 & 0.40 \\
0 & 0 & 1 & 0 & 0.45 \\
0 & 0 & 0 & 1 & 0.30 \\
0 & 0 & 0 & 0 & 0 \\
0 & 0 & 0 & 0 & 0
\end{array}\right].
$$

Therefore, $a = 0.20$, $b = 0.40$, $c = 0.45$, and $d = 0.30$. Thus, A is healthy, B and C are bone, and D is tumorous.

(d) As we saw in part (c), the six equations reduced to four independent equations. We need only four beams, correctly chosen, to obtain a solution. The four beams must pass through all four cells and must lead to independent equations. One such choice would be beams 1, 2, 3, and 6. Another choice would be beams 1, 2, 4, and 5.

55. Find the eigenvalues and eigenvectors.

$$
M = \left[\begin{array}{cc} 0.3 & 0.2 \\ 0.3 & 0.8 \end{array}\right]
$$

$$
\begin{aligned}
0 &= \det(M - \lambda I) \\
&= \left[\begin{array}{cc} 0.3 - \lambda & 0.2 \\ 0.3 & 0.8 - \lambda \end{array}\right] \\
&= (0.3 - \lambda)(0.8 - \lambda) - 0.2(0.3) \\
&= \lambda^2 - 1.1\lambda + 0.18
\end{aligned}
$$

Solve using the quadratic formula.

$$
\begin{aligned}
\lambda &= \frac{-(-1.1) \pm \sqrt{(-1.1)^2 - 4(0.18)}}{2} \\
&= \frac{1.1 \pm \sqrt{0.49}}{2} \\
&= \frac{1.1 \pm 0.7}{2} \\
&= 0.9, 0.2
\end{aligned}
$$

Find the eigenvector corresponding to $\lambda = 0.9$.

$$
0 = (M - \lambda I)X
$$

$$
\begin{aligned}
\left[\begin{array}{c} 0 \\ 0 \end{array}\right] &= \left[\begin{array}{cc} 0.3 - 0.9 & 0.2 \\ 0.3 & 0.8 - 0.9 \end{array}\right] \left[\begin{array}{c} x_1 \\ x_2 \end{array}\right] \\
&= \left[\begin{array}{cc} -0.6 & 0.2 \\ 0.3 & -0.1 \end{array}\right] \left[\begin{array}{c} x_1 \\ x_2 \end{array}\right]
\end{aligned}
$$

The augmented matrix for this system is

$$
\left[\begin{array}{cc|c} -0.6 & 0.2 & 0 \\ 0.3 & -0.1 & 0 \end{array}\right]
$$

A solution to this system is

$$\begin{bmatrix} x_1 \\ x_2 \end{bmatrix} = \begin{bmatrix} 1 \\ 3 \end{bmatrix}.$$

The eigenvector corresponding to $\lambda = 0.2$ contains a negative value so it cannot be considered as a population.
(a) Since the initial population is 10,000 a multiple of the eigenvector is used.

$$\begin{bmatrix} x_1 \\ x_2 \end{bmatrix} = k \begin{bmatrix} 1 \\ 3 \end{bmatrix} = \begin{bmatrix} k \\ 3k \end{bmatrix}$$

$$k + 3k = 10{,}000$$
$$4k = 10{,}000$$
$$k = 2{,}500$$

(b) Therefore, the initial population is

$$\begin{bmatrix} 2{,}500 \\ 3(2{,}500) \end{bmatrix} \text{ or } \begin{bmatrix} 2{,}500 \\ 7{,}500 \end{bmatrix}$$

with a growth factor of 0.9.

57. (a) Each row gives the fraction of each class of the herd that will become members of the class represented by that row in the next year.

(b) Let $V = \begin{bmatrix} k \\ 0 \\ 0 \\ 0 \\ 0 \\ 0 \end{bmatrix}$

$$MV = \begin{bmatrix} 0.95 & 0 & 0.75 & 0 & 0 & 0 \\ 0 & 0.95 & 0 & 0.75 & 0 & 0 \\ 0 & 0 & 0 & 0 & 0.6 & 0 \\ 0 & 0 & 0 & 0 & 0 & 0.6 \\ 0 & 0.48 & 0 & 0 & 0 & 0 \\ 0 & 0.42 & 0 & 0 & 0 & 0 \end{bmatrix} \begin{bmatrix} k \\ 0 \\ 0 \\ 0 \\ 0 \\ 0 \end{bmatrix} = \begin{bmatrix} 0.95k \\ 0 \\ 0 \\ 0 \\ 0 \\ 0 \end{bmatrix} = 0.95V$$

Therefore, $\lambda = 0.95$. This tells us that if the herd consists of only adult males, then next year it will consist of only adult males with population decreasing by a factor of 0.95.

(c) Let $X =$ population and
$Q =$ harvest.

Next year's herd $-$ harvest $=$ next year's herd without harvest

$$MX - Q = X$$
$$MX - X = Q$$
$$(M - I)X = Q$$
$$X = (M - I)^{-1}Q$$

(d) $X = (M - I)^{-1}Q$

$$= \begin{bmatrix} 0.95-1 & 0 & 0.75 & 0 & 0 & 0 \\ 0 & 0.95-1 & 0 & 0.75 & 0 & 0 \\ 0 & 0 & -1 & 0 & 0.6 & 0 \\ 0 & 0 & 0 & -1 & 0 & 0.6 \\ 0 & 0.48 & 0 & 0 & -1 & 0 \\ 0 & 0.42 & 0 & 0 & 0 & -1 \end{bmatrix}^{-1} \begin{bmatrix} 100 \\ 100 \\ 10 \\ 10 \\ 0 \\ 0 \end{bmatrix}$$

$$= \begin{bmatrix} -0.05 & 0 & 0.75 & 0 & 0 & 0 \\ 0 & -0.05 & 0 & 0.75 & 0 & 0 \\ 0 & 0 & -1 & 0 & 0.6 & 0 \\ 0 & 0 & 0 & -1 & 0 & 0.6 \\ 0 & 0.48 & 0 & 0 & -1 & 0 \\ 0 & 0.42 & 0 & 0 & 0 & -1 \end{bmatrix}^{-1} \begin{bmatrix} 100 \\ 100 \\ 10 \\ 10 \\ 0 \\ 0 \end{bmatrix}$$

$$= \begin{bmatrix} -20.0 & 31.0791 & -15.0 & 23.3094 & -9.0 & 13.9856 \\ 0 & 7.1942 & 0 & 5.3957 & 0 & 3.2374 \\ 0 & 2.0719 & -1.0 & 1.5540 & -0.6 & 0.9324 \\ 0 & 1.8129 & 0 & 0.3597 & 0 & 0.2158 \\ 0 & 3.4532 & 0 & 2.5899 & -1.0 & 1.5540 \\ 0 & 3.0216 & 0 & 2.2662 & 0 & 0.3597 \end{bmatrix} \begin{bmatrix} 100 \\ 100 \\ 10 \\ 10 \\ 0 \\ 0 \end{bmatrix} \approx \begin{bmatrix} 1,191 \\ 773 \\ 213 \\ 185 \\ 371 \\ 325 \end{bmatrix}$$

(e) Let $X =$ this year's herd.

Next year's herd $-$ harvest $=$ next year's herd with 10% increase

$$MX - Q = 1.1X$$
$$MX - 1.1X = Q$$
$$(M - 1.1I)X = Q$$
$$X = (M - 1.1I)^{-1}Q$$

(f) $X = (M - 1.1I)^{-1}Q$

$$= \begin{bmatrix} 0.95-1.1 & 0 & 0.75 & 0 & 0 & 0 \\ 0 & 0.95-1.1 & 0 & 0.75 & 0 & 0 \\ 0 & 0 & -1.1 & 0 & 0.6 & 0 \\ 0 & 0 & 0 & -1.1 & 0 & 0.6 \\ 0 & 0.48 & 0 & 0 & -1.1 & 0 \\ 0 & 0.42 & 0 & 0 & 0 & -1.1 \end{bmatrix}^{-1} \begin{bmatrix} 100 \\ 100 \\ 10 \\ 10 \\ 0 \\ 0 \end{bmatrix}$$

$$= \begin{bmatrix} -6.6667 & 192.00 & -4.5455 & 130.9091 & -2.4793 & 71.4050 \\ 0 & 161.33 & 0 & 110 & 0 & 60 \\ 0 & 38.40 & -0.9091 & 26.1818 & -0.4959 & 14.2810 \\ 0 & 33.60 & 0 & 22 & 0 & 12 \\ 0 & 70.40 & 0 & 48 & -0.9091 & 26.1818 \\ 0 & 61.60 & 0 & 42 & 0 & 22 \end{bmatrix} \begin{bmatrix} 100 \\ 100 \\ 10 \\ 10 \\ 0 \\ 0 \end{bmatrix}$$

$$= \begin{bmatrix} 19,797 \\ 17,233 \\ 4,093 \\ 3,580 \\ 7,520 \\ 6,580 \end{bmatrix}$$

59. (a) $\begin{bmatrix} 1 \\ 1 \end{bmatrix} x + \begin{bmatrix} 0 \\ 2 \end{bmatrix} y = \begin{bmatrix} 1 \\ 2 \end{bmatrix}$

$$\begin{bmatrix} x \\ x \end{bmatrix} + \begin{bmatrix} 0 \\ 2y \end{bmatrix} = \begin{bmatrix} 1 \\ 2 \end{bmatrix}$$

$$\begin{bmatrix} x \\ x + 2y \end{bmatrix} = \begin{bmatrix} 1 \\ 2 \end{bmatrix}$$

Since corresponding elements must be equal, $x = 1$ and $x + 2y = 2$. Substituting $x = 1$ in the second equation gives $y = \frac{1}{2}$. Note that $x = 1$ and $y = \frac{1}{2}$ are the values that balance the equation.

(b) $\qquad x\mathrm{CO}_2 + y\mathrm{H}_2 + z\mathrm{CO} = \mathrm{H}_2\mathrm{O}$

$$\begin{bmatrix} 1 \\ 0 \\ 2 \end{bmatrix} x + \begin{bmatrix} 0 \\ 2 \\ 0 \end{bmatrix} y + \begin{bmatrix} 1 \\ 0 \\ 1 \end{bmatrix} z = \begin{bmatrix} 0 \\ 2 \\ 1 \end{bmatrix}$$

$$\begin{bmatrix} x \\ 0 \\ 2x \end{bmatrix} + \begin{bmatrix} 0 \\ 2y \\ 0 \end{bmatrix} + \begin{bmatrix} z \\ 0 \\ z \end{bmatrix} = \begin{bmatrix} 0 \\ 2 \\ 1 \end{bmatrix}$$

$$\begin{bmatrix} x + z \\ 2y \\ 2x + z \end{bmatrix} = \begin{bmatrix} 0 \\ 2 \\ 1 \end{bmatrix}$$

Since corresponding elements must be equal, $x + z = 0, 2y = 2$, and $2x + z = 1$. Solving $2y = 2$ gives $y = 1$. Solving the system $\begin{cases} x + z = 0 \\ 2x + z = 1 \end{cases}$ gives $x = 1$ and $z = -1$. Thus, the values that balance the equation are $x = 1$, $y = 1$, and $z = -1$.

61. $\frac{1}{2}W_1 + \frac{\sqrt{2}}{2}W_2 = 150$ (1)

$\frac{\sqrt{3}}{2}W_1 - \frac{\sqrt{2}}{2}W_2 = \quad 0$ (2)

Adding equations (1) and (2) gives

$$\left(\frac{1}{2} + \frac{\sqrt{3}}{2} \right) W_1 = 150.$$

Multiply by 2.

$$(1 + \sqrt{3})W_1 = 300$$

$$W_1 = \frac{300}{1 + \sqrt{3}} \approx 110$$

From equation (2),

$$\frac{\sqrt{3}}{2}W_1 = \frac{\sqrt{2}}{2}W_2$$

$$W_2 = \frac{\sqrt{3}}{\sqrt{2}}W_1.$$

Substitute $\frac{300}{1+\sqrt{3}}$ from above for W_1.

$$W_2 = \frac{\sqrt{3}}{\sqrt{2}} \cdot \frac{300}{1 + \sqrt{3}} = \frac{300\sqrt{3}}{(1 + \sqrt{3})\sqrt{2}} \approx 134$$

Therefore, $W_1 \approx 110$ lb and $W_2 \approx 134$ lb.

63. Use a table to organize the information.

	Standard	Extra Large	Time Available
Hours Cutting	$\frac{1}{4}$	$\frac{1}{3}$	4
Hours Shaping	$\frac{1}{2}$	$\frac{1}{3}$	6

Let $x =$ the number of standard paper clips (in thousands),

and $y =$ the number of extra large paper clips (in thousands).

The given information leads to the system

$$\frac{1}{4}x + \frac{1}{3}y = 4$$
$$\frac{1}{2}x + \frac{1}{3}y = 6.$$

Solve this system by any method to get $x = 8$, $y = 6$. The manufacturer can make 8 thousand (8,000) standard and 6 thousand (6,000) extra large paper clips.

65. Let $x_1 =$ the number of blankets,

$x_2 =$ the number of rugs, and

$x_3 =$ the number of skirts.

The given information leads to the system

$$24x_1 + 30x_2 + 12x_3 = 306 \quad (1)$$
$$4x_1 + 5x_2 + 3x_3 = 59 \quad (2)$$
$$15x_1 + 18x_2 + 9x_3 = 201. \quad (3)$$

Simplify equations (1) and (3).

$$\frac{1}{6}R_1 \to R_1 \quad 4x_1 + 5x_2 + 2x_3 = 51 \quad (4)$$
$$4x_1 + 5x_2 + 3x_3 = 59 \quad (2)$$
$$\frac{1}{3}R_3 \to R_3 \quad 5x_1 + 6x_2 + 3x_3 = 67 \quad (5)$$

Solve this system by the Gauss-Jordan method. Write the augmented matrix and use row operations.

$$\begin{bmatrix} 4 & 5 & 2 & | & 51 \\ 4 & 5 & 3 & | & 59 \\ 5 & 6 & 3 & | & 67 \end{bmatrix}$$

$$\begin{matrix} -1R_1 + R_2 \to R_2 \\ -4R_3 + 5R_1 \to R_3 \end{matrix} \quad \begin{bmatrix} 4 & 5 & 2 & | & 51 \\ 0 & 0 & 1 & | & 8 \\ 0 & 1 & -2 & | & -13 \end{bmatrix}$$

Interchange the second and third rows.

$$\begin{bmatrix} 4 & 5 & 2 & | & 51 \\ 0 & 1 & -2 & | & -13 \\ 0 & 0 & 1 & | & 8 \end{bmatrix}$$

$-5R_2 + R_1 \to R_1$
$$\begin{bmatrix} 4 & 0 & 12 & | & 116 \\ 0 & 1 & -2 & | & -13 \\ 0 & 0 & 1 & | & 8 \end{bmatrix}$$

$-12R_3 + R_1 \to R_1$
$2R_3 + R_2 \to R_2$
$$\begin{bmatrix} 4 & 0 & 0 & | & 20 \\ 0 & 1 & 0 & | & 3 \\ 0 & 0 & 1 & | & 8 \end{bmatrix}$$

$\frac{1}{4}R_1 \to R_1$
$$\begin{bmatrix} 1 & 0 & 0 & | & 5 \\ 0 & 1 & 0 & | & 3 \\ 0 & 0 & 1 & | & 8 \end{bmatrix}$$

The solution of the system is $x = 5$, $y = 3$, $z = 8$.
5 blankets, 3 rugs, and 8 skirts can be made.

67. (a) High $\begin{bmatrix} 3{,}170 \\ 2{,}360 \\ 1{,}800 \end{bmatrix}$
Medium
Coated

(b) $\begin{bmatrix} x \\ y \\ z \end{bmatrix}$

(c) $\begin{bmatrix} 10 & 5 & 8 \\ 12 & 0 & 4 \\ 0 & 10 & 5 \end{bmatrix} \begin{bmatrix} x \\ y \\ z \end{bmatrix} = \begin{bmatrix} 3{,}170 \\ 2{,}360 \\ 1{,}800 \end{bmatrix}$

(d) $\begin{bmatrix} x \\ y \\ z \end{bmatrix} = \begin{bmatrix} 10 & 5 & 8 \\ 12 & 0 & 4 \\ 0 & 10 & 5 \end{bmatrix}^{-1} \begin{bmatrix} 3{,}170 \\ 2{,}360 \\ 1{,}800 \end{bmatrix}$

$= \begin{bmatrix} -0.154 & 0.212 & 0.0769 \\ -0.231 & 0.192 & 0.2154 \\ 0.462 & -0.385 & -0.231 \end{bmatrix} \begin{bmatrix} 3{,}170 \\ 2{,}360 \\ 1{,}800 \end{bmatrix}$

$= \begin{bmatrix} 150 \\ 110 \\ 140 \end{bmatrix}$

Chapter 10 Test

[10.1]

1. Solve the system using the Gauss-Jordan method.

$$3x + y = 11$$
$$x - 2y = -8$$

2. Solve the system using the Gauss-Jordan method.

$$x + 2y + 3z = 5$$
$$2x - y + z = 5$$
$$x + y + z = 2$$

3. Use the Gauss-Jordan method to find all solutions of the following system, given that all variables must be nonnegative integers.

$$2x + 2y - z = 30$$
$$3x + 2y - 2z = 41$$
$$x + 4y + z = 27$$

4. Use a system of equations to solve the following problem. Solve the system by the method of your choice.

An investor has $80,000 and she would like to earn 7.675% per year by investing it in mutual funds, bonds, and certificates of deposit. She wants to invest $4000 more in bonds than the total investment in mutual funds and certificates of deposit. Mutual funds pay 10% per year, bonds pay 7% per year, and certificates of deposit pay 5% per year. How much should she invest in each of the three?

[10.2]

5. Find the values of the variables.

$$\begin{bmatrix} 6-x & 2y+1 \\ 3m & 5p+2 \end{bmatrix} = \begin{bmatrix} 8 & 10 \\ -5 & 3p-1 \end{bmatrix}$$

[10.2–10.3]

6. Given the following matrices, perform the indicated operations, if possible.

$$A = \begin{bmatrix} 1 & 2 & -1 \\ 0 & 1 & 1 \\ 1 & 0 & 1 \end{bmatrix} \qquad B = \begin{bmatrix} 1 & -2 \\ 1 & 1 \\ 0 & 1 \end{bmatrix} \qquad C = \begin{bmatrix} 2 & 1 & 3 \\ 0 & 4 & 1 \\ 1 & 1 & 1 \end{bmatrix}$$

(a) AB (b) $2A - C$ (c) $A + 2B$

[10.4]

Find the inverse of each matrix which has an inverse.

7. $A = \begin{bmatrix} 1 & 0 & -1 \\ 2 & 1 & 1 \\ 1 & 1 & 5 \end{bmatrix}$ 8. $B = \begin{bmatrix} 2 & -1 & 1 \\ 0 & 2 & 4 \\ 2 & 1 & 5 \end{bmatrix}$

9. For $A = \begin{bmatrix} 1 & 0 & 1 \\ 1 & 1 & 1 \\ 2 & 1 & 3 \end{bmatrix}$, $A^{-1} = \begin{bmatrix} 2 & 1 & -1 \\ -1 & 1 & 0 \\ -1 & -1 & 1 \end{bmatrix}$.

   Use this inverse to solve the equation $AX = B$, where $B = \begin{bmatrix} 1 \\ 2 \\ -1 \end{bmatrix}$.

10. Solve the system using the inverse of the coefficient matrix.

$$x - 2y = 3$$
$$x + 3y = 5$$

11. Solve the system by using the inverse of the coefficient matrix. Use a graphing calculator.

$$0.103x - 0.247y + 0.489z = 0.936$$
$$-0.218x + 0.379y + 0.702z = 0.863$$
$$0.315x - 0.742y - 0.913z = -0.768$$

[10.5]

In Exercises 12–14 find the eigenvalues and corresponding eigenvectors.

12. $\begin{bmatrix} 3 & 1 \\ 4 & 3 \end{bmatrix}$ 13. $\begin{bmatrix} 1 & 3 \\ 2 & 6 \end{bmatrix}$ 14. $\begin{bmatrix} 0 & 1 & -2 \\ 2 & 1 & 0 \\ 4 & -2 & 5 \end{bmatrix}$

15. Find the repeated eigenvalue and eigenvector.
$$\begin{bmatrix} 3 & -4 \\ 4 & -5 \end{bmatrix}$$

16. Find the complex eigenvalues and corresponding eigenvectors.
$$\begin{bmatrix} 1 & -1 \\ 4 & 1 \end{bmatrix}$$

Chapter 10 Test Answers

1. $(2, 5)$

2. $(1, -1, 2)$

3. $(11, 4, 0)$, $(13, 3, 2)$, $(15, 2, 4)$, $(17, 1, 6)$, $(19, 0, 8)$

4. $26,000 in mutual funds; $42,000 in bonds, $12,000 in certificates of deposit

5. $x = -2$, $y = \frac{9}{2}$, $m = -\frac{5}{3}$, $p = -\frac{3}{2}$

6. (a) $\begin{bmatrix} 3 & -1 \\ 1 & 2 \\ 1 & -1 \end{bmatrix}$ (b) $\begin{bmatrix} 0 & 3 & -5 \\ 0 & -2 & 1 \\ 1 & -1 & 1 \end{bmatrix}$ (c) Not possible

7. $A^{-1} = \begin{bmatrix} \frac{4}{3} & -\frac{1}{3} & \frac{1}{3} \\ -3 & 2 & -1 \\ \frac{1}{3} & -\frac{1}{3} & \frac{1}{3} \end{bmatrix}$

8. B^{-1} does not exist.

9. $X = \begin{bmatrix} 5 \\ 1 \\ -4 \end{bmatrix}$

10. $\left(\frac{19}{5}, \frac{2}{5} \right)$

11. $(-2.061, -1.683, 1.498)$

12. $1, 5$; $\begin{bmatrix} 1 \\ -2 \end{bmatrix}$, $\begin{bmatrix} 1 \\ 2 \end{bmatrix}$

13. $0, 7$; $\begin{bmatrix} 3 \\ -1 \end{bmatrix}$, $\begin{bmatrix} 1 \\ 2 \end{bmatrix}$

14. $1, 2, 3$; $\begin{bmatrix} 0 \\ 2 \\ 1 \end{bmatrix}$, $\begin{bmatrix} 1 \\ 2 \\ 0 \end{bmatrix}$, $\begin{bmatrix} 1 \\ 1 \\ -1 \end{bmatrix}$

15. -1; $\begin{bmatrix} 1 \\ 1 \end{bmatrix}$

16. $1 + 2i, 1 - 2i$; $\begin{bmatrix} 1 \\ -2i \end{bmatrix}$, $\begin{bmatrix} 1 \\ 2i \end{bmatrix}$

DIFFERENTIAL EQUATIONS

11.1 Solutions of Elementary and Separable Differential Equations

1. $\dfrac{dy}{dx} = -2x + 3x^2$

$$y = \int (-2x + 3x^2)\, dx$$

$$= \frac{-2x^2}{2} + \frac{3x^3}{3} + C$$

$$= -x^2 + x^3 + C$$

3. $3x^3 - 2\dfrac{dy}{dx} = 0$

Solve for $\dfrac{dy}{dx}$.

$$\frac{dy}{dx} = \frac{3x^3}{2}$$

$$y = \frac{3}{2}\int x^3\, dx = \frac{3}{2}\left(\frac{x^4}{4}\right) + C = \frac{3x^4}{8} + C$$

5. $y\dfrac{dy}{dx} = x$

Separate the variables and take antiderivatives.

$$\int y\, dy = \int x\, dx$$

$$\frac{y^2}{2} = \frac{x^2}{2} + K$$

$$y^2 = \frac{2}{2}x^2 + 2K$$

$$y^2 = x^2 + C$$

7. $\dfrac{dy}{dx} = 2xy$

$$\int \frac{dy}{y} = \int 2x\, dx$$

$$\ln |y| = \frac{2x^2}{2} + C$$

$$\ln |y| = x^2 + C$$
$$e^{\ln |y|} = e^{x^2 + C}$$
$$y = \pm e^{x^2 + C}$$
$$y = \pm e^{x^2} \cdot e^C$$
$$y = ke^{x^2}$$

9. $\dfrac{dy}{dx} = 3x^2 y - 2xy$

$$\frac{dy}{dx} = y(3x^2 - 2x)$$

$$\int \frac{dy}{y} = \int (3x^2 - 2x)\, dx$$

$$\ln |y| = \frac{3x^3}{3} - \frac{2x^2}{2} + C$$

$$e^{\ln |y|} = e^{x^3 - x^2 + C}$$
$$y = \pm(e^{x^3 - x^2})e^C$$
$$y = ke^{(x^3 - x^2)}$$

11. $\dfrac{dy}{dx} = \dfrac{y}{x},\ x > 0$

$$\int \frac{dy}{y} = \int \frac{dx}{x}$$

$$\ln |y| = \ln x + C$$
$$e^{\ln |y|} = e^{\ln x + C}$$
$$y = \pm e^{\ln x} \cdot e^C$$
$$y = Me^{\ln x}$$
$$y = Mx$$

13. $\dfrac{dy}{dx} = y - 5$

$$\int \frac{dy}{y - 5} = \int dx$$

$$\ln |y - 5| = x + C$$
$$e^{\ln |y-5|} = e^{x + C}$$
$$y - 5 = \pm e^x \cdot e^C$$
$$y - 5 = Me^x$$
$$y = Me^x + 5$$

15. $\dfrac{dy}{dx} = y^2 e^x$

$$\int y^{-2}\, dy = \int e^x\, dx$$

$$-y^{-1} = e^x + C$$

$$-\frac{1}{y} = e^x + C$$

$$y = \frac{-1}{e^x + C}$$

17.
$$\frac{dy}{dx} = \frac{\cos x}{\sin y}$$

$$\int \sin y \, dy = \int \cos x \, dx$$

$$-\cos y = \sin x + C$$
$$\cos y = -\sin x - C$$

19. $\dfrac{dy}{dx} + 2x = 3x^2$; $y = 2$ when $x = 0$

$$\frac{dy}{dx} = 3x^2 - 2x$$

$$y = \frac{3x^3}{3} - \frac{2x^2}{2} + C$$

$$y = x^3 - x^2 + C$$

Since $y = 2$ when $x = 0$,

$$2 = 0 - 0 + C$$
$$C = 2.$$

Thus,

$$y = x^3 - x^2 + 2.$$

21. $2\dfrac{dy}{dx} = 4xe^{-x}$; $y = 42$ when $x = 0$

$$\frac{dy}{dx} = 2xe^{-x}$$

Use the table of integrals or integrate by parts.

$$y = 2(-x - 1)e^{-x} + C$$

Since $y = 42$ when $x = 0$,

$$42 = 2(0 - 1)(1) + C$$
$$42 = -2 + C$$
$$C = 44.$$

Thus,

$$y = -2xe^{-x} - 2e^{-x} + 44.$$

23. $\dfrac{dy}{dx} = \dfrac{x^2}{y}$; $y = 3$ when $x = 0$.

$$\int y \, dy = \int x^2 \, dx$$

$$\frac{y^2}{2} = \frac{x^3}{3} + C$$

$$y^2 = \frac{2}{3}x^3 + 2C$$

$$y^2 = \frac{2}{3}x^3 + k$$

Since $y = 3$ when $x = 0$,

$$9 = 0 + k$$
$$k = 0.$$

So $y^2 = \dfrac{2}{3}x^3 + 9$.

25. $(2x + 3)y = \dfrac{dy}{dx}$; $y = 1$ when $x = 0$.

$$\int (2x + 3) \, dx = \int \frac{dy}{y}$$

$$\frac{2x^2}{2} + 3x + C = \ln|y|$$

$$e^{x^2 + 3x + C} = e^{\ln|y|}$$

$$y = (e^{x^2 + 3x})(\pm e^C)$$

$$y = ke^{x^2 + 3x}$$

Since $y = 1$ when $x = 0$.

$$1 = ke^{0+0}$$
$$k = 1.$$

So $y = e^{x^2 + 3x}$.

27. $\dfrac{dy}{dx} = 4x^3 - 3x^2 + x$; $y = 0$ when $x = 1$.

$$y = \int (4x^3 - 3x^2 + x) \, dx$$

$$= x^4 - x^3 + \frac{x^2}{2} + C$$

Substitute.

$$0 = 1 - 1 + \frac{1}{2} + C$$

$$-\frac{1}{2} = C$$

$$y = x^4 - x^3 + \frac{x^2}{2} - \frac{1}{2}$$

29. $\dfrac{dy}{dx} = \dfrac{y^2}{x}$; $y = 5$ when $x = e$.

$$\int y^{-2} \, dy = \int \frac{dx}{x}$$

$$-y^{-1} = \ln|x| + C$$

$$-\frac{1}{y} = \ln|x| + C$$

$$y = \frac{-1}{\ln|x| + C}$$

Since $y = 5$ when $x = e$,

$$5 = \frac{-1}{\ln e + C}$$

$$5 = \frac{-1}{1 + C}$$

$$5 + 5C = -1$$
$$5C = -6$$

$$C = -\frac{6}{5}.$$

So
$$y = \frac{-1}{\ln|x| - \frac{6}{5}}$$
$$= \frac{-5}{5\ln|x| - 6}.$$

31. $\frac{dy}{dx} = (y-1)^2 e^{x-1}$; $y = 2$ when $x = 1$.

$$\frac{dy}{(y-1)^2} = e^{x-1}\, dx$$

$$\int (y-1)^{-2}\, dy = \int e^{x-1}\, dx$$

$$\frac{(y-1)^{-1}}{-1} = e^{x-1} + C$$

$$-\frac{1}{y-1} = e^{x-1} + C$$

$$-(y-1) = \frac{1}{e^{x-1}+C}$$

$$-y+1 = \frac{1}{e^{x-1}+C}$$

$$1 - \frac{1}{e^{x-1}+C} = y$$

$$y = \frac{e^{x-1}+C}{e^{x-1}+C} - \frac{1}{e^{x-1}+C}$$

$$y = \frac{e^{x-1}+C-1}{e^{x-1}+C}.$$

$y = 2$, when $x = 1$.

$$2 = \frac{e^0+C-1}{e^0+C}$$

$$2 = \frac{C}{1+C}$$

$$2 + 2C = C$$

$$C = -2$$

$$y = \frac{e^{x-1}-3}{e^{x-1}-2}.$$

33. $\frac{dy}{dx} = \frac{k}{N}(N-y)y$

(a) $\frac{N\,dy}{(N-y)y} = k\,dx$

Since $\frac{1}{y} + \frac{1}{N-y} = \frac{N}{(N-y)y}$,

$$\int \frac{dy}{y} + \int \frac{dy}{N-y} = k\,dx$$

$$\ln\left|\frac{y}{N-y}\right| = kx + C$$

$$\frac{y}{N-y} = Ce^{kx}.$$

For $0 < y < N$, $Ce^{kx} > 0$.
For $0 < N < y$, $Ce^{kx} < 0$.
Solve for y.

$$y = \frac{Ce^{kx}N}{1+Ce^{kx}} = \frac{N}{1+C^{-1}e^{-kx}}$$

Let $b = C^{-1} > 0$ for $0 < y < N$.

$$y = \frac{N}{1+be^{-kx}}$$

Let $-b = C^{-1} < 0$ for $0 < N < y$.

$$y = \frac{N}{1-be^{-kx}}$$

(b) For $0 < y < N$; $t = 0$, $y = y_0$.

$$y_0 = \frac{N}{1+be^0} = \frac{N}{1+b}$$

Solve for b.

$$b = \frac{N-y_0}{y_0}$$

(c) For $0 < N < y$; $t = 0$, $y = y_0$.

$$y_0 = \frac{N}{1-be^0} = \frac{N}{1-b}$$

Solve for b.

$$b = \frac{y_0-N}{y_0}$$

35. (a) $0 < y_0 < N$ implies that $y_0 > 0$, $N > 0$, and $N - y_0 > 0$.

Therefore,

$$b = \frac{N-y_0}{y_0} > 0.$$

Also, $e^{-kx} > 0$ for all x, which implies that $1 + be^{-kx} > 1$.

(1) $y(x) = \frac{N}{1+be^{-kx}} < N$ since $1 + be^{-kx} > 1$.

(2) $y(x) = \frac{N}{1+be^{-kx}} > 0$ since $N > 0$ and $1 + be^{-kx} > 0$.

Combining statements (1) and (2), we have

$$0 < \frac{N}{1+be^{-kx}} = y(x)$$

$$= \frac{N}{1+be^{-kx}} < N$$

or $0 < y(x) < N$ for all x.

(b) $\lim_{x\to\infty} \frac{N}{1+be^{-kx}} = \frac{N}{1+b(0)} = N$

$$\lim_{x\to-\infty} \frac{N}{1+be^{-kx}} = 0$$

Note that as $x \to -\infty$, $1 + be^{-kx}$ becomes infinitely large.

Therefore, the horizontal asymptotes are $y = N$ and $y = 0$.

(c) $y'(x) = \dfrac{(1 + be^{-kx})(0) - N(-kbe^{-kx})}{(1 + be^{-kx})^2}$

$= \dfrac{Nkbe^{-kx}}{(1 + be^{-kx})^2} > 0$ for all x.

Therefore, $y(x)$ is an increasing function.

(d) To find $y''(x)$, apply the quotient rule to find the derivative of $y'(x)$. The numerator of $y''(x)$ is

$y''(x) = (1 + be^{-kx})^2(-Nk^2be^{-kx})$
$\qquad - Nkbe^{-kx}[-2kbe^{-kx}(1 + be^{-kx})]$
$\quad = -Nkbe^{-kx}(1 + be^{-kx})$
$\qquad \cdot [k(1 + be^{-kx}) - 2kbe^{-kx}]$
$\quad = -Nkbe^{-kx}(1 + be^{-kx})(k - kbe^{-kx}),$

and the denominator is

$$[(1 + be^{-kt})^2]^2 = (1 + be^{-kx})^4.$$

Thus,

$$y''(x) = \frac{-Nkbe^{-kx}(k - kbe^{-kx})}{(1 + be^{-kx})^3}.$$

$y''(x) = 0$ when

$k - kbe^{-kx} = 0$
$be^{-kx} = 1$
$e^{-kx} = \dfrac{1}{b}$
$-kx = \ln\left(\dfrac{1}{b}\right)$
$x = -\dfrac{\ln\left(\frac{1}{b}\right)}{k}$
$= \dfrac{\ln\left(\frac{1}{b}\right)^{-1}}{k} = \dfrac{\ln b}{k}.$

When $x = \frac{\ln b}{k}$,

$y = \dfrac{N}{1 + be^{-k\left(\frac{\ln b}{k}\right)}} = \dfrac{N}{1 + be^{(-\ln b)}}$

$= \dfrac{N}{1 + be^{\ln(1/b)}} = \dfrac{N}{1 + b\left(\frac{1}{b}\right)} = \dfrac{N}{2}.$

Therefore, $\left(\frac{\ln b}{k}, \frac{N}{2}\right)$ is a point of inflection.

(e) To locate the maximum of $\frac{dy}{dx}$, we must consider, from part (d),

$$\frac{d}{dx}\left(\frac{dy}{dx}\right) = \frac{-Nkbe^{-kx}(k - kbe^{-kx})}{(1 + be^{-kx})^3}.$$

Since $y''(x) > 0$ for $x < \dfrac{\ln b}{k}$ and

$y''(x) < 0$ for $x > \dfrac{\ln b}{k}$,

we know that $x = \frac{\ln b}{k}$ locates a relative maximum of $\frac{dy}{dx}$.

37. (a)

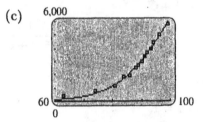

The data seems appropriate for a logistics function.

(b) $y = 11,074/(1 + 151,378e^{-0.1219x})$

(c)

6,000

60 ⌞_____⌟ 100
0

The logistic equation fits the data well.

(d) 11,074

39. $\dfrac{dN}{dt} = mN + i$

(a) $\qquad \dfrac{dN}{dt} = m\left(N + \dfrac{i}{m}\right)$

$\dfrac{1}{N + \frac{i}{m}}\,dN = m\,dt$

$\displaystyle\int \dfrac{dN}{N + \frac{i}{m}} = \int m\,dt$

$\ln\left(N + \dfrac{i}{m}\right) = mt + C$

$N + \dfrac{i}{m} = e^{mt+C}$

$N + \dfrac{i}{m} = e^{mt} \cdot e^C$

$N = e^{mt} \cdot e^C - \dfrac{i}{m}$

Since $N(0) = N_o$,

$$N_o = e^{m(0)} \cdot e^C - \frac{i}{m}$$

$$e^C = N_o + \frac{i}{m}.$$

Substituting,

$$N = \left(N_o + \frac{i}{m} \right) e^{mt} - \frac{i}{m}.$$

(b) Since $N_o = 0$,

$$N = \left(0 + \frac{i}{m} \right) e^{mt} - \frac{i}{m}$$

$$= \frac{i}{m}(e^{mt} - 1)$$

$$N(8) = \frac{i}{m}(e^{8m} - 1)$$

$$mN(8) = i(e^{8m} - 1)$$

$$i = \frac{mN(8)}{e^{8m} - 1}$$

(c) $m = \ln F_{sd}$

$$= \ln(0.709)$$

$$\approx -0.344$$

$$i = \frac{mN(8)}{e^{8m} - 1}$$

$$= \frac{\ln(0.709) \cdot 4.5}{e^{8\ln(0.709)} - 1}$$

$$\approx 1.65$$

41. (a) $\dfrac{dI}{dW} = 0.088(2.4 - I)$

Separate the variables and take antiderivatives.

$$\int \frac{dI}{2.4 - I} = \int 0.088 dW$$

$$- \ln |2.4 - I| = 0.088W + k$$

Solve for I.

$$\ln |2.4 - I| = -0.088W - k$$
$$|2.4 - I| = e^{-0.088W - k} = e^{-k}e^{-0.088W}$$
$$I - 2.4 = Ce^{-0.088W}, \text{ where } C = \pm e^{-k}.$$
$$I = 2.4 + Ce^{-0.088W}$$

Since $I(0) = 1$, then

$$1 = 2.4 + Ce^0$$
$$C = 1 - 2.4 = -1.4.$$

Therefore, $I = 2.4 - 1.4e^{-0.088W}$.

(b) Note that as W gets larger and larger $e^{-0.088W}$ approaches 0, so

$$\lim_{W \to \infty} I = \lim_{W \to \infty} (2.4 - 1.4e^{-0.088W})$$
$$= 2.4 - 1.4(0) = 2.4,$$

so I approaches 2.4.

43. (a) $\dfrac{dw}{dt} = k(C - 17.5w)$

C being constant implies that the calorie intake per day is constant.

(b) pounds/day $= k$ (calories/day)

$$\frac{\text{pounds/day}}{\text{calories/day}} = k$$

The units of k are pounds/calorie.

(c) Since 3,500 calories is equivalent to 1 pound, $k = \frac{1}{3,500}$ and

$$\frac{dw}{dt} = \frac{1}{3,500}(C - 17.5w).$$

(d) $\dfrac{dw}{dt} = \dfrac{1}{3,500}(C - 17.5w)$; $w = w_0$ when $t = 0$.

$$\frac{3,500}{C - 17.5w} dw = dt$$

$$\frac{3,500}{-17.5} \int \frac{-17.5}{C - 17.5w} dw = \int dt$$

$$-200 \ln |C - 17.5w| = t + k$$
$$\ln |C - 17.5w| = -0.005t - 0.005k$$
$$|C - 17.5w| = e^{-0.005t - 0.005k}$$
$$|C - 17.5w| = e^{-0.005t} \cdot e^{-0.005k}$$
$$C - 17.5w = e^{-0.005M} e^{-0.005t}$$
$$-17.5w = -C + e^{-0.005M} e^{-0.005t}$$

$$w = \frac{C}{17.5} - \frac{e^{-0.005M}}{17.5} e^{-0.005t}$$

(e) Since $w = w_0$ when $t = 0$,

$$w_0 = \frac{C}{17.5} - \frac{e^{-0.005M}}{17.5} \quad (1)$$

$$w_0 - \frac{C}{17.5} = -\frac{e^{-0.005M}}{17.5}$$

$$\frac{e^{-0.005M}}{17.5} = \frac{C}{17.5} - w_0.$$

Therefore,

$$w = \frac{C}{17.5} - \left(\frac{C}{17.5} - w_0 \right) e^{-0.005t}$$

$$w = \frac{C}{17.5} + \left(w_0 - \frac{C}{17.5} \right) e^{-0.005t}.$$

45. (a)

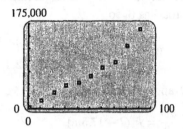

The points suggest that a logistic function is appropriate.

(b) A calculator with a logistic regression function determines that

$$y = \frac{484{,}900}{1 + 28.32e^{-0.02893x}}$$

best fits the data.

(c)

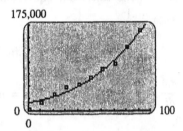

The model does produce appropriate y-values for the given x-values, particularly for the more recent years.

(d) As x gets larger and larger, $e^{-0.02893x}$ approaches 0, so y approaches

$$\frac{484{,}900}{1 + 28.32 \cdot 0} = \frac{484{,}900}{1} = 484{,}900.$$

Therefore, according to this model, the limiting size of the Jewish population in Toronto is 484,900.

47. $\dfrac{dy}{dt} = ky$

First separate the variables and integrate.

$$\frac{dy}{y} = k\,dt$$

$$\int \frac{dy}{y} = \int k\,dt$$

$$\ln|y| = kt + C_1$$

Solve for y.

$$|y| = e^{kt+C_1} = e^{C_1}e^{kt}$$
$$y = Ce^{kt}, \text{ where } C = \pm e^{C_1}.$$
$$y(0) = 35.5, \text{ so } 35.5 = Ce^0 = C, \text{ and }$$
$$y = 35.5e^{kt}.$$

Solve for k. Since $y(50) = 60.6$, then

$$60.6 = 35.5e^{50k}$$

$$e^{50k} = \frac{60.6}{35.5}$$

$$50k = \ln\left(\frac{60.6}{35.5}\right)$$

$$k = \frac{\ln\left(\frac{60.6}{35.5}\right)}{50} \approx 0.01070,$$

so $y = 35.5e^{0.01070t}$.

49. (a) $\dfrac{dy}{dt} = ky$

$$\int \frac{dy}{y} = \int k\,dt$$

$$\ln|y| = kt + C$$
$$e^{\ln|y|} = e^{kt+C}$$
$$y = \pm(e^{kt})(e^C)$$
$$y = Me^{kt}$$

If $y = 1$ when $t = 0$ and $y = 5$ when $t = 2$, we have the system of equations

$$1 = Me^{k(0)}$$
$$5 = Me^{2k}.$$

$$1 = M(1)$$
$$M = 1$$

Substitute.

$$5 = (1)e^{2k}$$
$$e^{2k} = 5$$
$$2k \ln e = \ln 5$$
$$k = \frac{\ln 5}{2}$$
$$\approx 0.8$$

(b) If $k = 0.8$ and $M = 1$,

$$y = e^{0.8t}.$$

When $t = 3$,

$$y = e^{0.8(3)}$$
$$= e^{2.4}$$
$$\approx 11.$$

(c) When $t = 5$,

$$y = e^{0.8(5)}$$
$$= e^4$$
$$\approx 55.$$

(d) When $t = 10$,

$$y = e^{0.8(10)}$$
$$= e^8$$
$$\approx 2{,}981.$$

51. (a) $\dfrac{dy}{dt} = -0.05y$

(b) $\displaystyle\int \dfrac{dy}{y} = -\int 0.05\,dt$

$$\ln|y| = -0.05t + C$$
$$e^{\ln|y|} = e^{-0.05+C}$$
$$y = \pm e^{-0.05t} \cdot e^C$$
$$y = Me^{-0.05t}$$

(c) Since $y = 90$ when $t = 0$.

$$90 = Me^0$$
$$M = 90.$$

So $y = 90e^{-0.05t}$.

(d) At $t = 10$,

$$y = 90e^{-0.05(10)}$$
$$\approx 55.$$

After 10 months, about 55 g are left.

11.2   Linear First-Order Differential Equations

1. $\dfrac{dy}{dx} + 2y = 5$

$$I(x) = e^{2\int dx} = e^{2x}$$

Multiply each term by e^{2x}.

$$e^{2x}\dfrac{dy}{dx} + 2e^{2x}y = 5e^{2x}$$
$$D_x\left(e^{2x}y\right) = 5e^{2x}$$

Integrate both sides.

$$e^{2x}y = \int 5e^{2x}\,dx$$
$$= \dfrac{5}{2}e^{2x} + C$$
$$y = \dfrac{5}{2} + Ce^{-2x}$$

3. $\dfrac{dy}{dx} + xy = 3x$

$$I(x) = e^{\int x\,dx} = e^{x^2/2}$$

$$e^{x^2/2}\dfrac{dy}{dx} + xe^{x^2/2}y = 3xe^{x^2/2}$$

$$D_x\left(e^{x^2/2}y\right) = 3xe^{x^2/2}$$

$$e^{x^2/2}y = \int 3xe^{x^2/2}\,dx$$

$$= 3e^{x^2/2} + C$$
$$y = 3 + Ce^{-x^2/2}$$

5. $x\dfrac{dy}{dx} - y - x = 0; \ x > 0$

$$\dfrac{dy}{dx} - \dfrac{1}{x}y = 1$$

$$I(x) = e^{-\int 1/x\,dx}$$
$$= e^{-\ln x} = \dfrac{1}{x}$$

$$\dfrac{1}{x}\dfrac{dy}{dx} - \dfrac{1}{x^2}y = \dfrac{1}{x}$$

$$D_x\left(\dfrac{1}{x}y\right) = \dfrac{1}{x}$$

$$\dfrac{y}{x} = \int \dfrac{1}{x}\,dx$$

$$\dfrac{y}{x} = \ln x + C$$
$$y = x\ln x + Cx$$

7. $2\dfrac{dy}{dx} - 2xy - x = 0$

$$\dfrac{dy}{dx} - xy = \dfrac{x}{2}$$

$$I(x) = e^{-\int x\,dx} = e^{-x^2/2}$$

$$e^{-x^2/2}\dfrac{dy}{dx} - xe^{-x^2/2}y = \dfrac{x}{2}e^{-x^2/2}$$

$$D_x\left(e^{-x^2/2}y\right) = \dfrac{x}{2}e^{-x^2/2}$$

$$e^{-x^2/2}y = \int \dfrac{x}{2}e^{-x^2/2}\,dx$$

$$= \dfrac{-1}{2}e^{-x^2/2} + C$$

$$y = -\dfrac{1}{2} + Ce^{x^2/2}$$

9. $x\dfrac{dy}{dx} + 2y = x^2 + 3x; \; x > 0$

$$\frac{dy}{dx} + \frac{2}{x}y = x + 3$$

$$I(x) = e^{\int 2/x \, dx} = e^{2 \ln x} = x^2$$

$$x^2\frac{dy}{dx} + 2xy = x^3 + 3x^2$$

$$D_x\left(x^2 y\right) = x^3 + 3x^2$$

$$x^2 y = \int (x^3 + 3x^2)\, dx$$

$$= \frac{x^4}{4} + x^3 + C$$

$$y = \frac{x^2}{4} + x + \frac{C}{x^2}$$

11. $y - x\dfrac{dy}{dx} = x^3; \; x > 0$

$$\frac{dy}{dx} - \frac{y}{x} = -x^2$$

$$I(x) = e^{-\int 1/x \, dx} = e^{-\ln x} = x^{-1}$$

$$\frac{1}{x}\frac{dy}{dx} - \frac{y}{x^2} = -x$$

$$D_x\left(\frac{1}{x}y\right) = -x$$

$$\frac{y}{x} = \int -x\, dx$$

$$= \frac{-x^2}{2} + C$$

$$y = \frac{-x^3}{2} + Cx$$

13. $\dfrac{dy}{dx} + y \cot x = x$

$$I(x) = e^{\int \cot x \, dx} = e^{\ln \sin x} = \sin x$$

$$\sin x\frac{dy}{dx} + y\cos x = x\sin x$$

$$D_x(y \sin x) = x \sin x$$
$$y \sin x = \int x \sin x \, dx$$
$$= \sin x - x\cos x + C$$

$$y = 1 - x\cot x + \frac{C}{\sin x}$$

15. $\dfrac{dy}{dx} + y = 2e^x; \; y = 100$ when $x = 0.$

$$I(x) = e^{\int dx} = e^x$$

$$e^x\frac{dy}{dx} + ye^x = 2e^{2x}$$

$$D_x\left(e^x y\right) = 2e^{2x}$$

$$e^x y = \int 2e^{2x}\, dx$$

$$= e^{2x} + C$$
$$y = e^x + Ce^{-x}$$

Since $y = 100$ when $x = 0,$

$$100 = e^0 + Ce^0$$
$$100 = 1 + C$$
$$C = 99.$$

Therefore,

$$y = e^x + 99e^{-x}.$$

17. $\dfrac{dy}{dx} - xy - x = 0; \; y = 10$ when $x = 1.$

$$\frac{dy}{dx} - xy - x = 0$$

$$\frac{dy}{dx} - xy = x$$

$$I(x) = e^{-\int x \, dx} = e^{-x^2/2}$$

$$e^{-x^2/2}\frac{dy}{dx} - xe^{-x^2/2}y = xe^{-x^2/2}$$

$$D_x\left(e^{-x^2/2}y\right) = xe^{-x^2/2}$$

$$e^{-x^2/2}y = \int xe^{-x^2/2}\, dx$$

$$e^{-x^2/2}y = -e^{-x^2/2} + C$$
$$y = -1 + Ce^{x^2/2}$$

Since $y = 10$ when $x = 1,$

$$10 = -1 + Ce^{1/2}$$
$$11 = Ce^{1/2}$$

$$C = \frac{11}{e^{1/2}}.$$

Therefore,

$$y = -1 + \frac{11}{e^{1/2}}\left(e^{x^2/2}\right)$$

$$= -1 + 11e^{x^2/2 - 1/2}$$
$$= -1 + 11e^{(x^2-1)/2}.$$

19. $x\dfrac{dy}{dx} + 5y = x^2$; $y = 12$ when $x = 2$.

$$x\dfrac{dy}{dx} + 5y = x^2$$

$$\dfrac{dy}{dx} + \dfrac{5}{x}y = x$$

$$I(x) = e^{\int 5/x\,dx} = e^{5\ln x} = x^5$$

$$x^5\dfrac{dy}{dx} + 5x^4 y = x^6$$

$$D_x\left(x^5 y\right) = x^6$$

$$x^5 y = \int x^6\,dx$$

$$x^5 y = \dfrac{x^7}{7} + C$$

$$y = \dfrac{x^2}{7} + \dfrac{C}{x^5}$$

Since $y = 12$, when $x = 2$,

$$12 = \dfrac{4}{7} + \dfrac{C}{32}$$

$$\dfrac{80}{7} = \dfrac{C}{32}$$

$$C = \dfrac{2{,}560}{7}.$$

Therefore,

$$y = \dfrac{x^2}{7} + \dfrac{2{,}560}{7x^5}.$$

21. $x\dfrac{dy}{dx} + (1+x)y = 3$; $y = 50$ when $x = 4$

$$\dfrac{dy}{dx} + \left(\dfrac{1+x}{x}\right)y = \dfrac{3}{x}$$

$$I(x) = e^{\int (1+x)\,dx/x}$$

$$= e^{\int (1/x)\,dx + dx}$$

$$= e^{(\ln x) + x}$$

$$= e^{\ln x} \cdot e^x$$

$$= xe^x$$

$$xe^x\dfrac{dy}{dx} + (1+x)e^x y = 3e^x$$

$$D_x\left(xe^x y\right) = 3e^x$$

$$xe^x y = \int 3e^x\,dx$$

$$xe^x y = 3e^x + C$$

$$y = \dfrac{3}{x} + \dfrac{C}{xe^x}$$

Since $y = 50$ when $x = 4$,

$$50 = \dfrac{3}{4} + \dfrac{C}{4e^4}$$

$$\dfrac{197}{4} = \dfrac{C}{4e^4}$$

$$C = 197e^4.$$

Therefore,

$$y = \dfrac{3}{x} + \dfrac{197e^4}{xe^x}$$

$$= \dfrac{3}{x} + \dfrac{197}{x}e^{4-x}$$

$$= \dfrac{3 + 197e^{4-x}}{x}.$$

23. (a) $\dfrac{dC}{dt} = -kC + D(t)$

$$\dfrac{dC}{dt} + kC = D(t)$$

$$I = e^{\int k\,dt} = e^{kt}$$

$$e^{kt}\dfrac{dC}{dt} + e^{kt}kC = e^{kt}D(t)$$

$$D_t\left(e^{kt}C\right) = e^{kt}D(t)$$

$$e^{kt}C = \int e^{kt}D(t)d(t)$$

$$= \int_0^t e^{kt}D(y)dy$$

$$C(t) = e^{-kt}\int_0^t e^{ky}D(y)dy + C_2$$

If $C(0) = 0$,

$$C(0) = e^0\int_0^0 e^{ky}D(y)dy + C_2$$

$$0 = 0 + C_2$$

Therefore,

$$C(t) = e^{-kt}\int_0^t e^{ky}D(y)dy.$$

(b) Let $D(y) = D$, a constant.

$$C(t) = e^{-kt}\int_0^t e^{ky}D\,dy$$

$$= De^{-kt}\int_0^t e^{ky}\,dy$$

$$= De^{-kt}\left(\dfrac{1}{k}\right)\left(e^{kt} - e^{k(0)}\right)$$

$$C(t) = \dfrac{D(1 - e^{-kt})}{k}$$

25. $\dfrac{dy}{dx} = cy - py^2$

  (a) Let $y = \frac{1}{z}$ and $\dfrac{dy}{dx} = -\dfrac{z'}{z^2}$.

$$-\frac{z'}{z^2} = c\left(\frac{1}{z}\right) - p\left(\frac{1}{z^2}\right)$$

$$z' = -cz + p$$

$$z' + cz = p$$

$$I(x) = e^{\int c\, dx} = e^{cx}$$

$$D_x\left(e^{cx} \cdot z\right) = \int p e^{cx}\, dx$$

$$e^{cx} \cdot z = \frac{p}{c}e^{cx} + k$$

$$z = \frac{p}{c} + ke^{-cx}$$

$$= \frac{p + kce^{-cx}}{c}$$

  Therefore,

$$y = \frac{c}{p + kce^{-cx}}.$$

  (b) Let $z(0) = \frac{1}{y_0}$.

$$\frac{1}{y_0} = \frac{p + kce^0}{c} = \frac{p + kc}{c}$$

$$\frac{c}{y_0} = p + kc$$

$$kc = \frac{c}{y_0} - p = \frac{c - py_0}{y_0}$$

$$k = \frac{c - py_0}{cy_0}$$

$$y = \frac{c}{p + \left(\frac{c - py_0}{cy_0}\right)ce^{-cx}}$$
$$\text{From part (a)}$$

$$= \frac{cy_0}{py_0 + (c - py_0)e^{-cx}}$$

  (c) $\displaystyle\lim_{x \to \infty} y = \lim_{x \to \infty}\left(\frac{cy_0}{py_0 + (c - py_0)e^{-cx}}\right)$

$$= \frac{cy_0}{py_0 - 0}$$

$$= \frac{c}{p}$$

27. $\dfrac{dy}{dt} = 0.02y + e^t$; $y = 10{,}000$ when $t = 0$.

$$\frac{dy}{dt} - 0.02y = e^t$$

$$I(t) = e^{\int -0.02\, dt} = e^{-0.02t}$$

$$e^{-0.02t}\frac{dy}{dt} - 0.02e^{-0.02t}y = e^{-0.02t} \cdot e^t$$

$$D_t\left(e^{-0.02t}y\right) = e^{0.98t}$$

$$e^{-0.02t}y = \int e^{0.98t}\, dt$$

$$= \frac{e^{0.98t}}{0.98} + C$$

$$y = \frac{e^t}{0.98} + Ce^{0.02t}$$

$$10{,}000 = \frac{1}{0.98} + C$$

$$C \approx 9{,}999$$

$$y \approx \frac{e^t}{0.98} + 9{,}999e^{0.02t}$$

$$= 1.02e^t + 9{,}999e^{0.02t}$$

29. $\dfrac{dy}{dt} = 0.02y - t$; $y = 10{,}000$ when $t = 0$.

$$\frac{dy}{dt} - 0.02y = -t$$

$$I(t) = e^{\int -0.02dt} = e^{-0.02t}$$

$$e^{-0.02t}\frac{dy}{dt} - 0.02e^{-0.02t}y = -te^{-0.02t}$$

$$D_t\left(e^{-0.02t}y\right) = -te^{-0.02t}$$

$$e^{-0.02t}y = \int -te^{-0.02t}\, dt$$

Integration by parts:

Let $u = -t$ $dv = e^{-0.02t}\, dt$

$\quad\; du = -dt$ $v = \dfrac{e^{-0.02t}}{-0.02}$

$$e^{-0.02t} = \frac{te^{-0.02t}}{0.02} - \int \frac{e^{-0.02t}}{0.02}\, dt$$

$$e^{-0.02t}y = \frac{te^{-0.02t}}{0.02} + \frac{e^{-0.02t}}{0.0004} + C$$

$$y = 50t + 2{,}500 + Ce^{0.02t}$$

$$10{,}000 = 2{,}500 + C$$

$$C = 7{,}500$$

$$y = 50t + 2{,}500 + 7{,}500e^{0.02t}$$

31. $\dfrac{dT}{dt} = -k(T - T_0)$

$$\frac{dT}{dt} = -kT + kT_0$$

$$\frac{dT}{dt} + kT = kT_0$$

$$I(t) = e^{\int k\,dt} = e^{kt}$$

Multiply both sides by e^{kt}.

$$e^{kt}\frac{dT}{dt} + ke^{kt}T = kT_0e^{kt}$$

$$D_t\left(Te^{kt}\right) = kT_0e^{kt}$$

$$Te^{kt} = \int kT_0e^{kt}\,dt$$

$$Te^{kt} = T_0e^{kt} + C$$
$$T = T_0 + Ce^{-kt}$$
$$T = Ce^{-kt} + T_0$$

33. If $T = ce^{-kt} + T_0$,

$$\lim_{t\to\infty} T = \lim_{t\to\infty} \left(ce^{-kt} + T_0\right)$$
$$= c(0) + T_0 \text{ since } k > 0$$
$$= T_0.$$

Thus, the temperature approaches T_0 according to Newton's law of cooling. We would expect the temperature of the object to approach the temperature of the surrounding medium.

35. Refer to Exercise 27.

$$T = Ce^{-kt} + T_0$$

(a) In this problem, $T_0 = 10$, $C = 88.6$, and $k = 0.24$.
Therefore,

$$T = 88.6e^{-0.24t} + 10.$$

(b)

(c) The graph shows the most rapid decrease in the first few hours which is just after death.

(d) If $t = 4$,

$$T = 88.6e^{-0.24(4)} + 10$$
$$T \approx 43.9.$$

The temperature of the body will be 43.9°F after 4 hours.

(e)
$$40 = 88.6e^{-0.24t} + 10$$
$$88.6e^{-0.24t} = 30$$

$$e^{-0.24t} = \frac{30}{88.6}$$

$$-0.24t = \ln\left(\frac{30}{88.6}\right)$$

$$t = \frac{\ln\left(\frac{30}{88.6}\right)}{-0.24}$$

$$t \approx 4.5$$

The body will reach a temperature of 40°F in 4.5 hours.

11.3   Euler's Method

Note: In each step of the calculations shown in this section, all digits should be kept in your calculator as you proceed through Euler's method. Do not round intermediate results.

1. $\dfrac{dy}{dx} = x^2 + y^2$; $y(0) = 1$, $h = 0.1$; find $y(0.5)$.

$$g(x, y) = x^2 + y^2$$
$$x_0 = 0; \ y_0 = 1$$
$$g(x_0, y_0) = 0 + 1 = 1$$

$$x_1 = 0.1; \ y_1 = 1 + 1(0.1) = 1.1$$
$$g(x_1, y_1) = (0.1)^2 + (1.1)^2$$
$$= 1.22$$

$$x_2 = 0.2; \ y_2 = 1.1 + 1.22(0.1)$$
$$= 1.222$$
$$g(x_2, y_2) = (0.2)^2 + (1.222)^2$$
$$= 1.533$$

$$x_3 = 0.3; \ y_3 = 1.222 + 1.533(0.1)$$
$$= 1.375$$
$$g(x_3, y_3) = (0.3)^2 + (1.375)^2$$
$$= 1.981$$

$$x_4 = 0.4; \ y_4 = 1.375 + (1.981)(0.1)$$
$$= 1.573$$
$$g(x_4, y_4) = (0.4)^2 + (1.573)^2$$
$$= 2.635$$

$$x_5 = 0.5; \ y_5 = 1.573 + (2.635)(0.1)$$
$$= 1.837$$

These results are tabulated as follows.

x_i	y_i
0	1
0.1	1.1
0.2	1.222
0.3	1.375
0.4	1.573
0.5	1.837

$y(0.5) = 1.837$

3. $\dfrac{dy}{dx} = 1 + y = g(x, y); \ y(0) = 2,$

$h = 0.1$; find $y(0.6)$.

$x_0 = 0; \ y_0 = 2$
$\quad g(x_0, y_0) = 1 + 2 = 3$

$x_1 = 0.1; \ y_1 = 2 + 3(0.1) = 2.3$
$\quad g(x_1, y_1) = 1 + 2.3 = 3.3$

$x_2 = 0.2; \ y_2 = 2.3 + 3.3(0.1) = 2.63$
$\quad g(x_2, y_2) = 1 + 2.63 = 3.63$

$x_3 = 0.3; \ y_3 = 2.63 + 3.63(0.1) = 2.993$
$\quad g(x_3, y_3) = 1 + 2.993 = 3.993$

$x_4 = 0.4; \ y_4 = 2.993 + 3.993(0.1) = 3.3923$
$\quad g(x_4, y_4) = 1 + 3.3923 = 4.3923$

$x_5 = 0.5; \ y_5 = 3.3923 + 4.392(0.1) = 3.8315$
$\quad g(x_5, y_5) = 1 + 3.8315 = 4.8315$

$x_6 = 0.6; \ y_6 = 3.8315 + 4.8315(0.1)$
$\qquad = 4.31465$

x_i	y_i
0	2
0.1	2.3
0.2	2.63
0.3	2.993
0.4	3.3923
0.5	3.8315
0.6	4.31465

$y(0.6) \approx 4.315$

5. $\dfrac{dy}{dx} = x + \sqrt{y} = g(x, y); \ y(0) = 1.$

$h = 0.1$; find $y(0.4)$.

$x_0 = 0; \ y_0 = 1$
$\quad g(x_0, y_0) = 0 + 1 = 1$

$x_1 = 0.1; \ y_1 = 1 + 1(0.1) = 1.1$
$\quad g(x_1, y_1) = 0.1 + \sqrt{1.1} = 1.149$

$x_2 = 0.2; \ y_2 = 1.1 + 1.149(0.1)$
$\qquad = 1.215$

$\quad g(x_2, y_2) = 0.2 + \sqrt{1.215} = 1.302$
$x_3 = 0.3; \ y_3 = 1.215 + 1.302(0.1)$
$\qquad = 1.345$

$\quad g(x_3, y_3) = 0.3 + \sqrt{1.345} = 1.460$
$x_4 = 0.4; \ y_4 = 1.345 + 1.460(0.1)$
$\qquad = 1.491$

x_i	y_i
0	1
0.1	1.1
0.2	1.215
0.3	1.345
0.4	1.491

$y(0.4) \approx 1.491$

7. $\dfrac{dy}{dx} = x\sqrt{1 + y^2}; \ y(1) = 1, h = 0.1$; find $y(1.5)$.

$g(x, y) = x\sqrt{1 + y^2}$

$x_0 = 1; \ y_0 = 1$
$g(x_0, y_0) = 1 \cdot \sqrt{1 + 1} = \sqrt{2}$

$x_1 = 1.1; \ y_1 = y_0 + g(x_0, y_0)0.1 = 1 + \dfrac{\sqrt{2}}{10} \approx 1.1414$

$g(x_1, y_1) = 1.1\sqrt{1 + \left(1 + \dfrac{\sqrt{2}}{10}\right)^2} \approx 1.6693$

$x_2 = 1.2; \ y_2 = y_1 + g(x_1, y_1)0.1 \approx 1.3083$
$g(x_2, y_2) = 1.2\sqrt{1 + y_2^2} \approx 1.9761$
$x_3 = 1.3; \ y_3 = y_2 + g(x_2, y_2)0.1 \approx 1.5060$
$g(x_3, y_3) = 1.3\sqrt{1 + y_3^2} \approx 2.3501$
$x_4 = 1.4; \ y_4 = y_3 + g(x_3, y_3)0.1 \approx 1.7410$
$g(x_4, y_4) = 1.4\sqrt{1 + y_4^2} \approx 2.8108$
$x_5 = 1.5; \ y_5 = y_4 + g(x_4, y_4)0.1 \approx 2.0220$

These results are tabulated as follows.

x_i	y_i
1	1
1.1	1.141
1.2	1.308
1.3	1.506
1.4	1.741
1.5	2.022

So, rounded to three decimal places,

$$y(1.5) \approx 2.022.$$

9. $\dfrac{dy}{dx} = y + \cos x = g(x,y); \; y(0) = 0, h = 0.1;$

find y(0.5).

$x_0 = 0; \; y_0 = 0$
$\qquad g(x_0, y_0) = 0 + 1.0 = 1.0$

$x_1 = 0.1; \; y_1 = 0 + 1.0(0.1)$
$\qquad\qquad = 0.1$
$\qquad g(x_1, y_1) = 0.1 + \cos(0.1) = 1.095$

$x_2 = 0.2; \; y_2 = 0.1 + 1.095(0.1)$
$\qquad\qquad = 0.210$
$\qquad g(x_2, y_2) = 0.210 + \cos(0.2) = 1.190$

$x_3 = 0.3; \; y_3 = 0.210 + 1.190(0.1)$
$\qquad\qquad = 0.329$
$\qquad g(x_3, y_3) = 0.329 + \cos(0.3) = 1.284$

$x_4 = 0.4; \; y_4 = 0.329 + 1.284(0.1)$
$\qquad\qquad = 0.457$
$\qquad g(x_4, y_4) = 0.457 + \cos(0.4) = 1.378$

$x_5 = 0.5; \; y_5 = 0.457 + 1.378(0.1)$
$\qquad\qquad = 0.595$

x_i	y_i
0	0.000
0.1	0.100
0.2	0.210
0.3	0.328
0.4	0.457
0.5	0.595

$y(0.5) \approx 0.595$

11. $\dfrac{dy}{dx} = -4 + x = g(x,y); \; y(0) = 1,$
$\quad h = 0.1$, find y(0.4).

$x_0 = 0; \; y_0 = 1$
$\qquad g(x_0, y_0) = -4 + 0 = -4$

$x_1 = 0.1; \; y_1 = 1 + (-4)(0.1) = 0.6$
$\qquad g(x_1, y_1) = -4 + 0.1 = -3.9$

$x_2 = 0.2; \; y_2 = 0.6 + (-3.9)(0.1) = 0.21$
$\qquad g(x_2, y_2) = -4 + 0.2 = -3.8$

$x_3 = 0.3; \; y_3 = 0.21 + (-3.8)(0.1)$
$\qquad\qquad = -0.17$
$\qquad g(x_3, y_3) = -4 + 0.3 = -3.7$

$x_4 = 0.4; \; y_4 = -0.17 + (-3.7)(0.1)$
$\qquad\qquad = -0.540$

x_i	y_i
0	1
0.1	0.6
0.2	0.21
0.3	-0.17
0.4	-0.540

$y(0.4) \approx -0.540$
Actual solution:

$$\frac{dy}{dx} = -4 + x$$

$$y = -4x + \frac{x^2}{2} + C$$

At $y(0) = 1$,

$$1 = -4(0) + \frac{0}{2} + C$$

$$C = 1.$$

Therefore,

$$y = -4x + \frac{x^2}{2} + 1$$

$$y(0.4) = -4(0.4) + \frac{(0.4)^2}{2} + 1$$

$$= -0.520.$$

13. $\frac{dy}{dx} = x^2$; $y(0) = 2$, $h = 0.1$, find $y(0.5)$.

$x_0 = 0$; $y_0 = 2$
$g(x_0, y_0) = 0^2 = 0$
$x_1 = 0.1$; $y_1 = 2 + 0(0.1) = 2$
$g(x_1, y_1) = (0.1)^2 = 0.01$

$x_2 = 0.2$; $y_2 = 2 + 0.01(0.1) = 2.001$
$g(x_2, y_2) = (0.2)^2 = 0.04$

$x_3 = 0.3$; $y_3 = 2.001 + 0.04(0.1)$
$\qquad = 2.005$
$g(x_3, y_3) = (0.3)^2 = 0.09$

$x_4 = 0.4$; $y_4 = 2.005 + 0.09(0.1)$
$\qquad = 2.014$
$g(x_4, y_4) = (0.4)^2 = 0.16$

$x_5 = 0.5$; $y_5 = 2.014 + 0.16(0.1)$
$\qquad = 2.030$

x_i	y_i
0	2
0.1	2
0.2	2.001
0.3	2.005
0.4	2.014
0.5	2.030

$y(0.5) \approx 2.030$

Actual solution:

$$\frac{dy}{dx} = x^2$$

$$y = \frac{x^3}{3} + C.$$

At $y(0) = 2$,

$$2 = \frac{0}{3} + C$$

$$C = 2.$$

Therefore,

$$y = \frac{x^3}{3} + 2.$$

$$y(0.5) = \frac{(0.5)^3}{3} + 2 = 2.042.$$

15. $\frac{dy}{dx} = 2xy$; $y(1) = 1$, $h = 0.1$, find $y(1.6)$.

$$g(x, y) = 2xy$$

$x_0 = 1$; $y_0 = 1$
$g(x_0, y_0) = 2(1)(1) = 2$

$x_1 = 1.1$; $y_1 = 1 + 2(0.1) = 1.2$
$g(x_1, y_1) = 2(1.1)(1.2) = 2.64$

$x_2 = 1.2$; $y_2 = 1.2 + 2.64(0.1)$
$\qquad = 1.464$
$g(x_2, y_2) = 2(1.2)(1.464) \approx 3.514$

$x_3 = 1.3$; $y_3 = 1.464 + 3.514(0.1)$
$\qquad = 1.815$
$g(x_3, y_3) = 2(1.3)(1.815)$
$\qquad = 4.720$

$x_4 = 1.4$; $y_4 = 1.815 + 4.720(0.1)$
$\qquad = 2.287$
$g(x_4, y_4) = 2(1.4)(2.287)$
$\qquad = 6.404$

$x_5 = 1.5$; $y_5 = 2.287 + 6.404(0.1)$
$\qquad = 2.9274$
$g(x_5, y_5) = 2(1.5)(2.9274)$
$\qquad = 8.782$

$x_6 = 1.6$; $y_6 = 2.927 + 8.7822(0.1)$
$\qquad = 3.806$

x_i	y_i
1	1
1.1	1.2
1.2	1.464
1.3	1.815
1.4	2.287
1.5	2.927
1.6	3.806

$y(1.6) \approx 3.806$

Actual solution:

$$\int \frac{dy}{y} = \int 2x \, dx$$

$$\ln |y| = x^2 + C$$

$$|y| = e^{x^2 + C}$$

$$y = ke^{x^2}.$$

At $y(1) = 1$,

$$1 = ke^1 = ke$$

$$k = \frac{1}{e}.$$

Therefore,

$$y = \frac{1}{e}(e^{x^2})$$
$$= e^{x^2-1}$$
$$y(1.6) = e^{(1.6)^2-1}$$
$$= 4.759.$$

17. $\frac{dy}{dx} = ye^x$; $y(0) = 1, h = 0.1$, find $y(0.3)$

$$g(x,y) = ye^x$$
$$x_0 = 0;$$
$$y_0 = 1$$
$$g(x_0, y_0) = 1 \cdot e^0 = 1$$

$$x_1 = 0.1;\ y_1 = y_0 + g(x_0, y_0)h$$
$$= 1 + 1(0.1) = 1.1$$
$$g(x_1, y_1) = 1.1e^{0.1} \approx 1.216$$

$$x_2 = 0.2;\ y_2 = y_1 + g(x_1, y_1)h$$
$$= 1.1 + 1.216(0.1) \approx 1.2216$$
$$g(x_2, y_2) = 1.2216e^{0.2} \approx 1.492$$

$$x_3 = 0.3;\ y_3 = y_2 + g(x_2, y_2)h$$
$$= 1.2216 + 1.492(0.1)$$
$$\approx 1.371$$

x_i	y_i
0	1
0.1	1.1
0.2	1.22
0.3	1.371

So, $y(0.3) \approx 1.371$
Actual solution: separate variables.

$$\frac{dy}{y} = e^x\, dx$$

Integrate.

$$\int \frac{dy}{y} = \int e^x + c$$
$$\ln|y| = e^x + c$$
$$|y| = e^{e^x+c} = e^c e^{e^x}$$
$$y = ke^{e^x}, \text{ where } k = \pm e^c$$

At $y(0) = 1$, $1 = ke^{e^0} = ke$, so $k = \frac{1}{e}$.

Therefore,

$$y = \frac{1}{e}e^{e^x} = e^{e^x-1},$$

so

$$y(0.3) = e^{e^{0.3}-1} \approx 1.419.$$

19. $\frac{dy}{dx} = ye^x$; $y(0) = 1, h = 0.05$, find $y(0.3)$.

x_i	y_i
0	1.0000000
0.05	1.0500000
0.10	1.1051917
0.15	1.1662630
0.20	1.2340132
0.25	1.3093746
0.30	1.3934381

So, $y(0.3) \approx 1.393$.
Actual solution:

$$y(0.3) \approx 1.1419 \text{ (from problem 17)}$$

$$\text{error} = 1.393 - 1.419 = -0.0026$$

Problem 17:

$$\text{error} = 1.371 - 1.419 = -0.048$$

The error with $h = 0.05$ is about half the error with $h = 0.1$.

21. $\frac{dy}{dx} = \sqrt[3]{x}$, $y(0) = 0$

Using the program for Euler's method in *The Graphing Calculator Manual* supplement, the following values are obtained:

x_i	y_i	$y(x_i)$	$y_i - y(x_i)$
0	0	0	0
0.2	0	0.08772053	−0.08772053
0.4	0.11696071	0.22104189	−0.10408118
0.6	0.26432197	0.37954470	−0.11522273
0.8	0.43300850	0.55699066	−0.12398216
1.0	0.61867206	0.75000000	−0.13132794

23. $\frac{dy}{dx} = 1 - y$, $y(0) = 0$

Using the program for Euler's method in *The Graphing Calculator Manual*, the following values are obtained:

x_i	y_i	$y(x_i)$	$y_i - y(x_i)$
0	0	0	0
0.2	0.2	0.1812692	0.0187308
0.4	0.36	0.32967995	0.03032005
0.6	0.488	0.45118836	0.03681164
0.8	0.5904	0.55067104	0.03972896
1.0	0.67232	0.63212056	0.04019944

25. $\frac{dy}{dx} = \sqrt[3]{x};\ y(0) = 0$

See the solution in Exercise 21 for plotting points for approximation of this equation.

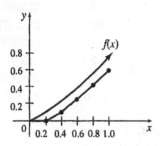

27. $\frac{dy}{dx} = 1 - y;\ y(0) = 0$

See the solution in Exercise 23 for plotting points. for approximations of this equation.

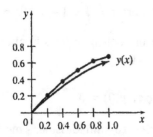

29. $\frac{dy}{dx} = y^2;\ y(0) = 1$

(a)

x_i	y_i
0	1
0.2	1.2
0.4	1.488
0.6	1.9308288
0.8	2.676448771
1.0	4.109124376

Thus, $y(1.0) \approx 4.109$.

(b) $\frac{dy}{dx} = y^2;\ y = 1$ when $x = 0$

$$\frac{dy}{dx} = y^2$$

$$\frac{1}{y^2}\,dy = dx$$

$$\int \frac{1}{y^2}\,dy = \int dx$$

$$-\frac{1}{y} = x + C$$

When $x = 0,\ y = 1$.

$$-\frac{1}{1} = 0 + C$$

$$C = -1$$

$$-\frac{1}{y} = x - 1$$

$$-1 = (x - 1)y$$

$$y = \frac{-1}{x - 1}$$

$$y = \frac{1}{1 - x}$$

As x approaches 1 from the left, y approaches ∞.

31. Let $y = $ the number of algae (in thousands) at time.

$$y \le 100;\ y(0) = 3$$

(a) $\frac{dy}{dt} = 0.01y(100 - y)$

$$= y - 0.01y^2$$

(b) Find $y(2);\ h = 0.5$.

x_i	y_i
0	3
0.5	4.455
1	6.583265
1.5	9.6582
2	14.0209

Therefore, $y(2) \approx 14$, so about 14,000 algae are present when $t = 2$.

33. $\frac{dy}{dt} = 0.05y - 0.1y^{1/2};\ y(0) = 60,\ h = 1;$ find $y(6)$.

x_i	y_i
0	60
1	62.22541
2	64.54785
3	66.97182
4	69.50205
5	72.14347
6	74.90127

Therefore, $y(6) \approx 75$, so about 75 insects are present after 6 weeks.

35. (a) Using Method 2 described after Example 1 of the text, store $\frac{dy}{dt}$ for the function variable Y_1 with $k = 0.5$ and $m = 4$. That is, Y_1 should equal $0.5(P - P^2)^{1.5}$. Store 5 for H (use keystrokes $5 \rightarrow H$), to $-h = -5$ for T ($-5 \rightarrow T$), and $p_0 = 0.1$ for $P (0.1 \rightarrow P)$. Next enter the keystrokes $T + H \rightarrow T$: $P + Y_1 H \rightarrow P$. Each time the ENTER key is pressed, the subsequent values for t_i will be stored into T and the corresponding values for p_i will appear on the screen. This summarized in the table below.

t_i	p_i
0	0.1
5	0.1675
10	0.297678
15	0.536660
20	0.846644
25	0.963605
30	0.980024

(b) By continuing to press the ENTER key, it appears that the values for p_i are approaching 1.

11.4   Linear Systems of Differential Equations

1.
$$\frac{dx}{dt} = 3x_1 - 2x_2$$
$$\frac{dx_2}{dt} = x_1$$

The system can be represented in matrix form as

$$\begin{bmatrix} \dfrac{dx_1}{dt} \\ \dfrac{dx_2}{dt} \end{bmatrix} = \begin{bmatrix} 3 & -2 \\ 1 & 0 \end{bmatrix} \begin{bmatrix} x_1 \\ x_2 \end{bmatrix}$$

Find the eigenvalues.

$$0 = \det(M - \lambda I)$$
$$= \det \begin{bmatrix} 3 - \lambda & -2 \\ 1 & 0 - \lambda \end{bmatrix}$$
$$= (3 - \lambda)(0 - \lambda) - (-2)(1)$$
$$= \lambda^2 - 3\lambda + 2$$
$$= (\lambda - 1)(\lambda + 2)$$
$$\lambda = 1 \text{ and } \lambda = 2$$

For $\lambda = 1$,

$$(M - \lambda I)X = \begin{bmatrix} 2 & -2 \\ 1 & -1 \end{bmatrix} \begin{bmatrix} x \\ y \end{bmatrix} = \begin{bmatrix} 0 \\ 0 \end{bmatrix}.$$

One solution to this system is $\begin{bmatrix} 1 \\ 1 \end{bmatrix}$.

For $\lambda = 2$,

$$(M - \lambda I)X = \begin{bmatrix} 1 & -2 \\ 1 & -2 \end{bmatrix} \begin{bmatrix} x \\ y \end{bmatrix} = \begin{bmatrix} 0 \\ 0 \end{bmatrix}.$$

One solution to this system is $\begin{bmatrix} 2 \\ 1 \end{bmatrix}$. The matrix of eigenvectors is

$$P = \begin{bmatrix} 1 & 2 \\ 1 & 1 \end{bmatrix}.$$

$$P^{-1} = \begin{bmatrix} -1 & 2 \\ 1 & -1 \end{bmatrix} \quad \text{and} \quad P^{-1}MP = \begin{bmatrix} 1 & 0 \\ 0 & 2 \end{bmatrix}.$$

Now let $X = PY$ and multiply both sides of the differential equation by P^{-1}, yielding

$$\frac{dY}{dt} = P^{-1}MPY$$

or,

$$\begin{bmatrix} \dfrac{dy_1}{dt} \\ \dfrac{dy_2}{dt} \end{bmatrix} = \begin{bmatrix} 1 & 0 \\ 0 & 2 \end{bmatrix} \begin{bmatrix} y_1 \\ y_2 \end{bmatrix}$$

or,

$$\frac{dy_1}{dt} = y_1, \text{ and } \frac{dy_2}{dt} = 2y_2.$$

Integrate these to yield

$$y_1 = C_1 e^t \text{ and } y_2 = C_2 e^{2t}$$

Finally, calculate $X = PY$.

$$\begin{bmatrix} x_1 \\ x_2 \end{bmatrix} = \begin{bmatrix} 1 & 2 \\ 1 & 1 \end{bmatrix} \begin{bmatrix} C_1 e^t \\ C_2 e^{2t} \end{bmatrix}$$
$$= \begin{bmatrix} C_1 e^t + 2C_2 e^{2t} \\ C_1 e^t + C_2 e^{2t} \end{bmatrix}$$

or,

$$x_1 = C_1 e^t + 2C_2 e^{2t}, \ x_2 = C_1 e^t + C_2 e^{2t}.$$

3. $\dfrac{dx_1}{dt} = 3x_1 + 2x_2$

$\dfrac{dx_2}{dt} = 3x_1 + 8x_2$

The system can be represented in matrix form as

$$\begin{bmatrix} \dfrac{dx_1}{dt} \\[2mm] \dfrac{dx_2}{dt} \end{bmatrix} = \begin{bmatrix} 3 & 2 \\ 3 & 8 \end{bmatrix} \begin{bmatrix} x_1 \\ x_2 \end{bmatrix}$$

Find the eigenvalues.

$0 = \det(M - \lambda I)$

$= \det \begin{bmatrix} 3-\lambda & 2 \\ 3 & 8-\lambda \end{bmatrix}$

$= (3-\lambda)(8-\lambda) - 2(3)$

$= \lambda^2 - 11\lambda + 18$

$= (\lambda - 2)(\lambda - 9)$

$\lambda = 2$ and $\lambda = 9$

For $\lambda = 2$,

$$(M - \lambda I)X = \begin{bmatrix} 1 & 2 \\ 3 & 6 \end{bmatrix} \begin{bmatrix} x \\ y \end{bmatrix} = \begin{bmatrix} 0 \\ 0 \end{bmatrix}.$$

One solution to this system is $\begin{bmatrix} 2 \\ -1 \end{bmatrix}$.

For $\lambda = 9$,

$$(M - \lambda I)X = \begin{bmatrix} -6 & 2 \\ 3 & -1 \end{bmatrix} \begin{bmatrix} x \\ y \end{bmatrix} = \begin{bmatrix} 0 \\ 0 \end{bmatrix}.$$

One solution to this system is $\begin{bmatrix} 1 \\ 3 \end{bmatrix}$. The matrix of eigenvectors is

$$P = \begin{bmatrix} 2 & 1 \\ -1 & 3 \end{bmatrix}.$$

$$P^{-1} = \frac{1}{7}\begin{bmatrix} 3 & -1 \\ 1 & 2 \end{bmatrix} \quad \text{and} \quad P^{-1}MP = \begin{bmatrix} 2 & 0 \\ 0 & 9 \end{bmatrix}.$$

Now let $X = PY$ and multiply both sides of the differential equation by P^{-1}, yielding

$$\frac{dY}{dt} = P^{-1}MPY$$

or,

$$\begin{bmatrix} \dfrac{dy_1}{dt} \\[2mm] \dfrac{dy_2}{dt} \end{bmatrix} = \begin{bmatrix} 2 & 0 \\ 0 & 9 \end{bmatrix} \begin{bmatrix} y_1 \\ y_2 \end{bmatrix}$$

or,

$$\frac{dy_1}{dt} = 2y_1, \text{ and } \frac{dy_2}{dt} = 9y_2.$$

Integrate these to yield

$$y_1 = C_1 e^{2t} \text{ and } y_2 = C_2 e^{9t}.$$

Finally, calculate $X = PY$

$$\begin{bmatrix} x_1 \\ x_2 \end{bmatrix} = \begin{bmatrix} 2 & 1 \\ -1 & 3 \end{bmatrix} \begin{bmatrix} C_1 e^{2t} \\ C_2 e^{9t} \end{bmatrix}$$

$$= \begin{bmatrix} 2C_1 e^{2t} + C_2 e^{9t} \\ -C_1 e^{2t} + 3C_2 e^{9t} \end{bmatrix}$$

or,

$$x_1 = 2C_1 e^{2t} + C_2 e^{9t}, \ x_2 = -C_1 e^{2t} + 3C_2 e^{9t}.$$

5. $\dfrac{dx_1}{dt} = 3x_1$

$\dfrac{dx_2}{dt} = 5x_1 + 2x_2$

$\dfrac{dx_3}{dt} = 2x_1 + 4x_2 - x_3$

The system can be represented in matrix form as

$$\begin{bmatrix} \dfrac{dx_1}{dt} \\[2mm] \dfrac{dx_2}{dt} \\[2mm] \dfrac{dx_3}{dt} \end{bmatrix} = \begin{bmatrix} 3 & 0 & 0 \\ 5 & 2 & 0 \\ 2 & 4 & -1 \end{bmatrix} \begin{bmatrix} x_1 \\ x_2 \\ x_3 \end{bmatrix}$$

Find the eigenvalues.

$0 = \det(M - \lambda I)$

$= \det \begin{bmatrix} 3-\lambda & 0 & 0 \\ 5 & 2-\lambda & 0 \\ 2 & 4 & -1-\lambda \end{bmatrix}$

$= (3-\lambda)[(2-\lambda)(-1-\lambda) - 0(4)] - 0 + 0$

$= (3-\lambda)(2-\lambda)(-1-\lambda)$

$\lambda = 3, 2,$ and -1

For $\lambda = 3$,

$$(M - \lambda I)X = \begin{bmatrix} 0 & 0 & 0 \\ 5 & -1 & 0 \\ 2 & 4 & -4 \end{bmatrix} \begin{bmatrix} x_1 \\ x_2 \\ x_3 \end{bmatrix} = \begin{bmatrix} 0 \\ 0 \\ 0 \end{bmatrix}.$$

One solution to this system is $\begin{bmatrix} 2 \\ 10 \\ 11 \end{bmatrix}$.

For $\lambda = 2$,

$$(M - \lambda I)X = \begin{bmatrix} 1 & 0 & 0 \\ 5 & 0 & 0 \\ 2 & 4 & -3 \end{bmatrix} \begin{bmatrix} x_1 \\ x_2 \\ x_3 \end{bmatrix} = \begin{bmatrix} 0 \\ 0 \\ 0 \end{bmatrix}.$$

One solution to this system is $\begin{bmatrix} 0 \\ 3 \\ 4 \end{bmatrix}$.

For $\lambda = -1$,

$$(M - \lambda I)X = \begin{bmatrix} 4 & 0 & 0 \\ 5 & 3 & 0 \\ 2 & 4 & 0 \end{bmatrix} \begin{bmatrix} x_1 \\ x_2 \\ x_3 \end{bmatrix} = \begin{bmatrix} 0 \\ 0 \\ 0 \end{bmatrix}.$$

One solution to this system is $\begin{bmatrix} 0 \\ 0 \\ 1 \end{bmatrix}$. The matrix of eigenvectors is

$$P = \begin{bmatrix} 2 & 0 & 0 \\ 10 & 3 & 0 \\ 11 & 4 & 1 \end{bmatrix}$$

$$P^{-1} = \frac{1}{6} \begin{bmatrix} 3 & 0 & 0 \\ -10 & 2 & 0 \\ 7 & -8 & 6 \end{bmatrix} \quad \text{and}$$

$$P^{-1}MP = \begin{bmatrix} 3 & 0 & 0 \\ 0 & 2 & 0 \\ 0 & 0 & -1 \end{bmatrix}$$

Now let $X = PY$ and multiply both sides of the differential equation by P^{-1}, yielding

$$\frac{dY}{dt} = P^{-1}MPY$$

or,

$$\begin{bmatrix} \dfrac{dy_1}{dt} \\ \dfrac{dy_2}{dt} \\ \dfrac{dy_3}{dt} \end{bmatrix} = \begin{bmatrix} 3 & 0 & 0 \\ 0 & 2 & 0 \\ 0 & 0 & -1 \end{bmatrix} \begin{bmatrix} y_1 \\ y_2 \\ y_3 \end{bmatrix}$$

or,

$$\frac{dy_1}{dt} = 3y_1, \frac{dy_2}{dt} = 2y_2 \text{ and } \frac{dy_3}{dt} = -y_3.$$

Integrate these to yield

$$y_1 = C_1 e^{3t}, y_2 = C_2 e^{2t} \text{ and } y_3 = C_3 e^{-t}.$$

Finally, calculate $X = PY$.

$$\begin{bmatrix} x_1 \\ x_2 \\ x_3 \end{bmatrix} = \begin{bmatrix} 2 & 0 & 0 \\ 10 & 3 & 0 \\ 11 & 4 & 1 \end{bmatrix} \begin{bmatrix} C_1 e^{3t} \\ C_2 e^{2t} \\ C_3 e^{-t} \end{bmatrix}$$

$$= \begin{bmatrix} 2C_1 e^{3t} \\ 10C_1 e^{3t} + 3C_2 e^{2t} \\ 11C_1 e^{3t} + 4C_2 e^{2t} + C_3 e^{-t} \end{bmatrix}$$

or,

$$x_1 = 2C_1 e^{3t}, \ x_2 = 10C_1 e^{3t} + 3C_2 e^{2t},$$
$$x_3 = 11C_1 e^{3t} + 4C_2 e^{2t} + C_3 e^{-t}.$$

7. $\dfrac{dx_1}{dt} = 4x_1 - 2x_2 + e^t$

   $\dfrac{dx_2}{dt} = -3x_1 + 9x_2 + e^{2t}$

$$\begin{bmatrix} \dfrac{dx_1}{dt} \\ \dfrac{dx_2}{dt} \end{bmatrix} = \begin{bmatrix} 4 & -2 \\ -3 & 9 \end{bmatrix} \begin{bmatrix} x_1 \\ x_2 \end{bmatrix} + \begin{bmatrix} e^t \\ e^{2t} \end{bmatrix}$$

Find the eigenvalues.

$$0 = \det(M - \lambda I)$$
$$= \det \begin{bmatrix} 4 - \lambda & -2 \\ -3 & 9 - \lambda \end{bmatrix}$$
$$= (4 - \lambda)(9 - \lambda) - (-2)(-3)$$
$$= \lambda^2 - 13\lambda + 30$$
$$= (\lambda - 3)(\lambda - 10)$$
$$\lambda = 3, 10$$

For $\lambda = 3$,

$$(M - \lambda I)X = \begin{bmatrix} 1 & -2 \\ -3 & 6 \end{bmatrix} \begin{bmatrix} x \\ y \end{bmatrix} = \begin{bmatrix} 0 \\ 0 \end{bmatrix}.$$

One solution to this system is $\begin{bmatrix} 2 \\ 1 \end{bmatrix}$.

For $\lambda = 10$,

$$(M - \lambda I)X = \begin{bmatrix} -6 & -2 \\ -3 & -1 \end{bmatrix} \begin{bmatrix} x \\ y \end{bmatrix} = \begin{bmatrix} 0 \\ 0 \end{bmatrix}.$$

One solution to this system is $\begin{bmatrix} 1 \\ -3 \end{bmatrix}$. The matrix of eigenvectors is

$$P = \begin{bmatrix} 2 & 1 \\ 1 & -3 \end{bmatrix}.$$

$$P^{-1} = \frac{1}{7} \begin{bmatrix} 3 & 1 \\ 1 & -2 \end{bmatrix} \quad \text{and} \quad P^{-1}MP = \begin{bmatrix} 3 & 0 \\ 0 & 10 \end{bmatrix}$$

Now let $X = PY$ and multiply both sides of the differential equation by P^{-1}, yielding

$$\frac{dY}{dt} = P^{-1}MPY + P^{-1} \begin{bmatrix} e^t \\ e^{2t} \end{bmatrix}$$

or

$$\begin{bmatrix} \dfrac{dy_1}{dt} \\ \dfrac{dy_2}{dt} \end{bmatrix} = \begin{bmatrix} 3 & 0 \\ 0 & 10 \end{bmatrix} \begin{bmatrix} y_1 \\ y_2 \end{bmatrix} + \frac{1}{7} \begin{bmatrix} 3 & 1 \\ 1 & -2 \end{bmatrix} \begin{bmatrix} e^t \\ e^{2t} \end{bmatrix}$$

or,

$$\frac{dy_1}{dt} = 3y_1 + \frac{3e^t}{7} + \frac{e^{2t}}{7}$$

$$\frac{dy_2}{dt} = 10y_2 + \frac{e^t}{7} - \frac{2e^{2t}}{7}$$

Multiply the first equation by an integrating factor $e^{\int -3dt} = e^{-3t}$.

$$\frac{dy_1}{dt} - 3y_1 = \frac{3e^t}{7} + \frac{e^{2t}}{7}$$

$$e^{-3t}\frac{dy_1}{dt} - 3e^{-3t}y_1 = \frac{3e^{-2t}}{7} + \frac{e^{-t}}{7}$$

$$D_t(e^{-3t}y_1) = \frac{3e^{-2t}}{7} + \frac{e^{-t}}{7}$$

$$e^{-3t}y_1 = \int \left(\frac{3e^{-2t}}{7} + \frac{e^{-t}}{7} \right) dt = -\frac{3e^{-2t}}{14} - \frac{e^{-t}}{7} + C_1$$

$$y_1 = -\frac{3e^t}{14} - \frac{3e^{2t}}{7} + C_1 e^{3t}$$

Multiply the second differential equation by an integrating factor $e^{\int -10dt} = e^{-10t}$.

$$\frac{dy_2}{dt} - 10y_2 = \frac{e^t}{7} - \frac{2e^{2t}}{7}$$

$$e^{-10t}\frac{dy_2}{dt} - 10e^{-10t}y_2 = \frac{e^{-9t}}{7} - \frac{2e^{-8t}}{7}$$

$$D_t(e^{-10t}y_2) = \frac{e^{-9t}}{7} - \frac{2e^{-8t}}{7}$$

$$e^{-10t}y_2 = \int \left(\frac{e^{-9t}}{7} - \frac{2e^{-8t}}{7} \right) dt$$

$$= -\frac{e^{-9t}}{63} + \frac{e^{-8t}}{28} + C_2$$

$$y_2 = -\frac{e^t}{63} + \frac{e^{2t}}{28} + C_2 e^{10t}$$

Finally, calculate $X = PY$.

$$\begin{bmatrix} x_1 \\ x_2 \end{bmatrix} = \begin{bmatrix} 2 & 1 \\ 1 & -3 \end{bmatrix} \begin{bmatrix} -\dfrac{3e^t}{14} - \dfrac{3e^{2t}}{7} + C_1 e^{3t} \\ -\dfrac{e^t}{63} + \dfrac{e^{2t}}{28} + C_2 e^{10t} \end{bmatrix}$$

$$= \begin{bmatrix} 2\left(-\dfrac{3e^t}{14} - \dfrac{3e^{2t}}{7} + C_1 e^{3t} \right) + \left(-\dfrac{e^t}{63} + \dfrac{e^{2t}}{28} + C_2 e^{10t} \right) \\ 1\left(-\dfrac{3e^t}{14} - \dfrac{3e^{2t}}{7} + C_1 e^{3t} \right) - 3\left(-\dfrac{e^t}{63} + \dfrac{e^{2t}}{28} + C_2 e^{10t} \right) \end{bmatrix}$$

$$= \begin{bmatrix} -\dfrac{4}{9}e^t - \dfrac{23}{28}e^{2t} + 2C_1 e^{3t} + C_2 e^{10t} \\ -\dfrac{1}{6}e^t - \dfrac{15}{28}e^{2t} + C_1 e^{3t} - 3C_2 e^{10t} \end{bmatrix}$$

or

$$x_1 = 2C_1 e^{3t} + C_2 e^{10t} - \frac{4}{9}e^t - \frac{23}{28}e^{2t},$$

$$x_2 = C_1 e^{3t} - 3C_2 e^{10t} - \frac{1}{6}e^t - \frac{15}{28}e^{2t}.$$

9. From exercise 1, the general solution is

$$x_1 = C_1 e^t + 2C_2 e^{2t}$$
$$x_2 = C_1 e^t + C_2 e^{2t}.$$

Setting $x_1(0) = 7$ and $x_2(0) = 5$ gives

$$7 = C_1 + 2C_2$$
$$5 = C_1 + C_2.$$

Solving this system gives

$$C_1 = 3, C_2 = 2.$$

Therefore,

$$x_1 = 3e^t + 4e^{2t}$$
$$x_2 = 3e^t + 2e^{2t}.$$

11. From exercise 3, the general solution is

$$x_1 = 2C_1 e^{2t} + C_2 e^{9t}$$
$$x_2 = -C_1 e^{2t} + 3C_2 e^{9t}.$$

Setting $x_1(0) = 3$ and $x_2(0) = 23$ gives

$$3 = 2C_1 + C_2$$
$$23 = -C_1 + 3C_2.$$

Solving this system gives

$$C_1 = -2, C_2 = 7.$$

Therefore,

$$x_1 = -4e^{2t} + 7e^{9t}$$
$$x_2 = 2e^{2t} + 21e^{9t}.$$

13. From exercise 5, the general solution is

$$x_1 = 2C_1 e^{3t}$$
$$x_2 = 10C_1 e^{3t} + 3C_2 e^{2t}$$
$$x_3 = 11C_1 e^{3t} + 4C_2 e^{2t} + C_3 e^{-t}.$$

Setting $x_1(0) = 6$ and $x_2(0) = 15$ and $x_3(0) = -1$ gives

$$6 = 2C_1$$
$$15 = 10C_1 + 3C_2$$
$$-1 = 11C_1 + 4C_2 + C_3.$$

Solving this system gives

$$C_1 = 3, C_2 = -5, C_3 = -14.$$

Therefore,

$$x_1 = 6e^{3t}$$
$$x_2 = 30e^{3t} - 15e^{2t}$$
$$x_3 = 33e^{3t} - 20e^{2t} - 14e^{-t}.$$

15. (a) $\dfrac{dx_1}{dt} = x_2 - 2$

$$\frac{dx_2}{dt} = -x_1 + 2$$

$$\begin{bmatrix} \dfrac{dx_1}{dt} \\ \dfrac{dx_2}{dt} \end{bmatrix} = \begin{bmatrix} 0 & 1 \\ -1 & 0 \end{bmatrix} \begin{bmatrix} x_1 \\ x_2 \end{bmatrix} + \begin{bmatrix} -2 \\ 2 \end{bmatrix}$$

Find the eigenvalues.

$$0 = \det(F - \lambda I)$$
$$= \det \begin{bmatrix} 0 - \lambda & 1 \\ -1 & 0 - \lambda \end{bmatrix}$$
$$= (0 - \lambda)(0 - \lambda) - 1(-1)$$
$$0 = \lambda^2 + 1$$

(b) $\lambda^2 + 1 = 0$
$$\lambda^2 = -1$$
$$\lambda = \pm\sqrt{-1}$$
$$= \pm i$$

(c) For $\lambda = -i$,

$$(M - \lambda I)X = \begin{bmatrix} i & 1 \\ -1 & i \end{bmatrix} \begin{bmatrix} x \\ y \end{bmatrix} = \begin{bmatrix} 0 \\ 0 \end{bmatrix}.$$

The augmented matrix for this system is

$$\begin{bmatrix} i & 1 & | & 0 \\ -1 & i & | & 0 \end{bmatrix}.$$

$$-iR_2 \to R_2 \quad \begin{bmatrix} i & 1 & | & 0 \\ i & 1 & | & 0 \end{bmatrix}$$

A solution to this system is

$$\begin{bmatrix} x \\ y \end{bmatrix} = \begin{bmatrix} 1 \\ -i \end{bmatrix}.$$

For $\lambda = i$,

$$(M - \lambda I)X = \begin{bmatrix} -i & 1 \\ -1 & -i \end{bmatrix} \begin{bmatrix} x \\ y \end{bmatrix} = \begin{bmatrix} 0 \\ 0 \end{bmatrix}.$$

The augmented matrix for this system is

$$\begin{bmatrix} -i & 1 & | & 0 \\ -1 & -i & | & 0 \end{bmatrix}.$$

$$iR_2 \to R_2 \quad \begin{bmatrix} -i & 1 & | & 0 \\ -i & 1 & | & 0 \end{bmatrix}$$

A solution to this system is

$$\begin{bmatrix} x \\ y \end{bmatrix} = \begin{bmatrix} 1 \\ i \end{bmatrix}.$$

The matrix of eigenvectors is

$$P = \begin{bmatrix} 1 & 1 \\ i & -i \end{bmatrix}.$$

(d) $PP^{-1} = \begin{bmatrix} 1 & 1 \\ i & -i \end{bmatrix} \begin{bmatrix} \frac{1}{2} & -\frac{i}{2} \\ \frac{1}{2} & \frac{i}{2} \end{bmatrix}$

$$= \begin{bmatrix} 1\left(\frac{1}{2}\right) + 1\left(\frac{1}{2}\right) & 1\left(-\frac{i}{2}\right) + 1\left(\frac{i}{2}\right) \\ i\left(\frac{1}{2}\right) - i\left(\frac{1}{2}\right) & i\left(-\frac{i}{2}\right) - i\left(\frac{i}{2}\right) \end{bmatrix}$$

$$= \begin{bmatrix} 1 & 0 \\ 0 & 1 \end{bmatrix} = I$$

(e) $P^{-1}MP = \begin{bmatrix} i & 0 \\ 0 & -i \end{bmatrix}$

Now let $X = PY$ and multiply both sides of the differential equation by P^{-1}, yielding

$$\frac{dY}{dt} = P^{-1}MPY + P^{-1} \begin{bmatrix} -2 \\ 2 \end{bmatrix}$$

or,

$$\begin{bmatrix} \dfrac{dy_1}{dt} \\ \dfrac{dy_2}{dt} \end{bmatrix} = \begin{bmatrix} i & 0 \\ 0 & -i \end{bmatrix} \begin{bmatrix} y_1 \\ y_2 \end{bmatrix} + \begin{bmatrix} \frac{1}{2} & -\frac{i}{2} \\ \frac{1}{2} & \frac{i}{2} \end{bmatrix} \begin{bmatrix} -2 \\ 2 \end{bmatrix}$$

$$= \begin{bmatrix} iy_1 \\ -iy_2 \end{bmatrix} + \begin{bmatrix} -1 - i \\ -1 + i \end{bmatrix}$$

or,

$$\frac{dy_1}{dt} = iy_1 + (-1 - i)$$
$$\frac{dy_2}{dt} = -iy_2 + (-1 + i).$$

(f) Multiply the first differential equation by an integrating factor $e^{\int -i\,dt} = e^{-it}$.

$$e^{-it}\frac{dy_1}{dt} - ie^{-it}y_1 = e^{-it}(-1-i)$$

$$D_t(e^{-it}y_1) = e^{-it}(-1-i)$$

$$e^{-it}y_1 = \int e^{-it}(-1-i)dt$$

$$= (-1-i)\int e^{-it}dt$$

$$= (-1-i)\left(\frac{1}{-i}e^{-it} + C\right)$$

$$= (-1-i)(ie^{-it} + C)$$

$$y_1 = (-1-i)(i + Ce^{it})$$

$$= (-1-i)Ce^{it} + (1-i)$$

Let $C_1 = (-1-i)C$.

$$y_1 = C_1 e^{it} + (1-i)$$

Multiply the second differential equation by an integrating factor $e^{\int i\,dt} = e^{it}$.

$$e^{it}\frac{dy_2}{dt} + ie^{it}y_2 = e^{it}(-1+i)$$

$$D_t(e^{it}y_2) = e^{it}(-1+i)$$

$$e^{it}y_2 = \int e^{it}(-1+i)dt$$

$$= (-1-i)\int e^{it}dt$$

$$= (-1-i)\left(\frac{1}{-i}e^{-it} + C\right)$$

$$= (-1+i)(-ie^{it} + C)$$

$$y_2 = (-1+i)(-i + Ce^{-it})$$

$$= (-1+i)Ce^{-it} + (1+i)$$

Let $C_2 = (-1+i)C$.

$$y_2 = C_2 e^{-it} + (1+i)$$

(g) Calculate $X = PY$.

$$\begin{bmatrix} x_1 \\ x_2 \end{bmatrix} = \begin{bmatrix} 1 & 1 \\ i & -i \end{bmatrix}\begin{bmatrix} C_1 e^{it} + (1-i) \\ C_2 e^{-it} + (1+i) \end{bmatrix}$$

$$= \begin{bmatrix} C_1 e^{it} + C_2 e^{-it} + 2 \\ iC_1 e^{2t} - iC_2 e^{-it} + 2 \end{bmatrix}$$

or

$$x_1 = C_1 e^{it} + C_2 e^{-it} + 2$$

$$x_2 = iC_1 e^{it} - iC_2 e^{-it} + 2.$$

(h) Setting $x_1(0) = 2$ and $x_2(0) = 3$ gives

$$2 = C_1 + C_2 + 2$$

$$3 = iC_1 - iC_2 + 2.$$

Simplifying

$$0 = C_1 + C_2$$

$$1 = iC_1 - iC_2.$$

The solution to this system is

$$C_1 = \frac{-i}{2}, C_2 = \frac{i}{2}.$$

(i) Therefore,

$$x_1 = \frac{-i}{2}e^{it} + \frac{i}{2}e^{-it} + 2$$

$$= \frac{-i}{2}(\cos t + i\sin t) + \frac{i}{2}(\cos t - i\sin t) + 2$$

$$= \frac{-i}{2}\cos t + \frac{1}{2}\sin t + \frac{i}{2}\cos t + \frac{1}{2}\sin t + 2$$

$$x_1 = \sin t + 2$$

$$x_2 = \frac{1}{2}e^{it} + \frac{1}{2}e^{-it} + 2$$

$$= \frac{1}{2}(\cos t + i\sin t) + \frac{1}{2}(\cos t - i\sin t) + 2$$

$$= \frac{1}{2}\cos t + \frac{i}{2}\sin t + \frac{1}{2}\cos t - \frac{i}{2}\sin t + 2$$

$$x_2 = \cos t + 2.$$

(j) The answer is biologically reasonable since x_1 and x_2 are real-valued.

11.5   Nonlinear Systems of Differential Equations

1. $\dfrac{dx_1}{dt} = 4x_1 - 2x_1 x_2$

$\dfrac{dx_2}{dt} = 3x_2 - x_1 x_2$

Find the equilibrium point.

$$\frac{dx_1}{dt} = 0 = 4x_1 - 2x_1 x_2 \quad , \quad \frac{dx_2}{dt} = 0 = 3x_2 - x_1 x_2$$

$$4x_1 = 2x_1 x_2 \qquad , \qquad 3x_2 = x_1 x_2$$

$$2 = x_2 \qquad , \qquad 3 = x_1$$

region	1	2	3	4
dx_1/dt	$-$	$-$	$+$	$+$
dx_2/dt	$-$	$+$	$+$	1

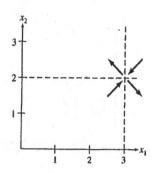

3. $\dfrac{dx_1}{dt} = 3x_1x_2 - x_1x_2^2$

$\dfrac{dx_2}{dt} = 2x_1^2x_2 - 2x_1x_2$

Find the equilibrium point.

$\dfrac{dx_1}{dt} = 0 = 3x_1x_2 - x_1x_2^2 \quad , \quad \dfrac{dx_2}{dt} = 0 = 2x_1^2x_2 - 2x_1x_2$

$3x_1x_2 = x_1x_2^2 \qquad , \qquad 2x_1^2x_2 = 2x_1x_2$

$3 = x_2 \qquad\qquad , \qquad\qquad x_1 = 1$

region	1	2	3	4
dx_1/dt	−	−	+	+
dx_2/dt	+	−	−	+

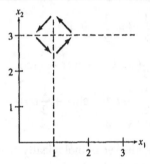

5. (a) $\dfrac{dx_1}{dt} = x_1^2 + x_1x_2 - 2x_1$

$\dfrac{dx_2}{dt} = 2x_1x_2 - x_2^2 - x_2$

Find the equilibrium point.

$\dfrac{dx_1}{dt} = 0 = x_1^2 + x_1x_2 - 2x_1$

$0 = x_1 + x_2 - 2$

$x_2 = -x_1 + 2 \quad \text{eq.(1)}$

$\dfrac{dx_2}{dt} = 0 = 2x_1x_2 - x_2^2 - x_2$

$0 = 2x_1 - x_2 - 1$

$x_2 = 2x_1 - 1 \quad \text{eq.(2)}$

Solving equations (1) and (2) simultaneously gives

$$x_1 = 1, x_2 = 1.$$

(b) Plot equations (1) and (2) and test each region.

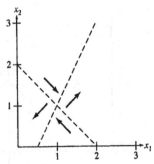

7. (a) $\dfrac{dx_1}{dt} = x_1^2x_2 + x_1x_2^2 - 3x_1x_2$

$\dfrac{dx_2}{dt} = 2x_1x_2^2 - x_1^2 - 3x_1x_2$

Find the equilibrium point.

$\dfrac{dx_1}{dt} = 0 = x_1^2x_2 + x_1x_2^2 - 3x_1x_2$

$0 = x_1 + x_2 - 3$

$x_2 = -x_1 + 3 \quad \text{eq.(1)}$

$\dfrac{dx_2}{dt} = 0 = 2x_1x_2^2 - x_1^2x_2 - 3x_1x_2$

$0 = 2x_2 - x_1 - 3$

$x_2 = \dfrac{x_1}{2} + \dfrac{3}{2} \quad \text{eq.(2)}$

Solving equations (1) and (2) simultaneously gives

$$x_1 = 1, x_2 = 2.$$

(b) Plot equations (1) and (2) and test each region.

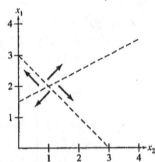

9. (a) $\dfrac{dx_1}{dt} = -2x_1 + 3x_1x_2$

$\dfrac{dx_2}{dt} = 3x_2 - 2x_1x_2$

$\dfrac{dx_2}{dx_1} = \dfrac{\frac{dx_2}{dt}}{\frac{dx_1}{dt}} = \dfrac{3x_2 - 2x_1x_2}{-2x_1 + 3x_1x_2}$

$= \dfrac{x_2(3 - 2x_1)}{x_1(-2 + 3x_2)}$

Separating variables yields

$$\dfrac{-2 + 3x_2}{x_2}dx_2 = \dfrac{3 - 2x_1}{x_1}dx_1$$

or,

$$\int\left(\dfrac{-2}{x_2} + 3\right)dx_2 = \int\left(\dfrac{3}{x_1} - 2\right)dx_1$$

$$-2\ln x_2 + 3x_2 = 3\ln x_1 - 2x_1 + C.$$

Use the initial conditions $x_2 = 2$ when $x_1 = 1$ to find C.

$$-2\ln 2 + 3(2) = 3\ln 1 - 2(1) + C$$
$$-\ln 4 + 6 = 0 - 2 + C$$
$$C = -\ln 4 + 8$$

The desired equation is

$$3\ln x_1 - 2x_1 + 2\ln x_2 - 3x_2 = \ln 4 - 8.$$

(b) $\dfrac{dx_1}{dt} = 0 = 3x_2 - 2x_1 x_2$

$$x_2 = \frac{3}{2}$$

$$\frac{dx_2}{dt} = 0 = -2x_1 + 3x_1 x_2$$

$$x_1 = \frac{2}{3}$$

(c)

The figure shows that the variations in the populations are cyclic. We know from Example 1 or from a phase plane diagram that the solution moves about this cycle in a clockwise directions.

11. $\dfrac{dy_1}{dt} = k_1 y_1 (1 - F_1 - y_1 - by_2)$

$$\frac{dy_2}{dt} = k_2 y_2 \left(1 - F_2 - \frac{y_2}{y_1}\right)$$

$$\frac{dy_1}{dt} = 0 = k_1 y_1 (1 - F_1 - y_1 - by_2)$$

$$y_1 = 1 - F_1 - by_2$$

$$\frac{dy_2}{dt} = 0 = k_2 y_2 \left(1 - F_2 - \frac{y_2}{y_1}\right)$$

$$1 = F_2 + \frac{y_2}{y_1}$$

$$y_1 = F_2 y_1 + y_2$$

$$y_1 - F_2 y_1 = y_2$$

Solve using the substitution method take out of subscript position.

$$y_1 = 1 - F_1 - b(y_1 - F_2 y_1)$$
$$= 1 - F_1 - by_1 + bF_2 y_1$$

$$y_1(1 + b - bF_2) = 1 - F_1$$

$$y_1 = \frac{1 - F_1}{1 + b(1 - F_2)}$$

$$y_2 = y_1 - F_2 y_1$$
$$= y_1(1 - F_2)$$

$$y_2 = \frac{(1 - F_1)(1 - F_2)}{1 + b(1 - F_2)}$$

13. $\dfrac{dx_1}{dt} = a - bx_1 - \beta x_1 x_2 + \gamma x_3$

$$\frac{dx_2}{dt} = \beta x_1 x_2 - (b + \alpha + v)x_2$$

$$\frac{dx_3}{dt} = vx_2 - (\beta + \gamma)x_3$$

Set the derivatives to zero.

$$a - bx_1 - \beta x_1 x_2 + \gamma x_3 = 0 \quad (1)$$
$$\beta x_1 x_2 - (b + \alpha + v)x_2 = 0 \quad (2)$$
$$vx_2 - (\beta + \gamma)x_3 = 0 \quad (3)$$

Begin by solving equation (2) for x_1.

$$\beta x_1 x_2 - (b + \alpha + v)x_2 = 0$$
$$x_2[\beta x_1 - (b + \alpha + v)] = 0$$

$$x_2 = 0 \quad \text{or} \quad \beta x_1 - (b + \alpha + v) = 0$$

$$x_1 = \frac{b + \alpha + v}{\beta}$$

Notice that, since all constants are positive, x_1 must be positive.

Now solve equation (3) for x_3 in terms of x_2.

$$vx_2 - (\beta + \gamma)x_3 = 0$$
$$-(\beta + \gamma)x_3 = -vx_2$$

$$x_3 = \frac{v}{\beta + \gamma} x_2$$

Replace x_1 and x_3 in equation (1) with the expressions just found.

$$a - bx_1 - \beta x_1 x_2 + \gamma x_3 = 0$$

$$a - b\left(\frac{b + \alpha + v}{\beta}\right) - \beta\left(\frac{b + \alpha + v}{\beta}\right)x_2 + \gamma\left(\frac{v}{\beta + \gamma}x_2\right) = 0$$

$$a - \frac{b(b + \alpha + v)}{\beta} - (b + \alpha + v)x_2 + \frac{\gamma v}{\beta + \gamma}x_2 = 0$$

$$\left[-(b + \alpha + v) + \frac{\gamma v}{\beta + \gamma}\right]x_2 = -\left[a - \frac{b(b + \alpha + v)}{\beta}\right]$$

Multiply both sides of this equation by the LCD, $\beta(\beta + \gamma)$.

$$[-\beta(\beta + \gamma)(b + \alpha + v) + \beta\gamma v]x_2 = -[a\beta(\beta + \gamma) - b(\beta + \gamma)(b + \alpha + v)]$$

$$x_2 = \frac{a\beta(\beta + \gamma) - b(\beta + \gamma)(b + \alpha + v)}{\beta(\beta + \gamma)(b + \alpha + v) - \beta\gamma v}$$

$$= \frac{(\beta + \gamma)[a\beta - b(b + \alpha + v)]}{\beta(\beta + \gamma)(b + \alpha + v) - \beta\gamma v}$$

Since all constants are positive, for x_2 and x_3 to be positive, the numerator of the last rational expression must be positive.

$$(\beta + \gamma)[a\beta - b(b + \alpha + v)] > 0$$
$$a\beta - b(b + \alpha + v) > 0$$
$$a\beta > b(b + \alpha + v)$$

$$\frac{a}{b} > \frac{b + \alpha + v}{\beta}$$

11.6   Applications of Differential Equations

1. (a) Let $y =$ the number of individuals infected. The differential equation is

$$\frac{dy}{dt} = a(N - y)y.$$

The solution is Equation 12 in Example 4, which is

$$y = \frac{N}{1 + (N - 1)e^{-aNt}}, \text{ where } a = \frac{k}{N}.$$

The number of individuals uninfected at time t is

$$y = N - \frac{N}{1 + (N - 1)e^{-aNt}}$$

$$= \frac{N + N(N - 1)e^{-aNt} - N}{1 + (N - 1)e^{-aNt}}$$

$$= \frac{N(N - 1)}{N - 1 + e^{aNt}}.$$

Now substitute $N = 5{,}000$ and $a = 0.00005$.

$$y = \frac{5{,}000(5{,}000 - 1)}{5{,}000 - 1 + e^{(0.00005)(5{,}000)t}}$$

$$= \frac{24{,}995{,}000}{4{,}999 + e^{0.25t}}$$

(b) $t = 30$

$$y = \frac{24,995,000}{4,999 + e^{0.25(30)}} = 3,672$$

(c) $t = 50$

$$\frac{24,995,000}{4,999 + e^{0.25(50)}} = 91$$

(d) From Example 4,

$$t_m = \frac{\ln(N-1)}{aN}$$

$$= \frac{\ln(5,000 - 1)}{(0.00005)(5,000)}$$

$$= 34.$$

The maximum infection rate will occur on the 34th day.

3. (a) The differential equation is

$$\frac{dy}{dt} = a(N - y)y.$$

$y_0 = 50$; $y = 300$ when $t = 10$, $N = 10,000$.

The solution is Equation 11 in Example 4, which is

$$y = \frac{N}{1 + be^{-kt}},$$

where $b = \frac{N - y_0}{y_0}$ and $k = aN$.
Since $y_0 = 50$ and $N = 10,000$,
$b = \frac{10,000 - 50}{50} = 199$; $k = 10,000a$.

Therefore,

$$y = \frac{10,000}{1 + 199e^{-10,000at}}.$$

$y = 300$ when $t = 10$.

$$300 = \frac{10,000}{1 + 199e^{-10,000(10)a}}$$

$$300 + 300(199)e^{-100,000a} = 10,000$$

$$e^{-100,000a} = \frac{9,700}{300(199)}$$

$$= 0.1624791$$

$$a = \frac{\ln(0.1624791)}{-100,000}$$

$$= 0.000018$$

$k = 10,000a = 10,000(0.000018) = 0.18$
Therefore,

$$y = \frac{10,000}{1 + 199e^{-0.18t}} \quad \text{or} \quad \frac{10,000e^{0.18t}}{e^{0.18t} + 199}.$$

(b) Half the community is $y = 5,000$. Find t for $y = 5,000$.

$$5,000 = \frac{10,000}{1 + 199e^{-0.18t}}$$

$$5,000 + 5,000(199)e^{-0.18t} = 10,000$$

$$e^{-0.18t} = \frac{5,000}{5,000(199)} = 0.005$$

$$t = \frac{\ln(0.005)}{-0.18}$$

$$= 29.44$$

Half the community will be infected in about 29 days.

5. (a) $\dfrac{dy}{dt} = -ay + b(f - y)Y$

$a = 1$, $b = 1$, $f = 0.5$, $Y = 0.01$;
$y = 0.02$ when $t = 0$.

$$\frac{dy}{dt} = -y + 1(0.5 - y)(0.01)$$

$$= -1.010y + 0.005.$$

$$\int \frac{dy}{-1.010y + 0.005} = \int dt$$

$$\frac{1}{-1.010} \ln|-1.010y + 0.005| = t + C_2$$

$$\ln|-1.010y + 0.005| = -1.010t + C_1$$

$$|-1.010y + 0.005| = e^{-1.010t + C_1}$$

$$= e^{C_1}e^{-1.010t}$$

$$-1.010y + 0.005 = Ce^{-1.010t}$$

$$y = 0.005 - 0.990Ce^{-1.010t}$$

Since $y = 0.02$ when $t = 0$,

$$0.02 = 0.005 - 0.990Ce^0$$

$$-0.990C = 0.015.$$

Therefore,

$$y = 0.005 + 0.015e^{-1.010t}.$$

(b) $\dfrac{dY}{dt} = -AY + B(F - Y)y$

$A = 1,\ B = 1,\ y = 0.1,\ F = 0.03;$
$Y = 0.01$ when $t = 0.$

$$\frac{dY}{dt} = -Y + 1(0.03 - Y)(0.1)$$
$$= -1.1Y + 0.003$$
$$\frac{dY}{-1.1Y - 0.003} = dt$$

$$-\frac{1}{1.1}\ln|-1.1Y + 0.003| = t + C_2$$
$$\ln|-1.1Y + 0.003| = -1.1t + C_1$$
$$|-1.1Y + 0.003| = e^{-1.1t + C_1}$$
$$= e^{C_1}e^{-1.1t}$$
$$-1.1Y + 0.003 = Ce^{-1.1t}$$

$$Y = \frac{C}{-1.1}e^{-1.1t} - \frac{0.003}{-1.1}$$
$$= 0.909Ce^{-1.1t} + 0.00273$$

Since $Y = 0.01$ when $t = 0,$

$$0.01 = -0.909Ce^0 + 0.00273$$
$$-0.909C = 0.00727.$$

Therefore,

$$Y = 0.00727e^{-1.1t} + 0.00273.$$

7. (a) $\dfrac{dy}{dt} = a(N - y)y$

$y_0 = 5;\ y = 15$ when $t = 3;\ N = 50$

The solution to this differential equation is Equation 11 in Example 4, which is

$y = \dfrac{N}{1 + be^{-kt}}$ where $b = \dfrac{N - y_0}{y_0}$ and

$k = aN.$

Since $y = 5$ and $N = 50,$

$$b = \frac{50 - 5}{5} = 9;\ k = 50a.$$

$$y = \frac{50}{1 + 9e^{-50at}}.$$

$y = 15$ when $t = 3,$ so

$$15 = \frac{50}{1 + 9e^{-50a}}$$
$$15 + 135e^{-150a} = 50$$
$$e^{-150a} = \frac{35}{135}$$
$$= \frac{7}{27}$$
$$-150a = \ln\frac{7}{27}$$
$$= -1.350$$
$$-50a = -0.45.$$

Therefore,

$$y = \frac{50}{1 + 9e^{-0.45t}}.$$

(b) When $y = 30,$

$$30 = \frac{50}{1 + 9e^{-0.45t}}$$
$$30 + 270e^{-0.45t} = 50$$
$$e^{-0.45t} = \frac{20}{270} = \frac{2}{27}$$
$$t = -\frac{1}{0.45}\ln\frac{2}{27} = 5.78.$$

In about 6 days, 30 employees have heard the rumor.

9. (a) $\dfrac{dy}{dt} = kye^{-at};\ a = 0.1;\ y = 5$ when $t = 0;$
$y = 15$ when $t = 3.$

$$\int\frac{dy}{y} = k\int e^{-0.1t}\,dt$$
$$\ln|y| = -10ke^{-0.1t} + C_1$$
$$|y| = e^{-10ke^{-0.1t} + C_1}$$
$$= e^{C_1}e^{-10ke^{-0.1t}}$$
$$y = Ce^{-10ke^{-0.1t}}$$

Since $y = 5$ when $t = 0,$

$$5 = Ce^{-10k}$$
$$C = 5e^{10k}.$$

Since $y = 15$ when $t = 3,$

$$15 = Ce^{-10ke^{-0.3}} = Ce^{-7.41k}$$
$$C = 15e^{7.41k}.$$

Solve the system

$$C = 5e^{10k}$$
$$C = 15e^{7.41k}.$$

$$5e^{10k} = 15e^{7.41k}$$
$$e^{10k} = 3e^{7.41k}$$

Take natural logarithms on both sides.

$$10k \ln e = \ln 3 + 7.41k \ln e$$
$$2.59k = \ln 3$$

$$k = \frac{1}{2.59} \ln 3 = 0.424$$

$$C = 5e^{10(0.424)} = 347$$

Therefore,

$$y = 347e^{-10(0.424)e^{-0.1t}}$$
$$= 347e^{-4.24e^{-0.1t}}$$

(b) If $y = 30$,

$$30 = 347e^{-4.24e^{-0.1t}}$$

$$e^{-4.24e^{-0.1t}} = \frac{30}{347} = 0.0865$$

$$-4.24e^{-0.1t} \ln e = \ln 0.0865$$

$$e^{-0.1t} = -\frac{1}{4.24} \ln 0.0865$$

$$= 0.5773$$

$$-0.1t \ln e = \ln 0.5773$$

$$t = -10 \ln 0.5773$$

$$= 5.493.$$

30 employees have heard the rumor in about 5.5 days.

11. Let y = the amount of salt present at time t.

(a)
$$\frac{dy}{dt} = \text{(rate of salt in)}$$
$$\qquad - \text{(rate of salt out)}$$
rate of salt in $= (3 \text{ gal/min})(2 \text{ lb/gal})$
$$= 6 \text{ lb/min}$$

rate of salt out $= \left(\frac{y}{V} \text{ lb/gal}\right)(2 \text{ gal/min})$

$$= \left(\frac{2y}{V} \text{ lb/min}\right)$$

$$\frac{dy}{dt} = 6 - \frac{2y}{V}; \ y(0) = 20 \text{ lb}$$

$$\frac{dV}{dt} = \text{(rate of liquid in)}$$
$$\qquad - \text{(rate of liquid out)}$$

$$= 3 \text{ gal/min} - 2 \text{ gal/min}$$
$$= 1 \text{ gal/min}$$

$$\frac{dV}{dt} = 1$$

$$V = t + C_1$$

When $t = 0$, $V = 100$. Thus,

$$C_1 = 100.$$
$$V = t + 100.$$

Therefore,

$$\frac{dy}{dt} = 6 - \frac{2y}{t+100}$$

$$\frac{dy}{dt} + \frac{2}{t+100}y = 6.$$

$$I(t) = = e^{\int 2\,dt/(t+100)}$$
$$= e^{2 \ln|t+100|}$$
$$= (t+100)^2$$

$$\frac{dy}{dt}(t+100)^2 + 2y(t+100) = 6(t+100)^2$$

$$D_t\left[y(t+100)^2\right] = 6(t+100)^2$$

$$y(t+100)^2 = 6\int(t+100)^2\,dt$$

$$y(t+100)^2 = 2(t+100)^3 + C$$

$$y = 2(t+100) + \frac{C}{(t+100)^2}$$

Since $t = 0$ when $y = 20$,

$$20 = 2(100) + \frac{C}{100^2}$$

$$C = -1,800,000.$$

$$y = 2(t+100) - \frac{1,800,000}{(t+100)^2}$$

$$= \frac{2(t+100)^3 - 1,800,000}{(t+100)^2}.$$

(b) $t = 1 \text{ hr} = 60 \text{ min}$

$$y = \frac{2(160)^3 - 1,800,000}{(160)^2} = 249.69$$

After 1 hr, about 250 lb of salt are present.

(c) As time increases, salt concentration continues to increase.

13. Let $y = $ the amount of salt present at time t minutes.

(a) $\dfrac{dy}{dt} = $ (rate of salt in)

$\qquad\qquad - $ (rate of salt out)

rate of salt in $= 0$

rate of salt out $= \left(\dfrac{y}{V} \text{ lb/gal}\right)(2 \text{ gal/min})$

$$= \dfrac{2y}{V} \text{ lb/min}$$

$$\dfrac{dy}{dt} = -\dfrac{2y}{V}; \ y(0) = 20$$

$$\dfrac{dV}{dt} = \text{(rate of liquid in)}$$

$$- \text{(rate of liquid out)}$$

$$= 2 \text{ gal/min} - 2 \text{ gal/min} = 0$$

$$\dfrac{dV}{dt} = 0$$

$$V = C_1$$

When $t = 0$, $V = 100$, so $C_1 = 100$.

Therefore,

$$\dfrac{dy}{dt} = -\dfrac{2y}{100} = -0.02y$$

$$\dfrac{dy}{y} = -0.02\,dt$$

$$\ln|y| = -0.02t + C_1$$
$$|y| = e^{-0.02t + C_1} = e^{C_1}e^{-0.02t}$$
$$= Ce^{-0.02t}.$$

Since $t = 0$ when $y = 20$,

$$20 = Ce^0$$
$$C = 20.$$
$$y = 20e^{-0.02t}.$$

(b) $t = 1 \text{ hr} = 60 \text{ min}$

$$y = 20e^{-0.02(60)} = 6.024$$

After 1 hr, about 6 lb of salt are present.

(c) As time increases, salt concentration continues to decrease.

15. Let $y = $ amount of the chemical at time t.

(a) $\dfrac{dy}{dt} = $ (rate of chemical in)

$\qquad\qquad - $ (rate of chemical out)

rate of chemical in

$$= (2 \text{ liters/min})(0.1 \text{ g/liter})$$
$$= 0.2 \text{ g/min}$$

rate of chemical out

$$= \left(\dfrac{y}{V} \text{ g/liter}\right)(1 \text{ liter/min})$$

$$= \dfrac{y}{V} \text{ g/liter}$$

$$\dfrac{dy}{dt} = 0.2 - \dfrac{y}{V}; \ y(0) = 5$$

$$\dfrac{dV}{dt} = \text{(rate of liquid in)}$$

$$- \text{(rate of liquid out)}$$

$$= 2 \text{ liter/min} - 1 \text{ liter/min}$$

$$= 1 \text{ liter/min}$$

$$\dfrac{dV}{dt} = 1$$

$$V = t + C_1$$

When $t = 0$, $V = 100$, so $C_1 = 100$.

$$V = t + 100$$

Therefore,

$$\dfrac{dy}{dt} = 0.2 - \dfrac{y}{t + 100}$$

$$\dfrac{dy}{dt} + \dfrac{1}{t + 100} \cdot y = 0.2$$

$$I(t) = e^{\int dt/(t+100)} = e^{\ln|t+100|}$$
$$= t + 100$$

$$\dfrac{dy}{dt}(t + 100) + y = 0.2(t + 100)$$

$$D_x\,(t + 100)y = 0.2(t + 100)$$

$$(t + 100)y = \int 0.2(t + 100)\,dt$$

$$(t + 100)y = 0.1(t + 100)^2 + C$$

$$y = 0.1(t + 100) + \dfrac{C}{t + 100}$$

$t = 0$, $y = 5$

$$5 = 0.1(100) + \dfrac{C}{100}$$

$$500 = 1,000 + C$$
$$C = -500$$

Therefore,

$$y = 0.1(t + 100) + \frac{-500}{t + 100}$$

$$= \frac{0.1(t + 100)^2 - 500}{t + 100}.$$

(b) When $t = 30$ min,

$$y = \frac{0.1(130)^2 - 500}{130} = 9.154,$$

After 30 min, about 9.2 g of chemical are present.

Chapter 11 Review Exercises

5. $y \dfrac{dy}{dx} = x + y$

Since you cannot separate the variables, that is, rewrite the equation in the form $g(y)dy = f(x)dx$ where g is a function of y alone and f is a function of x alone, then the equation is not separable. Since you cannot re-write the equation in the form $\frac{dy}{dx} + P(x)y = Q(x)$, then the equation is not a line first-order differential equation. Therefore, the equation is neither linear nor separable.

7. $\sqrt{x}\dfrac{dy}{dx} = \dfrac{1 + e^x}{y}$

Since you can rewrite the equation in the form $y\,dy = \frac{1+e^x}{\sqrt{x}}\,dx$, then the equation is separable, but since it cannot be rewritten in the form $\frac{dy}{dx} + P(x)y = Q(x)$, then the equation is not linear.

9. $\dfrac{dy}{dx} + x = xy$

Since you can rewrite the equation in the form $\frac{dy}{dx} + (-x)y = -x$, then the equation is linear. Since the equation can be rewritten in the form $\frac{dy}{y-1} = x\,dx$, then it is also separable. Therefore, it is both linear and separable.

11. $x\dfrac{dy}{dx} + y = e^x(1 + y)$

Since the equation can be rewritten in the form

$$\frac{dy}{dx} + y\left(\frac{1 - e^x}{x}\right) = \frac{e^x}{x},$$

then the equation is linear. Since the equation cannot be rewritten in the form $g(y)dy = f(x)dx$, then the equation is not separable.

13. $\dfrac{dy}{dx} = 2x^3 + 6x$

$$dy = (2x^3 + 6x)\,dx$$

$$y = \frac{x^4}{2} + 3x^2 + C$$

15. $\dfrac{dy}{dx} = 4e^x$

$$dy = 4e^x\,dx$$

$$y = 4e^x + C$$

17. $\dfrac{dy}{dx} = \dfrac{3x + 1}{y}$

$$y\,dy = (3x + 1)\,dx$$

$$\frac{y^2}{2} = \frac{3x^2}{2} + x + C_1$$

$$y^2 = 3x^2 + 2x + C$$

19. $\dfrac{dy}{dx} = \dfrac{2y + 1}{x}$

$$\frac{dy}{2y + 1} = \frac{dx}{x}$$

$$\frac{1}{2}\left(\frac{2\,dy}{2y + 1}\right) = \frac{dx}{x}$$

$$\frac{1}{2}\ln|2y + 1| = \ln|x| + C_1$$

$$\ln|2y + 1|^{1/2} = \ln|x| + \ln k$$

Let $\ln k = C_1$

$$\ln|2y + 1|^{1/2} = \ln k\,|x|$$

$$|2y + 1|^{1/2} = k\,|x|$$

$$2y + 1 = k^2\,x^2$$

$$2y + 1 = Cx^2$$

$$2y = Cx^2 - 1$$

$$y = \frac{Cx^2 - 1}{2}$$

21. $\dfrac{dy}{dx} + 5y = 12$

$$I(x) = e^{\int 5\,dx} = e^{5x}$$

$$e^{5x}\frac{dy}{dx} + 5e^{5x}y = 12e^{5x}$$

$$D_x\left(e^{5x}y\right) = 12e^{5x}$$

$$e^{5x}y = \frac{12}{5}e^{5x} + C$$

$$y = \frac{12}{5} + Ce^{-5x}$$

23. $3\dfrac{dy}{dz} + xy - x = 0$

$$\frac{dy}{dz} + \frac{x}{3}y = \frac{x}{3}$$

$$I(x) = e^{\int x/3 \, dx} = e^{x^2/6}$$

$$e^{x^2/6}\frac{dy}{dz} + \frac{x}{3}e^{x^2/6}y = \frac{x}{3}e^{x^2/6}$$

$$D_x\left(e^{x^2/6}y\right) = \frac{x}{3}e^{x^2/6}$$

$$e^{x^2/6}y = \int \frac{x}{3}e^{x^2/6} \, dx$$

$$= e^{x^2/6} + C$$

$$y = 1 + Ce^{-x^2/6}$$

25. $\dfrac{dy}{dx} = 2x\cos x^2;\ y = 1$ when $x = 0$.

$$dy = 2x\cos x^2 dx$$
$$y = \sin x^2 + C$$
$$1 = \sin(0)^2 + C$$
$$1 = 0 + C$$
$$1 = C$$
$$y = \sin x^2 + 1$$

27. $\dfrac{dy}{dx} = 5(e^{-x} - 1);\ y = 17$ when $x = 0$.

$$dy = 5(e^{-x} - 1)\, dx$$
$$y = 5(-e^{-x} - x) + C$$
$$= -5e^{-x} - 5x + C$$
$$17 = -5e^0 - 0 + C$$
$$17 = -5 + C$$
$$22 = C$$
$$y = -5e^{-x} - 5x + 22$$

29. $(5 - 2x)y = \dfrac{dy}{dx};\ y = 2$ when $x = 0$.

$$(5 - 2x)dx = \frac{dy}{y}$$

$$5x - x^2 + C = \ln|y|$$
$$e^{5x-x^2+C} = y$$
$$e^{5x-x^2} \cdot e^C = y$$
$$Me^{5x-x^2} = y$$
$$Me^0 = 2$$
$$M = 2$$
$$y = 2e^{5x-x^2}$$

31. $\dfrac{dy}{dx} = \dfrac{1 - 2x}{y + 3};\ y = 16$ when $x = 0$.

$$(y + 3)\, dy = (1 - 2x)\, dx$$
$$\frac{y^2}{2} + 3y = x - x^2 + C$$
$$\frac{16^2}{2} + 3(16) = 0 + C$$
$$176 = C$$
$$\frac{y^2}{2} + 3y = x - x^2 + 176$$
$$y^2 + 6y = 2x - 2x^2 + 352$$

33. $\dfrac{dy}{dx} = 4x^3 + 2;\ y = 3$ when $x = 1$.

$$dy = (4x^3 + 2)\, dx$$

$$\int dy = \int (4x^3 + 2)\, dx$$
$$y = x^4 + 2x + C$$

Since $y = 3$ when $x = 1$,

$$3 = (1)^4 + 2(1) + C$$
$$3 = 3 + C$$
$$C = 0$$

The particular solution is $y = x^4 + 2x$.

35. $\sqrt{x}\dfrac{dy}{dx} = xy;\ y = 4$ when $x = 1$.

$$\frac{1}{y}\, dy = \frac{x}{\sqrt{x}}\, dx$$
$$\int \frac{1}{y}\, dy = \int x^{1/2}\, dx$$
$$\ln|y| = \frac{x^{3/2}}{\frac{3}{2}} + C_1$$
$$\ln|y| = \frac{2}{3}x^{3/2} + C_1$$
$$|y| = e^{2/3x^{3/2} + C_1}$$
$$|y| = e^{C_1}e^{2/3x^{3/2}}$$
$$y = \pm e^{C_1}e^{2/3x^{3/2}}$$
$$y = Ce^{2/3x^{3/2}}$$

Since $y = 4$ when $x = 1$,

$$4 = Ce^{2/3(1)^{3/2}}$$
$$C = \frac{4}{e^{2/3}}$$
$$C \approx 2.054$$
$$y = 2.054e^{(2/3)x^{3/2}}.$$

39. $\dfrac{dy}{dx} = e^x + y$; $y(0) = 1, h = 0.2$; find $y(0.6)$.

$g(x, y) = e^x + y$
$x_0 = 0; \; y_0 = 1$
$\quad g(x_0, y_0) = e^0 + 1 = 2$

$x_1 = 0.2; \; y_1 = y_0 + g(x_0, y_0)h$
$\qquad = 1 + 2(0.2) = 1.4$
$\quad g(x_1, y_1) = e^{0.2} + 1.4 \approx 2.6214$

$x_2 = 0.4; \; y_2 = y_1 + g(x_1, y_1)h$
$\qquad = 1.4 + 2.6214(0.2)$
$\qquad \approx 1.9243$
$\quad g(x_2, y_2) = e^{0.4} + 1.9243 \approx 3.4161$

$x_3 = 0.6; \; y_3 = y_2 + g(x_2, y_2)h$
$\qquad = 1.9243 + 3.4161(0.2)$
$\qquad \approx 2.6075$

x_i	y_i
0	1
0.2	1.4
0.4	1.9243
0.6	2.608

So, $y(0.6) \approx 2.608$.

41. $\dfrac{dy}{dx} = 3 + \sqrt{y}$, $y(0) = 0$, $h = 0.2$, find $y(1)$.

x_i	y_i
0	0
0.2	0.6
0.4	1.354919
0.6	2.187722
0.8	3.083541
1	4.034741

Therefore, $y(1) \approx 4.035$.

43. $\dfrac{dx_1}{dt} = 2x_1 + 7x_2 + t$

$\dfrac{dx_2}{dt} = 7x_1 + 2x_2 + 2t$

$$\begin{bmatrix} \dfrac{dx_1}{dt} \\[2mm] \dfrac{dx_2}{dt} \end{bmatrix} = \begin{bmatrix} 2 & 7 \\ 7 & 2 \end{bmatrix} \begin{bmatrix} x_1 \\ x_2 \end{bmatrix} + \begin{bmatrix} t \\ 2t \end{bmatrix}$$

Find the eigenvalues.

$0 = \det(M - \lambda I)$

$\quad = \det \begin{bmatrix} 2 - \lambda & 7 \\ 7 & 2 - \lambda \end{bmatrix}$

$\quad = (2 - \lambda)(2 - \lambda) - 7(7)$
$\quad = \lambda^2 - 4\lambda - 45$
$\quad = (\lambda - 9)(\lambda + 5)$
$\lambda = -5, 9$

For $\lambda = -5$,

$$(M - \lambda I)X = \begin{bmatrix} 7 & 7 \\ 7 & 7 \end{bmatrix} \begin{bmatrix} x \\ y \end{bmatrix} = \begin{bmatrix} 0 \\ 0 \end{bmatrix}.$$

One solution to this system is $\begin{bmatrix} 1 \\ -1 \end{bmatrix}$.

For $\lambda = 9$,

$$(M - \lambda I)X = \begin{bmatrix} -7 & 7 \\ 7 & -7 \end{bmatrix} \begin{bmatrix} x \\ y \end{bmatrix} = \begin{bmatrix} 0 \\ 0 \end{bmatrix}.$$

One solution to this system is $\begin{bmatrix} 1 \\ 1 \end{bmatrix}$. The matrix of eigenvectors is

$$P = \begin{bmatrix} 1 & 1 \\ -1 & 1 \end{bmatrix}.$$

$$P^{-1} = \frac{1}{2} \begin{bmatrix} 1 & -1 \\ 1 & 1 \end{bmatrix} \text{ and } P^{-1}MP = \begin{bmatrix} -5 & 0 \\ 0 & 9 \end{bmatrix}.$$

Now let $X = PY$ and multiply both sides of the differential equation by P^{-1}, yielding

$$\frac{dY}{dt} = P^{-1}MPY + P^{-1} \begin{bmatrix} t \\ 2t \end{bmatrix}$$

or,

$$\begin{bmatrix} \dfrac{dy_1}{dt} \\[2mm] \dfrac{dy_2}{dt} \end{bmatrix} = \begin{bmatrix} -5 & 0 \\ 0 & 9 \end{bmatrix} \begin{bmatrix} y_1 \\ y_2 \end{bmatrix} + \frac{1}{2} \begin{bmatrix} 1 & -1 \\ 1 & 1 \end{bmatrix} \begin{bmatrix} t \\ 2t \end{bmatrix}$$

$$\frac{dy_1}{dt} = -5y_1 + \frac{t}{2} - t = -5y_1 - \frac{t}{2}$$

$$\frac{dy_2}{dt} = 9y_2 + \frac{t}{2} + t = 9y_2 + \frac{3t}{2}$$

Multiply the first equation by an integrating factor, $e^{\int 5\,dt} = e^{5t}$.

$$\frac{dy_1}{dt} + 5y_1 = -\frac{t}{2}$$

$$e^{5t}\frac{dy_1}{dt} + 5e^{5t}y_1 = \frac{-te^{5t}}{2}$$

$$D_t(e^{5t}y_1) = \frac{-te^{5t}}{2}$$

The antiderivative of $\frac{-te^{5t}}{2}$ is found by using integration by parts.

$$e^{5t}y_1 = -\frac{e^{5t}}{10}\left(t - \frac{1}{5}\right) + C_1$$

$$y_1 = -\frac{1}{10}\left(t - \frac{1}{5}\right) + C_1 e^{-5t}$$

Multiply the second differential equation by an integrating factor $e^{\int -9\,dt} = e^{-9t}$.

$$\frac{dy_2}{dt} - 9y_2 = \frac{3t}{2}$$

$$e^{-9t}\frac{dy_2}{dt} - 9e^{-9t}y_2 = \frac{3t}{2}e^{-9t}$$

$$D_t(e^{-9t}y_2) = \frac{3t}{2}e^{-9t}$$

The antiderivative of $\frac{3te^{-9t}}{2}$ is found by using integration by parts.

$$e^{-9t}y_2 = \frac{3e^{-9t}}{-18}\left(t + \frac{1}{9}\right) + C_2$$

$$y_2 = -\frac{1}{6}\left(t + \frac{1}{9}\right) + C_2 e^{9t}$$

Finally, calculate $X = PY$.

$$\begin{bmatrix} x_1 \\ x_2 \end{bmatrix} = \begin{bmatrix} 1 & 1 \\ -1 & 1 \end{bmatrix} \begin{bmatrix} -\frac{1}{10}\left(t - \frac{1}{5}\right) + C_1 e^{-5t} \\ -\frac{1}{6}\left(t + \frac{1}{9}\right) + C_2 e^{9t} \end{bmatrix}$$

$$= \begin{bmatrix} C_1 e^{-5t} + C_2 e^{9t} - \frac{4t}{15} + \frac{1}{675} \\ -C_1 e^{-5t} + C_2 e^{9t} - \frac{t}{15} - \frac{26}{675} \end{bmatrix}$$

$$x_1 = C_1 e^{-5t} + C_2 e^{9t} - \frac{4t}{15} + \frac{1}{675}$$

$$x_2 = -C_1 e^{-5t} + C_2 e^{9t} - \frac{t}{15} - \frac{26}{675}$$

45. (a) Set the derivatives to zero.

$$4x_1 - x_1^2 - x_1 x_2 = 0$$
$$x_2 = 4 - x_1 \quad (1),$$

$$2x_1^2 - x_1 x_2 - 2x_1 = 0$$
$$x_2 = 2x_1 - 2 \quad (2),$$

The solution to this system is

$$x_1 = 2, x_2 = 2.$$

(b) Graph equations (1) and (2) and test each region.

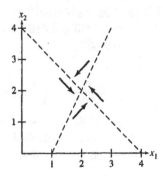

47. $\frac{dy}{dt} = \frac{-10}{1 + 5t}$; $y = 50$ when $t = 0$.

$$y = -2\ln(1 + 5t) + C$$
$$50 = -2\ln 1 + C$$
$$= C$$
$$y = 50 - 2\ln(1 + 5t)$$

(a) If $t = 24$,

$$y = 50 - 2\ln[1 + 5(24)]$$
$$\approx 40 \text{ insects.}$$

(b) If $y = 0$,

$$50 = 2\ln(1 + 5t)$$
$$1 + 5t = e^{25}$$
$$t = \frac{e^{25} - 1}{5}$$
$$\approx 1.44 \times 10^{10} \text{ hours}$$
$$\approx 6 \times 10^8 \text{ days}$$
$$\approx 1.6 \text{ million years.}$$

49.
$$\frac{dx_1}{dt} = 0.2x_1 - 0.5x_1x_2$$

$$\frac{dx_2}{dt} = -0.3x_2 + 0.4x_1x_2$$

$$\frac{dx_2}{dx_1} = \frac{\frac{dx_2}{dt}}{\frac{dx_1}{dt}}$$

$$= \frac{-0.3x_2 + 0.4x_1x_2}{0.2x_1 - 0.5x_1x_2}$$

$$= \frac{x_2(-0.3 + 0.4x_1)}{x_1(0.2 - 0.5x_2)}$$

$$\frac{0.2 - 0.5x_2}{x_2} dx_2 = \frac{-0.3 + 0.4x_1}{x_1} dx_1$$

$$\left(\frac{0.2}{x_2} - 0.5\right) dx_2 = \left(\frac{-0.3}{x_1} + 0.4\right) dx_1$$

$$0.2 \ln x_2 - 0.5x_2 = -0.3 \ln x_1 + 0.4x_1 + C$$
$$0.3 \ln x_1 + 0.2 \ln x_2 - 0.4x_1 - 0.5x_2 = C$$

Both growth rates are 0 if

$$0.2x_1 - 0.5x_1x_2 = 0 \text{ and } -0.3x_2 + 0.4x_1x_2 = 0.$$

If $x_1 \neq 0$ and $x_2 \neq 0$, we have

$$0.2 - 0.5x_2 = 0 \text{ and } -0.3 + 0.4x_1 = 0, \text{ so}$$

$x_1 = \frac{3}{4}$ units and $x_2 = \frac{2}{5}$ units.

51. Let $y =$ the amount in parts per million (ppm) of smoke at time t.
When $t = 0$, $y = 20$ ppm, $V = 15,000 \text{ ft}^3$ cu ft.

rate of smoke in $= 5$ ppm,
rate of smoke out
$$= (1,200 \text{ ft}^3/\text{min})(\frac{y}{V} \text{ ppm/ft}^3)$$

Rate of air in = rate of air out, so $V = 15,000 \text{ ft}^3$ for all t.

$$\frac{dy}{dt} = 5 - \frac{1,200y}{15,000} = 5 - \frac{2y}{25}$$

$$= \frac{125 - 2y}{25}$$

$$\frac{1}{125 - 2y} dy = \frac{dt}{25}$$

$$-\frac{1}{2} \ln(125 - 2y) = \frac{t}{25} + C$$

$$-\frac{1}{2} \ln(125 - 2(20)) = C$$

$$C = -\frac{1}{2} \ln 85$$

If $y = 10$,

$$-\frac{1}{2} \ln[125 - 2(10)] = \frac{t}{25} - \frac{1}{2} \ln 85$$

$$\ln 105 = \ln 85 - \frac{2t}{25}$$

$$t = \frac{25}{2}[\ln 85 - \ln 105],$$

which is negative.
It is impossible to reduce y to 10 ppm.

53. $y = \dfrac{N}{1 + be^{-kx}}$; $y = y_i$ when $x = x_i$,

$i = 1, 2, 3.$

x_1, x_2, x_3 are equally spaced:

$x_3 = 2x_2 - x_1$, so $x_1 + x_3 = 2x_2$, or $x_2 = \dfrac{x_1 + x_3}{2}$.

Show $N = \dfrac{\frac{1}{y_1} + \frac{1}{y_3} - \frac{2}{y_2}}{\frac{1}{y_1y_3} - \frac{1}{y_2^2}}$.

Let

$$A = \frac{1}{y_1} + \frac{1}{y_3} - \frac{2}{y_2}$$

$$= \frac{1 + be^{-kx_1}}{N} + \frac{1 + be^{-kx_3}}{N}$$

$$\quad - \frac{2(1 + be^{-kx_2})}{N}$$

$$= \frac{1}{N}(1 + be^{-kx_1} + 1 + be^{-kx_3} - 2 - 2be^{-kx_2})$$

$$= \frac{b}{N}[e^{-kx_1} + e^{-kx_3} - 2e^{-kx_2}].$$

Let

$$B = \frac{1}{y_1y_3} - \frac{1}{y_2^2}$$

$$= \frac{(1 + be^{-kx_1})(1 + be^{-kx_3})}{N^2} - \frac{(1 + be^{-kx_2})^2}{N^2}$$

$$= \frac{1}{N^2}[1 + be^{-kx_1} + be^{-kx_3} + b^2e^{-k(x_1+x_3)}$$

$$\quad - 1 - 2be^{-kx_2} - b^2e^{-2kx_2}]$$

$$= \frac{b}{N^2}[e^{-kx_1} + e^{-kx_3} + be^{-k(2x_2)}$$

$$\quad - 2e^{-kx_2} - be^{-2kx_2}]$$

$$= \frac{b}{N^2}[e^{-kx_1} + e^{-kx_3} - 2e^{-kx_2}].$$

Clearly, $\frac{A}{B} = N$.

Hence,

$$N = \frac{\frac{1}{y_1} + \frac{1}{y_3} - \frac{2}{y_2}}{\frac{1}{y_1 y_3} - \frac{1}{y_2^2}}.$$

55. From Exercise 53,

$$N = \frac{\frac{1}{40} + \frac{1}{204} - \frac{2}{106}}{\frac{1}{40(204)} - \frac{1}{(106)^2}} \approx 329.$$

(a) $y_0 = 40$,

$$b = \frac{N - y_0}{y_0} = \frac{329 - 40}{40}$$

$$\approx 7.23.$$

If 1920 corresponds to $x = 5$ decades, then

$$106 = \frac{329}{1 + 7.23e^{-5k}}$$

$$1 + 7.23e^{-5k} = \frac{329}{106}$$

$$e^{-5k} = \frac{1}{7.23}\left(\frac{329}{106} - 1\right)$$

so

$$k = -\frac{1}{5}\ln\left[\frac{1}{7.23}\left(\frac{329}{106} - 1\right)\right]$$

$$\approx 0.25.$$

(b) $y = \dfrac{329}{1 + 7.23e^{-0.25x}}$

In 1990, $x = 12$. If $x = 12$,

$$y = \frac{329}{1 + 7.23e^{-3}} \approx 242 \text{ million.}$$

The predicated population is 242 million while the actual value is 249 million.

(c) In 2000, $x = 13$, so

$$y = \frac{329}{1 + 7.23e^{-0.25(13)}}$$

$$\approx 257 \text{ million.}$$

In 2050, $x = 18$, so

$$y = \frac{329}{1 + 7.23e^{-0.25(18)}}$$

$$\approx 305 \text{ million.}$$

57. (a) $\dfrac{dx}{dt} = 1 - kx$

Separate the variables and integrate.

$$\int \frac{dx}{1 - kx} = \int dt$$

$$-\frac{1}{k}\ln|1 - kx| = t + C_1$$

Solve for x.

$$\ln|1 - kx| = -kt - kC_1$$
$$|1 - kx| = e^{-kC_1}e^{-kt}$$

$$1 - kx = Me^{-kt}, \text{ where } M = \pm e^{-kC_1}.$$

$$x = -\frac{1}{k}(Me^{-kt} - 1) = \frac{1}{k} + Ce^{-kt},$$

where $C = -\dfrac{M}{k}$.

(b) Write the linear first-order differential equation in the linear form

$$\frac{dx}{dt} + kx = 1.$$

The integrating factor is $I(t) = e^{\int k\,dt} = e^{kt}$. Multiply both sides of the differential equation by $I(t)$.

$$\frac{dx}{dt}e^{kt} + kxe^{kt} = e^{kt}$$

Replace the left side of this equation by

$$D_t(xe^{kt}) = \frac{dx}{dt}e^{kt} + xke^{kt}.$$
$$D_t(xe^{kt}) = e^{kt}$$

Integrate both sides with respect to t.

$$xe^{kt} = \int e^{kt}\,dt = \frac{1}{k}e^{kt} + C$$

Solve for x.

$$x = \frac{1}{k} + Ce^{-kt}.$$

(c) Since $k > 0$, then as t gets larger and larger, Ce^{-kt} approaches 0, so $\lim\limits_{t \to \infty} \left(\frac{1}{k} + Ce^{-kt} \right) = \frac{1}{k}$.

59.
$$\frac{dT}{dt} = k(T - T_F)$$

$$\frac{dT}{dt} = k(T - 300°)$$

$$T(0) = 40°$$
$$T(1) = 150°$$

$$\frac{dT}{T - 300} = k\,dt$$

$$\ln |T - 300| = kt + C_1$$
$$|T - 300| = e^{kt + C_1}$$
$$T - 300 = Ce^{kt}$$
$$T = Ce^{kt} + 300$$

Since $T(0) = 40°$,

$$40 = Ce^0 + 300$$
$$C = -260.$$

Therefore,

$$T = -260e^{kt} + 300.$$

At $T(1) = 150°$,

$$150 = -260e^k + 300$$
$$-150 = -260e^k$$

$$\frac{15}{26} = e^k$$

$$\ln \left(\frac{15}{26} \right) = k \ln e$$

$$k = \ln \left(\frac{15}{26} \right)$$

$$k = -0.55.$$

Therefore,

$$T = -260e^{-0.55t} + 300.$$

At $t = 2$,

$$T = -260e^{-0.55(2)} + 300$$
$$= 213°.$$

61. (a)
$$\frac{dv}{dt} = G^2 - K^2 v^2$$

$$dv = (G^2 - K^2 v^2)\,dt$$

$$\frac{1}{G^2 - K^2 v^2}\,dv = dt$$

$$\int \frac{1}{G^2 - K^2 v^2}\,dv = \int dt$$

$$\frac{1}{K^2} \int \frac{1}{\left(\frac{G}{K} \right)^2 - v^2} = \int dt$$

Use entry 7 from the table of integrals.

$$\frac{1}{K^2} \cdot \frac{1}{2\frac{G}{K}} \ln \left| \frac{\frac{G}{K} + v}{\frac{G}{K} - v} \right| = t + C_1$$

$$\frac{1}{2GK} \ln \left| \frac{G + Kv}{G - Kv} \right| = t + C_1$$

$$\ln \left| \frac{G + Kv}{G - Kv} \right| = 2GKt + C_2$$

Since $v < \frac{G}{K}$, $Kv < G$ and $\frac{G + Kv}{G - Kv}$ is positive.

$$\ln \left(\frac{G + Kv}{G - Kv} \right) = 2GK(t) + C_2$$

When $t = 0$, $v = 0$, so

$$\ln \left(\frac{G + 0}{G - 0} \right) = 2GK(0) + C_2$$

$$\ln 1 = C_2$$
$$C_2 = 0.$$

Thus,

$$\ln \left(\frac{G + Kv}{G - Kv} \right) = 2GKt$$

$$\frac{G + Kv}{G - Kv} = e^{2GKt}$$

$$G + Kv = Ge^{2GKt} - Kve^{2GKt}$$
$$Kv + Kve^{2GKt} = Ge^{2GKt} - G$$
$$vK(e^{2GKt} + 1) = G(e^{2GKt} - 1)$$

$$v = \frac{G(e^{2GKt} - 1)}{K(e^{2GKt} + 1)}$$

$$v = \frac{G}{K} \cdot \frac{e^{2GKt} - 1}{e^{2GKt} + 1}$$

(b) $\lim\limits_{t\to\infty} v = \lim\limits_{t\to\infty} \left(\dfrac{G}{K} \cdot \dfrac{e^{2GKt} - 1}{e^{2GKt} + 1} \right)$

$\quad\quad \lim\limits_{t\to\infty} v = \dfrac{G}{K} \; \lim\limits_{t\to\infty} \dfrac{e^{2GKt} - 1}{e^{2GKt} + 1}$

$\quad\quad \lim\limits_{t\to\infty} v = \dfrac{G}{K} \cdot 1 = \dfrac{G}{K}$

A falling object in the presence of air resistance has a limiting velocity, $\frac{G}{k}$.

(c) $\lim\limits_{t\to\infty} v = \dfrac{G}{k}$

$\quad\quad 88 = \dfrac{G}{k}$

$\quad\quad k = \dfrac{G}{88}$

$\quad\quad Gk = \dfrac{G^2}{88}$

$\quad\quad Gk = \dfrac{32}{88}$

$\quad 2\,Gk = \dfrac{64}{88}$

$\quad 2\,Gk \approx 0.727$

$\quad\quad v = \dfrac{G}{k} \dfrac{e^{2Gkt} - 1}{e^{2Gkt} + 1}$

$\quad\quad v = 88 \dfrac{e^{0.727t} - 1}{e^{0.727t} + 1}$

Chapter 11 Test

[11.1]

Find the general solutions for the following differential equations.

1. $\dfrac{dy}{dx} = 3x^2 + 4x - 5$ **2.** $\dfrac{dy}{dx} = 5e^{2x}$ **3.** $\dfrac{dy}{dx} = \dfrac{2x}{x^2 + 5}$

Find particular solutions for the following differential equations.

4. $\dfrac{dy}{dx} = 4x^3 + 2x^2 + 1$; $y = 4$ when $x = 1$ **6.** $\dfrac{dy}{dx} = 4\left(e^{-x} - 1\right)$; $y = 5$ when $x = 0$

5. $\dfrac{dy}{dx} = \dfrac{1}{3x + 1}$; $y = 4$ when $x = 3$

7. After use of an insecticide, the rate of decline of an insect population is given by

$$\frac{dy}{dt} = \frac{-8}{1 + 4t},$$

where t is the number of hours after the insecticide is applied. If there were 40 insects initially, how many are left after 20 hours?

Find general solutions for the following differential equations. (Some solutions may give y implicitly.)

8. $\dfrac{dy}{dx} = \dfrac{2x + 1}{y - 1}$ **9.** $\dfrac{dy}{dx} = \dfrac{3y}{x + 1}$ **10.** $\dfrac{dy}{dx} = \dfrac{e^x - x}{y + 1}$

Find particular solutions of the following differential equations.

11. $\sqrt{y}\dfrac{dy}{dx} = xy$; $y = 4$ when $x = 2$ **12.** $\dfrac{dy}{dx} = e^y \cdot x^3$; $y = 0$ when $x = 0$

[11.2]

Find the general solution for each differential equation.

13. $y' + 4y = 12$ **14.** $y' - xy = 2x$

Find the particular solution for each differential equation.

15. $y' + 3y = 10$; $y = 5$ when $x = 0$ **16.** $x^2\dfrac{dy}{dx} + 2x^3y + 2x^3 = 0$; $y = 4$ when $x = 0$

17. $e^x y' + e^x y = x + 1$; $y = \dfrac{2}{e}$ when $x = 1$

[11.3]

Use Euler's method to approximate the indicated function value for $y = f(x)$ to three decimal places using $h = 0.1$.

18. $y' = 2x - y^{-1}$; $f(0) = 1$; find $f(0.3)$.   **19.** $y' = xy$; $f(0) = 1$; find $f(0.4)$.

20. Let $y = f(x)$ and $y' = \frac{x^2}{2} + 3$ with $f(0) = 0$. Use Euler's method with $h = 0.1$ to approximate $f(0.3)$ to three decimal places. Then solve the differential equation and find $f(0.3)$ to three decimal places. Also, find $y_3 - f(x_3)$.

21. In Euler's method, large errors may occur. What can be done to reduce the error?

[11.4]

For Exercises 22–23, solve the system of differential equations.

22. $\dfrac{dx_1}{dt} = 4x_1 - 2x_2$ **23.** $\dfrac{dx_1}{dt} = x_1 + x_2 + 2e^t$

$\quad\ \dfrac{dx_2}{dt} = x_1 + x_2$ $\qquad\ \dfrac{dx_2}{dt} = 4x_1 + x_2 - e^t$

Find the particular solution for the initial value problem from Exercise 22, given $x_1(0) = 4$ and $x_2(0) = 1$.

[11.5]

25. Sketch a phase plane diagram of the system.

$$\frac{dx_1}{dt} = x_1 - x_1 x_2$$

$$\frac{dx_2}{dt} = x_2 - x_1 x_2$$

26. **(a)** Find the equilibrium point, and **(b)** sketch a phase plane diagram. Note that the lines where $\frac{dx_1}{dt} = 0$ and $\frac{dx_2}{dt} = 0$ are neither vertical nor horizontal.

$$\frac{dx_1}{dt} = x_1^2 x_2 + x_1 x_2^2 - 6x_1 x_2$$

$$\frac{dx_2}{dt} = 2x_1 x_2^2 - x_1^2 x_2 - 3x_1 x_2$$

[11.6]

27. Explain why differential equations are useful in problem solving.

28. Find an equation relating x to y given the following equations, which describe the interaction of two competing species and their growth rates.

$$\frac{dx}{dt} = 6x - 4xy$$

$$\frac{dy}{dt} = -8y + 12xy$$

Find the values of x and y for which both growth rates are 0.

29. In a population of size N, the rate at which an influenza epidemic spreads is given by

$$\frac{dy}{dt} = a\,(N - y) \cdot y,$$

where y is the number of people infected and a is a constant.

(a) If 40 people in a community of 10,000 people are infected at the beginning of an epidemic, and 100 people are infected 5 days later, write an equation for the number of people infected after t days.

(b) How many people are infected after 10 days?

30. A tank presently contains 100 gallons of a solution of dissolved salt and water, which is kept uniform by stirring. While pure water is allowed to flow into the tank at a rate of 5 gallons per minute, the mixture flows out of the tank at a rate of 3 gallons per minute. How much salt will remain in the tank after t minutes if 10 pounds of salt are in the mixture originally?

31. A tank filled with a salt solution has solution flowing in and out. The number of pounds y of salt in the tank is given by

$$y = 25\left(1 - e^{-0.02t}\right)$$

where t is the time in hours. Will there ever be 25 pounds of salt in the solution? Explain.

Chapter 11 Test Answers

1. $y = x^3 + 2x^2 - 5x + C$

2. $y = \frac{5}{2}e^{2x} + C$

3. $y = \ln\left(x^2 + 5\right) + C$

4. $y = x^4 + \frac{2}{3}x^3 + x + \frac{4}{3}$

5. $y = \frac{1}{3}\ln|3x + 1| + 4 - \frac{1}{3}\ln 10$

6. $y = -4e^{-x} - 4x + 9$

7. 31

8. $\frac{y^2}{2} - y = x^2 + x + C$

9. $y = C(x + 1)^3$

10. $\frac{y^2}{2} + y = e^x - \frac{x^2}{2} + C$

11. $y^{1/2} = \frac{x^2}{4} + 1$

12. $e^{-y} = 1 - \frac{x^4}{4}$ or $y = -\ln\left|1 - \frac{x^4}{4}\right|$

13. $y = 3 + Ce^{-4x}$

14. $y = Ce^{x^2/2} - 2$

15. $y = \frac{10}{3} + \frac{5}{3}e^{-3x}$

16. $y = 5e^{-x^2} - 1$

17. $y = \frac{x^2}{2}e^{-x} + xe^{-x} + \frac{1}{2}e^{-x}$

18. 0.725

19. 1.061

20. 0.903; $y = \frac{1}{6}x^3 + 3x$; 0.905; -0.002

21. The error can be reduced by using more subintervals of smaller width.

22. $x_1 = C_1 e^{3t} + C_2 e^{2t}$
    $x_2 = \frac{1}{2}C_1 e^{3t} + C_2 e^{2t}$

23. $x_1 = C_1 e^{3t} - C_2 e^{-t} + \frac{1}{4}e^t$
    $x_2 = 2C_1 e^{3t} + 2C_2 e^{-t} - 2e^t$

24. $x_1 = 6e^{3t} - 2e^{2t}$
    $x_2 = 3e^{3t} - 2e^{2t}$

25.

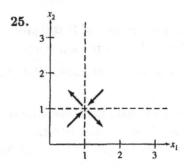

26. (a) (3,3)   (b)

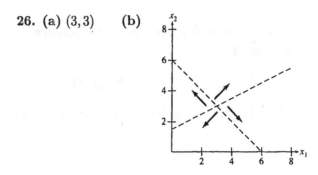

27. Many problems involve rates of change which in turn lead to differential equations.

28. $3\ln y - 2y = -4\ln x + 6x + C$; $x = \frac{2}{3}$, $y = \frac{3}{2}$

29. (a) $y = \frac{10{,}000}{1 + 249e^{-0.184t}}$ (b) 247

30. $y = 10 \cdot \left(\frac{50}{t + 50}\right)^{3/2}$

31. As $t \to \infty$, $e^{-0.02t} \to 0$, so $\lim\limits_{t \to \infty} y = \lim\limits_{t \to \infty} 25\left(1 - e^{-0.02t}\right) = 25$. The limiting value is 25. Theoretically, there will never be 25 lb of salt in the solution. However, as time increases the amount of salt will approach 25 lb.

PROBABILITY

12.1 Sets

1. $3 \in \{2, 5, 7, 9, 10\}$

The number 3 is not an element of the set, so the statement is false.

3. $\{3, 7, 12, 14\} = \{3, 7, 12, 14, 0\}$

Two sets are equal only if they contain exactly the same elements. Since 0 is an element of the second set but not the first, the statement is false.

5. $0 \in \emptyset$

The empty set has no elements. The statement is false.

In Exercises 7–13,

$$A = \{2, 4, 6, 8, 10, 12\},$$
$$B = \{2, 4, 8, 10\},$$
$$C = \{4, 8, 12\},$$
$$D = \{2, 10\},$$
$$E = \{6\},$$
$$\text{and} \quad U = \{2, 4, 6, 8, 10, 12, 14\}.$$

7. Since every element of A is also an element of U, A is a subset of U, written $A \subseteq U$.

9. A contains elements that do not belong to B, namely 2, 6, and 12, so A is not a subset of B, written $A \nsubseteq B$.

11. Since 0 is an element of $\{0, 2\}$, but is not an element of C, $\{0, 2\} \nsubseteq C$.

13. Since B has 4 elements, it has 2^4 or 16 subsets. There are exactly 16 subsets of B.

15. Since $\{7, 9\}$ is the set of elements belonging to both sets, which is the intersection of the two sets, we write

$$\{5, 7, 9, 19\} \cap \{7, 9, 11, 15\} = \{7, 9\}.$$

17. Since $\emptyset$ contains no elements, there are no elements belonging to both sets. Thus, the intersection is the empty set, and we write

$$\{3, 5, 9, 10\} \cap \emptyset = \emptyset.$$

19. $\{1, 2, 4\}$ is the set of elements belonging to both sets, and $\{1, 2, 4\}$ is also the set of elements in the first set or in the second set or possibly both. Thus,

$$\{1, 2, 4\} \cap \{1, 2, 4\} = \{1, 2, 4\}$$

and

$$\{1, 2, 4\} \cup \{1, 2, 4\} = \{1, 2, 4\}$$

are both true statements.

21. $X \cap Y$, the intersection of X and Y, is the set of elements belonging to both X and Y. Thus,

$$X \cap Y = \{2, 3, 4, 5\} \cap \{3, 5, 7, 9\}$$
$$= \{3, 5\}.$$

23. X', the complement of X, consists of those elements of U that are not in X. Thus,

$$X' = \{7, 9\}.$$

25. First find $X \cup Z$.

$$X \cup Z = \{2, 3, 4, 5\} \cup \{2, 4, 5, 7, 9\}$$
$$= \{2, 3, 4, 5, 7, 9\}$$

Now find $Y \cap (X \cup Z)$.

$$Y \cap (X \cup Z) = \{3, 5, 7, 9\} \cap \{2, 3, 4, 5, 7, 9\}$$
$$= \{3, 5, 7, 9\} \text{ or } Y$$

27. $(X \cap Y') \cup Z' = (\{2, 3, 4, 5\} \cap \{2, 4\}) \cup \{3\}$
$$= \{2, 4\} \cup \{3\}$$
$$= \{2, 3, 4\}$$

29. $A = \{2, 4, 6, 8, 10, 12\},$
$B = \{2, 4, 8, 10\},$
$C = \{4, 8, 12\},$
$D = \{2, 10\},$
$E = \{6\},$
$U = \{2, 4, 6, 8, 10, 12, 14\}$

A pair of sets is disjoint if the two sets have no elements in common. The pairs of these sets that are disjoint are B and E, C and E, D and E, and C and D.

31. $B \cap A'$ is the set of all elements in B *and* not in A.

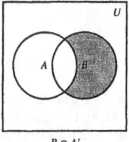

$B \cap A'$

33. $A' \cap B'$ is the set of all elements not in A *and* not in B.

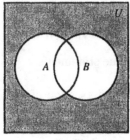

$A' \cap B'$

35. $(A \cap B) \cup B'$

First find $A \cap B$, the set of elements in A *and* in B. Now combine this region with B', the set of all elements not in B. For the union, we want those elements in $A \cap B$ *or* B', or both.

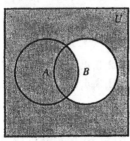

$(A \cap B) \cup B'$

37. The notation $n(A)$ represents the number of elements in set A.

39. $(A \cap C') \cup B$

First find $A \cap C'$, the region in A *and* not in C.

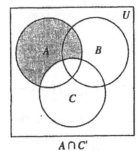

$A \cap C'$

For the union, we want the region in $(A \cap C')$ *or* in B, or both.

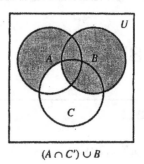

$(A \cap C') \cup B$

41. $A' \cap (B \cap C)$

First find A', the region not in A.

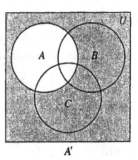

A'

Then find $B \cap C$, the region where B and C overlap.

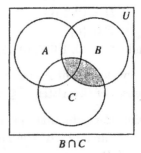

$B \cap C$

Now intersect these regions.

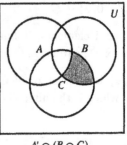

$A' \cap (B \cap C)$

43. $n(A \cup B) = n(A) + n(B) - n(A \cap B)$
$$= 5 + 8 - 4$$
$$= 9$$

45. $n(A \cup B) = n(A) + n(B) - n(A \cap B)$
$$20 = n(A) + 7 - 3$$
$$20 = n(A) + 4$$
$$16 = n(A)$$

47. $(A \cup B)' = A' \cap B'$

For $(A \cup B)'$, first find $A \cup B$.

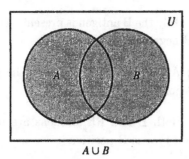

$A \cup B$

Now find $(A \cup B)'$, the region outside $A \cup B$.

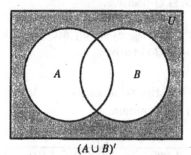

$(A \cup B)'$

For $A' \cap B'$, first find A' and B' individually.

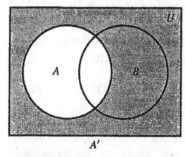

A'

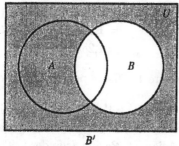

B'

Then $A' \cap B'$ is the region where A' and B' overlap, which is the entire region outside $A \cup B$ (the same result as in the second diagram). Therefore,

$$(A \cup B)' = A' \cap B'.$$

49. $A \cap (B \cup C) = (A \cap B) \cup (A \cap C)$

First find A and $B \cup C$ individually.

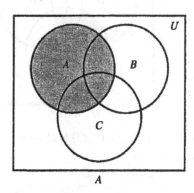

A

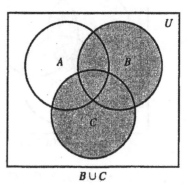

$B \cup C$

Then $A \cap (B \cup C)$ is the region where the above two diagrams overlap.

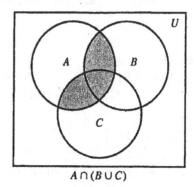

$A \cap (B \cup C)$

Next find $A \cap B$ and $A \cap C$ individually.

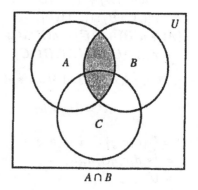

$A \cap B$

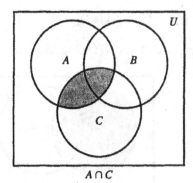

$A \cap C$

Then $(A \cap B) \cup (A \cap C)$ is the union of the above two diagrams.

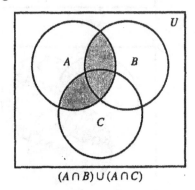

$(A \cap B) \cup (A \cap C)$

The Venn diagram for $A \cap (B \cup C)$ is identical to the Venn diagram for $(A \cap B) \cup (A \cap C)$, so conclude that

$$A \cap (B \cup C) = (A \cap B) \cup (A \cap C).$$

51. Prove

$n(A \cup B \cup C)$
$= n(A) + n(B) + n(C) - n(A \cap B) - n(A \cap C)$
$\quad - n(B \cap C) + n(A \cap B \cap C)$

$n(A \cup B \cup C)$
$= n[A \cup (B \cup C)]$
$= n(A) + n(B \cup C) - n[A \cap (B \cup C)]$
$= n(A) + n(B) + n(C) - n(B \cap C)$
$\quad - n[(A \cap B) \cup (A \cap C)]$
$= n(A) + n(B) + n(C) - n(B \cap C)$
$\quad - \{n(A \cap B) + n(A \cap C)$
$\quad - n[(A \cap B) \cap (A \cap C)]\}$
$= n(A) + n(B) + n(C) - n(B \cap C) - n(A \cap B)$
$\quad - n(A \cap C) + n(A \cap B \cap C)$

53. $U = \{s, d, c, g, i, m, h\}$ and $N = \{s, d, c, g\}$, so

$$N' = \{i, m, h\}.$$

55. $N \cup O = \{s, d, c, g\} \cup \{i, m, h, g\}$
$\quad = \{s, d, c, g, i, m, h\}$
$\quad = U$

57. **(a)** The blood has the A antigen but is Rh negative and has no B antigen. This blood type is A-negative.

(b) Both A and B antigens are present and the blood is Rh negative. This blood type is AB-negative.

(c) Only the B antigen is present. This blood type is B-negative.

(d) All antigens are present. The blood type is AB-positive.

(e) Both B and Rh antigens are present. This blood type is B-positive.

(f) Only the Rh antigen is present. This blood type is O-positive.

(g) No antigens at all are present. This blood type is O-negative.

59. Extend the table to include totals for each row and each column.

	W	B	I	A	Total
F	1,009,509	132,410	4,591	13,696	1,160,206
M	986,884	144,110	5,985	17,060	1,154,039
Total	1,996,393	276,520	10,576	30,756	2,314,245

(a) $n(F)$ is the total for the first row in the table. Thus, there are 1,160,206 people in the set F.

(b) $n(F \cap (I \cup A)) = n(F \cap I) + n(F \cap A) = 4{,}591 + 13{,}696 = 18{,}287$

There are 18,287 people in the set $F \cap (I \cup A)$.

(c) $n(M \cup B) = n(M) + n(B) - n(M \cap B)$
$\qquad = 1{,}154{,}039 + 276{,}520 - 144{,}110$
$\qquad = 1{,}286{,}449$

There are 1,286,449 people in the set $M \cup B$.

(d) $W' \cup I' \cup A'$ is the universe since each person is either *not* white, *not* American Indian, or *not* Asian or Pacific Islander. Thus, there are 2,314,245 in the set $W' \cup I' \cup A'$.

(e) The set $F \cap (I \cup A)$ consists of females who are either American Indian or Asian or Pacific Islander.

61. $n(G \cup B) = n(G) + n(B) - n(G \cap B) = 61.3 + 32.5 - 4.9 = 88.9$

There are 88.9 million people in the set $G \cap B$.

63. $n(F \cap (B \cup H)) = n(F \cap B) = 25.7$

There are 25.7 million people in the set $F \cap (B \cup H)$.

65. $A' \cap C' = B \cup D \cup E$ since the people who are both *not* in the set A and *not* in the set C are the people who are in one of the sets B, D, or E. The set $G' \cap (A' \cap C')$ consists of people in either F or H and also in either B, D, or E. Thus,

$$n(G' \cap (A' \cap C')) = n(F \cap B) + n(F \cap D)$$
$$+ n(F \cap E) + n(H \cap B)$$
$$+ n(H \cap D) + n(H \cap E)$$
$$= 25.7 + 7.6 + 1.5 + 1.9 + 0.8 + 0.2$$
$$= 37.7.$$

There are 37.7 million people in the set $G' \cap (A' \cap C')$.

67. Since the three sets are disjoint, the percentage who lived with either both parents, their father only, or neither parent is $74 + 5 + 3 = 82$. Since the set of children who lived with their mother only is the complement of this seet and since 100 percent represents the total, the percentage who lived with their mother only is $100 - 82 = 18$.

69. $V \cap J = \{$General Electric Inc., Intel Corp., Cisco Systems Inc., Microsoft Corp., Exxon Mobil Corp.$\} \cap \{$China Telecom Ltd., Cisco Systems Inc., Nokia Oyj, Nortel Networks, NTT Mobile Corp.$\}$
$= \{$Cisco Systems Inc.$\}$

71. $(J \cup F)' = (\{$China Telecom Ltd., Cisco Systems Inc., Nokia Oyj, Nortel Networks, NTT Mobile Corp.$\} \cup \{$General Electric Inc., Microsoft Corp., Cisco Systems Inc, Home Depot Inc., Intel Corp.$\})'$
$= \{$China Telecom Ltd., Cisco Systems Inc., Nokia Oyj, Nortel Networks, NTT Mobile Corp., General Electric Inc., Microsoft Corp., Home Depot Inc., Intel Corp.$\})'$
$= \{$Exxon Mobil Corp.,Citigroup Inc., Tyco International$\}$

73. Let A be the set of trucks that carried early peaches, B be the set of trucks that carried late peaches, and C be the set of trucks that carried extra late peaches. We are given the following information.

$$n(A) = 34$$
$$n(B) = 61$$
$$n(C) = 50$$
$$n(A \cap B) = 25$$
$$n(B \cap C) = 30$$
$$n(A \cap C) = 8$$
$$n(A \cap B \cap C) = 6$$
$$n(A' \cap B' \cap C') = 9$$

Start with $A \cap B \cap C$.
We know that $n(A \cap B \cap C) = 6$.
Since $n(A \cap B) = 25$, the number in $A \cap B$ but not in C is $25 - 6 = 19$.
Since $n(B \cap C) = 30$, the number in $B \cap C$ but not in A is $30 - 6 = 24$.
Since $n(A \cap C) = 8$, the number in $A \cap C$ but not in B is $8 - 6 = 2$.
Since $n(A) = 34$, the number in A but not in B or C is $34 - (19 + 6 + 2) = 7$.
Since $n(B) = 61$, the number in B but not in A or C is $61 - (19 + 6 + 24) = 12$.
Since $n(C) = 50$, the number in C but not in A or B is $50 - (24 + 6 + 2) = 18$.
Since $n(A' \cap B' \cap C') = 9$, the number outside $A \cup B \cup C$ is 9.

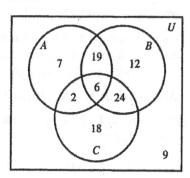

(a) From the Venn diagram, 12 trucks carried only late peaches.

(b) From the Venn diagram, 18 trucks carried only extra late peaches.

(c) From the Venn diagram, $7 + 12 + 18 = 37$ trucks carried only one type of peach.

(d) From the Venn diagram, $6 + 2 + 19 + 24 + 7 + 12 + 18 + 9 = 97$ trucks went out during the week.

77. The number of subsets of a set with 51 elements (50 states plus the District of Columbia) is

$$2^{51} \approx 2.252 \times 10^{15}.$$

79. The number of subsets of a set containing 9 elements is $2^9 = 512$.

12.2   Introduction To Probability

3. The sample space is the set of the twelve months, {January, February, March, ..., December}.

5. Let "surgery" be the outcome "have surgery," "medicine" be "treat with medicine," and "wait" be "wait six months and see."

{Surgery, medicine, wait}

7. Let h = heads and t = tails for the coin; the die can display 6 different numbers. There are 12 possible outcomes in the sample space, which is the set

$$\{(h,1),\ (t,1),\ (h,2),\ (t,2),\ (h,3),\ (t,3),$$
$$(h,4),\ (t,4),\ (h,5),\ (t,5),\ (h,6),\ (t,6)\}.$$

11. Use the first letter of each name. The sample space is the set

{AB, AC, AD, AE, BC, BD, BE, CD, CE, DE}.

(a) One of the committee members must be Chinn. This event is {AC, BC, CD, CE}.

(b) Alam, Bartolini, and Chinn may be on any committee; Dickson and Ellsberg may not be on the same committee. This event is

{AB, AC, AD, AE, BC, BD, BE, CD, CE}.

(c) Both Alam and Chinn are on the committee. This event is {AC}.

13. Let w = wrong; c = correct.

$S = \{www, wwc, wcw, cww, ccw, cwc, wcc, ccc\}$

(a) The student gets three answers wrong. This event is written $\{www\}$.

(b) The student gets exactly two answers correct. Since either the first, second, or third answer can be wrong, this can happen in three ways. The event is written $\{ccw, cwc, wcc\}$.

(c) The student gets only the first answer correct. The second and third answers must be wrong. This event is written $\{cww\}$.

15. "Getting a 2" is the event $E = \{2\}$, so $n(E) = 1$.

If all the outcomes in a sample space S are equally likely, then the probability of an event E is

$$P(E) = \frac{n(E)}{n(S)}.$$

In this problem,

$$P(E) = \frac{n(E)}{n(S)} = \frac{1}{6}.$$

17. "Getting a number less than 5" is the event $E = \{1, 2, 3, 4\}$, so $n(E) = 4$.

$$P(E) = \frac{4}{6} = \frac{2}{3}.$$

19. "Getting a 3 or a 4" is the event $E = \{3, 4\}$, so $n(E) = 2$.

$$P(E) = \frac{2}{6} = \frac{1}{3}.$$

In Exercises 21–25, the sample space contains all 52 cards in the deck, so $n(S) = 52$.

21. Let E be the event "a 9 is drawn." There are four 9's in the deck, so $n(E) = 4$.

$$P(9) = P(E) = \frac{n(E)}{n(S)} = \frac{4}{52} = \frac{1}{13}$$

23. Let F be the event "a black 9 is drawn." There are two black 9's in the deck, so $n(F) = 2$.

$$P(\text{black } 9) = P(F) = \frac{n(F)}{n(S)} = \frac{2}{52} = \frac{1}{26}$$

25. Let R be the event "drawing a black 7 or a red 8." There are two black 7's and two red 8's.

$$n(R) = 4$$
$$P(R) = \frac{4}{52} = \frac{1}{13}$$

For Exercises 27–29, the sample space consists of all the marbles in the jar. There are $2 + 3 + 5 + 8 = 18$ marbles, so $n(S) = 18$.

27. There are 3 orange marbles.

$$P(\text{orange}) = \frac{3}{18} = \frac{1}{6}$$

29. There are 8 marbles which are orange or yellow.

$$P(\text{orange or yellow}) = \frac{8}{18} = \frac{4}{9}$$

33. A person can be married and over 30 at the same time. No, these events are not mutually exclusive.

35. In one roll of a die, it is impossible to get both a 4 and an odd number at the same time. Yes, these events are mutually exclusive.

37. P(second die is 5 or the sum is 10)
$$= P(\text{second die is 5}) + P(\text{sum is 10})$$
$$- P(\text{second die is 5 and sum is 10})$$
$$= \frac{6}{36} + \frac{3}{36} - \frac{1}{36}$$
$$= \frac{8}{36}$$
$$= \frac{2}{9}$$

39. Assume that each box is equally likely to be drawn from and that within each box each marble is equally likely to be drawn. If Laurie does not redistribute the marbles, then the probability of winning the Porsche is $\frac{1}{2}$, since the event of a pink marble being drawn is equivalent to the event of choosing the first of the two boxes.

If however, Laurie puts 49 of the pink marbles into the second box with the 50 blue marbles, the probability of a pink marble being drawn increases to $\frac{74}{99}$. The probability of the first box being chosen is $\frac{1}{2}$, and the probability of drawing a pink marble from this box is 1. The probability of the second box being chosen is $\frac{1}{2}$, and the probability of drawing a pink marble from this box is $\frac{49}{99}$. Thus, the probability of drawing a pink marble is $\frac{1}{2} \cdot 1 + \frac{1}{2} \cdot \frac{49}{99} = \frac{74}{99}$.

41. (a) Less than a 4 would be an ace, a 2, or a 3. There are a total of 12 aces, 2's, and 3's in a deck of 52, so
$$P(\text{ace or 2 or 3}) = \frac{12}{52} = \frac{3}{13}.$$

(b) There are 13 diamonds plus three 7's in other suits, so
$$P(\text{diamond or 7}) = \frac{16}{52} = \frac{4}{13}.$$

Alternatively, using the union rule for probability,

$P(\text{diamond}) + P(7) - P(\text{7 of diamonds})$
$$= \frac{13}{52} + \frac{4}{52} - \frac{1}{52} = \frac{16}{52} = \frac{4}{13}.$$

(c) There are 26 black cards plus 2 red aces, so
$$P(\text{black or ace}) = \frac{28}{52} = \frac{7}{13}.$$

(d) There are 13 hearts plus 3 additional jacks in other suits, so
$$P(\text{heart or jack}) = \frac{16}{52} = \frac{4}{13}.$$

(e) There are 26 red cards plus 6 black face cards, so
$$P(\text{red or face card}) = \frac{32}{52} = \frac{8}{13}.$$

43. (a) There are 3 uncles plus 2 cousins out of 10, so
$$P(\text{uncle or cousin}) = \frac{5}{10} = \frac{1}{2}.$$

(b) There are 3 uncles, 2 brothers, and 2 cousins, for a total of 7 out of 10, so
$$P(\text{male or cousin}) = \frac{7}{10}.$$

(c) There are 2 aunts, 2 cousins, and 1 mother, for a total of 5 out of 10, so
$$P(\text{female or cousin}) = \frac{5}{10} = \frac{1}{2}.$$

45. Let E be the event "a 5 is rolled."
$$P(E) = \frac{1}{6} \text{ and } P(E') = \frac{5}{6}.$$
The odds in favor of rolling a 5 are
$$\frac{P(E)}{P(E')} = \frac{\frac{1}{6}}{\frac{5}{6}} = \frac{1}{5},$$
which is written "1 to 5."

47. Let E be the event "a 1, 2, 3, or 4 is rolled." Here $P(E) = \frac{4}{6} = \frac{2}{3}$ and $P(E') = \frac{1}{3}$. The odds in favor of E are
$$\frac{P(E)}{P(E')} = \frac{\frac{2}{3}}{\frac{1}{3}} = \frac{2}{1},$$
which is written "2 to 1."

49. The statement is not correct.

Assume two dice are rolled.

$P(\text{you win}) = P(\text{first die is greater than second})$
$$= \frac{15}{36}$$
$$= \frac{5}{12}$$

$P(\text{other player wins}) = 1 - P(\text{you win})$
$$= 1 - \frac{5}{12}$$
$$= \frac{7}{12}$$

51. $P(\text{You will eat out today.}) = \dfrac{1}{1+2} = \dfrac{1}{3}$

$P(\text{Bottled water will be tap water}) = \dfrac{1}{1+4} = \dfrac{1}{5}$

$P(\text{Earth will be struck by meteor.}) = \dfrac{1}{1+9000} = \dfrac{1}{9001}$

$P(\text{You will go to Disney World.}) = \dfrac{1}{1+9} = \dfrac{1}{10}$

$P(\text{You'll regain weight}) = \dfrac{9}{9+10} = \dfrac{9}{19}$

53. It is not possible to establish an exact probability for this event, so this is an empirical probability.

55. It is not possible to establish an exact probability for this event, so this is an empirical probability.

59. This assignment is not possible because one of the probabilities is -0.08, which is not between 0 and 1. A probability cannot be negative.

61. The answers that are given are theoretical. Using the Monte Carlo method with at least 50 repetitions on a graphing calculator should give values close to these.

(a) 0.2778   **(b)** 0.4167

63. The answers that are given are theoretical. Using the Monte Carlo method with at least 50 repetitions on a graphing calculator should give values close to these.

(a) 0.15625   **(b)** 0.3125

65. E: person smokes
F: person has a family history of heart disease
G: person is overweight

(a) $E \cup F$ occurs when E or F or both occur, so $E \cup F$ is the event "person smokes or has a family history of heart disease, or both."

(b) $E' \cap F$ occurs when E does not occur and F does occur, so $E' \cap F$ is the event "person does not smoke and has a family history of heart disease."

(c) $F' \cup G'$ is the event "person does not have a family history of heart disease or is not overweight, or both."

67. $P(C) = 0.039$, $P(M \cap C) = 0.035$,
$P(M \cup C) = 0.491$

Place the given information in a Venn diagram by starting with 0.035 in the intersection of the regions for M and C.

Since $P(C) = 0.039$,

$$0.039 - 0.035 = 0.004$$

goes inside region C, but outside the intersection of C and M. Thus,

$$P(C \cap M') = 0.004.$$

Since $P(M \cup C) = 0.491$,

$$0.491 - 0.035 - 0.004 = 0.452$$

goes inside region M, but outside the intersection of C and M. Thus, $P(M \cap C') = 0.452$. The labeled regions have probability

$$0.452 + 0.035 + 0.004 = 0.491.$$

Since the entire region of the Venn diagram must have probability 1, the region outside M and C, or $M' \cap C'$, has probability

$$1 - 0.491 = 0.509.$$

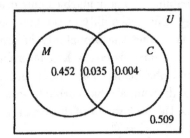

(a) $P(C') = 1 - P(C)$
$= 1 - 0.039$
$= 0.961$

(b) $P(M) = 0.452 + 0.035$
$= 0.487$

(c) $P(M') = 1 - P(M)$
$= 1 - 0.487$
$= 0.513$

(d) $P(M' \cap C') = 0.509$

(e) $P(C \cap M') = 0.004$

(f) $P(C \cup M')$
$= P(C) + P(M') - P(C \cap M')$
$= 0.039 + 0.513 - 0.004$
$= 0.548$

69. (a) Now red is no longer dominant, and RW or WR results in pink, so

$$P(\text{red}) = P(RR) = \frac{1}{4}.$$

(b) $P(\text{pink}) = P(RW) + P(WR)$

$$= \frac{1}{4} + \frac{1}{4} = \frac{1}{2}$$

(c) $P(\text{white}) = P(WW) = \frac{1}{4}$

71. The total population for 2000 is 275,400, and the total for 2025 is 338,300.

(a) $P(\text{Hispanic in 2000}) = \dfrac{32,500}{275,400}$

$$\approx 0.118$$

(b) $P(\text{Hispanic in 2025}) = \dfrac{56,900}{338,300}$

$$\approx 0.168$$

(c) $P(\text{African-American in 2000}) = \dfrac{33,500}{275,400}$

$$\approx 0.122$$

(d) $P(\text{African-American in 2025}) = \dfrac{44,700}{338,300}$

$$\approx 0.132$$

73. (a) $P(\text{III Corps}) = \dfrac{22,083}{70,076} \approx 0.32$

(b) $P(\text{lost in battle}) = \dfrac{22,557}{70,076} \approx 0.32$

(c) $P(\text{I Corps lost in battle}) = \dfrac{7,661}{20,706}$

$$\approx 0.37$$

(d) $P(\text{I Corps not lost in battle}) = \dfrac{20,706 - 7,661}{20,706}$

$$\approx 0.63$$

$P(\text{II Corps not lost in battle}) = \dfrac{20,666 - 6,603}{20,666}$

$$\approx 0.68$$

$P(\text{III Corps not lost in battle}) = \dfrac{22,083 - 8,007}{22,083}$

$$\approx 0.64$$

$P(\text{Calvary not lost in battle}) = \dfrac{6,621 - 286}{6,621}$

$$\approx 0.96$$

The Calvary had the highest probability of not being lost in battle.

(e) $P(\text{I Corps loss}) = \dfrac{7,661}{20,706} \approx 0.37$

$P(\text{II Corps loss}) = \dfrac{6,603}{20,666} \approx 0.32$

$P(\text{III Corps loss}) = \dfrac{8,007}{22,083} \approx 0.36$

$P(\text{Calvary loss}) = \dfrac{286}{6,621} \approx 0.04$

I Corps had the highest probability of loss.

75. (a) Divide each entry in the original table by 198,420.

	A	B	C	D
O	0.052	0.060	0.039	0.004
E	0.306	0.270	0.213	0.045
M	0.003	0.003	0.003	0

(b) This probability is the sum of the first column.

$$P(A) = 0.052 + 0.306 + 0.003 = 0.361$$

(c) $P(O \cap (C \cup D)) = P(O \cap C) + P(O \cap D)$

$$= 0.039 + 0.004$$
$$= 0.043$$

(d) $P(A \cup B) = P(A) + P(B)$

$$= 0.361 + (0.060 + 0.270 + 0.003)$$
$$= 0.694$$

(e) $P(E \cup (C \cup D))$

$$= P(E) + P(C \cup D) - P(E \cap (C \cup D))$$
$$= 0.834 + 0.304 - 0.258$$
$$= 0.88$$

77. (a) $P(\text{at least some intolerance of Fascists})$
$$= P(\text{not very much}) + P(\text{somewhat})$$
$$+ P(\text{extremely intolerant of Facists})$$

$$= \frac{20.7}{100} + \frac{43.1}{100} + \frac{22.9}{100} = \frac{86.7}{100} = 0.867$$

(b) $P(\text{at least some intolerance of Communists})$
$$= P(\text{not very much}) + P(\text{somewhat})$$
$$+ P(\text{extremely intolerant of Communists})$$

$$= \frac{33.0}{100} + \frac{34.2}{100} + \frac{17.1}{100} = \frac{84.3}{100}$$

$$= 0.843$$

12.3 Conditional Probability; Independent Events; Bayes' Theroem

1. Let A be the event "the number is 2" and B be the event "the number is odd."

   The problem seeks the conditional probability $P(A|B)$. Use the definition

   $$P(A|B) = \frac{P(A \cap B)}{P(B)}.$$

   Here, $P(A \cap B) = 0$ and $P(B) = \frac{1}{2}$. Thus,

   $$P(A|B) = \frac{0}{\frac{1}{2}} = 0.$$

3. Let A be the event "the number is even" and B be the event "the number is 6." Then

   $$P(A|B) = \frac{P(A \cap B)}{P(B)} = \frac{\frac{1}{6}}{\frac{1}{6}} = 1.$$

5. Let A be the event "sum of 6" and B be the event "double." 6 of the 36 ordered pairs are doubles, so $P(B) = \frac{6}{36} = \frac{1}{6}$. There is only one outcome, 3-3, in $A \cap B$ (that is, a double with a sum of 6), so $P(A \cap B) = \frac{1}{36}$. Thus,

   $$P(A|B) = \frac{\frac{1}{36}}{\frac{1}{6}} = \frac{6}{36} = \frac{1}{6}.$$

7. Use a reduced sample space. After the first card drawn is a heart, there remain 51 cards, of which 12 are hearts. Thus,

   $$P(\text{heart on 2nd}|\text{heart on 1st}) = \frac{12}{51} = \frac{4}{17}.$$

9. Use a reduced sample space. After the first card drawn is a jack, there remain 51 cards, of which 11 are face cards. Thus,

   $$P(\text{face card on 2nd}|\text{jack on 1st}) = \frac{11}{51}.$$

11. P(a jack and a 10)
    $= P$(jack followed by 10)
    $\quad + P$(10 followed by jack)

    $$= \frac{4}{52} \cdot \frac{4}{51} + \frac{4}{52} \cdot \frac{4}{51}$$

    $$= \frac{16}{2,652} + \frac{16}{2,652}$$

    $$= \frac{32}{2,652} \approx 0.012$$

13. P(two black cards)
    $= P$(black on 1st)
    $\quad \cdot P$(black on 2nd|black on 1st)

    $$= \frac{26}{52} \cdot \frac{25}{51}$$

    $$= \frac{650}{2,652} \approx 0.245$$

19. At the first booth, there are three possibilities: shaker 1 has heads and shaker 2 has heads; shaker 1 has tails and shaker 2 has heads; shaker 1 has heads and shaker 2 has tails. We restrict ourselves to the condition that at least one head has appeared. These three possibilities are equally likely so the probability of two heads is $\frac{1}{3}$.

   At the second booth we are given the condition of one head in one shaker. The probability that the second shaker has one head is $\frac{1}{2}$.

   Therefore, you stand the best chance at the second booth.

21. No, these events are not independent.

23. Since A and B are independent events.

    $$P(A \cap B) = P(A) \cdot P(B) = \frac{1}{4} \cdot \frac{1}{5} = \frac{1}{20}.$$

    Thus,

    $$P(A \cup B) = P(A) + P(B) - P(A \cap B)$$
    $$= \frac{1}{4} + \frac{1}{5} - \frac{1}{20}$$
    $$= \frac{2}{5}.$$

25. Use Bayes' theorem with two possibilities M and M'.

    $$P(M|N) = \frac{P(M) \cdot P(N|M)}{P(M) \cdot P(N|M) + P(M') \cdot P(N|M')}$$

    $$= \frac{0.4(0.3)}{0.4(0.3) + 0.6(0.4)}$$

    $$= \frac{0.12}{0.12 + 0.24}$$

    $$= \frac{0.12}{0.36} = \frac{12}{36} = \frac{1}{3}$$

27. Using Bayes' theorem,

$P(R_1|Q)$

$$= \frac{P(R_1) \cdot P(Q|R_1)}{P(R_1) \cdot P(Q|R_1) + P(R_2) \cdot P(Q|R_2) + P(R_3) \cdot P(Q|R_3)}$$

$$= \frac{0.05(0.40)}{0.05(0.40) + 0.6(0.30) + 0.35(0.60)}$$

$$= \frac{0.02}{0.41} = \frac{2}{41}.$$

29. Using Bayes' theorem,

$P(R_3|Q)$

$$= \frac{P(R_3) \cdot P(Q|R_3)}{P(R_1) \cdot P(Q|R_1) + P(R_2) \cdot P(Q|R_2) + P(R_3) \cdot P(Q|R_3)}$$

$$= \frac{0.35(0.60)}{0.05(0.40) + 0.6(0.30) + 0.35(0.60)}$$

$$= \frac{0.21}{0.41} = \frac{21}{41}.$$

31. We first draw the tree diagram and determine the probabilities as indicated below.

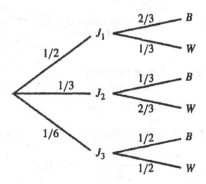

We want to determine the probability that if a white ball is drawn, it came from the second jar. This is $P(J_2|W)$. Use Bayes' theorem.

$P(J_2|W)$

$$= \frac{P(J_2) \cdot P(W|J_2)}{P(J_2) \cdot P(W|J_2) + P(J_1) \cdot P(W|J_1) + P(J_3) \cdot P(W|J_3)}$$

$$= \frac{\frac{1}{3} \cdot \frac{2}{3}}{\frac{1}{3} \cdot \frac{2}{3} + \frac{1}{2} \cdot \frac{1}{3} + \frac{1}{6} \cdot \frac{1}{2}}$$

$$= \frac{\frac{2}{9}}{\frac{2}{9} + \frac{1}{6} + \frac{1}{12}}$$

$$= \frac{\frac{2}{9}}{\frac{17}{36}} = \frac{8}{17}.$$

33. The sample space is

$$\{PW, WP, PP, WW\}.$$

The event "purple" is $\{PW, WP, PP\}$, and the event "mixed" is $\{PW, WP\}$.

$$P(\text{mixed}|\text{purple}) = \frac{n(\text{mixed and purple})}{n(\text{purple})}$$

$$= \frac{2}{3}.$$

Use the following tree diagram for Exercises 35 and 37.

1st child	2nd child	3rd child	Branch	Probability
			1	1/8
			2	1/8
			3	1/8
			4	1/8
			5	1/8
			6	1/8
			7	1/8
			8	1/8

35. $P(3 \text{ girls}|3\text{rd is a girl})$

$$= \frac{P(3 \text{ girls and 3rd is a girl})}{P(3\text{rd is a girl})}$$

$$= \frac{\frac{1}{8}}{\frac{1}{2}}$$

$$= \frac{1}{4}$$

37. $P(\text{all girls}|\text{at least 1 girl})$

$$= \frac{P(\text{all girls and at least 1 girl})}{P(\text{at least 1 girl})}$$

$$= \frac{n(\text{all girls and at least 1 girl})}{n(\text{at least 1 girl})}$$

$$= \frac{1}{7}$$

39. By the product rule for independent events, two events are independent if the product of their probabilities is the probability of their intersection.

$$0.75(0.4) = 0.3$$

Therefore, the given events are independent.

41. $P(C) = 0.039$, the total of the C row

43. $P(M \cup C) = P(M) + P(C) - P(M \cap C)$
$$= 0.487 + 0.039 - 0.035$$
$$= 0.491$$

45. $P(C|M) = \dfrac{P(C \cap M)}{P(M)}$

$$= \dfrac{0.035}{0.487}$$

$$\approx 0.072$$

47. (a) From the table,

$$P(C \cap D) = 0.0008 \text{ and}$$
$$P(C) \cdot P(D) = 0.0400(0.0200) = 0.0008.$$

Since $P(C \cap D) = P(C) \cdot P(D)$, C and D are independent events; color blindness and deafness are independent events.

49. (a) $P(\text{forecast of rain}|\text{rain}) = \dfrac{66}{80}$

$$= 0.825$$
$$\approx 83\%$$

$$P(\text{forecast of no rain}|\text{no rain}) = \dfrac{764}{920}$$

$$\approx 0.83$$
$$\approx 83\%$$

(b) $P(\text{rain}|\text{forecast of rain}) = \dfrac{66}{222} \approx 0.30$

(c) $P(\text{no rain}|\text{forecast of no rain}) = \dfrac{764}{778}$

$$\approx 0.98$$

51. (a) $P(N|T)$

$$= \dfrac{P(N) \cdot P(T|N)}{P(N) \cdot P(T|N) + P(N') \cdot P(T|N')}$$

$$= \dfrac{0.89(0.10)}{0.89(0.10) + 0.11(0.75)}$$

$$= \dfrac{0.0890}{0.1715} = \dfrac{178}{343} \approx 0.519$$

(b) $P(T'|N) = 0.90$, so $P(T|N) = 0.10$ (since either T or T' must occur).

$$P(N) = 1 - P(N') = 0.89$$

Using Bayes' theorem,

$$P(N'|T) = \dfrac{P(N') \cdot P(T|N')}{P(N') \cdot P(T|N') + P(N) \cdot P(T|N)}$$

$$= \dfrac{0.11(0.75)}{0.11(0.75) + 0.89(0.10)}$$

$$= \dfrac{0.0825}{0.1715} = \dfrac{165}{343} \approx 0.481.$$

53. Let H be the event "has hepatitis" and T be the event "positive test." First set up a tree diagram.

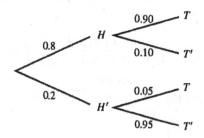

Using Bayes' theorem,

$$P(H|T) = \dfrac{P(H) \cdot P(T|H)}{P(H) \cdot P(T|H) + P(H') \cdot P(T|H')}$$

$$= \dfrac{0.8(0.90)}{0.8(0.90) + 0.2(0.05)}$$

$$= \dfrac{0.72}{0.73} = \dfrac{72}{73} \approx 0.986.$$

55.

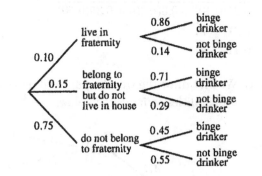

(a) $P(\text{binge drinker})$
$$= 0.10(0.86) + 0.15(0.71) + 0.75(0.45)$$
$$= 0.53$$

(b) $P(\text{lives in fraternity}|\text{binge drinker})$

$$= \dfrac{0.10(0.86)}{0.53} \approx 0.1623$$

57. Draw the tree diagram.

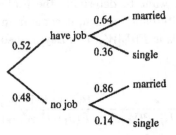

(a) $P(\text{married}) = P(\text{job and married})$
$$+ P(\text{no job and married})$$
$$= 0.52(0.64) + 0.48(0.86)$$
$$= 0.3328 + 0.4128 = 0.7456$$

(b) $P(\text{job and single}) = 0.52(0.36) = 0.1872$

59. $P(\text{between 35 and 44}|\text{never married}) = \dfrac{0.232(0.187)}{0.133(0.874) + 0.205(0.397) + 0.232(0.187) + 0.287(0.075) + 0.142(0.038)}$

$$= \dfrac{0.043384}{0.267932}$$

$$\approx 0.162$$

61. $P(\text{between 45 and 64}|\text{never married}) = \dfrac{0.284(0.062)}{0.123(0.775) + 0.194(0.298) + 0.219(0.121) + 0.284(0.062) + 0.181(0.047)}$

$$= \dfrac{0.017608}{0.205751}$$

$$\approx 0.086$$

63. Draw a tree diagram.

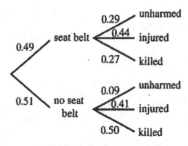

$$P(\text{not wearing seat belt}|\text{unharmed}) = \dfrac{0.51(0.09)}{0.49(0.29) + 0.51(0.09)} \approx 0.244$$

65. Let V represent "person votes" and "65" represent "65 or over." Draw the tree diagram.

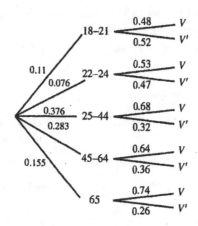

$$P(65|V) = \dfrac{0.155(0.74)}{0.11(0.48) + 0.076(0.53) + 0.376(0.68) + 0.283(0.64) + 0.155(0.74)}$$

$$= \dfrac{0.1147}{0.64458}$$

$$\approx 0.178$$

12.4   Discrete Random Variables; Applications to Decision Making

1. Let x denote the number of heads observed. Then x can take on 0, 1, 2, 3, or 4 as values. The probabilities are as follows.

$$P(x = 0) = \binom{4}{0}\left(\frac{1}{2}\right)^0\left(\frac{1}{2}\right)^4 = \frac{1}{16}$$

$$P(x = 1) = \binom{4}{1}\left(\frac{1}{2}\right)^1\left(\frac{1}{2}\right)^3 = \frac{4}{16} = \frac{1}{4}$$

$$P(x = 2) = \binom{4}{2}\left(\frac{1}{2}\right)^2\left(\frac{1}{2}\right)^2 = \frac{6}{16} = \frac{3}{8}$$

$$P(x = 3) = \binom{4}{3}\left(\frac{1}{2}\right)^3\left(\frac{1}{2}\right)^1 = \frac{4}{16} = \frac{1}{4}$$

$$P(x = 4) = \binom{4}{4}\left(\frac{1}{2}\right)^4\left(\frac{1}{2}\right)^0 = \frac{1}{16}$$

Therefore, the probability distribution is as follows.

Number of Heads	0	1	2	3	4
Probability	$\frac{1}{16}$	$\frac{1}{4}$	$\frac{3}{8}$	$\frac{1}{4}$	$\frac{1}{16}$

3. Use the probabilities that were calculated in Exercise 1. Draw a histogram with 5 rectangles, corresponding to $x = 0$, $x = 1$, $x = 2$, $x = 3$, and $x = 4$. $P(x \le 2)$ corresponds to

$$P(x = 0) + P(x = 1) + P(x = 2),$$

so shade the first 3 rectangles in the histogram.

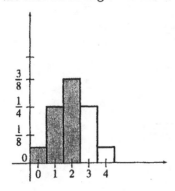

5. $E(x) = 2(0.1) + 3(0.4) + 4(0.3) + 5(0.2)$
$$= 3.6$$

$$Var(x) = (2 - 3.6)^2(0.1) + (3 - 3.6)^2(0.4)$$
$$+ (4 - 3.6)^2(0.3) + (5 - 3.6)^2(0.2)$$
$$= 0.84$$

$$\sigma = \sqrt{0.84}$$
$$\approx 0.917$$

7. $E(z) = 9(0.14) + 12(0.22) + 15(0.36)$
$$+ 18(0.18) + 21(0.10)$$
$$= 14.64$$

$$Var(x) = (9 - 14.64)^2(0.14) + (12 - 14.64)^2(0.22)$$
$$+ (15 - 14.64)^2(0.36) + (18 - 14.64)^2(0.18$$
$$+ (21 - 14.64)^2(0.10)$$
$$= 12.1104$$

$$\sigma = \sqrt{12.1104}$$
$$= 3.48$$

9. It is possible (but not necessary) to begin by writing the histogram's data as a probability distribution, which would look as follows.

x	1	2	3	4
$P(x)$	0.2	0.3	0.1	0.4

The expected value of x is

$$E(x) = 1(0.2) + 2(0.3) + 3(0.1) + 4(0.4)$$
$$= 2.7.$$

11. The expected value of x is

$$E(x) = 6(0.1) + 12(0.2) + 18(0.4) + 24(0.2)$$
$$+ 30(0.1)$$
$$= 18.$$

13. Using the data from Example 4, the expected winnings for Mary are

$$E(x) = -0.8\left(\frac{1}{4}\right) + 0.8\left(\frac{1}{4}\right) + 0.8\left(\frac{1}{4}\right) + (-0.8)\left(\frac{1}{4}\right)$$
$$= 0.$$

Yes, it is still a fair game if Mary tosses and Donna calls.

17. Let x represent the number of offspring. We have the following probability distribution.

x	0	1	2	3	4
$P(x)$	0.31	0.21	0.19	0.17	0.12

$$E(x) = 0(0.31) + 1(0.21) + 2(0.19) + 3(0.17) + 4(0.12)$$
$$= 1.58$$

19. With the new contamination probability, the probabilities for the branches becomes

Branch	1	2	3	4
Outcome (x)	0	0.1	1	1
$P(x)$	0.024	0.0066	0.0017	0.0009

Branch	5	6	7
Outcome (x)	0	1	0
$P(x)$	0.005	0.9131	0.0531

The expected outcome for transfusion is

$$E_1 = 0(0.024) + 0.1(0.0066) + 1(0.0017) + 1(0.0009)$$
$$+ 0(0.005) + 1(0.9131) + 0(0.0531)$$
$$\approx 0.9164$$

The expected outcome for no transfusion remains unchanged at 0.865.

Since the value of the outcome increases blood transfusions should be given after the blood has been screened.

21. (a) Let x represent the amount of damage in millions of dollars. For seeding, the expected value is

$$E(x) = 0.038(335.8) + 0.143(191.1) + 0.392(100)$$
$$+ 0.255(46.7) + 0.172(16.3)$$
$$\approx \$94.0 \text{ million.}$$

For not seeding, the expected value is

$$E(x) = 0.054(335.8) + 0.206(191.1) + 0.480(100)$$
$$+ 0.206(46.7) + 0.054(16.3)$$
$$\approx \$116.0 \text{ million.}$$

(b) Seed, since the total expected damage is less with that option.

23. Below is the probability distribution of x, which stands for the person's net winnings.

x	\$99	\$39	−\$1
$P(x)$	$\frac{1}{500} = 0.002$	$\frac{2}{500} = 0.004$	$\frac{497}{500} = 0.994$

The expected value of the person's winnings is

$$E(x) = 99(0.002) + 39(0.004) + (-1)(0.994)$$
$$\approx -\$0.64 \quad \text{or} \quad -64\cent.$$

Since the expected value of the winnings is not 0, this is not a fair game.

25. There are $18 + 20 = 38$ possible outcomes. In 18 cases you win a dollar and in 20 you lose a dollar; hence,

$$E(x) = 1\left(\frac{18}{38}\right) + (-1)\left(\frac{20}{38}\right)$$
$$= -\frac{1}{19}, \text{ or about } -5.3\cent.$$

27. You have one chance in a thousand of winning \$500 on a \$1 bet for a net return of \$499. In the 999 other outcomes, you lose your dollar.

$$E(x) = 499\left(\frac{1}{1,000}\right) + (-1)\left(\frac{999}{1,000}\right)$$
$$= \frac{-500}{1,000} = -50\cent$$

29. At any one restaurant, your expected winnings are

$$E(x) = 100,000\left(\frac{1}{176,402,500}\right) + 25,000\left(\frac{1}{39,200,556}\right)$$
$$+ 5,000\left(\frac{1}{17,640,250}\right) + 1,000\left(\frac{1}{1,568,022}\right)$$
$$+ 100\left(\frac{1}{288,244}\right) + 5\left(\frac{1}{7,056}\right) + 1\left(\frac{1}{588}\right)$$
$$= 0.00488.$$

Going to 25 restaurants gives you expected earnings of $25(0.00488) = 0.122$. Since you spent \$1, you lose $87.8\cent$ on the average, so your expected value is $-87.8\cent$.

31. (a) Expected value of a two-point conversion:

$$E(x) = 2(0.37) = 0.74$$

Expected value of an extra-point kick:

$$E(x) = 1(0.94) = 0.94$$

(b) Since the expected value of an extra-point kick is greater than the expected value of a two-point conversion, the extra-point kick will maximize the number of points scored over the long run.

Chapter 12 Review Exercises

In Exercises 1-5,

$$U = \{a, b, c, d, e, f, g\},$$
$$K = \{c, d, f, g\},$$
$$\text{and} \quad R = \{a, c, d, e, g\}.$$

1. K has 4 elements, so it has $2^4 = 16$ subsets.

3. $K \cap R$ (the intersection of K and R) is the set of all elements belonging to both set K and set R.

$$K \cap R = \{c, d, g\}$$

5. $(K \cap R)' = \{a, b, e, f\}$ since these elements are in U but not in $K \cap R$. (See Exercise 15.)

7. $A \cup B'$ is the set of all elements which belong to A or do not belong to B, or both.

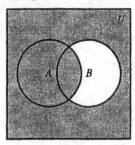

$$A \cup B'$$

9. $(A \cap B) \cup C$

First find $A \cap B$.

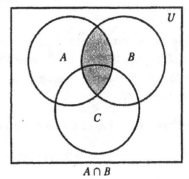

$$A \cap B$$

Now find the union of this region with C.

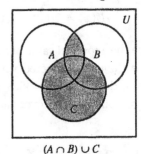

$$(A \cap B) \cup C$$

11. The sample space for rolling a die is

$$S = \{1, 2, 3, 4, 5, 6\}.$$

13. There are 16 possibilities.

$$S = \{hhhh, \ hhht, \ hhth, \ hthh, \ thhh, \\
hhtt, \ htht, \ htth, \ thht, \ tthh, \\
thth, \ httt, \ thtt, \ ttht, \ ttth, \ tttt\}$$

15. There are 3 face cards in each suit (jack, queen, and king) and there are 4 suits, so there are $3 \cdot 4 = 12$ face cards out of the 52 cards. Thus,

$$P(\text{face card}) = \frac{12}{52} = \frac{3}{13}.$$

17. There are 4 queens of which 2 are red, so

$$P(\text{red}|\text{queen}) = \frac{n(\text{red and queen})}{n(\text{queen})}$$

$$= \frac{2}{4} = \frac{1}{2}.$$

25. Let C be the event "a club is drawn." There are 13 clubs in the deck, so $n(C) = 13$, $P(C) = \frac{13}{52} = \frac{1}{4}$, and $P(C') = 1 - P(C) = \frac{3}{4}$. The odds in favor of drawing a club are

$$\frac{P(C)}{P(C')} = \frac{\frac{1}{4}}{\frac{3}{4}} = \frac{1}{3},$$

which is written "1 to 3."

27. The sum is 8 for each of the 5 outcomes 2-6, 3-5, 4-4, 5-3, and 6-2. There are 36 outcomes in all in the sample space.

$$P(\text{sum is 8}) = \frac{5}{36} \approx 0.139$$

29. $P(\text{sum is no more than 5})$

$$= P(2) + P(3) + P(4) + P(5)$$

$$= \frac{1}{36} + \frac{2}{36} + \frac{3}{36} + \frac{4}{36}$$

$$= \frac{10}{36} = \frac{5}{18}$$

$$\approx 0.278$$

31. Consider the reduced sample space of the 11 outcomes in which at least one die is a four. Of these, 2 have a sum of 7, 3-4 and 4-3. Therefore,

$P(\text{sum is 7}|\text{at least one die is a 4})$

$$= \frac{2}{11} \approx 0.182.$$

33. $P(E) = 0.51$, $P(F) = 0.37$, $P(E \cap F) = 0.22$

(a) $P(E \cup F) = P(E) + P(F) - P(E \cap F)$
$$= 0.51 + 0.37 - 0.22$$
$$= 0.66$$

(b) Draw a Venn diagram.

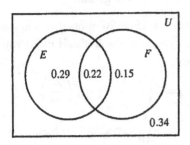

$E \cap F'$ is the portion of the diagram that is inside E and outside F.

$$P(E \cap F') = 0.29$$

(c) $E' \cup F$ is outside E or inside F, or both.

$$P(E' \cup F) = 0.22 + 0.15 + 0.34 = 0.71.$$

(d) $E' \cap F'$ is outside E and outside F.

$$P(E' \cap F') = 0.34$$

35. The probability that the ball came from box B, given that it is red, is

$P(B|\text{red})$

$$= \frac{P(B) \cdot P(\text{red}|B)}{P(B) \cdot P(\text{red}|B) + P(A) \cdot P(\text{red}|A)}$$

$$= \frac{\frac{5}{8}\left(\frac{2}{5}\right)}{\frac{5}{8}\left(\frac{2}{5}\right) + \frac{3}{8}\left(\frac{5}{6}\right)}$$

$$= \frac{4}{9} \approx 0.444.$$

37. (a) There are $n = 36$ possible outcomes. Let x represent the sum of the dice, and note that the possible values of x are the whole numbers from 2 to 12. The probability distribution is as follows.

x	2	3	4	5	6
$P(x)$	$\frac{1}{36}$	$\frac{2}{36} = \frac{1}{18}$	$\frac{3}{36} = \frac{1}{12}$	$\frac{4}{36} = \frac{1}{9}$	$\frac{5}{36}$

x	7	8	9	10	11	12
$P(x)$	$\frac{6}{36} = \frac{1}{6}$	$\frac{5}{36}$	$\frac{4}{36} = \frac{1}{9}$	$\frac{3}{36} = \frac{1}{12}$	$\frac{2}{36} = \frac{1}{18}$	$\frac{1}{36}$

(b) The histogram consists of 11 rectangles.

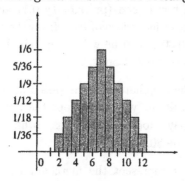

(c) The expected value is

$$E(x) = 2\left(\frac{1}{36}\right) + 3\left(\frac{2}{36}\right) + 4\left(\frac{3}{36}\right) + 5\left(\frac{4}{36}\right)$$

$$+ 6\left(\frac{5}{36}\right) + 7\left(\frac{6}{36}\right) + 8\left(\frac{5}{36}\right) + 9\left(\frac{4}{36}\right)$$

$$+ 10\left(\frac{3}{36}\right) + 11\left(\frac{2}{36}\right) + 12\left(\frac{1}{36}\right)$$

$$= \frac{252}{36} = 7.$$

39. The probability that corresponds to the shaded region of the histogram is the total of the shaded areas, that is,

$$1(0.1) + 1(0.3) + 1(0.2) = 0.6.$$

41. Let x represent the number of girls. The probability distribution is as follows.

x	0	1	2	3	4	5
$P(x)$	$\frac{1}{32}$	$\frac{5}{32}$	$\frac{10}{32}$	$\frac{10}{32}$	$\frac{5}{32}$	$\frac{1}{32}$

The expected value is

$$E(x) = 0\left(\frac{1}{32}\right) + 1\left(\frac{5}{32}\right) + 2\left(\frac{10}{32}\right) + 3\left(\frac{10}{32}\right)$$

$$+ 4\left(\frac{5}{32}\right) + 5\left(\frac{1}{32}\right)$$

$$= \frac{80}{32} = 2.5 \text{ girls.}$$

43. (a)

	N_2	T_2
N_1	$N_1 N_2$	$N_1 T_2$
T_1	$T_1 N_2$	$T_1 T_2$

Since the four combinations are equally likely, each has probability $\frac{1}{4}$.

(b) $P(\text{two trait cells}) = P(T_1 T_2) = \dfrac{1}{4}$

(c) $P(\text{one normal cell and one trait cell})$
$= P(N_1 T_2) + P(T_1 N_2)$

$$= \frac{1}{4} + \frac{1}{4}$$

$$= \frac{1}{2}$$

(d) $P(\text{not a carrier and does not have the disease})$
$= P(N_1 N_2)$

$$= \frac{1}{4}$$

45. Let $C =$ the set of viewers who watch situation comedies;

$G =$ the set of viewers who watch game shows;

and $M =$ the set of viewers who watch movies.

We are given the following information.

$n(C) = 20$
$n(G) = 19$
$n(M) = 27$
$n(C \cap G) = 5$
$n(G \cap M) = 8$
$n(C \cap M) = 10$
$n(C \cap G \cap M) = 3$
$n(C' \cap G' \cap M') = 6$

Start with $C \cap G \cap M$. $n(C \cap G \cap M) = 3$
Since $n(C \cap G) = 5$, the number in $C \cap G$, but not in M, is 2.
Since $n(G \cap M) = 8$, the number in $G \cap M$, but not in C, is 5.
Since $n(C \cap M) = 10$, the number in $C \cap M$, but not in G, is 7.
Since $n(C) = 20$, the number in C, but not in G or M, is $20 - (2 + 3 + 7) = 8$.
Since $n(G) = 19$, the number in G, but not in C or M, is $19 - (2 + 3 + 5) = 9$.
Since $n(M) = 27$, the number in M, but not in C or G, is $27 - (7 + 3 + 5) = 12$.
Since $n(C' \cap G' \cap M') = 6$, the number outside $C \cup G \cup M$ is 6.

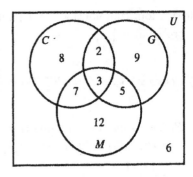

From the Venn diagram, we see that the number of viewers who

(a) were interviewed is

$$3 + 2 + 5 + 7 + 8 + 9 + 12 + 6 = 52;$$

(b) watch comedies and movies but not game shows is 7;

(c) watch only movies is 12;

(d) watch comedies and game shows but not movies is 2.

47. $P(\text{earthquake}) = \dfrac{9}{9 + 1} = \dfrac{9}{10} = 0.90$

49. (a) $P(\text{double miss}) = 0.05(0.05) = 0.0025$

(b) $P(\text{specific silo destroyed})$
$= 1 - P(\text{double miss})$
$= 1 - 0.0025$
$= 0.9975$

(c) $P(\text{all ten destroyed}) = (0.9975)^{10} \approx 0.9753$

(d) $P(\text{at least one survived})$
$= 1 - P(\text{none survived})$
$= 1 - P(\text{all ten destroyed})$
$= 1 - 0.9753$
$= 0.0247 \quad \text{or} \quad 2.47\%$

This does not agree with the quote of a 5% chance that at least one would survive.

(e) The events that each of the two bombs hit their targets are assumed to be independent. The events that each silo is destroyed are assumed to be independent.

51. If a box is good (probability 0.9) and the merchant samples an excellent piece of fruit from that box (probability 0.80), then he will accept the box and earn a $200 profit on it.

If a box is bad (probability 0.1) and he samples an excellent piece of fruit from the box (probability 0.30), then he will accept the box and earn a −$1,000 profit on it.

If the merchant ever samples a nonexcellent piece of fruit, he will not accept the box. In this case he pays nothing and earns nothing, so the profit will be $0.

Let x represent the merchant's earnings. Note that

$$0.9(0.80) = 0.72,$$
$$0.1(0.30) = 0.03,$$
$$\text{and } 1 - (0.72 + 0.03) = 0.25.$$

The probability distribution is as follows.

x	200	−1000	0
$P(x)$	0.72	0.03	0.25

The expected value when the merchant samples the fruit is

$$E(x) = 200(0.72) + (-1,000)(0.03) + 0(0.25)$$
$$= 144 - 30 + 0$$
$$= \$114.$$

We must also consider the case in which the merchant does not sample the fruit. Let x again represent the merchant's earnings. The probability distribution is as follows.

x	200	$-1,000$
$P(x)$	0.9	0.1

The expected value when the merchant does not sample the fruit is

$$E(x) = 200(0.9) + (-1,000)(0.1)$$
$$= 180 - 100$$
$$= \$80.$$

Combining these two results, the expected value of the right to sample is $\$114 - \$80 = \$34$, which corresponds to choice (c).

Chapter 12 Test

[12.1]

1. Write *true* or *false* for each statement.

   (a) $3 \in \{1, 5, 7, 9\}$ **(b)** $\{1, 3\} \not\subset \{0, 1, 2, 3, 4\}$ **(c)** $\emptyset \subset \{2\}$

   (d) A set of 6 distinct elements has exactly 64 subsets.

2. Let $U = \{1, 2, 3, 4, 5, 6, 7, 8, 9\}$, $A = \{1, 3, 4, 5\}$, $B = \{2, 4, 5\}$, and $C = \{1, 3, 5, 7\}$.
   List the members of each of the following sets, using set braces.

   (a) $A \cap B'$ **(b)** $A \cap (B \cup C')$

3. Draw a Venn diagram and shade the region that represents $A \cap (B \cup C')$.

4. Draw a Venn diagram and shade the region that represents $(A \cap B) \cup C'$.

5. Draw a Venn diagram and fill in the number of elements in each region given
   that $n(U) = 25, n(A) = 11, n(B \cap A') = 9$, and $n(A \cap B) = 6$.

6. A survey of 70 children obtained the following results:

   > 32 play soccer;
   > 29 play basketball;
   > 13 play tennis only;
   > 6 play all three sports;
   > 18 play soccer and basketball;
   > 15 play soccer and tennis;
   > 10 play basketball and tennis;
   > 14 play none of the three sports.

   Use a Venn diagram to answer the following questions.

   (a) How many children play basketball only?

   (b) How many children play tennis and basketball, but not soccer?

   (c) How many children play soccer and tennis, but not basketball?

   (d) How many children play exactly one of the three sports?

[12.2]

7. A single fair die is rolled. Find the probabilities of the following events.

   (a) Getting a 4   **(b)** Getting an even number

   (c) Getting a number less than 5   **(d)** Getting a number greater than 1

   (e) Getting any number except 4 or 5   **(f)** Getting a number less than 8

8. A card is drawn from a well-shuffled deck of 52 cards. Find the probability of drawing each of the following.

   (a) An ace

   (b) A black ace

   (c) A red card

9. Suppose that for events A and B, $P(A) = 0.4$, $P(B) = 0.3$ and $P(A \cup B) = 0.68$. Find each of the following probabilities.

   (a) $P(A \cap B)$ **(b)** $P(A')$

   (c) $P(A \cap B')$ **(d)** $P(A' \cap B')$

10. An urn contains 4 red, 3 blue, and 2 yellow marbles. A single marble is drawn.

   (a) Find the odds in favor of drawing a red marble.

   (b) Find the probability that a red or a blue marble is drawn.

[12.3]

11. For events E and F, $P(E) = 0.4$, $P(F) = 0.5$, and $P(E \cup F) = 0.8$. Find

   (a) $P(E|F)$ **(b)** $P(F|E')$.

12. Three cards are drawn without replacement from a standard deck of 52.

   (a) What is the probability that all three are spades?

   (b) What is the probability that all three are spades, given that the first card drawn is a spade?

13. The probability of passing the University of Waterloo's physical fitness test is 0.3. If you fail the first time, your chances of passing on the second try drop to 0.1. Draw a tree diagram and compute the probability that a person will pass on the first or second try.

14. If two cards are drawn without replacement from an ordinary deck, find the probability that two aces are drawn.

Decide whether each of the matrices in Exercises 15 and 16 could be a probability vector.

15. $\begin{bmatrix} \frac{2}{3} & \frac{1}{2} \end{bmatrix}$ **16.** $\begin{bmatrix} 0 & 1 \end{bmatrix}$

[12.4]

17. A nickel, a dime, and a quarter are tossed simultaneously. Let the random variable x denote the number of heads observed in the experiment, and prepare a probability distribution for this experiment.

18. Two dice are rolled, and the total number of points is recorded. Find the expected value.

19. There is a game called Double or Nothing, and it costs $1 to play. If you draw the ace of spades, you are paid $2, but you are paid nothing if you draw any other card. Is this a fair game?

20. At a large university, 62% of the students enrolled are female. A sample of 3 students are selected and the number of female students is noted.

   (a) Give the probability distribution for the number of females.

   (b) Sketch its histogram.

   (c) Find the expected value.

Chapter 12 Test Answers

1. (a) False (b) False (c) True (d) True

2. (a) $\{1,3\}$ (b) $\{4,5\}$

3.
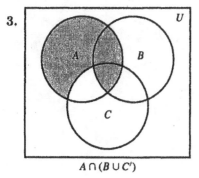
$A \cap (B \cup C')$

4.
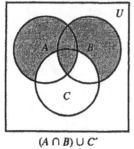
$(A \cap B) \cup C'$

5.
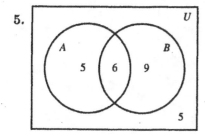

6. (a) 7 (b) 4 (c) 9 (d) 25

7. (a) $\frac{1}{6}$ (b) $\frac{1}{2}$ (c) $\frac{2}{3}$ (d) $\frac{5}{6}$ (e) $\frac{2}{3}$ (f) 1

8. (a) $\frac{1}{13}$ (b) $\frac{1}{26}$ (c) $\frac{1}{2}$

9. (a) 0.02 (b) 0.6 (c) 0.38 (d) 0.32

10. (a) 4 to 5 (b) $\frac{7}{9}$

11. (a) $\frac{1}{5}$ (b) $\frac{2}{3}$

12. (a) $\left(\frac{13}{52}\right)\left(\frac{12}{51}\right)\left(\frac{11}{50}\right) \approx 0.013$
    (b) $\left(\frac{12}{51}\right)\left(\frac{11}{50}\right) \approx 0.052$

13.
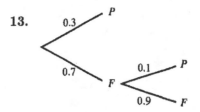

$P(\text{pass}) = 0.3 + 0.07 = 0.37$

14. 0.0045   15. No   16. Yes

17.

x	0	1	2	3
$P(x)$	$\frac{1}{8}$	$\frac{3}{8}$	$\frac{3}{8}$	$\frac{1}{8}$

18. 7

19. No, it is not a fair game.

20. **(a)**

Number of Females	0	1	2	3
Probability	0.0549	0.2686	0.4382	0.2383

(b) 0

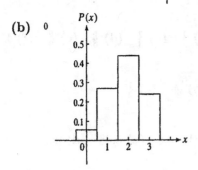

(c) 1.86

PROBABILITY AND CALCULUS

13.1 Continuous Probability Models

1. $f(x) = \frac{1}{9}x - \frac{1}{18}$; $[2, 5]$

Show that condition 1 holds.

Since $2 \le x \le 5$,

$$\frac{2}{9} \le \frac{1}{9}x \le \frac{5}{9}$$
$$\frac{1}{6} \le \frac{1}{9}x - \frac{1}{18} \le \frac{1}{2}.$$

Hence, $f(x) \ge 0$ on $[2, 5]$.

Show that condition 2 holds.

$$\int_2^5 \left(\frac{1}{9}x - \frac{1}{18}\right) dx = \frac{1}{9}\int_2^5 \left(x - \frac{1}{2}\right) dx$$
$$= \frac{1}{9}\left(\frac{x^2}{2} - \frac{1}{2}x\right)\Big|_2^5$$
$$= \frac{1}{9}\left(\frac{25}{2} - \frac{5}{2} - \frac{4}{2} + 1\right)$$
$$= \frac{1}{9}(8 + 1)$$
$$= 1$$

Yes, $f(x)$ is a probability density function.

3. $f(x) = \frac{1}{21}x^2$; $[1, 4]$

Since $x^2 \ge 0$, $f(x) \ge 0$ on $[1, 4]$.

$$\frac{1}{21}\int_1^4 x^2\, dx = \frac{1}{21}\left(\frac{x^3}{3}\right)\Big|_1^4$$
$$= \frac{1}{21}\left(\frac{64}{3} - \frac{1}{3}\right) = 1$$

Yes, $f(x)$ is a probability density function.

5. $f(x) = 4x^3$; $[0, 3]$

$$4\int_0^3 x^3\, dx = 4\left(\frac{x^4}{4}\right)\Big|_0^3$$
$$= 4\left(\frac{81}{4} - 0\right)$$
$$= 81 \ne 1$$

No, $f(x)$ is not a probability density function.

7. $f(x) = \frac{x^2}{16}$; $[-2, 2]$

$$\frac{1}{16}\int_{-2}^2 x^2\, dx = \frac{1}{16}\left(\frac{x^3}{3}\right)\Big|_{-2}^2$$
$$= \frac{1}{16}\left(\frac{8}{3} + \frac{8}{3}\right)$$
$$= \frac{1}{3} \ne 1$$

No, $f(x)$ is not a probability density function.

9. $f(x) = \frac{5}{3}x^2 - \frac{5}{90}$; $[-1, 1]$

Let $x = 0$. Then $f(x) = f(0) = -\frac{5}{90} < 0$.

So $f(x) < 0$ for at least one x-value in $[-1, 1]$.
No, $f(x)$ is not a probability density function.

11. $f(x) = kx^{1/2}$; $[1, 4]$

$$\int_1^4 kx^{1/2}\, dx = \frac{2}{3}kx^{3/2}\Big|_1^4$$
$$= \frac{2}{3}k(8 - 1)$$
$$= \frac{14}{3}k$$

If $\frac{14}{3}k = 1$,

$$k = \frac{3}{14}.$$

Notice that $f(x) = \frac{3}{4}x^{1/2} \ge 0$ for all x in $[1, 4]$.

13. $f(x) = kx^2$; $[0, 5]$

$$\int_0^5 kx^2\, dx = k\frac{x^3}{3}\Big|_0^5$$
$$= k\left(\frac{125}{3} - 0\right)$$
$$= k\left(\frac{125}{3}\right)$$

If $k\left(\frac{125}{3}\right) = 1$,

$$k = \frac{3}{125}.$$

Notice that $f(x) = \frac{3}{125}x^2 \ge 0$ for all x in $[0, 5]$.

15. $f(x) = kx$; $[0, 3]$

$$\int_0^3 kx\,dx = k\frac{x^2}{2}\Big|_0^3$$

$$= k\left(\frac{9}{2} - 0\right)$$

$$= \frac{9}{2}k$$

If $\frac{9}{2}k = 1$,

$$k = \frac{2}{9}.$$

Notice that $f(x) = \frac{2}{9}x \geq 0$ for all x in $[0, 3]$.

17. $f(x) = kx$; $[1, 5]$

$$\int_1^5 kx\,dx = k\frac{x^2}{2}\Big|_1^5$$

$$= k\left(\frac{25}{2} - \frac{1}{2}\right)$$

$$= 12k$$

If $12k = 1$,

$$k = \frac{1}{12}.$$

Notice that $f(x) = \frac{1}{12}x \geq 0$ for all x in $[1, 5]$.

19. The total area under the graph of a probability density function always equals 1.

23. $f(x) = \frac{1}{2}(1+x)^{-3/2}$; $[0, \infty)$

$$\frac{1}{2}\int_0^\infty (1+x)^{-3/2}\,dx$$

$$= \lim_{a \to \infty} \frac{1}{2}\int_0^a (1+x)^{-3/2}\,dx$$

$$= \lim_{a \to \infty} \frac{1}{2}(1+x)^{-1/2}\left(\frac{-2}{1}\right)\Big|_0^a$$

$$= \lim_{a \to \infty} [-(1+a)^{-1/2} + 1]$$

$$= \lim_{a \to \infty} \left(\frac{-1}{\sqrt{1+a}} + 1\right)$$

$$= 0 + 1 = 1$$

Since $x \geq 0$, $f(x) \geq 0$.
$f(x)$ is a probability density function.

(a) $P(0 \leq x \leq 2)$

$$= \frac{1}{2}\int_0^2 (1+x)^{-3/2}\,dx$$

$$= -(1+x)^{-1/2}\Big|_0^2$$

$$= -3^{-1/2} + 1$$

$$\approx 0.4226$$

(b) $P(1 \leq x \leq 3)$

$$= \frac{1}{2}\int_1^3 (1+x)^{-3/2}\,dx$$

$$= -(1+x)^{-1/2}\Big|_1^3$$

$$= -4^{-1/2} + 2^{-1/2}$$

$$\approx 0.2071$$

(c) $P(x \geq 5)$

$$= \frac{1}{2}\int_5^\infty (1+x)^{-3/2}\,dx$$

$$= \lim_{a \to \infty} \frac{1}{2}\int_5^a (1+x)^{-3/2}\,dx$$

$$= \lim_{a \to \infty} [-(1+x)^{-1/2}]\Big|_5^a$$

$$= \lim_{a \to \infty} [-(1+a)^{-1/2} + 6^{-1/2}]$$

$$= \lim_{a \to \infty} \left(\frac{-1}{\sqrt{1+a}} + 6^{-1/2}\right)$$

$$\approx 0 + 0.4082$$

$$= 0.4082$$

25. $f(x) = \frac{1}{2}e^{-x/2}$; $[0, \infty)$

$$\frac{1}{2}\int_0^\infty e^{-x/2}\,dx$$

$$= \lim_{a \to \infty} \frac{1}{2}\int_0^a e^{-x/2}\,dx$$

$$= \lim_{a \to \infty} \frac{1}{2}\left(\frac{-2}{1}e^{-x/2}\right)\Big|_0^a$$

$$= \lim_{a \to \infty} -e^{-x/2}\Big|_0^a$$

$$= \lim_{a \to \infty} \left(\frac{-1}{e^{a/2}} + 1\right)$$

$$= 0 + 1$$

$$= 1$$

$f(x) > 0$ for all x.
$f(x)$ is a probability density function.

(a) $P(0 \le x \le 1) = \dfrac{1}{2} \displaystyle\int_0^1 e^{-x/2}\, dx$

$$= -e^{-x/2}\Big|_0^1$$

$$= \dfrac{-1}{e^{1/2}} + 1$$

$$\approx 0.3935$$

(b) $P(1 \le x \le 3) = \dfrac{1}{2} \displaystyle\int_1^3 e^{-x/2}\, dx$

$$= -e^{-x/2}\Big|_1^3$$

$$= \dfrac{-1}{e^{3/2}} + \dfrac{1}{e^{1/2}}$$

$$\approx 0.3834$$

(c) $P(x \ge 2) = \dfrac{1}{2} \displaystyle\int_2^\infty e^{-x/2}\, dx$

$$= \lim_{a \to \infty} \dfrac{1}{2} \int_2^a e^{-x/2}\, dx$$

$$= \lim_{a \to \infty} \left(-e^{-x/2}\right)\Big|_2^a$$

$$= \lim_{a \to \infty} \left(\dfrac{-1}{e^{a/2}} + \dfrac{1}{e}\right)$$

$$\approx 0.3679$$

27. $f(x) = \dfrac{1}{2\sqrt{x}};\ [1,\ 4]$

(a) $P(3 \le x \le 4) = \displaystyle\int_3^4 \left(\dfrac{1}{2\sqrt{x}}\right) dx$

$$= \dfrac{1}{2} \int_3^4 x^{-1/2}\, dx$$

$$= \dfrac{1}{2}(2)x^{1/2}\Big|_3^4$$

$$= 2 - 3^{1/2} \approx 0.2679$$

(b) $P(1 \le x \le 2) = \displaystyle\int_1^2 \left(\dfrac{1}{2\sqrt{x}}\right) dx$

$$= \dfrac{1}{2}(2)x^{1/2}\Big|_1^2$$

$$= 2^{1/2} - 1 \approx 0.4142$$

(c) $P(2 \le x \le 3) = \displaystyle\int_2^3 \left(\dfrac{1}{2\sqrt{x}}\right) dx$

$$= \dfrac{1}{2}(2)x^{1/2}\Big|_2^3$$

$$= 3^{1/2} - 2^{1/2} \approx 0.3178$$

29. (a) $P(0 \le x \le 150)$

$$= \int_0^{150} 1.185 \cdot 10^{-9} x^{4.5222} e^{-0.049846x}\, dx$$

$$\approx 0.8131$$

(b) $P(100 \le x \le 200)$

$$= \int_{100}^{200} 1.185 \cdot 10^{-9} x^{4.5222} e^{-0.049846x}\, dx$$

$$\approx 0.4901$$

31. (a)

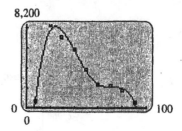

Of the types of functions available using the regression feature of a graphing utility, a polynomial function best matches the data.

(b) The function

$$N(x) = -0.00351465x^4 + 0.792884x^3 - 61.7955x^2$$
$$+ 1{,}814.54x - 9{,}709.20$$

provided by the calculator models this data well, as illustrated by the following graph.

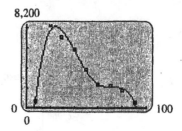

(c) Using the integration feature on our calculator, we find that

$$\int_{6.8}^{91.5} N(x) \approx 343{,}795.$$

So,

$$S(x) = \dfrac{1}{343{,}795} N(x)$$

$$= \dfrac{1}{343{,}795}(-0.00351465x^4 + 0.792884x^3$$

$$- 61.7955x^2 + 1{,}814.54x - 9{,}709.20)$$

will be a probability density function for $[6.8, 91.5]$, because

$$\int_{6.8}^{91.5} S(x)dx = 1, \text{ and } S(x) \geq 0$$

for all x in $[6.8, 91.5]$.

(d) Again using the integration feature on our calculator,

$$P(x < 25) = \int_{6.8}^{25} S(x)dx \approx 0.300$$

$$P(45 < x < 65) = \int_{45}^{65} S(x)dx \approx 0.179$$

$$P(x > 75) = \int_{75}^{91.5} S(x)dx \approx 0.072$$

From the table the actual probabilities are

$$P(x < 25) = \frac{630 + 8,173}{32,409} \approx 0.272$$

$$P(45 < x < 65) = \frac{3,872 + 2,390}{32,409} \approx 0.193$$

$$P(x > 75) = \frac{1,740 + 555}{32,409} \approx 0.071$$

The probabilities are closer where the curve better fits the data.

33. $f(t) = \dfrac{1}{960}e^{-t/960}$

(a) $P(t < 365) = \displaystyle\int_{0}^{365} \frac{1}{960}e^{-t/960}dt$

$$= -e^{-t/960}\Big|_{0}^{365}$$

$$= -e^{-365/960} + 1$$

$$\approx 0.32$$

(b) $P(t > 960) = \displaystyle\int_{960}^{\infty} \frac{1}{960}e^{-t/960}dt$

$$= \lim_{b \to \infty} \int_{960}^{b} e^{-t/960}dt$$

$$= \lim_{b \to \infty} (-e^{-t/960})\Big|_{960}^{b}$$

$$= \lim_{b \to \infty} (-e^{-b/960} + e^{-1})$$

$$= 0 + e^{-1}$$

$$\approx 0.37$$

35. $f(x) = \dfrac{8}{7(x-2)^2}$; $[3, 10]$

(a) $P(3 \leq x \leq 4) = \dfrac{8}{7}\displaystyle\int_{3}^{4} (x-2)^{-2}\, dx$

$$= -\frac{8}{7}(x-2)^{-1}\Big|_{3}^{4}$$

$$= -\frac{8}{7}\left(\frac{1}{2} - 1\right)$$

$$= -\frac{8}{7}\left(-\frac{1}{2}\right)$$

$$= \frac{4}{7}$$

$$\approx 0.5714$$

(b) $P(5 \leq x \leq 10) = \dfrac{8}{7}\displaystyle\int_{5}^{10} (x-2)^{-2}\, dx$

$$= -\frac{8}{7}(x-2)^{-1}\Big|_{5}^{10}$$

$$= \frac{8}{7}\left(\frac{1}{8} - \frac{1}{3}\right)$$

$$= -\frac{8}{7}\left(-\frac{5}{24}\right)$$

$$= \frac{5}{21}$$

$$\approx 0.2381$$

37. $f(x) = 3x^{-4}$; $[1, \infty)$

(a) $P(1 \leq x \leq 2) = \displaystyle\int_{1}^{2} 3x^{-4}\, dx$

$$= -x^{-3}\Big|_{1}^{2}$$

$$= 1 - \frac{1}{8}$$

$$= \frac{7}{8}$$

$$= 0.875$$

(b) $P(3 \leq x \leq 5) = \displaystyle\int_{3}^{5} 3x^{-4}\, dx$

$$= -x^{-3}\Big|_{3}^{5}$$

$$= \frac{1}{27} - \frac{1}{125}$$

$$\approx 0.029$$

(c) $P(x \geq 3) = \displaystyle\int_3^\infty 3x^{-4}\, dx$

$= \displaystyle\lim_{b \to \infty} \int_3^b 3x^{-4}\, dx$

$= \displaystyle\lim_{b \to \infty} (-x^{-3})\Big|_3^b$

$= \displaystyle\lim_{b \to \infty} \left(\frac{1}{27} - \frac{1}{b^3}\right)$

$= \dfrac{1}{27}$

≈ 0.037

39. $f(x) = \dfrac{1}{11}\left(1 + \dfrac{3}{\sqrt{x}}\right);\ [4, 9]$

(a) $P(6 \leq x \leq 9) = \displaystyle\int_6^9 \frac{1}{11}(1 + 3x^{-1/2})dx$

$= \dfrac{1}{11}(x + 6x^{1/2})\Big|_6^9$

$= \dfrac{1}{11}(9 + 18 - 6 - 6\sqrt{6})$

$= \dfrac{1}{11}(21 - 6\sqrt{6})$

≈ 0.5730

(b) $P(4 \leq x \leq 5) = \displaystyle\int_4^5 \frac{1}{11}(1 + 3x^{-1/2})dx$

$= \dfrac{1}{11}(x + 6x^{1/2})\Big|_4^5$

$= \dfrac{1}{11}(5 + 6\sqrt{5} - 4 - 12)$

$= \dfrac{1}{11}(6\sqrt{5} - 11)$

≈ 0.2197

(c) $P(4 \leq x \leq 7) = \displaystyle\int_4^7 \frac{1}{11}(1 + 3x^{-1/2})dx$

$= \dfrac{1}{11}(x + 6x^{1/2})\Big|_4^7$

$= \dfrac{1}{11}(7 + 6\sqrt{7} - 4 - 12)$

$= \dfrac{1}{11}(6\sqrt{7} - 9)$

≈ 0.6250

13.2   Expected Value and Variance of Continuous Random Variables

1. $f(x) = \dfrac{1}{4};\ [3, 7]$

$E(x) = \mu = \displaystyle\int_3^7 \frac{1}{4}x\, dx$

$= \dfrac{1}{4}\left(\frac{x^2}{2}\right)\Big|_3^7$

$= \dfrac{49}{8} - \dfrac{9}{8}$

$= 5$

$\mathrm{Var}(x) = \displaystyle\int_3^7 (x - 5)^2 \left(\frac{1}{4}\right) dx$

$= \dfrac{1}{4} \cdot \dfrac{(x - 5)^3}{3}\Big|_3^7$

$= \dfrac{8}{12} + \dfrac{8}{12}$

$= \dfrac{4}{3} \approx 1.33$

$\sigma \approx \sqrt{\mathrm{Var}(x)} = \sqrt{\dfrac{4}{3}}$

≈ 1.15

3. $f(x) = \dfrac{x}{8} - \dfrac{1}{4};\ [2, 6]$

$\mu = \displaystyle\int_2^6 x\left(\frac{x}{8} - \frac{1}{4}\right) dx$

$= \displaystyle\int_2^6 \left(\frac{x^2}{8} - \frac{x}{4}\right) dx$

$= \left(\dfrac{x^3}{24} - \dfrac{x^2}{8}\right)\Big|_2^6$

$= \left(\dfrac{216}{24} - \dfrac{36}{8}\right) - \left(\dfrac{8}{24} - \dfrac{4}{8}\right)$

$= \dfrac{208}{24} - 4$

$= \dfrac{26}{3} - 4$

$= \dfrac{14}{3}$

≈ 4.67

Use the alternative formula to find

$$\text{Var}(x) = \int_2^6 x^2 \left(\frac{x}{8} - \frac{1}{4}\right) dx - \left(\frac{14}{3}\right)^2$$

$$= \int_2^6 \left(\frac{x^3}{8} - \frac{x^2}{4}\right) dx - \frac{196}{9}$$

$$= \left(\frac{x^4}{32} - \frac{x^3}{12}\right)\Big|_2^6 - \frac{196}{9}$$

$$= \left(\frac{1,296}{32} - \frac{216}{12}\right) - \left(\frac{16}{32} - \frac{8}{12}\right) - \frac{196}{9}$$

$$\approx 0.89.$$

$$\sigma = \sqrt{\text{Var}(x)}$$

$$\approx \sqrt{0.89}$$

$$\approx 0.94$$

5. $f(x) = 1 - \dfrac{1}{\sqrt{x}}; \ [1, \ 4]$

$$\mu = \int_1^4 x(1 - x^{-1/2})dx$$

$$= \int_1^4 (x - x^{1/2})dx$$

$$= \left(\frac{x^2}{2} - \frac{2x^{3/2}}{3}\right)\Big|_1^4$$

$$= \frac{16}{2} - \frac{16}{3} - \frac{1}{2} + \frac{2}{3}$$

$$= \frac{17}{6}$$

$$\approx 2.83$$

$$\text{Var}(x) = \int_1^4 x^2(1 - x^{-1/2})dx - \left(\frac{17}{6}\right)^2$$

$$= \int_1^4 (x^2 - x^{3/2})dx - \frac{289}{36}$$

$$= \left(\frac{x^3}{3} - \frac{2x^{5/2}}{5}\right)\Big|_1^4 - \frac{289}{36}$$

$$= \frac{64}{3} - \frac{64}{5} - \frac{1}{3} + \frac{2}{5} - \frac{289}{36}$$

$$\approx 0.57$$

$$\sigma \approx \sqrt{\text{Var}(x)} \approx 0.76$$

7. $f(x) = 4x^{-5}; \ [1, \ \infty)$

$$\mu = \int_1^\infty x(4x^{-5})dx$$

$$= \lim_{a \to \infty} \int_1^a 4x^{-4}\, dx$$

$$= \lim_{a \to \infty} \left(\frac{4x^{-3}}{-3}\right)\Big|_1^a$$

$$= \lim_{a \to \infty} \left(\frac{-4}{3a^3} + \frac{4}{3}\right)$$

$$= \frac{4}{3} \approx 1.33$$

$$\text{Var}(x) = \int_1^\infty x^2(4x^{-5})dx - \left(\frac{4}{3}\right)^2$$

$$= \lim_{a \to \infty} \int_1^a 4x^{-3}\, dx - \frac{16}{9}$$

$$= \lim_{a \to \infty} \left(\frac{4x^{-2}}{-2}\right)\Big|_1^a - \frac{16}{9}$$

$$= \lim_{a \to \infty} \left(\frac{-2}{a^2} + 2\right) - \frac{16}{9}$$

$$= 2 - \frac{16}{9} = \frac{2}{9} \approx 0.22$$

$$\sigma = \sqrt{\text{Var}(x)} = \sqrt{\frac{2}{9}} \approx 0.47$$

11. $f(x) = \dfrac{\sqrt{x}}{18}; \ [0, \ 9]$

(a) $\ E(x) = \mu = \displaystyle\int_0^9 \frac{x\sqrt{x}}{18}\, dx$

$$= \int_0^9 \frac{x^{3/2}}{18}\, dx$$

$$= \frac{2x^{5/2}}{90}\Big|_0^9 = \frac{x^{5/2}}{45}\Big|_0^9$$

$$= \frac{243}{45} = \frac{27}{5} = 5.40$$

(b) $\ \text{Var}(x) = \displaystyle\int_0^9 \frac{x^2\sqrt{x}}{18}\, dx - \left(\frac{27}{5}\right)^2$

$$= \int_0^9 \frac{x^{5/2}}{18}\, dx - \left(\frac{27}{5}\right)^2$$

$$= \frac{x^{7/2}}{63}\Big|_0^9 - \left(\frac{27}{5}\right)^2$$

$$= \frac{2187}{63} - \left(\frac{27}{5}\right)^2 \approx 5.55$$

(c) $\sigma = \sqrt{\text{Var}(x)} \approx 2.36$

(d) $P(5.40 < x \le 9)$

$$= \int_{5.4}^{9} \frac{x^{1/2}}{18}\, dx$$

$$= \frac{x^{3/2}}{27}\bigg|_{5.4}^{9}$$

$$= \frac{27}{27} - \frac{(5.4)^{1.5}}{27} \approx 0.54$$

(e) $P(5.40 - 2.36 \le x \le 5.40 + 2.36)$

$$= \int_{3.04}^{7.76} \frac{x^{1/2}}{18}\, dx$$

$$= \frac{x^{3/2}}{27}\bigg|_{3.04}^{7.76}$$

$$= \frac{7.76^{3/2}}{27} - \frac{3.04^{3/2}}{27}$$

$$\approx 0.60$$

13. $f(x) = \frac{1}{2}x;\ [0,\ 2]$

(a) $E(x) = \mu = \int_{0}^{2} \frac{1}{2}x^2\, dx = \frac{x^3}{6}\bigg|_{0}^{2}$

$$= \frac{8}{6} = \frac{4}{3} \approx 1.33$$

(b) $\text{Var}(x) = \int_{0}^{1} \frac{1}{2}x^3\, dx - \frac{16}{9}$

$$= \frac{x^4}{8}\bigg|_{0}^{2} - \frac{16}{9}$$

$$= 2 - \frac{16}{9} = \frac{2}{9} \approx 0.22$$

(c) $\sigma = \sqrt{\text{Var}(x)} = \sqrt{\frac{2}{9}} \approx 0.47$

(d) $P\left(\frac{4}{3} < x \le 2\right) = \int_{4/3}^{2} \frac{x}{2}\, dx$

$$= \frac{x^2}{4}\bigg|_{4/3}^{2}$$

$$= 1 - \frac{16}{36} \approx 0.56$$

(e) $P\left(\frac{4}{3} - 0.47 \le x \le \frac{4}{3} + 0.47\right)$

$$= \int_{0.86}^{1.8} \frac{x}{2}\, dx = \frac{x^2}{4}\bigg|_{0.86}^{1.8}$$

$$= \frac{1.8^2}{4} - \frac{0.86^2}{4} \approx 0.63$$

15. $f(x) = \frac{1}{4};\ [3,\ 7]$

(a) $m = \text{median}:\ \int_{3}^{m} \frac{1}{4}\, dx = \frac{1}{2}$

$$\frac{1}{4}x\bigg|_{3}^{m} = \frac{1}{2}$$

$$\frac{m}{4} - \frac{3}{4} = \frac{1}{2}$$

$$m - 3 = 2$$

$$m = 5$$

(b) $E(x) = \mu = 5$ (from Exercise 1)

$$P(x = 5) = \int_{5}^{5} \frac{1}{4}\, dx = 0$$

17. $f(x) = \frac{x}{8} - \frac{1}{4};\ [2,\ 6]$

(a) $m = \text{median}:$

$$\int_{2}^{m} \left(\frac{x}{8} - \frac{1}{4}\right) dx = \frac{1}{2}$$

$$\left(\frac{x^2}{16} - \frac{x}{4}\right)\bigg|_{2}^{m} = \frac{1}{2}$$

$$\frac{m^2}{16} - \frac{m}{4} - \frac{1}{4} + \frac{1}{2} = \frac{1}{2}$$

$$m^2 - 4m - 4 + 8 = 8$$

$$m^2 - 4m - 4 = 0$$

$$m = \frac{4 \pm \sqrt{16 + 16(1)}}{2}$$

Reject $\frac{4 - \sqrt{32}}{2}$ since it is not in $[2,\ 6]$.

$$m = \frac{4 + \sqrt{32}}{2} \approx 4.83$$

(b) $E(x) = \mu = \frac{14}{3}$ (from Exercise 3)

$$P\left(\frac{14}{3} \le x \le 4.83\right)$$

$$= \int_{4.67}^{4.83} \left(\frac{x}{8} - \frac{1}{4}\right) dx$$

$$= \left(\frac{x^2}{16} - \frac{x}{4}\right)\bigg|_{4.67}^{4.83}$$

$$= \frac{4.83^2}{16} - \frac{4.83}{4} - \frac{4.67^2}{16} + \frac{4.67}{4}$$

$$\approx 0.055$$

19. $f(x) = 4x^{-5}$; $[1, \infty)$

(a) m = median:

$$\int_1^m 4x^{-5}\, dx = \frac{1}{2}$$

$$\frac{4x^{-4}}{-4}\Big|_1^m = \frac{1}{2}$$

$$-m^{-4} + 1 = \frac{1}{2}$$

$$1 - \frac{1}{m^4} = \frac{1}{2}$$

$$2m^4 - 2 = m^4$$

$$m^4 = 2$$

$$m = \sqrt[4]{2} \approx 1.19$$

(b) $E(x) = \mu = \dfrac{4}{3}$ (From Exercise 7)

$$P(1.19 \le x \le \tfrac{4}{3}) \approx \int_{1.19}^{1.33} 4x^{-5}\, dx$$

$$\approx -x^{-4}\Big|_{1.19}^{1.33}$$

$$\approx -\frac{1}{1.33^4} + \frac{1}{1.19^4}$$

$$\approx 0.18$$

21. $f(x) = \dfrac{1}{(\ln 20)x}$; $[1, 20]$

(a) $\mu = \displaystyle\int_1^{20} x \cdot \frac{1}{(\ln 20)x}\, dx$

$$= \int_1^{20} \frac{1}{\ln 20}\, dx$$

$$= \frac{x}{\ln 20}\Big|_1^{20}$$

$$= \frac{19}{\ln 20} \approx 6.34 \text{ seconds}$$

(b) $\text{Var}(x) = \displaystyle\int_1^{20} x^2 \cdot \frac{1}{(\ln 20)x}\, dx - \mu^2$

$$= \int_1^{20} \frac{x}{\ln 20}\, dx - \mu^2$$

$$= \frac{x^2}{2\ln 20}\Big|_1^{20} - (6.34)^2$$

$$= \frac{399}{2\ln 20} - (6.34)^2$$

$$\approx 26.40$$

$$\sigma \approx \sqrt{26.40}$$

$$\approx 5.14 \text{ sec}$$

(c) $P(6.34 - 5.14 < x < 6.34 + 5.14)$
$$= P(1.2 < x < 11.48)$$

$$= \int_{1.2}^{11.48} \frac{1}{(\ln 20)x}\, dx$$

$$= \frac{\ln x}{\ln 20}\Big|_{1.2}^{11.48}$$

$$= \frac{1}{\ln 20}(\ln 11.48 - \ln 1.2)$$

$$\approx 0.75$$

23. $f(x) = \dfrac{1}{2\sqrt{x}}$; $[1, 4]$

(a) $\mu = \displaystyle\int_1^4 x \cdot \frac{1}{2\sqrt{x}}\, dx$

$$= \int_1^4 \frac{x^{1/2}}{2}\, dx$$

$$= \frac{x^{3/2}}{3}\Big|_1^4$$

$$= \frac{1}{3}(8 - 1)$$

$$= \frac{7}{3} \approx 2.33 \text{ cm}$$

(b) $\text{Var}(x) = \displaystyle\int_1^4 x^2 \cdot \frac{1}{2\sqrt{x}}\, dx - \left(\frac{7}{3}\right)^2$

$$= \int_1^4 \frac{x^{3/2}}{2}\, dx - \frac{49}{9}$$

$$= \frac{x^{5/2}}{5}\Big|_1^4 - \frac{49}{9}$$

$$= \frac{1}{5}(32 - 1) - \frac{49}{9}$$

$$\approx 0.76$$

$$\sigma = \sqrt{\text{Var}(x)}$$

$$\approx 0.87 \text{ cm}$$

(c) $P(x > 2.33 + 2(0.87))$
$$= P(x > 4.07)$$
$$= 0$$

The probability is 0 since two standard deviations falls out of the given interval $[1, 4]$.

25. $f(x) = \dfrac{105}{4x^2}$ for $[15, 35]$

(a) $\mu = \displaystyle\int_{15}^{35} x \cdot \dfrac{105}{4x^2}\, dx$

$= \dfrac{105}{4} \displaystyle\int_{15}^{35} \dfrac{1}{x}\, dx$

$= \dfrac{105}{4} \ln x \Big|_{15}^{35}$

$= \dfrac{105}{4}(\ln 35 - \ln 15)$

$= \dfrac{105}{4} \ln\left(\dfrac{35}{15}\right)$

$= \dfrac{105}{4} \ln\left(\dfrac{7}{3}\right)$

≈ 22.2

(b) $\text{Var}(x) = \displaystyle\int_{15}^{35} x^2 \dfrac{105}{4x^2}\, dx - \left[\dfrac{105}{4} \ln\left(\dfrac{7}{3}\right)\right]^2$

$= \dfrac{105}{4} \displaystyle\int_{15}^{35} dx - \left[\dfrac{105}{4} \ln\left(\dfrac{7}{3}\right)\right]^2$

$= \dfrac{105}{4} x \Big|_{15}^{35} - \left[\dfrac{105}{4} \ln\left(\dfrac{7}{3}\right)\right]^2$

$= \dfrac{105}{4}(35 - 15) - \left[\dfrac{105}{4} \ln\left(\dfrac{7}{3}\right)\right]^2$

$= \dfrac{105}{4}(20) - \left[\dfrac{105}{4} \ln\left(\dfrac{7}{3}\right)\right]^2$

$= 525 - \left[\dfrac{105}{4} \ln\left(\dfrac{7}{3}\right)\right]^2$

$\sigma = \sqrt{525 - \left[\dfrac{105}{4} \ln\left(\dfrac{7}{3}\right)\right]^2}$

≈ 5.51

(c) One standard deviation below the mean is $22.24 - 5.51 = 16.73$.

$P(x \le 16.73) = \displaystyle\int_{15}^{16.73} \dfrac{105}{4x^2\, dx}$

$= \dfrac{105}{4}\left(-\dfrac{1}{x}\right)\Big|_{15}^{16.73}$

$= \dfrac{105}{4}\left(-\dfrac{1}{16.73} + \dfrac{1}{15}\right)$

≈ 0.18

27. $p(x, t) = \dfrac{e^{-x^2/(4Dt)}}{\int_0^L e^{-u^2/(4Dt)}\, du}$

Letting $t = 12$, $L = 6$, and $D = 38.3$,

$p(x, 12) = \dfrac{e^{-x^2(4\cdot38.3\cdot12)}}{\int_0^6 e^{-u^2/(4\cdot38.3\cdot12)}\, du}$

$= \dfrac{e^{-x^2/1{,}838.4}}{\int_0^6 e^{-u^2/(1{,}838.4)}\, du}$

$\approx \dfrac{1}{5.9611} e^{-x^2/1{,}838.4}.$

The integral in the denominator was evaluated using the integration feature on the calculator.

$E(x) = \displaystyle\int_0^L x \cdot p(x, 12)\, dx$

$= \dfrac{1}{5.9611} \displaystyle\int_0^6 x e^{-x^2/1{,}838.4}\, dx$

$= \dfrac{1}{5.9611} \cdot \left(-\dfrac{1}{2}\right) \cdot 1{,}838.4 e^{-x^2} \Big|_0^6$

≈ 2.99

The expected recapture distance is 2.99 m.

29. $f(t) = \dfrac{1}{960} e^{-t/960}$; $[0, \infty)$

$E(t) = \displaystyle\int_0^\infty \dfrac{t}{960} e^{-t/960}\, dt$

$= \lim_{b \to \infty} \displaystyle\int_0^b \dfrac{t}{960} e^{-t/960}\, dt$

Integrating by parts, choose

$$dv = e^{-t/960}\, dt \text{ and } u = \dfrac{t}{960}.$$

Then

$$v = -960 e^{-t/960} \text{ and } du = \dfrac{1}{960}\, dt.$$

So,

$$\int u\, dv = uv - \int v\, du$$

$\displaystyle\int_0^b \dfrac{t}{960} e^{-t/960}\, dt = -t e^{-t/960} \Big|_0^b - \displaystyle\int_0^b -e^{-t/960}\, dt$

$= -t e^{-t/960} \Big|_0^b - 960 e^{-t/960} \Big|_0^b$

$= -(t + 960) e^{-t/960} \Big|_0^b$

$= -(b + 960) e^{-b/960} + 960$

Thus,

$$E(t) = \lim_{b \to \infty} \int_0^b \frac{t}{960} e^{-t/960} dt$$

$$= \lim_{b \to \infty} [-(b + 960)e^{-b/960} + 960]$$

$$= 960 \text{ days}$$

$$\text{Var}(t) = \int_0^\infty \frac{t^2}{960} e^{-t/960} dt - 960^2$$

$$= \frac{1}{960} \left[\lim_{b \to \infty} \int_0^b t^2 e^{-t/960} dt \right] - 960^2$$

Use the column method for integration by parts.
Let $u = t^2$ and $dv = e^{-t/960} dt$.

D		I
t^2	$+$	$e^{-t/960}$
$2t$	$-$	$-960e^{-t/960}$
2	$+$	$960^2 e^{-t/960}$
0		$-960^3 e^{-t/960}$

$$\int_0^b t^2 e^{-t/960} dt$$

$$= -960t^2 e^{-t/960} - 2t(960^2 e^{-t/960})$$
$$+ 2(-960^3 e^{-t/960}) \Big|_0^b$$

$$= -960e^{-t/960}(t^2 + 2(960)t + 2(960^2)) \Big|_0^b$$

$$= -960e^{-b/960}(b^2 + 2(960)b + 2(960^2)) + 2(960^3)$$

So,

$$\text{Var}(t) = \frac{1}{960} \left[\lim_{b \to \infty} \int_0^b t^2 e^{-t/960} dt \right] - 960^2$$

$$= \frac{1}{960} \{ \lim_{b \to \infty} [-960e^{-b/960}(b^2 + 2(960)b$$
$$+ 2(960^2)) + 2(960^3)] \} - 960^2$$

$$= \frac{1}{960} \cdot 2(960^3) - 960^2$$

$$= 960^2$$

$$\sigma = \sqrt{\text{Var}(t)} = \sqrt{960^2} = 960 \text{ days}$$

31. $f(x) = \frac{1}{11} \left(1 + \frac{3}{\sqrt{x}} \right)$; $[4, 9]$

    (a) From Exercise 6, $\mu \approx 6.41$ yr.

    (b) $\sigma \approx 1.45$ yr

    (c) $P(x > 6.41)$

$$= \int_{6.41}^9 \frac{1}{11}(1 + 3x^{-1/2}) dx$$

$$= \frac{1}{11}(x + 6x^{1/2}) \Big|_{6.41}^9$$

$$= \frac{1}{11}[9 + 18 - 6.41 - 6(6.41)^{1/2}]$$

$$\approx 0.49$$

13.3   Special Probability Density Functions

1. $f(x) = \frac{5}{4}$ for $[4, 4.8]$

    This is a uniform distribution:
    $a = 4$, $b = 4.8$.

    (a) $\mu = \frac{1}{2}(4.8 + 4) = \frac{1}{2}(8.8)$

$$= 4.4 \text{ cm}$$

    (b) $\sigma = \frac{1}{\sqrt{12}}(4.8 - 4)$

$$= \frac{1}{\sqrt{12}}(0.8)$$

$$\approx 0.23 \text{ cm}$$

    (c) $P(4.4 < x < 4.4 + 0.23)$

$$= \int_{4.4}^{4.63} \frac{5}{4} dx$$

$$= \frac{5}{4}x \Big|_{4.4}^{4.63} = \frac{5}{4}(4.63 - 4.4)$$

$$\approx 0.29$$

3. $f(t) = 0.03e^{-0.03t}$ for $[0, \infty)$

    This is an exponential distribution:
    $a = 0.03$.

    (a) $\mu = \frac{1}{0.03} \approx 33.33$ yr

    (b) $\sigma = \frac{1}{0.03} \approx 33.33$ yr

(c) $P(33.33 < x < 33.33 + 33.33)$

$$= \int_{33.33}^{66.66} 0.03e^{-0.03t} \, dt$$

$$= -e^{-0.03t}\Big|_{33.33}^{66.66}$$

$$\doteq \frac{-1}{e^{0.03(66.66)}} + \frac{1}{e^{0.03(33.33)}}$$

$$\approx 0.23$$

5. $f(x) = e^{-t}$ for $[0, \infty)$

This is an exponential distribution:
$a = 1$.

(a) $\mu = \dfrac{1}{1} = 1$ day

(b) $\sigma = \dfrac{1}{1} = 1$ day

(c) $P(1 < x < 1 + 1) = \displaystyle\int_{1}^{2} e^{-t} \, dt = -e^{-t}\Big|_{1}^{2}$

$$= -\frac{1}{e^2} + \frac{1}{e}$$

$$\approx 0.23$$

In Exercises 7-13, use the table in the Appendix for areas under the normal curve.

7. $z = 3.50$

Area to the left of $z = 3.50$ is 0.9998. Given mean $\mu = z - 0$, so area to left of μ is 0.5.
Area between μ and z is

$$0.9998 - 0.5 = 0.4998.$$

Therefore, this area represents 49.98% of total area under normal curve.

9. Between $z = 1.28$ and $z = 2.05$

Area to left of $z = 2.05$ is 0.9798 and area to left of $z = 1.28$ is 0.8997.

$$0.9798 - 0.8997 = 0.0801$$

Percent of total area $= 8.01\%$

11. Since $10\% = 0.10$, the z-score that corresponds to the area of 0.10 to the left of z is -1.28.

13. 18% of the total area to the right of z means $1 - 0.18$ of the total area is to the left of z.

$$1 - 0.18 = 0.82$$

The closest z-score that corresponds to the area of 0.82 is 0.92.

19. Let m be the median of the exponential distribution $f(x) = ae^{-ax}$ for $[0, \infty)$.

$$\int_{0}^{m} ae^{-ax} \, dx = 0.5$$

$$-e^{-ax}\Big|_{0}^{m} = 0.5$$

$$-e^{-am} + 1 = 0.5$$

$$0.5 = e^{-am}$$

$$-am = \ln 0.5$$

$$m = -\frac{\ln 0.5}{a}$$

or $-am = \ln \dfrac{1}{2}$

$$-am = -\ln 2$$

$$m = \frac{\ln 2}{a}$$

21. The area that is to the left of x is

$$A = \int_{-\infty}^{x} \frac{1}{\sigma\sqrt{2\pi}} e^{-\frac{(t-\mu)^2}{2\sigma^2}} \, dt.$$

Let $u = \frac{t-\mu}{\sigma}$. Then $du = \frac{1}{\sigma} \, dt$ and $dt = \sigma \, du$.
If $t = x$,

$$u = \frac{x - \mu}{\sigma} = z.$$

As $t \to -\infty$, $\mu \to -\infty$.
Therefore,

$$A = \int_{-\infty}^{z} \frac{1}{\sigma\sqrt{2\pi}} e^{(-1/2)u^2} \sigma \, du$$

$$= \frac{\sigma}{\sigma} \int_{-\infty}^{z} \frac{1}{\sqrt{2\pi}} e^{-u^2/2} \, du$$

$$= \int_{-\infty}^{z} \frac{1}{\sqrt{2\pi}} e^{-u^2/2} \, du.$$

This is the area to the left of z for the standard normal curve.

In Exercises 23 and 25, use Simpson's rule with $n = 100$ or the integration feature on a graphing calculator to approximate the integrals. Answers may vary slightly from those given here depending on the method that is used.

23. **(a)** $\displaystyle\int_{0}^{50} 0.5e^{-0.5x} \, dx \approx 1.00000$

(b) $\displaystyle\int_{0}^{50} 0.5xe^{-0.5x} \, dx \approx 1.99999$

(c) $\displaystyle\int_{0}^{50} 0.5x^2 e^{-0.5x} \, dx = 8.00003$

25. $\int_{-\infty}^{\infty} \frac{1}{\sqrt{2\pi}} e^{-x^2/2}\, dx$

$\approx \int_{-4}^{4} \frac{1}{\sqrt{2\pi}} e^{-x^2/2}\, dx$

(a) $\mu = 1.75 \times 10^{-14} \approx 0$

(b) $\sigma = 0.999433 \approx 1$

27. For a uniform distribution,

$f(x) = \frac{1}{b-a}$ for [a, b].

$f(x) = \frac{1}{36-20} = \frac{1}{16}$ for [20, 36]

(a) $\mu = \frac{1}{2}(20+36) = \frac{1}{2}(56)$

$= 28$ days

(b) $P(30 < x \le 36)$

$= \int_{30}^{36} \frac{1}{16}\, dx = \frac{1}{16}x \Big|_{30}^{36}$

$= \frac{1}{16}(36-30)$

$= 0.375$

29. We have an exponential distribution, with a = 1.
$f(t) = e^{-t}$, $[0, \infty)$

(a) $\mu = \frac{1}{1} = 1$ hr

(b) $P(t < 30$ min$)$

$= \int_{0}^{0.5} e^{-t}\, dt$

$= -e^{-t} \Big|_{0}^{0.5}$

$1 - e^{-0.5} \approx 0.39$

31. $f(x) = ae^{-ax}$ for $[0, \infty)$

Since $\mu = 25$ and $\mu = \frac{1}{a}$,

$a = \frac{1}{25} = 0.04.$

This, $f(x) = 0.04e^{-0.04x}$.

(a) We must find t such that
$P(x \le t) = 0.90$.

$\int_{0}^{t} 0.04e^{-0.04x}\, dx = 0.90$

$-e^{-0.04x} \Big|_{0}^{t} = 0.90$

$-e^{-0.04t} + 1 = 0.90$

$0.10 = -e^{-0.04t}$

$-0.04t = \ln 0.10$

$t = \frac{\ln 0.10}{-0.04}$

$t \approx 57.56$

The longest time within which the predator will be 90% certain of finding a prey is approximately 58 min.

(b) $P(x \ge 60)$

$= \int_{60}^{\infty} 0.04e^{-0.04x}\, dx$

$= \lim_{b \to \infty} \int_{60}^{b} 0.04e^{-0.04x}\, dx$

$= \lim_{b \to \infty} (-e^{-0.04x}) \Big|_{60}^{b}$

$= \lim_{b \to \infty} [-e^{-0.04b} + e^{-0.04(60)}]$

$= 0 + e^{-2.4}$

≈ 0.0907

The probability that the predator will have to spend more than one hour looking for a prey is approximately 0.09.

33. For an exponential distribution, $f(x) = ae^{-ax}$ for $[0, \infty)$.
Since $\mu = \frac{1}{a} = 11.5$, $a = \frac{1}{11.5}$

(a) $P(x \ge 20) = \int_{20}^{\infty} \frac{1}{11.5} e^{-x/11.5}\, dx$

$= \lim_{b \to \infty} \int_{20}^{b} \frac{1}{11.5} e^{-x/11.5}\, dx$

$= \lim_{b \to \infty} \left(-e^{-x/11.5} \Big|_{20}^{b} \right)$

$= \lim_{b \to \infty} (-e^{-b/11.5} + e^{-20/11.5})$

$= e^{-20/11.5}$

≈ 0.18

(b) $P(10 \leq x \leq 20) = \int_{10}^{20} \frac{1}{11.5} e^{-x/11.5} dx$

$$= -e^{-x/11.5} \Big|_{10}^{20}$$

$$= -e^{-20/11.5} + e^{-10/11.5}$$

$$\approx 0.24$$

35. Uniform distribution on [32, 44]

$$f(x) = \frac{1}{44 - 32} = \frac{1}{12} \text{ for } [32, 44]$$

(a) $\mu = \frac{1}{2}(32 + 44) = 38$ inches

(b) $P(38 < x < 40)$

$$= \int_{38}^{40} \frac{1}{12} dx$$

$$= \frac{x}{12} \Big|_{38}^{40}$$

$$= \frac{1}{6} \approx 0.17$$

37. For an exponential distribution, $f(x) = ae^{-ax}$ for $[0, \infty)$. Since $a = \frac{1}{609.5}$, $f(x) = \frac{1}{609.5} e^{-x/609.5}$.

(a) The expected number of days is

$\mu = \frac{1}{a} = 609.5$. The standard deviation is

$\sigma = \frac{1}{a} = 609.5$.

(b) $P(x > 365) = \int_{365}^{\infty} \frac{1}{609.5} e^{-x/609.5} dx$

$$= 1 - \int_{0}^{365} \frac{1}{609.5} e^{-x/609.5} dx$$

$$= 1 + \left(e^{-x/609.5} \Big|_{0}^{365} \right)$$

$$= 1 + \left(e^{-365/609.5} - 1 \right)$$

$$= e^{-365/609.5}$$

$$\approx 0.55$$

39. For a uniform distribution,

$$f(x) = \frac{1}{b - a} \text{ for } [a, b].$$

Thus, we have

$$f(x) = \frac{1}{85 - 10} = \frac{1}{75}$$

for [10, 85].

(a) $\mu = \frac{1}{2}(10 + 85) = \frac{1}{2}(95)$

$$= 47.5 \text{ thousands}$$

Therefore, the agent sells \$47,500 in insurance.

(b) $P(50 < x < 85) = \int_{50}^{85} \frac{1}{75} dx$

$$= \frac{x}{75} \Big|_{50}^{85}$$

$$= \frac{85}{75} - \frac{50}{75}$$

$$= \frac{35}{75} = 0.47$$

41. (a) Since we have an exponential distribution with $\mu = 4.25$,

$$\mu = \frac{1}{a} = 4.25$$

$$a = 0.235.$$

Therefore, $f(x) = 0.235 e^{-0.235x}$ on $[0, \infty)$.

(b) $P(x > 10)$

$$= \int_{10}^{\infty} 0.235 e^{-0.235x} dx$$

$$= \lim_{a \to \infty} \int_{10}^{a} 0.235 e^{-0.235x} dx$$

$$= \lim_{a \to \infty} \left(-e^{-0.235x} \right) \Big|_{10}^{a}$$

$$= \lim_{a \to \infty} \left(-\frac{1}{e^{0.235a}} + \frac{1}{e^{2.35}} \right)$$

$$= \frac{1}{e^{2.35}} = 0.095$$

43. (a) $\mu = 2.5$, $\sigma = 0.2$, $x = 2.7$

$$z = \frac{2.7 - 2.5}{0.2} = 1$$

Area to the right of $z = 1$ is

$$1 - 0.8413 = 0.1587.$$

Probability $= 0.1587$

(b) Within 1.2 standard deviations of the mean is the area between $z = -1.2$ and $z = 1.2$.
Area to left of $z = 1.2 = 0.8849$
Area to the left of $z = -1.2 = 0.1151$

$$0.8849 - 0.1151 = 0.7698$$

Probability $= 0.7698$

Chapter 13 Review Exercises

1. In a probability function, the y-values (or function values) represent probabilities.

3. A probability density function f for $[a,\ b]$ must satisfy the following two conditions:

(1) $\int_a^b f(x)\,dx = 1$;

(2) $f(x) \geq 0$ for all x in the interval $[a,\ b]$.

5. $f(x) = \sqrt{x} \geq 0;\ [4,\ 9]$

$$\int_4^9 x^{1/2}\,dx = \frac{2}{3}x^{3/2}\Big|_4^9$$
$$= \frac{2}{3}(27 - 8)$$
$$= \frac{38}{3} \neq 1$$

$f(x)$ is not a probability density function.

7. $f(x) = e^{-x};\ [0,\ \infty)$

$$\int_0^\infty e^{-x}\,dx = \lim_{b \to \infty} \int_0^b e^{-x}\,dx$$
$$= \lim_{b \to \infty} -e^{-x}\Big|_0^b$$
$$= \lim_{b \to \infty} \left(1 - \frac{1}{e^b}\right) = 1$$

$f(x) > 0$ for all x.
$f(x)$ is a probability density function.

9. $f(x) = kx^2;\ [0,\ 3]$

$$\int_0^3 kx^2\,dx = \frac{kx^3}{3}\Big|_0^3$$
$$= 9k$$

Since $f(x)$ is a probability density function,

$$9k = 1$$
$$k = \frac{1}{9}.$$

11. $f(x) = \frac{1}{10}$ for $[10,\ 20]$

(a) $P(10 \leq x \leq 12)$
$$= \int_{10}^{12} \frac{1}{10}\,dx$$
$$= \frac{x}{10}\Big|_{10}^{12}$$
$$= \frac{1}{5}$$
$$= 0.2$$

(b) $P\left(\frac{31}{2} \leq x \leq 20\right)$
$$= \int_{31/2}^{20} \frac{1}{10}\,dx$$
$$= \frac{x}{10}\Big|_{31/2}^{20}$$
$$= 2 - \frac{31}{20}$$
$$= \frac{9}{20}$$
$$= 0.45$$

(c) $P(10.8 \leq x \leq 16.2)$
$$= \int_{10.8}^{16.2} \frac{1}{10}\,dx$$
$$= \frac{x}{10}\Big|_{10.8}^{16.2}$$
$$= 0.54$$

13. The distribution that is tallest or most peaked has the smallest standard deviation. This is the distribution pictured in graph (b).

15. $f(x) = \frac{2}{9}(x - 2);\ [2,\ 5]$

$$\mu = \int_2^5 \frac{2x}{9}(x - 2)\,dx$$
$$= \int_2^5 \frac{2}{9}(x^2 - 2x)\,dx$$
$$= \frac{2}{9}\left(\frac{x^3}{3} - x^2\right)\Big|_2^5$$
$$= \frac{2}{9}\left(\frac{125}{3} - 25 - \frac{8}{3} + 4\right)$$
$$= 4$$